Trends in Intelligent Systems and Computer Engineering

Lecture Notes in Electrical Engineering

Volume 6

Trends in Intelligent Systems and Computer Engineering
Oscar Castillo, Li Xu, and Sio-Iong Ao
ISBN 978-0-387-74934-1, 2008

Recent Advances in Industrial Engineering and Operations Research
Alan H. S. Chan, and Sio-Iong Ao
ISBN 978-0-387-74903-7, 2008

Advances in Communication Systems and Electrical Engineering
Xu Huang, Yuh-Shyan Chen, and Sio-Iong Ao
ISBN 978-0-387-74937-2, 2008

Time-Domain Beamforming and Blind Source Separation
Julien Bourgeois, and Wolfgang Minker
ISBN 978-0-387-68835-0, 2007

Digital Noise Monitoring of Defect Origin
Telman Aliev
ISBN 978-0-387-71753-1, 2007

Multi-Carrier Spread Spectrum 2007
Simon Plass, Armin Dammann, Stefan Kaiser, and K. Fazel
ISBN 978-1-4020-6128-8, 2007

Oscar Castillo • Li Xu • Sio-Iong Ao
Editors

Trends in Intelligent Systems and Computer Engineering

Springer

Editors

Oscar Castillo
Tijuana Institute of Technology
Department of Computer Science
P.O. Box 4207
Chula Vista CA 91909
USA

Li Xu
Zhejiang University
College of Electrical Engineering
Department of Systems Science and Engineering
Yu-Quan Campus
310027 Hangzhou
People's Republic of China

Sio-Iong Ao
IAENG Secretariat
37–39 Hung To Road
Unit 1, 1/F
Hong Kong
People's Republic of China

ISBN: 978-0-387-74934-1 e-ISBN: 978-0-387-74935-8
DOI: 10.1007/978-0-387-74935-8

Library of Congress Control Number: 2007935315

Printed on acid-free paper

9 8 7 6 5 4 3 2 1

springer.com

Preface

A large international conference, Intelligent Systems and Computer Engineering, was held in Hong Kong, March 21–23, 2007, under the International MultiConference of Engineers and Computer Scientists (IMECS) 2007. The IMECS 2007 is organized by the International Association of Engineers (IAENG), a nonprofit international association for engineers and computer scientists. The IMECS conferences serve as good platforms for the engineering community to meet with each other and to exchange ideas. The conferences also strike a balance between theoretical and application development. The conference committees have been formed with over two hundred committee members who are mainly research center heads, faculty deans, department heads, professors, and research scientists from over thirty countries. The conferences are truly international meetings with a high level of participation from many countries. The response that we have received for the multiconference is excellent. There have been more than one thousand one hundred manuscript submissions for the IMECS 2007. All submitted papers have gone through the peer review process and the overall acceptance rate is 58.46%.

This volume contains revised and extended research articles on intelligent systems and computer engineering written by prominent researchers participating in the multiconference IMECS 2007. There is huge demand, not only for theories but also applications, for the intelligent systems and computer engineering in the society to meet the needs of rapidly developing top-end high technologies and to improve the increasing high quality of life. Topics covered include automated planning, expert systems, machine learning, fuzzy systems, knowledge-based systems, computer systems organization, computing methodologies, and industrial applications. The papers are representative of these subjects. The book offers state-of-the-art tremendous advances in intelligent systems and computer engineering and also serves as an excellent reference work for researchers and graduate students working with intelligent systems and computer engineering.

Sio Iong Ao, Oscar Castillo, and Li Xu

July 2007

Hong Kong, Mexico, and China

Contents

Contributors

Mehreen Afzal
National University of Science and Technology, Pakistan, mehreenafzal00@hotmail.com

Suneeta Agarwal
Computer Science & Engineering Department, Motilal Nehru National Institute of Technology, Allahabad, India, sunneta@mnnit.ac.in

Morteza Analoui
Computer Engineering Department, Iran University of Science and Technology, Narmak, Tehran 16846, Iran, analoui@iust.ac.ir

Behrouz Minaei Bidgoli
Department of Computer Science and Engineering, Michigan State University, East Lansing, MI 48824, USA, minaeibi@cse.msu.edu

Henrik Boström
Skövde Cognition and Artificial Intelligence Lab, School of Humanities and Informatics, University of Skövde, SE-541 28 Skövde, Sweden, henrik.bostrom@his.se

Martin Brown
School of Electronic and Electrical Engineering, The University of Manchester, Manchester M60 1QD, UK, martin.brown@manchester.ac.uk

Gary Bundell
Centre for Intelligent Information Processing Systems, School of Electrical, Electronic and Computer Engineering, University of Western Australia, Crawley, WA 6009, Australia, bundell@ee.uwa.edu.au

B. Eng. Hua Chang
Heinz Nixdorf Institute, University of Paderborn, Fuerstenallee 11, 33102 Paderborn, Germany, Hua.Chang@hni.uni-paderborn.de

Hsueh-Wei Chang
Environmental Biology, Kaohsiung, changhw@kmu.edu.tw

Chiu-Hsiung Chen
Department of Computer Sciences and Information Engineering, China University of Technology, HuKou Township 303, Taiwan, Republic of China, chchchen@cute.edu.tw

Ken Chen
School of Computer & Information Engineering, Hunan Agricultural University, Changsha 410128, China, Chenk@hunau.edu.cn

Yu Chen
Institute of Electronic Information Engineering, College of Information Science and Engineering, Northeastern University, Shenyang, China, chenyuren.001@163.com

Yu-Huei Cheng
Kaohsiung University, yuhuei.cheng@gmail.com

Yu-Min Cheng
Chroma Corporation, Taiwan, Republic of China, nick.cheng@chroma.com.tw

Chiao-Yu Chuang
Department of Electronic Engineering, National Kaohsiung University of Applied Sciences, Kaohsiung, Taiwan 807, Republic of China

Li-Yeh Chuang
University, Kaohsiung, Taiwan

Gero Decker
Hasso-Plattner-Institute for Software Systems Engineering, Prof.-Dr.-Helmert-Str. 2-3, 14482 Potsdam, Germany, gero.decker@hpi.uni-potsdam.de

Tzimeas Dimitrios
Department of Computer Science and Informatics, University College of Dublin, Dublin, Ireland

Walter Dosch
University of Lübeck, Institute of Software Technology and Programming Languages, Lübeck, Germany, http://www.isp.uni-luebeck.de

Yang-Hsin Fan
Department of Electronic Engineering, Institute of Computer and Communication Engineering, National Taipei University of Technology, Taipei, Taiwan, Republic of China
Information System Section of Library, National Taitung University, Taitung, Taiwan, Republic of China, yhfan@nttu.edu.tw

Jürgen Gausemeier
Heinz Nixdorf Institute, University of Paderborn, Fuerstenallee 11, 33102 Paderborn, Germany, Juergen.Gausemeier@hni.uni-paderborn.de

Mohammad Ghodsi
Computer Engineering Department, Sharif University of Technology, Tehran, Iran
IPM School of Computer Science, Tehran, Iran, ghodsi@sharif.edu

Henrik Grimm
Centre for Intelligent Automation, University of Skövde, Sweden

Ashutosh Gupta
Computer Science & Engineering Department, Institute of Engineering and Rural Technology, Allahabad, India, ashutosh333@rediffmail.com

Akira Hara
Graduate School of Information Sciences, Hiroshima City University, 3-4-1, Ozuka-higashi, Asaminami-ku, Hiroshima 731-3194, Japan, ahara@its.hiroshima-cu.ac.jp

Fei He
School of Electronic and Electrical Engineering, The University of Manchester, Manchester M60 1QD, UK
Department of Electronic Engineering, City University of Hong Kong, Hong Kong, eefeihe@cityu.edu.hk

Rong-Shue Hsiao
Department of Electronic Engineering, Institute of Computer and Communication Engineering, National Taipei University of Technology, Taipei, Taiwan, Republic of China, rsh@ntut.edu.tw

Larry Hughes
Department of Electrical and Computer Engineering, Dalhousie University, Halifax, Nova Scotia, Canada, larry.hughes@dal.ca

Yi-Hui
Department of Electronic Engineering, National Kaohsiung University of Applied Sciences, Kaohsiung, Taiwan 807, Republic of China

Lin Ming-Hung
Department of Electrical Engineering, Yuan-Ze University 135, Far-East Rd., Chung-Li, Tao-Yuan, 320, Taiwan, Republic of China, cml@saturn.yzu.edu.tw

Jeong-Hoon Hwang
Human-Robot Interaction Research Center, Korea Advanced Institute of Science and Technology, 373-1 Guseong-dong, Yuseong-gu, Daejeon 305-701, Korea

R. N. Ibrahim
Department of Mechanical Engineering, Monash University, Wellington Rd., Clayton 3800, Australia, Raafat.Ibrahim@eng.monash.edu.au

Takumi Ichimura
Graduate School of Information Sciences, Hiroshima City University, 3-4-1, Ozuka-higashi, Asaminami-ku, Hiroshima 731-3194, Japan, ichimura@its.hiroshima-cu.ac.jp

Masumi Ishikawa
Kyushu Institute of Technology, 2-4 Hibikino, Wakamatsu, Kitakyushu 808-0196, Japan, ishikawa@brain.kyutech.ac.jp

Sun Junping
Graduate School of Computer and Information Sciences, Nova Southeastern University, Fort Lauderdale, FL 33314, USA

B. Kalpana
Department of Computer Science, Avinashilingam University for Women, Coimbatore, India

Kunihiko Kaneko
Graduate School of Information Science and Electrical Engineering, Kyushu University, Fukuoka, Japan, kaneko@db.is.kyushu-u.ac.jp

Thashmee Karunaratne
Department of Computer and Systems Sciences, Stockholm University/Royal Institute of Technology, Forum 100, SE-164 40 Kista, Sweden, si-thk@dsv.su.se

M. L. Kulthon Kasemsan
Science Program in Information Technology (MSIT), Faculty of Information Technology, Rangsit University, Pathumtani, Thailand 12000, kasemsan@rangsit.rsu.ac.th

Yoshiaki Kurosawa
Graduate School of Information Sciences, Hiroshima City University, 3-4-1, Ozuka-higashi, Asaminami-ku, Hiroshima 731-3194, Japan, kurosawa@its.hiroshima-cu.ac.jp

Behrouz T. Ladani
The University of Isfahan, Iran

Kang-Woo Lee
School of Media, Soongsil University, Sangdo-dong 511, Dongjak-gu, Seoul 156-743, South Korea, kangwooster@gmail.com

Trong-Yen Lee
Department of Electronic Engineering, Institute of Computer and Communication Engineering, National Taipei University of Technology, Taipei, Taiwan, Republic of China, tylee@ntut.edu.tw

Konstantina Lepinioti
School of Design, Engineering and Computing, Bournemouth University, UK, tlepinio@bournemouth.ac.uk

Chih-Min Lin
Department of Electrical Engineering, Yuan-Ze University 135, Far-East Rd., Chung-Li, Tao-Yuan, 320, Taiwan, Republic of China, cml@saturn.yzu.edu.tw

Ming-Hsien Lin
Kaohsiung University, iamminghsien@gmail.com

Cheng-Yuan Liou
Department of Computer Science and Information Engineering, National Taiwan University, Taipei, Taiwan, Republic of China

Junfang Liu
College of Information Science and Engineering, Northeastern University, Shenyang 110004, China

P. B. Lochert
Mechanical Engineering, Monash University, Wellington Rd, Clayton 3800, Australia, plochert@bigpond.net.au

Heitor S. Lopes
CPGEI, Universidade Tecnológica Federal do Paraná (UTFPR), Av. 7 de setembro, 3165 - Curitiba - Paraná, Brazil, hslopes@cpgei.cefetpr.br

Yingliang Lu
Graduate School of Information Science and Electrical Engineering, Kyushu University, Fukuoka, Japan, riku@db.is.kyushu-u.ac.jp

Ka Lok Man
Centre for Efficiency-Oriented Languages (CEOL), Department of Computer Science, University College Cork (UCC), Cork, Ireland

Eleni Mangina
Department of Computer Science and Informatics, University College of Dublin, Dublin, Ireland

Sukanya Manna
Department of Computer Science and Information Engineering, National Taiwan University, Taipei, Taiwan, Republic of China

Akifumi Makinouchi
Department of Information and Network Engineering, Kurume Institute of Technology, Fukuoka, Japan, akifumi@cc.kurume-it.ac.jp

Lena Mashayekhy
The University of Isfahan, Iran

Ashraf Masood
National University of Science and Technology, Pakistan

Stephen McKearney
School of Design, Engineering and Computing, Bournemouth University, UK, smckearn@bournemouth.ac.uk

Tsuyoshi Morimoto
Electronics and Computer Science Department, Fukuoka University, 8-19-1 Nanakuma, Jonan-ku, Fukuoka 814-0180, Japan, morimoto@tlsun.tl.fukuoka-u-ac.jp

R. Nadarajan
Department of Mathematics and Computer Applications, PSG College of Technology, Coimbatore, India

Mohammad A. Nematbakhsh
The University of Isfahan, Iran

Flávio Neves Junior
CPGEI, Universidade Tecnológica Federal do Paraná (UTFPR), Av. 7 de setembro, 3165 - Curitiba - Paraná, Brazil, hslopes@cpgei.cefetpr.br

Amos Ng
Centre for Intelligent Automation, University of Skövde, Sweden

Shinichi Oeda
Department of Information and Computer Engineering, Kisarazu National College of Technology, Kisarazu, Japan

Serhan Ozdemir
Mechanical Engineering Department, Izmir Institute of Technology, Izmir 35430, Turkey, serhanozdemir@iyte.edu.tr

Kaveh Pashaei
Electrical and Computer Engineering Faculty, School of Engineering, University of Tehran, Tehran, Iran, k.pashaei@ece.ut.ac.ir

Anna Person
Centre for Intelligent Automation, University of Skövde, Sweden

Farzad Peyravi
Electrical and Computer Engineering Faculty, School of Engineering, University of Tehran, Tehran, Iran, f.peyravi@ece.ut.ac.ir

Loonis Pierre
Universite de La Rochelle, Laboratoire Informatique Image Interaction, Avenue Michel Crepeau 17000 La Rochelle, France, ploonis@univ-lf.fr

Jörg Preißinger
Institut für Informatik, Technische Universität München, Boltzmannstr. 385748 Garching bei München, Germany, preissin@in.tum.de

Monruthai Radeerom
Science Program in Information Technology (MSIT), Faculty of Information Technology, Rangsit University, Pathumtani, Thailand 12000, mradeerom@yahoo.com

Imran Raza
Department of Computer Science, COMSATS Institute of Information Technology, Lahore, Pakistan, iraza@ciitlahore.edu.pk

Muhammad Hassan Raza
Department of Engineering Mathematics and Internetworking, Dalhousie University, Halifax, Nova Scotia, Canada, hraza@dal.ca

Mohammad Hossein Rezvani
Computer Engineering Department, Iran University of Science and Technology, Narmak, Tehran 16846, Iran, rezvani@iust.ac.ir

Sara Salehi
Computer Engineering Department, Azad University, Tehran-South Branch, Iran, sarasalehi@gmail.com

Tout Samir
Department of Computer Science, Eastern Michigan University, Ypsilanti, MI 48197, USA

Hassan Sayyadi
Computer Engineering Department, Sharif University of Technology, Tehran, Iran, sayyadi@ce.sharif.edu

Michel P. Schellekens
Centre for Efficiency-Oriented Languages (CEOL), Department of Computer Science, University College Cork (UCC), Cork, Ireland, m.schellekens@cs.ucc.ie

Marcel Seelig
Hasso-Plattner-Institute for Software Systems Engineering, Prof.-Dr.-Helmert-Str. 2-3, 14482 Potsdam, Germany, marcel.seelig@hpi.uni-potsdam.de

Roberto Selow
Electrical Engineering Department, Centro Universitário Positivo, Rua Prof. Pedro Viriato Parigot de Souza, 5300 - Curitiba - Paraná, Brazil, rselow@unicenp.edu.br

Idris Abu Seman
Malaysia Palm Oil Board No. 6, Persiaran Institusi, Bandar Baru Bangi, 43000 Kajang, Malaysia, idris@mpob.gov.my

Hakki Erhan Sevil
Mechanical Engineering Department, Izmir Institute of Technology, Turkey, erhansevil@iyte.edu.tr

Hadi Shahriar Shahhoseini
Electrical Engineering Department, Iran University of Science and Technology, Narmak, Tehran 16844, Iran, hshsh@iust.ac.ir

Mohsen Sharifi
Iran University of Science and Technology, Computer Engineering Department, Tehran 16846-13114, Iran, msharifi@iust.ac.ir

Yue Shen
School of Computer & Information Engineering, Hunan Agricultural University, Changsha 410128, China, shenyue@hunau.edu.cn

Mohsen Soryani
Iran University of Science and Technology, Computer Engineering Department, Tehran 16846-13114, Iran, soryani@iust.ac.ir

Chonawat Srisa-an
Science Program in Information Technology (MSIT), Faculty of Information Technology, Rangsit University, Pathumtani, Thailand 12000, chonawat@rangsit.rsu.ac.th

Annette Stümpel
University of Lübeck, Institute of Software Technology and Programming Languages, Lübeck, Germany

Te-Jen Su
Department of Electronic Engineering, National Kaohsiung University of Applied Sciences, Kaohsiung, Taiwan 807, Republic of China, sutj@cc.kuas.edu.tw

Machi Suka
Department of Preventive Medicine, St. Marianna University School of Medicine, Kawasaki, Japan

Mazliham Mohd Su'ud
Universiti Kuala Lumpur, Sek 14, Jalan Teras Jernang 43650 Bandar Baru Bangi, Selangor, Malaysia, mazliham@tm.net.my
Universite de La Rochelle, Laboratoire Informatique Image Interaction, Avenue Michel Crepeau 17000 La Rochelle, France

Fattaneh Taghyareh
Electrical and Computer Engineering Faculty, School of Engineering, University of Tehran, Tehran, Iran, ftaghiyar@ut.ac.ir

K. Tahera
Mechanical Engineering, Monash University, Wellington Road, Clayton 3800, Australia, KhadizaTahera@eng.monash.edu.au

Shin-ya Takahashi
Electronics and Computer Science Department, Fukuoka University, 8-19-1 Nanakuma, Jonan-ku, Fukuoka 814-0180, Japan, takahasi@tlsun.tl.fukuoka-u-ac.jp

Jun Takezawa
Department of Emergency and Intensive Care Medicine, Nagoya University Graduate School of Medicine, Nagoya, Japan

Chia-Chun Tsai
Department of Computer Science and Information Engineering, Nanhua University, Chia-Yi, Taiwan, Republic of China, chun@mail.nhu.edu.tw

Wen-Pin Tsai
Department of Electronic Engineering, National Kaohsiung University of Applied Sciences, Kaohsiung, Taiwan 807, Republic of China

Anna Wang
Institute of Electronic Information Engineering, College of Information Science and Engineering, Northeastern University, Shenyang, China, wanganna@mail.neu.edu.cn

Sverdlik William
Department of Computer Science, Eastern Michigan University, Ypsilanti, MI 48197, USA

Jie Wu
414# mailbox, North Eastern University, Shen Yang, Liao Ning, China 110004, wan-ganna@mail.neu.edu.cn

Ronghui Wu
College of Computer & Communication, Hunan University, Changsha 410082, China

Cheng Xu
College of Computer & Communication, Hunan University, Changsha 410082, China

Cheng-San Yang
Hospital, Taiwan

Cheng-Hong Yang
National Kaohsiung University, chyang@cc.kuas.edu.tw

Lam Fat Yeung
Department of Electronic Engineering, City University of Hong Kong, Hong Kong, eelyeung@cityu.edu.hk

Katsumi Yoshida
Department of Preventive Medicine, St. Marianna University School of Medicine, Kawasaki, Japan

Fei Yu
School of Computer & Information Engineering, Hunan Agricultural University, Changsha 410128, China, yufei@hunau.edu.cn
Jiangsu Provincial Key Laboratory of Computer Information Processing Technology, Suzhou University, Suzhou 2150063, China, hunanyufei@126.com
College of Computer & Communication, Hunan University, Changsha 410082, China, yufei@hunau.net

Hong Zhang
Kyushu Institute of Technology, 2-4 Hibikino, Wakamatsu, Kitakyushu 808-0196, Japan, zhang@brain.kyutech.ac.jp

Xinhua Zhang
414# mailbox, North Eastern University, Shen Yang, Liao Ning, China 110004, wan-ganna@mail.neu.edu.cn

Weiqun Zheng
Centre for Intelligent Information Processing Systems, School of Electrical, Electronic and Computer Engineering, University of Western Australia, Crawley, WA 6009, Australia, zheng@ee.uwa.edu.au

Chapter 1
A Metamodel-Assisted Steady-State Evolution Strategy for Simulation-Based Optimization

Anna Persson, Henrik Grimm, and Amos Ng

1.1 Introduction

Evolutionary algorithms (EAs) have proven to be highly useful for optimization of real-world problems due to their powerful ability to find near-optimal solutions of complex problems [8]. A variety of successful applications of EAs has been reported for problems such as engineering design, operational planning, and scheduling. However, in spite of the great success achieved in many applications, EAs have also encountered some challenges. The main weakness of using EAs in real-world optimization is that a large number of simulation evaluations are needed before an acceptable solution can be found. Typically, an EA requires thousands of simulation evaluations and one single evaluation may take a couple of minutes to hours of computing time. This poses a serious hindrance to the practical application of EAs in real-world scenarios, and to address this problem the incorporation of computationally efficient metamodels has been suggested, so-called metamodel-assisted EAs [11]. The purpose of metamodels is to approximate the relationship between the input and output variables of a simulation by computationally efficient mathematical models. If the original simulation is represented as

$$y = f(x)$$

then a metamodel is an approximation of the form

$$y' = f'(x)$$

such that $y' = y + \varepsilon$, where ε represents the approximation error. By adopting metamodels in EAs, the computational burden of the optimization process can be greatly reduced because the computational cost associated with running a metamodel is negligible compared to a simulation run [11].

This chapter presents a new metamodel-assisted EA for optimization of computationally expensive simulation-optimization problems. The proposed algorithm is

Oscar Castillo et al. (eds.), *Trends in Intelligent Systems and Computer Engineering.*

basically an evolution strategy inspired by concepts from genetic algorithms. For maximum parallelism and increased efficiency, the algorithm uses a steady-state design. The chapter describes how the algorithm is successfully applied to optimize two real-world problems in the manufacturing domain. The first problem considered is about optimal buffer allocation in a car engine production line, and the second problem considered is about optimal production scheduling in a manufacturing cell for aircraft engines. In both problems, artificial neural networks (ANNs) are used as the metamodel.

In the next section, background information of EAs is presented and some examples of combining EAs and ANNs are given.

1.2 Background

This section describes the fundamentals of EAs and ANNs and presents some examples of combining these two techniques.

1.2.1 Evolutionary Algorithms

EAs define a broad class of different optimization techniques inspired by biological mechanisms. Generally, EAs are recognized by a genetic representation of solutions, a population-based solution approach, an iterative evolutionary process, and a guided random search. In evolving a population of solutions, EAs apply biologically inspired operations for selection, crossover, and mutation. The solutions in the initial population are usually generated randomly, covering the entire search space. During each generation, some solutions are selected to breed offspring for the next generation of the population. Either a complete population is bred at once (generational approach), or one individual at a time is bred and inserted into the population (steady-state approach).

The solutions in the population are evaluated using a simulation (Fig. 1.1). The EA feeds a solution to the simulation, which measures its performance. Based on the evaluation feedback given from the simulation, possibly in combination with previous evaluations, the EA generates a new set of solutions. The evaluation of

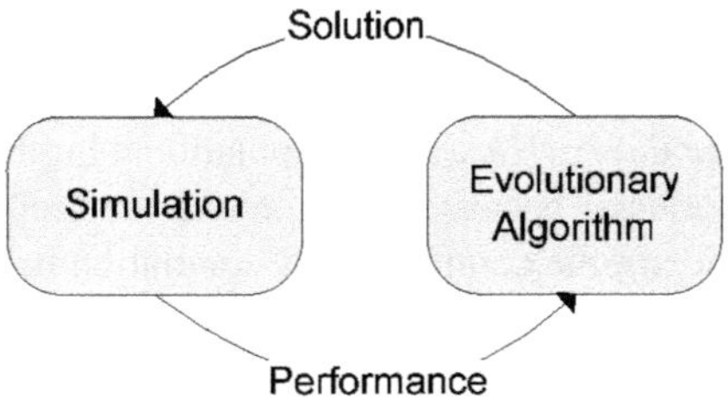

Fig. 1.1 Evaluation of solutions using a simulation model

solutions continues until a user-defined termination criterion has been fulfilled. This criterion may, for example, be that (a) a solution that satisfies a certain fitness level has been found, (b) the evaluation process has repeated for a certain number of times, or (c) that the best solutions in the last n evaluations have not changed (convergence has been reached).

Two well-defined EAs have served as the basis for much of the activity in the field: evolution strategies and genetic algorithms, which are described in the following.

Evolution strategies (ESs) are a variant of EAs founded in the middle of the 1960s. In an ES, λ offspring are generated from μ parents ($\lambda \neq \mu$) [1]. The selection of parents to breed offspring is random-based and independent of the parents' fitness values. Mutation of offspring is done by adding a normally distributed random value, where the standard deviation of the normal distribution usually is self-adaptive. The μ out of the λ generated offspring having the best fitness are selected to form the next generation of the population.

Genetic algorithms (GAs) became widely recognized in the early 1970s [4]. In a GA, μ offspring are generated from μ parents. The parental selection process is fitness-based and individuals with high fitness have a higher probability to be selected for breeding the next generation of the population. Different methods exist for the selection of parents. One example is tournament selection, in which a few individuals are chosen at random and the one with the best fitness is selected as the winner. In this selection method individuals with worse fitness may also be selected, which prevents premature convergence. A common approach is that the best individuals among the parents are carried over to the next generation unaltered, a strategy known as elitism.

1.2.2 Combining Evolutionary Algorithms and Artificial Neural Networks

The use of metamodels was first proposed to reduce the limitations of time-consuming simulations. Traditionally, regression and response surface methods have been two of the most common metamodeling approaches. In recent years, however, ANNs have gained increased popularity as this technique requires fewer assumptions and less precise information about the systems being modeled when compared with traditional techniques [3]. The first work providing the foundations for developing ANN metamodels for simulation was done. Both of these studies yielded results that indicated the potential applications of ANNs as metamodels for discrete-event and continuous simulation, particularly when saving computational cost is important.

In general terms, an ANN is a nonlinear statistical data modeling method used to model complex relationships between inputs and outputs. Originally, the inspiration for the technique was from the area of neuroscience and the study of neurons as information processing elements in the central nervous system. ANNs have universal

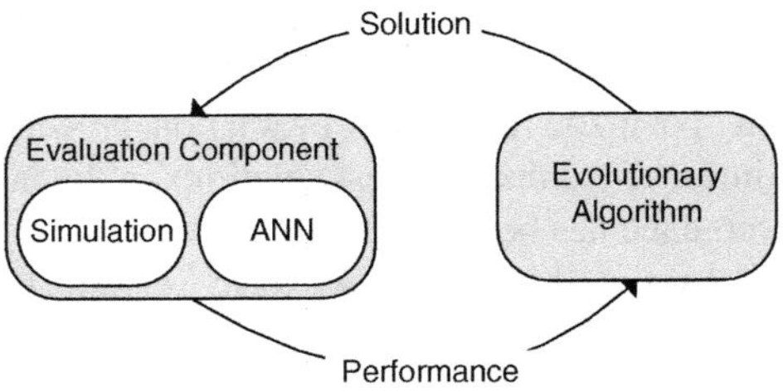

Fig. 1.2 Evaluation of solutions using both a simulation model and a metamodel

approximation characteristics and the ability to adapt to changes through training. Instead of only following a set of rules, ANNs are able to learn underlying relationships between inputs and outputs from a collection of training examples, and to generalize these relationships to previously unseen data. These attributes make ANNs very suitable to be used as surrogates for computationally expensive simulation models.

There exist several different approaches of using ANNs as simulation surrogates. The most straightforward approach is to first train the ANN using historical data and then completely replace the simulation with the ANN during the optimization process. These approaches can, however, only be successful when there is a small discrepancy between the outputs from the ANN and the simulation. Due to lack of data and the high complexity of real-world problems, it is generally difficult to develop an ANN with sufficient approximation accuracy that is globally correct and ANNs often suffer from large approximation errors which may introduce false optima [6]. Therefore, most successful approaches instead alternate between the ANN and the simulation model during optimization (Fig. 1.2).

In conjunction with EAs, ANNs have proven to be very useful for reducing the time consumption of optimizations. Most work within this area has focused on GAs, but there are also a few reports of combining ANNs with ESs. Some examples of this work are presented in the following.

Most work in combining ANNs and EAs is focused on GAs. Bull [2] presents an approach where an ANN is used in conjunction with a GA to optimize a theoretical test problem. The ANN is first trained with a number of initial samples to approximate the simulation and the GA then uses the ANN for evaluations. In every 50 generations, the best individual in the population is evaluated using the simulation. This individual then replaces the sample representing the worst fitness in the training dataset and the ANN is retrained. The author found that the combination of GAs and ANNs has great potential, but that one must be careful so that the optimization is not misled by the ANN when the fitness landscape of the modelled system is complex.

Jin et al. [6] propose another approach for using ANNs in combination with GAs. The main idea of this approach is that the frequency at which the simulation is used and the ANN is updated is determined by the estimated accuracy of the ANN. The authors introduce the concept of evolution control and propose two control methods: controlled individuals and controlled generations. With controlled individuals, part of the individuals in a population is chosen and evaluated using the simulation.

The controlled individuals can be chosen either randomly or according to their fitness values. With controlled generations, the whole population of N generations are evaluated with the simulation in every M generations $(N \leq M)$. Online learning of the ANN is applied after each call to the simulation when new training data are available. The authors carry out empirical studies to investigate the convergence properties of the implemented evolution strategy on two benchmark problems. They find that correct convergence occurs with both control mechanisms.

A third approach of combining ANNs and GAs is presented by Khu et al. [7]. The authors propose a strategic and periodic scheme of updating the ANN to ensure that it is constantly relevant as the search progresses. In the suggested approach, the whole population is first evaluated using the ANN and the best individuals in the population are then evaluated using the simulation. The authors implement an ANN and a GA for hydrological model calibration and show that there is a significant advantage in using ANNs for water and environmental system design.

Hüsken et al. [5] present an approach of combining ANNs and ESs. The authors propose an approach in which λ offspring are generated from μ parents and evaluated using the ANN $(\lambda > \mu)$. The ANN evaluations are the basis for the preselection of $s\,(0 < s < \lambda)$ individuals to be simulated. Of the s simulated individuals, the μ individuals having the highest simulation fitness form the next generation of the population. The authors apply their proposed algorithm to optimize an example problem in the domain of aerodynamic design and experiment on different ANN architectures. Results from the study show that structurally optimized networks exhibit better performance than standard networks.

1.3 A New Metamodel-Assisted Steady-State Evolution Strategy

In this chapter an optimization algorithm based on an ES and inspired by concepts from GA is proposed. The algorithm uses a steady-state design, in which one individual at a time is bred and inserted into the population (as opposed to generational approaches in which a whole generation is created at once). The main reason for choosing a steady-state design is that it has a high degree of parallelism, which is a very important efficiency factor when simulation evaluations are computationally expensive.

The implementation details of the algorithm are presented with pseudocode in Fig. 1.3. An initial population of μ solutions is first generated and evaluated using the simulation. The simulated samples are used to construct a metamodel (e.g., an ANN). Using crossover and mutation, λ offspring are generated from parents in the population chosen using the GA concept of tournament selection. The offspring are evaluated using the metamodel and one individual is selected to be simulated, again using tournament selection. When the individual has been simulated, the simulation input–output sample is used to train the metamodel online. Before the simulated individual is inserted into the population, one of the μ solutions already in the population is removed. Similar to the previous selection processes, the individual to

```
population ← Generate_Initial_Population( )
for each individual in population
    Simulation_Evaluation(individual)
    Update_Metamodel(individual)
end
while (not Stop_Optimization( )) do
    offspring ← Ø
    repeat λ times
        parent1 ← Select_For_Reproduction(population)
        parent2 ← Select_For_Reproduction(population)
        individual ← Crossover(parent1, parent2)
        Mutate(individual)
        Metamodel_Evaluation(individual)
        offspring.Add(individual)
    end
    replacement_individual ← Select_For_Replacement(offspring)
    Simulation_Evaluation(replacement_individual)
    Update_Metamodel(replacement_individual)
    population.Remove(Select_Individual_For_Removal(population))
    population.Add(replacement_individual)
end
```

Fig. 1.3 Pseudocode of proposed algorithm

be replaced is chosen using tournament selection. In the replacement strategy, the GA concept of elitism is used; that is, the individual in the population having the highest fitness is always preserved.

To make use of parallel processing nodes, several iterations of the optimization loop are executed in parallel.

1.4 Real-World Optimization

This section describes how the algorithm described in the previous section has been implemented in the optimization of two real-world problems in the manufacturing domain.

1.4.1 Real-World Optimization Problems

1.4.1.1 Buffer Allocation Problem

The first problem considered is about finding optimal buffer levels in a production line at the engine manufacturer Volvo Cars Engines, Skövde, Sweden. The Volvo Cars factory is responsible for supplying engine components for car engines to assembly plants and the specific production line studied in this chapter is responsible for the cylinder blocks. Production is largely customer order-driven and several models are built on the same production line, which imposes major demands on flexibility. As a way to achieve improved production in the cylinder block line, the management team wants to optimize its buffer levels. It is desirable to find a configuration of the buffer levels that maximizes the overall throughput of the line, while simultaneously minimizing the lead time of cylinder blocks. To analyze the system and perform optimizations, a detailed simulation model of the line has been developed using the QUEST software package.

For the scenario considered here, 11 buffers are subject to optimization and a duration corresponding to a two-week period of production is simulated. As the production line is complex and the simulation model is very detailed, one single simulation run for a period of this length takes about two hours to complete. Because there is a high degree of stochastic behavior in the production line due to unpredictable machine breakdowns, the simulation of each buffer level configuration is replicated five times and the average output of the five replications is taken as the simulation result. The optimization objective is described by

$$\left(\sum_{i \in C} w_1 i_{lead\_time}\right) / num\_cylinderblocks - w_2 throughput$$

where C is the set of all cylinder blocks and w_n is the weighted importance of an objective. The goal of the optimization is to minimize the objective function value.

1.4.1.2 Production Scheduling Problem

The second problem considered is a production scheduling problem at Volvo Aero (Sweden). The largest activity at Volvo Aero is development and production of advanced components for aircraft engines and gas turbines. Nowadays, more than 80% of all new commercial aircraft with more than 100 passengers are equipped with engine components from Volvo Aero. Volvo Aero also produces engine components for space rockets. As a partner of the European space program, they develop rocket engine turbines and combustion chambers.

At the Volvo Aero factory studied in this chapter, a new manufacturing cell has recently been introduced for the processing of engine components. The highly automated cell comprises multiple operations and is able to process several component types at the same time. After a period of initial tests, full production is now to be started in the cell. Similar to other manufacturing companies, Volvo Aero continuously strives for competitiveness and cost reduction, and it is therefore important that the new cell is operated as efficiently as possible.

To aid production planning, a simulation model of the cell has been built using the SIMUL8 software package. The simulation model provides a convenient way to perform what-if analyses of different production scenarios without the need of experimenting with the real system. Besides what-if analyses, the simulation model can also be used for optimization of the production. We describe how the simulation model has been used to enhance the production by optimization of the scheduling of components to be processed in the cell. For the production to be as efficient as possible, it is interesting to find a schedule that is optimal with respect to maximal utilization in combination with minimal shortage, tardiness, and wait-time of components. The optimization objective is described by

$$\sum_{i \in P} \left(w_s i_{shortage} + w_t i_{tardiness} \right) - w_u utilisation$$

where P is the set of all products and w is the weighted importance of an objective. The goal of the optimization is to minimize the objective function value.

1.4.2 Optimization Parameters

The population comprises 20 individuals (randomly initiated). From the parent population, 15 offspring are generated by performing a one-point crossover between two solutions (with a probability of 0.5) selected using tournament selection, that is, taking the better of two randomly chosen solutions. Each value in a created offspring is mutated using a Gaussian distribution with a deviation that is randomly selected from the interval (0,10).

1.4.3 Metamodel

For each of the two optimization problems, a fast metamodel of the simulation model is constructed by training an ANN to estimate fitness as a function of input parameters (buffer levels and planned lead-times, respectively). The ANN has a feedforward architecture with two hidden layers (Fig. 1.4). When the optimization

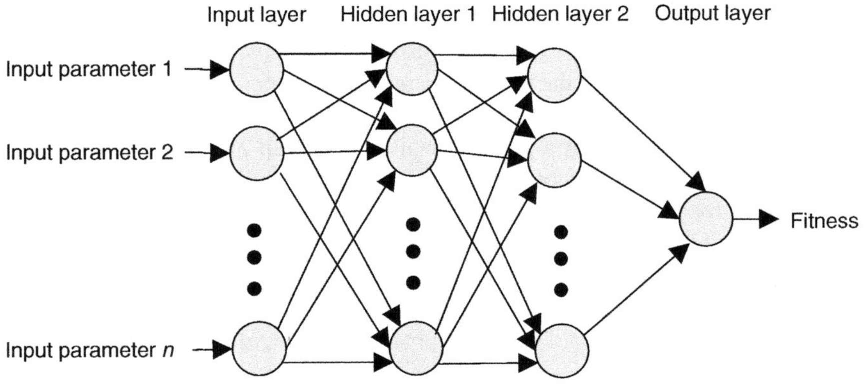

Fig. 1.4 Conceptual illustration of ANN

starts the ANN is untrained and after each generation, the newly simulated samples are added to the training dataset and the ANN is trained with the most recent samples (at most 500) using continuous training. To avoid overfitting, 10% of the training data is used for cross-validation. The training data is linearly normalized to values between 0 and 1. If any of the new samples has a lower or higher value than any earlier samples, renormalization of the data is performed and the weights of the ANN are reset.

1.4.4 Platform

The optimization has been realized using the OPTIMIZE platform, which is an Internet-based parallel and distributed computing platform that supports multiple users to run experiments and optimizations with different deterministic/stochastic simulation systems [10]. In the platform various EAs, ANN-based metamodels, deterministic/stochastic simulation systems, and a corresponding database management system are integrated in a parallel and distributed fashion and made available to users through Web services technology.

1.5 Results

This section presents the results of the proposed algorithm applied to the two real-world optimization problems described in the previous section. For an indication of the performance of the proposed algorithm, a standard steady-state ES not using a metamodel is also implemented for the two optimization problems. This algorithm uses the same representation, objective function, and mutation operator as the proposed metamodel-assisted algorithm.

In Fig. 1.5, results from the buffer allocation optimization are shown. In this experiment, 100 simulations have been performed (where each simulation is the average result of five replications). Figure 1.6 shows results from the production scheduling problem. In this experiment, 1000 simulations have been performed and the presented result is the average of 10 replications of the optimization.

As Figs. 1.5 and 1.6 show, the proposed metamodel-assisted algorithm converges significantly faster than the standard ES for both optimization problems, which indicates the potential of using a metamodel.

1.6 An Improved Offspring Selection Procedure

A possible enhancement of the proposed algorithm would be an improved offspring selection procedure. In the selection of the next offspring to be inserted into the population, a number of different approaches have been proposed in the literature.

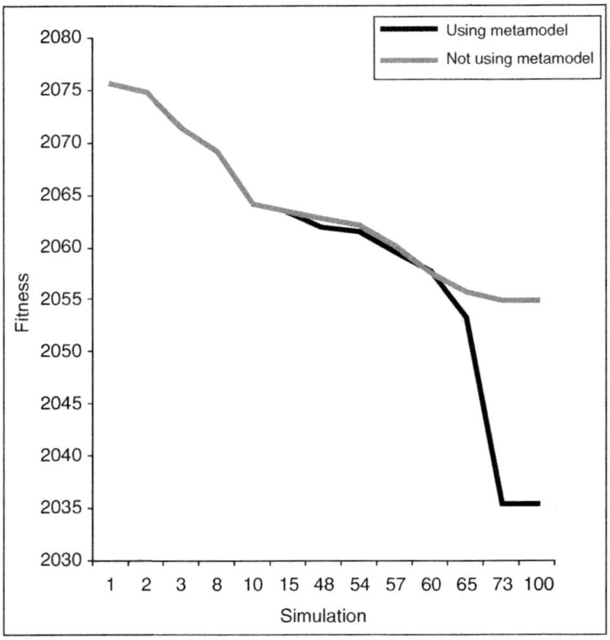

Fig. 1.5 Optimization results for buffer allocation problem

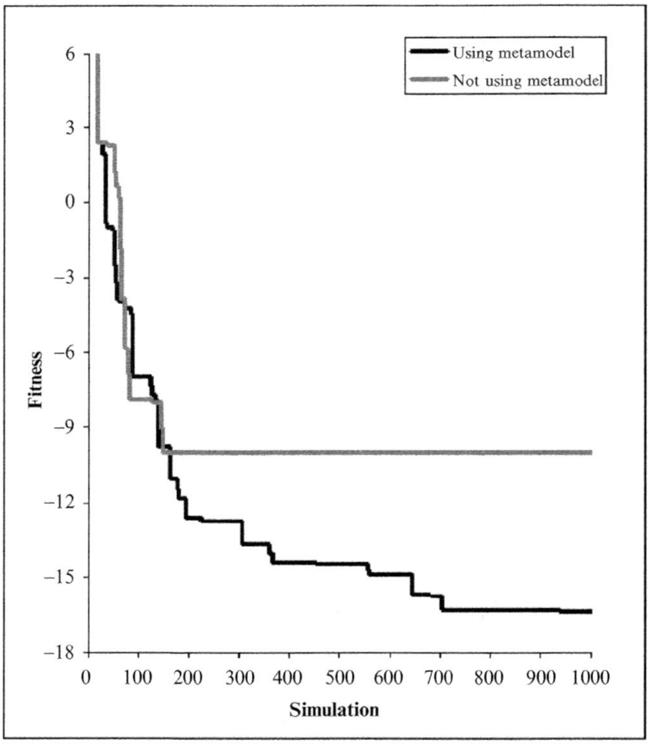

Fig. 1.6 Optimization results for production scheduling problem

The most common approach is to simply select the offspring having the best metamodel fitness. Metamodels in real-world optimization problems are, however, often subject to estimation errors and when these uncertainties are not accounted for, a premature and suboptimal convergence may occur on complex problems with many misleading local optima [12]. Poor solutions might be kept for the next generation and the good ones might be excluded. Optimization without taking the uncertainties into consideration is therefore likely to perform badly [9]. Although this is a well-known problem, the majority of existing metamodel-assisted EAs do not account for metamodel uncertainties.

We suggest a new offspring selection procedure that is aware of the uncertainty in metamodel estimations. In this procedure, the probability of each offspring having the highest simulation fitness among all offspring is quantified and taken into account when selecting the offspring to be inserted into the population. This means that a higher confidence in the potential of an offspring will increase the chances that it is selected.

1.6.1 Overall Selection Procedure

First of all, each offspring is evaluated using the metamodel and assigned a metamodel fitness value. The accuracy of the metamodel is then measured and its estimation error is expressed through an error probability distribution. This distribution, in combination with the metamodel fitness values, is used to calculate the probability of each offspring having the highest simulation fitness (the formulas used for the calculation are presented in the next section). Based on these probabilities, one offspring is chosen using roulette wheel selection to be simulated and inserted into the population.

1.6.2 Formulas for Probability Calculation

The metamodel error is represented by a probability distribution e. This distribution is derived from a list of differences between metamodel fitness value and simulation fitness value for samples in a test set. Based on e, the offspring probabilities are calculated using two functions: f and F.

The function f is a probability distribution over x of the simulation output given a metamodel output o, according to Eq. 1.1.

$$f(o,x) = e(x-o) \tag{1.1}$$

The function F is a cumulative probability distribution for a given metamodel output o, representing the probability that the simulated output would be less than the value of x (in case of a maximization problem), according to Eq. 1.2.

$$F(o,x) = \int_{-\infty}^{x} f(o,t)dt \tag{1.2}$$

Based on the two functions f and F, the probability of an offspring a having the highest simulation fitness among all offspring is calculated according to Eq. 1.3,

$$p(a) = \int_{-\infty}^{\infty} \left(f(a_o,x) \prod^{y \in O, y \neq a} F(y_o,x)dx \right) \tag{1.3}$$

where a_o is the output of offspring a, O is the set of all offspring, and y_o is the output of offspring y.

When probabilities for all offspring have been calculated, one of the offspring is selected using roulette wheel selection based on the probabilities.

1.7 Summary

This chapter presents a new metamodel-assisted EA for the optimization of computationally expensive problems. The algorithm is basically a hybrid of two common EAs: evolution strategies (ESs) and genetic algorithms (GAs). The proposed algorithm is based on a steady-state design, in which one individual at a time is bred and inserted into the population (as opposed to generational approaches in which a whole generation is created at once). A steady-state design is used because it supports a high degree of parallelism, which is a very important efficiency factor when simulations are computationally expensive.

The proposed algorithm is successfully applied to optimize two real-world problems in the manufacturing domain. The first problem considered is about finding optimal buffer levels in a car engine production line, and the second problem considered is about optimal production scheduling in a manufacturing cell for aircraft engines. In both problems, an ANN is used as the metamodel.

Results from the optimization show that the algorithm is successful in optimizing both real-world problems. A comparison with a corresponding algorithm not using a metamodel indicates that the use of metamodels may be very efficient in simulation-based optimization of complex problems.

A possible enhancement of the algorithm in the form of an improved offspring selection procedure that is aware of uncertainties in metamodel estimations is also discussed in the chapter. In this procedure, the probability of each offspring having the highest simulation fitness among all offspring is quantified and taken into consideration when selecting the offspring to be inserted into the population.

References

1. Beyer, H.G., Schwefel, H.P. (2002) Evolution strategies—A comprehensive introduction. *Natural Computing* 1(1), pp. 3–52.
2. Bull, L. (1999) On model-based evolutionary computation. *Software Computing* (3), pp. 76–82.
3. Fonseca, D.J., Navaresse, D.O., Moynihan, G.P. (2003) Simulation metamodeling through artificial neural networks. *Engineering Applications of Artificial Intelligence* 16(3), pp. 177–183.
4. Holland, J.H. (1975) *Adaptation in Natural and Artificial Systems*, University of Michigan Press, Ann Arbor.
5. Hüsken, M., Jin, Y., Sendhoff, B. (2005) Structure optimization of neural networks for evolutionary design optimization. *Source Soft Computing—A Fusion of Foundations, Methodologies and Applications* 9(1), pp. 21–28.
6. Jin, Y., Olhofer, M., Sendhoof, B. (2002) A framework for evolutionary optimization with approximate fitness functions. *IEEE Transactions on Evolutionary Computation* 6(5), pp. 481–494.
7. Khu, S.T., Savic, D., Liu, Y., Madsen, H. (2004) A fast evolutionary-based metamodelling approach for the calibration of a rainfall-runoff model. In: *Proceedings of the First Biennial Meeting of the International Environmental Modelling and Software Society*, pp. 147–152, Osnabruck, Germany.
8. Laguna, M., Marti, R. (2002) Neural network prediction in a system for optimizing simulations. *IEEE Transactions* (34), pp. 273–282.
9. Lim, D., Ong, Y.-S., Lee, B.-S. (2005) Inverse multi-objective robust evolutionary design optimization in the presence of uncertainty. In: *Proceedings of the 2005 Workshops on Genetic and Evolutionary Computation*, pp. 55–62, Washington, DC.
10. Ng, A., Grimm, H., Lezama, T., Persson, A., Andersson, M., Jägstam, M. (2007) Web services for metamodel-assisted parallel simulation optimization. In: *Proceedings of The IAENG International Conference on Internet Computing and Web Services (ICICWS'07)*, March 21–23, pp. 879–885, Hong Kong.
11. Ong, Y.S., Nair, P.B., Keane, A.J., Wong, K.W. (2004) Surrogate-assisted evolutionary optimization frameworks for high-fidelity engineering design problems. In: *Knowledge Incorporation in Evolutionary Computation*, pp. 307–332, Springer, New York.
12. Ulmer, H., Streichert, F., Zell, A. (2003) Evolution strategies assisted by Gaussian processes with improved pre-selection criterion. In: *Proceedings of IEEE Congress on Evolutionary Computation (CEC'03)*, December 8–12, 2003, pp. 692–699, Canberra, Australia.

Chapter 2
Automatically Defined Groups for Knowledge Acquisition from Computer Logs and Its Extension for Adaptive Agent Size

Akira Hara, Yoshiaki Kurosawa, and Takumi Ichimura

2.1 Introduction

Recently, a large amount of data is stored in databases through the advance of computer and network environments. To acquire knowledge from the databases is important for analyses of the present condition of the systems and for predictions of coming incidents. The log file is one of the databases stored automatically in computer systems. Unexpected incidents such as system troubles as well as the histories of daily service programs' actions are recorded in the log files. System administrators have to check the messages in the log files in order to analyze the present condition of the systems. However, the descriptions of the messages are written in various formats according to the kinds of service programs and application software. It may be difficult to understand the meaning of the messages without the manuals or specifications. Moreover, the log files become enormous, and important messages are liable to mingle with a lot of insignificant messages. Therefore, checking the log files is a troublesome task for administrators.

Log monitoring tools such as SWATCH [1], in which regular expressions for representing problematic phrases are used for pattern matching, are effective for detecting well-known typical error messages. However, various programs running in the systems may be open source software or software companies' products, and they may have been newly developed or upgraded recently. Therefore, it is impossible to detect all the problematic messages by the predefined rules. In addition, in order to cope with illegal use by hackers, it is important to detect unusual behavior such as the start of the unsupposed service program, even if the message does not correspond to the error message. To realize this system, the error-detection rules depending on the environment of the systems should be acquired adaptively by means of evolution or learning.

Genetic programming (GP) [2] is one of the evolutionary computation methods, and it can optimize the tree structural programs. Much research on extracting rules from databases by GP has been done in recent years. In the research [3–5], the tree structural program in a GP individual represents an IF-THEN rule. In order to acquire multiple rules, we had previously proposed an outstanding method that united GP with cooperative problem-solving by multiple agents. We called this method automatically defined groups (ADG) [6, 7]. By using this method, we had developed the rule extraction algorithm from the database [8–12]. In this system, two or more rules hidden in the database, and respective rules' importance can be acquired by cooperation of agents. However, we meet a problematic situation when the database has many latent rules. In this case, the number of agents runs short for search and for evaluation of each rule because the number of agents is fixed in advance. In order to solve this problem, we have improved ADG so that the method can treat the variable number of agents. In other words, the number of agents increases adaptively according to the acquired rules.

In Sect. 2.2, we explain the algorithm of ADG, and the application to rule extraction from classified data. In Sect. 2.3, we describe how to extract rules from log files by ADG, and show a preliminary experiment using a centralized control server for many client computers. In Sect. 2.4, we describe an issue in the case where we apply the rule-extracting algorithm to a large-scale log file, and then we propose the ADG with variable agent size for solving the problem. We also show the results of experiments using the large-scale log files. In Sect. 2.5, we describe conclusions and future work.

2.2 Rule Extraction by ADG

2.2.1 Automatically Defined Groups

In the field of data processing, to cluster the enormous data and then to extract common characteristics from each cluster of data are important for knowledge acquisition. In order to accomplish this task, we adopt a multiagent approach, in which agents compete with one another for their share of the data, and each agent generates a rule for the assigned data; the former corresponds to the clustering of data, and the latter corresponds to the rule extraction in each cluster. As a result, all rules are extracted by multiagent cooperation. However, we do not know how many rules subsist in the given data and how data should be allotted to each agent. Moreover, as we prepare abundant agents, the number of tree structural programs increases in an individual. Therefore, search performance declines.

In order to solve these problems, we have proposed an original evolutionary method, automatically defined groups. The method is an extension of GP, and it optimizes both the grouping of agents and the tree structural program of each group in the process of evolution. By grouping multiple agents, we can prevent the increase of search space and perform an efficient optimization. Moreover, we can easily analyze agents' behavior group by group. Respective groups play different roles from one another for cooperative problem-solving. The acquired group structure is utilized

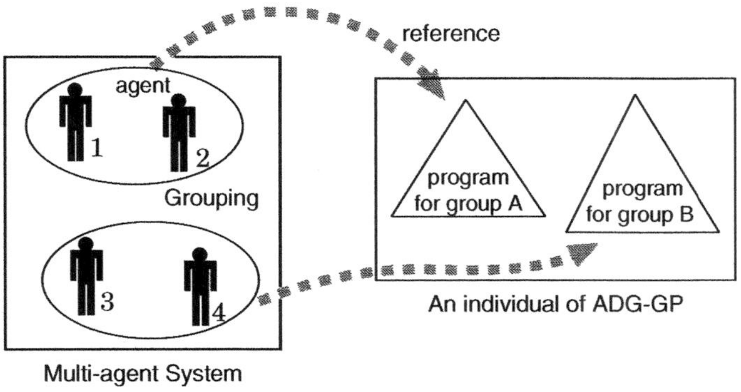

Fig. 2.1 Concept of automatically defined groups

for understanding how many roles are needed and which agents have the same role. That is, the following three points are automatically acquired by using ADG.

- How many groups (roles) are required to solve the problem?
- To which group does each agent belong?
- What is the program of each group?

In the original ADG, each individual consists of a predefined number of agents. The individual maintains multiple trees, each of which functions as a specialized program for a distinct group as shown in Fig. 2.1. We define a group as the set of agents referring to the same tree for the determination of their actions. All agents belonging to the same group use the same program.

Generating an initial population, agents in each GP individual are divided into several groups at random. Crossover operations are restricted to corresponding tree pairs. For example, a tree referred to by agent 1 in an individual breeds with a tree referred to by agent 1 in another individual. This breeding strategy is called restricted breeding [13–15]. In ADG, we also have to consider the sets of agents that refer to the trees used for the crossover. The group structure is optimized by dividing or unifying the groups according to the inclusion relationship of the sets.

The concrete processes are as follows. We arbitrarily choose an agent for two parental individuals. A tree referred to by the agent in each individual is used for crossover. We use T and T' as expressions of these trees, respectively. In each parental individual, we decide a set $A(T)$, the set of agents that refer to the selected tree T. When we perform a crossover operation on trees T and T', there are the following three cases.

(a) If the relationship of the sets is $A(T) = A(T')$, the structure of each individual is unchanged.
(b) If the relationship of the sets is $A(T) \supset A(T')$, the division of groups takes place in the individual with T, so that the only tree referred to by the agents in

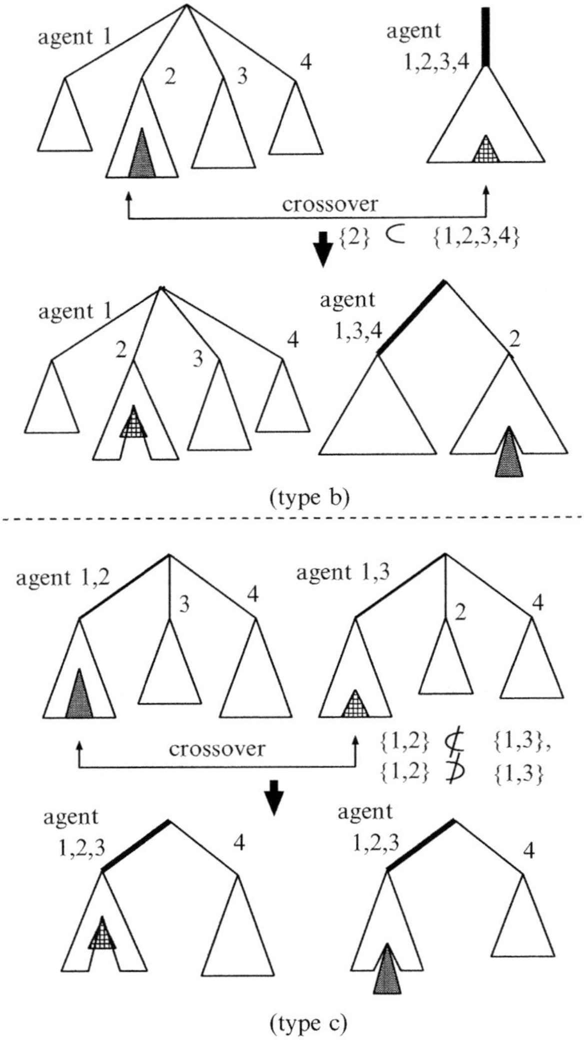

Fig. 2.2 Examples of crossover

$A(T) \cap A(T')$ can be used for crossover. The individual which maintains T' is unchanged. Figure 2.2 (type b) indicates an example of this type of crossover.

(c) If the relationship of the sets is $A(T) \not\supset A(T')$ and $A(T) \not\subset A(T')$, the unification of groups takes place in both individuals so that the agents in $A(T) \cup A(T')$ can refer to an identical tree. Figure 2.2 (type c) shows an example of this crossover.

We expect that the search works efficiently and the adequate group structure is acquired by using this method.

2.2.2 Rule Extraction from Classified Data

In some kinds of databases, each datum is classified into the positive or negative case (or more than two categories). For example, patient diagnostic data in hospitals are classified into some categories according to their diseases. It is an important task to extract characteristics for a target class. However, even if data belong to the same class, all the data in the class do not necessarily have the same characteristics. A part of a dataset might show a different characteristic. It is possible to apply ADG to rule extraction from such classified data. In ADG, multiple tree structural rules are generated evolutionally, and each rule represents the characteristic of a subset in the same class of data. Figure 2.3 shows a concept of rule extraction using ADG. Each agent group extracts a rule for the divided subset, and the rules acquired by multiple groups can cover all the data in the target class. Moreover, when agents are grouped, the load of each agent and predictive accuracy of its rule are considered. As a result, a lot of agents come to belong in the group with the high use-frequency and high-accuracy rule. In other words, we can regard the number of agents in each group as the important degree of the rule. Thus, two or more rules and the important degree of respective rules can be acquired at the same time. This method was applied to medical data and the effectiveness has been verified [8–11].

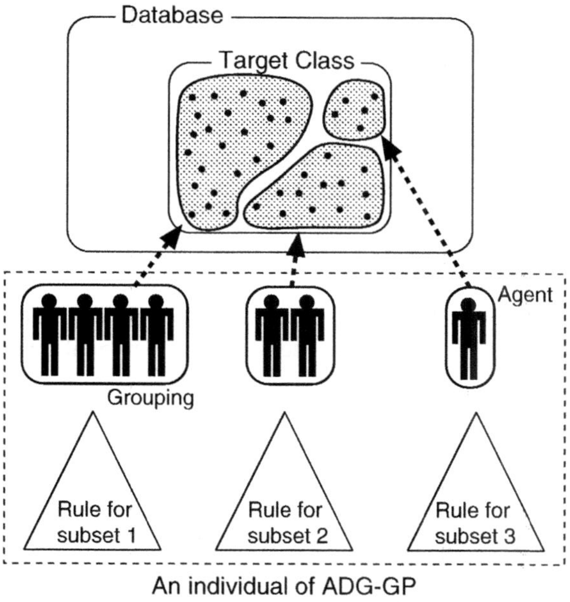

Fig. 2.3 Rule extraction using ADG

2.3 Knowledge Acquisition from Log Files by ADG

2.3.1 How to Extract Rules from Unclassified Log Messages

We apply the rule extraction method using ADG to detect trouble in computer systems from log files. In order to use the method described in the previous section, we need supervised information for its learning phase. In other words, we have to classify each message in the log files into two classes: normal message class and abnormal message class indicating system trouble. However, this is a difficult task because complete knowledge for computer administration is needed and log data are of enormous size. In order to classify log messages automatically into the appropriate class, we consider a state transition pattern of computer system operation. We focus on the following two different states and make use of the difference of the states as the supervised information.

1. Normal state. This is the state in the period of stable operation of the computer system. We assume that the administrators keep good conditions of various system configurations in this state. Therefore, frequently observed messages (e.g., "Successfully access," "File was opened," etc.) are not concerned with the error messages. Of course, some insignificant warning messages (e.g., "Short of paper in printer," etc.) may sometimes appear.
2. Abnormal state. This is the state in the period of unstable operation of the computer system. The transition to the abnormal state may happen due to hardware trouble such as hard disk drive errors, or by restarting service programs with new configurations in the current system. Moreover, some network security attacks may cause the unstable state. In this state, many error messages (e.g., "I/O error," "Access denied," "File not found," etc.) are included in the log files. Of course, the messages observed in the normal state also appear in the abnormal state.

The extraction of rules is performed by using log files in the respective states. First, we define the base period of the normal state, which seems to be stable, and define the testing period, which might be in the abnormal state. Then we prepare the two databases. One is composed of log messages in the normal state period, and the other is composed of log messages in the abnormal state period. By evolutionary computations, we can find rules, which respond to the messages appearing only in the abnormal state.

For knowledge representation to detect a remarkable problematic case, we use the logical expressions, which return true only to such problematic messages. The tagging procedure using regular expressions as described in [16] was used for the preprocessing to the log files and the representation of the rules. Figure 2.4 shows an illustration of the preprocessing. Each message in the log files is separated into several fields (e.g., daemon name field, host name field, comment field, etc.) by the preprocessing, and each field is tagged. Moreover, words that appear in the log messages are registered in the word lists for respective tags beforehand.

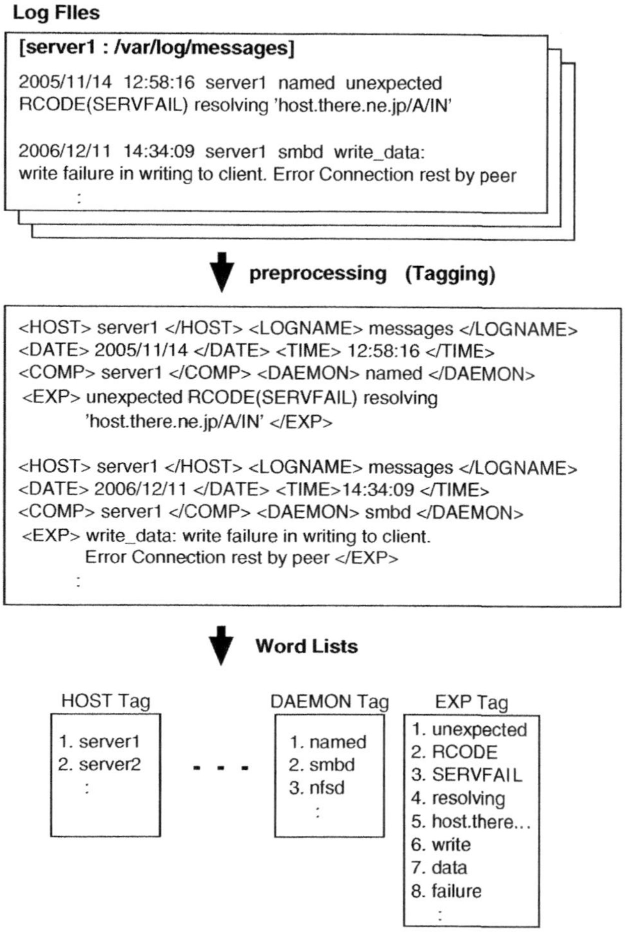

Fig. 2.4 Preprocessing to log files

The rule is made by the conjunction of multiple terms, each of which judges whether the selected word is included in the field of the selected tag. The following expression is an example of the rule.

```
(and (include <DAEMON> 3)(include <EXP> 8))
```

We assume that the word "nfsd" is registered third in the word list for the <DAEMON> tag, and the word "failure" is registered eighth in the word list for the <EXP> tag. For example, this rule returns true to the message including the following strings.

```
<DAEMON>nfsd</DAEMON> <EXP>Warning:access failure</EXP>
```

Multiple trees in an individual of ADG represent the respective logical expressions. Each message in the log files is input to all trees in the individual. Then, calculations are performed to determine whether the message satisfies each logical expression. The input message is regarded as the problematic case if one or more trees in the individual return true. In contrast, the input message is not regarded as the problematic case if all trees in the individual return false. Therefore, all the rules should return false to the messages that appear in both states.

The fitness is calculated based on the accuracy for error detection and load balancing among agents. High accuracy for error detection means that the rules detect as many messages as possible in the abnormal state and react to as few messages as possible in the normal state. The concept of each agent's load arises from the viewpoint of cooperative problem-solving by multiple agents. The load is calculated from the adopted frequency of each group's rule and the number of agents in each group. The adopted frequency of each rule is counted when the rule returns true to the messages in the abnormal state log. If multiple trees return true for a message, the frequency of the tree with more agents is counted. When the agent a belongs to the group g, the load of the agent w_a is defined as follows,

$$w_a = \frac{f_g}{n^g_{\text{Agent}}}, \tag{2.1}$$

where n^g_{Agent} represents the number of agents that belong to group g, and f_g represents the adopted frequency of g. For balancing every agent's load, the variance of the loads V_w as shown in (2.2) should be minimized.

$$V_w = \frac{1}{N_{\text{Agent}}} \left\{ \sum_{i=1}^{N_{\text{Agent}}} (\bar{w} - w_i)^2 \right\}, \tag{2.2}$$

$$\bar{w} = \frac{1}{N_{\text{Agent}}} \sum_{i=1}^{N_{\text{Agent}}} w_i, \tag{2.3}$$

where N_{Agent} represents the number of agents in the individual. By load balancing, more agents are allotted to the group that has a greater frequency of adoption. On the other hand, the number of agents in the less-adopted group becomes small. Therefore, the number of agents of respective rules indicates how general each rule is for the detection of problematic messages. Moreover, when usual messages in the normal state are judged to be problematic messages through a mistake of a rule, it is considered that the number of agents who support the rule should be small. To satisfy the requirements mentioned above, we maximize the fitness f defined as follows.

$$f = \frac{H_{\text{Abn}}/N_{\text{Abn}}}{H_{\text{Norm}}/N_{\text{Norm}}} - \beta \frac{\sum_{N_{\text{Norm}}} \text{fault\_agent}}{H_{\text{Norm}} \times N_{\text{Agent}}} - \delta V_w. \tag{2.4}$$

In this equation, N_{Abn} and N_{Norm} represent the number of messages in the abnormal state and normal state, respectively. H_{Abn} and H_{Norm} represent the frequency that one or more trees in the individual return true for abnormal state logs and normal state logs, respectively. fault_agent represents the number of agents who support the wrong rule, when the rule returns true for messages in the normal state. Therefore, the second term represents the average rate of agents who support the wrong rules when misrecognition occurs. By this term, the allotment of agents to a rule with more misrecognition will be restrained. By the third term, load balancing of agents will be achieved. In addition, in order to inhibit the redundant division of groups, the fitness value is modified according to the number of groups, G $(G \geq 1)$, in the individual as follows,

$$f \leftarrow \gamma^{G-1} \times f \quad (0 < \gamma < 1), \tag{2.5}$$

where γ represents the discount rate for the fitness. This equation means that the fitness is penalized according to the increase of G.

By evolution, one of the multiple trees learns to return true for problematic messages that appear only in the abnormal state logs, and all trees learn to return false for normal messages that appear both in the normal and abnormal state logs. Moreover, agents are allotted to respective rules according to the adopted frequency and the low rate of misrecognition. Therefore, the rule with more agents is the typical and accurate error-detection rule, and the rule with less agents is a specialized rule for the rare case.

2.3.2 Preliminary Experiment

In order to examine the effectiveness of the rule extraction method, we apply the method to the log files in an actual computer environment. As the actual log files, the logs in the centralized control server for many client computers are used. The server can apply management operations such as boot or shutdown to client machines all at once. The numbers of messages included in log files, N_{Norm} and N_{Abn}, are 2992 and 728, respectively.

The parameter settings are as follows: population size is 300. The number of agents in each individual at initial population is 50. The respective weights in (2.4) and (2.5) are $\beta = 0.0001$, $\delta = 0.0001$, and $\gamma = 0.9999$. These parameter values were determined by preliminary trials.

As the result of a tagging procedure using regular expressions, six kinds of tags (HOST, LOGNAME, SORT, FUNC, EXP, and DATA) are attached to the messages in the log files. When we make word lists for respective tags, the largest word list size is 81 for the EXP tag. Figure 2.5 illustrates the generated word lists for respective tags.

Table 2.1 shows GP functions and terminals for these experiments. We impose constraints on the combination of these symbols, such as strongly typed genetic programming [17]. For example, terminal symbols do not enter directly in the arguments of the and function. Crossovers and mutations that break the constraints are not performed. In addition, the sizes of word lists for respective tags are different from one another. Therefore, the value in the second argument of the `include` function may exceed the size of the word list for the corresponding tag. In that case, the remainder in dividing the value by the word list size is used for the selection of a word.

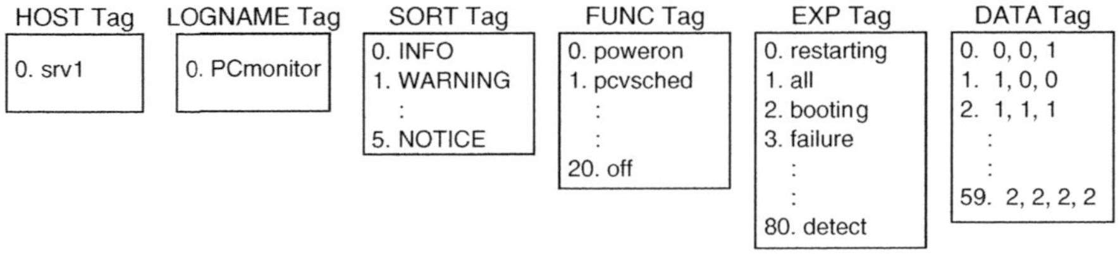

Fig. 2.5 Word lists for the centralized control server's log

Table 2.1 GP functions and terminals

Symbol	#Args	Functions
`and`	2	arg0 ∧ arg1
`include`	2	If Tag arg0 includes arg1 (Word) then T else F
<HOST>,<EXP>,...	0	Tag name
`0,... ,N-1`	0	Select corresponding word from word list. N is the number of words in list.

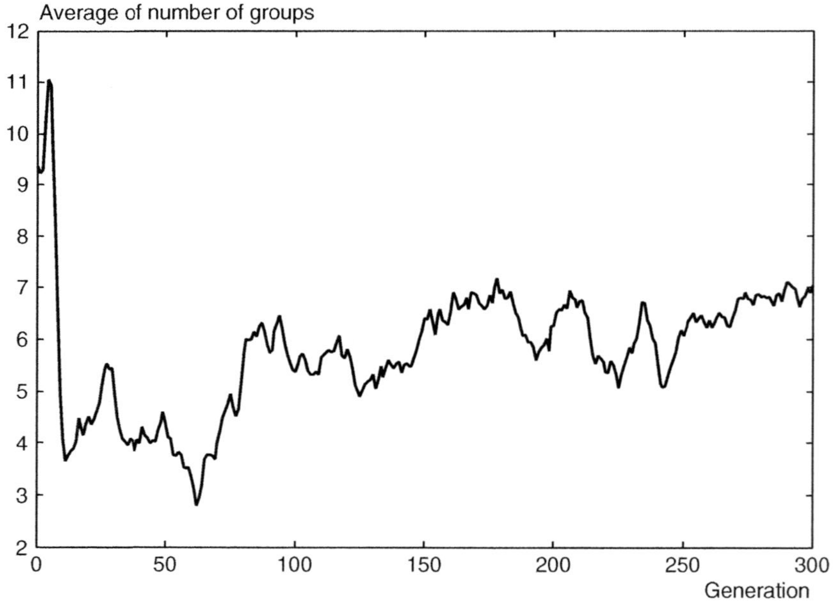

Fig. 2.6 Change of the average number of groups

As the result of a evolutionary optimization, multiple rules, which respond to the messages appearing only in the abnormal state, were acquired successfully. Figure 2.6 illustrates the change of the average number of groups. The number of groups corresponds to the number of extracted rules. As a result, 50 agents in the best individual were divided into eight groups. The best individual detected 157

```
Rule1 (42agents): (and (and (include <LOGNAME> PCmonitor)
           (include <HOST> srv1))(and (include <EXP> booting)
           (include <SORT> INFO)))

Rule2 ( 2agents): (include <DATA> 2,0,2,2,2,2,2,2,2,2,2,2,2,2,2,
                   2,2,2,2,2,2,2,2,2,2,2,2,2,2,2,2,2,2,2,2,2)

Rule3 ( 1agent)  : (include <EXP> Ftrans)

Rule4 ( 1agent)  : (and (include <LOGNAME> PCmonitor)
          (include <EXP> dst=/usr/aw/maya7.0/scripts/startup))

Rule5 ( 1agent)  : (include <DATA> 2,2,2,2,2,2,2,2,2,2,2,2,2,2,2,
                   2,2,2,2,2,2,2,2,2,2,0,2,0,2,2,2,2,2,2,2,2)

Rule6 ( 1agent)  : (and (include <EXP> rexec)
                        (include <LOGNAME> PCmonitor))

Rule7 ( 1agent)  : (include <EXP> Rexec)

Rule8 ( 1agent)  : (include <DATA> 2,2,2,2,2,2,2,2,2,2,2,2,2,2,2,
                   2,2,2,2,2,2,2,2,2,2,0,2,0,2,0,2,0,2,0,2,0)
```

Fig. 2.7 Acquired rules for the centralized control server

```
<HOST>srv1</HOST> <LOGNAME>PCmonitor</LOGNAME>
<SORT>INFO[0]</SORT>
<FUNC>ftrans(470).boot_request_ftrans</FUNC>
<EXP>pc300 Already booting</EXP>

<HOST>srv1</HOST> <LOGNAME>PCmonitor</LOGNAME>
<SORT>INFO[0]</SORT><FUNC>poweron(31968).main</FUNC>
<EXP>322ALL</EXP> <DATA>2,0,2,2,2,2,2,2,2,2,2,2,2,2,2,
2,2,2,2,2,2,2,2,2,2,2,2,2,2,2,2,2,2,2,2,2</DATA>
```

Fig. 2.8 Examples of detected messages by the acquired rules

messages as problematic ones from the abnormal state log. On the other hand, the best individual detected no messages in the normal log.

Figure 2.7 shows the acquired rules in the best individuals. These rules are arranged according to the number of agents, and the second arguments of the `include` function are replaced by the corresponding words. Figure 2.8 shows examples of detected messages by the acquired rules. For example, the word "2, 0, 2, ..., 2" in the second rule in Fig. 2.7 represents a part of an output message, which is returned when performing the "power on" operation for all the client machines. According to an system administration expert, the values delimited by a comma correspond to the conditions of respective machines, and the value "0" means the occurrence of a failure of the operation. Thus, the proposed method can extract error-detection rules without expertise by utilizing the characteristic that such error messages do not appear in the normal state.

In addition, the first rule detects a large part of problematic messages (136 messages) from the abnormal state log. That is, the first rule can be considered as the typical error-detection rule for the environment. As shown in Fig. 2.7, the great mass of agents is assigned to the search of the rule. Thus, the number of agents in each group represents the important degree of the corresponding rule.

2.4 ADG with Variable Agent Size

2.4.1 Issue in Applying ADG to Large-Scale Logs

As mentioned in the previous section, we can extract problematic messages from actual log files by ADG without expertise on administration. As another experiment, we applied the method to large-scale log files. The log files used in the experiment were picked up by a network file server for many client computers. The numbers of messages included in log files N_{Norm} and N_{Abn} are 32,411 and 17,561, respectively. As a result, 44 rules were extracted by ADG using 50 agents as shown in Fig. 2.9. This result indicates an issue of this method. Only one agent was allotted for most rules because the number of prepared agents is insufficient. Even the rule with the maximum number of agents has only three agents. It was impossible to understand minutely the difference of the importance of the rules.

The above results show that the capability of evaluating the important degree of each rule is correlated with the number of agents in an individual. Each agent group extracts one rule. Therefore, the maximum of extractable rules is equal to the number of agents. We have to use as many agents as the supposed rules at least. Moreover, the rules' importance can be judged by the number of agents. Therefore, to evaluate rules in detail, more agents are needed so that the number of agents can exceed the extracted number of rules sufficiently. In short, the problem is that the number of agents is not sufficient to manage the rules. However, it is impossible to estimate the extracted number of rules because it is difficult to set a large enough number of agents beforehand.

2.4.2 ADG with Variable Agent Size

In order to solve the problem on the number of agents, we set that the number of agents dynamically increases to be more than multiples of the number of the acquired rules. The procedures for increasing agents are as follows.

In the best individual of each generation t, we find N^t_{Rules}, the number of rules that return true for problematic messages. When the number of agents in each individual

```
Rule1 ( 3agents): (include <DAEMON> smbd)
Rule2 ( 3agents): (include <EXP> race)
Rule3 ( 2agents): (include <EXP> nrpc)
Rule4 ( 1agent) : (include <EXP> NOTLIB)
 :
 :
Rule42( 1agent) : (include <DAEMON> gdm(pam_unix))
Rule43( 1agent) : (include <EXP Connection)
Rule44( 1agent) : (include <EXP> I/O)
```

Fig. 2.9 Acquired rules for a large-scale log files

at the generation t is N^t_{Agents}, the condition for increasing agents is expressed as follows,

$$N^t_{\text{Agents}} < kN^t_{\text{Rules}} \quad (k \geq 1.0), \tag{2.6}$$

where k is the parameter for controlling the agent size. When this condition is satisfied, the number of agents in each individual is incremented by one. The flow of the evolutionary process is shown below.

(a) Initialization of individuals.
(b) Fitness evaluation of each individual.
(c) Genetic operations
(Selection + Elitist Strategy, Crossover, Mutation).
(d) Operation for increasing agents.
(We find the number of rules and agents in the best individual. If the condition for increasing agents is satisfied, one agent is added to respective individuals.)
(e) If termination condition is not satisfied, return to (b).

When the number of extracted rules increases to N_{Rules}, the number of agents finally reaches kN_{Rules} by the above operations.

2.4.3 Experiments for Large-Scale Logs

2.4.3.1 Comparison Among Variable/Fixed Agent Size Methods

We apply the proposed method to rule extraction from large-scale log files, where the problem concerning the number of agents previously occurred as described in Sect. 2.4.1. We set the number of agents in each individual at initial population at 50, and set the parameter k in the condition for increasing agents at 3.0. For comparison with fixed large size of agents, we also perform another experiment using ADG with fixed 200 agents in each individual.

The parameter settings are as follows: population size is 300. The respective weights in (2.4) and (2.5) are $\beta = 0.001$, $\delta = 0.01$, and $\gamma = 0.9999$. These parameter values were determined by preliminary experiments.

As a result, agents in the best individual were divided into 72 groups by the proposed method. That is, we could get 72 rules. The number of agents was 216 at the last generation. Figure 2.10 shows the best fitness curves by the conventional ADG [16] using fixed 50 or 200 agents and the proposed ADG with variable agent size. We can see from this figure that the search of any method converged by 1000 generations, and the proposed method got better fitness value than the conventional fixed agent size methods. Figure 2.11 shows the change of the number of extracted rules and agents by the proposed method. We can see from this figure that 50 agents are enough for search upto 127 generations, but after the generation the number of agents increases according to the number of rules so as not to be short.

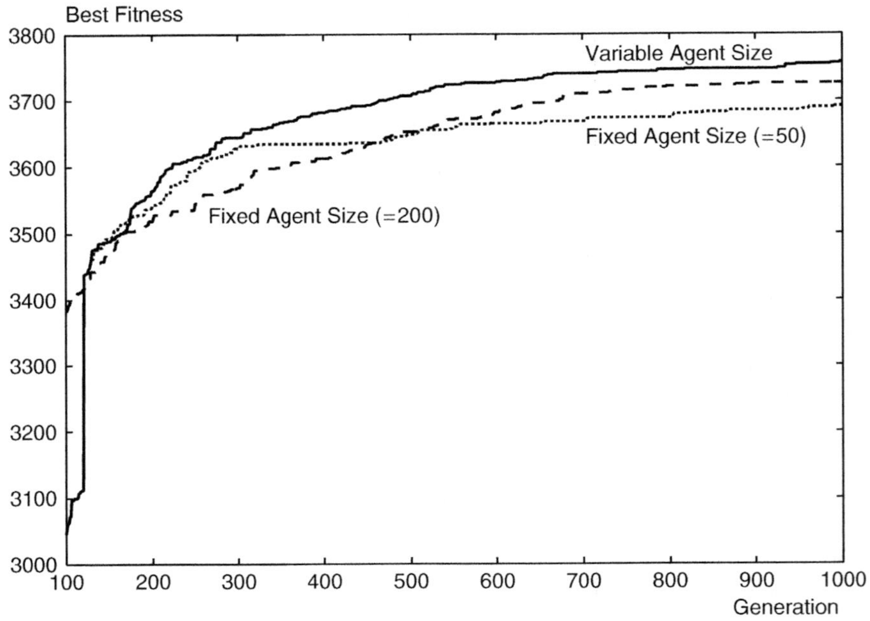

Fig. 2.10 Comparison of the best fitness curves

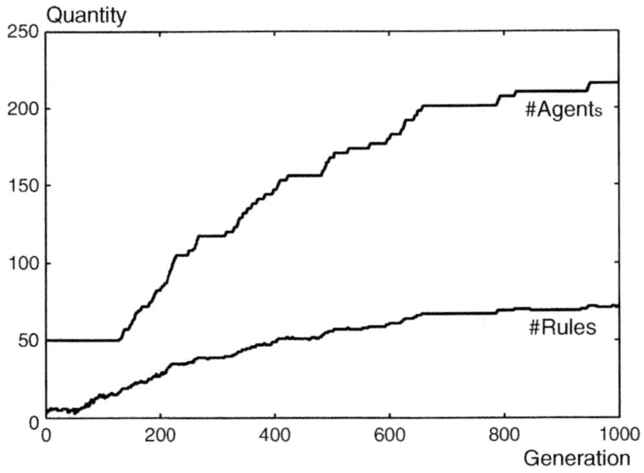

Fig. 2.11 The number of extracted rules and agents

When the number of the extracted rules converged at about the 950th generation, the number of agents also converged. Table 2.2 shows some of the acquired rules by the conventional method and proposed method. Respective rules correspond to the tree structural programs in the best individual. Table 2.2 also shows the number of agents of each rule. These rules are arranged according to the number of agents.

Table 2.2 Some acquired rules and the number of agents

ID	Rule	#Agents fixed (50)	#Agents variable
1	(Include DAEMON smbd)	3	15
2	(Include EXP race)	3	12
3	(Include EXP nrpc)	2	14
4	(Include EXP NOTLB)	1	9
⋮			
42	(Include DAEMON gdm(pam_unix))	1	2
43	(Include EXP Connection)	1	1
44	(Include EXP I/O)	1	1
⋮			
72	(Include EXP Journalled)	–	1

```
<HOST> fsv </HOST> <LOGNAME> messages </LOGNAME>
<DATE>2006/04/19</DATE> <TIME>14:15:37</TIME>
<COMP> fsv </COMP> <DAEMON> smbd </DAEMON>
<EXP> decode_pac_date: failed to verify PAC server
 signature </EXP>

<HOST> fsv </HOST> <LOGNAME> messages </LOGNAME>
<DATE>2006/04/19</DATE> <TIME>14:34:09</TIME>
<COMP> fsv </COMP> <DAEMON> smbd </DAEMON>
<EXP> write_data: write failure in writing to
client. Error Connection reset by peer </EXP>

<HOST> fsv </HOST> <LOGNAME> messages </LOGNAME>
<DATE>2006/04/19</DATE> <TIME>16:43:30</TIME>
<COMP> fsv </COMP> <DAEMON> kernel </DAEMON>
<EXP> I/O error: dev 08:f0, sector 0 </EXP>
```

Fig. 2.12 Log messages detected by the acquired rules

In the conventional method, the number of agents for each rule ranges from one to three, and most rules (rule ID 4, 5, ..., and 44) have only one agent. Therefore, we cannot list the rules in order of importance. In the proposed method, the number of agents for each rule ranges widely from 1 to 15 by the increase of agents. This result shows that the proposed method is useful for the minute evaluation of the importance of the respective rules.

Furthermore, the number of acquired rules by the proposed method is 72. On the other hand, the number of rules by the conventional method is 44. That is, we can get more rules by using a variable number of agents. This result indicates that the search performance of the proposed method becomes better with the increase of agents, and the new rule can be acquired. Therefore, the proposed method shows a higher fitness value than the conventional method as shown in Fig. 2.10. Examples of log messages detected by the acquired rules are shown in Fig. 2.12.

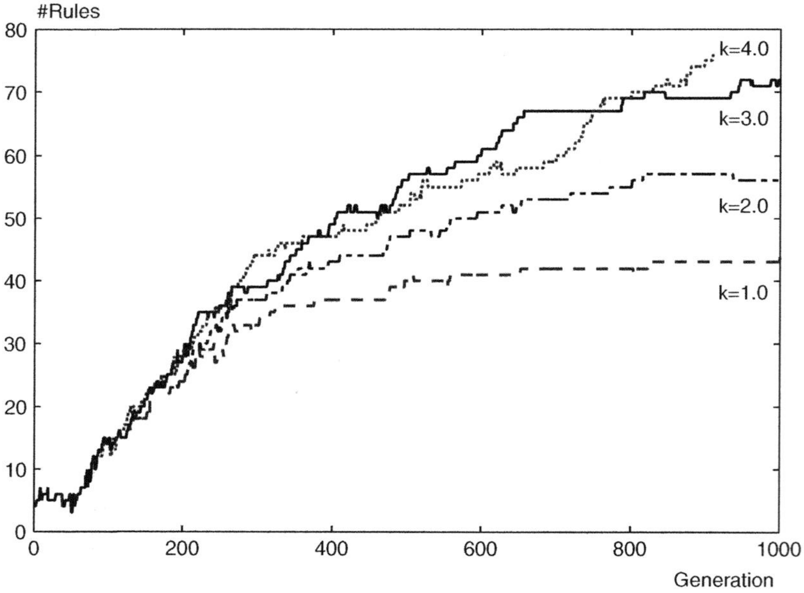

Fig. 2.13 The number of extracted rules in various k

2.4.3.2 Effect of Parameter for Agent Size

In order to examine the effect of the parameter k used in condition (2.6) for increasing agents, we perform experiments with various values of k. Figure 2.13 shows the number of extracted rules of four kinds of k ($k = 1.0, 2.0, 3.0$, and 4.0). We can see from this figure that 44 rules, 56 rules, 72 rules, and 75 rules were extracted, respectively. In the case of $k = 1.0$, the condition (2.6) for increasing agents had not been satisfied in the evolutionary process, and the number of agents remains 50 for the initial individual. As the value of k increases from 1.0 to 3.0, the number of rules also increases. That is, the more agents are used for evolutionary search, the more rules can be acquired. However, the number of acquired rules in $k = 4.0$ is almost the same as that in $k = 3.0$. To determine the adequate value of k adaptively is one of the challenges for improving the method.

2.5 Conclusions and Future Work

In this research, the mechanism where the number of agents increases in proportion to the number of discovered rules was introduced to the ADG method. As a result, two good effects were observed. One effect is that it becomes possible to evaluate the importance of respective rules in detail, and the other is that the number of extracted rules increases. In these experiments, we set that the number of agents should be more than three times the number of extracted rules. We have to examine an appropriate criterion for increasing the number of agents. Moreover, in the present fitness function, agents are allotted to respective rules from the viewpoints of load balancing of agents and from the viewpoint of the decrease of agents who support the wrong rules. As a result, the number of agents becomes an index of the importance for each rule, in which both the frequency of use and accuracy are considered. However, when log information is treated, not only the occurrence frequency but also the degree of the influence on computer systems becomes important. We have to investigate the way to introduce other viewpoints (e.g., risk or urgency level, etc.) into the fitness function, so that the number of agents can become a more profitable index.

References

1. SWATCH: The Simple WATCHer of Logfiles. (2007) http://swatch.sourceforge.net/
2. J.R. Koza (1992) *Genetic Programming – On the Programming of Computers by Means of Natural Selection*. The MIT Press
3. C.C. Bojarczuk, H.S. Lopes, A.A. Freitas (2000) Genetic programming for knowledge discovery in chest-pain diagnosis. *IEEE Engineering in Medicine and Biology*. Vol. 19, No. 4, pp. 38–44
4. C.C. Bojarczuk, H.S. Lopes, A.A. Freitas (2003) An innovative application of a constrained-syntax genetic programming system to the problem of predicting survival of patients. In: *Proceedings of Euro GP 2003*. pp. 11–21
5. L. Hirsch, M. Saeedi, R. Hirsch (2005) Evolving rules for document classification. In: *Proceedings of Euro GP 2005*. pp. 85–95
6. A. Hara, T. Nagao (1999) Emergence of cooperative behavior using ADG; Automatically defined groups. In: *Proceedings of the 1999 Genetic and Evolutionary Computation Conference*. pp. 1039–1046
7. A. Hara, T. Nagao (2002) Construction and analysis of stock market model using ADG; Automatically defined groups. *International Journal of Computational Intelligence and Applications (IJCIA)*. Vol. 2, No. 4, pp. 433–446
8. A. Hara, T. Ichimura, K. Yoshida (2005) Discovering multiple diagnostic rules from coronary heart disease database using automatically defined groups. *Journal of Intelligent Manufacturing*. Vol. 16, No. 6, pp. 645–661
9. A. Hara, T. Ichimura, T. Takahama, Y. Isomichi (2004) Discovery of cluster structure and the clustering rules from medical database using ADG; Automatically defined groups. In: T. Ichimura and K. Yoshida (eds) *Knowledge-Based Intelligent Systems for Healthcare*. pp. 51–86, CRC Press
10. T. Ichimura, S. Oeda, M. Suka, A. Hara, K.J. Mackin, K. Yoshida (2005) Knowledge discovery and data mining in medicine. In: N. Pal and L.C. Jain (eds) *Advanced Techniques in Knowledge Discovery and Data Mining*. pp. 177–210, Springer
11. A. Hara, T. Ichimura, T. Takahama, Y. Isomichi (2005) Extraction of risk factors by multi-agent voting model using automatically defined groups. In: *Proceedings of the Ninth Conference on Knowledge-Based Intelligent Information and Engineering Systems (KES'2005)*. Vol. 3, pp. 1218–1224
12. A. Hara, T. Ichimura, T. Takahama, Y. Isomichi (2003) Extraction of rules by heterogeneous agents using automatically defined groups. In: *Proceedings of the Seventh Conference on Knowledge-Based Intelligent Information and Engineering Systems (KES'2003)*. Vol. 2, pp. 1405–1411

13. S. Luke, L. Spector (1996) Evolving teamwork and coordination with genetic programming. In: *Genetic Programming 1996: Proceedings of the First Annual Conference*. pp. 150–156
14. H. Iba (1996) Emergent cooperation for multiple agents using genetic programming. In: *Parallel Problem Solving from Nature IV. Proceedings of the International Conference on Evolutionary Computation*. pp. 32–41
15. H. Iba (1997) Multiple-agent learning for a robot navigation task by genetic programming. In: *Genetic Programming 1997: Proceedings of the Second Annual Conference*. pp. 195–200
16. Y. Kurosawa, A. Hara, T. Ichimura, Y. Kawano (2006) Extraction of error detection rules without supervised information from log files using automatically defined groups. In: *Proceedings of The 2006 IEEE International Conference on System, Man and Cybernetics*. pp. 5314–5319
17. T. Haynes, R. Wainwright, S. Sen, D. Schoenefeld (1995) Strongly typed genetic programming in evolving cooperation strategies. In: *Genetic Algorithms: Proceedings of the Sixth International Conference (ICGA95)*. pp. 271–278

Chapter 3
Robust Hybrid Sliding Mode Control for Uncertain Nonlinear Systems Using Output Recurrent CMAC

Chih-Min Lin, Ming-Hung Lin, and Chiu-Hsiung Chen

3.1 Introduction

In recent years, the sliding-mode control (SMC) theory has been widely and successfully applied in different systems such as control systems, power systems, and biped robotics, among others [12, 13]. For control systems the salient features of SMC techniques are fast convergence, external disturbance rejection, and strong robustness [12]. However, the uncertainty bound in SMC may not be easily obtained; a large switching control gain is always chosen in order to guarantee system stability. Unfortunately, this causes high-frequency control chattering which may result in unforeseen instability and deteriorates system performance. The uncertainty bound can be estimated via an adaptive algorithm or intelligent approximation tool, for example, neural networks or fuzzy systems. A number of works that proposed an uncertainty bound estimator to reduce control chattering in SMC have been reported in [8, 15, 17].

In the above approaches, the NN-based adaptive algorithms have been incorporated into the sliding-mode controllers which combine the advantages of the sliding-mode control with robust characteristics and the online tuning ability of adaptive algorithms; so that the stability, convergence, and robustness of the system can be ameliorated. However, the parameters of the neural network need to be adjusted for every neuron at each iteration; the above literature suffers complex computational loading. This will restrict the practical applications of these systems. Moreover, they only concern the single-input single-output control systems [8, 15].

Since the introduction of the cerebellar model articulation controller (CMAC) by Albus [1], these biologically inspired models have been widely used. CMAC represents one kind of associative memory technique, which can be used in control system and function approximation. CMAC is classified as a nonfully connected perceptronlike associative memory network with overlapping receptive fields. One advantage of CMAC, compared with multilayer neural networks, is its excellent learning characteristics. CMAC has been adopted widely for the closed-loop control of complex dynamical systems because of its simple computation, good generalization capability, and fast learning property. It has been already validated that CMAC can approximate a nonlinear function over a domain of interest to any desired accuracy [7].

The advantages of using CMAC over NN in many practical applications have been presented in recent literature [4–6, 14]. However, the conventional CMAC uses local constant binary receptive-field basis functions. The disadvantage is that its output is constant within each quantized state and the derivative information is not preserved. For acquiring the derivative information of input and output variables, Chiang and Lin developed a CMAC network with a nonconstant differentiable Gaussian receptive-field basis function, and provided the convergence analyses of this network [3]. Based on this concept, some researchers have utilized the CMACs with a Gaussian receptive-field basis function to control nonlinear systems [9, 10]. However, the major drawback of these CMACs is that their application domain is limited to static problems due to their inherent network structure.

In order to resolve the static CMAC problem, eliminate the chattering phenomenon, and preserve the main advantage of SMC, a robust hybrid sliding-mode control (RHSMC) system is proposed to control the multi-input multi-output (MIMO) uncertain nonlinear systems. In this system a dynamic output recurrent cerebellar model articulation controller (ORCMAC), which includes the delayed recurrent units from the output space to input space, is used to estimate the unknown modeling uncertainties.

This hybrid control system consists of three parts. The first one is a main controller that contains a dynamic ORCMAC modeling uncertainty estimator. The second one is a compensation controller; it is used to compensate for the difference between modeling uncertainty and its estimated value. In this compensation controller, a bound estimation mechanism is utilized to observe the approximation error bound so that the chattering phenomenon of the control effort can be eliminated. The other is an H_∞ robust controller that is designed to attenuate the effect of external disturbances to a prescribed level. The Lyapunov stability theory and H_∞ control technique are utilized to derive the control algorithms so that the stability of the system can be guaranteed and robust tracking performance can be achieved. Finally, the proposed design method is applied to control a MIMO nonlinear nine-link biped robot to illustrate its effectiveness.

This study is organized as follows. Problem formulation is described in Sect. 3.2. The proposed RHSMC scheme is constructed in Sect. 3.3. In Sect. 3.4, the simulation results of the proposed RHSMC for a nonlinear biped robot system are presented. Conclusions are drawn in Sect. 3.5.

3.2 Problem Formulation

Consider an nth-order multi-input multi-output uncertain nonlinear system expressed in the following form.

$$\begin{cases} \boldsymbol{x}^{(n)}(t) = \boldsymbol{F}(\boldsymbol{X}(t)) + \boldsymbol{G}(\boldsymbol{X}(t))\boldsymbol{u}(t) + \boldsymbol{d}(t) \\ \boldsymbol{y} = \boldsymbol{x}(t) \end{cases} \tag{3.1}$$

where $\boldsymbol{x}(t) \in \Re^m$ is a state vector, and $\boldsymbol{F}(\boldsymbol{X}(t)) \in \Re^m$ and $\boldsymbol{G}(\boldsymbol{X}(t)) \in \Re^{m\times m}$ are nonlinear uncertain functions assumed to be bounded; $\boldsymbol{u}(t) \in \Re^m$ and $\boldsymbol{y} \in \Re^m$ are control inputs and system outputs, respectively, in which m is the number of system inputs and outputs; $\boldsymbol{d}(t) \in \Re^m$ is an unknown but bounded external disturbance; and $\boldsymbol{X}(t) = [\boldsymbol{x}^T(t), \dot{\boldsymbol{x}}^T(t), \cdots, \boldsymbol{x}^{(n-1)T}(t)]^T \in \Re^{mn}$ is a state vector of the system which is assumed to be available for measurement. For the system (3.1) to be controllable, it is required that $det\,(\boldsymbol{G}(\boldsymbol{X}(t))) \neq 0$ for all $\boldsymbol{X}(t)$ in a certain controllability region $U_c \in \Re^{mn}$.

Assuming that all the parameters of the system are well known and the external disturbances are absent, the nominal model of the uncertain nonlinear system (1) can be represented as

$$\boldsymbol{x}^{(n)}(t) = \boldsymbol{F}_n(\boldsymbol{X}(t)) + \boldsymbol{G}_n(\boldsymbol{X}(t))\boldsymbol{u}(t) \tag{3.2}$$

where $\boldsymbol{F}_n(\boldsymbol{X}(t))$ and $\boldsymbol{G}_n(\boldsymbol{X}(t))$ are the nominal values of $\boldsymbol{F}(\boldsymbol{X}(t))$ and $\boldsymbol{G}(\boldsymbol{X}(t))$, respectively, and assume $\boldsymbol{G}_n^{-1}(\boldsymbol{X}(t))$ exists. If the modeling uncertainty occurs and external disturbance is included, then the system (3.2) can be modified as

$$\begin{aligned} \boldsymbol{x}^{(n)}(t) &= \boldsymbol{F}_n(\boldsymbol{X}(t)) + \Delta\boldsymbol{F}(\boldsymbol{X}(t)) + [\boldsymbol{G}_n(\boldsymbol{X}(t)) + \Delta\boldsymbol{G}(\boldsymbol{X}(t))]\boldsymbol{u}(t) + \boldsymbol{d}(t) \\ &= \boldsymbol{F}_n(\boldsymbol{X}(t)) + \boldsymbol{G}_n(\boldsymbol{X}(t))\boldsymbol{u}(t) + \alpha \;\; \alpha(\boldsymbol{X}(t)) + \boldsymbol{d}(t) \end{aligned} \tag{3.3}$$

where $\Delta\boldsymbol{F}(\boldsymbol{X}(t))$ and $\Delta\boldsymbol{G}(\boldsymbol{X}(t))$ denote the unknown uncertainties of $\boldsymbol{F}(\boldsymbol{X}(t))$ and $\boldsymbol{G}(\boldsymbol{X}(t))$, respectively; $\alpha(\boldsymbol{X}(t))$ is called the modeling uncertainty and is defined as

$$\alpha(\boldsymbol{X}(t)) = \Delta\boldsymbol{F}(\boldsymbol{X}(t)) + \Delta\boldsymbol{G}(\boldsymbol{X}(t))\boldsymbol{u}(t).$$

Define the tracking error as

$$\boldsymbol{e} \underline{\underline{\Delta}} \boldsymbol{y}_d - \boldsymbol{y} \in \Re^m \tag{3.4}$$

and then the system tracking error vector is defined as

$$\boldsymbol{E} \underline{\underline{\Delta}} [\boldsymbol{e}^T, \dot{\boldsymbol{e}}^T, \cdots, \boldsymbol{e}^{(n-1)T}]^T \in \Re^m \tag{3.5}$$

The control objective is to find a suitable control law so that the state vector $\boldsymbol{X}$(t) can track a specified bounded reference trajectory $\boldsymbol{Y}_d = [\boldsymbol{y}_d^T, \dot{\boldsymbol{y}}_d^T, \cdots, \boldsymbol{y}_d^{(n-1)T}]^T \in \Re^{mn}$.

Define a sliding surface

$$\boldsymbol{S}(t) \underline{\underline{\Delta}} \boldsymbol{e}^{(n-1)} + \boldsymbol{k}_1 \boldsymbol{e}^{(n-2)} + \cdots + \boldsymbol{k}_n \int_0^t \boldsymbol{e}(\tau)d\tau \tag{3.6}$$

where $\boldsymbol{k}_i = diag(\boldsymbol{k}_{i1}, \boldsymbol{k}_{i2}, \ldots, \boldsymbol{k}_{im}) \in \Re^{m\times m}$ is a nonzero positive constant diagonal matrix, which should be chosen to guarantee the stability of the sliding surface.

3.3 Robust Hybrid Sliding-Mode Control System Design

A robust hybrid sliding-mode control (RHSMC) system shown in Fig. 3.1 is proposed to control the MIMO uncertain nonlinear systems. This hybrid control system is composed of three controllers and is defined as

$$\boldsymbol{u}_{RHSMC}(t) = \boldsymbol{u}_m(t) + \boldsymbol{u}_c(t) + \boldsymbol{u}_h(t) \tag{3.7}$$

where $\boldsymbol{u}_m(t)$ is the main controller that contains an ORCMAC modeling uncertainty estimator to approximate online the unknown modeling uncertainty $\boldsymbol{\alpha}$. $\boldsymbol{u}_c(t)$ is a compensation controller used to dispel the approximation error between modeling uncertainty and its estimated value. In this compensation controller, an error-bound estimation mechanism is utilized to adjust the approximation error bound so that the control chattering can be reduced. $\boldsymbol{u}_h(t)$ represents an H_∞ controller that is designed to achieve robust tracking performance and to attenuate the effect of external disturbance to a prescribed level.

3.3.1 ORCMAC Modeling Uncertainty Estimator

An ORCMAC is proposed and shown in Fig. 3.2, in which ΔT denotes a time delay. This ORCMAC is composed of input space, association memory space, receptive-field space, weight memory space, output space, and recurrent weights.

The signal propagation and the basic function in each space are described as follows.

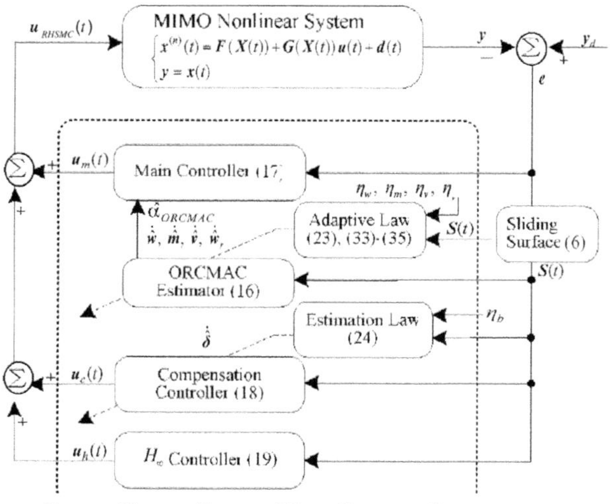

Fig. 3.1 The block diagram of the RHSMC feedback control system

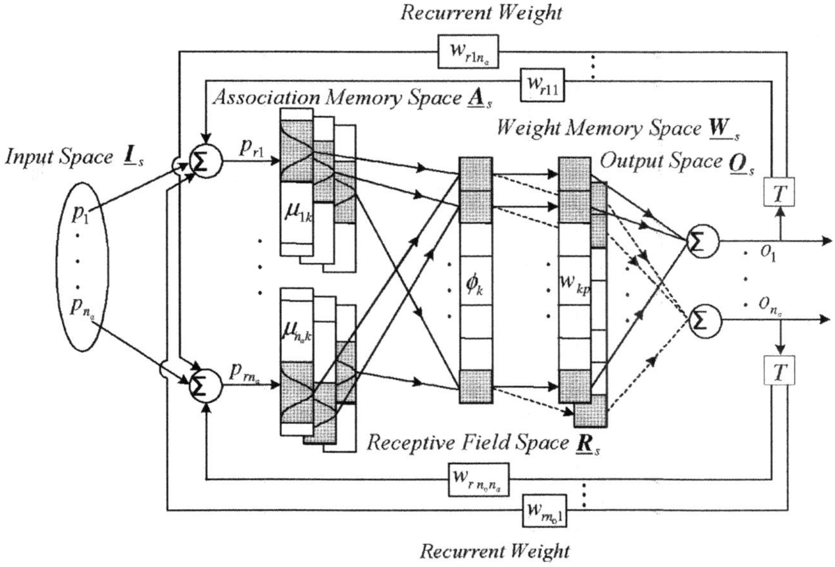

Fig. 3.2 Architecture of an ORCMAC

1. Input space $\underline{I}_s$: For a given $\boldsymbol{I} = [I_1, \cdots, I_j, \cdots, I_{na}]^T \in \Re^{na}$, each input state variable I_j must be quantized into discrete regions (called elements) according to given control space. The number of elements n_e is termed a resolution.
2. Association memory space $\underline{A}_s$: Several elements can be accumulated as a block, the number of blocks is usually greater than or equal to two. $\underline{A}_s$ denotes an association memory space. The inputs of every block $\boldsymbol{I}_r = [I_{r1}, \cdots, I_{rj}, \cdots, I_{rna}]^T \in \Re^{na}$ are represented as

$$\boldsymbol{I}_r = \boldsymbol{I} + \boldsymbol{w}_r \boldsymbol{O}(t - \Delta T) \tag{3.8}$$

where

$$\boldsymbol{w}_r = [\boldsymbol{w}_{r1}, \cdots, \boldsymbol{w}_{rp}, \cdots, \boldsymbol{w}_{rna}] = \begin{bmatrix} w_{r11} & \cdots & w_{rp1} & \cdots & w_{rn_o1} \\ \vdots & \ddots & \vdots & \vdots & \vdots \\ w_{r1j} & \cdots & w_{rpj} & \cdots & w_{rn_oj} \\ \vdots & \vdots & \vdots & \ddots & \vdots \\ w_{r1n_a} & \cdots & w_{rpn_a} & \cdots & w_{rn_on_a} \end{bmatrix} \in \Re^{n_a \times n_o}$$

and w_{rpj} is the recurrent weight from the output space into input space, $\boldsymbol{O}$ represents the output of ORCMAC, n_o denotes the number of the output space and $\boldsymbol{O}(t - \Delta T)$ represents the value of $\boldsymbol{O}$ through delay time ΔT. It is clear that the input of ORCMAC contains the memory term $\boldsymbol{O}(t - \Delta T)$, which stores the past information of the network and presents a dynamic mapping. This is the apparent difference between the proposed ORCMAC and the conventional CMAC. In this

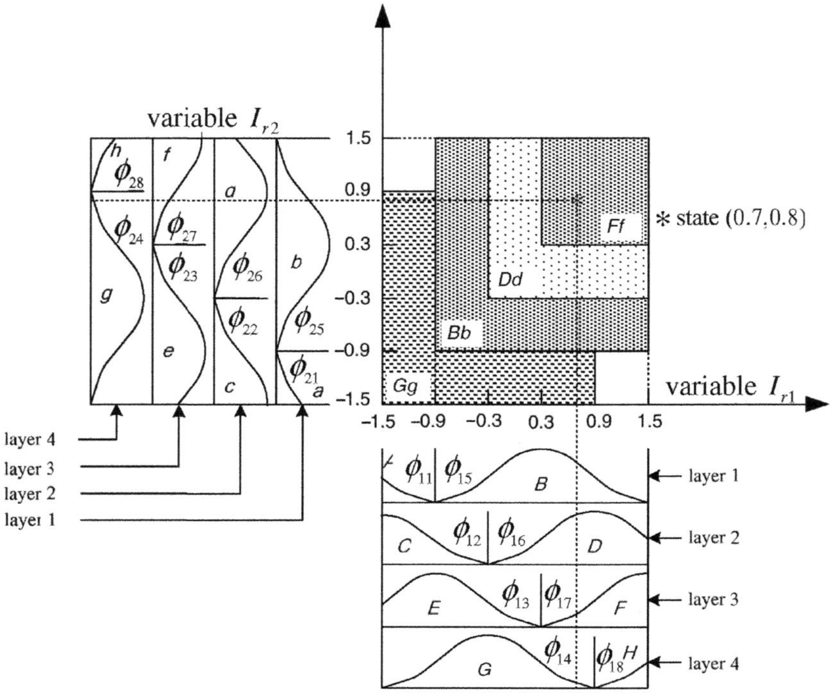

Fig. 3.3 A two-dimensional ORCMAC with $n_f = 4$ and $n_e = 5$

space, each block performs a receptive-field basis function; the Gaussian function is adopted here as the receptive-field basis function, which can be represented as

$$\phi_{jk}(I_{rj}) = exp\left[\frac{-(I_{rj} - m_{jk})^2}{\sigma_{jk}^2}\right], \text{ for } k = 1, 2, \cdots, n_b \tag{3.9}$$

where $\phi_{jk}(I_{rj})$ represents the kth block of the jth input I_{rj} with the mean m_{jk} and variance σ_{jk}. Figure 3.3 depicts the schematic diagram of a two-dimensional ORCMAC with $n_e = 5$ and $n_f = 4$ (n_f is the number of elements in a complete block), in which I_{r1} is divided into blocks A and B, and I_{r2} is divided into blocks a and b. By shifting each variable one element, different blocks will be obtained. For instance, blocks C and D for I_{r1}, and blocks c and d for I_{r2} are possible shifted elements. Each block in this space has two adjustable parameters m_{jk} and σ_{jk}.

3. Receptive-field space $\underline{\boldsymbol{R}}_s$: Areas formed by blocks, named Aa and Bb are called receptive-fields. The kth multidimensional receptive-field function is defined as

$$\gamma_k(\boldsymbol{I}_r, \boldsymbol{m}_k, \boldsymbol{v}_k, \boldsymbol{w}_r) = \prod_{j=1}^{n_a} \phi_{jk} = exp\left[\sum_{j=1}^{n_a} \frac{-(I_{rj} - m_{jk})^2}{\sigma_{jk}^2}\right] \text{ for } k = 1,2,\cdots,n_R \tag{3.10}$$

where $\boldsymbol{m}_k = [m_{1k}, m_{2k}, \cdots, m_{n_ak}]^T \in \Re^{n_a}$ and $\boldsymbol{v}_k = [\sigma_{1k}, \sigma_{2k}, \cdots, \sigma_{n_ak}]^T \in \Re^{n_a}$. The multidimensional receptive-field functions can be expressed in a vector notation as

$$\boldsymbol{\Gamma}(\boldsymbol{I}_r, \boldsymbol{m}, \boldsymbol{v}, \boldsymbol{w}_r) = [\gamma_1, \cdots, \gamma_k, \cdots, \gamma_{n_R}]^T \tag{3.11}$$

where $\boldsymbol{m} = [\boldsymbol{m}_1^T, \boldsymbol{m}_2^T, \cdots, \boldsymbol{m}_k^T, \cdots, \boldsymbol{m}_{n_R}^T]^T \in \Re^{n_a n_R}$ and $\boldsymbol{v} = [\boldsymbol{v}_1^T, \boldsymbol{v}_2^T, \cdots, \boldsymbol{v}_k^T, \cdots, \boldsymbol{v}_{n_R}^T]^T \in \Re^{n_a n_R}$.

In the ORCMAC scheme, no receptive-field is formed by the combination of different layers such as "*A*, *B*" and "*c*, *d*." Therefore, *Cc* and *Dd* are new receptive-fields resulting from different blocks. With this kind of quantization and receptive-field composition, each state is covered by n_f (less than or equal to n_f) different receptive-fields. If the input falls within the *k*th receptive-field, this field becomes active. Nearby inputs can activate one or more of the same n_f weights, which can produce similar outputs. This correlation provides a very useful property of ORCMAC, namely, local generalization.

4. Weight memory space $\underline{\boldsymbol{W}}_s$: Each location of $\underline{\boldsymbol{R}}_s$ to a particular adjustable value in the weight memory space can be expressed as

$$\boldsymbol{w} = [\boldsymbol{w}_1, \cdots, \boldsymbol{w}_p, \cdots, \boldsymbol{w}_{n_o}] = \begin{bmatrix} w_{11} & \cdots & w_{1p} & \cdots & w_{1n_o} \\ \vdots & \ddots & \vdots & \vdots & \vdots \\ w_{k1} & \cdots & w_{kp} & \cdots & w_{kn_o} \\ \vdots & \vdots & \vdots & \ddots & \vdots \\ w_{n_R1} & \cdots & w_{n_Rp} & \cdots & w_{n_Rn_o} \end{bmatrix} \tag{3.12}$$

where $\mathbf{w}_p = [w_{1p}, \cdots, w_{kp}, \cdots, w_{n_Rp}]^T \in \Re^{n_R}$, and w_{kp} denotes the connecting weight value of the *p*th output associated with the *k*th receptive-field. The behavior of storing weight information in ORCMAC is similar to that of the cerebellum of a human, which distributes and stores information on different cell layers.

5. Output space $\underline{\boldsymbol{O}}_s$: The output computation of ORCMAC is the algebraic sum of the activated weights in the weight memory, and is expressed as

$$O_p = \boldsymbol{w}_p^T \boldsymbol{\Gamma}(\boldsymbol{I}_r, \boldsymbol{m}, \boldsymbol{v}, \boldsymbol{w}_r) = \sum_{k=1}^{n_R} w_{kp} \gamma_k, \text{ for } p = 1, 2, \cdots, n_o \tag{3.13}$$

The outputs of ORCMAC can be also expressed in a vector notation as

$$\boldsymbol{O} = [O_1, \cdots, O_p, \cdots, O_{no}]^T = \mathbf{w}^T \Gamma \tag{3.14}$$

In the two-dimensional case shown in Fig. 3.3, the output of ORCMAC is the sum of the value in receptive-fields *Bb, Dd, Ff,* and *Gg*, where the input state is (0.7, 0.8).

3.3.2 Robust Hybrid Sliding-Mode Control

In this study, an ORCMAC modeling uncertainty estimator is designed to approximate the unknown modeling uncertainty. The outputs of ORCMAC are the estimated values of modeling uncertainties. By the universal approximation theorem [16], there exists an optimal ORCMAC estimator α^*_{ORCMAC} to approach the modeling uncertainty $\alpha(x(t))$ such that

$$\alpha(x(t)) = \alpha^*_{ORCMAC}(\boldsymbol{I}_r|\boldsymbol{w}^*) + \boldsymbol{\Delta} = \boldsymbol{w}^{*T}\boldsymbol{\Gamma} + \boldsymbol{\Delta} \tag{3.15}$$

where $\boldsymbol{\Delta} = [\Delta_1, \ldots, \Delta_i, \ldots, \Delta_m]^T \in \Re^m$ denotes an approximation error and $\boldsymbol{w}^*$ is the optimal constant parameter matrix of $\boldsymbol{w}$. The absolute value of Δ_i is assumed to be less than a small positive constant δ_i (i.e., $|\Delta_i| < \delta_i$). However, it is difficult to determine this approximation error bound δ_i, so that an estimation law of this bound is derived in the following. Also, the optimal estimator is unobtainable, thus an online ORCMAC modeling uncertainty estimator is defined as

$$\hat{\alpha}_{ORCMAC}(\boldsymbol{I}_r|\hat{\mathbf{w}}) = [O_1, \cdots, O_p, \cdots, O_{no}]^T = \hat{\mathbf{w}}^T\Gamma \tag{3.16}$$

The proposed hybrid control law is composed of three controllers and is defined as (3.7) where the main controller is given as

$$\boldsymbol{u}_m(t) = \boldsymbol{G}_n^{-1}(\boldsymbol{X}(t))[-\boldsymbol{F}_n(\boldsymbol{X}(t)) + \boldsymbol{y}_d^{(n)} + \boldsymbol{K}^T\boldsymbol{E} - \hat{\alpha}_{ORCMAC}] \tag{3.17}$$

the compensation controller is given as

$$\boldsymbol{u}_c(t) = \boldsymbol{G}_n^{-1}(\boldsymbol{X}(t))\hat{\boldsymbol{\delta}}sgn(\boldsymbol{S}(t)) \tag{3.18}$$

and the H_∞ controller is given as

$$\boldsymbol{u}_h(t) = \boldsymbol{G}_n^{-1}(\boldsymbol{X}(t))[(2\boldsymbol{R}^2)^{-1}(\boldsymbol{R}^2 + \mathbf{I})\boldsymbol{S}(t)] \tag{3.19}$$

In the main controller, the modeling uncertainty is estimated by an ORCMAC. In the compensation controller, $\hat{\boldsymbol{\delta}} = diag(\hat{\delta}_1, \ldots, \hat{\delta}_i, \ldots, \hat{\delta}_m) \in \Re^{m\times m}$ is the estimated value of approximation error bound. In the H_∞ controller, $\boldsymbol{R} = diag(r_1, \ldots, r_i, \ldots, r_m) \in \Re^{m\times m}$ is a specified attenuation constant diagonal matrix for the disturbances.

Substituting (3.7) and (3.17–3.19) into (3.3), yields

$$\begin{aligned}\mathbf{e}^{(n)} + \mathbf{K}^T\mathbf{E} &= [\hat{\mathbf{w}}^T\boldsymbol{\Gamma} - (\mathbf{w}^{*T}\boldsymbol{\Gamma} + \boldsymbol{\Delta}) - \hat{\boldsymbol{\delta}}sgn(\mathbf{S}(t))] - [(2\mathbf{R}^2)^{-1}(\mathbf{R}^2 + \mathbf{I})\mathbf{S}(t) + \mathbf{d}] \\ &= -[\tilde{\mathbf{w}}^T\boldsymbol{\Gamma} + \hat{\boldsymbol{\delta}}sgn(\mathbf{S}(t)) + \boldsymbol{\Delta} + (2\mathbf{R}^2)^{-1}(\mathbf{R}^2 + \mathbf{I})\mathbf{S}(t) + \mathbf{d}] \\ &= \dot{\mathbf{S}}(t)\end{aligned} \tag{3.20}$$

where $\tilde{\boldsymbol{w}} = \boldsymbol{w}^* - \hat{\boldsymbol{w}}$. In case of the existence of $\boldsymbol{d}$, consider a specified H_∞ tracking performance [2]

$$\sum_{i=1}^{m}\int_{0}^{T} s_i^2(t)dt \le \sum_{i=1}^{m} s_i^2(0) + \frac{1}{\eta_w}\sum_{i=1}^{m}\tilde{\boldsymbol{w}}_i^T(0)\tilde{\boldsymbol{w}}_i(0) + \frac{1}{\eta_b}\sum_{i=1}^{m}\tilde{\delta}_i^2(0) + \sum_{i=1}^{m} r_i^2 \int_0^T d_i^2(t)dt \tag{3.21}$$

where η_w and η_b are positive constants; r_i is a prescribed attenuation constant; $\tilde{\delta}_i - \delta_i - \hat{\delta}_i$; and choose $n_o = m$. If the system starts with initial conditions $s_i(0)=0$, $\tilde{\boldsymbol{w}}_i(0) = 0$, $\tilde{\delta}_i(0) = 0$, the H_∞ tracking performance in (3.21) can be rewritten as

$$\sup_{d_i \in L_2[0,T]} \sum_{i=1}^{m}\left(\frac{||s_i||}{||d_i||} \le r_i\right) \tag{3.22}$$

where $\|s_i\|^2 = \int_0^T s_i^2(t)dt$ and $\|d_i\|^2 = \int_0^T d_i^2(t)dt$. This shows that r_i is an attenuation level between the disturbance d_i and system output s_i. If $r_i = \infty$, this is the case of minimum error tracking control without disturbance attenuation. Then, the following theorem can be stated and proven.

Theorem 3.1. *Consider the nth-order MIMO uncertain nonlinear systems represented by (3.1). The hybrid control law is designed as (3.7). The main controller $\boldsymbol{u}_m(t)$ is given in (3.17), which contains a modeling uncertainty estimator given in (3.16), and the adaptation algorithm of ORCMAC is given as (3.23). The compensation controller $\boldsymbol{u}_c(t)$ is designed as (3.18), where the bound estimation algorithm is presented in (3.24). The H_∞ controller is given in (3.19). Then, the desired robust tracking performance in (3.21) can be achieved for the specified attenuation levels r_i, $i = 1, 2, \ldots, m$.*

$$\dot{\hat{\boldsymbol{w}}}_i = -\eta_w s_i(t)\boldsymbol{\Gamma} \quad \text{where} \quad i = 1, 2, \ldots, m \tag{3.23}$$

$$\dot{\hat{\delta}}_i = \eta_b |s_i(t)| \quad \text{where} \quad i = 1, 2, \ldots, m \tag{3.24}$$

Proof. A Lyapunov function candidate is defined as

$$V_{RHSMC}(\boldsymbol{S}(t), \tilde{\boldsymbol{w}}, \tilde{\boldsymbol{\delta}}) = \frac{1}{2}\boldsymbol{S}^T(t)\boldsymbol{S}(t) + \frac{1}{2\eta_w}tr(\tilde{\boldsymbol{w}}^T\tilde{\boldsymbol{w}}) + \frac{1}{2\eta_b}tr(\tilde{\boldsymbol{\delta}}^T\tilde{\boldsymbol{\delta}}) \tag{3.25}$$

where the estimation error matrix is defined as $\tilde{\boldsymbol{\delta}} = \boldsymbol{\delta} - \hat{\boldsymbol{\delta}}$. Taking the derivative of the Lyapunov function with respect to time and using (3.20), yields

$$\begin{aligned}
\dot{V}_{RHSMC}(\boldsymbol{S}(t), \tilde{\boldsymbol{w}}, \tilde{\boldsymbol{\delta}}) &= \boldsymbol{S}^T(t)\dot{\boldsymbol{S}}(t) + \frac{1}{\eta_w}tr(\tilde{\boldsymbol{w}}^T\dot{\tilde{\boldsymbol{w}}}) + \frac{1}{\eta_b}tr(\tilde{\boldsymbol{\delta}}^T\dot{\tilde{\boldsymbol{\delta}}}) \\
&= -\boldsymbol{S}^T(t)[\tilde{\boldsymbol{w}}^T\boldsymbol{\Gamma} + \hat{\boldsymbol{\delta}}sgn(\boldsymbol{S}(t)) + \boldsymbol{\Delta} + (2\boldsymbol{R}^2)^{-1}(\boldsymbol{R}^2 - \mathbf{I})\boldsymbol{S}(t) + \boldsymbol{d}] \\
&\quad + \frac{1}{\eta_w}\sum_{i=1}^{m}\tilde{\boldsymbol{w}}_i^T\dot{\tilde{\boldsymbol{w}}}_i + \frac{1}{\eta_b}\sum_{i=1}^{m}\tilde{\delta}_i\dot{\tilde{\delta}}_i \\
&\quad - \boldsymbol{S}^T(t)\tilde{\boldsymbol{w}}^T\boldsymbol{\Gamma} - \boldsymbol{S}^T(t)\hat{\boldsymbol{\delta}}sgn(\boldsymbol{S}(t)) - \boldsymbol{S}^T(t)\boldsymbol{\Delta} + \frac{1}{\eta_w}\sum_{i=1}^{m}\tilde{\boldsymbol{w}}_i^T\dot{\tilde{\boldsymbol{w}}}_i \\
&\quad + \frac{1}{\eta_b}\sum_{i=1}^{m}\tilde{\delta}_i\dot{\tilde{\delta}}_i + \boldsymbol{S}^T(t)[-(2\boldsymbol{R}^2)^{-1}(\boldsymbol{R}^2 + \mathbf{I})\boldsymbol{S}(t) - \boldsymbol{d}]
\end{aligned}$$

$$
\begin{aligned}
&= -\sum_{i=1}^{m}\left[s_i(t)\tilde{\boldsymbol{w}}_i^T\boldsymbol{\Gamma} - \frac{1}{\eta_w}\tilde{\boldsymbol{w}}_i^T\dot{\hat{\boldsymbol{w}}}_i + \hat{\delta}_i|s_i(t)| + s_i(t)\Delta_i + \frac{1}{\eta_b}\tilde{\delta}_i\dot{\hat{\delta}}_i\right] \\
&\quad + \sum_{i=1}^{m}\left[-\frac{r_i^2+1}{2r_i^2}s_i^2(t) - s_i(t)d_i\right] \\
&= \sum_{i=1}^{m}\left[\tilde{\boldsymbol{w}}_i^T\left(\frac{1}{\eta_w}\dot{\hat{\boldsymbol{w}}}_i - s_i(t)\boldsymbol{\Gamma}\right) - \hat{\delta}_i|s_i(t)| - s_i(t)\Delta_i - \frac{1}{\eta_b}(\delta_i - \hat{\delta}_i)\dot{\hat{\delta}}_i\right] \\
&\quad + \sum_{i=1}^{m}\left[-\frac{r_i^2+1}{2r_i^2}s_i^2(t) - s_i(t)d_i\right] \qquad (3.26)
\end{aligned}
$$

From (3.23) and (3.24), Eq. (3.26) can be rewritten as

$$
\begin{aligned}
\dot{V}_{RHSMC}(\boldsymbol{S}(t),\tilde{\boldsymbol{w}},\tilde{\boldsymbol{\delta}}) &\le \sum_{i=1}^{m}\left[-|s_i(t)|(\delta_i - |\Delta_i|) - \frac{r_i^2+1}{2r_i^2}s_i^2(t) - s_i(t)d_i\right] \\
&\le \sum_{i=1}^{m}\left[-\frac{r_i^2+1}{2r_i^2}s_i^2(t) - s_i(t)d_i\right] \\
&= \sum_{i=1}^{m}\left[-\frac{s_i^2(t)}{2} - \frac{1}{2}\left(\frac{s_i(t)}{r_i} + r_i d_i\right)^2 + \frac{1}{2}r_i^2 d_i^2\right] \\
&\le \sum_{i=1}^{m}\left[-\frac{s_i^2(t)}{2} + \frac{1}{2}r_i^2 d_i^2\right] \qquad (3.27)
\end{aligned}
$$

Assuming $d_i \in L_2[0,T]$, $\forall T \in [0,\infty)$, integrating the above equation from $t=0$ to $t=T$, yields

$$
V_{RHSMC}(T) - V_{RHSMC}(0) \le \sum_{i=1}^{m}\left[-\frac{1}{2}\int_0^T s_i^2(t)dt + \frac{1}{2}r_i^2\int_0^T d_i^2(t)dt\right] \qquad (3.28)
$$

Because $V_{RHSMC}(T) \ge 0$, the above inequality implies the following inequality

$$
\frac{1}{2}\sum_{i=1}^{m}\int_0^t s_i^2(t)dt \le V_{RHSMC}(0) + \frac{1}{2}\sum_{i=1}^{m}r_i^2\int_0^T d_i^2(t)dt \qquad (3.29)
$$

Using (3.25), the above inequality is equivalent to the following.

$$
\sum_{i=1}^{m}\int_0^T s_i^2(t)dt \le \sum_{i=1}^{m}s_i^2(0) + \frac{1}{\eta_w}\sum_{i=1}^{m}\tilde{\boldsymbol{w}}_i^T(0)\tilde{\boldsymbol{w}}_i(0) + \frac{1}{\eta_b}\sum_{i=1}^{m}\tilde{\delta}_i^2(0) + \sum_{i=1}^{m}r_i^2\int_0^T d_i^2(t)dt \qquad (3.30)
$$

Thus the proof is completed.

3.3.3 Online Parameter Learning

Selection of parameters for the recurrent weights and means and variances of the receptive-field basis functions will significantly affect the performance of ORCMAC. Inappropriate recurrent weights and receptive-field basis functions will degrade the learning performance. In order to train ORCMAC effectively, an online parameter training methodology, which is derived using the gradient descent method, is proposed such that it can automatically adjust the recurrent weights and means and variances of the receptive-field basis functions. This training scheme increases the learning speed of ORCMAC. First, the adaptive law shown in (3.30) can be rewritten as

$$\dot{\hat{w}}_{ki} = -\eta_w s_i(t)\gamma_k \tag{3.31}$$

According to the gradient descent method, the adaptive law of the weights can be also represented as follows.

$$\dot{\hat{w}}_{ki} = -\eta_w \frac{\partial V_{RHSMC}}{\partial u_{RHSMC_i}} \frac{\partial u_{RHSMC_i}}{\hat{w}_{ki}} = -\eta_w \frac{\partial V_{RHSMC}}{\partial u_{RHSMC_i}} \gamma_k \tag{3.32}$$

Thus, the Jacobin of the controlled system $\partial V_{RHSMC}/\partial u_{RHSMC_i} = s_i(t)$.

Consider the receptive-field basis functions, the adaptive laws of mean $\hat{m}_{jk}$, variance $\hat{\sigma}_{jk}$, and recurrent weight $\hat{w}_{rpj}$ can be derived via the gradient descent method as

$$\begin{aligned}\dot{\hat{m}}_{jk} &= -\eta_m \sum_{i=1}^{m} \frac{\partial V_{RHSMC}}{\partial u_{RHSMC_i}} \frac{\partial u_{RHSMC_i}}{\partial \gamma_k} \frac{\partial \gamma_k}{\partial \phi_{jk}} \frac{\partial \phi_{jk}}{\partial \hat{m}_{jk}} \\ &= -\eta_m \sum_{i=1}^{m} s_i(t)\hat{w}_{ki}\gamma_k \frac{2(I_{rj} - \hat{m}_{jk})}{\hat{\sigma}_{jk}^2}\end{aligned} \tag{3.33}$$

$$\begin{aligned}\dot{\hat{\sigma}}_{jk} &= -\eta_v \sum_{i=1}^{m} \frac{\partial V_{RHSMC}}{\partial u_{RHSMC_i}} \frac{\partial u_{RHSMC_i}}{\partial \gamma_k} \frac{\partial \gamma_k}{\partial \phi_{jk}} \frac{\partial \phi_{jk}}{\partial \hat{\sigma}_{jk}} \\ &= -\eta_v \sum_{i-1}^{m} s_i(t)\hat{w}_{ki}\gamma_k \frac{2(I_{rj} - \hat{m}_{jk})^2}{\hat{\sigma}_{jk}^3}\end{aligned} \tag{3.34}$$

$$\begin{aligned}\dot{\hat{w}}_{rpj} &= -\eta_r \sum_{k=1}^{n_R} \sum_{i=1}^{m} \frac{\partial V_{RHSMC}}{\partial u_{RHSMC_i}} \frac{\partial u_{RHSMC_i}}{\partial \gamma_k} \frac{\partial \gamma_k}{\partial \phi_{jk}} \frac{\partial \phi_{jk}}{\partial I_{rj}} \frac{\partial I_{rj}}{\partial \hat{w}_{rpj}} \\ &= \eta_r \sum_{k=1}^{n_R} \sum_{i=1}^{m} s_i(t)\hat{w}_{ki}\gamma_k \frac{2(I_{rj} - \hat{m}_{jk})}{\hat{\sigma}_{jk}^2} O_p(t - \Delta T)\end{aligned} \tag{3.35}$$

The derivation of the adaptation laws (3.33–3.35) can help to overcome the inappropriate selection of the recurrent weights, means, and variances of the receptive-field basis functions. These adaptive laws will not destroy the stability property presented in Theorem 3.1, in as much as the maximum output value of the receptive-field basis functions is limited to unity.

3.4 Simulation Results

In this section, the proposed RHSMC control system is applied to control a nonlinear nine-link biped robot.

3.4.1 A Nine-Link Biped Robot

Consider a nine-link biped robot as shown in Fig. 3.4 and assume this system is subjected to nonlinear faults with the dynamic system presented as follows [11].

$$\ddot{\boldsymbol{q}} = \boldsymbol{M}^{-1}(\boldsymbol{q})\left(\boldsymbol{\tau} - \boldsymbol{C}(\boldsymbol{q},\dot{\boldsymbol{q}})\dot{\boldsymbol{q}} - \boldsymbol{g}(\boldsymbol{q}) + \lambda(t-t_0)\bar{\boldsymbol{f}}_t(\boldsymbol{q},\dot{\boldsymbol{q}})\right) \tag{3.36}$$

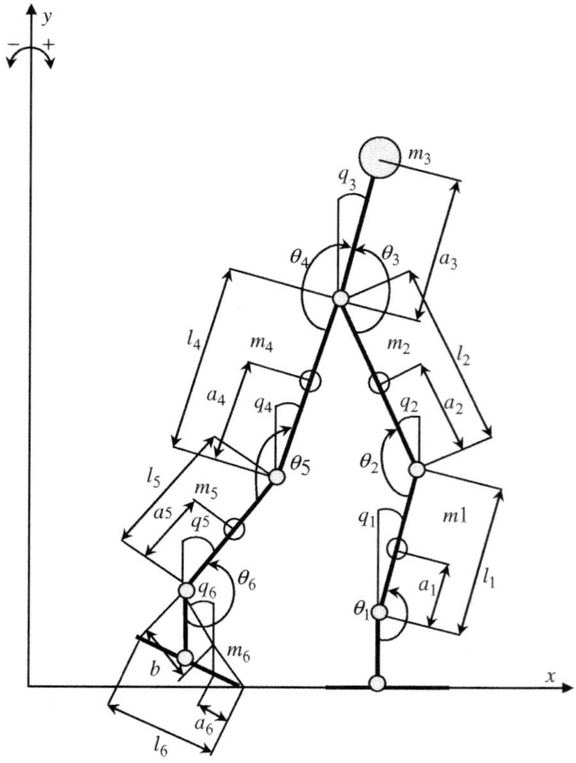

Fig. 3.4 Nine-link biped robot

where $q,\ddot{q},\ddot{q} \in \Re^6$ are vectors of joint positions, velocities, and accelerations, respectively; $M(q) = \{\delta_{ij} cos(q_i - q_j)\} \in \Re^{6\times 6}$ is the inertia matrix, $\tau \in \Re^6$ is the input torque vector, and $C(q,\ddot{q}) = \{\delta_{ij} sin(q_i - q_j)\} \in \Re^{6\times 6}$ is the Coriolis/centripetal matrix; $g(q) = \{-h_i sin(q_i)\} \in \Re^6$ is the gravitational force; where δ_{ij} and h_i are some system parameters. Unknown vector $\bar{f}_t(q,\ddot{q}) \in \Re^6$ stands for the change in the biped robot due to a fault.

Without loss of generality and for the sake of convenience of analysis, the fault of the biped robot dynamics is represented as the changes in the dynamics and is presented as

$$\bar{f}_t(q,\ddot{q}) = M(q) f_t(q,\ddot{q}) \tag{3.37}$$

where $f_t(q,\ddot{q}) \in \Re^6$. Then the robot dynamic equation (3.37) can be rewritten as

$$\ddot{q} = M^{-1}(q)\,[\tau - C(q,\dot{q})\dot{q} - g(q)] + \lambda(t-t_0)\,f_t(q,\dot{q}) \tag{3.38}$$

The nonlinear parameters of biped robot are given as follows.

$\delta_{11} = m_1 a_1^2 + (m_2+m_3+m_4+m_5+m_6) l_1^2 + I_1$

$\delta_{22} = m_2 a_2^2 + (m_3+m_4+m_5+m_6) l_2^2 + I_2$

$\delta_{33} = m_3 a_3^2 + I_3$

$\delta_{44} = m_4 (l_4 - a_4)^2 + (m_5+m_6) a_4^2 + I_4$

$\delta_{55} = m_5 (l_5 - a_5)^2 + m_6 l_5^2 + I_5$

$\delta_{66} = m_6 b^2 + I_5$

$\delta_{12} = m_2 l_1 a_2 + (m_3+m_4+m_5+m_6) l_1 l_2$

$\delta_{13} = m_3 l_1 a_3$

$\delta_{14} = -m_4 l_1 (l_4 - a_4) - (m_5+m_6) l_1 l_4$

$\delta_{15} = -m_5 l_1 (l_5 - a_5) - m_6 l_1 l_5$

$\delta_{16} = -m_6 l_1 b$

$\delta_{23} = m_3 l_2 a_3$

$\delta_{24} = -m_4 l_2 (l_4 - a_4) - (m_5+m_6) l_2 l_4$

$\delta_{25} = -m_5 l_2 (l_5 - a_5) - m_6 l_2 l_5$

$\delta_{26} = -m_6 l_2 b$

$\delta_{34} = \delta_{35} = \delta_{36} = 0$

$\delta_{45} = m_5 l_4 (l_5 - a_5) + m_6 l_4 l_5$

$\delta_{46} = m_6 l_4 b$

$\delta_{56} = m_6 l_5 b$

$\delta_{ij} = \delta_{ji}$ for $i = 1,2,\ldots,6$ and $j = 1,2,\ldots,6$

$h_1 = (m_1 a_1 + m_2 l_1 + m_3 l_1 + m_4 l_1 + m_5 l_1 + m_6 l_1) g$

$$h_2 = (m_2a_2 + m_3l_2 + m_4l_2 + m_5l_2 + m_6l_2)g$$
$$h_3 = m_3a_3g$$
$$h_4 = (m_4a_4 - m_4l_4 - m_5l_4 - m_6l_4)g$$
$$h_5 = (m_5a_5 - m_5l_5 - m_6l_5)g$$
$$h_6 = -m_6bg$$

The desired reference trajectories are offline planed as $\boldsymbol{q}_d, \dot{\boldsymbol{q}}_d$, and $\ddot{\boldsymbol{q}}_d$. In order to demonstrate the efficiency of the proposed RHSMC, two simulation cases for nonlinear faults and modeling uncertainties are simulated for the biped robotic system. In Case 1, consider a nonlinear fault due to a tangle of complex factors that is assumed to manifest itself as a nonlinear change in the biped robotic system; in Case 2, in order to study the robustness of controller and the fault-tolerant control ability, the system uncertainties containing parameter variations, exogenous disturbance, and a fault with a 75% change in the mass of link 1 and link 5 are simulated in the biped robotic system. These simulation cases are addressed as follows.

Case 1: A fault with the nonlinear change in link 1 and link 2 occurs at the sixth sec with the following failure function.

$$\boldsymbol{f}_{\mathrm{t}}(\boldsymbol{q}, \dot{\boldsymbol{q}}) = \begin{bmatrix} 75q_1^2 + 100q_1^2\dot{q}_2^2 + 7q_2 + 17 \\ 100q_1q_2 + 25 \\ 0 \\ 0 \\ 0 \\ 0 \end{bmatrix} \tag{3.39}$$

Case 2: The biped robotic system has system uncertainties and has a fault with 75% of the changes in the mass of link 1 and link 5 occurring at the sixth sec. The system uncertainties are given as

$$\boldsymbol{f}_{sd}(\dot{\boldsymbol{q}}) = \begin{bmatrix} 0.5sign(\dot{q}_1) + 2\dot{q}_1 \\ 0.5sign(\dot{q}_2) + 2\dot{q}_2 \\ 0.5sign(\dot{q}_3) + 2\dot{q}_3 \\ 0.5sign(\dot{q}_4) + 2\dot{q}_4 \\ 0.5sign(\dot{q}_5) + 2\dot{q}_5 \\ 0.5sign(\dot{q}_6) + 2\dot{q}_6 \end{bmatrix}, \quad \boldsymbol{\tau}_d(t) = \begin{bmatrix} exp(-0.1t) \\ exp(-0.1t) \\ exp(-0.1t) \\ exp(-0.1t) \\ exp(-0.1t) \\ exp(-0.1t) \end{bmatrix} \tag{3.40}$$

where $\boldsymbol{f}_{sd}(\dot{\boldsymbol{q}}) \in \Re^6$ is a vector containing the unknown static and dynamic friction terms, and $\boldsymbol{\tau}_d(t) \in \Re^6$ is a vector representing external disturbance. Then, the biped robot dynamic equation becomes:

$$\ddot{\boldsymbol{q}} = \boldsymbol{M}^{-1}(\boldsymbol{q})\,[\tau + \tau_d - \boldsymbol{C}(\boldsymbol{q}, \dot{\boldsymbol{q}})\dot{\boldsymbol{q}} - \boldsymbol{G}(\boldsymbol{q}) - \boldsymbol{f}_{sd}] + \lambda(t - t_0)\boldsymbol{f}_{\mathrm{t}}(\boldsymbol{q}, \dot{\boldsymbol{q}}) \tag{3.41}$$

The control objective is to let the biped robot joint positions follow the reference trajectories under the occurrence of system failures and modeling uncertainties. In the simulations, the initial joint positions are:

$$\begin{aligned} \boldsymbol{q}(0) &= [q_1(0)\; q_2(0)\; q_3(0)\; q_4(0)\; q_5(0)\; q_6(0)]^T \\ &= [0.37 - 1\; 0.75 - 0.15 - 0.56\; 3.85]^T\, rad. \end{aligned}$$

An ORCMAC used in this study is characterized as follows.

- Number of input state variables: $n_a = 6$.
- Number of elements for each state variable: $n_e = 5$ (elements).
- Generalization: $n_f = 4$ (elements/block).
- Number of receptive-fields for each state variable:
- $n_b = 2(\text{receptive-fields/layer}) \times 4(\text{layer}) = 8(\text{receptive-fields})$.
- Receptive-field basis functions: $\mu_{ik} = exp[-(p_{rik} - c_{ik})^2/v_{ik}^2]$ for $i = 1,2,\cdots,6$ and $k = 1,2,\cdots,8$. The input spaces of input signals are normalized within $\{[-1.5,\ 1.5][-1.5,\ 1.5][-1.5,\ 1.5][-1.5,\ 1.5][-1.5,\ 1.5][-1.5,\ 1.5]\}$.
- The initial means of the Gaussian functions are divided equally and are set as $c_{i1} = -2.1$, $c_{i2} = -1.5$, $c_{i3} = -0.9$, $c_{i4} = -0.3$, $c_{i5} = 0.3$, $c_{i6} = 0.9$, $c_{i7} = 1.5$, $c_{i8} = 2.1$, and the initial variances are set as $v_{ik} = 0.5$ for $i = 1,2,\cdots,6$ and $k = 1,2,\cdots,8$. The weights of are $\boldsymbol{w}$ initialized from zero.
- For ORCMAC the learning rates are chosen as $\eta_w = 1$ and $\eta_c = \eta_v = \eta_r = 0.01$.

For comparison, simulations are performed by using a nominal control, which is the main control in (3.17) without the uncertainty estimator, and the proposed RHSMC for the biped robot. The simulation results of nominal control and the proposed RHSMC for cases 1 are shown in Figs. 3.5 and 3.6, respectively. Figures 3.5a and 3.6a show the tracking performance of the joint angle of each link. Figures 3.5b and 3.6b show the angular velocity of each link. Figures 3.5c and 3.6c show the applied torque of each link. Figures 3.5d and 3.6d show the tracking error of each link.

The simulation results of nominal control and the proposed RHSMC for case 2 are depicted in Figs. 3.7 and 3.8, respectively. Figures 3.7a and 3.8a show the tracking performance of the joint angle of each link. Figures 3.7b and 3.8b show the angular velocity of each link. Figures 3.7c and 3.8c show the applied torque of each link. Figures 3.7d and 3.8d show the tracking error of each link. The simulation results show that without the uncertainty estimator, the nominal controller cannot effectively deal with the system uncertainties and faults; however, the proposed RHSMC control system can favorably track the reference trajectories, even with the system uncertainties and faults.

3.5 Conclusions

This study has successfully demonstrated that the robust hybrid sliding-mode control (RHSMC) system can achieve favorable tracking performance for nonlinear systems with modeling uncertainties and external disturbances. All parameters in the RHSMC system are tuned based on the Lyapunov stability theory and gradient descent method, thus the stability of the system can be guaranteed. Finally, the proposed RHSMC for a multi-input multi-output nonlinear nine-link biped robot is

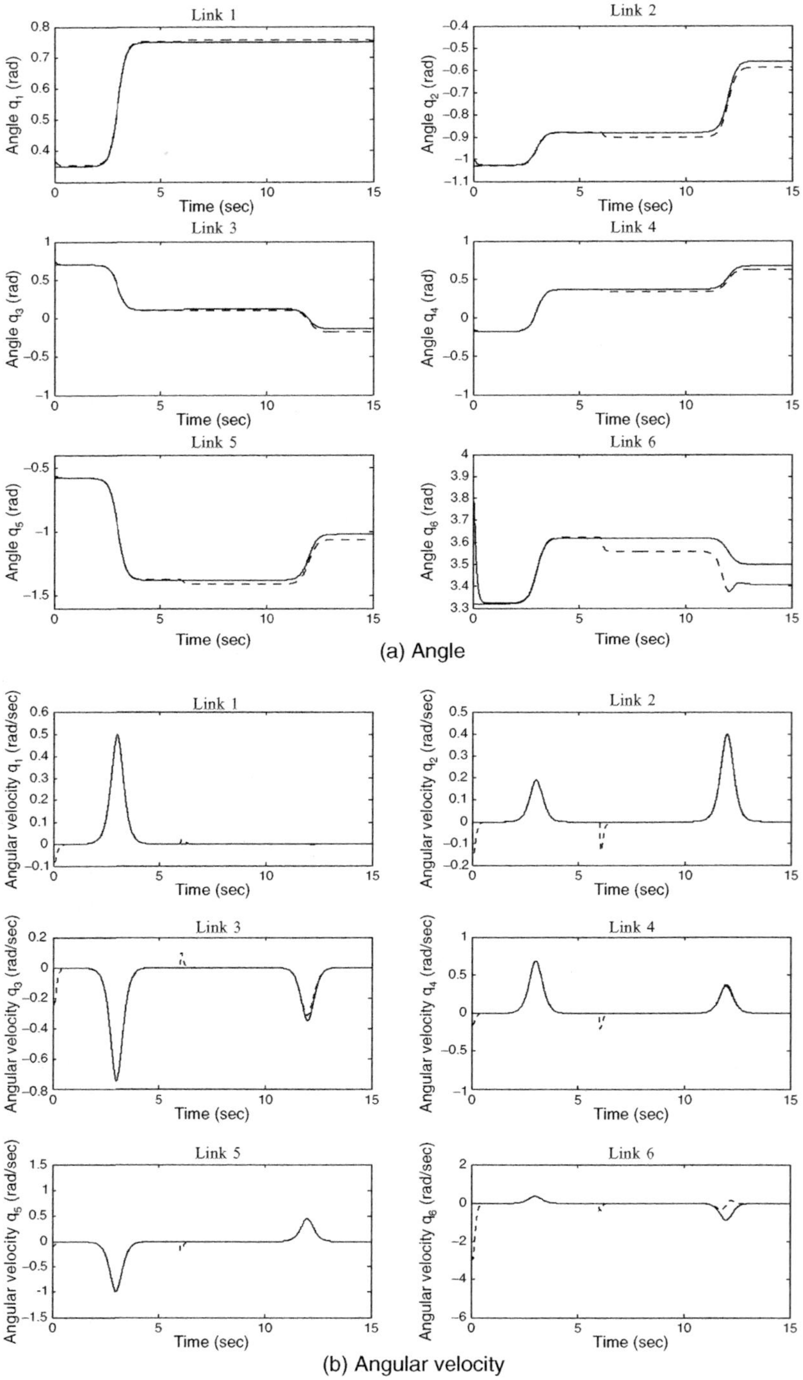

Fig. 3.5 The simulation results of biped robot using nominal control for case 1 (—reference trajectory, —system trajectory)

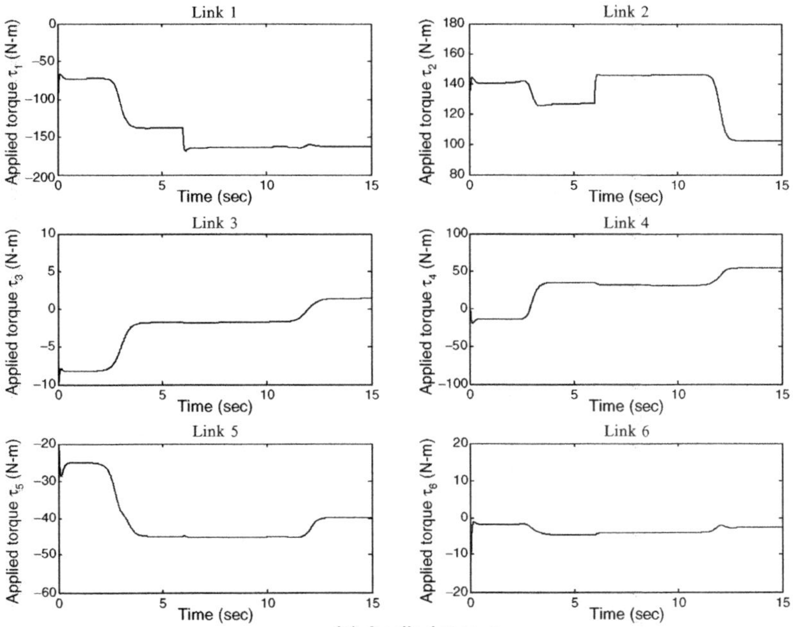

(c) Applied torque

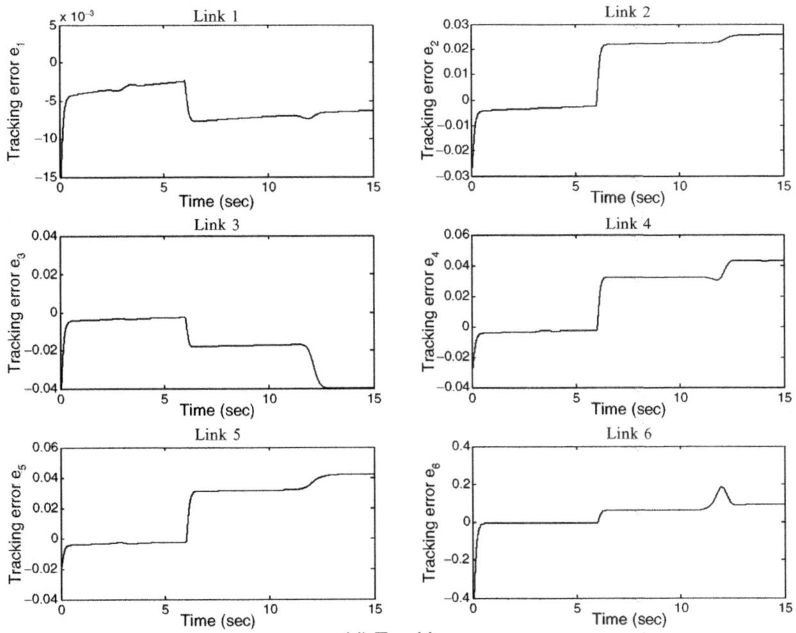

(d) Tracking error

Fig. 3.5 Continued

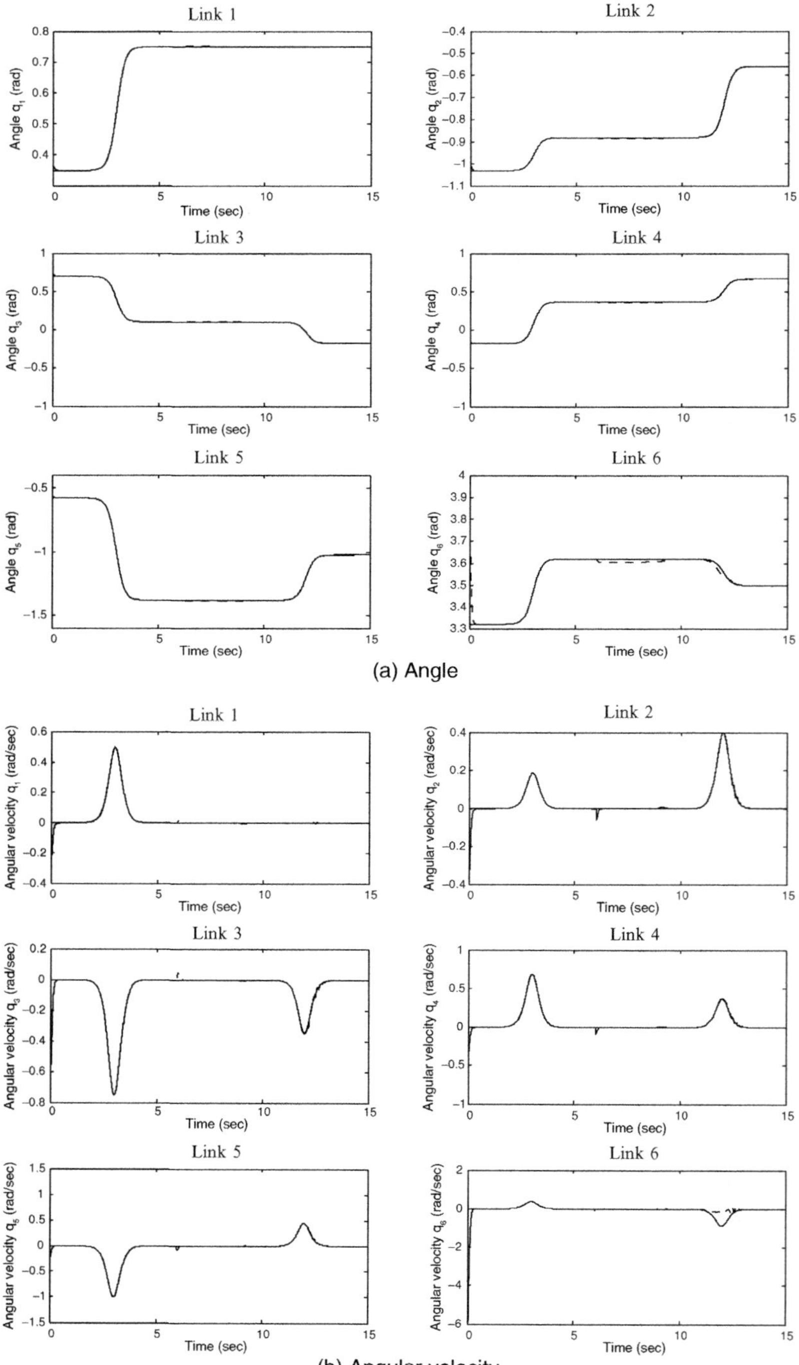

Fig. 3.6 The simulation results of biped robot using RHSMC for case 1 (—reference trajectory, —system trajectory)

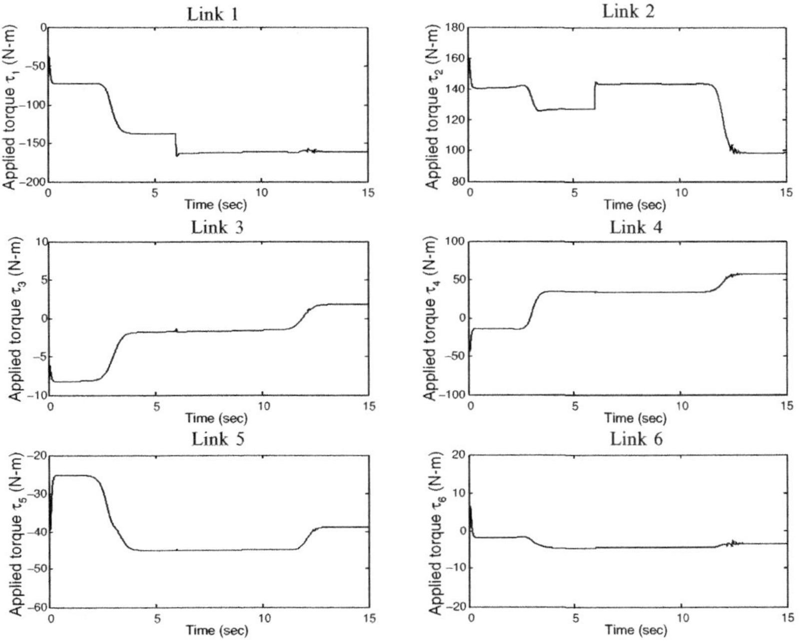

(c) Applied torque

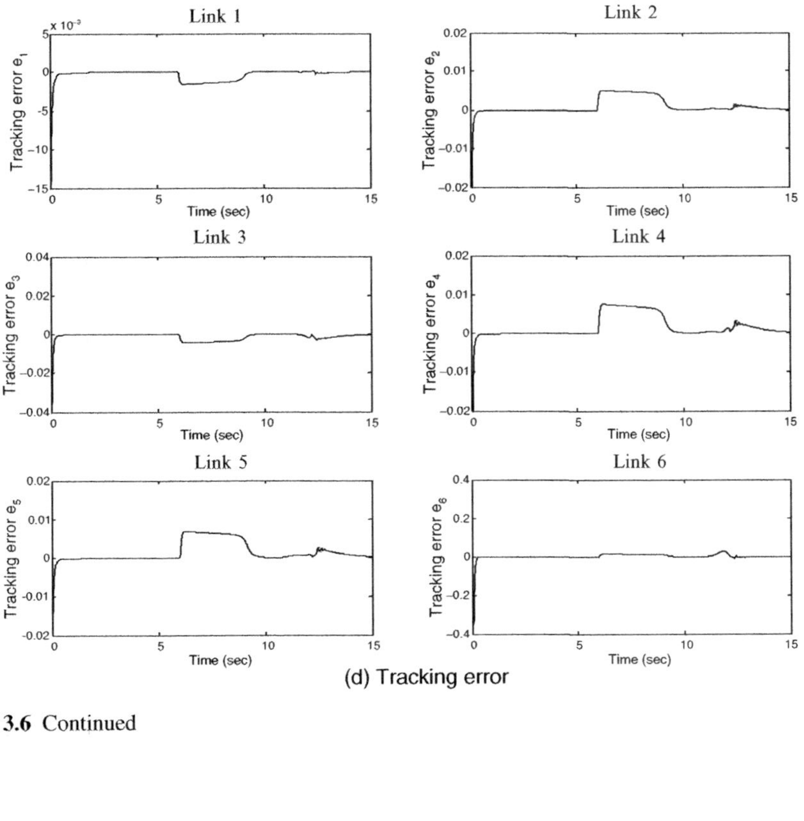

(d) Tracking error

Fig. 3.6 Continued

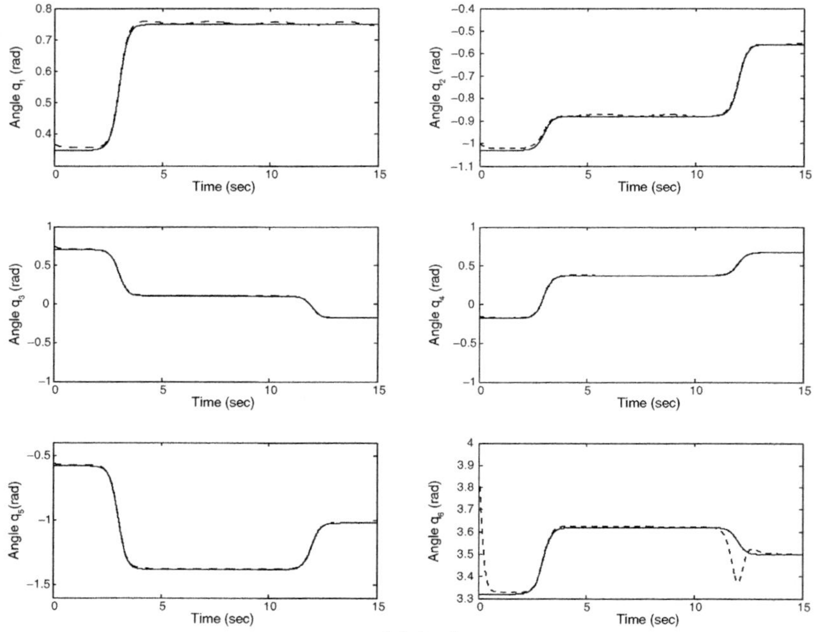

(a) Angle

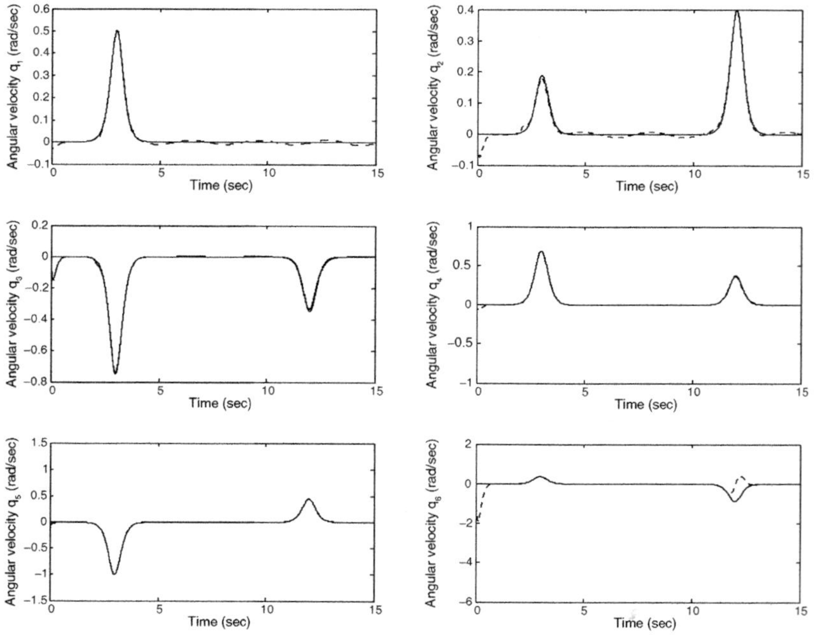

(b) Angular velocity

Fig. 3.7 The simulation results of biped robot using nominal control for case 1 (—reference trajectory, —system trajectory)

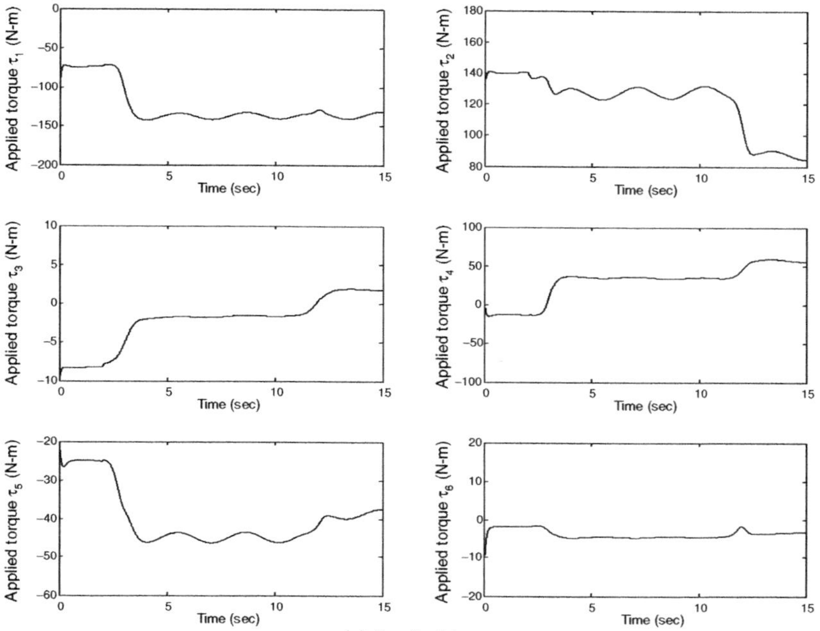

(c) Applied torque

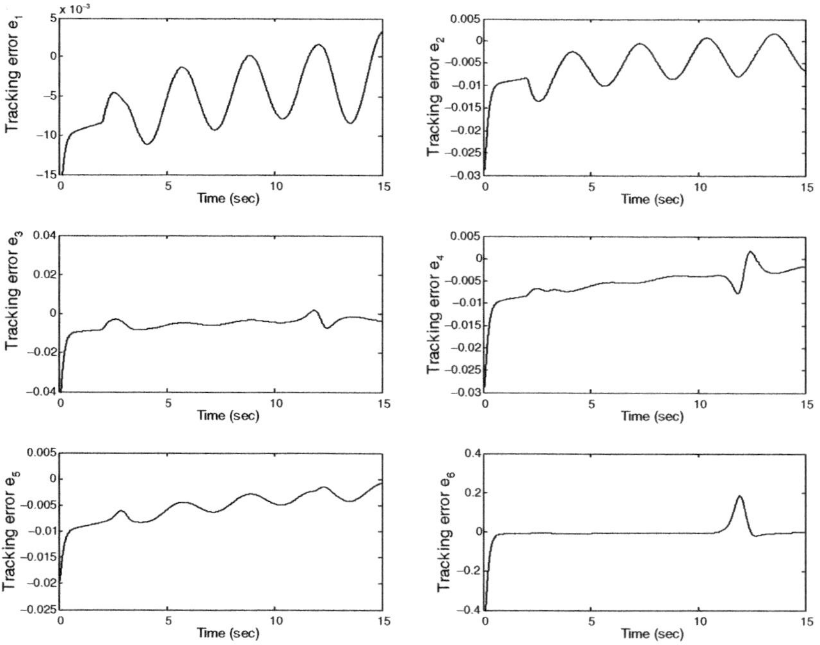

(d) Tracking error

Fig. 3.7 Continued

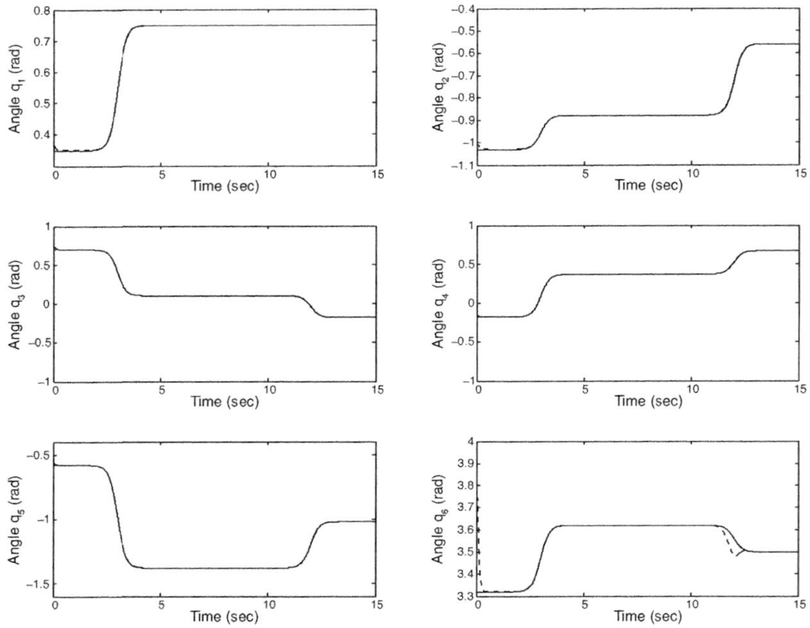

(a) Angle

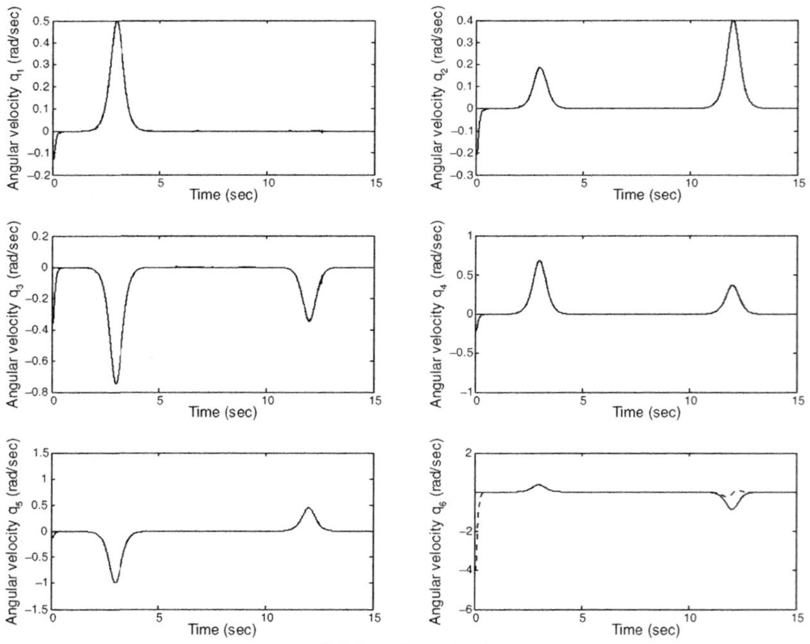

(b) Angular velocity

Fig. 3.8 The simulation results of biped robot using RHSMC for case 2 (—reference trajectory, —system trajectory)

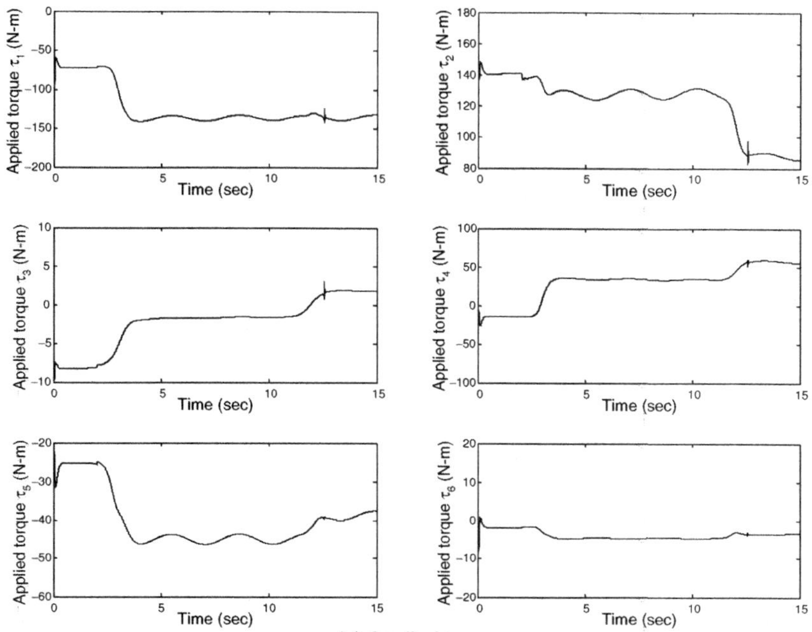

(c) Applied torque

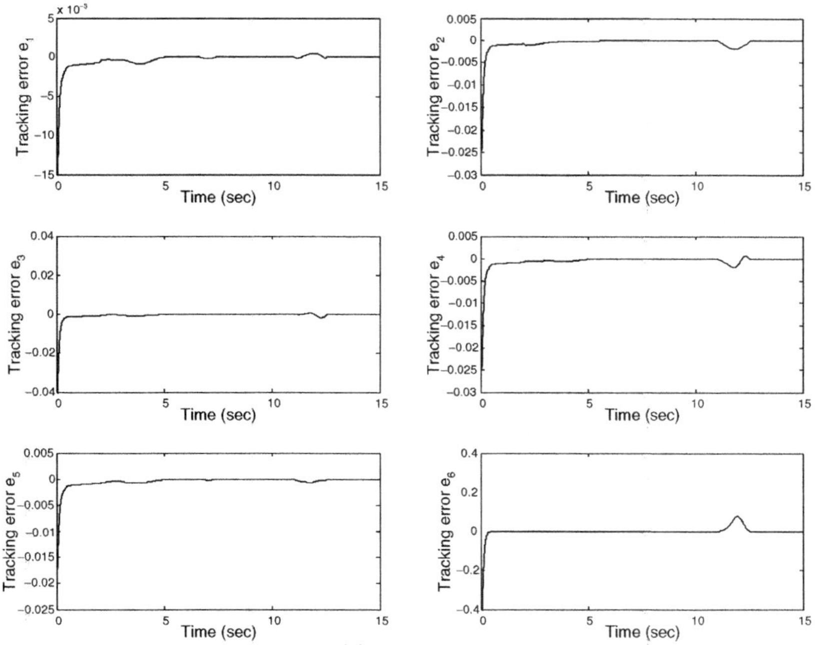

(d) Tracking error

Fig. 3.8 Continued

presented to illustrate the effectiveness of the proposed control scheme. The simulation results show that the effect of modeling uncertainty, approximation error, and external disturbance are efficiently attenuated, and the chattering of control effort is significantly reduced by using the proposed control approach.

References

1. Albus J S (1975) Data storage in the cerebellar model articulation controller (CMAC), *J. Dyn. Syst. Measurement Contr.*, vol. 97, no. 3, pp. 228–233.
2. Chen B S, Lee C H, Chang Y C (1996) H_∞ tracking design of uncertain nonlinear SISO systems: Adaptive fuzzy approach, *IEEE Trans. Fuzzy Syst.*, vol. 4, no. 1, pp. 32–43.
3. Chiang C T, Lin C S (1996) CMAC with general basis functions, *Neural Netw.*, vol. 9, no. 7, pp. 1199–1211.
4. Gonzalez-Serrano F J, Figueiras-Vidal A R, Artes-Rodriguez A (1998) Generalizing CMAC architecture and training, *IEEE Trans. Neural Netw.*, vol. 9, no. 6, pp. 1509–1514.
5. Jan J C, Hung S L (2001) High-order MS_CMAC neural network, *IEEE Trans. Neural Netw.*, vol. 12, no. 3, pp. 598–603.
6. Kim Y H, Lewis F L (2000) Optimal design of CMAC neural-network controller for robot manipulators, *IEEE Trans. Syst. Man Cybern. C*, vol. 30, no. 1, pp. 22–31.
7. Lane S H, Handelman D A, Gelfand J J (1992) Theory and development of higher-order CMAC neural networks, *IEEE Control Syst. Mag.*, vol. 12, no. 2, pp. 23–30.
8. Lin C M, Hsu C F (2003) Neural network hybrid control for antilock braking systems, *IEEE Trans. Neural Netw.*, vol. 14, no. 2, pp. 351–359.
9. Lin C M, Peng Y F (2004) Adaptive CMAC-based supervisory control for uncertain nonlinear systems, *IEEE Trans. Syst. Man Cybern. B*, vol. 34, no. 2, pp. 1248–1260.
10. Lin C M, Peng Y F (2005) Missile guidance law design using adaptive cerebellar model articulation controller, *IEEE Trans. Neural Netw.*, vol. 16, no. 3, pp. 636–644.
11. Liu Z, Li C (2003) Fuzzy neural networks quadratic stabilization output feedback control for biped robots via H_∞ approach, *IEEE Trans. Syst. Man Cybern. B*, vol. 33, no. 1, pp. 67–84.
12. Slotine J J E, Li W P (1991) *Applied Nonlinear Control*. Englewood Cliffs, NJ: Prentice-Hall.
13. Utkin V I (1992) *Sliding Modes in Control and Optimization*. New York: Springer-Verlag.
14. Wai R J, Lin C M, Peng Y F (2003) Robust CMAC neural network control for LLCC resonant driving linear piezoelectric ceramic motor, *IEE Proc. Control Theory Appl.*, vol. 150, no. 3, pp. 221–232.
15. Wai R J, Lin F J (1999) Fuzzy neural network sliding-model position controller for induction servo motor driver, *IEE Proc. Electr. Power Appl.*, vol. 146, no. 3, pp. 297–308.
16. Wang L X (1994) *Adaptive Fuzzy Systems and Control: Design and Stability Analysis*. Englewood Cliffs, NJ: Prentice-Hall.
17. Yoo B, Ham W (1998) Adaptive fuzzy sliding mode control of nonlinear system, *IEEE Trans. Fuzzy Syst.*, vol. 6, no. 2, pp. 315–321.

Chapter 4
A Dynamic GA-Based Rhythm Generator

Tzimeas Dimitrios and Mangina Eleni

4.1 Introduction

Musical problems such as composition, harmonization, and arrangement are widely popular in the world of programming as they fulfill various criteria [3], including within mathematical laws of harmony and symmetry and aesthetic definitions of "pleasant" and "beautiful." There are many examples of GAs being able to cope with this challenge as they can explore large search spaces with minimal requirements [1]. However, the main bottleneck of the algorithm is the design of the fitness function. This is the critical part of the algorithm that decides which candidate solutions will survive and evolve to the next population [6]. The difficulty lies in characterization of what is "good" or "bad."

Interactive GAs solve this problem by asking the user to become the one who will take these decisions, but this dramatically increases the running time and makes the applications exhausting for the user [20]. Automated GAs give a description of which output is "pleasing" or "preferred" in the definition of the fitness function and provide a convergence based on these choices [9]. However, the algorithm may locate a local maximum or can be disoriented, giving an output contrary to the user's preferences.

The strategic choice of the software developer will decide the quality of his music system. Automated GAs tend to give better solutions when the problem and the requested solution are well defined whereas they fail in more general creativity problems where it is difficult, if not impossible, to define an aesthetically accepted solution. In these cases, interactive GAs replace the fitness function with a human evaluator who orients the exploration of the search space according to his stylistic preferences [10].

The rest of the chaper presents relevant work and the *SENEgaL* system. Two representative music systems, which generate rhythms based on evolutionary computing, are introduced analyzing their main advantages and disadvantages. The critical damped oscillator (CDO) fitness function is described in the next section, illustrating how it can be implemented for GA-based music problems to overcome

Oscar Castillo et al. (eds.), *Trends in Intelligent Systems and Computer Engineering.*

the difficulties of designing the fitness function. The algorithmic design of *SENEgaL*, the prototype music system which generates rhythms from western Africa is presented in detail. Certain focus is given in the musical education interface of *SENEgaL* and the analysis of its evaluation experiment. Finally, the conclusions of our work and a short description of the future development of this system are provided.

4.1.1 Related Work

Intelligent music systems composing rhythms are not popular in the area of GAs. The most significant ones appeared during the last decade whereas systems composing melodies, synthesized sounds, or harmonization have been under development from the 1980s since GAs were widely applied [3]. CONGA [20] and Sbeat [23] are the most representative systems of this kind.

4.1.1.1 CONGA

CONGA stands for "Composition in Genetic Approach" and enables users to compose their own rhythmic patterns [20]. The core of the system is based on the combination of interactive evolutionary computing (IEC) and artificial neural networks (ANN) [2, 12]. Randomly generated rhythmic patterns are evaluated by the user in a binary scaling (good/bad) and the algorithm, guided by these decisions, evolves towards better/preferred patterns following a simple GA [6]. A genetic programming (GP) [13] technique is employed to arrange the preferred patterns in music bars. In order to decrease the total time, ANN evaluation assistance "learns" the user's rhythmic preferences and automatically gives a fitness value for known rhythmic patterns.

The main advantage of this method is that the system adapts to the musical style of the user, creating the opportunity to compose new rhythmic patterns. This attribute makes CONGA a really useful composition tool. On the other hand, running time is a significant problem as the user has to continually listen to and evaluate each candidate solution until a satisfying solution is reached. GAs have a great exploration ability [6] but it may take several hundreds of generations to "unlock" from a local maximum and explore the fitness landscape of a higher peak. This waiting time increases the psychological burden imposed on users.

4.1.1.2 Sbeat

Sbeat is an advanced prototype system, which not only constructs rhythmic patterns, but also arranges melodies of two additional musical instruments in a friendly graphic environment [23].The method is based on the idea of simulated breeding [22], where the user selects his favorite individuals to advance in the next generation and produce new offspring. This technique disables the stochastic selection aiming to reduce the user's operations to assign the fitness values.

The system gives a "pleasant" and musically complete output as it arranges a musical piece for a full drum kit and two additional instruments. This trio could be any combination such as drums, acoustic wood bass, and acoustic nylon guitar. The sounds are provided by the general MIDI (GM) library [4]. The main disadvantage of the method is the running time of the system and the further limitations that occur in order to decrease it. The output is set to be maximum sixteen beats, which is a four 4/4 bar piece. In addition the population is limited to 20–30 individuals, this obliges the user to reset and reinitialize the whole population in order to restart the exploration of the search space.

4.2 The Critical Damped Oscillator Fitness Function

An alternative strategy which avoids the long running time of IEC music systems is the critical damped oscillator fitness function [21]. It is an automated fitness rater employed in a simple GA, where the user has to define her music targets and the system dynamically allocates the fitness values based on these decisions. The CDO fitness function is inspired by the concept of the typical damped oscillator [19] which is described by Eq. 4.1.

$$\sum F = -c\dot{x} - kx \tag{4.1}$$

where k is a positive constant and c is the friction constant/decay parameter.

The CDO model sets the fitness values in a similar way, aiming to "balance" the appearances of the predefined schemes acting as a "spring force" that prevents schemes from dominating the chromosome. Figure 4.1 illustrates this behavior in a simple melodic problem where the system has to compose a melody according to the following predefined targets/ratios.

1. Tonal notes: Melody must consist of notes belonging to the same tonality (C major in this example) with a ratio of 48 (out of chromosome size of 56).
2. Allowed jumps: The melody must not have melodic jumps greater than one octave with a ratio of 45.
3. C (Do) on-beat: The first on-beat note of every two bars must be a C (Do) in order to give the feeling of a cadence with a ratio of 6.
4. D (Re) off-beat: The off-beat note preferred to be a D (Re) with a ratio of 6.
5. Motif appearances: It is preferable for a predecided melodic motif to appear in the melody in order to give a melodic consistency with a ratio of 10.

The left column of graphs in Fig. 4.1 illustrates the fitness values for each of the targets and the right column shows the appearance of the corresponding scheme/target. The dotted line represents the desired ratio that is set for each target separately. The fitness values are given a real number at the end of each generation according to how far the present appearances are from the ratio and how many more appearances occurred than in the previous generation. In this way the mathematical modeling of the critical damped oscillator in (4.1) is adhered to.

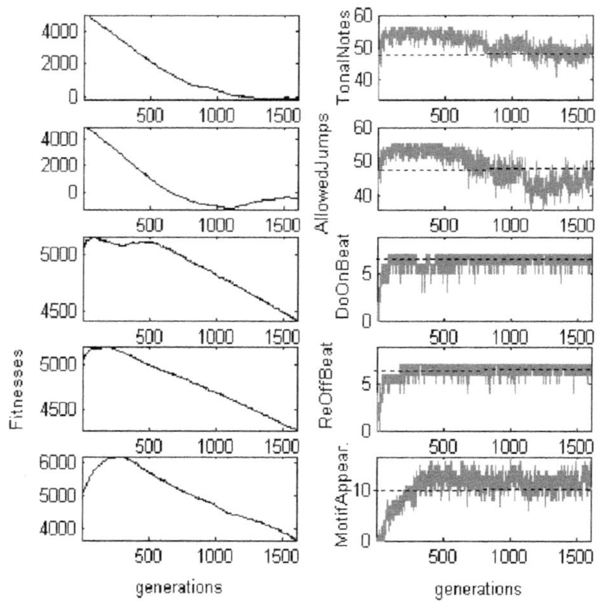

Fig. 4.1 The CDO fitness values allocation in a simple melodic problem

The success of the method is based on the idea that the fitness values increase when a target moves away from the ratio and decreases, respectively (4.1), when it "accelerates"; that is, its appearances tend to exceed them. Consequently, the algorithm dynamically decides the fitness values and effectively balances targets of different natures achieving multiobjectivity in a simple GA.

The main advantage of the technique is that it avoids continuous interaction with the user making the system much faster and more flexible as it is now free of any limitations that are imposed upon an interactive system. It can handle populations' and individuals' length (melody length in this example) of any desired size.

4.2.1 SENEgaL

SENEGaL is GA-based music system which enables the user to generate and simulate rhythms from the area of West Africa. The variety and the musical richness of these rhythms provide a great number of choices and makes the system an interesting music composition "game." The present development of the system includes the following basic rhythms [5, 7].

1. Gahu: A traditional rhythm from Ghana
2. Linjen: A popular West African rhythm often prefered by American djembe players
3. Nokobe: A traditional rhythm of the Ewe people of Ghana
4. Kaki Lambe: A traditional Senegalese rhythm
5. Fanga: A Liberian welcome rhythm

Each one of these rhythms consists of different rhythmic patterns played on two or more instruments (percussion). The user can follow the suggested arrangement of the output or even choose his own combination of percussion from the general MIDI (GM) library [4]. Table 4.1 provides a list of the arrangement choices. Musical analysis of each of these rhythms allowed for the individual categorization of each of the instruments' rhythmic patterns which, in turn, make up the final dancing beat. Figure 4.2 illustrates "Linjen" and its basic rhythmic patterns for each instrument.

Table 4.1 The GM percussion kit

High Q	Mid Tom 1	Low Timbale
Slap	High Tom 2	Open Hi Hat
Scratch Push	Crash Cymbal 1	Low Conga
Scratch Pull	High Tom 1	High Timbale
Sticks	Ride Cymbal 1	Low Timbale
Square Click	Chinese Cymbal	High Timbale
Metronome Click	Ride Bell	Low Timbale
Metronome Bell	Tambourine	High Agogo
Kick Drum 2	Splash Cymbal	Low Agogo
Kick Drum 1	Cowbell	Cabasa
Side Stick	Crash Cymbal 2	Maracas
Snare Drum 1	Vibra Slap	Short Hi WhistleLong Low
Hand Clap	Ride Cymbal 2	Whistle Short Guiro
Snare Drum 2	High Bongo	High Timbale
Low Tom 2 R	Low Bongo	Low Timbale
Closed Hi Hat	Mute High Conga	
Low Tom 1	Open High Conga	
Pedal Hi Hat	Low Conga	
Mid Tom 2	Open Hi Hat	

Fig. 4.2 Basic rhythmic patterns of Linjen

Low Tom follows pattern E, whereas High Tom follows the A, B, C, or D pattern. These patterns are the targets for the CDO fitness function described in Sect. 4.3.

4.2.1.1 Method

SENEGaL follows the evolution of a simple GA, that is, selection–recombination–mutation. The evaluation of the population is automated and the fitness values are calculated dynamically at the end of each generation according to the CDO fitness function, as described in Sect. 4.3. In this case the targets are defined by the analysis of the rhythms, which provide the schemas that are encouraged to appear. The ratio can be adjusted by the user according to how strict an output is desired. If the ratio is 100% of the length of the chromosome, then the output will be a simulation of the actual African rhythm. Otherwise, the output will provide other interesting patterns, giving a more innovative sound.

We decided upon an integer phenotypic data representation which describes the candidate solutions and improves the performance of the system thus making it effectively fast. Each note is represented by an integer number which corresponds to the number of the minimum value in the rhythm. In Fig. 4.3, the minimum value is the sixteenth; hence, the quarter is represented by 4, the eighth by 2, and so on. The negative values, respectively, correspond to a rest. The chromosome stores the values of each instrument in different dimensions; that is, in the example of Fig. 4.3 the chromosome is a one-dimensional array, whereas in Linjen (Fig. 4.2) the chromosome has two dimensions storing separately the values for High Tom and Low Tom. Another advantage of this data representation is that it does not impose any limit on the length of the output. For instance two bars of 4/4 are represented by 10 integers, which means that the system can produce much longer outputs without any impact on the performance.

The GA parameters were based upon published work and the findings of empirical studies. The mutation rate was set to 0.025 and the crossover rate to 0.9, the crossover method to stochastic universal sampling, and the crossover technique to "multiple point," in order to achieve an extended exploration of the search space [6, 17]. The size of the population was set to 40 as larger populations did not show any more favorable results or improvements in performance. In addition, at the end of each generation an escape function was implemented in order to terminate the run when the target criteria were met.

Fig. 4.3 *SENEGaL* data representation

4.2.1.2 Results and Discussion

The system was set up in a flexible way so that it could provide rapid convergence. This strategy was not only chosen for its better performance but, as the user will most likely fail to sufficiently describe the targets in his first try, he will probably have to redefine them. Thus by having an average runtime of ~3 min., the user is given the opportunity to "play" the game of composing African rhythms over and over without experiencing any significant mental strain.

Figure 4.4 illustrates how the user has to define the target schemas and their ratios. A default setting is provided, so the output will be close to the actual African rhythm. In addition, the system is adjustable and the user can choose and experiment with extra target motifs or schemas and ratios. Based on this, under the user's guidance, the algorithm can explore other areas of the search space as well. If the total ratio per instrument is higher than 70–80% then the output will probably be a simulation of the rhythm, whereas total ratios less than 70% will generate a new rhythm with "African" influences as it will include patterns of the rhythm.

When the schemas and ratios are set for three instruments as in Fig. 4.4 (Low Tom, Whistle, High Tom) instead of a typical set of two instruments (Low Tom, High Tom) as in Fig. 4.2, *SENEgaL* generates "Gen_Linjen," the first three bars of which are provided in Fig. 4.5. The dotted circles note the appearances of the desired patterns mixed with other schemas that were allowed to evolve by the ratios (~80% in average). The output is musically pleasant and the option of arranging each output in a completely different combination of instruments invites the user to use her imagination and experiment with different rhythms, ratios, patterns, and sounds. Even arranging the same output with different MIDI instruments [4] can provide interesting compositions, which can be further used as part of a bigger musical arrangement.

Figure 4.6 illustrates how the fitness values are assigned by the CDO model. This figure represents the target appearances and the corresponding fitness values of "High Tom." The arrows indicate the CDO behavior. For instance, in generations

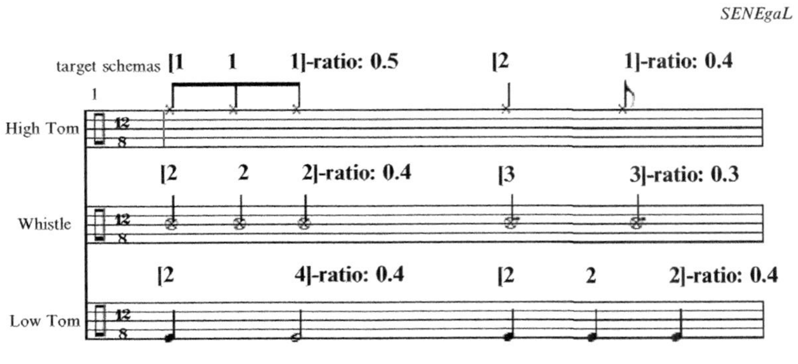

Fig. 4.4 The target schemas and their ratios in Linjen

Fig. 4.5 *SENEgaL* output for Linjen rhythm

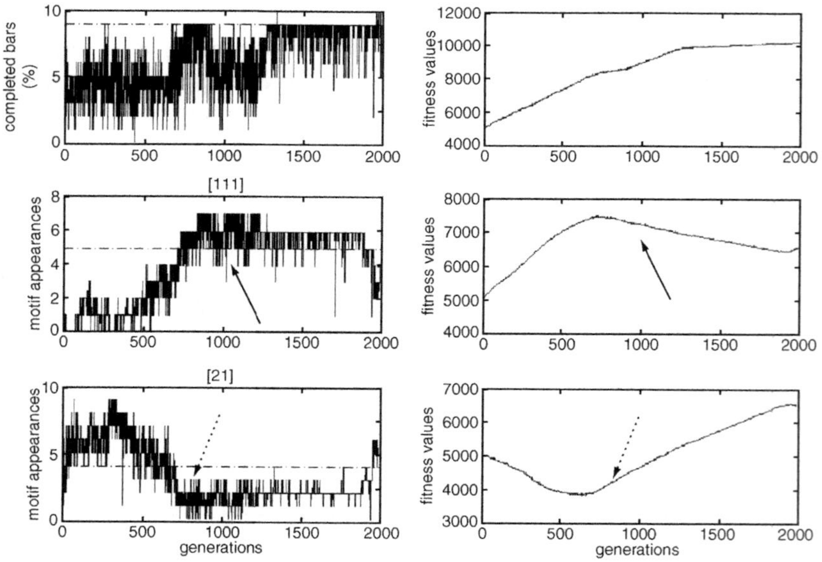

Fig. 4.6 The motif appearances and their fitness values of High Tom

0–500 motif $\{2\ 1\}$ exceeds the desired ratio (dotted line) tending to dominate the population. Therefore, the fitness values continue to decrease preventing this motif from surviving in the next generations. This behavior causes a rapid decrease in its appearances in generations 500–1000.

Crossing the desired ratio will change again the behavior of its fitness value which will start increasing in order to "force" the appearances to balance around the desired ratio, exactly as it happens with the critical damped oscillator. On the other hand, it is difficult for motif $\{1\ 1\ 1\}$ to increase in generations 0–500 as it is longer than motif $\{2\ 1\}$ which tends to dominate $\{1\ 1\ 1\}$ due to its length. In this case the fitness values of $\{1\ 1\ 1\}$ increase, constantly resulting in a successful convergence to the desired ratio around generation 700. This simultaneous dynamic allocation of the fitness values is the key of the success of the CDO fitness function which does not allow the dominance/extinction of any motif.

In this run the "escape function" was deactivated in order to show the behavior of the system up to 2000 generations. The actual convergence occurred around generation 900 when the motifs of the other two instruments (Whistle, Low Tom) converged as they were well guided by the CDO fitness function in exactly the same way. The target "completed bars" represent the musical arrangement of the appearances of each motif in a way that they complete a bar but do not cause rhythmic discontinuities or syncope. The level of continuity or discontinuity can also be adjusted by the user in order to obtain a result closer to his preferences. In the example provided the ratio was set to 90% which means the 9/10 bars will be fully completed.

4.3 Music Education Interface

The main characteristics of *SENEgaL* led to the idea of creating a user-friendly interface with the aim of developing a musical education tool in mind rather than a common GA-based rhythm generator. The difference between such a system and the previously presented systems, CONGA and Sbeat, is found in the goal of the application itself: to invite the user to participate in an interesting and amusing "game" which will increase her musical understanding of rhythms, percussion, and African music. In this "music game" the users, by passing through several steps, try to generate their own rhythms based on an African rhythm. These steps involve numerous choices varying from simple music settings to tutorials about percussion and historical background of African music. As further explained in the following sections, the user defines silently all the necessary CDO fitness function settings while taking musical decisions regarding the final result. In this way, he participates actively in the exploration of the fitness landscape and the evolution of chromosomes in the algorithm.

4.3.1 System Description

The interface is designed in an aesthetically related visual/audio environment. Colors, audio samples, and images are delicately used in order to create a friendly atmosphere (Fig. 4.7) and bring the user closer to the task that she is going to carry out [15, 16].

The principal task of the application is to create new rhythms or patterns, based on an African rhythm. More precisely, the main steps of the interface are the following.

- *Introduction*: The user receives a general overview of the system and the several tasks she is going to carry out.
- *Select African Rhythm*: The user is given the option to select which African rhythms she prefers. Each rhythm is supported by help files (music/historical background of the particular rhythm), music notation (PDF format), and audio samples (MIDI format). The current version supports three rhythms: Linjen, KakiLambe, and Gahu [15, 16]. Each rhythm is characterized by different instruments, patterns, and style.
- *Set tempo parameters and select instruments*: The user has to select which instruments are going to play in the system's final result. There are multiple supports available for this step as the main philosophy of the application is to attract users of all levels, from amateur to advanced. Hence, there are several ways to select the drum kit: (i) directly from the given list box, (ii) following an additional tutorial "help me choose instruments," or (iii) by pushing the "suggest!" button which automatically provides instrument suggestions. The selection of the in-

Fig. 4.7 Processing rhythms snapshot in *SENEgaL*

Fig. 4.8 Choosing instruments and tempo parameters in *SENEgaL*

struments is of high importance and the help windows guide the user towards a wide-frequency combination of percussion (bass, middle, and treble drum). However, the user is given the freedom to experiment with her own combinations of drums and create a new drum set. In the same step, as illustrated in Fig. 4.8, the user decides that the tempo parameters of the output should be the same as the three help options. On the top right corner of the window there is a group of help buttons related to the African rhythm the user chose in step 2, in order to recall music information and remind users of the main characteristics of this rhythm (audio sample, music notation, and music/historical background).

- *Choose motifs*: The user is given the main patterns that define the chosen African rhythm in music notation. These patterns are in fact the target schemas of the CDO fitness function. The user then sets the ratio of these patterns by adjusting a slider bar depending on how much she desires to replicate the actual African rhythm. In this way, the system receives the target ratios, necessary for the CDO fitness function. At this step the algorithm adapts to the aesthetic preferences of the user creating a fitness function based on the user's decision.
- *Final screen*: The algorithm takes about 3–4 mins. to locate a local optima of the fitness landscape that has been created by the user's preferences. The final screen gives the options of listening to the generated rhythm (MIDI format) and viewing it as a music notation sheet (PDF format). Without exiting or restarting the application the additional option of rearranging the drum set for the same output is given.
- *Rearranging the drum set*: The user can create a different drum kit to play the already generated rhythm by following the tutorial in Fig. 4.9. This instrument

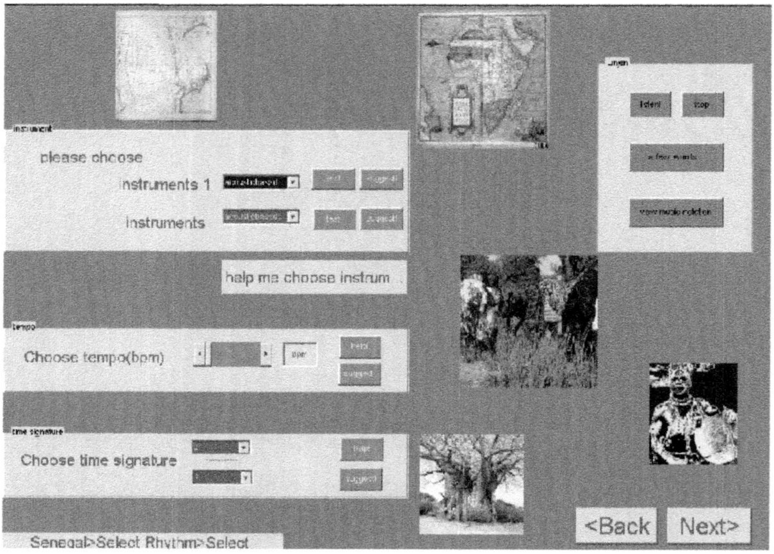

Fig. 4.9 Rearranging instruments in *SENEgaL*

changeover takes only a few seconds and gives the opportunity to explore the potential of the combination of all the supported percussion instruments of Table 4.1. In addition, all help options are activated in order to make this experimentation helpful and fun.

The average running time of the system is relatively short. Depending on the familiarity of the user and how fast he makes decisions about the rhythm, motif, instrument, and tempo settings it takes 5–10 mins. to achieve a complete output, following all the six steps described above. This attribute makes *SENEgaL* a light and fast application which attracts the user to reuse it in order to play with a different combination of rhythm, instrument, motif, and tempo parameters. In addition, the stochastic nature of the Gas [6, 8] results in a completely different output after each run. Even if the user has inserted exactly the same preferences the actual generated rhythm is different after each run; this is due to the ability of the GAs to explore different areas of the landscape and give multiple solutions [18].

One other main advantage of *SENEgaL*, compared to Sbeat and CONGA, is the ability to personalize the evolution of the population without using real-time interactive strategies. The preferences of the user define a dynamic CDO fitness function and the exploration of the search space takes place according to this definition. Instead of listening and evaluating each of the candidate solutions after the pass of each generation, the user is invited to play an entertaining music game which increases his musical understanding while trying to compose a new rhythm or set a new combination of instruments.

4.4 Evaluation

Students from University College of Dublin were asked to use *SENEgaL* in order to create their own rhythms. The experiment was conducted as follows.

1. Students were given an appointment to perform the experiment individually. The experiment took place in a controlled environment, with the same moderator running the experiment in the same room, on the same computer for 30 min.
2. The moderator would first give the student a sheet of information and would ask him or her to read it carefully. The sheet contained general information about the experiment and the description of the task that had to be done.
3. After the given time of 30 min. and the completion of the tasks, the student was asked to complete a questionnaire containing 24 items and 3 fields of personal data (age, sex, music familiarity).

4.4.1 Questionnaire

Usability is defined by ISO 9126 [11] as the capability of the software product to attract the user as well as to be understood, learned, and used when tested under specified conditions. The questionnaire consisted of 24 items based on Lewis' [14] factor analysis which describe the usability [16] of *SENEgaL* on four dimensions: usefulness, ease of use, ease of learning, and satisfaction. The majority of the items were kept in their originally proposed form and the rest were adapted to the nature of the problem. More specifically, usefulness was oriented towards measuring the ability of the system to increase the users' understanding of rhythm, instruments, and African musical tradition. All the items used were measured using a seven-point Likert-type scale [15].

4.4.2 Demographics

A total of 30 students in the University College of Dublin, participated in this study. The ages of the students ranged from 20 to 30 years old. Eighty percent were aged between 20 and 25 years old, 20% of them were between 25 and 30 years old, and none was more than 30 years old. Of the participants, 46.67% (14 subjects) were male and 53.33% (16 subjects) were female.

4.4.3 Perceived Computer Knowledge and Data Analysis

Figure 4.10 illustrates the distribution of music familiarity. The subjects belong to all categories symmetrically (mean: 3.93, STD: 1.82) and Fig. 4.11 provides the

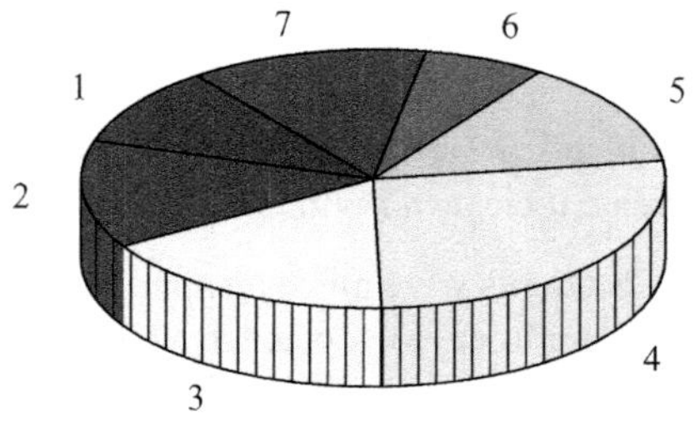

Fig. 4.10 The music familiarity distribution of the subjects

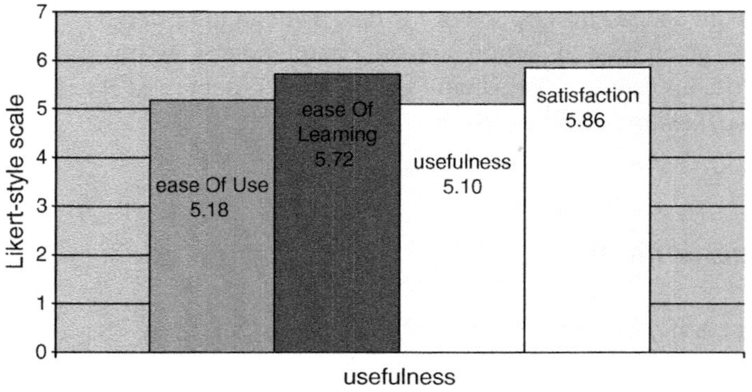

Fig. 4.11 Means of ease of use, ease of learning, usefulness, and satisfaction

results per dimension. As shown in Fig. 4.11, the system sustains a high level of usability in the dimensions of ease of use (mean: 5.18, STD: 1.49), ease of learning (mean: 5.72, STD: 1.49), usefulness (mean: 5.1, STD: 1.14), and satisfaction (mean: 5.86, STD: 1.26).

Usability is a factor of great importance to be taken into account especially when the software platform is accessed by non-IT experts. The effect of the high levels of usability can be confirmed by Fig. 4.12, where the system proves to have higher usability levels to subjects of low music familiarity. For instance, subjects of music familiarity 1 (very poor) have mean "ease of use" of 5.53 and standard deviation of 1.36, whereas subjects of music familiarity 7 (very good) have mean of 4.65 and standard deviation of 1.98. This fact shows that the system appears to be more attractive to users of poor music familiarity because it helps them to learn with greater ease (mean: 5.92, STD: 0.90) than advanced users (mean: 4.50, STD: 1.21).

The significantly high levels of usability in all categories of music familiarity illustrated in Figs. 4.12 and 4.13, show that *SENEgaL* is a user-friendly (ease to use, mean: 5.18, STD: 1.49), useful music education system (usefulness, mean: 5.1, STD: 1.14) that succeeds in bringing the user closer to the world of music and percussion instruments (ease of learning, mean: 5.72, STD: 1.49) in a pleasant manner (satisfaction, mean: 5.86, STD: 1.26).

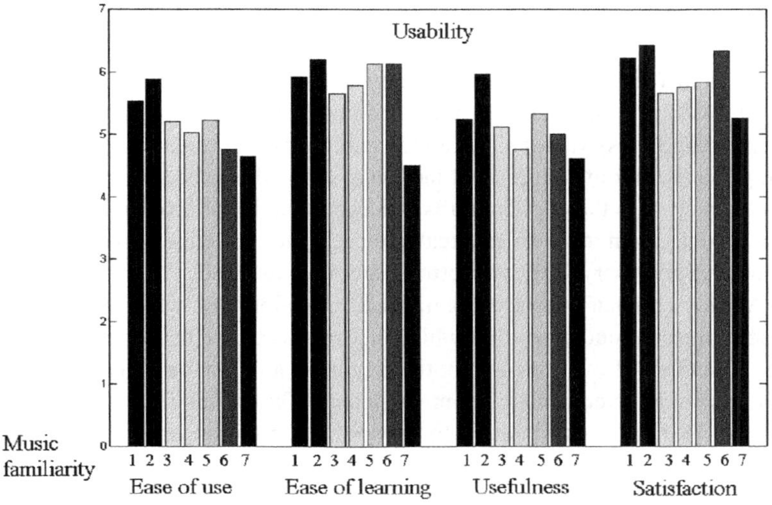

Fig. 4.12 Usability analysis by music familiarity

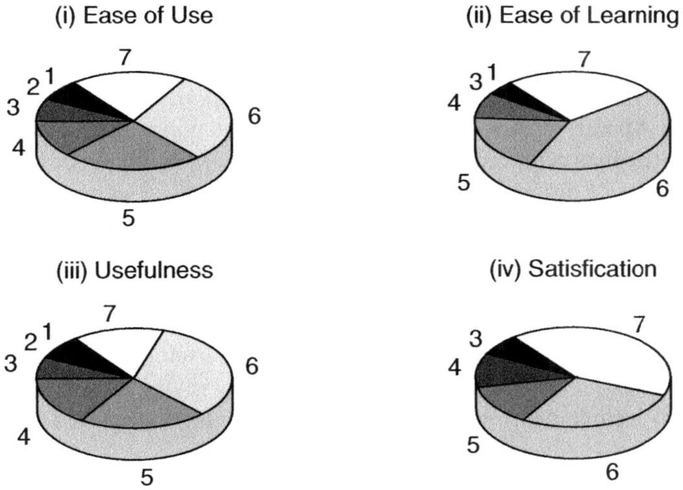

Fig. 4.13 Likert-style analysis of **(i)** ease of use, **(ii)** ease of learning, **(iii)** usefulness, and **(iv)** satisfaction

4.5 Conclusions and Future Work

In this chapter we discussed the application of GAs on evolutionary music systems and the main issues which arise from it. The comparison of interactive and automated GAs showed that the nature of the music problem is the principal factor that will determine which strategy will provide a satisfying output. We presented the CDO fitness function which is a specialized technique of designing a fitness function suitable for musical creativity problems. We analyzed the methods and the algorithmic design of *SENEgaL*, a new GA music system for generating music rhythms from Western Africa. The novelty of this work is based on the idea of the CDO fitness function, where the target schemas are assigned dynamically, and the corresponding fitness values simulate the behavior of a critical damped oscillation. *SENEgaL* is an automated fitness function system where the user has to define his or her target before the application is launched, making it flexible and rapid. Thus the user's obligation to listen and evaluate each one of the individuals in each generation, as is necessary with interactive systems, is avoided.

SENEgaL's human interface is a musical education tool, which aims to increase the user's musical understanding while she tries to create her own rhythms. It includes a plethora of instruments, motifs, rhythms, and other musical features making the system an interesting and amusing game. The system fulfills all the required algorithmic parameters and adapts to the aesthetic preferences of the user while he makes decisions regarding rhythms, patterns, and tempo settings. Furthermore the application is not exclusively for advanced musicians as it involves plenty of helpful files and tutorials which make each of the decisions and steps comprehensible even to amateur users. The experiment evaluating the usability of the system showed that it is a user-friendly, useful musical education system that succeeds in bringing the user closer to the world of music and percussion instruments in a pleasant manner.

Future work centers mainly on the launch of the application on the Internet. The database of African rhythms will be enlarged and categorized according to level of difficulty. Rhythms with more complicated instrument combinations and patterns will be placed in a different group to simpler rhythm cases. Hence:

1. *Amateur level*: Simple rhythms with maximum two instruments and two motifs per instrument. All "help" and "suggest" files/tutorials will be available.
2. *Intermediate level*: More complicated rhythms with up to three instruments and unlimited patterns per instrument. Only "help" files/tutorials will be available.
3. *Advanced level*: The users will have access to any rhythm and will be given the option to include their own motifs in addition to those suggested by *SENEgaL*. All "help" and "suggest" files/tutorials will be disabled.

The users will have to log in to the system and create a profile by giving a short description of their musical background. According to their musical understanding and their familiarity with the *SENEgaL* system they will be able to advance from one category to another. Additionally, the Web site will include plenty of music-related features: discussion forums, database of rhythms generated by the users, best-rated generated rhythms, video/audio tutorials about percussion techniques and performance, a musical theory encyclopedia, and useful links. Such a development will bring together users of completely different backgrounds and create a small musical community around *SENEgaL* making it a powerful tool in musical education.

References

1. Biles, JA (1994) GenJam: A genetic algorithm for generating jazz solos. In: *International Computer Music Conference (ICMC'94)*, Aarhus, Denmark, pp. 131–137.
2. Biles, JA, Anderson, PG, Loggi, LW (1996) Neural network fitness functions for a musical IGA. In: *International ICSC Symposia on Intelligent Industrial Automation And Soft Computing (IIA96/SOCO96)*, ICSC Academic Press, pp. B39–44.
3. Burton, AR, Vladimirova, T (1999) Generation of musical sequences with genetic techniques. *Computer Music Journal* 23:59–73.
4. De Furia, S (1988) *The MIDI book: Using MIDI and Related Interfaces*. Third Earth Productions, Pompton Lakes, NJ.
5. Dworsky, AL, Sansby, B (2000) *How to Play Djembe: West African Rhythms for Beginners*. Scb Distributors, Gardena, CA.
6. Goldberg, DE (1989) *Genetic Algorithms in Search, Optimization, and Machine Learning*. Wokingham Addison-Wesley, Reading, MA.
7. Hartigan, R (1995) *West African Rhythms for Drumset*. Warner Brothers, Burbank, CA.
8. Holland, JH (1975) *Adaptation in Natural and Artificial Systems*. The University of Michigan Press, Ann Arbor.
9. Horner, A, Goldberg, DE (1991) Genetic algorithms and computer-assisted music composition. In: *1991 International Computer Music Conference*, Montreal (Canada), pp. 479–482.
10. Horowitz, D (1994) Generating rhythms with genetic algorithms. In: *1994 International Computer Music Conference*, International Computer Music Association, Aarhus, Denmark, pp. 142–143.
11. ISO/IECTR9126 (2000) Software engineering–Product quality. International Standard.
12. Johanson, B, Poli, R (1998) GP-music: An interactive genetic programming system for music generation with automated fitness raters. In: *GP'98*, Helsinki, Morgan Kaufmann, pp. 181–186.
13. Koza, JR (1992) *Genetic Programming: On the Programming of Computers by Means of Natural Selection*. MIT Press, Cambridge, MA.
14. Lewis, JR (1995) IBM computer usability satisfaction questionnaires: Psychometric evaluation and instructions for use. *International Journal of Human-Computer Interaction* 7:57–78.
15. Likert, R (1932) A technique for the measurement of attitudes. *Archives of Psychology* 140:15–20.
16. Lund, AM (2001) Measuring usability with the USE questionnaire usability interface, *Society for Technical Communication* 8.
17. Mitchell, M, Forrest, S, Holland, JH (1992) The royal road function for genetic algorithms: Fitness landscapes and GA performance. In: *First European Conference on Artificial Life*, MIT Press, pp. 245–254.
18. Papadopoulos, G, Wiggins, G (1998) A genetic algorithm for the generation of jazz melodies. In: *STeP 98*, Jyvaskyla, Finland.
19. Serway, RA, Jewett, JW (2003) *Physics for Scientists and Engineers*. Brooks/Cole, Belmont, CA.
20. Tokui, N (2000) Music composition with interactive evolutionary computation. In: *Third International Conference on Generative Art*, Milan, pp. 215–226.
21. Tzimeas, D, Mangina, E (2007) The critical damped oscillator fitness function in music creativity problems. In: *International Joint Conference on Artificial Intelligence 2007 (AI and Music Workshop)*, Hyderabad, India, pp. 71–82.
22. Unemi, T (1999) SBART 2.4: Breeding 2D CG Images and movies and creating a type of collage. In: *Third International Conference on Knowledge-based Intelligent Information Engineering Systems, IEEE 99TH8410*, pp. 288–291.
23. Unemi, T, Nakada, E (2002) SBEAT3: A tool for multi-part music composition by simulated breeding. In: *Eighth International Conference on Artificial Life*, MIT Press, Sydney, pp. 410–413.

Chapter 5
Evolutionary Particle Swarm Optimization: A Metaoptimization Method with GA for Estimating Optimal PSO Models

Hong Zhang and Masumi Ishikawa

5.1 Introduction

Particle swarm optimization (PSO) is an algorithm for swarm intelligence based on stochastic and population-based adaptive optimization inspired by social behavior of bird flocks and fish swarms [5, 10].

In recent years, many variants of PSO have been proposed [11] and successfully applied in many research and application disciplines because of its intuitive understandability, ease of implementation, and the ability to efficiently solve large-scale and highly nonlinear optimization problems prevalent in social sciences, computer science, and complex engineering systems [9, 14, 15].

Although the original PSO is very simple with only a few parameters to adjust, it provides better performance in computing speed, computing accuracy, and memory size compared with other methods such as machine learning, neural network learning, and genetic computation. Needless to say, each parameter in PSO greatly affects the performance of PSO. However, how to determine appropriate values of parameters in PSO is yet to be found.

How to determine appropriate values of parameters in PSO can be regarded as metalevel optimization. Hence many researchers have paid much attention to this challenging problem. One approach to their determination is to try various values of parameters randomly to find a proper parameter set for PSO which handles many kinds of optimization problems reasonably well [2, 3]. Because this is exhaustive, the computing cost is heavy in high-dimensional parameter space.

The other approach is to implement a composite PSO (CPSO) [13], in which the differential evolution (DE) algorithm [16] handles the PSO heuristics in parallel during optimization. The instantaneous fitness of the global optimal particle is used in CPSO for evaluating the performance of PSO. Its experimental results indicated that CPSO could surpass the success ratio[1] of the original PSO for some

[1] Success refers to cases where particles reach a globally optimal solution. Success ratio is defined as the relative frequency of success.

Oscar Castillo et al. (eds.), *Trends in Intelligent Systems and Computer Engineering.*

benchmark test problems. As an expansion of the CPSO, optimized particle swarm optimization (OPSO) has been proposed and was applied to the training of artificial neural networks [12].

On the other hand, for specifically determining the values of parameters in PSO, Eberhart and Shi proposed using an inertia weight using a constriction factor [6]. These concludes that the best approach is to use the constriction factor while limiting the maximum velocity $v_{\max}$ of each particle within the dynamic range of the variable $x_{\max}$ in each dimension. However, the selection of the inertia parameter and maximum velocity are problem-dependent [18].

In order to systematically determine the values of parameters in PSO, this chapter proposes a metaoptimization method called evolutionary particle swarm optimization (EPSO). Specifically, EPSO estimates appropriate parameter values in PSO by a real-coded genetic algorithm with elitism strategy (RGA/E). This is a new metaoptimization method which is different from CPSO and OPSO. Firstly, the values of parameters in PSO are directly estimated by RGA/E in EPSO. The elitism strategy is performed to improve the performance and convergence behavior of EPSO. Secondly, a temporally cumulative fitness of the best particle is used in EPSO for effectively evaluating the performance of PSO. This is a key idea of this chapter for successfully accomplishing metaoptimization.

As is well known, PSO is a stochastic adaptive optimization algorithm. Sometimes particle swarm searches well by chance, even if parameter values are not appropriate for solving a given optimization problem. Because the temporally cumulative fitness proposed reflects the sum of instantaneous fitness, its variance is inversely proportional to the time span. Therefore, applying the fitness to EPSO is expected to systematically determine appropiate values of parameters in PSO by RGA/E for various optimization problems reasonably well, and greatly increase the success ratio of the original PSO for efficiently finding the global optimal solution corresponding to a given optimization problem.

To demonstrate the effectiveness of the proposed EPSO method, computer experiments on a two-dimensional optimization problem are carried out. We show experimental results, confirm the characteristics of dependency on initial conditions, and analyze the resulting PSO models.

The rest of the chapter is organized as follows. Section 5.2 briefly describes the original PSO and RGA/E. Section 5.3 presents the proposed EPSO method and a key idea about the temporally cumulative fitness that we used in the method. Section 5.4 discusses the results of computer experiments applied to a two-dimensional optimization problem and Sect. 5.5 gives conclusions.

5.2 The Original PSO and RGA/E

5.2.1 The Original PSO

The original PSO is modeled by particles with position and velocity in a high-dimensional space. Particles move in hyperspace and remember the best position that they have found. In general, members of a swarm communicate the information of good positions with each other and adjust their own position and velocity based on (1) a global best position that is known to all, and (2) local best positions that are known by neighboring particles.

Let x_k^i and v_k^i denote the position and velocity of the ith particle at time k, respectively. The position and velocity of ith particle is updated by

$$x_{k+1}^i = x_k^i + v_{k+1}^i \tag{5.1a}$$

$$v_{k+1}^i = c_0 v_k^i + c_1 r_1 (x_l^i - x_k^i) + c_2 r_2 (x_g - x_k^i), \tag{5.1b}$$

where c_0 is an inertial factor, c_1 is an individual confidence factor and c_2 is a swarm confidence factor, and $r_1, r_2 \in U[0,1]^2$ are random values. $x_l^i (= \arg \max_{k=1,2,\ldots} \{g(x_k^i)\}$, where $g(x_k^i)$ is the fitness value of the ith particle at time k.) is the local best position corresponding to the ith particle up to now; $x_g (= \arg \max_{i=1,2,\ldots} \{g(x_l^i)\})$ is the global best position of particles of the entire swarm, respectively. Note that a constant, $v_{\max}$, is used to arbitrarily limit the velocity of each particle and to improve the resolution of the search.

The procedure of the original PSO is described as follows.

Step 1. Initialize the position x_0^i and the velocity v_0^i of each particle randomly. Their range is domain-specific. Set counter $k = 1$ and the maximum number K of iterations in the search.

Step 2. Calculate the fitness value $g(x_k^i)$ of each particle x_k^i for determining the local best position x_l^i and the global best position x_g up to the current time.

Step 3. Update v_{k+1}^i and x_{k+1}^i by Eq. (5.1) for obtaining the new position and velocity of each particle. It is to be noted that v_{k+1}^i has an upper bound of velocity $v_{\max}$; then $v_{k+1}^i = v_{\max}$. If the current fitness value of x_l^i is larger than the previous one, replace x_l^i with the current one. If the current fitness value of x_g is larger than the previous one, replace x_g with the current one. If $k < K$ then $k = k+1$ and go to *Step* 2, else stop the search.

Each parameter (i.e., c_0, c_1 and c_2) in the original PSO has a role in exploration, respectively. How to determine them is to greatly affect the performance of PSO. So far, these values usually are determined by a designer based on a large number of experimental results. The original PSO uses a set of parameter values, $(c_0, c_1, c_2) = (1, 2, 2)$, for various optimization problems [10].

5.2.2 RGA/E

A real-coded genetic algorithm with elitism strategy RGA/E is applied to simulate the survival of the fittest among individuals over consecutive generation for

---

[2] $U[0,1]$ denotes that the probability distribution is a continuous uniform one. The support of the interval is defined by the two parameters 0 and 1 which are its minimum and maximum values.

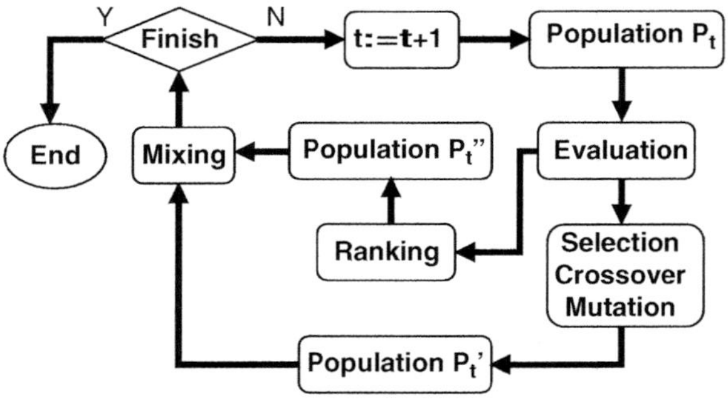

Fig. 5.1 Flowchart of RGA/E

solving real-valued optimization problems. The fitness function used in it, which is always problem-dependent, is defined over the genetic representation and measures the quality of the represented solution. Figure 5.1 depicts the flowchart of RGA/E.

For convenience, the individual w^i in a population is represented by a K-dimensional vector of real numbers:

$$w^i = (w^i_1, w^i_2, \ldots, w^i_K).$$

Concretely, the following genetic operations (i.e., selection, crossover, mutation, and rank algorithm) are used in RGA/E.

5.2.2.1 Roulette Wheel Selection

Roulette wheel selection is a very popular deterministic operator nowadays, but in most implementations it has random components. The probability of selecting the individual w^i operation can be expressed as follows.

$$p[w^i] = \frac{g(w^i)}{\sum_{i=1}^{M} g(w^i)},$$

where $g(\cdot)$ is the fitness value of the individual w^i, and M is the number of individuals in the population.

5.2.2.2 BLX-α Crossover

BLX-α crossover [7] reinitializes the values of offspring w^{ij} with values from a extended range given by the parents (w^i, w^j), where $w^{ij}_k = U(w^{ij\min}_k - I \times \alpha, w^{ij\max}_k + I \times \alpha)$[3] with $w^{ij\min}_k = \mathrm{Min}(w^i_k, w^j_k)$, $w^{ij\max}_k = \mathrm{Max}(w^i_k, w^j_k)$ and $I = w^{ij\max}_k - w^{ij\min}_k$. The parameter α is a predetermined constant.

---

[3] $U[a,b]$ denotes that the probability distribution is a continuous uniform one. The support of the interval is defined by the two parameters, a and b, which are its minimum and maximum values.

5.2.2.3 Random Mutation

For a randomly chosen gene k of an individual w^i, the allele w^i_k is added by a randomly chosen value, $\Delta w^i_k \in U[-\Delta w_b, \Delta w_b]$ [17]. Note that Δw_b is the given boundary value for a mutant variant in operation of the random mutation. Therefore, the allele $w^{i\,\prime}_k$ of the offspring $w^{i\prime}$ can be obtained as follows.

$$w^{i\,\prime}_k = w^i_k + \Delta w^i_k.$$

5.2.2.4 Ranking Algorithm

The elitism strategy is adopted for improving convergence behavior of genetic algorithms. For executing the strategy, the individuals in the current population P_t are sorted according to their fitness. Then the sorted population is referred to as a new one. Therefore, the index of individuals in the population will be used to determine that individuals are allowed to reproduce directly to the next generation by mixing operation. Here, s_n is defined as the predetermined number of the superior individuals from the population P_t''.

5.3 The Proposed EPSO Method

In this section, we describe the proposed EPSO method that efficiently estimates appropriate values of parameters in PSO corresponding to a given optimization problem by RGA/E and the temporally cumulative fitness of the best particle that we adopted.

Figure 5.2 illustrates the basic concept of EPSO. An iterative procedure of EPSO is composed of two parts: one is an outer loop using RGA/E. The other is an inner loop using PSO. For a given set of parameters provided by RGA/E, whereas PSO finds a global optimal solution, the resulting fitness value according to each particle

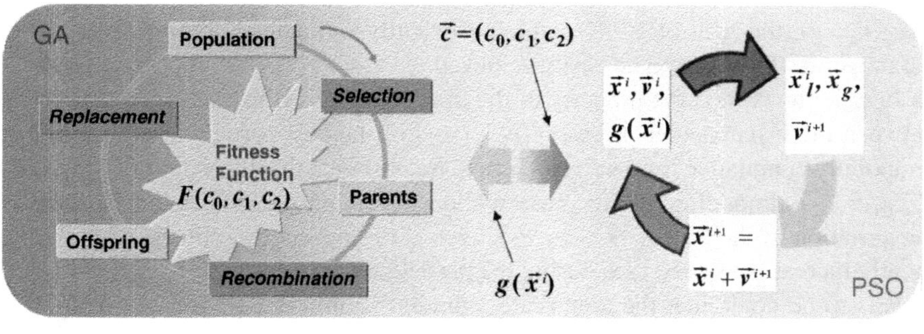

Fig. 5.2 Basic concept of the proposed EPSO method

is provided to RGA/E for online evolutionary computation. Owing to evolutionary computation, it is expected that the values of parameters in PSO, c_0, c_1, c_2, will be improved, and the corresponding fitness value generated by PSO will increase as iteration proceeds.

Usually, a fitness function $f(c_0,c_1,c_2)$ in RGA/E can be defined by

$$\begin{aligned} f(c_0,c_1,c_2) &= \max_{k=1,2,\dots} \left\{ \max_{i=1,2,\dots} \{g(x_k^i)|_{c_0,c_1,c_2}\} \right\} \\ &= \max_{k=1,2,\dots} \{g(x_k^b)|_{c_0,c_1,c_2}\} \\ &= g(x_g)|_{c_0,c_1,c_2}, \end{aligned} \tag{5.2}$$

where x_k^b is the best position in the entire swarm at time k. However, because the fitness $f(c_0,c_1,c_2)$ is an instantaneous one, it is unstable for evaluating the performance of PSO. Accordingly, it is not suitable for a reliable evaluation of the stochastic object.

The goal of EPSO is to find appropriate values of parameters in PSO for efficiently solving a given optimization problem. In order to change the temporary stability of the fitness as advantageously as possible, we propose to adopt a temporally cumulative fitness instead of the above instantaneous fitness in RGA/E for effectually evaluating the performance of PSO.

The goodness of a particle swarm search can be used by the best particle x_k^b as a practical measure. For eliminating the influence of the mentioned temporary instability of the fitness $f(c_0,c_1,c_2)$, we use the temporally cumulative fitness of the best particle for evaluating the performance of PSO. Specifically, the fitness function used in EPSO can be expressed by

$$F(c_0,c_1,c_2) = \sum_{k=1}^{K} g(x_k^b)|_{c_0,c_1,c_2}. \tag{5.3}$$

It is obvious that the temporally cumulative fitness $F(c_0,c_1,c_2)$ evaluates the entire dynamic process of the best particle x_k^b. Therefore, it can basically eliminate the influence from the instantaneous fitness $f(c_0,c_1,c_2)$, which can just be determined by an instantaneous value. Based on the property of the proposed fitness $F(c_0,c_1,c_2)$, the goal of EPSO, which efficiently estimates appropriate values of parameter in PSO, should be easily achieved.

In order to express the property of the fitness, Fig. 5.3 illustrates the relationship between the instantaneous fitness $f(c_0,c_1,c_2)$ of the best particle at time k and its temporally cumulative fitness $F(c_0,c_1,c_2)$. We observed that the values of the fitness $F(c_0,c_1,c_2)$ almost linearly increase without receiving the influence according to the variation of the fitness $f(c_0,c_1,c_2)$, even if the value of the fitness $f(c_0,c_1,c_2)$ greatly increases or decreases during the particle swarm searches.

It is to be noted that the temporally cumulative fitness $F(c_0,c_1,c_2)$ of the best particle is used to express the performance of the estimated PSO model as a swarm

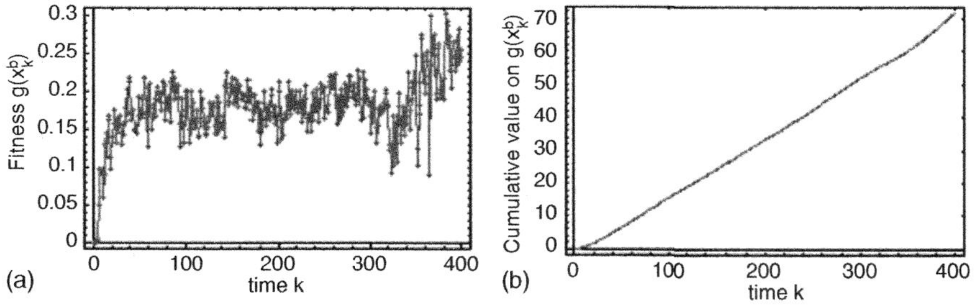

Fig. 5.3 The relationship between the instantaneous fitness $f(c_0, c_1, c_2)$ and the temporally cumulative fitness $F(c_0, c_1, c_2)$: **a** instantaneous fitness, $f(c_0, c_1, c_2)$, of the best particle at time k; **b** The temporally cumulative value of the instantaneous fitness $F(c_0, c_1, c_2)$ over time

representative. This is a new trial in evaluating a stochastic objective, and as far as we know, the fitness $F(c_0, c_1, c_2)$ has never been proposed, and it is applied for the first time to evaluating the performance of PSO by RGA/E.

5.4 Computer Experiments

5.4.1 Task and Parameters in Experiment

To demonstrate the effectiveness of EPSO, computer experiments with a two-dimensional optimization problem are carried out. Based on the obtained experimental results, we investigate the characteristics of dependency on different initial conditions, and analyze the resulting PSO models.

The search space of the optimization problem we used is 60×60 (i.e., $x_{\max} = 30$), the fitness value regarding the search environment in Fig. 5.4 is definded by

$$g(x) = 0.4\,e^{-\frac{(x_1-20)^2+(x_2-3)^2}{2\times 3^2}} + 0.25\,e^{-\frac{(x_1-15)^2+(x_2+20)^2}{2\times 4^2}} + 0.2\,e^{-\frac{(x_1+10)^2+(x_2+15)^2}{2\times 4^2}}$$
$$+ 0.2\,e^{-\frac{(x_1-10)^2+(x_2-20)^2}{2\times 4^2}} + 0.25\,e^{-\frac{(x_1-0)^2+(x_2+1)^2}{2\times 4^2}} + 0.05\,e^{-\frac{(x_1+20)^2+(x_2+5)^2}{2\times 6^2}}.$$

About the definition of initial conditions we used: particles start from a designated region (z_s, z_c). z_s denotes the area of the region, and z_c denotes the central coordinates of the region, respectively. In order to clarify how these conditions affect the performance of PSO, we use the following cases, Case 1: $z_{s_1} = 10\times 10$, $z_{c_1} = (-20, 20)$; Case 2: $z_{s_2} = 10\times 10$, $z_{c_2} = (0, 0)$; Case 3: $z_{s_3} = 30\times 30$, $z_{c_3} = (0, 0)$; Case 4: $z_{s_4} = 60\times 60$, $z_{c_4} = (0, 0)$ for solving the given optimization problem.

The computational cost for each iteration is proportional to the number of particles. Due to heavy computational cost for simultaneously executing RGA/E and PSO, EPSO should be carried out by using a small number of particles. To reduce

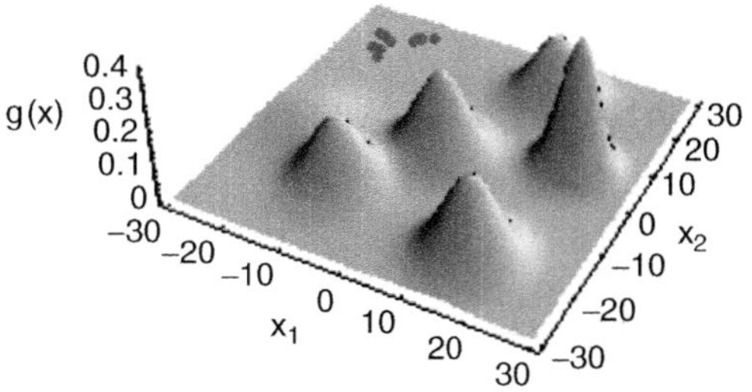

Fig. 5.4 Search environment

Table 5.1 Major parameters used in EPSO

Items	Parameters
The number of individuals	M = 100
The number of generation	G = 20
The number of superior individuals	s_n = 72
Roulette wheel selection	—
Probability of BLX-2.0 crossover	p_c = 1.0
Probability of random mutation	p_m = 1.0
Boundary value of the mutation	Δx_b = 1.5
The number of particle	P = 10
The number of iterations	K = 400
The maximum velocity	$v_{\max}$ = 20

the computational cost, we have to reduce the number of particles. On the other hand, the number of iterations might increase for a small number of particles; by taking the trade-off into account, we decide the number of particles is ten in our computer experiments for finding a global optimal solution [1].

Table 5.1 gives the major parameters used in EPSO for solving the given optimization problem shown in Fig. 5.4 in the following experiments. Note that because the search space of the given optimization problem is just two-dimensional, a global and effective search must be kept by increasing the diversity of individuals at each generation. So we set the parameters on the probability of crossover and mutation operators to be 1.0 in order do keep a high diversity of individuals in a population at each generation.

5.4.2 Experimental Results

Firstly, we observe the varying status of the values of parameters in PSO with implementing EPSO. For example, Fig. 5.5 shows the variations of the temporally cumulative fitnesses ($F(c_0, c_1, c_2)$), which are the superior six ranks among the entire population and corresponding to the parameter values of these individuals (c_0, c_1, c_2) at a search process (different generations).

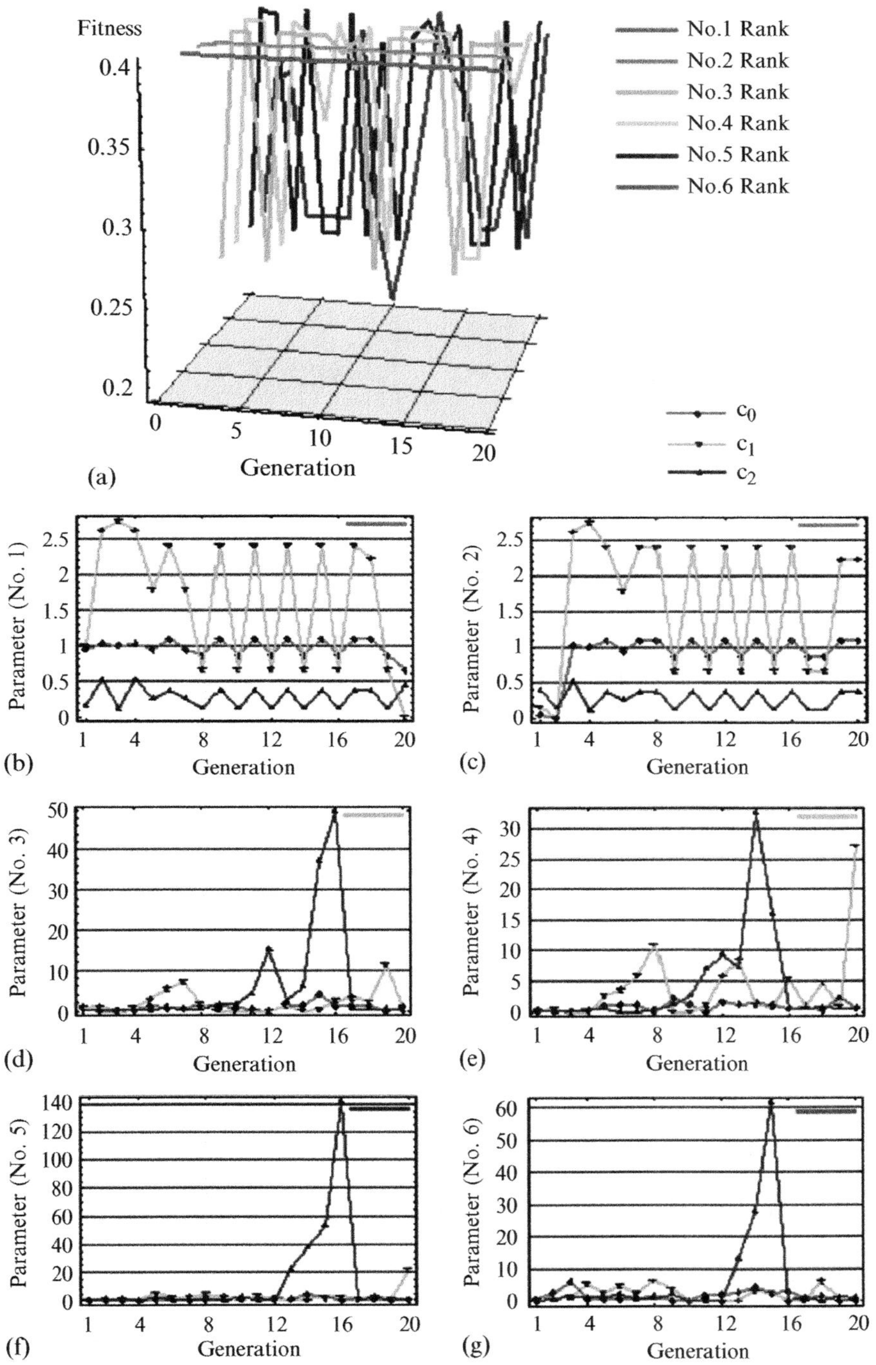

Fig. 5.5 The variations of the fitness which are the superior six ranks among the entire population and corresponding to the parameter values in a search process for Case 1: **a** the variations of the fitnesses with implementing EPSO; **b–g** The variations corresponding to parameter values c_0, c_1, c_2

Table 5.2 The resulting appropriate values of parameters in PSO by starting particles from the different initial conditions, that is, Cases 1 to 4.

Case	Parameter			Model	Success
	c_0	c_1	c_2	type	ratio (%)
	–	4.65 ± 0.0	2.22 ± 0.0	*b*	9.10
1	0.69 ± 0.2	–	5.45 ± 4.7	*c*	36.4
	1.01 ± 0.2	1.37 ± 0.8	1.62 ± 0.8	*d*	54.5
	–	–	11.6 ± 0.7	*a*	15.4
2	–	1.84 ± 0.0	7.86 ± 0.0	*b*	7.70
	0.87 ± 0.1	–	3.58 ± 3.2	*c*	30.8
	1.15 ± 0.3	1.65 ± 0.4	1.70 ± 0.6	*d*	46.1
	–	–	2.91 ± 1.9	*a*	23.5
3	–	0.66 ± 0.0	3.12 ± 0.0	*b*	5.90
	0.74 ± 0.4	–	1.31 ± 0.5	*c*	11.7
	1.05 ± 0.3	0.92 ± 0.6	1.66 ± 1.1	*d*	58.8
	–	4.96 ± 4.2	4.37 ± 2.1	*b*	16.7
4	0.81 ± 0.1	–	2.22 ± 1.1	*c*	33.3
	1.00 ± 0.2	1.91 ± 0.9	1.77 ± 0.9	*d*	50.0

As seen in Fig. 5.5(a), the result of the obtained highest fitness $F(\cdot) = 0.400008$ indicates that EPSO can find appropriate values of parameters in PSO for efficiently solving the given optimization problem, and the resulting parameter values of PSO arriving at the global optimal solution are not unique. This means that the proposed EPSO method is effective in estimating the optimal PSO model, and the obtained optimal PSO model is not the only one for solving the given optimization problem.

Table 5.2 indicates the resulting values of parameters in PSO for finding the global optimal solution by starting particles with the different initial conditions, that is, Cases 1 to 4.[4]

The "type" in Table 5.2 stands for the following velocity equations in PSO.

$$\begin{cases} type\ a: & v_{k+1}^i = c_2 r_2 (x_g - x_k^i) \\ type\ b: & v_{k+1}^i = c_1 r_1 (x_l^i - x_k^i) + c_2 r_2 (x_g - x_k^i) \\ type\ c: & v_{k+1}^i = c_0 v_k^i + c_2 r_2 (x_g - x_k^i) \\ type\ d: & v_{k+1}^i = c_0 v_k^i + c_1 r_1 (x_l^i - x_k^i) + c_2 r_2 (x_g - x_k^i) \end{cases}$$

Secondly, we observed that the estimated mean value of the parameter c_2 is never zero under any initial condition. It declares that the swarm confidence factor plays an important role in finding a global optimal solution. The fact perfectly agrees with the role of swarm communication in PSO; that is, the members of the swarm communicate the information of the best positions to each other.

We also observed that the mean values of parameters c_1 and c_2 become small and the mean value of parameter c_0 becomes bigger (approaches to 1.0) with extending the area of the square zone (comparison with the resulting statistical data

[4] Computing enviroment: Intel(R) Xeon(TM); CPU 3.40 GHz; memory 2GB RAM; computing tool: Mathematica 5.2; computing time: about 3 min per case.

of Cases 2–4). This means that the effect of individual confidence factor c_1 and swarm confidence factor c_2 becomes weaker and the effect of the inertia factor c_0 becomes stronger, respectively, for enhancing performance of PSO.

Thirdly, we implemented each resulting PSO model, respectively, for investigating convergence behaviors and characteristics. For example, Table 5.3 gives the obtained experimental data shown in Fig. 5.6 (with a box-and-whisker plot[5]) and

Table 5.3 The statistical data of the fitness by the resulting PSO models with different type (mean ± standard deviation) for Cases 1–4[a]

Items	Model				Case
	Type *a*	Type *b*	Type *c*	Type *d*	
Fitness	–	0.213 ± 0.149	0.218 ± 0.193	0.290 ± 0.121	1
PR	–	22.73%	36.36%	40.91%	
Fitness	0.287 ± 0.066	0.302 ± 0.073	0.362 ± 0.066	0.349 ± 0.073	2
PR	17.86%	21.43%	21.43%	39.29%	
Fitness	0.241 ± 0.043	0.260 ± 0.049	0.260 ± 0.049	0.362 ± 0.066	3
PR	5.0%	10.0%	10.0%	75.0%	
Fitness	–	0.347 ± 0.073	0.340 ± 0.094	0.384 ± 0.045	4
PR	–	27.91%	32.56%	39.53%	

[a]PR: The proportion ratio for success

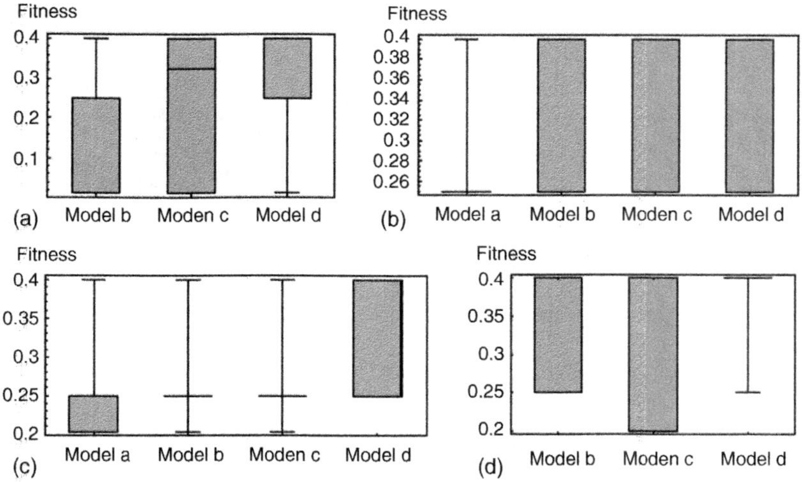

Fig. 5.6 Distribution of the fitness corresponding to each model for Cases 1 through 4

[5] A box-and-whisker plot is a histogramlike method of displaying data. The range is simply the difference between the maximum and minimum values in the data. The quartiles divide the data into quarters as the median divides the data into halves. There are three quartiles: $\{Q_1, Q_2, Q_3\}$. The first quartile Q_1 is the median of the lower part of the data, the second quartile Q_2 is the median of the entire set of the data, and the third quartile Q_3 is the median of the upper part of the data.

corresponding to the proportion ratio of the success by using the mean values of each parameter in each model for Cases 1 through 4.

Comparision with the resulting data of the proportion ratio of the success shown in Table 5.3, we observed that the type *d* of PSO models has better performance than those of other types of PSO models in finding the global optimal solution; that is, the mean value of the fitness is higher and the standard deviation value is smaller than those of others, respectively. This experimental result sufficiently confirms that the structure of the original PSO model is a rational and effective one for efficiently finding the global optimal solution.

As an example, Fig. 5.7 shows a search process of PSO with the resulting mean values of parameters in PSO (model *d*) by EPSO for Case 1. We can see that the result indicates that these particles arrived at the positions in the search space and distribution of these positions at the search process, and the effect of EPSO in exploration. Note that the pattern of the distribution is deeply dependent on the given optimization problem and the parameters used in the PSO.

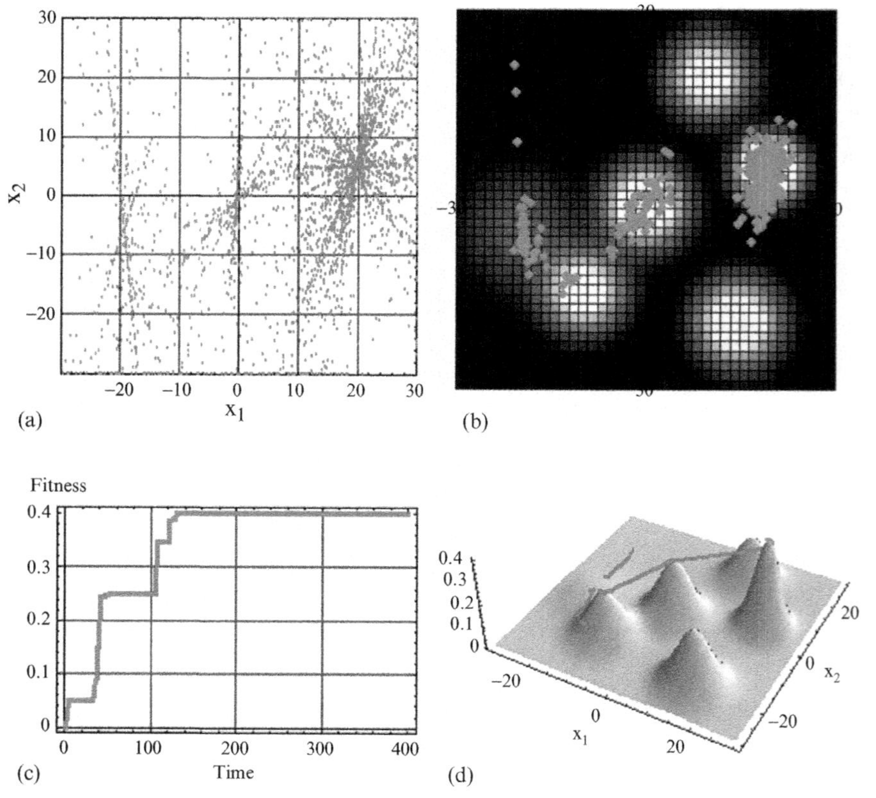

Fig. 5.7 A search process of PSO with the resulting average values of parameters by EPSO (model *d*) for Case 1: **a** the distribution of positions for all particles; **b** the distribution of positions for the best particle; **c** the variation of fitness for the global best particle; **d** the moving track of the global best particle in the search environment

5.4.3 Discussions

We are interested in finding optimal parameter sets that provide the PSO with reasonable exploration and convergence capability with applying a restriction for the velocity. It is obvious that the success ratio of using the structure of the original PSO (model *d*) is significantly bigger than those of other models (models *a*, *b*, and *c*) irrespective of various initial conditions. Because the value of parameter c_2 exists in each kind of velocity equation, the best search strategy in PSO for finding the global optimal solution is to follow the best particle which is nearest to the global optimal solution. And through implementing EPSO, the optimal structure of PSO models and their parameter values corresponding to the given optimization problem are simultaneously obtained.

However, what about the effect of EPSO in comparison with other methods? For answering the question, in the following experiments we compared the performances of EPSO, PSO, and RGA/E for the given optimization problem. Figure 5.8 illustrates the obtained experimental results for Case 1. The parameters in RGA/E shown in Table 5.1 were used.

As seen in Fig. 5.8, EPSO has the best performance in the frequency of fitness; that is, the particles found the global optimal solution (EPSO: 11, PSO: 3, RGA/E: 6). The obtained results and the goal of the use of the fitness $F(c_0, c_1, c_2)$ accord well together. From the viewpoint of finding the global optimal solution, the performance of RGA/E is inferior to EPSO and is superior to PSO. However, comparing the distributions of frequency of fitness, we observed that the search of EPSO or PSO relatively falls into local minima than those of RGA/E.

As contrasted with the above explaination, Table 5.4 gives the mean and standard deviation of the fitness values by EPSO, PSO, and RGA/E for Case 1. The mean values of the frequency distribution indicate that PSO is superior to EPSO and RGA/E.

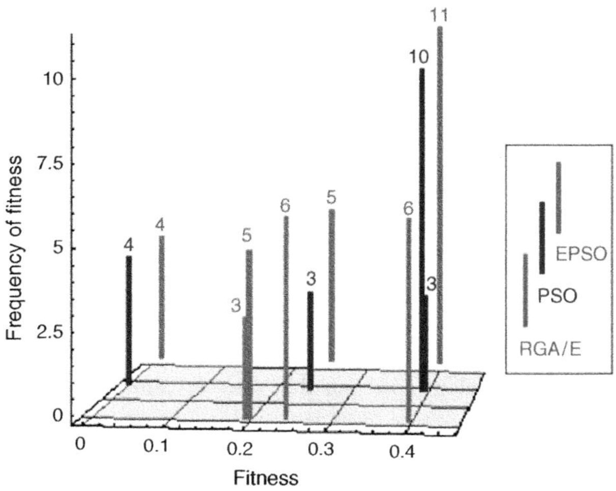

Fig. 5.8 Performance comparison among EPSO (red), PSO (blue), and RGA/E (green) for Case 1

Table 5.4 The statistical data of the fitness by EPSO, PSO, and RGA/E (mean ± standard deviation) for Case 1

	EPSO	PSO	RGA/E
Fitness	0.284 ± 0.149	0.299 ± 0.152	0.275 ± 0.083

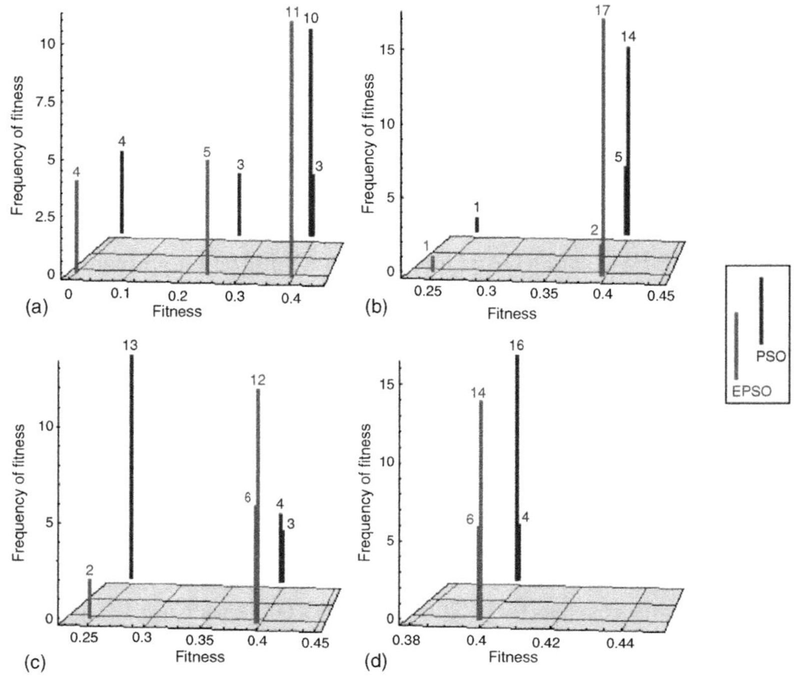

Fig. 5.9 Frequency distribution of fitness on EPSO and the original PSO. The fitness value for the global optimal solution is 0.400008. **a** Case 1; histogram with the values of parameters in the original PSO, histogram with the mean value of parameters in the PSO (type d) estimated by EPSO, and comparison with them; **b** Case 2; **c** Case 3; **d** Case 4

The reason is because the total number of the frequency of fitness arriving at the global optimal solution and near the global optimal solution for PSO is higher than that of EPSO and RGA/P. However, comparing with the number ratio (11:3) of the frequency of fitness reaching the global best position between EPSO and PSO, we found that EPSO is superior to PSO in the accuracy of statistics. This result sufficientlly reflects the characteristics of the temporally cumulative fitness $F(c_0, c_1, c_2)$ that we proposed in EPSO.

Of course, we should also point out that the performance of RGA/E depends upon given parameters too. Their parameter values also need to be optimized for fair comparison.

To further examine the effectiveness of EPSO, Fig. 5.9 illustrates the frequency distributions of the fitness by implementing the original PSO and by implementing

EPSO. We observed that the proposed EPSO method has superior performance for finding the global optimal solution than the original PSO has in Cases 1 through 4. The experimental results demonstrate the effectiveness of the proposed EPSO method, and indicate the variation of the performance of the estimated models according to different initial conditions. The results clearly exhibit that the temporally cumulative fitness $F(c_0, c_1, c_2)$ is effective, and suitable for finding appropriate values of parameters in PSO to efficiently solve the given optimization problem.

The use of the fitness $F(c_0, c_1, c_2)$ provides a useful way that systematically evaluates the performance of stochastic and population-based systems. Based on the property of the fitness, it is considered that the proposal is not only applicable to PSO, but also is applicable to other population-based adaptive optimization methods such as ant colony optimization (ACO) [4], genetic algorithms (GAs) [8], and so on. So it can be applied for efficiently estimating optimal models as a general fitness function of the metaoptimization method.

5.5 Conclusion

In this chaper, we have proposed a metaoptimization method, evolutionary particle swarm optimization, EPSO, that efficiently determines the values of parameters in PSO for finding a global optimal solution. It adopts a temporally cumulative fitness, which evaluates the varying position of the best particle at a search process, for evaluating the performance of PSO. This is the first proposal for estimating the appropriate values of parameters in PSO by a real-coded genetic algorithm with elitism strategy. Furthermore, it is considered that the fitness is not only applicable to evaluate PSO, but also extendable to various stochastic adaptive optimization algorithms with adjustable parameters for evaluating dynamic process.

Our experimental results show the effectiveness of the proposed method, and declare that the values of parameters in PSO are correctly estimated for finding the global optimal solution to the given two-dimensional optimization problem. The proposed EPSO method exhibits that it has a higher success ratio than those of the original PSO and RGA/E in exploration. Most importantly, the results suggest that the proposed method successfully provides a new paradigm for designing dynamic objective.

Even though better experimental results have been obtained, only a small-scale optimization problem with various initial conditions has so far been carried out. In order to enhance the efficiency and exactness in model selection and parameter identification, applications of the proposed EPSO method to complex and high-dimensional optimization problems are left for near-future studies.

Acknowledgment This research was partially supported by the 21st century COE (Center of Excellence) program (#19) granted to Kyushu Institute of Technology from the Ministry of Education, Culture, Sports, Science, and Technology (MEXT), Japan.

References

1. Bergh F and Engelbrecht AP (2001) Effects of swarm size on cooperative partical swarm optimisers. *Proceedings of the Genetic and Evolutionary Computation Conference* (GECCO2001), Morgan Kaufmann, San Francisco, CA, 892–899
2. Beielstein T, Parsopoulos KE, and Vrahatis MN (2002) Tuning PSO parameters through sensitivity analysis, Technical report of the Collaborative Research Center 531 Computational Intelligence CI-124/02, University of Dortmund
3. Carlisle A and Dozier G (2001) An off-the-shelf PSO. *Proceedings of the Workshop on Particle Swarm Optimization Indianapolis*, 1–6
4. Dorigo M, Maniezzo V, and Colorni A (1996) The ant system: Optimization by a colony of cooperating agents. *IEEE Transactions on Systems, Man, and Cybernetics-Part B,* 1:1–13
5. Eberhart RC and Kennedy J (1995) A new optimizer using particle swarm theory. *Proceedings of the Sixth International Symposium on Micro Machine and Human Science* Nagoya, Japan, 39–43
6. Eberhart RC and Shi Y (2000) Comparing inertia weights and constriction factors in particleswarm optimization. *Proceedings of the 2000 IEEE Congress on Evolutionary Computation* La Jolla, CA, 1:84–88
7. Eshelman LJ and Schaffer JD (1993) Real-coded genetic algorithms and interval-schemata. *Foundations of Genentic Algorithms* Morgan Kaufman, San Mateo, CA, 2:187–202
8. Goldberg DE (1989) *Genetic Algorithms in Search, Optimization, and Machine Learning.* Addison-Wesley, Boston
9. Gudise VG and Venayagamoorthy GK (2003) Evolving digital circuits using particle swarm. Neural Networks. *Proceedings of the International Joint Conference on Special Issue* 1:468–472
10. Kennedy J and Eberhart RC (1995) Particle swarm optimization. *Proceedings of the 1995 IEEE International Conference on Neural Networks* Piscataway, NJ, 1942–1948
11. Kennedy J (2006) In search of the essential particle swarm. *Proceedings of 2006 IEEE Congress on Evolutionary Computations*, Vancouver, BC, 6158–6165
12. Meissner M, Schmuker M, and Schneider G (2006) Optimized particle swarm optimization (OPSO) and its application to artificial neural network training. *BMC Bioinformatics* 7:125–135
13. Parsopoulos KE and Vrahatis MN (2002) Recent approaches to global optimization problems through particle swarm optimization. *Natural Computing* 1:235–306
14. Reyes-Sierra M and Coello Coello CA (2006) Multi-objective particle swarm optimizers: A Survey of the state-of-the-art. *International Journal of Computational Intelligence Research* 2(3):287–308
15. Spina R (2006) Optimisation of injection moulded parts by using ANN-PSO approach. *Journal of Achievements in Materials and Manufacturing Engineering* 15(1–2):146–152
16. Storn R and Price K (1997) Differential evolution—A simple and efficient heuristic for global optimization over continuous space. *Journal of Global Optimization* 11:341–359
17. Zhang H and Ishikawa M (2005) A hybrid real–coded genetic algorithm with local search. *Proceedings of the 12th International Conference on Neural Information Processing* (ICONIP2005) Taipei, Taiwan R.O.C, 732–737
18. Xie XF, Zhang WJ, and Yang ZL (2002) A disspative particle swarm optimization. *Proceedings of the IEEE Congress on Evolutionary Computation* (CEC20020) Honolulu, 1456–1461

Chapter 6
Human–Robot Interaction as a Cooperative Game

Kang Woo Lee and Jeong-Hoon Hwang

6.1 Introduction

As robots become important parts of our daily lives and interact with human users, human–robot interaction (HRI) is raised as an important issue in robotics as well as relevant academic areas, and researchers in those areas pay more attention to how robots can communicate and cooperate with humans. Nevertheless, there seem to be few theoretical frameworks that guide designing, evaluating, and explaining human–robot interaction. Much current research on the topic is rather technically oriented, and thus suffers from poverty of theory.

This does not simply originate from the current limitation of artificial intelligence (AI) technologies or lack of formalization of psychological studies. The studies on humans and robots have been independently carried on in psychology and robotics, and, no theoretical bridge that links two agents with heterogeneous abilities has been developed to explain the interaction process.

In this chapter we try to build a bridge between human and robot using a game-theoretic approach. Game theory has been developed to explain strategic decision making with respect to another agent's decision. It also has been widely applied to explain human behavior and design interaction processes not only in psychology, auction, military strategy, diplomatic policy, and law enforcement [2, 5, 8], but also computational biology, artificial intelligence, and so on. Game theory provides a formal framework to account for interaction activities in various situations. Especially, our research interests focus on communicative interaction and cooperative decision making that is described later.

We first consider some characteristics of human–robot interaction that illustrate the heterogeneous abilities of humans and robots, the triadic relationship, and interaction state transition of HRI. Secondly, we briefly review a basic introduction to game theory, and try to explain communicative games of HRI in terms of the game-theoretic perspective. Thirdly, we describe a cooperative decision-making

Oscar Castillo et al. (eds.), *Trends in Intelligent Systems and Computer Engineering*.

method in which cues from human users can be integrated with knowledge of a robotic system. Finally, a short conclusion follows at the end of this chapter.

6.2 Some Characteristics of Human–Robot Interaction

6.2.1 Heterogeneous Abilities of Humans and Robots

An important aspect of HRI is that it can be characterized as the interaction between more than two agents with different perceptual, cognitive, emotive, and executive abilities. That is, the artificial intelligence afforded a robot to think could be fundamentally different from a human's natural intelligence in the way that they perceive, recognize, and represent an object. A robot may perceive a face with respect to skin color and eigenvector space, whereas a human may perceive it in a way to process a global configuration and a relative relation from one feature, such as a nose, to another such as a mouth. That is, a user and robot may differently represent what they are watching even if both see the same object. Moreover, robots with AI are still not capable of satisfactorily carrying out real-world tasks. They are limited to solving a task that requires complex information processing abilities.

These qualitative and quantitative differences between two information processing systems may impose constraints on what can be interacted on, how the interaction can be accomplished, or what kinds of role robots may have to take. In this sense, Fong et al. [9] considered a robot as an "asker of question" because a robot usually performs poorly whenever its capabilities are inadequate or ill-suited for a task. On the other hand, a human user may take an assistant role to provide advice. Therefore, we may design the interaction system of a robot that allows a user's intervention throughout a task if the robot requests a human user's help. Similarly, the human–robot relation can be analogical to a parent–infant or teacher–student relation in the developmental robotic paradigm [7, 16]. A robot may continuously acquire new objects or skills and update its knowledge bases in a manner such that it communicates with and imitates human users or other robot mates.

6.2.2 Triadic Relationship of Human–Robot Interaction

Conventionally, AI has focused on the linkage between an intelligent system and its environment, whereas human–computer interaction (HCI) or human–machine interaction (HMI) has focused on the linkage between the human user and an artificial system. These linkages have been independently studied in those areas. However, as a robot moves in our daily environment and frequently interacts with ordinary people, a robotic system is asked not only to be intelligent but also to be communicative.

In AI, the establishment of a relationship between an agent and its environment concerns knowledge representation about the environment in which the agent operates. Given that how the agent interacts with its environment is based on its environmental knowledge, the knowledge constructs the links between the agent and its environment. However, knowledge about the environment is not the environment itself; rather, it represents the environment in various internalized forms.

In terms of the interaction between two agents, communication is defined as the interactive process of sharing information. In a simplistic form of communication, at least two communicators are involved in the process, sending and receiving information via symbolic mediates that construct themselves and are constructed from each communicator's knowledge.

Any message delivered by a sender contains information from his or her knowledge, and the received message is interpreted on the basis of the knowledge possessed by the receiver. If the message contains information that is not known to the receiver, the receiver makes a great effort to interpret the message. Therefore, the two agents may communicate easily if the communicative activities are based on their common knowledge, or on their shared ground.

Thus so far, interactions—agent–environment and agent–agent—have been considered in the two-way relationship. However, these two-way relationships cannot capture important aspects of HRI for the following reasons. First, robots with artificial intelligence remain incapable of satisfactorily carrying out real-world tasks. They are limited in perceptual, cognitive, executive, and emotional abilities, and frequently require human involvement as they interact with the environment. Second, as is the nature of HRI, agents have different intelligence capabilities originating from both quantitative and qualitative differences between two representation systems. Consequently, the same external object can be represented differently by each agent, or cannot perhaps be represented. This demands a triadic ground among those agents and external objects. Therefore, the two types of interactions are interwoven and the three components—human, robot, and environment—are interconnected in HRI.

6.2.3 State Transition of Interaction States

The uncertainty caused by two different systems forces them to minimize it through the interaction in which the states of the systems are evolved. That is, this may imply the interaction can be decomposed into states that represent how much common ground they share. Along this line, we proposed an interaction model that illustrates interaction states as well as state transition using information theory as shown in Figs. 6.1 and 6.2 [12, 13].

In the model, the three-way relationship between human knowledge, robot knowledge, and environmental objects is decomposed into various areas that are shared by the two agents, as well as those that are not shared. Each area is closely

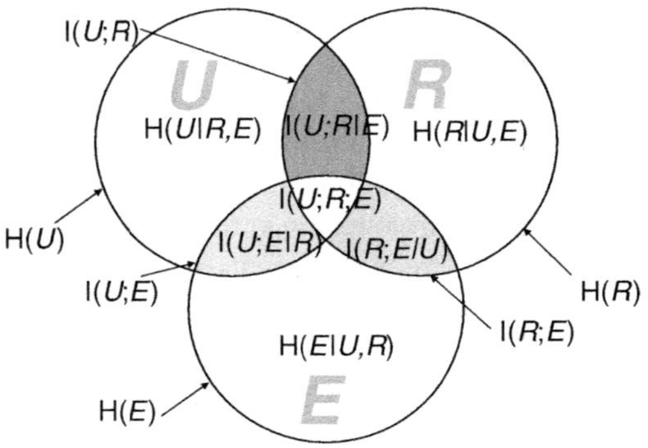

Fig. 6.1 The characters U, R, and E denote the knowledge possessed by a human, the knowledge possessed by a robot, and the environment. The knowledge possessed by agents can be described in probabilistic terms. The amount of the uncertainty of the random variables is defined as being taken over the average of the variable, and can be expressed as the entropy of the variables [6]

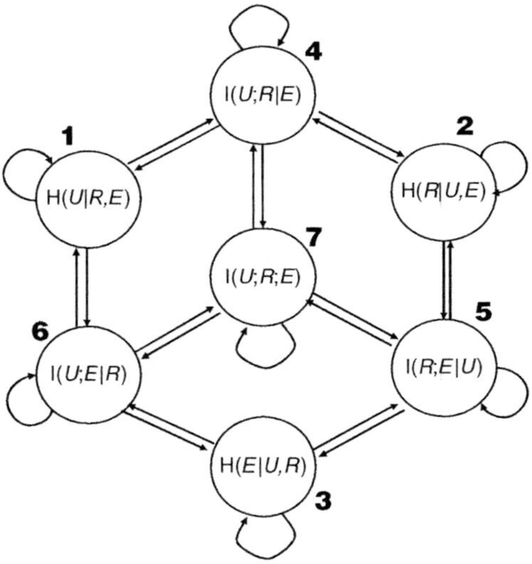

Fig. 6.2 State transition diagram of the HRI process model. Each region of Fig. 6.1 is mapped into each state of the state transition diagram

related to the patterns of interaction, as the interaction pattern varies according to what they commonly know or do not know.

Furthermore, it was assumed that the shared ground between two agents or between agents and environment can be established through interactions associated with reducing the uncertainty in each agent's knowledge. In order to efficiently communicate between agents, the knowledge states of the agents should converge into $\mathrm{I}(U;R;E)$, in which the knowledge of both agents is in the shared ground state. This implies that the interaction between two agents can be considered as a sequential process in which an agent's knowledge state evolves over time. The sequential process of the interaction is modeled as a state transition diagram with the proposed approach as shown in Fig. 6.2. The regions in the diagram of the three-way relationship are mapped to the knowledge states of each agent.

6.3 Interaction as a Game

In a game, the players can represent the participants or agents who are involved in decision making in a game. Each player has a set of actions available to make a move during the game. A player can be placed at a stage in which she is required to make a decision. The decision can be made on the basis of the state of the game at a state that represents the configuration of the game. A state of the game can be transferred to a new state through a state transition function that specifies the output probability of an action at a state.

A strategy is a plan of action that tells players what to do in each situation, and specifies what actions (moves) a player could take in coping with all possible counteractions of the co-player. A strategy can be deterministic or probabilistic depending on whether a random move is allowed. With a pure strategy where a decision is made deterministically, a game is played in the same way whenever a game is performed, whereas a game may reach different ends with a mixed strategy where a decision can be made randomly to some extent.

In a game, payoffs represent a player's motivation for the selection of a particular action. The payoffs may represent the amount of reward (profit), penalty (loss), or utility using real values or ordinal ranks. The payoff is also associated with the amount of satisfaction (or preference) derived from objects or events, so that for an agent who prefers "coffee" to "tea," it would be said to associate with high utility with coffee, whereas it would be said to associate low utility with tea. In this sense agents act so as to maximize (or minimize) their utility (penalty) [2, 17].

Strategic games can be divided into two classes, cooperative and noncooperative games according to whether players can make binding commitments. In a cooperative game, agents are allowed to choose their actions and jointly reach some desired outcomes through binding commitment, whereas in a noncooperative game, agents are self-interested and care about maximizing their profit (or minimizing their loss) regardless of the results for others, and thus independently make decisions. Therefore, a set of possible joint actions of a group are primitives in a cooperative game, but a set of possible individual agents are primitives in a noncooperative game.

In this respect, agents in a cooperative game may ask, "What strategic decision will lead to the best outcome for all of participants?" whereas agents in a noncooperative game may ask, "What is the rational decision to cope with other agents' choices?" The answers to the question are closely related to the solution in game theory, and usually refer to an equilibrium which is a stable state: forces in a system balance each other and remain at rest unless disturbed by an external force.

In conventional game theory a decision made by an agent is premised on the two assumptions: common knowledge and rationality. That is, any decision is bound not only to common knowledge that is shared by agents, but also rationality that drives an agent to choose an action generating maximum utility.

6.4 Communicative Game in Human–Robot Interaction

6.4.1 Twenty Questions Game

A major portion of human–robot interaction is concerned with communicative activities that are composed of sending and receiving signals in various forms including verbal, facial, gestural, and iconic forms. Basically, a communicative game consists of sending a message to a receiver and his response to that message [4, 10].

Suppose that communicative agents are in a particular state and have elements of some set M, in which a sender can observe the true state, but the receiver cannot. However, the receiver believes what that true state might be with probability distribution P over M. Let M be the set of the message where $M = \{m_1, \mathrm{K}, m_n\}$. This can be interpreted as the action set that the sender has, whereas the action set of a receiver can be denoted $A = \{a_1, \mathrm{K}, a_k\}$. In a possible cooperative game of HRI, a sender is supposed to send a message that can be easily understandable for the receiver, whereas the receiver is supposed to take an action corresponding to the message.

Consider the game Twenty Questions in which one of the two players explains about an object, and the other answers what it might be (see Fig. 6.3). If each

Fig. 6.3 Twenty Questions game. In the experiment subjects were asked to carry out a questioning and answering task in which one of a pair of human subject and robot explained about an object, and the other answered what it might be

message m delivered by a questioner contains one object feature such as color, shape, or function, then a responder has to decide what it might be among many others on the basis of the given messages. In this interaction the decision made by the receiver can be represented as the probability of an object over jointly distributed independent messages. Therefore, the probability of an object at a certain interaction stage n can be written as follows.

$$P(a_i|M_n) = \prod_{k=1}^{n} P(a_i|m_k) \tag{6.1}$$

where M_n is m_1, Λ, m_n. The conditional probability is sequentially updated through the interaction between a sender and receiver. Therefore, Eq. 6.1 can be rewritten as follows.

$$P(a_i|M_n) = P(a_i|M_{n-1}, m_n) \tag{6.2}$$

Based on Eq. 6.2, the receiver finds the object that has the maximal conditional probability in given messages, and sets the object as her belief of what the user indicated. If the belief is not what the sender meant, the receiver may set the conditional probability of the object 0 and replace the belief with the object that has the second maximal probability or request further information.

6.4.2 Common Knowledge—Shared Ground Between Human and Robot

Underlying the Twenty Questions game was the first assumption (common knowledge). That is, any game involving more than two persons—shaking hands, teaching, kissing—cannot be played without some amount of information or common ground [3]. Consider a simple situation where a human user asks a robot to bring a cup. In this simple statement, a vast amount of assumptions (what the robot can do, what it recognizes, and so on) are implicitly embedded, and a proper response from the robot is driven by actions corresponding to the assumptions.

According to Lewis [15] and Aumann [1] who introduced and formalized the concept of common knowledge, knowledge can be common among a group of agents if everyone has it, everyone knows that everyone has it, everyone knows that everyone knows that everyone has it, and so on. Therefore, it is recursively endless and thus is difficult to apply when explaining HRI. In our work, HRI was expressed in terms of a three-way interaction among a human, robot, and environment as shown in Fig. 6.1. The amount of information shared by the variables was expressed by conditional information as well as mutual information. Furthermore, we pointed out that the interaction patterns between human and robot are linked by how they share the knowledge of the environment.

For instance, the state $\mathrm{I}(U;E|R)$ means that the robot does not have the environmental knowledge that the human user possesses. In this case, the robot is required to update its knowledge through interaction with the human user, or the user is required to take different strategies to communicate with the robot. Therefore, we can easily expect that if human and robot have more in common, the interaction can be more efficient. In this aspect, how to establish the common ground between the two agents that have heterogeneous abilities is critical in HRI.

6.5 Cooperative Decision Making

In the communicative game Twenty Questions the decision made by a receiver is based on the messages delivered by a sender. Therefore, the performance of the task is determined by how much knowledge they share. That is, the performance of a robotic system depends on whether the built-in knowledge is constructed to deal with human knowledge. This is potentially problematic in as much as there are a huge number of objects and object variations, and thus no current robotic system cannot deal with every detail of the environmental objects.

In fact, it is impossible even for us to memorize the objects or events surrounding us. We often use information from other persons or devices that extends our cognitive abilities, and completes a task. Knowledge can be distributed across groups of people and external devices [11]. This means that it is not necessary to internalize every detail of an environmental object to construct its knowledge base, rather it is more important to deal cooperatively with knowledge from different systems.

In this regard we present a cooperative decision-making method in which the information from a human user is integrated with the knowledge of a robotic system. The potential benefits of the cooperative decision making are (1) avoiding combinational explosion, (2) increasing the reliability of the decision-making process, and more important, (3) establishing cooperative interaction between a human user and the robotic system.

6.5.1 A Simple Decision-Making Process

Consider a simple task in which a player decides whether a given message (M) describing a single feature belongs to signal or noise. Because the message contains an object feature, such as color, shape, and the like, that is distributed along some degree of the feature continuum, there is always uncertainty as to what it might be. The task for the player concerns two kinds of message types (signal and noise) and two possible actions, say "signal" or "noise" $\{S,N\} \in A$ so there are four possible outcomes of the message–response matrix—correct rejection, false alarm, miss, and hit—as presented in Table 6.1. For instance, the hit means that the player says "S" occurs when a signal S is presented.

If the player's decision is based on only a given message M, then the player's response to the message can be written as conditional probabilities of responses given the message M.

Table 6.1 Message–response matrix of simple decision making

Message types	Responses	
	S	N
Signal	Correct rejection (a)	False alarm (b)
Noise	Miss (c)	Hit (d)

The characters a, b, c, and d represent the amounts of cost or utilities corresponding to its decisions

$$P(A=S|M)=\frac{P(M|S)P(S)}{P(M)} \tag{6.3}$$

$$P(A=N|M)=\frac{P(M|N)P(N)}{P(M)}$$

One way to decide whether the message M is a signal s or noise n is to calculate the likelihood ratio:

$$L(M)=\frac{P(M|A=S)}{P(M|A=N)} \tag{6.4}$$

where each point of M along the feature dimensions has an associated value of the likelihood ratio. The likelihood ratio indicates the relative heights of the two distributions that cross at $L(M)=1.0$ if the prior probabilities $P(S)$ and $P(N)$ are equally distributed. Therefore, the optimal criterion that maximizes the number of correct responses can be found if one makes a decision in the following way. If $L(M)<1$, respond N, otherwise respond S. Similarly, if we consider the utilities when it makes a decision, then we have

$$\frac{P(M|A=S)}{P(M|A=N)} \underset{A=S}{\overset{A=N}{\gtrless}} \frac{(b-a)}{(c-d)} \tag{6.5}$$

Now, consider the message M that has n statistically independent components (or features). Then, the conditional probability of Eq. 6.3 can be written as the conditional probability of each response given the production of feature probabilities $P(m_i)$ as follows.

$$P(A=S|M)=\prod_{i=1}^{n}P(S|m_i) \tag{6.6}$$

$$P(A=N|M)=\prod_{i=1}^{n}P(N|m_i)$$

The likelihood ratio becomes

$$L(M) = \prod_{i=1}^{n} \frac{P(m_i|S)}{P(m_i|N)} \tag{6.7}$$

Equivalently, we can write the above equation in the log-likelihood form.

$$\ln L = \sum_{i=1}^{n} \ln \Lambda_i \tag{6.8}$$

where $\Lambda_i = P(m_i|S)/P(m_i|N)$.

6.5.2 Relationship Among Response, Robot's Knowledge, and Human User's Message

In a typical communicative task with a robot, the robot as a receiver is required to decide whether a message belongs to the class A (signal) or the class B (noise) based on its knowledge as well as the human user's message. For instance, the command, "Bring the cup," with a finger pointing gesture can be analyzed into two types of information: one belonging to the cup itself and the other belonging to the pointing gesture. The decision task for a robotic system is to find the cup indicated by the human user among many other objects.

If we assume that a human user and a robot share some knowledge about external objects, and the user names an object with finger pointing when he interacts with the robot about the object, then the object features are linked with the name of the object and the cue is linked with the pointing gesture. Therefore, the message from a human user can be decomposed into two types: the components belonging to the object properties T and the components belonging to cues C associated with the human's finger pointing. As a result, the conditional probability of Eq. 6.6 can be decomposed as follows.

$$P(A|X) = P(A|T,C) = \prod_{i=1}^{n1} P(A|x_i^T) \prod_{j=1}^{n2} P(A|x_j^C) \tag{6.9}$$

That is, the responses in a cooperative decision task can be represented as a conditional probability of A jointly given object features x^T associated with the robot's perceptual knowledge and cues x^C given by the human user.

The robot may strategically make a decision about possible candidate objects derived from the robot's sensors. That is, if a candidate object is strongly associated with the pointing cue, the candidate object is more likely to be a target. If a candidate object is weakly associated with the pointing cue, the candidate object is less likely to be a target. Similarly, a noncandidate object not associated with the pointing cue is more likely to be a nontarget, whereas a noncandidate object associated with the pointing cue is less likely to be a nontarget. Therefore, the robot

Table 6.2 The relationship between robot perceptual knowledge about objects and cues given by a human user

Robot's perception	Human's point cue	
	Pointed	Not pointed
Target	Matched	Not matched
Nontarget	Not matched	Matched

According to the relationship between the components, the consistency and inconsistency can be assigned to the matrix

may choose an object from the most likely candidate object to the less likely candidates in a descending manner. In this sense, the two components—what the robot senses and where the user indicates—are correlated. These relations between them are summarized in Table 6.2.

6.5.3 Model of Cooperative Decision Making

Based on the relation between cue and test stimulus, a model of cooperative decision can be described from Eq. 6.9 as follows.

$$P(A|X^T, X^C) = f(X^T, g(X^T, X^C)) \tag{6.10}$$

in which both inputs are denoted as the weighted sums, $X^T = \sum_{i=1}^{n1} x_i\, w_i$ and $X^C = \sum_{j=1}^{n2} x_j\, w_j$, respectively.

Therefore, the cooperative decision is the function of the two terms driven by the candidate object x^T and the interaction between the candidate object and cue $g(x^T, x^C)$ [14]. The decision is basically made by the perceptual knowledge of a robot if no cue given by a human user is available. If a cue from a human user is given to the robot, the decision is made by the perceptual knowledge of the robot as well as the interaction between the perceptual knowledge and the cue. To some extent, the second term can be considered as a gain that influences on the decision task.

Even though there are many ways to describe the interaction between two variables, the multiplicative operation provides an interesting description of the dynamic interaction between more than two variables. This multiplicative operator correlates two variables, and provides the measure of their consistency.

$$g(X^T, X^C) = X^T \exp(X^T X^C) \tag{6.11}$$

We now take the two terms into a sigmoid function, then the response function is:

$$P(A|X^T, X^C) = \frac{1}{1 + \exp(-(X^T + g(X^T, X^C)))} \tag{6.12}$$

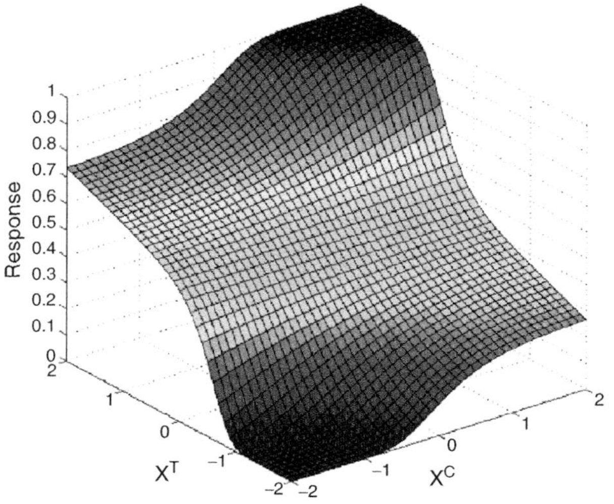

Fig. 6.4 The activation of the response function. The activation values were plotted over two inputs x^T and x^C. Interestingly, if two inputs agree, then the function obtains gains, but if two inputs disagree, then it pays costs

The response of the function was plotted in Fig. 6.4. The figure shows interesting properties that are desirable for cooperative decision making. First of all, when two input variables agree, then the function obtains gains (positive or negative gains). However, if two inputs disagree, then it pays costs. In other words, the objects that are jointly indicated by the robot's perceptual knowledge and a human user's cue are highly activated, otherwise less activated.

6.6 Conclusion

In summary, a game-theoretic approach for HRI has been developed to provide a plausible and formal account for the interaction. In particular, we propose a model of cooperative decision making for human–robot interaction. The model utilizes both the robot's knowledge and a human user's cue, and cooperatively makes a decision. Our research focus does not lie on the development of an artificial system that works alone, but on a cooperative system that interacts with a human user. In this regard our research is a distinctive attempt to link agents with heterogeneous abilities. However, no empirical study is presented in this chapter, therefore future work will extend our framework to experimental situations in which human subjects and robots interact and analyze interaction patterns based on our approach. We are also preparing to design experimental situations to show cooperative interaction between humans and robots. Finally, we expect that our approach will contribute to designing, analyzing, and evaluating the interaction process between human users and robotic systems.

References

1. Aumann R (1995) Backward induction and common knowledge of rationality, *Games and Economic Behavior* 8: 6–19.
2. Bierman HS, Fernandez L (1998) *Game Theory with Economic Applications*, Addison-Wesley, Reading, MA.
3. Clark HH, Brennan SE (1991) Grounding in communication. In LB Resnick, J Levine, SD Teasley (Eds), *Perspectives on Socially Shared Cognition*. APA.
4. Coiera E (2001) *Mediated Agent Interaction*. LNCS 2101, Springer, New York, pp. 1–15.
5. Colman AM (2003) Cooperation, psychological game theory, and limitations of rationality in social interaction. *The Behavioral and Brain Sciences*, 26: 139–153.
6. Cover TM, Thomas JA (1991) *Elements of Information Theory*. John Wiley & Sons, New York.
7. Dautehahn K, Billard A (1999) Studying robot social cognition within developmental psychology framework. In *Proceedings of the Third European Workshop on Advanced Mobile Robots, Switzerland*, pp. 187–199.
8. Downs A (1957) *An Economic Theory of Democracy*. Harper, New York.
9. Fong T, Thorpe C, Baur C (2003) Robot, asker of questions. *Robotics and Autonomous Systems*, 42: 235–243.
10. Hasida K (1996) Issues in communication game. In *Proceedings of the 16th Conference on Computational Linguistics*, pp. 531–536.
11. Hutchins E (1995) *Cognition in the Wild*. MIT Press, Cambridge, MA.
12. Hwang JH, Lee KW, Kwon DS (2006) A Formal model of sharing grounds for Human-Robot Interaction. In *Proceedings of 15th IEEE International Workshop on Robot and Human Interactive Communication*.
13. Hwang JH, Lee KW, Kwon DS (2006) A formal model of human-robot interaction and its application for measuring interactivity, *International Journal of Human Computer Interaction*.
14. Kay J, Phillips WA (1997) Activation functions, computational goals and learning rules for local processors with contextual guidance. *Neural Computation*, 9: 895–910.
15. Lewis D (1969) *Convention: A Philosophical Study*, Harvard University Press, Cambridge, MA.
16. Lungarella M, Metta G, Pfeifer R, Sandini G (2003) Developmental robotics: A survey. *Connection Science*, 15(4):151–190.
17. Osborne MJ, Rubinstein A (1994) *A Course in Game Theory*, MIT Press, Cambridge, MA.

Chapter 7
Swarm and Entropic Modeling for Landmine Detection Robots

Cagdas Bayram, Hakki Erhan Sevil, and Serhan Ozdemir

7.1 Introduction

Even at the dawn of the 21st century, landmines still pose a global threat. Buried just inches below the surface, combatants and noncombatants alike are all at risk of stepping on a mine. Their very nature is such that these furtive weapons do not discriminate, making it an urgent task to tackle the problem. According to the U.S. State Department [1], based on an estimate reported just a few years ago, there are well over 100 million anti-personnel mines around the world. The existence of these passive weapons causes a disruption in the development of already impoverished regions, as well as maiming or killing countless innocent passers-by. Since the ratification of the anti-personnel mine total ban treaty in 1997, their detection, removal, and elimination have become a top priority. Nevertheless, at the current rate, given the manpower and the man-hours that could be dedicated to the removal of these sleeping arms, it would take centuries. The concerns regarding the speed of removal and safety of the disposers eventually bring us to the discussion of the proposed method.

There are numerous efforts for utilizing robots for landmine detection and/or removal whereas nearly all research activity based on robotics seems to be focused on using sophisticated systems with costly hardware [2]. Even then, the speed that these robots can offer for mine detection is limited because the high associated cost limits the number of robots procured. Another shortcoming of a complex system is the difficulty of repair and maintenance in a harsh environment such as a minefield and also the possible catastrophic loss of the entire system due to an unexpected mine detonation.

As opposed to the idea of having a complete agent with state-of-the-art equipment, the goal may be accomplished by down-to-earth individuals working as a team, indirectly guided by a competent alpha agent. The task of the swarm is to autonomously sweep an area for mines as quickly as possible, as safely as possible. The swarm should be scalable and robust: loss or gain of members should not affect the behavior and reliability of the system, and obstacles or any other disturbances

Oscar Castillo et al. (eds.), *Trends in Intelligent Systems and Computer Engineering.*

should not affect the stability. Following these guidelines, the main objective of this study is to present an efficient autonomous navigation and detection method to guide a group of inexpensive robotic agents. To lower the cost of the agent, a minimal number of sensors, actuators, Ics, and other components should be used. In addition, the navigation method should fare well without needing very precise (and costly) sensors.

Nature already provided good solutions to manage groups of less able beings: fish schools, ant swarms, animal packs, bird flocks, and so on. With the growing desire of humans to create intelligent systems, these biosystems are being thoroughly inspected [3–10] and implemented [11–14] in various studies.

In this study a robotic agent is referred to as a drone, the group of robotic agents is referred to as a swarm, and the agent with mapping abilities is referred to as the alpha drone.

7.2 Desired Swarm Behavior and Drone Properties

Before going into the details of anything else, one should define the desired behavior of the swarm. The swarm should:

- Autonomously sweep a prescribed area.
- Exhibit swarming (collision avoidance, polarization, attraction to swarm mates).
- Designate the mine locations with an acceptable accuracy.
- Find all the mines in the swept area (high reliability).
- Be able to tolerate loss of members due unexpected situations.

At this point, equipping all members with advanced sensors and microcomputers will quite increase the cost, so it is decided to have two types of agents: drones and an alpha drone. Because our main interest is to have minimalist robotic agents that could be fielded in large numbers to speed up the mine cleaning process, a drone should:

- Have a unique identification number.
- Know and control its heading and speed.
- Have a means of wide-angle proximity detection (i.e., sonar array). These sensors need not be very precise. The behavior model should work for rough and noisy sensor readings.
- Have a means of detecting mines (i.e., metal detector).
- Have a means of wireless communication although it should consume low power, be inexpensive, and therefore low-range.
- Have the means of making simple preprogrammed decisions.
- Avoid stepping on mines.

The alpha drone's main task is to record the mine locations and indirectly control the drones by presenting them with a desired heading; the alpha drone should:

- Know its absolute location with good accuracy (i.e., using GPS).

- Know the boundaries of the area to be swept.
- Be able to indirectly force the drones to move in a direction.
- Collect landmine location data from drones and mark them on a map.
- Never step on a mine.
- Have all the necessary subsystems of a drone.

7.3 Drone Model

There are some proposed distributed behavior models for fish schools and bird flocks. Our particular interest is in the models proposed by Aoki [8], Huth and Wissel [9], Couzin et al. [10], and Reynolds [11]. To summarize, schooling and flocking was explained using three concentric zones: a zone of repulsion, zone of orientation, and zone of attraction. Also it was shown that the overall heading of the flock can be controlled by adding a migratory urge, which is simply a direction. The models not only explain the schooling phenomena to a good extent but also give a good tool to manage groups of robots. In a previous simulation work by authors based on these fish school models, it has been seen that the school tends to move in a hexagonal close-packed formation. This is an ideal formation pattern to be used in mine sweeping because there are no gaps left in a group of mine detectors. The authors began with these preliminary models, altered them to fit the world of mobile robots by translating the means of sensing and locomotion, and extended the model further.

Perhaps the most important problem of adapting these originator models to the world of robotics is the means of sensing. In the biological world, thanks to millions of years of evolution, even the simplest organism is equipped with highly precise and effective sensors. However, the robotic systems still have to utilize relatively poor sensors compared to those of biological organisms. Despite the advances in image processing and pattern recognition techniques, a full-blown visual sensor is still too costly or merely incompetent to deal with the complicated real world. The proposed model in this study is especially devised for robotic agents equipped with simple, readily available, and well-understood sensors such as infrared transceivers.

In this study we call an individual mobile robotic agent a drone. A drone is a simple entity, trying to find its way following the alpha drone's migration orders and mimicking other drones' movement while trying to survive. In our basic model, a drone has an array of near-range proximity sensors (possibly ultrasonic), a low-range wireless transmitter/receiver (possibly RF), differential locomotion (i.e., tracks), a simple microcontroller, a mine detector, a digital compass, and an attraction beacon (possibly an IR beacon with a certain frequency).

The drone model implemented in simulation is given in Fig. 7.1. Radii r_{PSA}, r_{MDS}, r_{AB}, and r_{WCS} are ranges for the proximity sensor array (PSA), mine detection sensor (MDS), attraction beacon (AB), and wireless communication system (WCS), respectively, where l_{MDS} represents the distance where the mine detection sensor is placed apart from the robot body. The hatched circle represents the robot body.

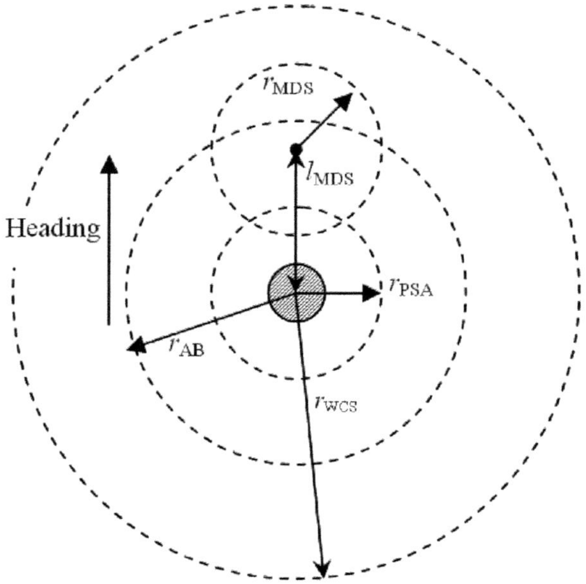

Fig. 7.1 Proposed drone model

The PSA is an active sensor array that gives true/false outputs within a certain angle resolution; a drone has a rough idea of the bearings of nearby objects. MDS gives an analog reading; in the case of a metal detector the output will be higher when a metallic object is closer and vice versa. WCS has two bidirectional channels, one being used for communication with the alpha drone (alpha channel) and the other with drones (beta channel). A drone also knows and can change its speed and heading. The attraction sensor array (ASA) is a passive sensor that detects other AB signals within a certain angle resolution; a drone has a rough idea of the bearings of other drones.

WCS broadcasts the following information from the beta channel in specific time intervals: a unique ID number, its speed, and heading. AB emits a unidirectional, "I am here," signal at specific time intervals. When MDS detects a mine, WCS broadcasts its ID along with a, "Mine detected," message and a life count from the alpha channel. All drones rebroadcast any message they receive from the alpha channel, coming both from drones and the alpha drone, after decreasing the life count by one. A message with zero life count is not broadcast.

This system is a simplified version of the packet routing method used in the Internet protocol and eliminates the possibility of unendingly broadcasting the same message over and over. It's important to understand that it may be difficult and impractical to precisely synchronize the "clocks" of the drones, hence the so-called WCS broadcasts will occur asynchronously. Another interesting point is that beta channel broadcasts may also be forwarded like alpha channel broadcasts, thus enabling a drone to know all members' current velocity. However this will result in much more crowded network traffic and may not be applicable in practice for large swarms. In simulation, both cases are considered.

7.4 Alpha Drone Model

The alpha drone is nothing more than a drone with two additional subsystems as a GPS and a means of knowing the relative position of drones in the flock. One such method is proposed by Wildermuth and Schneider [14] based on vision and pattern recognition. Also, if the wireless communication system is selected to operate on RF, triangulation techniques may be used to obtain the relative position data. The alpha drone has two main tasks: to present a general heading, the migratory urge, for the swarm, and to mark the mine locations on a map that is reported by drones and detected by the alpha drone itself. In addition, the alpha drone exhibits swarming as do other drones. Ultimately, the alpha drone requires more computational power and memory.

In case detecting the relative position of drones becomes too complicated or too slow, certain other approaches may be used. (1) Whenever a mine is detected, the alpha drone marks the place where it is currently located. The mine map generated will only give a density distribution of the minefield, without giving the actual co-ordinates. (2) All drones are equipped with GPS, which may increase the cost to undesirable levels.

7.5 Distributed Behavioral Rules and Algorithm

The behavior of drones can be divided into two categories: migrating and swarming, and mine detection and avoidance (see Fig. 7.2). These behavior modes are fused by a decision-making process. All inputs from subsystems are multiplied by weights and a resultant velocity request is generated. Finally, the velocity request is fed into the traction system to generate motion. The inputs are composed of PSA, MDS, and ASA readings, the heading imposed by the alpha drone, and an average of received velocity broadcasts by other drones.

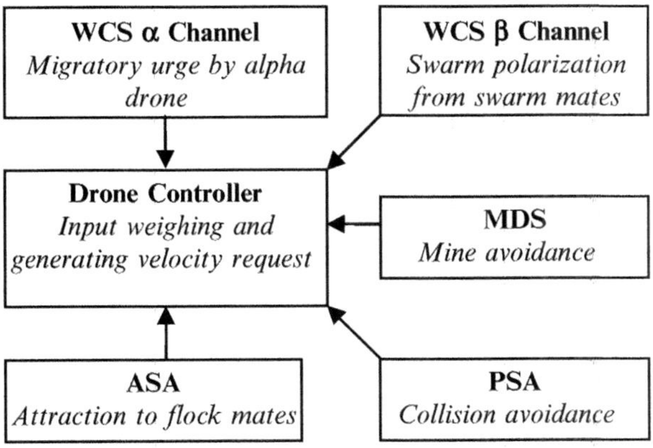

Fig. 7.2 Drone subsystems

Assume that a drone is able to fully perceive its surroundings, thus knowing the exact locations of obstacles. To exhibit basic collision avoidance, the drone should move in the opposite direction to the sum of unit position vectors (in its local coordinate frame) of obstacles.

In Fig. 7.3, O is the local coordinate frame for a drone; U_1 and U_2 are the unit vectors pointing towards obstacles. For the general case:

$$\hat{\mathbf{u}}_c = -\frac{\sum_{i=1}^{n} \mathbf{u}_i}{\left|\sum_{i=1}^{n} \mathbf{u}_i\right|} \tag{7.1}$$

where n is the number of obstacles, $\mathbf{u}_i$ is the unit vector pointing towards the ith obstacle, and $\hat{\mathbf{u}}_c$ is the unit vector pointing towards the required direction of motion to avoid collision.

In our model, a drone has a specific number of proximity sensors n_{PSA} that are placed symmetrically on a circle. Each sensor is assumed to cover an angle equal to $2\pi/n_{PSA}$. In addition, these sensors are not able to detect the distance to an obstacle but just provide an on/off signal if something is detected or not in a certain range.

Figure 7.4 shows a PSA with $n_{PSA} = 6$. For this particular example, each sensor covers an area of $2\pi/6 = 60°$. If an obstacle comes in PSA range within 0–59° then the first sensor is activated, within 60–129° the second sensor is activated, and so on. Because a sensor does not indicate the exact bearing of the obstacle, it is assumed that the obstacle is just in the middle of the sensor coverage. That is, if the first

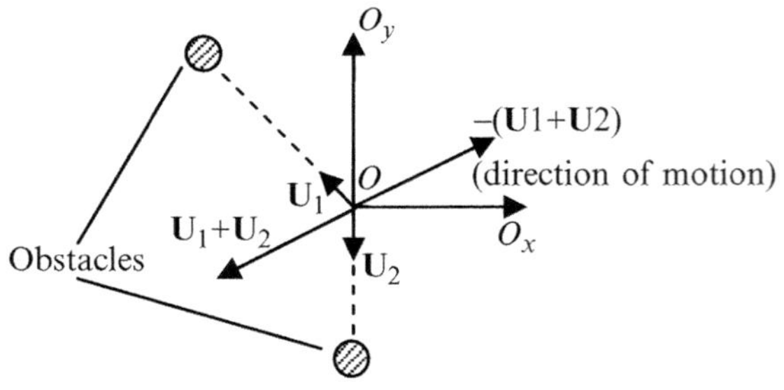

Fig. 7.3 Basic collision avoidance

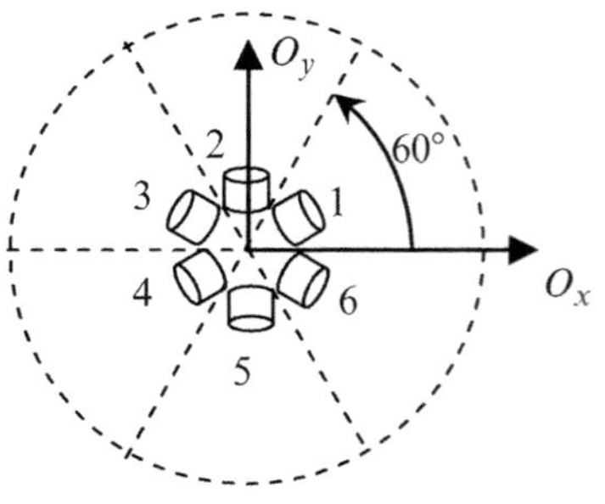

Fig. 7.4 PSA with six sensors

sensor is activated, we assume that the obstacle is at 30°, for the second sensor it is 90°, for the ith sensor it is

$$\phi_{obstacle} = \frac{2\pi}{n_{PSA}}(i-0.5)$$

Because we are dealing only with unit vectors in Eq. 7.1, it's enough to find the polar angle of $\hat{\mathbf{u}}_c$. This angle is our heading to avoid collision that is given by

$$\theta_c = -\text{ATAN2}\left[\sum_{i=1}^{n_{PSA}} \sin(q_i)\,,\ \sum_{i=1}^{n_{PSA}} \cos(q_i)\right]$$
$$q_i = \frac{2\pi\mu_i(i-0.5)}{n_{PSA}} \tag{7.2}$$

where μ_i is the respective sensor output as 1 or 0 (1 if the respective sensor detects an object, 0 otherwise), and θ_c is the collision avoidance heading request.

The attraction heading request is derived exactly as collision avoidance with a single exclusion. The minus sign is removed because we want the drone to move towards the other drones. Also note that ASA is a passive sensor and it only detects the signal emitted by other drones.

$$\theta_a = -\text{ATAN2}\left[\sum_{i=1}^{n_{ASA}} \sin(q_i)\,,\ \sum_{i=1}^{n_{ASA}} \cos(q_i)\right]$$
$$q_i = \frac{2\pi\mu_i(i-0.5)}{n_{ASA}} \tag{7.3}$$

where n_{ASA} is the number of sensors in attraction sensor array, μ_i is the respective sensor output as 1 or 0, and θ_a is the attraction heading request.

The heading request for migration is supplied by the alpha drone; because a drone knows its heading relative to true north, this migration direction is simply converted to a heading request.

The swarm polarization heading is generated by summing the broadcasts from other drones. The main problem is that the broadcasts are asynchronous. We have two solutions to this problem: use a fixed length array in memory to keep the incoming broadcasts; or use a dynamic array (stack) with a specified maximum size, add each incoming broadcast to the stack with a timestamp, and delete broadcasts that are older than a certain time. The second approach is used in simulation. The broadcasts of drones are in the same manner as the migration urge broadcast of the alpha drone, but in this case drones broadcast their actual heading in terms of compass directions.

$$\theta_p = \sum_{i=1}^{n} \frac{\theta_i}{n} \tag{7.4}$$

where n is the number of elements in polarization stack, and θ_i is the heading data in the ith stack element.

$$\begin{aligned}\theta_r = \text{ATAN2}(&\eta_m \text{S}\theta_m + \eta_p \text{S}\theta_p + \eta_a \text{S}\theta_a + \eta_c \text{S}\theta_c - \eta_l \mu_l, \\ &\eta_m \text{C}\theta_m + \eta_p \text{C}\theta_p + \eta_a \text{C}\theta_a + \eta_c \text{C}\theta_c)\end{aligned} \tag{7.5}$$

where C and S stands for cosine and sine; $\eta_m, \eta_p, \eta_a, \eta_c, \eta_l$ are the weights of importance for migration, polarization and attraction, collision avoidance, and mine avoidance, respectively; and $\theta_m, \theta_p, \theta_a, \theta_c$ are the heading requests generated by said behaviors. μ_l is the signal strength of MDS. Equation 7.5 is in fact scaling and addition of unit vectors describing behaviors. Another point is that, for example, by selecting η_l and η_c much bigger then the others, the system behavior shifts to hierarchical where survival supersedes all other rules. Only in the absence of mines or obstacles, do the other factors come into effect.

Now that the drone knows where to turn, it needs to know how fast it should go. The guidelines for speed selection can be given as: (1) the fewer drones you see around, go faster to catch up with the flock, and (2) try to move with the same speed as the other flock mates, which helps polarization. At this point, the same type of stack, which is used to store bearing broadcasts, is used to store velocity broadcasts.

$$v' = f(m)\left(\lambda\sum_{i=1}^{n}\frac{v_i}{n} + (1-\lambda)v_{\max}\right) \qquad f(m) = \left| \begin{matrix} m/a, \; m > a \\ 1, \; m \leq a \end{matrix} \right. \tag{7.6}$$

where $f(m)$ is pseudo-acceleration, n is the number of elements in the stack, v_i is the speed data in ith stack element, λ is the polarization parameter, $v_{\max}$ is the maximum attainable speed, m is the number of inactive sensors in ASA (i.e., sensors not detecting anything), and a is a limitation value that prevents too much speed loss for members near the center of the flock. Note that $1 \leq a <$ number of sensors in ASA and $a = 1$ means no speed loss limitation. Also note that $0 \leq \lambda < 1$, for $\lambda = 0$ speed matching will not occur.

7.6 Simulation Results

Simulations were carried out in two phases. In the first, interactions between two individuals are taken into consideration. The idea was whether a concept of swarm stability could be specified. Among any collection of individuals, the quality of being a swarm is inversely proportional to the distance between the particles, or agents. There is such a distance that the agents are not in a swarm formation any more but rather are acting freely. This first phase defines a "swarm stability" or swarm entropy, which is quantified in Tsallis entropy, whose details may be explained in a separate study. Agents normally roam apart from each other in search of food (a mine) so as to cover an area as fast as possible with the least likelihood of missing anything during the search. But this separation should not be too great in order not to lose the swarm behavior along with all the advantages that accompany it.

The definition of entropy, which was first discovered by Ludwig Boltzmann, can be given as a measure of disorder. There are many types of entropy definition in the literature. One of them is Tsallis entropy, first explained in 1988 [15]. Tsallis modified the mathematical expression of entropy definition in his study and defined a new parameter, q.

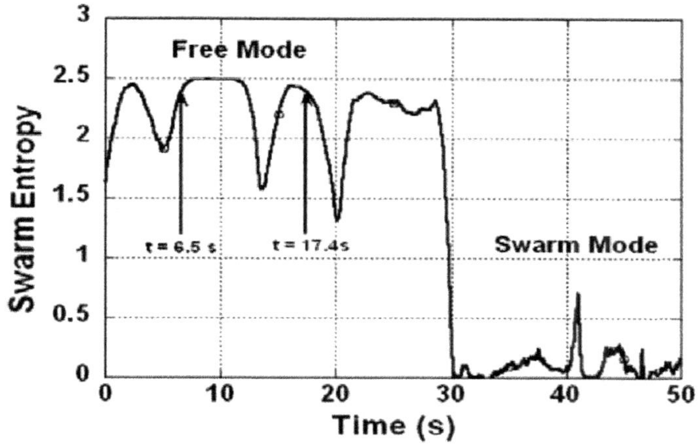

Fig. 7.5 Entropy changing with time for free and swarm modes

$$S_T = \frac{1 - \sum_i P_i^q}{1 - q} \tag{7.7}$$

During the roaming of the individuals, entropy fluctuates at around its highest, meaning the distance between the agents is rather far, and swarm stability is low; that is, the quality of remaining as a swarm may disappear should the particles get farther away. However, a sudden decrease in entropy may occur as in Fig. 7.5, when the individuals converge after the discovery of food (a mine).

Peaks in the entropy of a swarm mode were created by attractive and repulsive behavior of individuals. Close encounters are considered as risks of collision and quick reactions to avert it. Please note that such peaks are missing in the free mode (i.e., roaming a certain field).

In Fig. 7.6, at around 62.2 s, a repulsion may be seen because these two agents have moved too close, namely into the repulsive field, and at 62.7, they start to move back again.

In the second phase, the whole model is implemented in a computer program in an object-oriented fashion. Lengths are described in terms of "units." The following drone parameters are used for each simulation run (Fig. 7.7).

Maximum speed: 20 units/s	**Turning rate: 180°s**
Drone shape: disc	Diameter: 10 units

Four distinct swarm behaviors are observed. These are: high polarization (HP), balanced polarization (BP), low polarization (LP), and disarray. The first three behaviors have their uses where the disarray behavior indicates an unstable swarm, which is not desired.

High polarization means that the velocity (both speed and heading) of an individual drone is nearly the same as the swarm average in the absence of disturbances. The average speed of the swarm is maximized. The main disadvantage is that the swarm aggregates very slowly when it meets a disturbance (an obstacle or a mine).

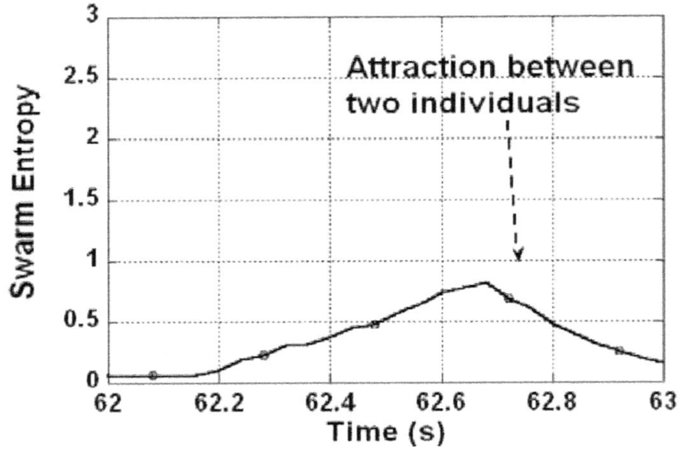

Fig. 7.6 Entropy variation where there is an attraction between two individuals

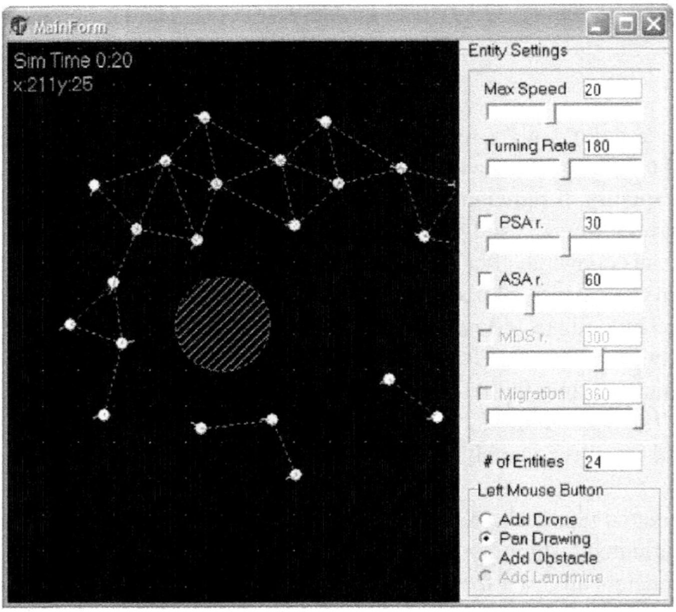

Fig. 7.7 Simulation screenshot: drones detected a mine. The "Landmine Detection Simulator" can be found at the author's Web site: http://www.iyte.edu.tr/~erhansevil/landmine.htm

Thus the mine detection reliability is decreased significantly. This behavior results in either a high migration weight μ_m, or a high polarization weight μ_p. It is an ideal swarm behavior for traversing mine-safe zones to go quickly to an objective area.

Balanced polarization means that the velocity of an individual drone is close to the neighboring drones but not necessarily close to the average swarm velocity. This offers high speed (although lower then HP) and high reliability. The swarm aggregates quickly after meeting disturbances. This behavior results in nearly equal μ_m and μ_p, and high μ_a. It is an ideal swarm behavior for most cases.

Low polarization means that the velocity of an individual drone is highly different from that of its neighbors. This happens when μ_a is high and μ_p is low and also μ_m is selected between. The only use for this behavior is that the swarm can find its way when there are too many obstacles, such as a labyrinth.

Disarray occurs if:

- μ_a is too low (swarm disintegrates).
- μ_l is too low (drones step on mines).
- μ_c is too low (drones collide with each other).
- μ_m is too low (swarm moves in a random direction).

Note that, by really unsuitable parameters, more than one symptom of disarray can be observed. Surprisingly, if the other parameters are chosen well, a low μ_p, even zero, does not lead to disarray.

Another important concept is efficiency. What should be the optimal number of drones to be used? It is observed that up to an optimum population, efficiency of the swarm increases. After that point, adding more drones does not improve the mine detection speed or performance much. This is mainly because too many drones form a useless bulk in the center of the swarm. However, the optimum number of drones also depends on the terrain (rough, smooth, etc.), landmine density, actual speed and turning rates of drones, sensor ranges, and swarm behavior.

7.7 Conclusion

A distributed behavioral model to guide a group of minimalist mobile robots is presented. The main point of interest for the model is that it is based on weighting sensor inputs and not on precedence-based rules. By chancing the weights, it is possible to shift the behavior of the swarm while all other physical parameters (such as sensor ranges) remain constant. The model is presented in a computer simulation that gave promising results.

It should be noted that the selection of weights changes the behavior of the swarm drastically and sometimes unexpectedly. To optimize the drone behavioral weights is the upcoming part of this study on which the authors are currently working.

References

1. US State Department (1998) *Hidden Killers*: The Global Landmine Crisis. Report released by the U.S. Department of State, Bureau of Political-Military Affairs, Office of Humanitarian Demining Programs, Washington, DC, September 1998.
2. Huang QJ, Nonami K (2003) Humanitarian mine detecting six-legged walking robot and hybrid neuro walking control with position/force control. *Mechatronics* 13: 773–790.

3. Dolby AS, Grubb TC (1998) Benefits to satellite members in mixed-species for aging groups: An experimental analysis. *Animal Behavior* 56: 501–509.
4. Adioui M, Treuil JP, Arino O (2003) Alignment in a fish school: A mixed Lagrangian–Eulerian approach. *Ecological Modelling* 167: 19–32.
5. Cale PG (2003) The influence of social behaviour, dispersal and landscape fragmentation on population structure in a sedentary bird. *Biological Conservation* 109: 237–248.
6. Smith VA, King AP, West MJ (2002) The context of social learning: Association patterns in a captive flock of brown-headed cowbirds. *Animal Behavior* 63: 23–35.
7. Green M, Alerstam T (2002) The problem of estimating wind drift in migrating birds. *Theoretical Biology* 218: 485–496.
8. Aoki I (1982) A simulation study on the schooling mechanism in fish. *Social Science Fish* 48: 1081–1088.
9. Huth A, Wissel C (1992) The simulation of the movement of fish schools. *Theoretical Biology* 156: 365–385.
10. Couzin D, Krause J, James R, Ruxton GD, Franks NR (2002) Collective memory and spatial sorting in animal groups. *Theoretical Biology* 218: 1–11.
11. Reynolds, Flocks CW (1987) Herds and schools: A distributed behavioral model. *Computational Graph* 21: 25–34.
12. Sugawara K, Sano M (1997) Cooperative acceleration of task performance: Foraging behavior of interacting multi-robots system. *Physica D* 100: 343–354.
13. Martin M, Chopard B, Albuquerque P (2002) Formation of an ant cemetery: Swarm intelligence or statistical accident? *FGCS* 18: 951–959.
14. Wildermuth D, Schneider FE (2003) Maintaining a common co-ordinate system for a group of robots based on vision. *Robotics and Autonomous Systems* 44: 209–217.
15. Tsallis C (1988) Possible generalization of Boltzmann–Gibbs statistics. *Journal of Statistical Physics* 52: 479–487.

Chapter 8
Iris Recognition Based on 2D Wavelet and AdaBoost Neural Network

Anna Wang, Yu Chen, Xinhua Zhang, and Jie Wu

8.1 Introduction

Biometrics refers to automatic identity authentication of a person on the basis of one's unique physiological or behavioral characteristics. To date, many biometric features have been applied to individual authentication. The iris, a kind of physiological feature with genetic independence, contains an extremely information-rich physical structure and unique texture pattern, and thus is highly complex enough to be used as a biometric signature. Statistical analysis reveals that irises have an exceptionally high degree-of-freedom up to 266 (fingerprints show about 78) [1], and thus are the most mathematically unique feature of the human body, more unique than fingerprints. Hence, the human iris promises to deliver a high level of uniqueness for authentication applications that other biometrics cannot match.

Indeed, Daugman's approach relies on the use of Gabor wavelets in order to process the image at several resolution levels. An iris code composed of binary vectors is computed this way and a statistical matcher (logical exclusive OR operator) analyzes basically the average Hamming distance between two codes (bit-to-bit test agreement) [2]. Some recent works follow this direction. Another approach, in the framework of iris verification, introduced by Wildes, consists of measuring the correlation between two images using different small windows of several levels of resolution [3]. Also, other methods for iris verification have been proposed, in particular relying on ICA [4].

The outline of this chapter is as follows. The method that uses a 2-D wavelet transform to obtain a low-resolution image and a Canny transform to localize pupil position is presented in Sect. 8.2. By the center of the pupil and its radius, we can acquire the iris circular ring. Section 8.3 adopts the Canny transform to extract iris texture in the iris circular ring as feature vectors and vertical projection to obtain a 1-D energy signal. The wavelet probabilistic neural network is a very simple classifier model that has been used as an iris biometric classifier and is introduced in Sect. 8.4. Two different extension techniques are used: wavelet packets versus Gabor

Oscar Castillo et al. (eds.), *Trends in Intelligent Systems and Computer Engineering.*

wavelets. The wavelet probabilistic neural network can compress the input data into a small number of coefficients and the proposed wavelet probabilistic neural network is trained by the AdaBoost algorithm. The experimental results acquired by the method are presented in this section. Finally, some conclusions and proposed future work can be found in Sect. 8.8.

8.2 Preprocessing

The iris image, as shown in Fig. 8.1, does not only contain abundant texture information, but also some useless parts, such as eyelid, pupil, and so on. We use a simple and efficient method to localize the iris. The steps are as follows.

1. A new image is the representation of the original image by 2-D wavelet, and its size is only a quarter of the original image. The wavelet coefficients are calculated by the formulas:

$$f(x,y) = \sum_k c_{j0}(k)\varphi_{j0,k} + \sum_{j=j0}^{\infty}\sum_k d_j(k)\phi_{j,k}(x) \tag{8.1}$$

$$c_{j0}(k) = <f(x), \varphi_{j0,k}(x)> = \int f(x)\varphi_{j0,k}(x)dx \tag{8.2}$$

$$d_j(k) = <f(x), \phi_{j,k}(x)> = \int f(x)\phi_{j,k}(x)dx \tag{8.3}$$

2. The edge of the pupil in the new image is detected by the Canny transform.

$$H_G = \int_{-W}^{W} G(-x)f(x)dx \tag{8.4}$$

$$H_n = n_0\left[\int_{-W}^{W} f^2(x)dx\right]^{1/2} \tag{8.5}$$

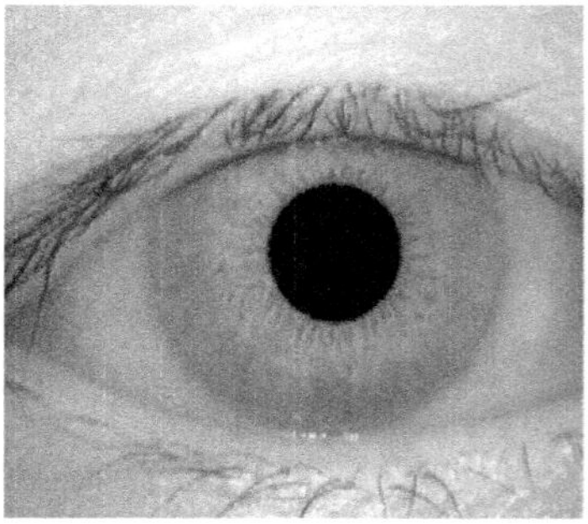

Fig. 8.1 An eye photo before processing

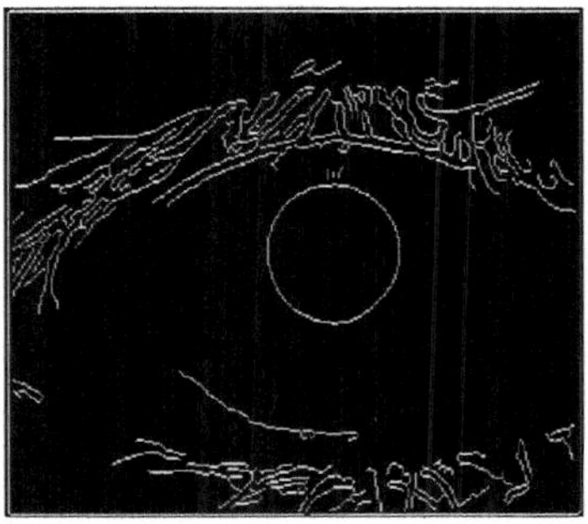

Fig. 8.2 Image after wavelet transform

3. The center coordinates and the radius of the pupil are determined by the Canny transform; the result is shown in Fig. 8.2.
4. When the center coordinates and the radius of the pupil are multiplied by two, the center coordinates and the radius of original pupil are obtained.
5. The iris circular ring is obtained by the position of original pupil.

We construct the wavelet transfer function and the scale transfer function as follows.

$$G(w) = -e^{-iw} \sin^2 \frac{w}{2} \tag{8.6}$$

$$H(w) = e^{-iw} \cos^4 \frac{w}{2} \tag{8.7}$$

We know the connection between the wavelet function and the wavelet transfer function:

$$\psi(w) = G(w)\psi(w) \tag{8.8}$$

By the same principle:

$$\phi(w) = H(w)\phi(w) \tag{8.9}$$

So we can get the wavelet function and the scale function:

$$\psi(x) = \begin{cases} -2x, & (-3/4 \le x < -1/4) \\ 3-4x, & (-1/4 \le x < 1/4) \\ 3-2x, & (1/4 \le x < 3/4) \\ 0, & (\text{other}) \end{cases} \tag{8.10}$$

$$\phi(x) = \begin{cases} x^2/2, & (0 \le x < 1) \\ 3/4-(x-3/2)^2, & (1 \le x < 2) \\ 1/2(x-3)^2, & (2 \le x < 3) \\ 0, & (\text{other}) \end{cases} \tag{8.11}$$

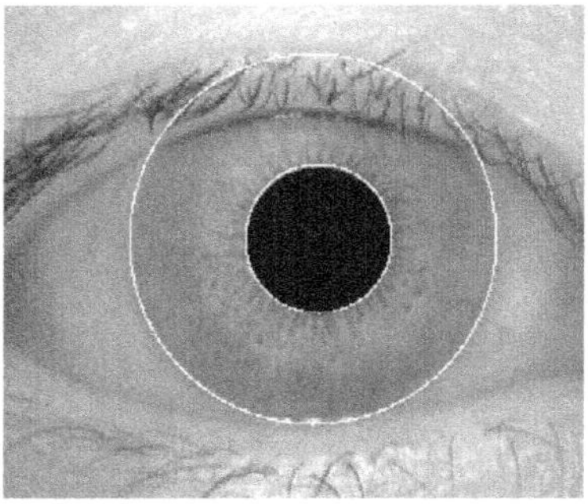

Fig. 8.3 The result of iris location

When the procedure has been done, we can localize the position of the pupil. But the position of the pupil is in the new image; the doubled center coordinates and radius of the pupil are the position of the pupil in original image. The result is shown in Fig. 8.3. When the center coordinates and the radius of the pupil in the original are obtained, the iris circular ring is extracted as features.

The more iris circular ring is extracted, the more information is used as a feature. The recognition performance is much better, but the efficiency is slightly affected [5]. In the next section, a detailed description of the iris feature extraction method is presented.

8.3 Iris Localization

8.3.1 Unwrapping

The purpose of the Canny transform is to extract the iris texture. The geometry of the iris is circular and most of its interesting textural features are extended in the radial and, to a lesser extent, the angular direction, therefore analysis is simplified by an adapted polar transform, suggested by Daugman [6]. The adaptation is motivated by the fact that the pupil and iris are not necessarily concentric; the pupil is often somewhat displaced towards the nose and downwards and the pupil diameter is not constant. This is amended by a transform that normalizes the distance between the pupil boundary and the outer iris boundary [7]. Such a transform is expressed by Eqs. 8.12 and 8.13, where (x_p, y_p) and (x_i, y_i) are a pair of coordinates on the pupil and iris border. The figure for Eqs. 8.12 and 8.13 defines the angle variable 0 and the radius variable r. The sketch map is shown in Fig. 8.4. The radius is normalized to the interval [0,1]. The unwrapped image is then histogram-equalized to increase the contrast of the texture. The results are shown in Figs. 8.5 to 8.7.

$$x(r,\theta) = rx_i(\theta) + (1-r)x_p(\theta) \tag{8.12}$$

$$(r,\theta) = ry_i(\theta) + (1-r)y_p(\theta) \tag{8.13}$$

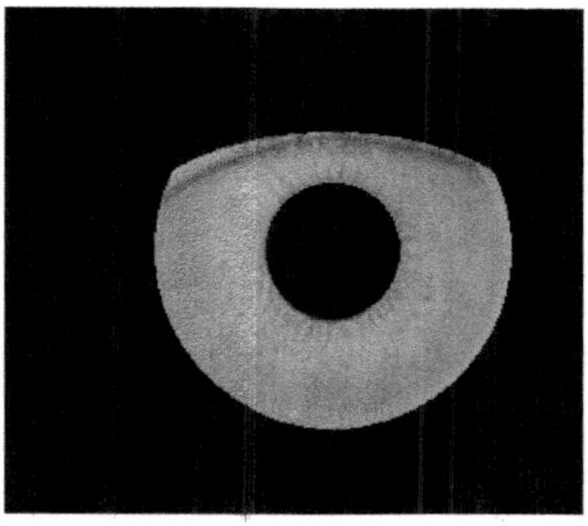

Fig. 8.4 Iris image

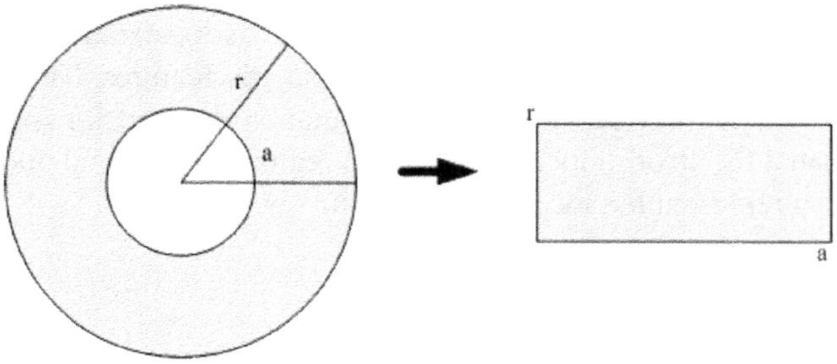

Fig. 8.5 Iris unwrapping principium

Fig. 8.6 Iris texture after unwrapping

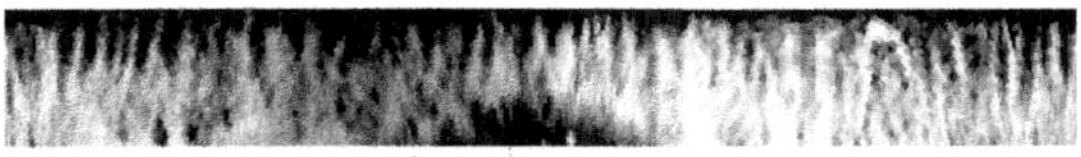

Fig. 8.7 Unwrapped iris texture after histogram equalization

8.3.2 *Vertical Projection*

To reduce system complexity, we adopt vertical projection to obtain a l-D energy profile signal. To exploit the benefits deriving from the concentrated energy, every row is accumulated as an energy signal. This method is evaluated on the CASIA iris databases [8].

Let X be an iris image of size $m \times n$; m is the number of iris circular rings, and n is the pixels of each iris circular ring.

$$X = [x_{1\times 1} \ldots x_{1\times n}; \ldots ; x_{m\times 1} \ldots x_{m\times n}] \tag{8.14}$$

After vertical projection, the l-D energy signal Y is obtained: $Y = [y_1, \ldots, y_n]$.

The m is much smaller than the n. Thus, the information of the iris texture after vertical projection is more abundant than the information of the iris texture after horizontal projection. Thus, we adopt the vertical projection to extract the 1-D energy signal [9].

8.4 Iris Feature Extraction

In this section, we use the adaptive method to facilitate iris pattern matching by fusing global features and local features. Both features are extracted from the log Gabor wavelet filter at different levels. The first one is the global feature that is invariant to the eye image rotation and the inexact iris localization. The statistics of texture features is used to represent the global iris features. The introduction of the global features has decreased the computation demand for local matching and compensated the error in localizing the iris region. The second one is the local feature that can represent iris local texture effectively.

8.4.1 Global Feature Extraction

The wavelet transform is used to obtain frequency information based on an pixel in an image. We are interested in calculating global statistics features in a particular frequency and orientation. To obtain the information, we must use nonorthogonal wavelets. We prefer to use log Gabor wavelets rather than Gabor wavelets. Log Gabor wavelet filters allow arbitrarily large bandwidth filters to be constructed while still maintaining a zero DC component in the even-symmetric filters. On the linear frequency scale, the log Gabor function has a transfer function of the form

$$G(f) = \exp\left(\frac{-(\log(f/f_0))^2}{2(\log(\sigma/f_0))^2}\right) \tag{8.15}$$

where f_0 represents the center frequency, and σ gives the bandwidth of filter.

If we let $I(x,y)$ denote the image and W_n^e and W_n^0 denote the even-symmetric and odd-symmetric wavelets at scale, we can think of the responses of each quadrature pair of filters as forming a response vector $[e_n(x,y), O_n(x,y)] = [I(x,y) * W_n^e + I(x,y) * W_n^0]$.

The amplitude of the transform at a given wavelet scale is given as

$$A_n(x,y) = \sqrt{e_n^2(x,y) + O_n^2(x,y)} \tag{8.16}$$

The phase is given by

$$\phi_n = \operatorname{atan2}(e_n(x,y), O_n(x,y)) \tag{8.17}$$

From $A_n(x,y)$ and ϕ_n we can obtain the amplitude and phase of the image. The statistical values of the amplitude are arrayed as the global feature to be classified with the weighting Euclidean distance. The system proposed includes the 32 global features including the mean and average absolute deviation of each image with four orientation and four frequency level Gabor wavelet filters.

8.4.2 Local Feature Extraction

The global feature represents the global characteristic of the iris image well. But the local difference can't availably reveal it and the recognition rate is affected with a different iris having similar global features. The global feature needs a local feature to perfect the recognition. This chapter encodes the iris image into binary code to match with the Hamming distance. Due to the texture, at the high-frequency levels the feature is strongly affected by noise. We extract the local iris feature at the intermediate levels.

For small data size and fast comparison, we can quantize the iris image into binary code with definite arithmetic. The local window is divided into $m \times n$ smaller subimages with a $p \times q$ pixel window. We calculate the image convolution with the log Gabor wavelet filter, which is a bandpass filter, and encode the amplitude into binary. The resulting code is called the local code. The D region that is part of the encoded region convoluted by the log Gabor filter is encoded into binary. We can define the regulation for encoding. The regulation is based on real numbers and the imaginary number sum of D1, D2, D3, and D4. If the real number sum of D1 is more than zero, the corresponding binary is low, whereas the corresponding binary is high. In addition, if the imaginary number sum of D1 is more than zero, the corresponding binary is low, whereas the corresponding binary is high. This proposed system applied 64-byte local features to fuse with global features.

8.5 Structure Local Interconnect Neural Network

We use the WPNN that combines the wavelet neural network and probabilistic neural network for an iris recognition classifier [10]. Figure 8.7 presents the architecture of a four-layer WPNN, which consists of a feature layer, wavelet layer, Gaussian layer, and decision layer (Figure 8.8). In the feature layer, $X_1,\dots,X_N$ are sets of feature vectors or input data, and N is the dimension of the data sets. The wavelet layer is a linear combination of several multidimensional wavelets. Each wavelet neuron is equivalent to a multidimensional wavelet, and the wavelet in the following form.

$$\phi_{a,b}(x) = \sqrt{a}\phi\left(\frac{x-b}{b}\right) a,b \in R \tag{8.18}$$

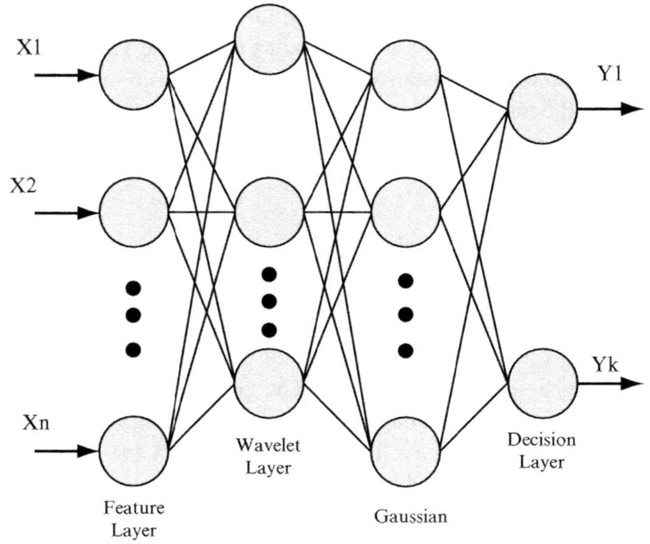

Fig. 8.8 The wavelet probabilistic neural network

It is a family of functions generated from one single function $p(x)$ by scaling and translation, which is localized in both the time space and the frequency space, is called a mother wavelet, and the parameters are named the scaling and translation factors, respectively [11]. In the Gaussian layer, the probability density function of each Gaussian neuron is of the following form.

$$f_i = \frac{1}{(2\pi)^{2/p} \times \sigma^p} \left(\frac{1}{n_i}\right) \sum_{i=1}^{n_a} \exp\left(\frac{-(X - S_j^i)^2}{2\sigma^2}\right) \tag{8.19}$$

where X is the feature vector, p is the dimension of the training set, n is the dimension of the input data, j is the jth data set, S_j^i is the training set, and σ is the smoothing factor of the Gaussian function.

When the input data are changed, we do not change the architecture of WPNN or train the factors. It is suitable for a biometric recognition classifier. Finally, the scaling factor, translation factor, and smoothing factor are randomly initialized at the beginning and are optimized by the AdaBoost algorithm [12].

8.6 Learning Algorithm

Boosting is one of the most effective learning methods for pattern classification in the past ten years. The goal of the boosting algorithm is to combine the output of many "weak" classifiers to generate the effective vote committee [13].

Consider a two-class question; the output variable code is $y \in [-1,1]$. Given a vector x to predict the subvariable, classifier $h(x)$ is generated in the prediction value in $[-1,1]$. The error rate of the training sample is

$$\overline{\text{err}} = \frac{1}{N}\sum_{i=1}^{N} I(y_i \neq h(x_i)) \tag{8.20}$$

A weak classifier $h(x_i)$ performs just slightly better than random guessing. The goal of boosting is to apply the algorithm continuously to the repeating modified data and give a weak classifier series $h_m(x), m = 1,2,\ldots,M$. Then a weighted majority vote combines all the predictions to get the final prediction:

$$H(x) = \text{sign}\left(\sum_{m=1}^{M} \alpha_m h_m(x)\right) \tag{8.21}$$

Here a_1, a_2, a_M are computed by the boosting algorithm and weight is assigned to the contribution of each $h_m(x)$, which can affect the accuracy of the classifiers in the series greatly.

AdaBoost algorithm

1. Initialize $w_1(i) = 1/N, i = 1,2,\ldots,N$ For $m = 1$ to M.
2. Choose weight $w_m(i)$ and classifier $h_m(x)$ to fit the training values.
3. Compute $\text{err}_m = \sum_{i=1}^{N} -w_m(i) y_i h_m(x_i)$.
4. Compute $\alpha_m = \frac{1}{2}\ln\left(\frac{1-\text{err}_m}{1+\text{err}_m}\right)$.
5. $w_m + 1 \leftarrow \frac{w_m(i)\exp(-\alpha_m y_1 h_m(x_1))}{\sum_i^N w_m(i)\exp(-\alpha_m y_i h_m(x_i))}$ $i = 1,2,\ldots,N$.
6. *OutputY*$(k) = \text{sign}(\sum_{m=1}^{M} \alpha_m X_m(k))$.

We use the AdaBoost algorithm to "boost" the fourth-level classifier and obtain very high accuracy.

8.7 Experiment Procedure and Its Results

In this section, we refer to our method of combining the low-complexity feature extraction algorithm with WPNN for iris recognition. The iris database used in the comparison is the CASIA iris database. The database contains 756 iris images acquired from 108 individuals (7 images per individual).

In the following experiments, a total of 324 iris images (three iris images of each person are extracted) was randomly selected as the training set and the remainder as the test set from the remaining images. Such procedure was carried out 100 times. The experimental platform is the AMD K7 Athlon 2.2 GHz processor, 512M DDRAM, Windows XP, and the software is MATLAB 6.5.

8.7.1 Evaluation of Iris Verification with the Proposed Method

In a real application, the iris verification experiment classifies an individual as either a genuine user (called an enrollee) or an impostor. Thus, the experiment has two types of recognition errors: it either falsely accepts an impostor or falsely rejects an enrollee. We define two types of error rate. The false acceptance rate (FAR) is the probability that an unauthorized individual is authenticated. The false rejection rate (FRR) is the probability that an authorized individual is inappropriately rejected [14].

$$\text{FAR} = \frac{\text{number} \cdot \text{of} \cdot \text{false} \cdot \text{acceptances}}{\text{number} \cdot \text{of} \cdot \text{impostor}}$$

$$\text{FRR} = \frac{\text{number} \cdot \text{of} \cdot \text{false} \cdot \text{rejections}}{\text{number} \cdot \text{of} \cdot \text{enrollee} \cdot \text{attempts}}$$

The performance of iris verification is estimated with the equal error rate (EER). The lower the EER value, the higher is the performance of the iris recognition. In most biometrics system, FRR is usually seen as a not very important problem, because it is the probability at which authentic enrolled users are rejected. Nevertheless, FAR is the most important error rate in the majority of biometrics systems, because it is the probability at which an unauthorized unenrolled person is accepted as an authentic user. Thus, a reasonable threshold is selected for adjusting the performance of the system. As shown in Table 8.1, the best result for a biometrics system is in a FAR of 0.0% and in a FRR of 30.57% as the threshold is 0.72. This suggests that we can obtain a lower FAR value, but the FRR will be sacrificed. Sometimes, we promote the FRR in the tolerated range for decreasing the FAR.

The results are shown in Table 8.2. In these experiments, the best EER is 3.34%, the average EER is 5.35%, and the recognition time is less than 1 ms per image. These results illustrate the superiority of the proposed method. These observations demonstrate that the iris recognition techniques can be suitable for low-power applications showing that the complexity of the proposed method is very low.

Table 8.1 The FAR and FRR of iris recognition

Threshold	FAR (%)	FRR (%)
0.62	3.25	3.2
0.63	1.98	3.3
0.64	1.22	5.57
0.65	0.71	7.73
0.66	0.36	9.25
0.67	0.21	11.41
0.68	0.12	14.85
0.69	0.04	17.8
0.70	0.03	20.08
0.71	0.01	27.17
0.72	0.00	30.57

Table 8.2 The recognition performance of our method

Method	Proposed
Average EER	5.35%
Best EER	3.34%
Recognition	<1 ms

Table 8.3 Comparison of previous method and proposed

Method	Proposed	Previous
Best EER	3.35%	4.37%

8.7.2 Evaluation on Iris Identification with Existing Methods

In Table 8.3, we show the performance of our new method is much better than that of our previous method. In the previous method, we adopt the Canny transform and vertical projection to extract iris texture in the iris circular ring as feature vectors. The 1-D discrete wavelet transform is used to reduce the dimensionality of the feature vector and the WPNN is adopted as the classifier. In this chapter, the proposed method combines a very simple feature extraction algorithm and WPNN, which can significantly outperform established iris recognition systems on standard datasets. Owing to the high efficiency and simplicity of the proposed method, it is very suitable for low-power applications or HW platforms having small amount of memory available (smartcard).

8.8 Conclusions

The chapter proposes a wavelet probabilistic neural network as an iris recognition classifier. The WPNN combines the wavelet neural network and probabilistic neural network. AdaBoost is used for adjusting the parameters of WPNN. We only need a few parameters as the weights of WPNN and the evaluation of WPNN is based on our experiment. From the simulation results described in experiments, it is clear that the proposed method has high efficiency. The complexity of the feature extraction method for iris recognition is very low. The proposed method is excellently effective, and achieves a considerable computational reduction while keeping good performance. We have proved the proposed method can achieve high performance and high efficiency in iris recognition. In future, we will further improve the recognition performance of iris recognition and apply it to embedded systems.

References

1. Y. Freund and R. Schapire, A decision-theoretic generalization of online learning, *Proceedings of the Ninth Annual Conference on Computational Learning Theory*, V32, N3, pp. 252–260, 8/02
2. A. Elmabrouk and A. Aggoun, Edge detection using local histogram analysis, *Electronic Letters*, V1, N12, pp. 11–30, 6/98
3. J. Daugman, How iris recognition works, *IEEE Transactions on Circuits and Systems for Video Technology*, V14, N1, pp. 21–31, 1/04
4. C. Rafael and M. Gonzalez, *Digital Image Processing*, 2nd Edition, Publishing House of Electronics Industry, 2005
5. P. Wildes, Iris recogition: An emerging biometric technology, *Proceedings of the IEEE*, V85, N9, pp. 1348–1363, 9/97
6. P. Wildes, A system for automated iris recognition, *Proceedings of the IEEE*, V1, N12, pp. 121–128, 12/94
7. L. Ma and T. Tan, Personal identification based on iris texture analysis, *IEEE Transactions on Pattern Analysis and Machine Intelligence*, V25, N12, pp. 1519–1533, 12/03
8. P.-F. Zhang, A novel iris recognition method based on feature fusion, *Proceedings of the Third International Conference on Machine Learning and Cybernetics*, Shanghai V26, N29, pp. 3661–3665, 8/04
9. Z. Sun and Y. Wang, Improve iris recognition accuracy via cascaded classifiers, *IEEE Transactions on Systems*, V35, N3, pp. 435–441, 8/05
10. Y. Wang and J.Q. Han, Iris recognition using independent component analysis, *Proceedings of the Fourth International Conference on Machine Learning and Cybernetics*, Guangzhou V18, N21, pp. 4487–4492, 8/05
11. W. Robert and B. Bradford, Effect of image compression on iris recognition, *IMTC 2005 Instrumentation and Measurement Technology Conference*, Ottawa V17, N19, pp. 2054–2058, 5/05
12. W. Yuan and Z. Lin, A rapid iris location method based on the structure of human eyes, *Proceedings of the 2005 IEEE Engineering in Medicine and Biology 27th Annual Conference*, Shanghai V1, N4, pp. 3020–3025, 9/05
13. A. Natalia and V. Manasi, Performance analysis of iris based identification system at the matching score level, *IEEE Transactions on Information Forensics and Security*, V1, N2, pp. 154–168, 6/06
14. C. Wang and S. Song, Iris segmentation based on shape from shading and parabolic template, *Proceedings of the Sixth World Congress on Intelligent Control and Automation*, Dalian V21, N23, pp. 10088–10091, 6/06

Chapter 9
An Improved Multiclassifier for Soft Fault Diagnosis of Analog Circuits

Anna Wang and Junfang Liu

9.1 Introduction

Fault diagnosis in analog circuits is far from the point of automation and relies heavily on the experience and intuition of engineers to develop diagnosis strategies. This requires the engineer to have some detailed knowledge of the circuit's experience in operational characteristics and experience in developing diagnosis strategies. And diagnosing and locating faults in analog circuits is becoming an increasingly difficult task, due to the growing complexity of electronic circuits and the spiralling number of applications that are characterized by the coexistence of both digital and analog parts. In fact, for analog systems, the lack of simple fault models and the presence of component tolerances, noise, and circuit nonlinearities make fault diagnosis automation and fault location procedures very complex. As a consequence, the automation of fault diagnosis and location for analog circuits is much more primitive than for the digital case, for which fully automated testing methodologies have been developed and are commonly used. All those also imply that in an analog fault diagnosis an extremely large and more expensive number of simulations with respect to the digital case are required. Thus the automation of fault diagnosis procedures has a development level much less advanced than that with respect to the digital case.

The engineering community began to look into analog circuit diagnosis problems in the mid-1970s, because analog systems were such an iterative process. During the past few years, there has been significant research on analog fault diagnosis at the system, board, and chip levels [1, 2]. Normally fault diagnosis schemes generally fall into two categories: estimation methods and pattern recognition methods. The estimation methods require mathematical process models that represent the real process satisfactorily, and a component is identified as a faulty one when the calculated value is beyond its tolerance range. If the model is complex, computation can easily become very time-consuming. Thus the application of the estimation methods

Oscar Castillo et al. (eds.), *Trends in Intelligent Systems and Computer Engineering.*

is very limited in practice. However, no mathematical model of the process is required in pattern recognition methods as the operation of the process is classified by matching the measurement data. Several neural network-based pattern recognition methods have been proposed in recent years that provide relative satisfactory results [3]. Such diagnostic systems make use of the neural network's ability to generalize as well as its ability to function without physical fault models or tolerance information. Nevertheless, generally, neural network-based pattern recognition usually needs a larger sample set and has low efficiency for a sample set in high-dimensional spaces. The development of SVM offers us a possibility to train nonlinear classifiers in high-dimensional spaces with good generalization ability using a small training set.

SVMs are the classifiers that were originally designed for binary classification. Practical classification applications are commonly multiclass problems. Forming a multiclass classifier by combining several binary classifiers is the usual approach; methods such as o-a-o, o-a-r, and DDAG SVM are all based on binary classification. A binary tree based on SVM is also a good way to solve multiclass problems.

In this chapter, we introduce the theory and algorithm of several common multiclassifiers based on SVM and put forward one kind of automatically generated unsupervised binary-tree multiclassifier based on SOMNN clustering and SVM classification aiming at the characteristics of fault diagnosis of an analog circuit with tolerances, noise, and circuit nonlinearities.

The chapter is organized as follows. In Sect. 9.2, the theory of several common multiclass SVMs are briefly described; in Sect. 9.3, the three binary tree structures are defined and a multiclassifier based on an unsupervised binary tree and SVM is introduced. Simulation experiments and results are presented in Sect. 9.4. Finally conclusions are given in Sect. 9.5.

9.2 Three Common Multiclassifiers Based on SVM

The traditional SVM only provides two-class classification arithmetic and it is important to extend it to multiclass classification. There are many multiclass classified methods that have been put forward before and are comparatively mature. The multiclass classification problem is commonly solved by a decomposition to solve several two-class problems for which the standard SVM is used. There are two important factors that affect classified accuracy and speed of SVM: one is distance between classes; the other is the number of training samples. The farther the distance is and the smaller the number is, the higher the accuracy and speed of SVM obtained. Generally we apply the three modes of o-a-r, o-a-o, and DDAG, which are introduced as follows.

9.2.1 One-Against-Rest Method [4]

The o-a-r method is the simplest strategy to extend SVM to the N-class pattern recognition problems. This strategy builds N binary SVM classifiers, one for each of the N classes. Input testing samples X to N two-class classifiers, respectively, and calculate the discriminant functions of every classifier. Then choose the class whose discriminant function is maximal as the one of testing data.

The shortcomings of this multiclass classification algorithm are as follows,

1. During the training phase, all the training samples must be trained; before determining that the test samples belong to one of N classes, all the N discriminant functions must be calculated. So when there is a larger training sample number N, the training and testing speeds are low.
2. When the testing samples do not belong to any of N classes (the red part in Fig. 9.1) or only belong to one of N classes (the blue part in Fig. 9.1a), there is a large nonseparable region. Figure 9.1 shows there is a three-class problem with nonseparable region.

9.2.2 One-Against-One Method [5]

In this mode, each two classes construct a child classifier of SVM. In this way, $N*(N-1)/2$ child classifiers of SVM will be obtained. Combine all $N*(N-1)/2$ classification functions to determine the class of the testing sample through the accumulation of predicting classification. In the testing, all testing data $N*(N-1)/2$ child classifiers of SVM and the class with the object samples which appear maximal times totally is the predicting classification using the voting method.

The shortcomings of this multiclass classification algorithm are as follows.

1. The number of two-class classifiers $N*(N-1)/2$ increases sharply with the increase of the class number N. In this way, there are so many calculations, which lead to train and test slowly. It is impossible realize real-time classification.

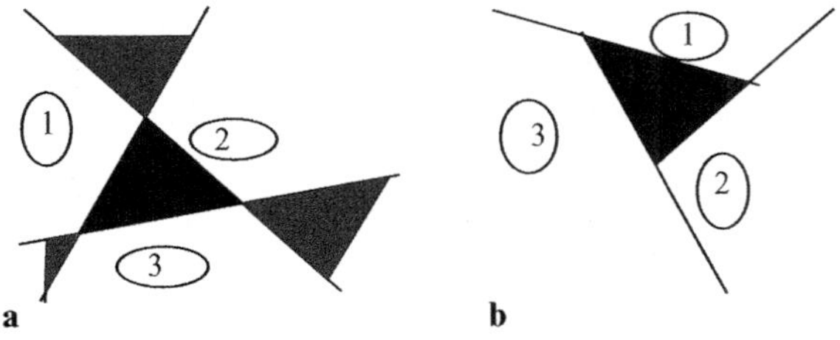

Fig. 9.1 **a** The nonseparable region of o-a-r; **b** the nonseparable region of o-a-o

2. In the testing classification, when the voting numbers of any two classes are the same, there is a part of the nonseparable region. Figure 9.1b (red part) shows there is a three-class problem with the nonseparable region.

9.2.3 DDAG Method [6]

In this method, there are the same classes except one at the decision nodes in a level; namely, there are many overlapping pattern classes at the level nodes except leaf nodes, so more support vectors are needed to separate two subgroups, which affect badly the training and testing speed. In this way, $N * (N-1)/2$ child classifiers of SVM will be obtained, without a nonseparable region. In addition, a different choice of root node will give birth to a different result for a decided class, which makes classification uncertain. The blue parts between class 1 and class 2 in Fig. 9.1a in the root nodes of SVM1-23 and that of SVM2-13 belong to class 1 and class 2, respectively.

9.3 Design and Analysis of Multiclassifiers Based on Binary Trees and SVM

9.3.1 Forming of the Binary Tree and Its Effect for Classified Performance

Because SVMs are based on two-class classification, multiclass classification based on the binary tree and SVM can be developed through combining its theory with the basic theory of the binary tree. The binary-tree classification method can transform a multiclass problem to a two-class problem effectively and decrease the error ratio.

The process of training is the process of forming the decision tree. Every node except the leaf node corresponds to a division at the top separating some classes from the others. The division at the top node of the binary tree should be determined first. Then new nodes are created to handle each of the resulting partitions and this process is repeated until only one class remains in the separated region. When the training process is finished, the feather space is divided into K regions (assuming K pattern classes) and there is no unclassifiable region in the feature space.

According to the forming process of the binary tree, there are many schemes to construct a strict binary tree with K leaf nodes according to different division methods. Figures 9.2 and 9.3 present two examples of the division of feature space and the corresponding binary tree for a decided classification problem. Different tree structures correspond to different divisions of feature space, and the classification performance of the classifier is closely related to the tree structure. The structure of the binary tree is closely related to the classification performance of the classifier

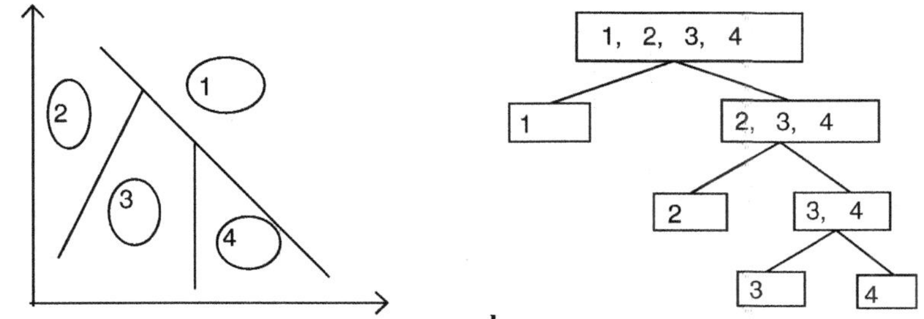

a b

Fig. 9.2 The division of feature space and the corresponding binary tree, Example 1: **a** the example of the division of feather space; **b** expression by binary tree

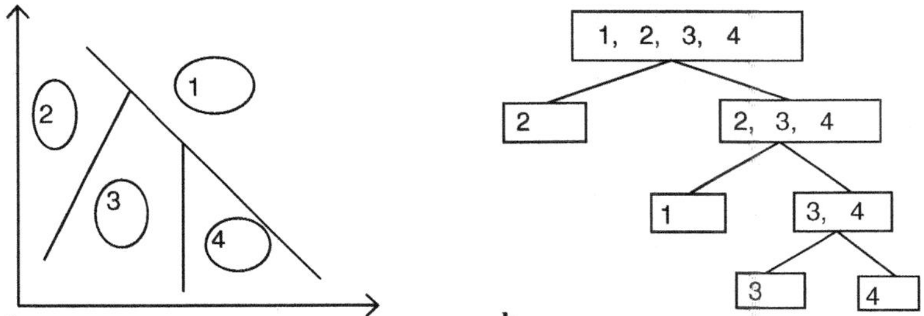

a b

Fig. 9.3 The division of feature space and the corresponding binary tree, Example 2: **a** Example 2 of the division of feather space; **b** expression by binary tree

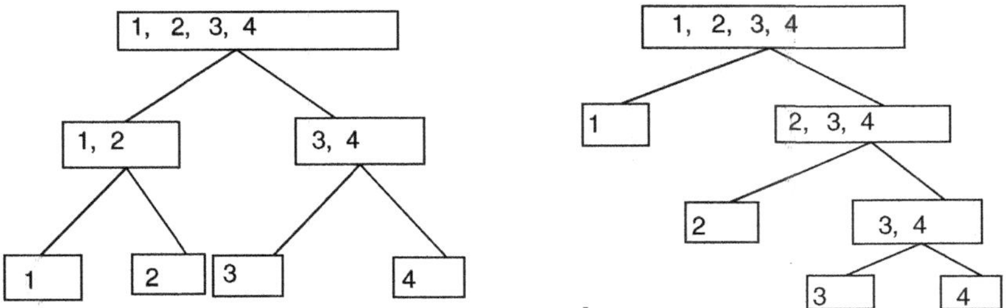

a b

Fig. 9.4 Structures of two peculiar binary trees: **a** symmetrical binary tree; **b** inclined binary tree

in this method; it is important to study how to determine the structure of the binary tree according to the training samples so that the classification error is minimized. In order to verify that the binary-tree structure of the proposed method based on similarity between pattern classes (the SOMNN clustering) has higher generalization ability, we define three kinds of binary-tree structures: symmetrical binary tree (SBT), inclined binary tree (IBT), and randomized binary tree (RBT). Figures 9.4a and b illustrate two peculiar binary-tree structures for the four-class classification: the SBT and IBT. Any binary-tree structures between the SBT and IBT are called the RBT. There is a little chance for the proposed unsupervised binary tree (UBT) to be in conformity with the SBT or IBT; in other words, the SBT and IBT are two extreme structures of the UBT. The SBT keeps perfect symmetry structure but neglects separability between classes, which gives rise to more support vectors, and there is further low classification accuracy if this structure is combined with SVM to form a multiclassifier. However, if the unsupervised binary tree based on SOMNN clustering conforms with the SBT, a classifier based on this structure and SVM will obtain the best classification performance, because there is the smallest number of average training samples with this structure. A large number of the training samples are trained repeatedly by a classifier based on the IBT and SVM, so that it will increase training and testing time, and further low classification speed.

9.3.2 Construction and Algorithms of Unsupervised Binary Tree

If the classification performance is not good at the upper node of the binary tree, the overall classification performance becomes worse. Therefore more separable classes should be separated at the upper nodes of the binary tree.

Based on the above reasons, we define a novel tree based on SOMNN clustering, namely, an unsupervised binary tree, which is completely data-driven. In this way, the SOMNN is first used to cluster all fault classes to generate two nodes, then new nodes are created to handle each of the resulting partitions and the process continues until all the classes at the node are separated. According to the preprocess results of SOMNN, the classes at each node are decided followed by the structure of an unsupervised tree. Then SVM are utilized to segment each decision node accurately. This structure relies on the training samples, so a specific structure is given in the experiment.

9.3.2.1 SOMNN Based on Kernels

The SOM is one of the most popular artificial neural network algorithms [7]. The SOM is based on unsupervised competitive learning, which means that the training is entirely based on separability between pattern classes and the neurons of the map compete with each other.

Based on the classical Kohonen formulations, we construct the kernel SOM algorithm. In this way, although we replace the inner product in pattern space with a kernel function, in essence, clustering still proceeds in the original pattern space. The difference is: Euclidean distance is no longer the measure between patterns and weight vectors; instead kernel functions are employed. Several common kernel functions based on the Mercer condition are as follows.

$$\text{Gauss RBF kernel function: } K(\mathbf{x},\mathbf{x}') = \exp((-\parallel \mathbf{x}-\mathbf{x}' \parallel)^2 \div \sigma^2) \qquad (9.1)$$

$$\text{Polynomial kernel function: } K(\mathbf{x},\mathbf{x}') = ((\mathbf{x}\cdot\mathbf{x}')+c)^d,\ (c \geq 0) \tag{9.2}$$

$$\text{Sigmoid kernel function: } K(\mathbf{x},\mathbf{x}') = \tanh(K(\mathbf{x},\mathbf{x}')+v),\ (k>0, v>0) \tag{9.3}$$

In this paper, we adopt the Gauss RBF kernel function.

The weight adjusting formula for nerve neuron j in the input space is:

$$\mathbf{w_j}(n+1) = \mathbf{w_j}(n) + \eta(n)h_{ji}(n)(\mathbf{x}(\mathbf{n}) - \mathbf{w_j}(n)),\ (\mathbf{x} \in R^N) \tag{9.4}$$

The weight adjusting formula for nerve neuron j in the feature space is:

$$\mathbf{w_j}(\mathbf{n+1}) = \mathbf{w_j}(\mathbf{n}) + \eta(n)h_{ji}(\phi(x(n) - \mathbf{w_j}(n))),\ (\mathbf{w_j}, \phi \in R^M) \tag{9.5}$$

where $N \ll M$ is small clearly than M, $\mathbf{wj}$ is the weight vector of nerve neuron j; i is the winner of the network; $\phi(x)$ is the mapping function; hji is the neighborhood function; and $\eta(n)$ is the learning rate. We define $\mathbf{wj}$ as

$$\mathbf{w_j}(\mathbf{n}) = \sum_{k=1}^{L}\left(a_{j,k}^{(n)}\phi(\mathbf{x_k})\right) \tag{9.6}$$

Substitute (9.5) into (9.6),

$$\begin{aligned}\sum_{k=1}^{L} a_{jk}^{(n+1)}\phi(\mathbf{x_k})\cdot\phi(\mathbf{x}_\iota) = (1-\eta(n)h_{ji}(n))\sum_{k=1}^{L} a_{jk}^{(n)}\phi(\mathbf{x_k})\cdot\phi(\mathbf{x}_\iota)\\ +\eta(n)h_{ji}(n)\phi(\mathbf{x})\cdot\phi(\mathbf{x}_\iota)\end{aligned} \tag{9.7}$$

Another expression of (9.7) is:

$$A_j^{(n+1)}K = (1-\eta(n)h_{ji}(n))A_j^{(n)}K + \eta(n)k \tag{9.8}$$

where

$$K = [k(\mathbf{x_j}), \mathbf{x}_\iota]_{(j,\iota)=1}^{L*L}, \qquad (j,\iota = 1,2,\ldots,L); \tag{9.9}$$

$$k(\mathbf{x_j}, \mathbf{x}_\iota) = \phi(\mathbf{x_j})\cdot\phi(\mathbf{x}_\iota); \tag{9.10}$$

$$A_{j()n} = \left[a_{j1}^{(n)}\right], \left[a_{j2}^{(n)}\right], \ldots, \left[a_{jL}^{(n)}\right]; \tag{9.11}$$

$$\mathbf{x}_\iota = [x_1, x_2, \ldots, x_L]; \tag{9.12}$$

$$\phi(\mathbf{x}_\iota) = [\phi(\mathbf{x_1}), \phi(\mathbf{x_2}), \ldots, \phi(\mathbf{x_L})]; \tag{9.13}$$

$$\mathbf{k} = [k(\mathbf{x}, \mathbf{x_1}), k(\mathbf{x}, \mathbf{x_2}), \ldots, \mathbf{k}(\mathbf{x}, \mathbf{x_L})] \tag{9.14}$$

The weight-adjusting formula for neuron j in the feature space becomes:

$$A_j^{(n+1)} = ((1-\eta(n)h_{ji}(n))A_j^{(n)} + \eta(n)h_{ji}(n)\mathbf{k} * K^{-1} \tag{9.15}$$

$$i(\mathbf{x_i}) = \operatorname{argmin}_j \left\|\phi\left(\mathbf{x_i} - A_j^{(n)}\right)\right\| \tag{9.16}$$

Due to

$$\left\|\phi(\mathbf{x}) - A_j^{(n)}\right\|^2 = k(\mathbf{x},\mathbf{x}) + A_j^{(n)} K^T A_j^{(n)T} - 2A_j^{(n)} k(\mathbf{x_k},\mathbf{x}) \tag{9.17}$$

Substitute (9.17) into (9.16);

$$i(\mathbf{x_i}) = \operatorname{argmin}_j \{k(\mathbf{x_i},\mathbf{x_i}) + A_j^{(n)} K^T A_j^{(n)T} - 2A_j^{(n)} \mathbf{k_i}^T\} \tag{9.18}$$

$$\text{where } \mathbf{k_i} = [k(\mathbf{x_1},\mathbf{x_i}), k(\mathbf{x_2},\mathbf{x_i}), \ldots, k(\mathbf{x_L},\mathbf{x_i})] \tag{9.19}$$

With the operating system for MATLAB 6.5, outputs of the active nerve neuron are 1 and outputs of neighborhood nerve neurons are 0.5.

9.3.2.2 Process of Formation of the Unsupervised Binary Tree

Step 1. The SOMNN-based kernel was first used to cluster for all training samples to generate two nodes; then new nodes were created to handle each of the clustering results and the process continued, until all the classes at the node were separated.

Step 2. According to each of the clustering results, ensure classes are associated with each node, construct binary-tree and select the training samples for each node.

Step 3. Every decision node is segmented accurately by SVM. All the nodes except leaf nodes correspond to a hyperplane separating some classes from the others. When the training process is finished, the feather space is divided into K regions and this division is unique.

9.4 Experiments and Analysis of Simulation Result

The sample circuit we have used to demonstrate fault diagnosis of nonlinear circuits with tolerances based on the asymmetry binary tree is shown in Fig. 9.5 shows a second-order filter with all components set to their nominal values.

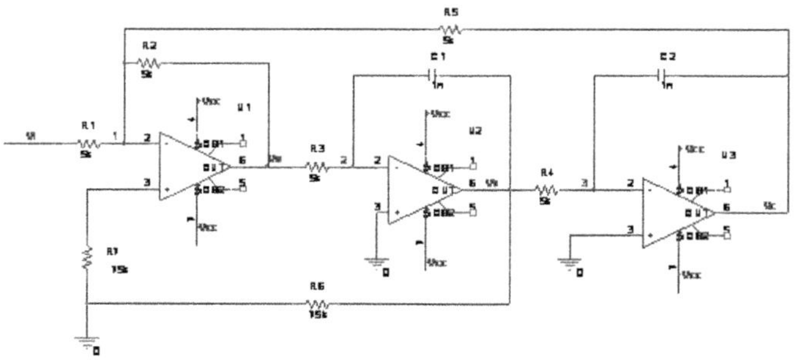

Fig. 9.5 The second-order filter

The faults associated with this circuit are assumed to be R2↑, R2↓, R5↑, R5↓, R6↑, R6 ↓, C1↑, C1 ↓, C2↑, C2↓. In this notation, ↑and↓ imply significantly higher or lower than nominal values. The fault components associated with the second-order filter are shown in Table 9.1. In order to generate training data for the fault classes, we set the value for faulty components and vary other resistors and capacitors within their standard tolerances of $5/100$ and $10/100$, respectively. Fault diagnosis of this circuit requires the neural network to recognize 11 pattern classes, 10 fault classes indicated above, and the NFT (fault-free) class. To generate training data for these fault classes, we can select V1, V2, V3, Va, Vb, and Vc, six node voltages that reflect the faults effectively. In Table 9.1, we describe the datasets that we used in our experiments involving the proposed algorithm for constructing an unsupervised binary tree. We implemented the algorithm for constructing unsupervised decision trees on a Pentium 4 machine with MATLAB 6.5. In order to test the performance of the proposed method, 400 samples are chosen as training samples and 200 as testing samples. We illustrate the partitions generated by the SOMNN on the synthetic dataset in Fig. 9.6. The number inside each region represents the node of the binary tree to which the corresponding subsets of data records

Table 9.1 A set of node voltages for all the fault classes

Fault Classes	Fault Elements	V1(v)	V2(v)	V3(v)	Va(v)	Vb(v)	Vc(v)
F0	NFT	−.0060	−.0737	0.0192	0.0314	0.0418	−.0178
F1	R2↑	−.0130	0.0000	0.0193	0.0350	0.0418	−.0195
F2	R2↓	−.0060	0.0000	0.0183	0.0504	0.0418	−.0196
F3	R5↑	−.0060	0.0000	0.0193	0.0472	0.0418	−.0203
F4	R5↓	−.0058	−.0725	0.0193	0.0363	0.0418	−.0155
F5	R6↑	−.0072	0.0000	0.0183	0.0412	0.0418	0.0378
F6	R6↓	−.0048	0.0737	0.0233	0.0424	0.0378	−.0215
F7	C1↑	−.0060	−.0883	0.0183	0.0540	−.0180	−.0179
F8	C1↓	−.0060	−.0883	0.0195	0.0540	.00380	−.0180
F9	C2↑	−.0060	0.0737	0.0379	0.0438	0.0338	−.0179
F10	C2↓	−.0060	0.0737	0.0163	0.0438	0.0338	−.0179

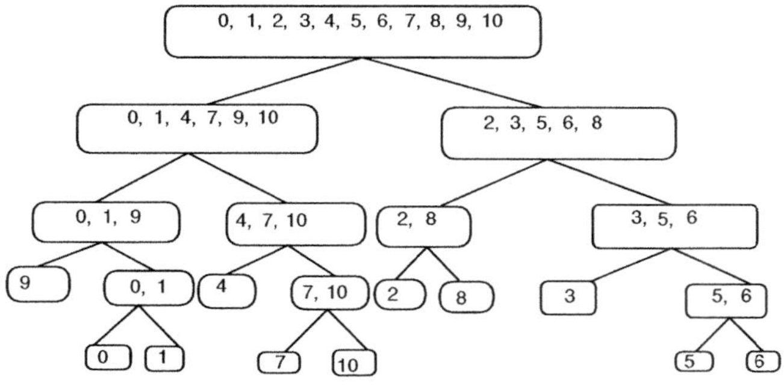

Fig. 9.6 The binary-tree structure based on training sample dataset

Table 9.2 Comparison of several multiclass classification methods

Methods	Training Samples	Training Sample	Time (s)	Accuracy (1/100)
o-a-r	400	200	4.18	92.67
o-a-o	400	200	1.73	93.33
DDAG	400	200	1.61	93.78
RBT	400	200	1.27	93.37
SBT	400	200	1.11	95.90
IBT	400	200	1.30	98.02
UBT	400	200	1.01	97.91

are allocated by the algorithm. The training of the SVM is under the same conditions when comparing the performance of the seven multiclass classification methods: o-a-r, o-a-o, DDAG based on SVM and RBT, SBT, IBT, UBT combined with SVM. Experiments have been made several times under the same conditions. The average correct classification rates and times for the testing data of the fault classes are given in Table 9.2. From Table 9.2 we can see that the average classification time of the multiclassifiers based on binary trees and SVM is 1.37 s, 1.11 s, 1.30 s, and 1.01 s, respectively, which is relatively shorter than the other methods, for there is no unclassifiable region in the feature space with those methods. Due to all the SVM at every decision node based on maximal separability between classes, the proposed multiclassifier obtains the shortest testing time. Although the average classification rate of the proposed method is slightly lower than that based on IBT and SVM by $0.11/100$, 0.29 s is saved. It is advisable that the proposed algorithm be applied to implement real-time fault diagnosis of the analog circuit at the cost of a little lower accuracy.

Classification results for this dataset also prove the performance improvement of the proposed multiclassifier.

9.5 Conclusion

In this chapter, we discussed several multiclassifiers based on SVM and analyzed their classification performance. Aiming at the characteristic of fault diagnosis of analog circuits with tolerances, we proposed a novel algorithm combining an unsupervised binary tree multiclassifier based on self-organizing map nerve network clustering and support vector machine classification. Separability between pattern classes was considered in constructing the binary tree, which reduced the number of support vectors when SVM were adopted to classify these classes, so that the algorithm obtains higher classified accuracy and speed.

References

1. Che M, Grellmann W, Seidler S (1987) *Selected Papers on Analog Fault Diagnosis*. New York: IEEE Press.
2. Liu R-W (1991) Testing and Diagnosis of Analog Circuit and Systems. New York: Van Nostrand.
3. Aminian F, Aminian M, Collin HW (2002) Analog fault diagnosis of actual circuit using neural network. *IEEE Transactions on Instrumentation and Measurement*, 51: 544–550.
4. Vapnik V (1995) *The Nature of Statistical Learning*. New York: Springer-Verlag.
5. Hsu C-W, Lin C-J (2002) A comparison of methods for multi-class support vector machines. *IEEE Transactions on Neural Networks*, 13(2): 415–425.
6. Vapnik V (2000) *Advances in Neural Information Systems*: 547–553, MIT Press.
7. Haykins S (2004) *Neural Netwoks: A Comprehensive Foundation [M]*. 2nd Edition: 321–347, Prentice Hall.

Chapter 10
The Effect of Background Knowledge in Graph-Based Learning in the Chemoinformatics Domain

Thashmee Karunaratne and Henrik Boström

10.1 Introduction

Typical machine learning systems often use a set of previous experiences (examples) to learn concepts, patterns, or relations hidden within the data [1]. Current machine learning approaches are challenged by the growing size of the data repositories and the growing complexity of those data [1, 2]. In order to accommodate the requirement of being able to learn from complex data, several methods have been introduced in the field of machine learning [2]. Based on the way the input and resulting hypotheses are represented, two main categories of such methods exist, namely, logic-based and graph-based methods [3]. The demarcation line between logic- and graph-based methods lies in the differences of their data representation methods, hypothesis formation, and testing as well as the form of the output produced.

Logic-based methods, which fall in the area of inductive logic programming [4], use logic programs to represent the input and the resulting hypotheses. Graph-based methods use graphs to encode input data and discover frequent or novel patterns in the form of subgraphs. These graphs contain labeled nodes which represent attributes of data entities and labeled edges that represent the relationship among the attributes of corresponding entities.

Logic-based methods use several search approaches to explore the hypothesis space. Common methods are depth-first, breadth-first, or heuristic-based greedy search. These methods often require strong restrictions on the search space in order to be efficient, something which is commonly referred to as the search bias. Graph-based methods use similar techniques for search tree traversal, but the main difference lies in that graph learning methods either look for most frequent subgraphs or novel patterns within the graph data, depending on the task. In doing so, the search tree is incrementally expanded node by node. Some methods explore the complete search space, but almost all graph-based methods use some pruning technique, such as considering supergraphs of infrequent graphs to also be infrequent,

Oscar Castillo et al. (eds.), *Trends in Intelligent Systems and Computer Engineering.*

which was first done in the a priori algorithm [5]. Washio and Motoda [6] provide a detailed description of existing subgraph search methods.

Apart from considering various types of algorithms in order to improve classification accuracy, considering various types of domain-specific knowledge (or background knowledge) is also a possible way of improving learning [7]. For logic-based methods, it is a well-established fact that adding relevant background knowledge may enhance the performance to a great extent. Several earlier studies have investigated the amount of improvement that can be achieved by introducing relevant background knowledge into ILP systems; for example, Srinivasan et al. [8] raise the question, "How does domain specific background information affect the performance of an ILP system?" In contrast to this, such studies have not been undertaken for graph-based methods, but we think that raising a similar question for these methods is highly relevant.

The main purpose of our study is to investigate the effect of incorporating background knowledge into graph learning methods. The ability of graph learning methods to obtain accurate theories with a minimum of background knowledge is of course a desirable property, but not being able to effectively utilize additional knowledge that is available and has been proven important is clearly a disadvantage. Therefore we examine how far additional, already available, background knowledge can be effectively used for increasing the performance of a graph learner. Another contribution of our study is that it establishes a neutral ground to compare classification accuracies of the two closely related approaches, making it possible to study whether graph learning methods actually would outperform ILP methods if the same background knowledge were utilized [9].

The rest of this chapter is organized as follows. The next section discusses related work concerning the contribution of background knowledge when learning from complex data. Section 10.3 provides a description of the graph learning method that is used in our study. The experimental setup, empirical evaluation, and the results from the study are described in Sect. 10.4. Finally, Sect. 10.5 provides conclusions from the experiments and points out interesting extensions of the work reported in this study.

10.2 Related Work

Studies reported in [8, 10, 11] are directly related to investigation of the effect of background knowledge in enhancing classification accuracy for logic-based methods. These studies have used different levels of background knowledge to illustrate the improvement in predictive accuracies that could be achieved by adding relevant background knowledge.

Existing logic-based approaches could be categorized into two main categories based on the way they construct the hypotheses. The first category contains so-called propositionalization approaches (e.g., [12, 13]), which generate feature sets from background knowledge to be used by classical attribute value learners. The second category contains so-called relational learning approaches, which discover sets of classification rules expressed in first-order logic. Examples of such approaches are PROGOL [14], FOIL [15], TILDE [16], and XRULES [17]. Both relational learning and propositionalization approaches utilize relevant additional background knowledge in the form of logic programs, and most of the approaches (e.g., [12, 14, 15]), use a hypothesis language that is restricted by user-defined mode declarations.

Graph-based learning methods use concepts borrowed from classical mathematical graph theory in order to represent input data and to construct the search mechanisms. Theoretically, learning from graphs is a search across the graph lattice for all possible (frequent or novel) subgraphs [18]. This graph lattice could either be a large single graph or a disconnected set of small graphs. Different search mechanisms are employed in order to traverse the search tree. Breadth-first and depth-first search have been used in early graph-based learning methods [19].

These approaches have limitations w.r.t. memory consumption due to the fact that typically a very large number of candidate subgraphs are generated in the middle of the lattice [19]. Most of the current algorithms use various measures in order to overcome the complexity of candidate generation. For example, SUBDUE [20] carries out a heuristic-based greedy search using minimum description length to compute support for candidate subgraphs. Subgraphs below a predefined minimum description length are removed from the set of candidate subgraphs. gSpan [21] defines a lexicographic order among graphs, which is a canonical labeling of input graphs, and maps each graph to a canonical label, called the unique minimumDFS. Methods that rely on the definition of kernel functions [22] have also proved successful in this context for effective learning, but at a computational cost [23].

Graph-based learning methods have been improved over time with respect to efficiency in graph encoding and subgraph search, but very little attention has been paid to the encoding of background knowledge for graph-based methods. Specifically for the chemoinformatics domain, almost all the current learning methods encode atom names into node labels and bond types into edge labels in order to transform a molecule into a graph. For example, a typical graph learning algorithm encodes a molecular fragment containing two carbon atoms sharing an aromatic bond as in Fig. 10.1, where c is the node label, 22 is the atomic value of the carbon atom, and 7 is the edge label given for the aromatic bond.

SUBDUE [9] has produced a detailed representation of additional background knowledge in the graph as shown in Fig. 10.2. The additional background

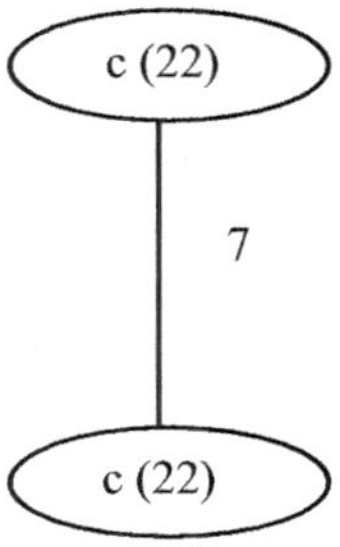

Fig. 10.1 General representation of atom–bond data in graph-based learning methods

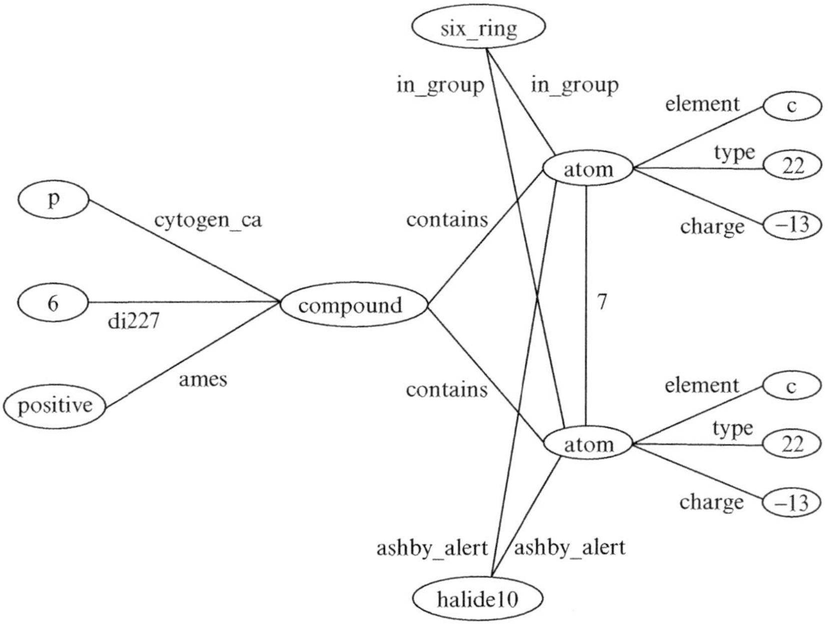

Fig. 10.2 Including all the available background information available for the same graph in Fig. 10.1, as given in [9]

information such as the charge (−13), and two subgroups, *Halide* and *Six-ring*, each of which the atoms are part of, are also included in the graph. Yet the potential gain acquired by this representation was not examined in that work. On the other hand, this representation allows graphs to grow exponentially with the number of nodes. Figure 10.2 corresponds to two nodes (atoms) and three pieces of additional background knowledge only, but a typical molecule contains at least 5–6 atoms and several structural relations. This complexity may be the reason why there has been no systematic investigation of effectively using available background knowledge to enhance classification accuracy of existing graph-based learning methods.

10.3 Graph Learning with DIFFER

DIFFER [24] is a graph propositionalization method that employs a canonical form of graph transformation called finger printing. DIFFER's graph transformation and subgraph search algorithms are simple, straightforward, and are not associated with any NP-complete graph matching problems as are most other graph-based methods. Most of the graph learning methods require solving an NP-complete problem either during discovery or classification. For example, subgraph search is a subgraph isomorphism problem which is NP-complete [19] and defining kernel functions is also NP-complete [23].

DIFFER [24] uses finger printing as the method of transforming a graph into a canonical form of it. In general terms, a finger print of a graph is nothing but a collection of distinct node–edge–node segments that are parts of the respective graph. For example, a triple $(c,\ o,\ 2)$ of a molecule from the domain of chemoinformatics would correspond to a double bond between a carbon and an oxygen atom. Therefore the fingerprint of a graph is defined as follows.

Definition: A finger print

Let *G(V, l, E, a)* be a graph, where V is a finite nonempty set of vertices, and l be a total function $l : V \rightarrow \Sigma_V$. E is a set of unordered pairs of distinct vertices called edges, and a is a total function such that $a : E \rightarrow \Sigma_E$. Let V_i and V_j be two nodes in the graph with labels l_i and l_j, respectively. Furthermore, let e be the label of the edge E between V_i and V_j. Then for all l_i and $l_j \in G$, define a set:

$$\begin{cases} (l_i, l_j, e) & if \quad e \neq \phi \\ (l_i, l_j) & otherwise \end{cases}$$

This set is called the *finger print* of G.

The finger prints are used for substructure search in such a way that for all pairs of examples, the intersection of their finger prints, which is referred to as the maximal common substructure, is formed, and ranked according to their frequency in the entire set of examples (i.e., the number of finger prints for which the maximal common substructure is a subset). The maximal common substructure search algorithm [24] searches for common elements among the fingerprints. In doing so the algorithm executes in a pairwise manner, that is, taking two fingerprints at a time. The algorithm is given in Fig. 10.3.

The set of *MaximalSubstructures* discovered by the maximal common substructure search algorithm is treated as the feature set that could be used for propositionalization. Yet all the *MaximalSubstructures* may not be equally important for the given learning task. Therefore irrelevant substructures have to be removed from

```
function PairWiseMaximalSubstructures((f1,...,fn); list of finger prints)
   j = 1
   while (j ≤ n1)
do
      k = j + 1
      while (k ≤ n)
      do
      add fj ∩ fk to MaximalSubstructures
      k++
   done
   j++
done
return MaximalSubstructures
```

Fig. 10.3 Maximal common substructure search algorithm

the set of *MaximalSubstructures* prior to use in the propositional learner. In order to detect the irrelevant features the set of *MaximalSubstructures* is ranked with respect to the frequency in the finger prints. High frequency reflects substructures that more frequently appear in the finger prints irrespective of the class labels, and low frequency reveals the infrequently appearing substructures in finger prints. Substructures that possess any of these properties do not contribute to unusual distributions over the classes, and therefore we could detect these irrelevant substructures and remove them from the feature set. A score is calculated for each substructure in order to come up with a measure of relevance of features. According to this definition the score S for a feature f is defined as

$$S = \frac{\text{number of finger prints containing } f}{\text{total number of finger prints}}$$

A suitable upper and lower threshold value to the score S is applied to the selection of most relevant features from the set of *MaximalSubstructures*.

An upper and lower threshold is applied to select the most informative substructures (features) for classification. The selected elements of the finger prints are used as (binary) features, allowing predictive models to be built by any standard attribute-value learner.

10.4 Empirical Evaluation

Several studies (e.g., [8, 10, 11]) have explored the effect of adding background knowledge in the domain of chemoinformatics, and background knowledge useful for enhancing predictive accuracy has been identified for several tasks. This domain is also chosen for this study together with earlier formulated background knowledge. The objective of the empirical evaluation is to study the effect on accuracy when incrementally adding background knowledge for tasks solved using the graph learning method DIFFER.

10.4.1 Background Knowledge for Chemoinformatics

In [8], four levels of chemical background knowledge are defined that are relevant for the mutagenesis dataset [25]. Most of the experiments that have been performed using this dataset concern one or more of these levels. In brief, the levels are (a more detailed description is given in Sect. 10.4.3):

1. The atoms that are present in the molecule are given as well as the element and type of each atom. Bonds between the atoms are also given with the type of each bond (single, double, etc.).
2. The structural indicators and global molecular properties for molecules as defined by [25].

3. The two-dimensional substructures that are associated with molecules such as benzene rings, nitro groups, and the like in addition to atom–bond data.
4. Three-dimensional molecular descriptions between the substructures within molecules in addition to background knowledge on level 3.

A similar description of background knowledge which is relevant for the carcinogenesis dataset is given in [26].

10.4.2 Datasets

We have chosen two datasets for this study: mutagenesis [25] and carcinogenesis [27]. We have selected these two datasets because they have been widely used in research and several studies have shown that the background information that is available for them is truly helpful in enhancing the predictive accuracies of the logic-based methods. The datasets are publicly available, thus any reader may compare the results obtained in this study with that of other machine learning algorithms. A brief description of the two datasets is given below.

10.4.2.1 Mutagenesis

The problem related to the mutagenesis dataset is to predict the mutagenicity of *Salmonella typhimurium TA98*, using a set of 230 aromatic and heteroaromatic nitro compounds. Because not all the compounds can be tested empirically, a machine learning method is required for prediction of whether a compound in this dataset is mutagenic. Debnath et al. [25] have recognized two subsets of this dataset: 188 compounds that could be fitted using linear regression, and 42 compounds that could not. As in most of the earlier studies, we consider only the regression-friendly dataset of 188 examples that contains 125 mutagenic and 63 nonmutagenic examples.

10.4.2.2 Carcinogenesis

The carcinogenesis dataset contains more than 330 chemicals that are tested for carcinogenicity [27]. Using rat and mouse strains (of both genders) as predictive surrogates for humans, levels of evidence of carcinogenicity are obtained from the incidence of tumors on long-term (two-year) exposure to the chemicals. The NTP [27] have assigned the following levels of evidence: clear evidence (CE), some evidence (SE), equivocal evidence (E), and no evidence (NE). Similar to most earlier studies, we have used the 298 training examples from the PTE dataset [28], with only the overall levels of positive activity, if CE or SE and negative otherwise, resulting in that the 298 training examples are divided into 162 positive examples and 136 negative examples.

10.4.3 Embedding Different Levels of Background Knowledge in DIFFER

The additional background knowledge could be viewed as novel relations among the existing elements in the structure, or new relations of existing elements with new entities or attributes. This additional knowledge could be included in graphs in two ways. One way is to define new edge labels using the additional relations and new nodes for additional entities or attributes. SUBDUE [9] has used this method as described in Fig. 10.2, for incorporating the 2-D descriptions of molecular substructures such as the *helide10* and *six_group*, and the global properties of the molecules such as *lumo, ames*, and so on.

Incorporating new knowledge as new nodes and edges is straightforward, yet this representation may end up with massive graphs. For example, the graph in Fig. 10.2 contains only two atoms, whereas a typical molecule in the chemoinformatics domain may contain about 20 atoms on average. This representation might then require several constraints due to computational demands, resulting in incomplete search or missing important relations.

The second approach to encoding additional background knowledge is to incorporate it as part of the existing node definition, which is exactly our approach. This approach allows expanding the node definitions in an appropriate manner to accommodate the additional information.

We introduce an extension to the graph representation in [24] by expanding the node and edge definitions by various forms of node and edge labels that enable incorporating different levels of background knowledge into graphs. We have used five different sets of labels that correspond to two different levels of background knowledge available in the chemoinformatics domain (levels 1 and 3 as discussed in Sect. 10.4.1). The following graph encodings are considered, where encodings D1, D2, and D3 belong to level 1 and D4 and D5 belong to level 3.

- *D1*: Each node is labeled with an atom name, its type, and a set of bonds by which the atom is connected to other atoms. The node definition for D1 can be represented by *(atom_name, atom_type, [bond type/s])*. For example, a node representing a carbon atom of type 22, which is connected to other atoms by an aromatic bond is labeled with *(c, 22, [7])*, where 7 denotes the aromatic bond. Figure 10.4 depicts a molecular segment containing two such carbon atoms. No edge label is associated with this representation (or all edges can be considered to have the same label *connected*).
- *D2*: The amount of information used for encoding is similar to D1, but the node and edge labels are different. Each node label in D2 is of the bonds by which the atom is connected to other atoms and it can be represented by *(atom_name, atom_type)*. The edges are labeled with the bond type by which two atoms are connected. For example, a node representing a carbon atom of type 22, which is connected to two other atoms by one single and one double bond is labeled with *(c, 22)*, and the edges to the nodes corresponding to the other atoms are labeled with *single* and *double*, respectively.

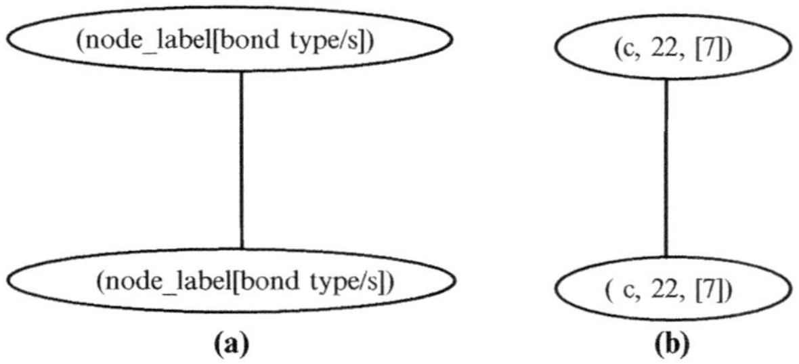

Fig. 10.4 a General definition for graph for node definition D1. **b** Corresponding representation of graph given in the example

- *D3*: Because DIFFER's finger prints are sets of triples as described in Sect. 10.3, duplicate triples are effectively removed. For example, if a molecule contains a six-ring, its encoding in the finger print will be (*c22, c22, aromatic*), because the six repetitions of the same label will be treated as one. However, the number of such similar bonds may be an additional piece of independent information for DIFFER. This information is encoded by extending the edge label with counts; that is, the edge label becomes *(bond_type, count)*. For example, the graph of the six-ring will then have an edge labeled *(aromatic, 6)*. The node labels are the same as for D2.
- *D4*: In addition to the background knowledge level 1, the atom's presence in complex structures such as benzene rings or nitro groups is also included in the node labels. Accordingly, the node label for D4 would be (*atom_name, atom_type, [list of complex structures]*). Hence this encoding includes background knowledge on level 3. For example, a carbon atom of type 22 that is part of a nitro group, a benzene group, and a 5-aromatic ring is labeled with (*c, 22, [nitro, benzene, 5-aromatic ring]*). The edge labels are the same as for D2.
- *D5*: The node labels are the same as for D4, but the edge labels contain counts as in D3.

10.4.4 Experimental Setup and Results

Feature sets are constructed using DIFFER for all five different graph encodings for the two datasets. The feature sets are given as input to a number of learning methods: random forests (RF), support vector machines (SVM), logistic regression, PART, and C4.5, as implemented in the WEKA data-mining toolkit [29]. All experiments are carried out using tenfold cross-validation. The results are summarized in Table 10.1, where the best learning method for each feature set is shown within parentheses.

The results for the different graph encodings reveal that the predictive accuracies for both datasets increase substantially when adding background knowledge. Also the inclusion of the number of repeated fragments in the finger print helped enhance the predictive accuracy. The difference in accuracy between the lowest and highest levels of background knowledge is significant according to McNemar's test [26]. Runtimes of DIFFER for different graph encodings are not reported here, inasmuch as measuring the runtimes was not our main objective, but the results reported in Table 10.1 were obtained within a few hours.

There is almost no difference in the accuracies of the graph encodings D1 and D2, reflecting the fact that these encodings do not differ in information content, but only in the formulation.

We have compared the predictive accuracies reported above with accuracies reported earlier for some standard logic-based and graph-based methods. The earlier studies include using the logic-based relational rule learner PROGOL [8, 26], the logic-based propositionalization method RSD [12], the graph-based frequent molecular fragments miner MolFea [31], and the graph-based relational data-mining method SUBDUE [9, 20]. The best accuracies reported by those methods are shown in Table 10.2, together with the results for DIFFER, where the predictive accuracies for all methods have been obtained by using tenfold cross-validation.

From Table 10.2, it can be seen that DIFFER outperforms all the other methods for the carcinogenesis dataset, and outperforms some and is outperformed by some for the mutagenesis dataset. One reason DIFFER works relatively better for carcinogenesis compared to mutagenesis may be that the latter contains smaller molecules with a smaller number of different atoms. The whole mutagenesis dataset is a combination of 8 different atoms, where the largest molecule contains only 40 atoms and 44 bonds. The carcinogenesis dataset has 18 different atoms with most of the molecules containing more than 80 atoms, which are involved in very complex structures. DIFFER is able to effectively deal with very large structures because it does not apply any constraints on the search for common substructures and hence involves no

Table 10.1 Performance of DIFFER with five different graph encodings

Data set	Accuracy				
	D1 (%, RF)	D2 (%, RF)	D3 (%, SVM)	D4 (%, SVM)	D5 (%, SVM)
Mutagenesis	80.61	80.61 (RF)	84.04	87.77	88.3
Carcinogenesis	61.25	62.1	68.73	71.03	75.0

Table 10.2 Best reported accuracies

Method	Mutagenesis (%)	Carcinogenesis (%)
DIFFER	88.3	75.0
PROGOL	88.0 [8]	72.0 [26]
RSD	92.0 [12]	61.4
MolFea	95.7 [31]	67.4 [31]
SUBDUE	81.6 [20]	61.54 [9]

search bias [24]. Furthermore, DIFFER follows a method where only the different atom–bond–atom relations are considered and therefore it performs comparatively well with datasets containing heterogeneous examples such as in carcinogenesis.

10.5 Concluding Remarks

The purpose of this study was to investigate the effect of adding background knowledge on the predictive accuracy of graph learning methods, which earlier had been studied only for logic-based approaches. Our study showed that graph learning methods may indeed gain in predictive accuracy by incorporating additional relevant background knowledge. Hence it can be concluded that the predictive performance of a graph learner is highly dependent on the way in which nodes and edges are formed. In the domain of chemoinformatics, we showed that the accuracy can be substantially improved by incorporating background knowledge concerning two-dimensional substructures for both tasks of predicting mutagenicity and carcinogenicity. Comparing the results obtained in this study with earlier reported results, one can conclude that even a quite simple graph learning approach, such as DIFFER, may outperform more elaborate approaches, such as frequent subgraph methods or kernel methods, if appropriate background knowledge is encoded in the graphs.

One area for future research is to study the effect of incorporating further background knowledge also on the remaining levels for the chemoinformatics domain, such as three-dimensional structural information corresponding to level 4. For DIFFER, this type of additional background knowledge could be encoded as extensions of current node and edge labels. Another possibility is to investigate the effect of encoding background knowledge as additional nodes and edges, as done in [9]. Another direction for future research is to study the effect of background knowledge also for other, more complex, graph learning methods, such as graph kernel methods.

References

1. Mitchell, T.M. (2006), *The Discipline of Machine Learning*, CMU-ML-06-108, School of Computer Science, Carnegie Mellon University, Pittsburgh.
2. Page, D. and Srinivasan A. (2003), ILP: A short look back and a longer look forward, *Journal of Machine Learning Research*, 4:415–430.
3. Ketkar, N., Holder, L., and Cook, D. (2005), Comparison of graph-based and logic-based MRDM, *ACM SIGKDD Explorations*, 7(2) (*Special Issue on Link Mining*).
4. Muggleton, S. and De Raedt L. (1994), Inductive logic programming: Theory and methods. *Journal of Logic Programming*.
5. Agrawal, R. and Srikant, R. (1994), Fast algorithms for mining association rules, *VLDB*, Chile, pp. 487–99.

6. Washio, T. and Motoda, H. (2003), State of the art of graph-based data mining. *SIGKDD Explorations*, 5(1):59–68 (*Special Issue on Multi-Relational Data Mining*).
7. Muggleton, S.H. (1991), Inductive logic programming. *New Generation Computing*, 8(4):295–318.
8. Srinivasan, A., King, R.D., and Muggleton, S. (1999), The role of background knowledge: Using a problem from chemistry to examine the performance of an ILP program, TR PRG-TR-08-99, Oxford.
9. Gonzalez, J., Holder, L.B., and Cook, D.J. (2001), Application of graph-based concept learning to the predictive toxicology domain, in *Proceedings of the Predictive Toxicology Challenge Workshop*.
10. Srinivasan, A., Muggleton, S.H., Sternberg, M.J.E., and King, R.D. (1995), The effect of background knowledge in inductive logic programming: A case study, PRG-TR-9-95, Oxford University Computing Laboratory.
11. Lodhi, H. and Muggleton, S.H. (2005), Is mutagenesis still challenging?, in *Proceedings of the 15th International Conference on Inductive Logic Programming*, ILP 2005, Late-Breaking Papers, pp. 35–40.
12. Lavrac, N., Zelezny, F., and Flach, P., (2002), RSD: Relational subgroup discovery through first-order feature construction, in *Proceedings of the 12th International Conference on Inductive Logic Programming (ILP'02)*, Springer-Verlag, New York.
13. Flach, P., and Lachiche, N. (1999), *1BC*: A first-order Bayesian classifier, in S. Daezeroski and P. Flach (Eds.), *Proceedings of the 9th International Workshop on Inductive Logic Programming*, pp. 92–103. Springer-Verlag, New York.
14. Muggleton, S. (1995), Inverse entailment and progol, *New Generation Computing*, 13 (3–4):245–286.
15. Quinlan, J.R. and Cameron-Jones, R.M. (1993), FOIL, in *Proceedings of the 6th European Conference on Machine Learning*, Lecture Notes in Artificial Intelligence, Vol. 667, pp. 3–20. Springer-Verlag, New York.
16. Blockeel, H. and De Raedt, L. Top-down induction of first-order logical decision trees, *Artificial Intelligence* (101)1–2:285–297.
17. Zaki, M.J. and Aggarwal, C.C. (2003), *XRules: An Effective Structural Classifier for XML Data* KDD, Washington, DC, ACM 316–325.
18. Cook, J. and Holder, L. (1994), Graph-based relational learning: Current and future directions, *JAIR*, 1:231–255.
19. Fischer, I. and Meinl, T. (2004), Graph based molecular data mining—An overview, in *IEEE SMC 2004 Conference Proceedings*, pp. 4578–4582.
20. Ketkar, N., Holder, L., and Cook, D. (2005), Qualitative comparison of graph-based and logic-based multi-relational data mining: A case study, in *Proceedings of the ACM KDD Workshop on Multi-Relational Data Mining*, August 2005.
21. Xifeng, Y. and Jiawei, H. (2002), "gSpan: Graph-based substructure pattern mining," in *Second IEEE International Conference on Data Mining (ICDM'02)*, p. 721.
22. Borgwardt, K.M. and Kriegel, H.P. (2005), Shortest-path kernels on graphs, *ICDM*, pp. 74–81.
23. Ramon, J. and Gaertner, T. (2003), Expressivity versus efficiency of graph kernels, in *Proceedings of the First International Workshop on Mining Graphs, Trees and Sequences*, pp. 65–74.
24. Karunaratne, T. and Boström, H. (2006), Learning from structured data by finger printing, in *Proceedings of 9th Scandinavian Conference of Artificial Intelligence*, Helsinki, Finland (to appear).
25. Debnath, A.K., Lopez de Compadre, R.L., Debnath, G., Shusterman, A.J., and Hansch, C. (1991), Structure-activity relationship of mutagenic aromatic and heteroaromatic nitro compounds: Correlation with molecular orbital energies and hydrophobicity, *JMC*, 34:786–797.
26. Srinivasan, A., King, R.D., Muggleton, S.H., and Sternberg, M.J.E. (1997), Carcinogenesis predictions using ILP, in *Proceedings of the 7th International Workshop on Inductive Logic Programming*.
27. US National Toxicology program, http://ntp.niehs.nih.gov/index.cfm?objectid = 32BA9724-F1F6-975E-7FCE50709CB4C932.

28. The predictive toxicology dataset, at ftp site: ftp://ftp.cs.york.ac.uk/pub/ML_GROUP/Datasets/carcinogenesis.
29. Witten, I.H. and Eibe, F. (2005), *Data Mining: Practical Machine Learning Tools and Techniques*, 2nd Edition, Morgan Kaufmann, San Mateo, CA.
30. Helma, C., Kramer, S., and De Raedt, L. (2002), The molecular feature miner molfea, molecular informatics: Confronting complexity, in *Proceedings of the Beilstein-Institut Workshop*, Bozen, Italy.

Chapter 11
Clustering Dependencies with Support Vectors

I. Zoppis and G. Mauri

11.1 Introduction

Experimental technologies in molecular biology (particularly oligonucleotide and cDNA arrays) now make it possible to simultaneously measure mRna-levels for thousand of genes [1]. One drawback is the difficulty in organizing this huge amount of data in functional structures (for instance, in cluster configurations); this can be useful to gain insight into regulatory processes or even for a statistical dimensionality reduction.

Most of the methods currently used, for example, to infer about gene coexpression, are performed by computing pairwise similarity or dissimilarity indices and then by clustering with one of the many available techniques [2–4]. Furthermore, in order to capture meaningful inferences over the course of phenotypic change, the problem is sometimes treated by evaluating time series data [5]; in this case the gene expression level is measured at a small number of points in time.

We argue that among all the inference models, kernel methods [6, 7] have become increasingly popular in genomics and computational biology [8]. This is due to their good performance in real-world applications and to their strong modularity which makes them suitable for a wide range of problems. These machine learning approaches provide an elegant way of dealing with nonlinear algorithms by reducing them to linear ones in a suitable feature space F (generally) nonlinearly related to the input dataset. In this class of algorithms we apply a clustering approach (SVC) [9] to identify groups of dependencies between pairs of genes in respect to some measure (i.e., kernel function) of their regulatory activity (activation or inhibition relationships).

In this chapter, we consider a simplified model based on mRna-data only, which is an effective gene-to-gene interaction structure. This can provide at least a starting point for hypotheses generation for further data mining[1].

---

[1] This approach might look too simplistic in view of the models which include metabolites, protein, and the like, but it can be thought of as a projection into the space of genes [10].

Oscar Castillo et al. (eds.), *Trends in Intelligent Systems and Computer Engineering.*

According to [11] we have performed our investigation by analyzing peak values of each expression profile. To this end, we define the expression of a gene as a set of peaks: we represent then the interaction between different genes through the interaction between their respective sets of peaks.

In the general case, SVC has the task to find a hypersphere with minimal radius R and center $\mathbf{a}$ which contains most of the data points (i.e., points from the training set). Novel test points are then identified as those that lie outside the boundaries of the hypersphere. As a byproduct of this algorithm, a set of contours that enclose the data is obtained. These contours can be interpreted as cluster boundaries and linkages between each pair of data items can be estimated.

There are two main steps in the SVC algorithm, namely SVM-training and cluster labeling. The SVM-training part is responsible for novelty model training: this is performed by fixing the kernel function to compare similar pairs of inputs. We deal with this task by evaluating similar interactions between pairs of genes: in other words, we use the kernel to measure the similarity between pairs of dependencies. The cluster labeling part checks the connectivity for each pair of points based on a "cut-off" criterion obtained from the trained SVMs. This is generally a critical issue from the time complexity point of view. However, when the label is based, for example, on appropriate proximity graphs [12], reduction in time can be obtained.

In order to use appropriate a priori knowledge that can avoid useless checks, we propose to consider a starting functional structure derived from the approximation of a combinatorial optimization problem, that is, MGRN [11]. This shortcut also gives us the advantage of being coherent with logical and biological considerations.

The chapter is organized as follows. In Sect. 11.2 we give a brief overview of the kernel methods and SVC algorithm. In Sect. 11.3 we address the MGRN problem and in Sect. 11.4 we apply our formulation to clustering the training set. In Sect. 11.5 we discuss the numerical results and finally, in Sect. 11.6, we conclude and discuss some directions for future work.

11.2 Kernel Methods

Kernel methods have been successful in solving different problems in machine learning. The idea behind these approaches is to map the input data (any set X) into a new feature (Hilbert) space $\mathcal{F}$ in order to find there some suitable hypothesis; so doing, complex relations in the input space can be simplified and more easily discovered. The feature map Φ in question is implicitly defined by a kernel function K which allows us to compute the inner product in $\mathcal{F}$ using only objects of the input space X, hence without carrying out the map Φ. This is sometimes referred as the kernel trick.

Definition 11.1 (Kernel function): *A kernel is a function* $K\colon X \times X \to \mathbb{R}$ *capable of representing through* $\Phi\colon X \to \mathcal{F}$ *the inner product of* $\mathcal{F}$; *that is,*

$$K(x,y) = <\Phi(x), \Phi(y)> . \qquad (11.1)$$

To assure that such equivalence exists, a kernel must satisfy Mercer's theorem [13]. Hence, under certain conditions (for instance, semidefiniteness of K), by fixing a kernel, one can ensure the existence of a mapping Φ and a Hilbert space $\mathcal{F}$ for which Eq. 11.1 holds. These functions can be interpreted as similarity measures of data objects into a (generally nonlinearly related) feature space (see, for instance, [6]).

11.2.1 Support Vector Clustering

Any linear algorithm that can be carried out in terms of the inner product Eq. 11.1 can be made nonlinear by substituting the kernel function K chosen a priori. The (kernel-based) clustering algorithm we use for our application is known as support vector clustering. It uses a Gaussian kernel function $K(x,y) = e^{-q||x-y||^2}$ to implicitly map data points in a high-dimensional feature space. In that space one looks for the minimal enclosing hypersphere; this hypersphere, when mapped back to data space, can separate into several components each enclosing a separate cluster of points. Here we briefly report basic ideas to formulate this method; for a complete treatment of the problem, see, for instance, [9].

11.2.1.1 SVM-Training Step

Given a set of input points $X = \{\mathbf{x}_i, i = 1,\dots,n\} \subset \mathrm{I\!R}^d$ and a nonlinear mapping $\mathbf{\Phi}$, the first step of SVC is to compute the smallest hypersphere containing $\{\mathbf{\Phi}(\mathbf{x}_i) : i = 1,\dots,n\}$ by solving the problem:

$$\begin{array}{rl} \min_{c,R,\xi} & R^2 + C\sum_1^n \xi_i \\ \text{s.t.} & ||\mathbf{\Phi}(\mathbf{x}_i) - \mathbf{a}||^2 \leq R^2 + \xi_i \\ & \xi_i > 0, i = 1,\dots,n, \end{array} \tag{11.2}$$

where $\mathbf{a}$ and R are the center and the radius of the hypersphere, ξ_i are slack variables that allow the possibility of outliers in the training set [9], C controls the trade-off between the volume and the errors ξ_i, and $||.||$ is the Euclidean norm.

Defining the Lagrangian and applying Karush–Kuhn–Tucker optimality conditions (KKT) the solution is obtained solving the dual problem:

$$\begin{array}{ll} \max_\beta & \sum_i \beta_i \mathbf{\Phi}(\mathbf{x}_i)\mathbf{\Phi}(\mathbf{x}_i) - \sum_{i,j} \beta_i \beta_j \mathbf{\Phi}(\mathbf{x}_i), \Phi(\mathbf{x}_j) \\ \text{s.t.} & 0 < \beta_i < C, \qquad \sum_i \beta_i = 1, \qquad i = 1,\dots,n. \end{array} \tag{11.3}$$

This procedure permits us to handle three different set of points in the training data

- Bounded support vectors (BSV): They are characterized by values $\beta_i = C$. These points lie outside the boundaries of the hypersphere and are treated as an exception (novel).

- Support vectors (SV): $0 < \beta_i < C$. They lie on cluster boundaries.
- The set of all other points that lie inside the boundaries.

Therefore when $C \geq 1$ no BSVs exist. Following kernel methods, the representation Eq. 11.1 can be adopted and Eq. 11.3 can be rewritten as

$$\begin{array}{ll} \max_\beta & \sum_i \beta_i K(\mathbf{x}_i,\mathbf{x}_i) - \sum_{i,j} \beta_i\beta_j K(\mathbf{x}_i,\mathbf{x}_j) \\ \text{s.t.} & 0 < \beta_i < C, \qquad \sum_i \beta_i = 1, \qquad i = 1,\dots,n. \end{array} \tag{11.4}$$

We notice that an explicit calculation of the feature map $\mathbf{\Phi}$ is not required but only the value of the inner product between mapped patterns is used.

The distance $R^2(\mathbf{x}) = ||\mathbf{\Phi}(x) - \mathbf{a}||$ from the center of the hypersphere to the image of a point $\mathbf{x}$ can be expressed [7] as

$$R^2(x) = K(\mathbf{x},\mathbf{x}) - 2\sum_j \beta_j K(\mathbf{x_j},\mathbf{x}) + \sum_{i,j} \beta_i\beta_j K(\mathbf{x}_i,\mathbf{x}_j). \tag{11.5}$$

The radius R of the hypersphere is the distance between the hypersphere center and the support vectors. A test point $\mathbf{x}$ is novel when $R(\mathbf{x}) > R$. Cluster boundaries can be approximated using data point images that lie on the surface of the minimal hypersphere, formally the set of points $\{\mathbf{x}_i | R(\mathbf{x}_i) = R\}$.

11.2.1.2 Cluster Labeling Step

This approach does not permit the assignment of points to clusters, but one can apply the following geometrical estimation: given a pair of inputs belonging to different clusters, any path connecting them (for instance, a segment) must exit from the hypersphere in the feature space. Such a path contains a point $\mathbf{y}$ with $R(\mathbf{y}) > R$. Therefore, by considering the adjacency matrix A of components

$$[A]_{i,j} = \begin{cases} 1 & \text{if } R(\mathbf{x}_i - \lambda(\mathbf{x}_j - \mathbf{x}_i)) \leq R, \forall\lambda \in [0,1] \\ 0 & \text{otherwise,} \end{cases} \tag{11.6}$$

clusters are then obtained from the connected components of A. Unfortunately, for practical purposes, when checking the line segment it is generally necessary to sample a number of points which contribute to creating a runtime versus accuracy trade-off.

11.3 The Optimization Problem

The time complexity of the cluster labeling part is generally a critical issue. One has to check the connectivity for each pair of points based on the decision criteria obtained from the trained SVMs: given M sampled points, on the segment connecting all pairs of n inputs, the algorithm takes $O(n^2M)$. In order to reduce this computational time, here we propose to start the clustering investigation on the structure obtained from the approximation of the MGRN problem [11].

This section is intended to approximate the subset of genes that act as the "true" regulatory elements by deleting the regulations which seem to be "spurious." The structure resulting from the optimization of the objective criteria of MGRN (i.e., a graph with vertices representing genes and edges representing activation or inhibition activities) can in fact be used as an appropriate a priori knowledge (hence, avoiding redundant checks on unrelated connections), as naturally suggested by the following logical and biological considerations.

- The genes involved in a particular pathway of a cell process are of two types: activation/inhibition. These types are mutually exclusive with very few exceptions. It turns out that one of the requirements that we must expect from the deleting procedure, is to output an activation/inhibition label for the nodes of the reference graph. A direct consequence is that the labeling of nodes must be consistent with the edge labels: by deleting an edge, the procedure simply has to prevent an activating (inhibiting) node to be the source of an inhibiting (activating) edge.
- The deleting procedure has to output a graph that achieves the maximum number of nodes with both activating/inhibiting incoming edges. If a gene can be both activated and inhibited, then it can be controlled. What the procedure must search for is a network in which the number of controlled elements are maximized.

These considerations give rise to the following problem on graphs.

> MGRN:[2] *Given a directed graph with A/I labeled edges, assign each vertex either an A or I label in order to maximize the number of vertices with both A and I labeled input edges, after deleting all edges whose label differs from its parent vertex.*

The above description can be formulated as follows (we work mainly as reported in [11]): for each vertex v_j of the graph representing the instance of the MGRN problem, we associate a Boolean expression C_j. Let us denote with C_j^+ and C_j^- the set of subscripts of the vertices that are connected to v_j labeled, respectively, as activator and inhibitor. For every vertex v_j, let x_j represent a Boolean variable which is true if and only if the v_j vertex is labeled as activator. Hence, we have $C_j = \left(\bigvee_{i\in C_j^+} x_i\right) \wedge \left(\bigvee_{i\in C_j^-} \neg x_i\right)$. As can be verified, C_j is satisfied if and only if the vertex v_j is controlled. Let z_j represent a Boolean variable which is true if and only if the vertex v_j is controlled. Then the following linear programming can be given.

$$\begin{array}{lll} \max_{\mathbf{x},\mathbf{z}} & \sum_j z_j & \\ \text{s.t.} & z_j \leq \sum_{i\in C_j^+} x_i & \text{for all } j \\ & z_j \leq \sum_{i\in C_j^-} (1-x_i) & \text{for all } j \\ & x_i \in \{0,1\} & \text{for all } i \\ & z_j \in \{0,1\} & \text{for all } j. \end{array} \tag{11.7}$$

[2] The problem is NP-hard [11], but through its relaxed version a performance approximation ratio of 1/2 can be achieved (see, e.g., [11, 14]).

Formulation (11.7) is similar to that of the MAXSAT problem [15]: if we want to maximize the number of controlled vertices we have to maximize $\sum_j z_j$. Consider a vertex v_j and the correspondent Boolean variable z_j. Whenever, for instance, we have $z_j = 1$, then the first two constraints in Eq. 11.7 become $\sum_{i \in C_j^+} x_i \geq 1$ and $\sum_{i \in C_j^-} (1 - x_i) \geq 1$. In order for these constraints to be simultaneously satisfied, at least one of the Boolean variables x_i associated with the vertices in C_j^+ must be assigned to 1 and at least one of the Boolean variables associated to the vertices in C_j^- must be assigned to 0. That is, the v_j vertex results controlled.

11.4 Method

Our cluster investigation begins in agreement with [11] by considering the expression profile of a gene $g \in S$ in the training set S, as the set $P_g = \{\mathbf{p}_i : i = 1, \ldots, n\}$ of n activation or inhibition peaks. Here we consider $\mathbf{p}_i \in \mathrm{I\!R}^3$ with component values given by $[\mathbf{p_i}]_{\text{start}} = x_{\text{start}}, [\mathbf{p_i}]_{\max} = x_{\max}, [\mathbf{p_i}]_{\text{end}} = x_{\text{end}}$; that is, each component being given by the start, max, and final temporal value of the expression profile. Specifically, peaks are represented by extracting from each profile any data points with value greater than the average expression. Hence, all consecutive points lying between two points with less than average expression value belong to the same peak.

Intuitively, peak $\mathbf{p}_i \in P_g$ should be considered as a good candidate to activate $\mathbf{p}_j \in P_t$ if the "leading edge" of $\mathbf{p}_i$ appears "slightly before" the "leading edge" of $\mathbf{p}_j$. Formally, given two genes g and t, this is expressed by the activation grade $\mathcal{A}\colon P_g \times P_t \rightarrow \mathrm{I\!R}$:

$$\mathcal{A}(\mathbf{p}_i, \mathbf{p}_j) = e^{-\alpha_1 (D_1 + D_2)/2}, \tag{11.8}$$

where $D_1 = [\mathbf{p}_j]_{\text{start}} - [\mathbf{p}_i]_{\text{start}} + 1$, $D_2 = [\mathbf{p}_j]_{\max} - [\mathbf{p}_i]_{\max}$, and $\alpha_1 \in \mathrm{I\!R}$.

Similarly, $\mathbf{p}_i \in P_g$ should be considered good to inhibit $\mathbf{p}_j \in P_t$ if its leading edge is after the leading edge of $\mathbf{p}_j$ and close enough to its trailing edge of $\mathbf{p}_j$ or more formally, the inhibition grade of $\mathbf{p}_i$ on $\mathbf{p}_j$ is:

$$\mathcal{I}(\mathbf{p}_i, \mathbf{p}_j) = e^{-\alpha_2 (D_1 + D_2)/2}, \tag{11.9}$$

where $D_1 = [\mathbf{p}_j]_{\max} - [\mathbf{p}_i]_{\text{start}} + 1$, $D_2 = [\mathbf{p}_j]_{\text{end}} - [\mathbf{p}_i]_{\max}$, and $\alpha_2 \in \mathrm{I\!R}$.

In order to decide whether a gene acts as activator or inhibitor we evaluate the values of $f\colon S \times S \rightarrow \{0, 1\}$ defined, for each pair of genes g, t, as

$$f(g,t) = \mathcal{H}\left(\sum_{\mathbf{p}_i \in P_g, \mathbf{p}_j \in P_t} \mathcal{A}(\mathbf{p}_i, \mathbf{p}_j) - \sum_{\mathbf{p}_i \in P_g, \mathbf{p}_j \in P_t} \mathcal{I}(\mathbf{p}_i, \mathbf{p}_j) \right), \tag{11.10}$$

where $\mathcal{H}$ is the Heaviside function $\mathcal{H}(x) = 1$ if $x > 0$ and 0 otherwise. Hence, whenever the overall inhibition grade is greater than the overall activation grade, the interaction between the pair of involved genes is assumed to be an inhibition (respectively, an activation in the opposite case).

The values assumed by Eq. 11.10 constitute all putative relationships to prune with the optimization Eq. 11.3. The approximation we obtain gives a direct graph whose vertices correspond to specific genes, which act as the regulatory elements, and edges representing activation or inhibition dependencies (labeled with 1 or 0).

Our goal is now to design clusters where these dependencies are homogeneous in respect to some measure K.

Because we have represented peaks as points in the Euclidean space, an interaction could occur when in such a space points are not far away from each other. Such an interaction can naturally be expressed by $\mathbf{p}_j - \mathbf{p}_i$, whereas by considering $\mathbf{c}_{g,t} = 1/N \sum_{\mathbf{p}_i \in P_g} \sum_{\mathbf{p}_j \in P_t} \mathbf{p}_j - \mathbf{p}_i$ we can give an idea of the overall average among N interactions of peaks of two different genes. In order to measure the local similarity between two different dependencies we finally endow the set $C = \{\mathbf{c}_{g,t} : g,t \in S\}$ with the kernel:

$$K(\mathbf{c}_{g,t}, \mathbf{c}_{s,v}) = e^{-q\|\mathbf{c}_{g,t} - \mathbf{c}_{s,v}\|^2} \tag{11.11}$$

with $g,t,s,v \in S$. As reported in Sect. 11.2 the kernel Eq. 11.11 is used to compute the minimal enclosing hypersphere for all the dependencies we found in our experiments.

11.5 Numerical Results

The objective of this section is mainly to compare the results of our application to other clustering approaches; specifically, we conduct numerical evaluations when searching for clusters which are as dense and separated as possible. Because the standard algorithms used for comparison, K-means and hierarchical clustering (applied with Euclidean distances), do not explicitly take into account the outliers, we decided to avoid the case also in the SVC procedure ($C = 1$). This choice prevents the full exploitation of the support vector abilities of clustering in a noisy environment [9].

We argue that even with the optimization shortcut defined in Sect. 11.3 the computational cost could remain high for large samples. In fact, we trained the SVM with 471 dependencies coming from a uniformly distributed sample of 50 genes of the budding yeast *Saccaromyce Cerevisiae* data [16]. For such illustrative examples, we considered only activation dependencies Eq. 11.8 with $D_1, D_2 > 0$.

In agreement with [11], first we filtered out both series whose maximum level was below a detection threshold ($\leq$200), and those whose patterns were expressed but without a significant variation over time: for instance, maximum and average expression level satisfied $(\text{MAX} - \text{AVG})/\text{AVG} \leq 0.1$. We proceeded as in [9] to set the free parameter q of SVC: first starting with a low value, when a single cluster appears, then increasing q to observe the formation of an increasing number of clusters (the Gaussian kernel fits the data with increasing precision). We stopped this iteration when the numbers of SVs were excessive; that is, a large fraction of the data turns into SVs (in our case, 271).

To get an idea of the cluster modularity we used the silhouette index

$$s_j(i) = \frac{b_j(i) - d_j(i)}{\max\{d_j(i), b_j(i)\}}, \tag{11.12}$$

where $d_j(i)$ is the average distance between the ith sample and all the samples included in the cluster C_j and $b_j(i)$ is the minimum average distance between the ith sample and all the samples clustered in C_k with $k \neq j$. Expression Eq. 11.12 measures how close each point in one cluster is to points in the neighboring clusters. It ranges from $+1$, indicating points that are very distant from neighboring clusters, through 0, indicating points that are not distinctly in one cluster or another, to -1, indicating points that are probably assigned to the wrong cluster. Figures 11.1 to 11.3 report silhouette values for different parameters q. Most points in the first clusters of the SVC have a large silhouette value, greater than 0.8 or around 0.8, indicating that clusters are somewhat separated from neighboring clusters. However, K-means and hierarchical clustering contain more points with low silhouette values or with negative values, indicating those clusters are not well separated. All the experiments gave us a better performance for the application with the SVC approach. Figure 11.4 reports the global silhouette value

$$\mathrm{GS} = \frac{1}{c} \sum_{j=1}^{c} S_j, \tag{11.13}$$

where c is the number of clusters and $S_j = 1/m_j \sum_{i=1}^{m_j} s_j(i)$ which characterizes the heterogeneity and isolation properties of each cluster.

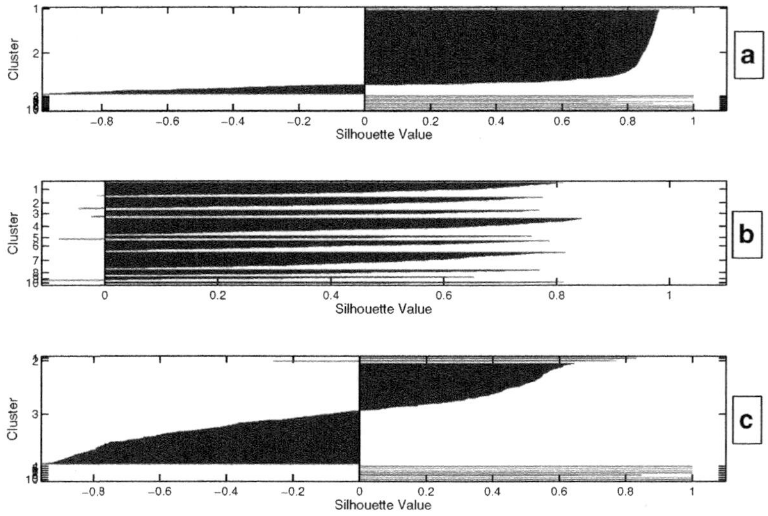

Fig. 11.1 Silhouette values for $q = 6, c = 10$: **a** SVC; **b** K-means; **c** hierarchical

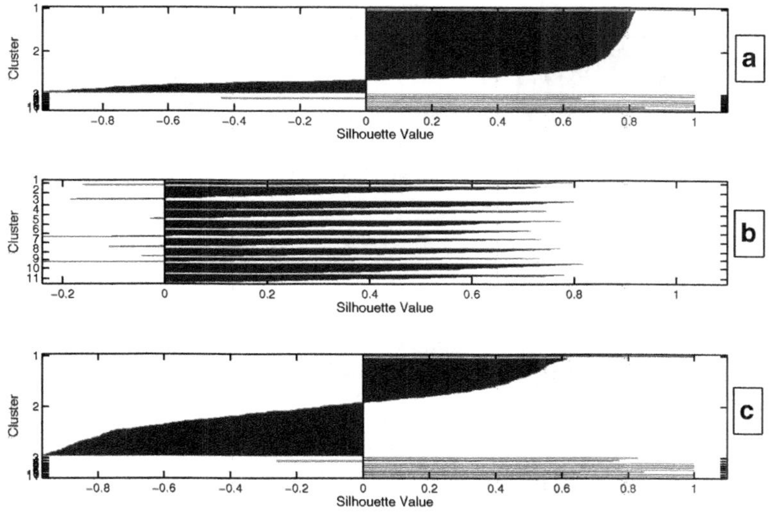

Fig. 11.2 Silhouette values for $q = 8, c = 11$: **a** SVC; **b** K-means; **c** hierarchical

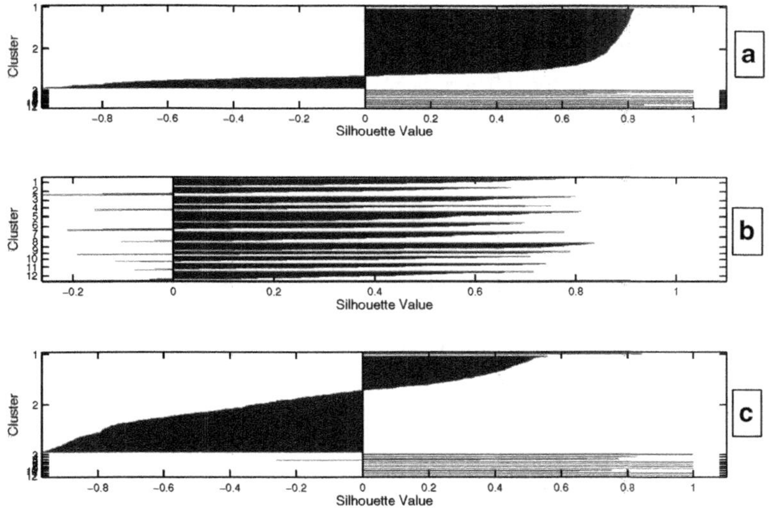

Fig. 11.3 Silhouette values for $q = 10, c = 12$: **a** SVC; **b** K-means; **c** hierarchical

11.6 Conclusions

In this contribution our intention was mainly to apply two efficient and theoretically well-founded machine learning ideas in order to cluster homogeneous dependencies in a complex system. First, we used SVC to group regulatory mechanisms such as a pair of gene-to-gene activities. Then, because this algorithm is computationally hard in its labeling part, we applied the MGRN problem to overcome useless checks when dealing with potential (spurious) dependencies. We dealt with data coming from microarray experiments (i.e., *Saccaromyce Cerevisiae*) and our preliminary

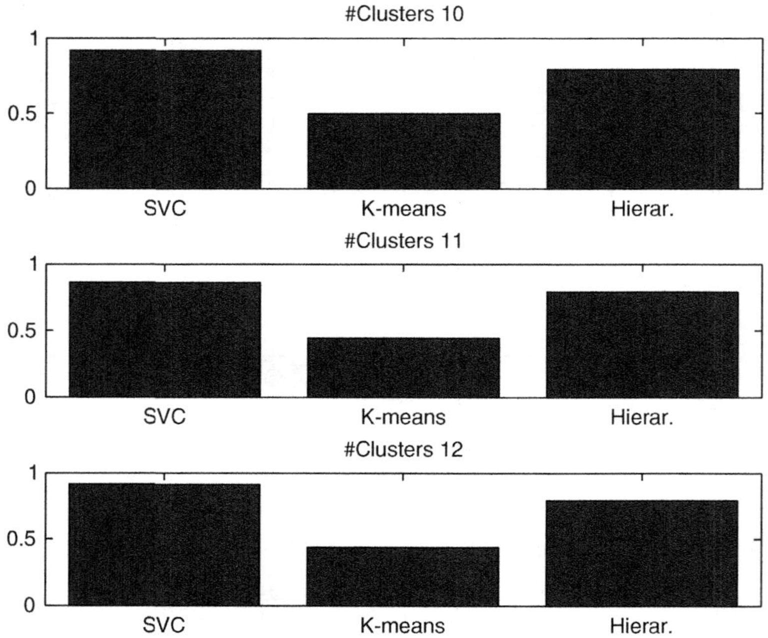

Fig. 11.4 Global silhouette for different number of clusters. #Cluster 10: SVC = 0.9195, K-means = 0.4971, Hierar = 0.7922; #Cluster 11: SVC = 0.8625, K-means = 0.4452, Hierar = 0.7922; #Cluster 12: SVC = 0.9145, K-means = 0.4391, Hierar = 0.7932

numerical results, evaluated on the basis of a quantitative index (silhouette value), encourage the use of this approach in respect to K-means and hierarchical clustering. From a biological point of view our application has been performed on the simplified assumption of a simple gene-to-gene interaction mechanism. This difficulty should be overcome in order to obtain better results, for example, by considering in our future work, to integrate both

- Information or heuristics in order to represent known a priori liability of transcription control by genes
- Information on transcription factors binding preferences to sequence motifs

Acknowledgements The research has been partially funded by Università di Milano Bicocca, FIAR 2006.

References

1. Eisen, M., Brown, P. (1999) Dna arrays for analysis of gene expression. *Methods in Enzymology* **303**: 179–205.
2. Bittner, M., Meltzer, P., Trent, J. (1999) Data analysis and integration: Of steps and arrows. *Nature Genetics* **22**: 213–215.

3. Chen, Y., Bittner, M.L., Rougherty, E.R. (1999) Issues associated with microarray data analysis and integration. *Nature Genetics* **22**: 213–215.
4. Heyer, L.J., Kruglyak, S., Yooseph, S. (1999) Exploring expression data: Identification and analysis of coexpressed genes. *Genome Research* **9**: 1106–1115.
5. Filkov, V., Skiena, S., Zhi, J. (2002) Analysis techniques for microarray time-series data. *Journal of Computational Biology* **9**: 317–330.
6. Shawe-Taylor, J., Cristianini, N. (2004) *Kernel Methods for Pattern Analysis*. Cambridge University Press, UK.
7. Schölkopf, B., Smola, A.J., Muller, K.R. (1999) *Advances in Kernel Method - Support Vector Learning*. Cambridge, MA: MIT Press.
8. Schölkopf, B., Tsuda, K., Vert, J.P. (2004) *Kernel Methods in Computational Biology*. Cambridge, MA: MIT Press.
9. Ben-Hur, A., Horn, D., Siegelmann, H.T., Vapnik, V. (2001) Support vector clustering. *Journal of Machine Learning Research* **2**: 125–137.
10. Gustafsson, M., Hörnquist, M., Lombardi, A. (2003) Large-scale reverse engineering by the lasso. *Proceedings of International Conference on Systems Biology*: 135–136.
11. Chen, T., Filkov, V., Skiena, S. (1999) Identifying gene regulatory networks from experimental data. *Proceedings of the 3rd Annual International Conference on Computational Molecular Biology*: 94–103.
12. Yang, J., Estivill-Castro, V., Chalup, S.K. (2002) Support vector custering trough proximity graph modelling. *Proceedings of 9th International Conference on Neural Information Processing* **2**: 898–903.
13. Courant, R., Hilbert, D. (1953) *Methods of Mathematical Physics*, vol. 1. New York: Interscience.
14. Pozzi, S., Della Vedova, G., Mauri, G. (2005) An explicit upper bound for the approximation ratio of the maximum gene regulatory network problem. *Proceedings on CMSB*, 3082: 1–8.
15. Cook, S. (1971) The complexity of theorem prouvem procedures. *Proceedings of the 3rd Symposium of the ACM on the Theory of Computing*: 151–158.
16. Cho, R., Campbell, M., Winzeler, E., Steinmetz, L., Conway, A., Wodicka, L., Wolfsberg, T., Gabrielian, A., Landsman, D., Lockhart, D., Davis, R. (1998) A genomic-wide transcriptional analysis of the mitotic cell cycle. *Molecular Cell* **2**: 65–73.

Chapter 12
A Comparative Study of Gender Assignment in a Standard Genetic Algorithm

K. Tahera, R. N. Ibrahim, and P. B. Lochert

12.1 Introduction

The genetic algorithm is a population-based heuristic search algorithm which has become a popular method for solving optimisation problems. The concept of the genetic algorithm was inspired by nature and was then successfully developed by John Holland in 1975. The basic concepts borrowed from nature are: randomness, fitness, inheritance, and creation of a new species. The genetic algorithm was developed based on the fact that successful matching of parents will tend to produce better offspring. This idea is supported by the building block theory [7]. The individuals in a population (i.e., the parents) are selected based on Darwin's law of natural selection and survival of the fittest. The genetic information of the parents is exchanged in the hope of producing improved offspring. Occasionally, a mutation operator randomly changes genes to produce new individuals. For a detailed review of the GA concept, see Haupt and Haupt [6].

Unlike natural processes, a standard genetic algorithm uses a population where each individual has the same gender (or has no gender) and any two individuals can be mated to cross over. Thus, this algorithm does not implement the concept of gender for reproduction. However, the nature is gender-specific and permits the reproduction of offspring only from opposite genders. To mimic nature more closely, only a few papers incorporate gender in a standard genetic algorithm. Allenson [1] incorporated gender where the gender of the offspring was decided according to the gender of the individual who was discarded from the population. Therefore, the number of males and females was kept constant throughout the algorithm. However, in nature, this constant phenomenon is rarely observed.

Lis and Eiben [9] developed a multisexual genetic algorithm for multiobjective optimisation. However, gender was used from a different perspective. Instead of having the traditional two categories of gender (male and female), they considered gender as an integer number which was equal to the number of optimisation criteria. Each individual was evaluated according to the optimisation criteria related

Oscar Castillo et al. (eds.), *Trends in Intelligent Systems and Computer Engineering.*

to its gender. During the recombination process, one individual from each gender was used to reproduce offspring. Drezner and Drezner [5] included randomness during the determination of the gender of the offspring. Thus, the number of males and females was not constant. However, the random determination of gender might lead to a population with a single gender, and thus regeneration might not occur.

To avoid the possibility of no regeneration, we introduce two approaches to assign gender to individuals. The first approach is "adaptive gender assignment" in which the gender of a new individual is decided based on gender density of the population. If the number of a particular gender is reduced below a threshold limit, the new individual's gender is assigned to that particular gender otherwise the gender is randomly determined. The other approach is "fitness gender assignment" in which the gender of an individual is assigned based on its fitness. The proposed algorithms were tested on a mechanical design optimisation problem.

12.2 The Gender Approach in a GA

Convergence of a genetic algorithm typically occurs when the individuals in a population are similar to one another and no better offspring are generated by selecting parents from the existing population members. As a consequence, the algorithm tends to perform better if the population is kept as diverse as possible for many generations. Therefore, procedures that delay the formation of a homogeneous population are more likely to result in better solutions. Mutation is one example of a diversity promoting process. The other example is to introduce the gender concept.

The pseudo-code of a gendered genetic algorithm with adaptive gender assignment and fitness gender assignment is given below:

STEP 1

An initial population of individuals (i.e., solutions) is randomly generated. A gender is assigned to each individual based on gender probability factor (γ). This factor ranges from $0 \leq \gamma \leq 1$ and it controls the number of males and females in the population. For example, if $\gamma = 0.5$ then half of the population is assigned as males and the other half as females. The gender probability factor also controls whether the genetic algorithm is gender-specific or gender-neutral. For $\gamma = 0$, all the population members are males. On the other hand, for $\gamma = 1$, all the population members are females. Thus in both cases, the algorithm converts to a gender-neutral standard genetic algorithm.

STEP 2

Pairs of population members are selected for mating to reproduce offspring. In the pair selection process it is ensured that the mating is between a male and a female. In a gendered genetic algorithm, male and female members are grouped in male and female pools, respectively. Then the first candidate is selected from the males' pool and the second candidate is selected from the females' pool. Thus, the mating of opposite genders is ensured. The selection from each of the gender's pools is done based on a tournament selection approach. In this approach, a small tournament size is considered (i.e., $k = 3$). A number of individuals, equal to tournament size, are chosen from each pool and they compete based on their fitness values. The fittest individual gets the chance to be selected for reproduction. The tournament size is important to control selection pressure and to keep diversity in the population. If the tournament size is larger, weak individuals have a smaller chance of being selected. On other hand, if the tournament size is smaller, weak individuals have a greater chance of being selected. Thus, the population becomes more diverse.

STEP 3

The crossover operation produces two children. The gender of the children, generated in a crossover operation, is determined based on either adaptive gender assignment or fitness gender assignment. In adaptive gender assignment, the gender of the child depends on constantly monitoring the number of males and females in the population of each generation. If any gender falls below a threshold limit (l), the gender of the child is assigned to that particular gender, otherwise the gender is randomly assigned as male or female. The threshold limit is determined by the following equation.

$$l = \frac{PopulationSize}{4}$$

In fitness gender assignment, genders of all individuals are reassigned after genetic operations. Thus the gender of the child is unassigned at this stage.

STEP 4

Occasionally with a small probability p_m, we alter the population of the children (i.e., the newly created points). We do not perform mutation in the final generation as we do not want to lose the best solution. We generate a random number r. If $r \leq p_m$ then we do the mutation. An individual is randomly selected and a random gene of the individual is changed with a newly created gene value. The mutation operator introduces the exploratory feature in the genetic process and leads the algorithm to search in the new region.

STEP 5

A new population is created for the next generation by replacing weaker individuals. The strategy adopted here is to store both parents and children in a temporary pool which is twice the population size. The best individuals equal to the population size are selected for the next generation. The gender of the child in a fitness gender assignment is applied at this stage. The gender of an individual changes in each generation based on its fitness value. The population is sorted from best to worst; the best half is assigned as males and the other half is assigned as females.

STEP 6

The process from Steps 2–6 continues until a termination criterion is met which is to run the algorithm for fixed number of generations.

STEP 7

The best solution found throughout the process is the result of the algorithm.

12.3 Application to Mechanical Design Problem

In mechanical design, sizing a mechanical system implies solving a problem of optimisation; this is called the optimal design problem. Optimal design problems generally involve interdependent discrete parameters whose values are taken from standardized tables, that is, lists of commercially available prefabricated sizes. The novel genetic algorithm is applied to the mechanical design optimisation problem of a pressure vessel as stated below.

12.3.1 Design of a Pressure Vessel

The following problem is taken from Kannan and Kramer [8]. A cylindrical vessel is capped at both ends by hemispherical heads, as shown in Fig. 12.1.

The objective is to minimise the total cost, including the cost of the material, forming and welding. There are four design variables: $T_S(x_1)$ (thickness of the shell), $T_h(x_2)$ (thickness of the head), $R(x_3)$ (inner radius), and $L(x_4)$ (length of the cylindrical section of the vessel, not including the head). T_S and T_h are integer multipliers of 0.0625 in which are the available thicknesses of rolled steel plates and R and L are continuous. The problem can be stated as follows.

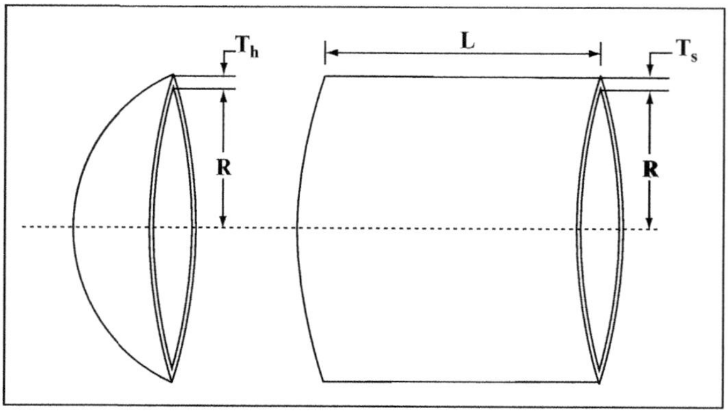

Fig. 12.1 Pressure vessel

Minimise:

$$f(x) = 0.6334x_1x_3x_4 + 1.7781x_2x_3^2 + 3.1661x_1^2x_4 + 19.84x_1^2x_3$$

Subject to

$$g_1(x) = -x_1 + 0.0193x_3 \leq 0$$
$$g_2(x) = -x_2 + 0.00954x_3 \leq 0$$
$$g_3(x) = -\pi x_3^2 x_4 - \frac{4}{3}\pi x_3^3 + 1296000 \leq 0$$
$$g_4(x) = x_4 - 240 \leq 0$$

This problem has been solved by Sandgren [10] using a branch-and-bound approach, by Kannan and Kramer [8] using an augmented Lagrangian multiplier approach, by Deb [4] using GeneAS (genetic adaptive search), and by Coello and Mpntes [3] using a genetic algorithm with a dominance-based tournament selection approach. Their results are shown in Table 12.1. The table shows that the design variables found by Kannan slightly violate a constraint.

The proposed genetic algorithm is applied to this problem. For comparison purpose, the results of five variants of genetic algorithms are presented. These are:

1. Gender neutral or standard genetic algorithm, GA-I.
2. Gendered genetic algorithm with constant gender assignment, GA-II. In this version, one of the children is assigned as male and the other child is assigned as female.
3. Gendered genetic algorithm with random gender assignment, GA-III. The gender of the children is assigned randomly.
4. Gendered genetic algorithm with adaptive gender assignment, GA-IV. If the number of a particular gender group falls below a threshold limit, then the gender of the child is assigned to that particular gender; otherwise it is randomly determined.

Table 12.1 Comparison of the results for optimisation of a pressure vessel

Design variables	Coello	Deb	Kannan	Sandgren
x1	0.8125000	0.9375000	1.1250000	1.1250000
x2	0.4375000	0.5000000	0.6250000	0.6250000
x3	42.0973980	48.3290000	58.2910000	47.7000000
x4	176.6540470	112.6790000	43.6900000	117.7010000
Constraints				
g1	−0.0000202	−0.0047503	0.0000163	−0.2043900
g2	−0.0358908	−0.0389413	−0.0689039	−0.1699420
g3	−546.5323390	−4175.9878717	−542.8693940	−467.3929114
g4	−63.3459530	−127.3210000	−196.3100000	−122.2990000
Objective Function				
f(x)	6059.9464093	6410.3811385	7198.0428258	8129.1036010

5. Gendered genetic algorithm with fitness gender assignment, GA-V. The population is sorted from best to worst and genders of all the individuals are reassigned according to fitness. The best half is assigned as males and the other half is assigned as females.

In all cases, the same initial population and parameter set (Table 12.2) is used. The gender of each individual is randomly assigned in the initial population ($\gamma = 0.5$).

Table 12.3 shows the results of the genetic algorithms. The comparison is done based on the solution quality. Due to the stochastic nature of genetic algorithms, hundreds of trials are conducted and the best of all these trials is considered as the optimum result.

It can be seen from Table 12.3 that the performances of genetic algorithms proposed in this chapter are better than the other variants of genetic algorithms. It could be argued that the performance of the genetic algorithm by Coello and Mpntes [3], as can be seen from Table 12.1, is even better than the proposed genetic algorithms. However, the proposed genetic algorithms attempt to implement a different version of GA using a gender-based approach. Coello and Mpntes [3] used dominance-based tournament selection in a standard genetic algorithm whereas the proposed algorithm uses normal tournament selection in a gender-based genetic algorithm.

Table 12.2 Genetic algorithm parameters

Population Size	Total Generation	Total Trials	Selection Type	Crossover Type	Crossover Rate	Mutation Rate
64	100	100	Tournament Selection	Single Point	0.8	0.08

Table 12.3 Comparison of the results for optimisation of a pressure vessel by using genetic algorithms

	Standard GA		Gendered GA		
Child's Gender Assignment	None	Constant	Random	Adaptive	Fitness
Design variables					
x1	0.8125	0.9375	0.875	0.875	0.8125
x2	0.5000	0.4375	0.4375	0.4375	0.4375
x3	41.9462	45.4622	44.5066	44.7145	41.9017
x4	179.6848	142.9886	150.7743	147.0379	179.1737
Constraints					
g1	−0.003	−0.060	−0.016	−0.012	−0.004
g2	−0.100	−0.004	−0.013	−0.011	−0.038
g3	−6896.281	−26556.724	−12080.005	−2593.073	−984.575
g4	−60.315	−97.011	−89.226	−92.962	−60.826
Objective Function					
f(x)	6300.735	6591.545	6236.988	6171.604	6085.772

In either case, the results in Tables 12.1 and 12.3 summarise that the genetic algorithm is a better optimisation algorithm than the other conventional algorithms in the mechanical design of a pressure vessel.

12.4 Conclusions

The motivation of introducing gender in a standard genetic algorithm is to increase diversity in the selection process. Instead of selecting individuals from a single group for mating as is seen in a standard genetic algorithm, a gendered genetic algorithm creates two groups and permits mating only between opposite groups. The selection from two groups helps to increase diversity in the selection process.

The gender assignment of offspring is an issue in a gendered genetic algorithm. Earlier research considered the gender assignment to be either constant or random. The constant phenomenon avoids the possibility of no regeneration. This chapter presents two approaches to gender assignment of offspring: adaptive gender assignment and fitness gender assignment. In adaptive gender assignment, the gender is assigned based on the gender density of the population. If a gender falls below a threshold limit, the offspring's gender is assigned to that particular gender. In fitness gender assignment, the population is sorted according to fitness and the best half is assigned as males and the other half is assigned as females. This strategy allows mating between good and poor individuals and thereby it maintains population diversity. It is noteworthy that in this approach the gender of an individual changes in each generation. Both of these approaches prevent the population from being a single gender.

The proposed algorithm is applied to a mechanical design problem. Different strategies for gender assignment are studied and it is seen that the adaptive gender assignment and fitness gender assignment provides better results.

Acknowledgements The authors will like to acknowledge the support received from CRCIEAM and the postgraduate scholarship to carry out the research work.

References

1. Allenson, R. (1992), Genetic algorithms with gender for multi-function optimisation, Technical Report EPCC-SS92-01, Edinburgh Parallel Computing Centre, Edinburgh, Scotland.
2. Belegundu, A.D. (1982), A study of mathematical programming methods for structural optimisation, PhD Thesis, Department of Civil and Environmental Engineering, University of Iowa, Iowa.
3. Coello, C.A.C. and Mpntes, E.M. (2002), Constraint handling in genetic algorithms through the use of dominance-based tournament selection, *Journal of Advanced Engineering Informatics*, 16: 193–203.

4. Deb, K. (1997), GeneAS: A robust optimal design technique for mechanical component design, In: Dasgupta, D., Michalewicz, Z., Editors, *Evolutionary Algorithms in Engineering Applications*, Berlin: Springer, pp. 497–514.
5. Drezner, T. and Drezner, Z. (2006), Gender specific genetic algorithms, *INFOR*, 44(2).
6. Haupt, R.L. and Haupt, E. (2004), *Practical Genetic Algorithms*, 2nd edition, New York: Wiley Interscience.
7. Holland, J. (1975), *Adaptation in Natural and Artificial Systems*, Ann Arbor: University of Michigan Press.
8. Kannan, B.K. and Kramer, S.N. (1994), An augmented Lagrange multiplier based method for mixed integer discrete continuous optimisation and its applications to mechanical design, *Journal of Mechanical Design Transactions ASME*, 116: 318–320.
9. Lis, J. and Eiben, A.E. (1997), A multi-sexual genetic algorithm for multi objective optimisation, in *IEEE International Conference on Evolutionary Computation*, pp. 59–64.
10. Sandgren, E. (1988), Nonlinear integer and discrete programming in mechanical design, in *Proceedings of the ASME Design Technology Conference*, Kissimmee, FL, pp. 95–105.

Chapter 13
PSO Algorithm for Primer Design

Ming-Hsien Lin, Yu-Huei Cheng, Cheng-San Yang, Hsueh-Wei Chang, Li-Yeh Chuang, and Cheng-Hong Yang

13.1 Introduction

In recent years, polymerase chain reactions (PCR) have been widely applied in medical science. The PCR technique allows a small amount of DNA to be amplified exponentially, thus ensuring that the amount of DNA is sufficient for DNA sequence analysis or gene therapy. It is important to choose a feasible primer pair to work quickly and efficiently.

Before conducting a PCR experiment, common primer design constraints have to be set in order to identify optimal primer pairs, which can selectively clip the desired DNA fragment. These constraints influence the success and efficiency of the PCR experiment. Commonly considered primer design constraints are the lengths of the primers, GC content, melting temperature, dimer, self-dimer, and hairpin. The melting temperature of primers should be within the range of 50–62°C, and the difference of the melting temperature of a primer pair should not exceed more than 5°C. The length of primers should be within 18–26 bps, and the difference of the primer pair length should not exceed 3 bps. The GC content of primers should be within 40–60%. Finally, the 3′ end of primers should be G or C whenever possible. In the next section, we show how these constraints are employed to objectively evaluate the fitness of each primer pair.

Recently, many kinds of primer design software were developed, but most of these do not allow the use of sequence accession numbers for primer design. Examples are Primer Design Assistant (PDA) [4] and GeneFisher [8]. The system we introduce in this chapter is based on a particle swarm optimization (PSO) algorithm. It incorporates the RefSeq database, which enables users to enter sequence accession numbers directly, or to copy/paste entire sequences in order to design primer sets. The software interface allows users to easily design fitting primer sets according to their needs.

The user-friendly interface allows (1) accession number input, (2) sequence input, and (3) input of primer constraints. The proposed PSO algorithm helps

Oscar Castillo et al. (eds.), *Trends in Intelligent Systems and Computer Engineering.*

in correctly and quickly identifying an optimal primer pair required for a specific PCR experiment. Finally, information about a feasible primer pair is graphically depicted on the user interface.

13.2 System and Method

13.2.1 System Design

Four modules were developed for the system; they are shown in Fig. 13.1. They are (1) sequence input module, (2) primer design constraints module, (3) PSO primer design module, and (4) output module. Through the input module users can input the RNA accession number, contig accession number, or a sequence as a PCR template sequence by the RNA accession number or contig accession number input will query the RefSeq database to obtain a mapping sequence. Through the primer design constraints module a user sets the desired primer. Through the PSO primer design module, a feasible primer can be designed. Finally, using the output module feasible primer set information is shown. The four modules are described below.

13.2.1.1 Sequence Input Module

This module allows the user three input ways. One is "RNA Accession Number" input whereby users can input the RNA accession number, such as NM_002372 (organism: human), NM_011065 (organism: mouse), or NM_031137 (organism: rat),

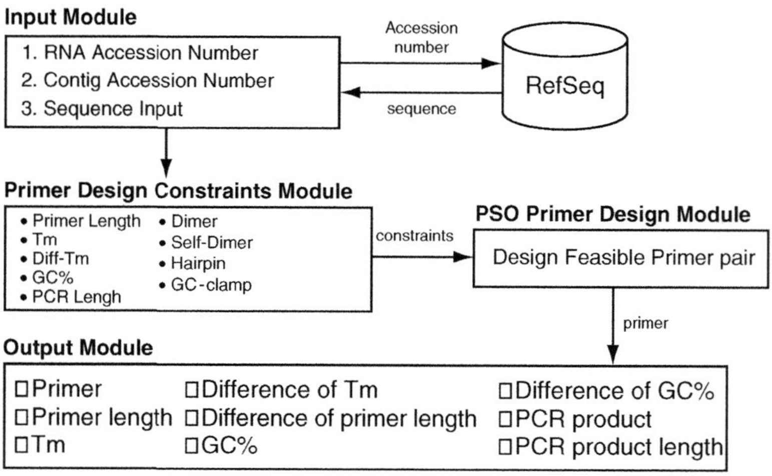

Fig. 13.1 System design modules

and so on, to do primer design and get a feasible primer set. Another is "Contig Accession Number" input whereby users can input the contig accession number, such as NT_079572 (organism: human), NT_060478 (organism: mouse), or NW_047416 (organism: rat), and so on, to do primer design and get a feasible primer set. And finally is "Sequence input" whereby users can paste a sequence directly to do primer design and get a feasible primer set. It has a simple and convenient input for users to design a primer.

13.2.1.2 Primer Design Constraints Module

This module provides basic primer design constraints, which include primer length, Tm, Diff-Tm, GC%, and PCR product. And it has four specific constraints: dimer check, self-dimer check, hairpin check, and GC-clamp check. All these constraints can be adjusted by users. Users can obtain the desired primer set by setting parameters of these constraints. If a user does not manually change the constraint parameters, the system will proceed by default.

13.2.1.3 PSO Primer Design Module

The PSO Primer Design Module is the core of this system. It mainly utilizes the PSO algorithm to implement. It employs the Sequence Input Module and Primer Design Constraints Module as input to design a feasible primer set. The PSO is described as follows.

Particle swarm optimization (PSO) is a population-based stochastic optimization technique, which was developed by Kennedy and Eberhart in 1995 [10]. PSO simulates the social behavior of organisms, such as birds in a flock or fish in a school, and describes an automatically evolving system. In PSO, each single candidate solution can be considered "an individual bird of the flock," that is, a particle in the search space. Each particle makes use of its own memory, as well as knowledge gained by the swarm as a whole to find the best (optimal) solution.

All of the particles have fitness values, which are evaluated by an optimized fitness function. They also have velocities which direct the movement of the particles. During movement, each particle adjusts its position according to its own experience, and according to the experience of a neighboring particle, thus making use of the best position encountered by itself and its neighbor. The particles move through the problem space by following a current of optimum particles. The process is then reiterated a predefined number of times or until a minimum error is achieved [11].

PSO was originally introduced as an optimization technique for real-number spaces, and has been successfully applied in many areas including function optimization, artificial neural network training, fuzzy system control, and other application problems. A comprehensive survey of the PSO algorithms and their applications can be found in Kennedy et al. [12].

13.2.1.4 Output Module

Finally, the output module shows the results of the primer design, and includes details about the primer length, primer Tm, difference of primer Tm, GC%, difference of GC%, as well as details about the PCR product, such as its length, and so on. The results are graphically displayed.

13.2.2 Database

The RefSeq database is employed. The system mainly builds human, mouse, and rat DNA data. Current database versions of the human and mouse genome are 36.1; the rat genome version is 3.1. In particular, mRNA and genomic DNA accession numbers are used to obtain available sequences for optimal primer design. The integrated database allows efficient sequence input by a user.

13.2.3 Particle Swarm Optimization for Primer Design

First of all, we define a vector to present the primer set, shown as follows.

$$P = (F_s,\ F_l,\ P_l,\ R_l) \tag{13.1}$$

F_s: the start index of the forward primer

F_l: the length of the forward primer

P_l: the PCR product length

R_l: the length of the reverse primer

We can calculate the reverse primer start index from P; that is,

$$R_s = F_s + P_l - R_l \tag{13.2}$$

R_s: the start index of the forward primer

Figure 13.2 shows the PSO primer design flowchart. The PSO primer design procedure is described below.

13.2.3.1 Initial Particle Swarm

To initialize a population, 100 particles $P = (F_s,\ F_l,\ P_l,\ R_l)$ are randomly generated and each particle is given a velocity (v). Initially, F_s is randomly generated in S, F_l and R_l are generated within the primer lengths set by the user, and P_l is randomly generated within the constraints of the PCR length set by the user. The velocity of

Fig. 13.2 PSO primer design flowchart

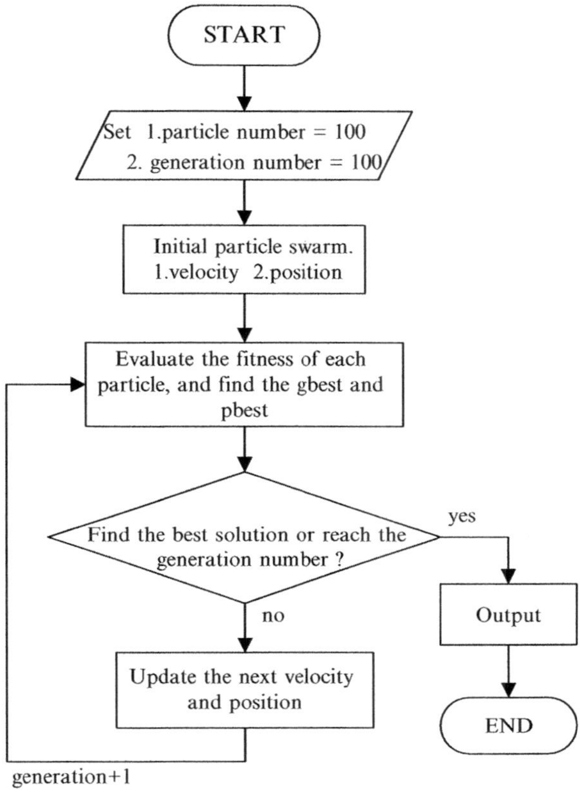

each particle is randomly generated within 0–1. Then the constriction factors (c1, c2) are set to 2 and the inertia weight (w) is set to 0.8. These values have been proven to yield good results in lower dimensions [8].

13.2.3.2 Fitness Evaluation

PSO requires a fitness function to evaluate the fitness of each particle in order to check whether the primers satisfy the design constraints. We use the primer design constraints as values for the fitness function.

Let P' indicate the primer, F_P indicate the forward primer, and R_P indicate the reverse primer. In a PCR experiment the feasible primer length is considered to be in the range of 18–26 bps; if a primer is longer, its specificity is higher; in this case a relatively high Tm is also required. A relatively short length will decrease the specificity. Hence, a primer that is neither too long or too short is suitable.

A length difference of 3 bps for the forward and reverse primer is considered optimal. $|F_P|$ and $|R_P|$ represent the number of nucleotides of the forward primer and the reverse primer, respectively.

The length (P) is used to check whether the length of a primer pair is within 18–26 bps; Δlength (P) is used to check whether the length difference of a primer pair exceeds 3 bps.

$$\text{Length}(P) = \begin{cases} 0, & \text{if } 18 \leq |F_P|, |R_P| \leq 26 \\ 1, & \text{other condition} \end{cases} \tag{13.3}$$

$$\Delta\text{Length}(P) = \begin{cases} 0, & \text{if the modulus of } (|F_P| - |R_P|) \leq 3 \\ 1, & \text{other condition} \end{cases} \tag{13.4}$$

$\text{P}_{\text{total}}(\text{G})$, $\text{P}_{\text{total}}(\text{C})$, $\text{P}_{\text{total}}(\text{A})$, and $\text{P}_{\text{total}}(\text{T})$ are denoted as the number of nucleotides G, A, C, and T of the primer, respectively. In this chapter, the value of the melting temperature of the primer is denoted Tm (P'), which uses the Wallace formula; it can be written as $\text{Tm}(P') = 4^*(\text{P}_{\text{total}}(\text{G}) + \text{P}_{\text{total}}(\text{C})) + 2^*(\text{P}_{\text{total}}(\text{A}) + \text{P}_{\text{total}}(\text{T}))$. Function $\text{Melt\_tm}(P')$ is used to check whether the melting temperature of a primer pair is between 52°C and 62°C, and $\Delta\text{Melt\_tm}(P)$ is used to check whether the difference of the melting temperature exceeds 5°C.

$$\text{Melt\_tm}(P') = \begin{cases} 0, & \text{if } 52 \leq \text{Tm}(F_P), \text{Tm}(R_P) \leq 62 \\ 1, & \text{other condition} \end{cases} \tag{13.5}$$

$$\Delta\text{Melt\_tm}(P) = \begin{cases} 0, & \text{if the absolute of } (\text{Tm}(F_P) - \text{Tm}(R_P)) \\ & \leq 5 \\ 1, & \text{if the absolute of } (\text{Tm}(F_p) - \text{Tm}(R_p)) \\ & > 5 \end{cases} \tag{13.6}$$

The GC ratio in the primer is denoted as $\text{GCratio}(P')$. The appropriate GC ratio in a primer should be in the range of 40–60%. The $\text{GCratio}(P')$ and $\text{GC\%}(P)$ are defined as follows.

$$\text{GC}_{\text{ratio}}(P') = \frac{\text{P}_{\text{total}}(\text{G}) + \text{P}_{\text{total}}(\text{C})}{|P'|} \tag{13.7}$$

$$\text{GC\%}(P) = \begin{cases} 0, & \text{if } 40 \leq (\text{GC}_{\text{ratio}}(F_P), \text{GC}_{\text{ratio}}(F_P)) \leq 60\% \\ 1, & \text{other condition} \end{cases} \tag{13.8}$$

In primer design primers that bind to any site on a sequence indiscriminately have to be avoided. Furthermore it should also be avoided that a forward primer complements the reverse primer or that one primer is a complement of itself. Dimer (P) is used to check whether the forward and reverse primer complement each other. Self-dimer (P) checks whether a primer pair is a complement of itself.

Dimer (P) and Self-dimer (P) are defined as follows.

$$\mathrm{Dimer}(P) = \begin{cases} 0, & \text{if it isn't a complement of } P \\ 1, & \text{if it is a complement of } P \end{cases} \tag{13.9}$$

$$\mathrm{Self\_Dimer}(P) = \begin{cases} 0, & \text{if it doesn't complement itself of } F_P \text{ and } R_P \\ 1, & \text{if it does complement itself of } F_P \text{ and } R_P \end{cases} \tag{13.10}$$

A primer should avoid complementing itself at the $3'$ end in the U form and the Hairpin (P) function is used as a check. The Hairpin (P) function can be written as follows.

$$\mathrm{Hairpin}(P) = \begin{cases} 0, & \text{if primer doesn't complement itself in the U form of } F_p \text{ and } R_p \\ 1, & \text{if primer complements itself in the U form of } F_p\text{and } R_p \end{cases} \tag{13.11}$$

GC_clamp(P) is used to check whether the terminal end of a primer is G or C; it is defined as follows.

$$\mathrm{GC\_clamp}(P) = \begin{cases} 0, & \text{if } 3' \text{ end of } F_P \text{ and } R_P \text{ is G or C} \\ 1, & \text{if } 3' \text{ end of } F_P \text{ and } R_P \text{ is A or T} \end{cases} \tag{13.12}$$

This constraint is used to judge whether the primer repeats in sequence to ensure the specificity of the primer. The PCR experiment might fail if the primer is not site-specific, and appears more than once in the sequence.

$$\mathrm{Unipair}(P) = \begin{cases} 0, & \text{if } F_P \text{ and } R_P \text{ appear in } S \text{ once} \\ 1, & \text{if } F_P \text{ and } R_P \text{ appear in } S \text{ more more than once} \end{cases} \tag{13.13}$$

The fitness of each particle is evaluated by the fitness function, which is constructed using the primer design constraints. A low fitness value means that the particle fits more constraints. The default fitness function is written as

$$\begin{aligned} \mathbf{Fitness}(\boldsymbol{P}) = {} & \mathbf{10}^{*}\,\mathbf{Tm}(\boldsymbol{P}) + \mathbf{5}^{*}\,\mathbf{GC\%}(\boldsymbol{P}) \\ & + 3^{*}\,(\mathrm{Length}(P) + \Delta\mathrm{Length}(Ppair) + \Delta\mathrm{Melt\_tm}(P) \\ & + \mathrm{GCclamp}(P) + \mathrm{Dimer}(P) + \mathrm{Self\text{-}dimer}(P) \\ & + \mathrm{Hairpin}(P) + \mathrm{Unipair}(P)) \end{aligned} \tag{13.14}$$

13.2.3.3 Updating of the Velocity and Position of the Next Generation of Each Particle

One of the characteristics of PSO is that each particle has a memory of its own best experience. Each particle can find its individual personal best position and velocity (pbest) and the global best position and velocity (gbest) by evaluation. With these reference values, each particle adjusts its direction in the next generation. If the particle fitness is better than in the previous generation pbest will be updated in the current generation. The memory-sharing property of PSO always allows an optimal fitness value to be found in the search space.

Equations 13.15 and 13.16 are the updating formulas for each particle.

$$v_i^{next} = w \times v_i^{now} + c_1 \times rand() \times (s_i^p - s_i^{now}) \\ + c_2 \times rand() \times (s_i^g - s_i^{now}) \tag{13.15}$$

$$s_i^{next} = s_i^{now} \times v_i^{next} \tag{13.16}$$

In (13.15) and (13.16) v_i^{next} is the updated velocity of a particle; v_i^{now} is the current velocity of a particle; c_1 and c_2 are constriction factors set at 2; the inertial weight w is set to 0.8; *rand()* is a number which is randomly generated within 0–1; s_i^p is the individual best position of a particle; s_i^g is the global best position of the particles; s_i^{now} is the current position of a particle; s_i^{next} is the updated position of a particle. A maximum and minimum $v_{\max}$ and $s_{\max}$ are set to a certain range; if the updated values of $v_{\max}$ and $s_{\max}$ are outside the range limits they will be set to the maximum or minimum value [5]. In this study, the $s_{\max}$ of P of F_{len}, R_{len} are limited within the range of the primer pair length set by the user in order to control the length of the primer.

13.2.3.4 Termination Condition

The algorithm is terminated when particles have achieved the best position: that is, their fitness value is 0, or the number of generations has reached 100.

13.3 Results and Discussion

Primer design has become an important issue over the last decade. The quality of primers always influences whether a PCR experiment is successful. Many primer design tools have been developed, but most of them are inefficient or have a complex interface, and will not result in optimal primers for use in a PCR experiments. Table 13.1 shows a comparison of the primer design tools.

PSO is based on the idea of collaborative behavior and swarming in biological populations. PSO shares many similarities with evolutionary computation techniques such as GAs. The performance of GAs has been shown to outperform SFS (sequential forward search), PTA (plus and take away), and SFFS (sequential forward floating search) [13]. Both PSO and GAs are population-based search approaches that depend on information-sharing among their population members to enhance the search processes by using a combination of deterministic and probabilistic rules.

Table 13.1 Comparison of primer design tools

Function	Proposed	Primer3	Genefisher [8]	PDA [4]
Sequence past panel	✓	✓	✓	✓
Accession number input	✓	×	×	×
The weight degree selected of each constraint	✓	×	×	×
Primer length	✓	✓	✓	✓
Tm	✓	✓	✓	×
Maximum of differential Tm	✓	✓	×	×
GC%	✓	✓	✓	×
Product size	✓	✓	✓	✓
Primer dimer check	✓	×	×	✓
Primer self-dimer check	✓	✓	×	×
Primer hairpin check	✓	✓	×	✓
GC-clamp check	✓	✓	✓	×
Visualized output	✓	✓	×	×

However, PSO does not include genetic operators such as crossover and mutation. The recognition and social model of interaction between particles is similar to crossover however, such as in Eq. 13.1 where the random parameters rand1 and rand2 will affect the speed of a particle, similarly to mutation in a GA. In fact, the only difference between them is that crossover and mutation in a GA is probabilistic (crossover rate and mutation rate), but the renewed particle in PSO should be processed at each iteration without any probability.

Compared with GAs, the information-sharing mechanism in PSO is considerably different. In GAs, the evolution is generated by using crossover and mutation in the same population. Chromosomes share information with each other, so the whole population moves as one group towards an optimal area. In the problem space, this model is similar to a search of only one area. Therefore, the drawback of this model is that it can easily become trapped in a local optimum. Although mutation is used, the probability usually is lower, limiting performance. In PSO, particles are uniformly distributed in the problem space, and only gbest gives out information to other particles. It is a one-way information-sharing mechanism. Evolution only looks for the best solution. In most cases all the particles tend to converge to the best solution quickly, even in the local version.

Compared to GAs, PSO has a more profound intelligent background and can be performed more easily [14]. Computation time used in PSO is shorter than in GAs [15]. The performance of PSO is affected by the parameter settings, inertia weight w, and the acceleration factors c_1 and c_2. However, if the proper parameter values are set, the results can easily be optimized. Proper adjustment of the inertia weight w and the acceleration factors c_1 and c_2 is very important. If the parameter adjustment is too small, the particle movement is too small. This scenario will also result in useful data, but is a lot more time-consuming. If the adjustment is excessive, particle movement will also be excessive, causing the algorithm to weaken early,

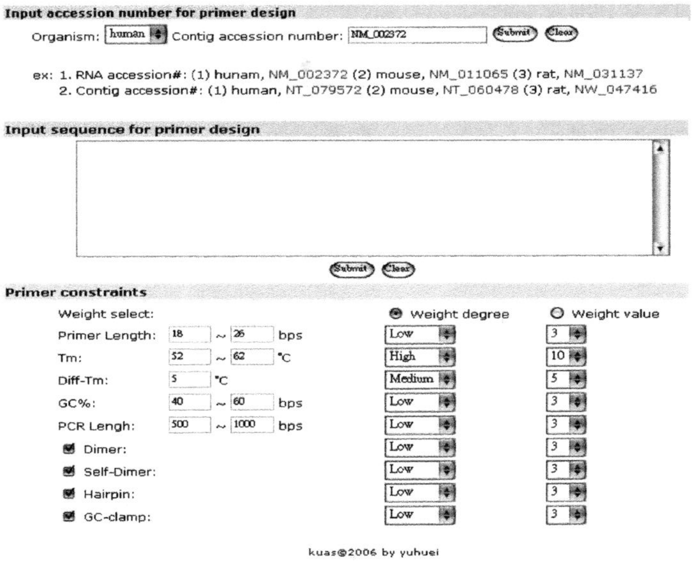

Fig. 13.3 System input interface

Table 13.2 Default values of constraints

	Range or check	Weight degree/value
Primer length	18–26 bps	Low/3
Tm	52–62°C	High/10
Diff-Tm	5°C	Medium/5
GC%	40–60%	Low/3
PCR length	500–1000 bps	Low/3
Dimer	Check	Low/3
Self-Dimer	Check	Low/3
Hairpin	Check	Low/3
GC-clamp	Check	Low/3

so that a useful feature set cannot be obtained. Hence, suitable parameter adjustment enables particle swarm optimization to increase the efficiency of feature selection.

Figure 13.3 is the system input interface. Default values of each constraint are shown in Table 13.2.

The weight degrees are set to the three values 10, 5, 3, respectively, which are marked as "High," "Medium," and "Low."

As an example, the sequence of NM_011065 was tested, and the results are shown in Figs. 13.4–13.6, as well as in Table 13.3.

Primer design has become an important issue over the last decade. The quality of primers always influences whether a PCR experiment is successful. In this chapter, we propose a PSO algorithm to design optimal primer pairs, which can be correctly and efficiently identified. The above results demonstrate that feasible primers could indeed be identified using this software system.

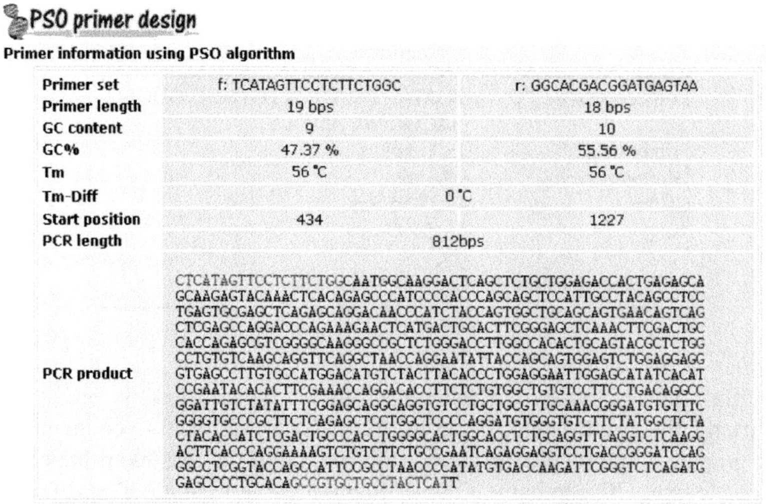

PSO primer design

Primer information using PSO algorithm

Primer set	f: TCATAGTTCCTCTTCTGGC	r: GGCACGACGGATGAGTAA
Primer length	19 bps	18 bps
GC content	9	10
GC%	47.37 %	55.56 %
Tm	56 °C	56 °C
Tm-Diff	0 °C	
Start position	434	1227
PCR length	812bps	
PCR product	CTCATAGTTCCTCTTCTGGCAATGGCAAGGACTCAGCTCTGCTGGAGACCACTGAGAGCA GCAAGAGTACAAACTCACAGAGCCCATCCCCACCCAGCAGCTCCATTGCCTACAGCCTCC TGAGTGCGAGCTCAGAGCAGGACAACCCATCTACCAGTGGCTGCAGCAGTGAACAGTCAG CTCGAGCCAGGACCCAGAAAGAACTCATGACTGCACTTCGGGAGCTCAAACTTCGACTGC CACCAGAGCGTCGGGGCAAGGGCCGCTCTGGGACCTTGGCCACACTGCAGTACGCTCTGG CCTGTGTCAAGCAGGTTCAGGCTAACCAGGAATATTACCAGCAGTGGAGTCTGGAGGAGG GTGAGCCTTGTGCCATGGACATGTCTACTTACACCCTGGAGGAATTGGAGCATATCACAT CCGAATACACACTTCGAAACCAGGACACCTTCTCTGTGGCTGTGTCCTTCCTGACAGGCC GGATTGTCTATATTTCGGAGCAGGCAGGTGTCCTGCTGCGTTGCAAACGGGATGTGTTTC GGGGTGCCCGCTTCTCAGAGCTCCTGGCTCCCCAGGATGTGGGTGTCTTCTATGGCTCTA CTACACCATCTCGACTGCCCACCTGGGGCACTGGCACCTCTGCAGGTTCAGGTCTCAAGG ACTTCACCCAGGAAAAGTCTGTCTTCTGCCGAATCAGAGGAGGTCCTGACCGGGATCCAG GGCCTCGGTACCAGCCATTCCGCCTAACCCCATATGTGACCAAGATTCGGGTCTCAGATG GAGCCCCTGCACAGCCGTGCTGCCTACTCATT	

Fig. 13.4 Output information of NM_011065 by PSO

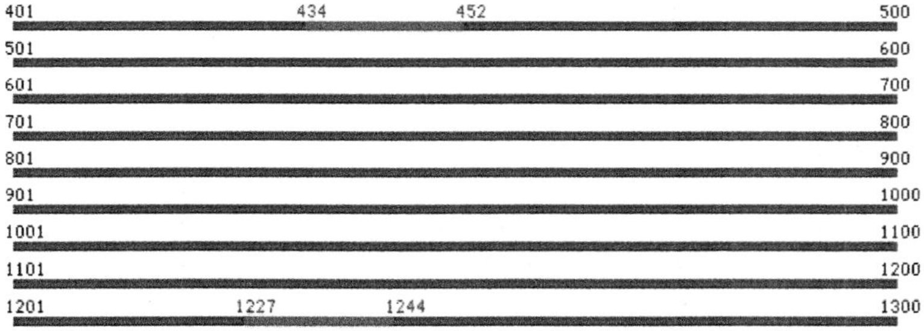

Fig. 13.5 Graphic depiction of the primer position in sequences

Fig. 13.6 Color coding to represent the PCR product, which can be clipped by the primer

13.4 Conclusion

In this study, we built the RefSeq database, which contains mRNA and genomic DNA data, and can be used to enter sequences through accession numbers or simply by pasting the sequence directly into the input interface. A user can individually set

Table 13.3 Primer information of NM_011065

	Forward/reverse primer
Primer set($5' \rightarrow 3'$)	TCATAGTTCCTCTTCTGGC/GGCAC-GACGGATGAGTAA
Primer length	19/18 bps
GC component	9/10 bps
GC%:	47.37/55.56%
Tm:	56/56°C
Tm-Diff	0°C
PCR product length	812 bps

a range for constraint criteria. Each constraint can be adjusted by weight degrees, which easily allows a feasible primer set for a PCR experiment to be identified.

The graphic output depiction shows information of the feasible primer set, such as primer length, GC content, GC%, PCR product, and length, Tm, difference of Tm of primers, and start position of primer in sequence. A color-coded graphic display shows the location of the primer set in the sequence. A feasible primer set can always be found using the PSO algorithm.

References

1. Liu, W.-T. Primer set selection in multiple PCR experiments, 2004, pp. 9–24.
2. Wu, J.-S., Lee, C., Wu, C.-C. and Shiue, Y.-L. Primer design using genetic algorithm, *Bioinformatics*, 2004, pp. 1710–1717.
3. Vallone, P.M. and Butler, J.M. AutoDimer: A screening tool for primer-dimer and hairpin structures, *BioTechniques* vol. 37, 2004, pp. 226–231.
4. Chen, S.H., Lin, C.Y., Cho, C.S., Lo, C.Z. and Hsiung, C.A. Primer design assistant (PDA): A web-based primer design tool, *Nucleic Acids Research*, vol. 31, no. 13, 2003, pp. 3751–3754.
5. Shi, Y. and Eberhart, R.C. Empirical study of particle swarm optimization, in: *Proceedings of the Evolutionary Computation 1999 Congress*, vol. 3, 1999, pp. 1945–1950.
6. Chen, H.-C., Chang, C.-J. and Liu, C.-H. Research of particle swarm optimization question, in: *First Conference on Taiwan Research Association and 2004 Technology and Management.*
7. Chang, H.-W. and Lin, C.-H. Introduction of polymerase chain reaction, *Nano-communication*, vol. 12, no 1, 2005, pp. 6–11.
8. Meyer, F., Schleiermacher, C. and Giegerich, R. (1995) Genefisher software support for the detection of postulated genes [Online]. Available: http://bibiserv.techfak.ni-bielefild.de/docs.gf_paper.html.
9. Shi, Y. and Eberhart, R.C. A modified particle swarm optimizer, in: *IEEE Proceedings of the Evolutionary Computation*, vol. 3, 1999, pp. 1945–1950.
10. Kennedy, J. and Eberhart, R.C. Particle swarm optimization, in: *Proceedings of the 1995 IEEE International Conference on Neural Networks*, Perth, Australia, vol. 4, 1995, pp. 1492–1948.
11. Kennedy, J. and Eberhart, R.C. A discrete binary version of particle swarm algorithm, *System, Man, and Cybernetics. 'Computational Cybernetics and Simulation,' 1997 IEEE International Conference*, vol. 5, Oct 12–15, 1997, pp. 4101–4108.
12. Kennedy, J., Eberhart, R. and Shi, Y. *Swarm Intelligence*. Morgan Kaufmann, San Francisco.
13. Oh, I.-S., Lee, J.-S. and Moon, B.-R. Hybrid genetic algorithms for feature selection, *IEEE Trans. Pattern Analysis and Machine Intelligence*, vol. 26, no. 11, Nov. 2004.

14. Shi, X.H., Liang, Y.C., Lee, H.P., Lu, C. and Wang, L.M. An improved GA and a novel PSO-GA-based hybrid algorithm. *Information Processing Letters*, vol. 93, no. 5, 2005, pp. 255–261.
15. Rahmat-Samii, Y. Genetic algorithm (GA) and particle swarm optimization (PSO) in engineering electromagnetics, in: *Proceedings of the Seventeenth International Conference on Applied Electromagnetics and Communications*, 2003, pp. 1–5.

Chapter 14
Genetic Algorithms and Heuristic Rules for Solving the Nesting Problem in the Package Industry

Roberto Selow, Flávio Neves, Jr., and Heitor S. Lopes

14.1 Introduction

The cutting/nesting problem in the packing industry can be stated as finding the maximum number of packages that can be arranged in a paper sheet of known size, in such a way as to minimize the loss of material. Figure 14.1 illustrates an example of six packages that will be drawn from a standard paper sheet and turned into a box. This problem is commonly found in many industrial areas that deal with cutting out shapes from raw stock, such as fabric, steel plate, paper, and so on.

An important factor in the search for the optimal solution for this problem is the number of parts that will be manipulated in the mounting settle. This is discussed later. There is a combinatorial explosion as the number of parts increases, leading to infeasible computational costs. For real-world problems, the number of parts is usually not larger than 20.

Genetic algorithms (GA) [10] have been used successfully in the last decades for several complex combinatorial problems and also for problems similar to the above-mentioned one [5, 12]. Therefore, the objective of this work is to propose a new method that uses genetic algorithms and heuristic rules to solve the problem.

14.2 Previous Work

The nesting/cutting problem of parts in a plane stock has been widely studied in the recent literature, inasmuch as it is a common problem found in several industries, such as packing, plating, clothing, and furniture, among others. According to Han and Na [7], there are two different versions of this problem, depending on the way the parts of the raw material are shaped: the first considers only rectangle like parts and the second, irregular-shaped parts.

Oscar Castillo et al. (eds.), *Trends in Intelligent Systems and Computer Engineering*.

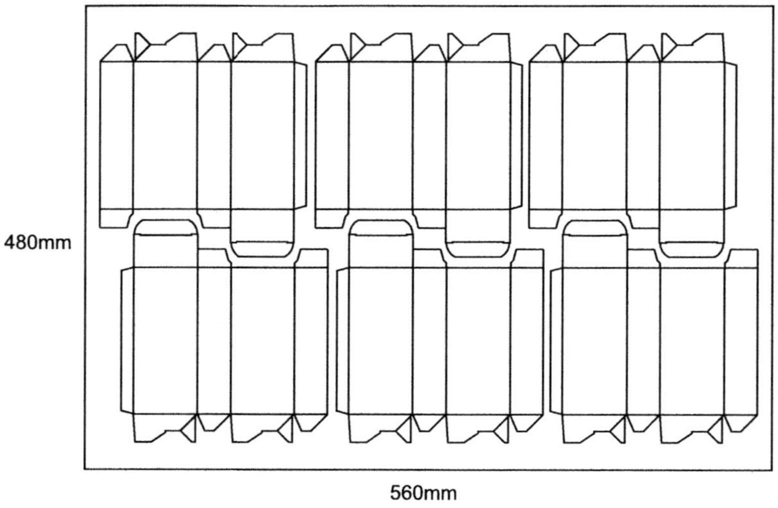

Fig. 14.1 Arrangement of six packages in a paper sheet

Possibly the works of Gilmore and Gomory [8, 9] were the seminal works in this area. They used rectangular parts and tackled the problem using linear programming techniques. They also have succeeded in working with problems with one, two, or three dimensions.

The problem was studied by Haims and Freeman [11] without the restriction of the number of parts to be cut in a sheet of raw material. The method they have developed consists in obtaining a rectangle, called a module, which encloses one or more irregular parts using the smallest possible area. Modules were then grouped in a sheet of material by means of dynamic programming. This algorithm requested that the rectangular module be positioned in one of the corners of the sheet. Later, Adamowicz and Albano [1] proposed an improvement in this algorithm, eliminating such limitation. This algorithm was used in the naval construction industry.

For a much more complex problem, dealing with irregular parts, Albano and Sapuppo [2] proposed a technique that uses heuristic search methods. Also using heuristic methods, Nee [14] has proposed an algorithm for the nesting problem in the steel plating industry.

The first use of genetic algorithms for this problem is, possibly, the work of Ismail and Hon [13]. Since then, several other authors proposed further improvements. For instance, Fujita and Gakkai [6] presented a hybrid approach using both a GA and a local minimization algorithm. The method presented by András et al. [3] is also based on GAs, but the combination of parts is represented in a tree.

More recently, Uday et al. [18] described a new approach for obtaining optimized arrangements of packages. The solution is based on a hybrid system that uses parallel genetic algorithms and a heuristic process based on the combination of contouring characteristics. Some topologies for communication between the subpopulations, as well as several migration policies, were tested in the experiments.

The simulations demonstrate that the proposed approach generates good results for this and other types of problems with large search space.

In the work of Chen et al. [4] some approaches for optimized arrangement layouts were proposed. Irregular flat shapes (convex and concave) were used. Genetic algorithms were among the techniques used.

14.3 Proposed Methodology

The proposed methodology for obtaining optimized layout package arrangements in a sheet of paper is presented in the next sections. Basically, the method requests the definition of the following: package representation, package operations, heuristic rules, search space encoding schemes, the fitness function, and adaptation of the fitness function.

14.3.1 Package Representation

The implementation of the problem for layout optimization requires creating a basic design for the packages, called E, based on a real package that, once cut out, will be folded into a box. See Fig. 14.2a for a detailed representation of an actual package. The model of the basic package is composed of a set of i rectangles (R_i) with given positions (P) and dimensions (D), as shown in Fig. 14.2b, and formalized by the expressions:

$$E = \{P, D\},\ P = (x_i, y_i) \text{ and } D = (l_i, h_i) \tag{14.1}$$

where:

x_i = horizontal coordinate of the origin of the ith rectangle referenced at its left corner.
y_i = vertical coordinate of the origin of the ith rectangle referenced at its lower corner.
l_i = horizontal dimension of the ith rectangle.
h_i = vertical dimension of the ith rectangle.

Therefore, the encoding of the basic package model shown in Fig. 14.2b is represented as

$$\begin{aligned}
P = \{&(x_1 = 7,\ y_1 = 0), (x_2 = 44,\ y_2 = 8), (x_3 = 63,\ y_3 = 0), (x_4 = 101,\ y_4 = 8), \\
&(x_5 = 0,\ y_5 = 16), (x_6 = 7,\ y_6 = 100), (x_7 = 44,\ y_7 = 100), (x_8 = 100,\ y_8 = 100\}\} \\
D = \{&(l_1 = 36,\ h_1 = 16), (l_2 = 19,\ h_2 = 8), (l_3 = 37,\ h_3 = 16), (l_4 = 19,\ h_4 = 8), \\
&(l_5 = 120,\ h_5 = 84),\ (l_6 = 36,\ h_6 = 27),\ (l_7 = 19,\ h_7 = 10), (l_8 = 19,\ h_8 = 10\}\})
\end{aligned}$$

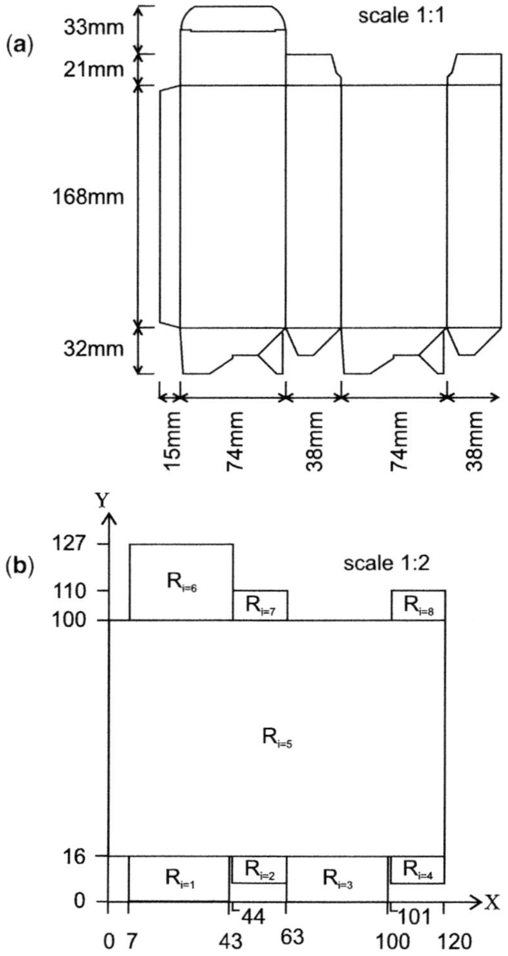

Fig. 14.2 **a** Real unfolded package sample. **b** Basic package model represented by rectangles

14.3.2 Basic Package Operations

In this work, we define two basic operations to be done with packages: rotation and displacement. The objective of such operations is to allow the movement of the packages within the search space (raw sheet of paper) in order to determine a setting without overlapping between pieces and with the smallest possible loss of material without.

14.3.2.1 Rotation of the Packages

Practical cases of the package industry have shown that, for most cases, the orientation of the packages in a good arrangement occurs according to angles of 0, 90, 180, and 270 degrees. In this work these angles are represented by ϕ.

14.3.2.2 Displacement of the Packages

In addition to rotation, it is also necessary to displace packages in a paper sheet to optimally arrange them. This is done by adding values x_a and y_a to all the horizontal (x_i) and the vertical (y_i) coordinates, respectively, that define the position of the rectangles. The new set of coordinates, which will define the new position of the packing, is called P_{new} and is presented as

$$P_{\text{new}} = \{(x_a + x_i), (y_a + y_i)\} \tag{14.2}$$

where:

x_a = horizontal displacement.
y_a = vertical displacement.

14.3.3 Heuristic Rules

To improve the performance of the GA, the creation of a heuristic approach is suggested, composed of a sequence of rules, defined by experts, which will guide the operations to be done over the set of the packages. The quality of the results obtained is not limited by these rules. On the contrary, they allow the organization of the packages in the setting and also reduce the search space for the algorithm. The proposed heuristic rules are presented below.

1. All the packages are aligned in columns.
2. Rotation is applied to all the packages of the same column.
3. The first column has its horizontal coordinate origin coincident with the origin of the search space.
4. Each column can move to the left within the horizontal search region (R_x), except the first one that cannot move horizontally.
5. The horizontal origin coordinate of each column is based on the origin of the column that is at its left.
6. Each column can be displaced within the vertical search region (R_y) above the horizontal axis.
7. The packages of each column can be displaced among themselves, within a 'between boxes' search region (R_{ec}).

The sizes of the R_x, R_y, and R_{ec} regions should be defined to allow the application of the rules presented above. Their values are integer numbers and obtained from the value of M_{me}, the larger dimension of a package (considering its height and width), as follows.

$$R_{xk} = [0..M_{me}] \tag{14.3}$$

$$R_{yk} = [0..M_{me}] \tag{14.4}$$

$$R_{\text{eck}} = [0..2M_{me}] \tag{14.5}$$

$$\phi_k = [0..3] \tag{14.6}$$

Table 14.1 Range of variables for the model of Fig. 14.2a, b

Horizontal search regions	Vertical search regions
$R_{x1} = 0$	$R_{y1} = [0..127]$
$R_{x2} = [0..127]$	$R_{y2} = [0..127]$
$R_{x3} = [0..127]$	$R_{y3} = [0..127]$
$R_{x4} = [0..127]$	$R_{y4} = [0..127]$
'Between boxes' search regions	Possible rotation angles
$R_{ec1} = [0..254]$	$\phi_1 = [0..3]$
$R_{ec2} = [0..254]$	$\phi_2 = [0..3]$
$R_{ec3} = [0..254]$	$\phi_3 = [0..3]$
$R_{ec4} = [0..254]$	$\phi_4 = [0..3]$

where:

R_{xk} = horizontal search region.
R_{yk} = vertical search region.
R_{eck} = between-boxes search regions.
ϕ_k = rotation angle (represented as integer numbers).
k = column index.

To illustrate the application of the previous definitions, Table 14.1 presents an example of values for R_{xk}, R_{yk}, and R_{eck} based on the package model shown in Fig. 14.2a, b, whose M_{me} value is 127. In this case, the layout has four columns, as shown below.

14.3.4 Search Space Encoding Schemes

Once the package representation and the possible operations over them are defined, it is necessary to encode the variables of the problem in the structure of a chromosome for the GA. First, the number of columns (K) that will compose the arrangement should be calculated. The genes that constitute the chromosome will be defined with these values. Each gene encodes a specific variable of a given arrangement representing values according to the R_{xk}, R_{yk}, R_{eck}, and ϕ_k previously defined limits.

The number of packages in each column (NEC) has to be large enough to fill the sheet in the vertical dimension and should be estimated by the user. The total horizontal and vertical dimensions of the paper sheet are *TFX* and *TFY*, respectively. Consequently, the package set must be contained within these dimensions. When the dimensions are defined, they must match the adopted scale (in this work, 1:2). Therefore, a chromosome of the GA can be represented by a string of integers, such that the encoded genes have the meaning given by expression (14.7) below. The range of each variable is shown in Table 14.1.

$$\text{Chromosome} = x_1\ y_1\ ec_1\ \phi_1 \quad \ldots \quad x_k\ y_k\ ec_k\ \phi_k \tag{14.7}$$

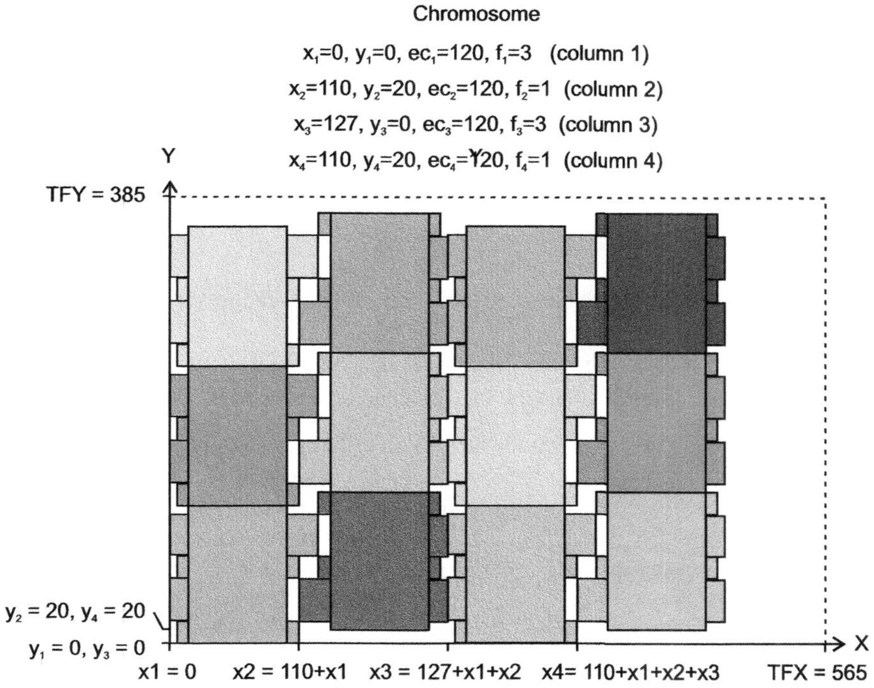

Fig. 14.3 Graphical representation of a chromosome

Following, there is an example based on the model previously presented. The layout of the packages is represented by the pieces of information contained in the chromosome as in Eq. 14.7. Next, the chromosome is decoded and then a graphical representation is created (Fig. 14.3).

14.3.5 Fitness Function

The quality of a solution s is evaluated by using a fitness function $f(s)(15)$. According to Selow [16], the fitness function for this problem is composed by two terms. The first term is the total area index $IA(s)$. It shows the total area occupied by the arrangement of packages. The second term is the index of overlapping packages $IS(s)$. It evaluates the amount of overlapping among the whole set of packages represented by the chromosome. The objective of the GA search is to find an arrangement with the smallest total area index $IA(s)$, and without overlapping between packages or $IS(s) = 0$.

14.3.5.1 Computation of the Total Area Index

The computation of the first term $IA(s)$ is based on the work of Han and Na [7]. In this work, the overall momentum of the package arrangement is minimized. In the present work, all packages in the arrangement have, by default, the same shape and dimensions. Therefore, a simplification was adopted: the area of the packages was not considered and we used, instead, the sum of the Euclidean distances between the origin of the search space and the origin of each package. This measure is represented by Eq. 14.8:

$$d_{kn} = \sqrt{xo_{k,n}^2 + yo_{k,n}^2} \tag{14.8}$$

where:

$d_{k,n}$ = Euclidean distance between the origin of the package and the origin of the search space.
$xo_{k,n}$ = horizontal coordinate of the origin of the package.
$yo_{k,n}$ = vertical coordinate of the origin of the package.

Recall that the maximum Euclidean distance that any package can assume is $d_{\max}$, defined by Eq. 14.9. This value is obtained considering the worst case where all packages are located in the opposite side of the origin on the paper sheet.

$$d_{\max} = \sqrt{TFX^2 + TFY^2} \tag{14.9}$$

Figure 14.4 illustrates the Euclidean distance for three packages out of the nine of the arrangement.

A given package of the arrangement is uniquely identified by $E_{k,n}$, where k and n correspond, respectively, to the column and the position in the column in which the package is set.

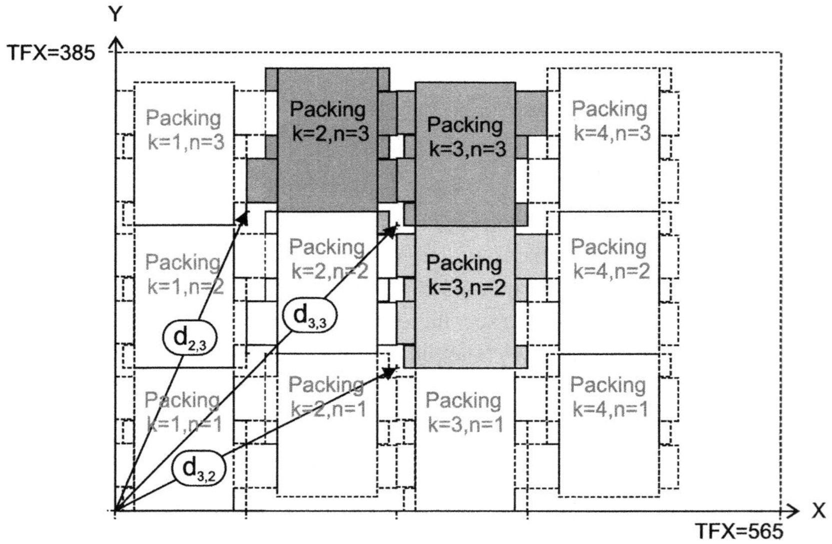

Fig. 14.4 Euclidean distances from the origin of the search space to the origin of three different packages

The normalization of $IA(s)$ is shown in Eq. 14.10. It is based on the worst case for a given arrangement, previously mentioned.

$$IA(s) = \left(\frac{\Sigma d_{k,n}}{K \cdot NEC \cdot d_{\max}} \right) \tag{14.10}$$

where:

$d_{k,n}$ = Euclidean distance from the origin of the package to the origin of the search space.
$d_{\max}$ = maximum Euclidean distance of a given package.
k = column index.
n = index of the position of the package in the column.
K = number of columns.
NEC = number of packages per column.
S = current solution under evaluation.

14.3.5.2 Computation of Overlapping Index

In order to compute $IS(s)$ it is necessary to determine the exact overlapping area of packages by means of computing the overlapping area of their compounding rectangles. Figure 14.5 presents the eight rectangles that compose a package and their encoding relative to a given arrangement. In this example, each rectangle is identified as $R_{k,n,i}$ where k is the index that identifies the column (in the arrangement) where the package is, n is the position of the package in the column, and i is the index of the rectangle.

Based on this encoding, the overlapping between two rectangles of different packages is represented as $S(R_{k,n,i},\ R_{k',n',i'})$, as shown in Fig. 14.6. The sum of all individual overlapping areas is represented by Eq. 14.11:

$$S(s) = \sum S(R_{k,n,i}, R_{k',n',i'}) \quad \text{such that } k \neq k' \vee n \neq n' \tag{14.11}$$

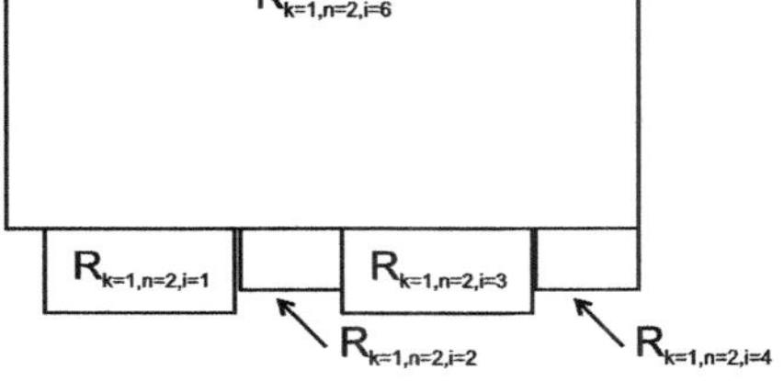

Fig. 14.5 Encoding example of the rectangles that compose the packing $k = 1,\ n = 2$

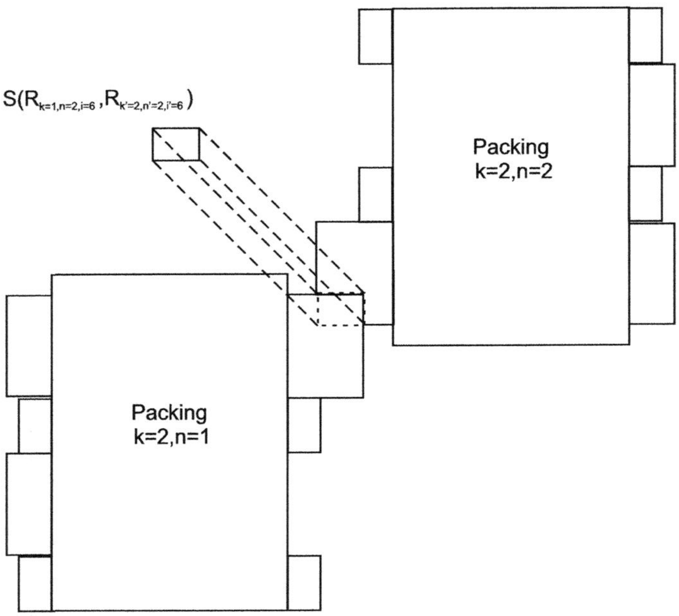

Fig. 14.6 Example of the overlapping between two packages in which $S(R_{k=1,n=2,i=6}, R_{k'=2,n'=2,i'=6})$

where:

$S(s)$ = total overlapping area.
k = column index.
n = position of the package in the column.
s = current solution under evaluation.

The total overlapping area $S(s)$ is normalized according to the maximum possible overlapping area of all packages ($S_{\max}$). To find such value, we consider that all packages of the arrangement are perfectly overlapped. In this case, $S_{\max}$ is obtained by Eq. 14.12:

$$S_{\max} = \left(\frac{AE \cdot (NEC \cdot K) \cdot (NEC \cdot K - 1)}{2} \right) \tag{14.12}$$

where:

$S_{\max}$ = maximum overlapping area of all packages.
AE = area of a package.
NEC = number of packages per column.
K = number of columns.

The equation that finally expresses the overlapping index of the arrangement of packages is:

$$IS(s) = \frac{S(s)}{S_{\max}} \tag{14.13}$$

14.3.6 Adaptation of the Fitness Function

To improve the performance of the GA, some studies suggest the use of some technique to modify dynamically the fitness function during the search [15, 17, 19]. In this work, we propose a term $IVD(s)$ which represents the index of dynamic variation, defined by Eq. 14.14.

$$IVD(s) = k_1 + k_2 \times g \tag{14.14}$$

where:

k_1, k_2 = arbitrary constants.
g = generation number.

Consequently, the final fitness function is shown in Eq. 14.15, in which the influence of both *IA* and *IS* indices increase along generations. By using the proposed fitness measure it is possible to discriminate between two different arrangements, favoring the one that is more compact.

$$f(s) = 1 - \left(\frac{IA(s)^{IVD(s)} + IS(s)^{IVD(s)}}{2} \right) \tag{14.15}$$

14.4 Experiments and Results

The proposed method was validated by comparison with the results obtained by the experts on the same problems. Three different models of packages frequently found in the industry were used in the tests, representing real-world situations. A total of 120 evaluations with different populations and generations was done including each of the test cases. The measure used to compare the performance of each arrangement was "paper area/package". Solutions with performance equal to or better than the best arrangements obtained by the experts were considered as successful solutions, otherwise, as failed solutions. Finally, the processing time for each test was measured, using a PC with Pentium III processor running at 866 MHz.

14.4.1 Results for Case 1

Figure 14.7 shows the arrangement of packages obtained by the experts for the first package model. This arrangement has $725\,\text{cm}^2/\text{pac}$.

Table 14.2 shows a summary of the results of the simulations for the case shown in the figure. The GA was run with different population sizes and number of generations. For each combination, ten independent runs with distinct random seeds were done. Each one of the ten runs had its time measured and the average time is presented.

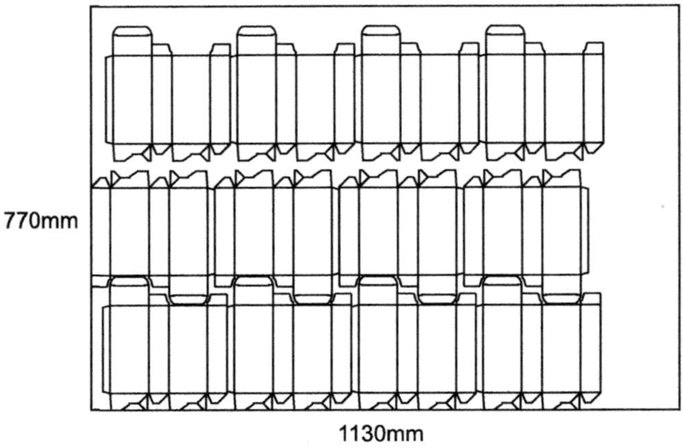

Fig. 14.7 Arrangement obtained by experts for the first case

Table 14.2 Summary of the results of simulations for the first case

Population size	Number of generations	Successful (%) solutions	Average running time for each simulation (s)
200	200	20	89
	400	20	178
	600	40	266
	800	70	356
400	200	20	177
	400	10	353
	600	70	530
	800	40	710
800	200	30	353
	400	70	706
	600	50	1070
	800	80	1414

An example of a successful arrangement for Case 1 was obtained with a population of 200 individuals during 600 generations. The arrangement obtained by GA had 725 cm^2/pac corresponding to a fitness of 0.687798828 (see Fig. 14.8). For this specific run, the curves of the maximum, average, and minimum fitness of the population are shown in Fig. 14.9.

14.4.2 Results for Case 2

For the second model of the package, the smallest paper area/packing that experts obtained was 544 cm^2/pac, as presented in Fig. 14.10.

Table 14.3 shows a summary of the results of the simulations for this case. The same simulations with the AG done for Case 1 were repeated here. An example of

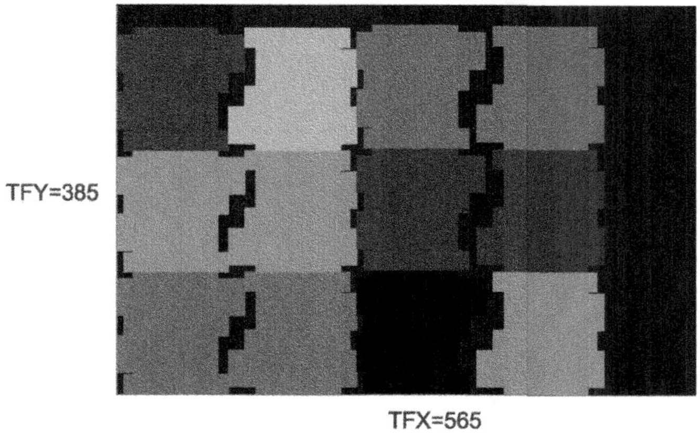

Fig. 14.8 Example of a successful arrangement found for first case

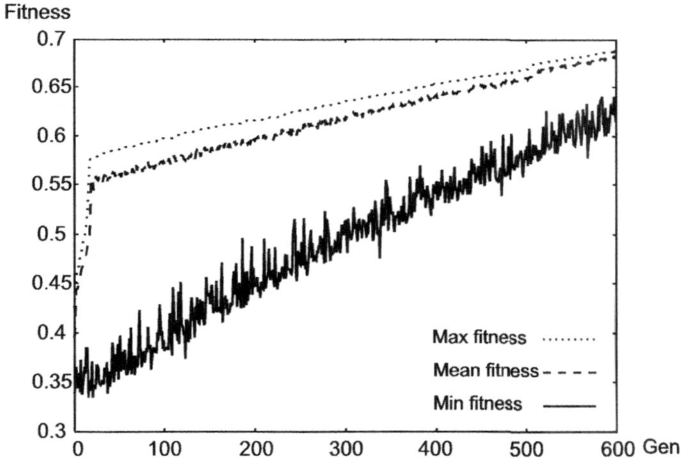

Fig. 14.9 Maximum, average, and minimum fitness during generations for the arrangement obtained in Fig.14.8

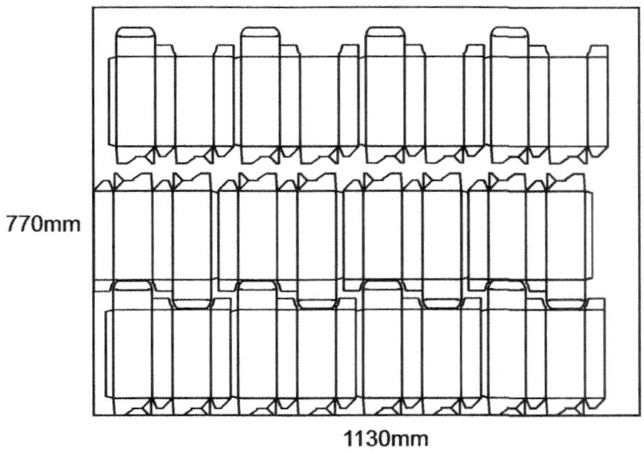

Fig. 14.10 Arrangement obtained by experts for the second case

Table 14.3 Summary of the results of simulations for the second case

Population size	Number of generations	Successful solutions (%)	Average running time for each simulation (s)
200	200	30	26
	400	50	52
	600	50	78
	800	70	103
400	200	30	51
	400	80	103
	600	60	155
	800	100	207
800	200	80	103
	400	100	205
	600	100	309
	800	100	412

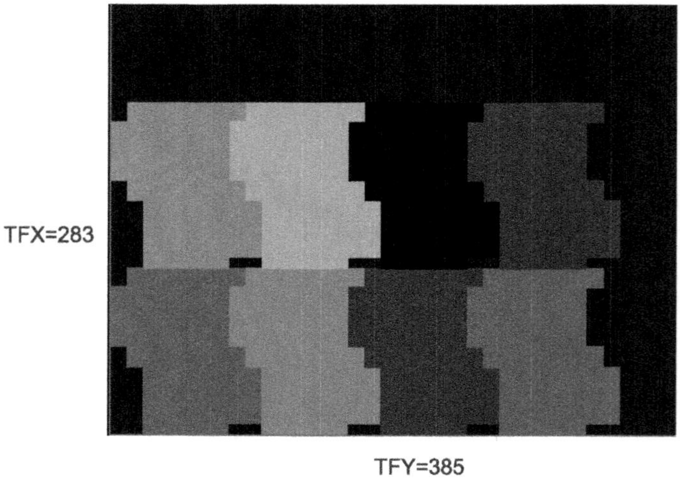

Fig. 14.11 Best arrangement found by the GA for the second case

a successful solution for Case 2 was obtained with a population of 200 individuals during 400 generations. The arrangement obtained by GA had $544\,\text{cm}^2/\text{pac}$ corresponding to a fitness of 0.705119103 (Fig. 14.11). For this solution, the curves of the maximum, average, and minimum fitness of the population are shown in Fig. 14.12.

14.4.3 Results for Case 3

For the third model of package, the smallest paper area/packing that experts obtained was $1450\,\text{cm}^2/\text{pac}$, as presented in Fig. 14.13.

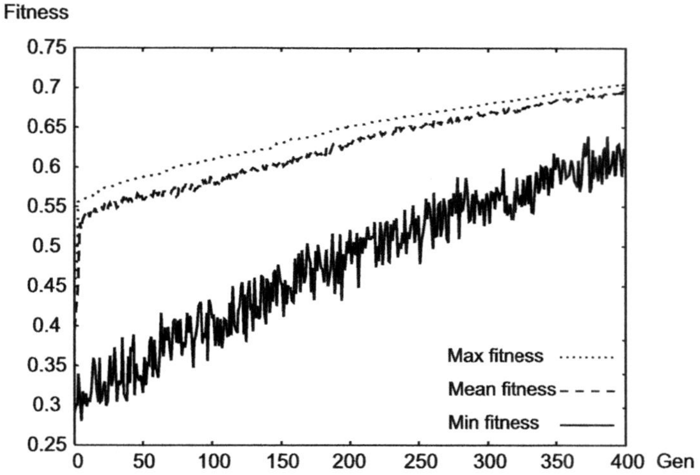

Fig. 14.12 Maximum, average, and minimum fitness during generations for the arrangement obtained in Fig. 14.11

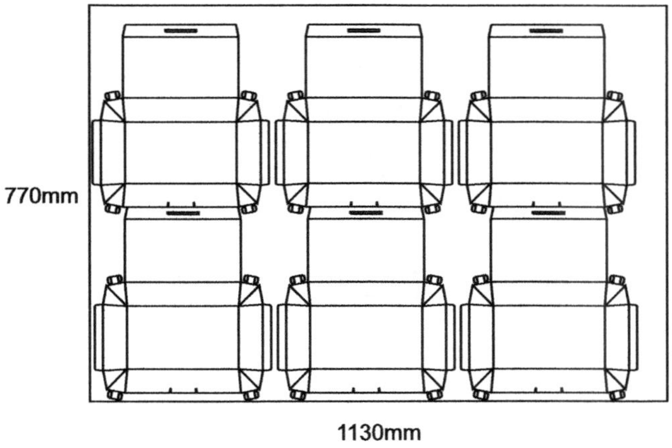

Fig. 14.13 Arrangement obtained by experts for the third case

Table 14.4 shows a summary of the results of the simulations for this case. The same simulations with the AG done for Case 1 are repeated here. An example of a successful solution for Case 3 was obtained with a population of 400 individuals during 400 generations. The arrangement obtained by GA had 1450 cm^2/pac corresponding to a fitness of 0.691554695 (Fig. 14.14). For this solution, the curves of the maximum, average, and minimum fitness of the population are shown in Fig. 14.15.

The performance of the proposed GA was compared with the experts for the three cases. Results regarding efficiency are summarized in Table 14.5, and those regarding the time necessary for obtaining solutions are shown in Table 14.6.

Table 14.4 Summary of the results of simulations for the third case

Population size	Number of generations	Successful solutions (%)	Average running time for each simulation (s)
200	200	10	17
	400	10	32
	600	60	48
	800	50	64
400	200	20	33
	400	20	65
	600	40	97
	800	40	129
800	200	0	65
	400	40	130
	600	20	194
	800	20	259

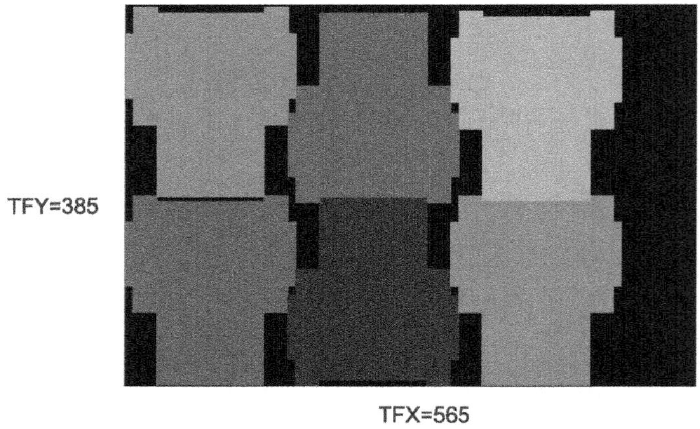

Fig. 14.14 Best arrangement found by the GA for the third case

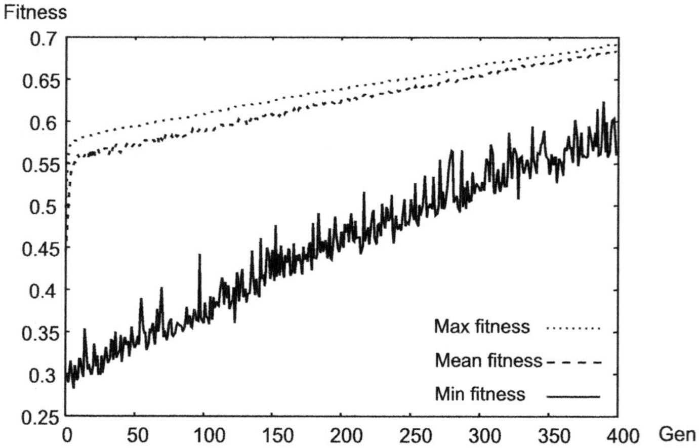

Fig. 14.15 Maximum, average, and minimum fitness during generations for the arrangement obtained in Fig. 14.14

Table 14.5 Comparison of efficiency between experts and the GA

		Case 1	Case 2	Case 3
GA	Total of arrangements evaluated	120	120	120
	Number of successful solutions	52 (43%)	85 (71%)	33 (28%)
Experts	Total of arrangements obtained	12	9	9
	Number of successful solutions	4 (33%)	5 (56%)	3 (33%)
Performance advantage of GA over experts		+10%	+15%	−5%

Table 14.6 Comparison of processing time between experts and the GA

		Case 1	Case 2	Case 3
GA	Average time per arrangement	517 s	150 s	94 s
Experts	Average time per arrangement	690 s	410 s	993 s
Performance advantage of GA over experts		33%	173%	956%

14.5 Conclusions

The good results obtained for the three real-world cases suggest that the proposed methodology presented here is feasible and efficient. These experiments led to the development of a GA-based software tool for optimizing package arrangements in the industry.

Before the use of heuristic rules in the system, two problems were verified. First, a large computational effort was spent moving packages from one edge of the paper sheet to the other. Because all the packages of the arrangement are equal, just a minor movement, that is, in a smaller region, would be enough. Second, we observed a lack of organization of the packages in the arrangement. From a practical point of view, this option is not adequate and can represent some problems in the production line. Therefore, the use of the heuristic rules indeed improved the performance of the GA.

The fitness curves in Figs. 14.9, 14.12 and 14.15 suggest that even better results could be obtained, because the difference among maximum, medium, and minimum fitness is still decreasing. But, it must be considered that it is not possible to get an arrangement with a number greater than $NEC.K/AE$, where AE represents the area of a package and NEC represents the number of packages per column. This was the reason for the stop criterion to be a limited number of generations.

Results obtained by the proposed GA, regarding efficiency, were better than those obtained by the experts, most of the time (Table 14.5). Possibly, this is due to the association of the efficiency of the GA as a global search method and the expert knowledge incorporated in the system by the heuristic rules. Regarding the time necessary for finding a good solution, again, the proposed GA had a great advantage when compared with human experts (Table 14.6). Overall, considering both efficiency and the processing time, the proposed methodology using the GA is very promising for application to real-world problems.

Future work will include the adaptation of the proposed methodology for dealing with packages of different shapes in the same arrangement.

References

1. M. Adamowicz, A. Albano, Nesting two-dimensional shapes in rectangular modules, *Computer Aided Design*, 1976, vol. 8, no. 1, pp. 27–33.
2. Albano, G. Sapuppo, Optimal allocation of two-dimensional irregular shapes using heuristic search methods, *IEEE Transactions on Systems, Man, and Cybernetics*, 1980, vol. 10, no. 5, pp. 242–248.
3. P. András, A. András, S. Zsuzsa, A genetic solution for the cutting stock problem, *Proceedings of the First On-Line Workshop on Soft Computing*, 1996, Nagoya University, pp. 87–92.
4. P. Chen, Z. Fu, A. Lim, B. Rodrigues, Two-dimensional packing for irregular shaped objects, *Hawaii International Conference on Information Sciences* (HICSS-36, Hawaii, USA), 2003.
5. P.C. Chu, J.E. Beasley, A genetic algorithm for the generalized assignment problem, *Computers in Operations Research*, 1997, vol. 24, no. 1, pp. 17–23.
6. K. Fujita, S. Gakkai, Approach for optimal nesting algorithm using genetic algorithm and local minimization algorithm, *Transactions of the Japanese Society of Mechanical Engineers*, 1993, part C, vol. 59, no. 564, pp. 2576–2583.
7. G.C. Han, S.J. Na, Two-stage approach for nesting in two-dimensional cutting problems using neural network and simulated annealing, *Proceedings of the Institution of Mechanical Engineering Part B Journal of Engineering Manufacture*, 1996, vol. 210, no. 6, pp. 509–519.
8. P.C. Gilmore, R.E. Gomory, Multistage cutting stock problems of two and more dimensons, *Operations Research*, 1965, vol. 13, pp. 94–120.
9. P.C. Gilmore, R.E. Gomory, The theory and computation of knapsack functions, *Operations Research*, 1966, vol. 14, no. 61, pp. 1045–1074.
10. D.E. Goldberg, *Genetic Algorithms in Search, Optimization, and Machine Learning*, Reading, MA: Addison-Wesley, 1989.
11. M.J. Haims, H. Freeman, Amultistage solution of the template layout problem, *IEEE Transactions on Systems Science and Cybernetics*, 1970, vol. 6, no. 2, pp. 145–151.
12. E. Hopper, B. Turton, A genetic algorithm for a 2D industrial packing problem, *Computers & Industrial Engineering*, 1999, vol. 37, pp. 375–378.
13. H.S. Ismail, K.K.B. Hon, New approaches for the nesting of two-dimensional shapes for press tool design, *International Journal of Production Research*, 1992, vol. 30, no. 4, pp. 825–837.
14. A.Y.C. Nee, A heuristic algorithm for optimum layout of metal stamping blanks, *Annals of CIRP*, 1984, vol. 33, no. 1, pp. 317–320.
15. V. Petridis, S. Kazarlis, A. Bazarlis, Varying fitness functions in genetic algorithm constrained optimization: The cutting stock and unit commitment problems, *IEEE Transactions on Systems, Man, and Cybernetics—Part B: Cybernetics*, 1998, vol. 28, no. 5, pp. 629–639.
16. R. Selow, Optimized arrangement of packages using genetic algorithms, M.Sc. Thesis, 2001, UTFPR, Brazil [in Portuguese].

17. W. Siedlecki, W. Sklanski, Constrained genetic optimization via dynamic reward-penalty balancing and its use in pattern recognition, *Proceedings of Third International Conference on Genetic Algorithms*, Ed., San Mateo, CA: Morgan Kaufmann, 1989, pp. 141–150.
18. Uday, E. Goodman, A. Debnath, Nesting of Irregular Shapes Using Feature Matching and Parallel Genetic Algorithms, *Genetic and Evolutionary Computation Conference Late-Breaking Papers*, E. Goodman, Ed., San Francisco: ISGEC Press, 2001, pp. 429–494.
19. H. Wang, Z. Ma, K. Nakayama, Effectiveness of penalty function in solving the subset sum problem, *Proceedings of Third IEEE Conference on Evolutionary Computation*, 1996, pp. 422–425.

Chapter 15
MCSA-CNN Algorithm for Image Noise Cancellation

Te-Jen Su, Yi-Hui, Chiao-Yu Chuang, and Wen-Pin Tsai

15.1 Introduction

Many optimization algorithms have been developed and adapted for several problems by intelligence computing. A new computational intelligence called the artificial immune system (AIS), which was inspired by the biological immune system, has attracted more and more interest in the last few years [1, 2, 4].

De Castro and Von Zuben [4] presented a clonal selection algorithm, which took into account the affinity maturation of the immune response, in order to solve complex problems, that is, learning and multimodal optimization. Among the clonal selection algorithms, mutation plays an important role in generating the next population. The mutation randomly modifies some antibody and is responsible for search space exploration. The reasonable and generally effective ways to improve the performance of the clonal selection algorithm is to allow its mutation probability to be self-adaption and diversified mutation operators.

In this chapter, we propose a modified clonal selection algorithm (MCSA), with an adaptive maturation strategy and novel clone framework to search approximate optimal solutions. We propose the pyramid framework with self-adaption mutation probability in clones, and perform different mutation operators of Gaussian mutation, swapping mutation, and multipoint mutation in the respective levels of the pyramid; next, a response mechanism is applied to avoid local search for optimization. Employing the above improvements, the MCSA enables a better capability for optimization.

The organization of this chapter is summarized as follows. In Sect. 15.2, the clonal selection algorithm is proposed, and the modified maturation strategy is applied in MCSA in Sect. 15.3. Another important role is described in Sect. 15.4. In Sect. 15.5, a hybrid MCSA and CNN for image noise cancellation are submitted. Finally, the simulation results and conclusions are drawn in Sects. 15.6 and 15.7, respectively.

Oscar Castillo et al. (eds.), *Trends in Intelligent Systems and Computer Engineering.*

15.2 Clonal Selection Algorithm

15.2.1 Immune System (IS)

The human immune system is a complex system of cells, molecules, and organs that represent an identification mechanism capable of perceiving and combating dysfunction from our own cells and the action of exogenous infectious microorganisms. The human immune system protects our bodies from infectious agents such as viruses, bacteria, fungi, and other parasites. Any molecule that can be recognized by the adaptive immune system is known as an antigen. The basic component of the immune system is the lymphocytes or the white blood cells.

Lymphocytes exist in two forms, B cells and T cells. These two types of cells are rather similar, but differ with relation to how they recognize antigens and by their functional roles. B cells are capable of recognizing antigens free in solution, whereas T cells require antigens to be presented by other accessory cells. Each of these has distinct chemical structures and produces many-shaped antibodies on its surfaces to kill the antigens. Antibodies are molecules attached primarily to the surface of B cells whose aim is to recognize and bind to antigens.

The immune system possesses several properties such as self/nonself-discrimination, immunological memory, positive/negative selection, immunological network, clonal selection, and learning which perform complex tasks.

15.2.2 Artificial Immune System (AIS)

The artificial immune system (AIS) is a set of advanced techniques that attempt to algorithmically imitate the natural behavior of the immune system and utilize the natural immune system as a metaphor for solving computational problems. AIS is the beneficial mechanisms extracted or gleaned from the immune system that can be used to solve particular problems, for example, misbehavior detection, identification, robotics, control, optimization problems, and so on.

The immune algorithm, which was proposed by Fukuda et al. [2], mathematically modeled immune diversity, network theory, and clonal selection as a multimodal function optimization problem. The guide of diversity and multiple solution vectors instituted are kept as the memory of the system.

15.2.3 Clonal Selection Algorithm (CSA)

The clonal selection algorithm that considered the affinity maturation of the immune response, presented by De Castro and Von Zuben [3–5], which makes it possible for immunological evolution to be used for engineering applications, such as pattern recognition, machine-learning, and multimodal and multiobjective function optimization.

15.3 Modified CSA

15.3.1 Immune System (IS)

In CSA, the whole population of clonal cells is mutated with equal probability; it is not adaptable in evolutionary generation. So we present a pyramid framework which divides whole clonal cells' population into three parts. The proportional quantity of antibodies in this pyramid framework is 1 : 2 : 3 from top to bottom as Fig. 15.1. The top of the pyramid means the best solutions with higher affinities of the whole population in the present generation; decreasing progressively, the bottom of the pyramid is the worst solution with lower affinities. By using this framework with three respective mutation operators Gaussian mutation, swapping mutation, and multipoint mutation (shown as follows), the MCSA could converge rapidly.

15.3.2 Gaussian Mutation

We implement Gaussian mutation operators [6] for CSA used to optimize numeric affinity functions and also implement self-adaption Gaussian mutation which allows the CSA to vary the mutation strength during the run; this gives further improvement on some of the applications.

We investigate the usefulness of Gaussian mutation in clonal selection used for numeric affinity function optimization. Self-adaption is a powerful tool that has been used to set the values of the algorithm parameters. Therefore, we use self-adaption to allow the CSA to control the variance for Gaussian mutation; that gives significant improvement for some applications, and changes of the variance over time are appropriate for the different types of applications that would be optimized.

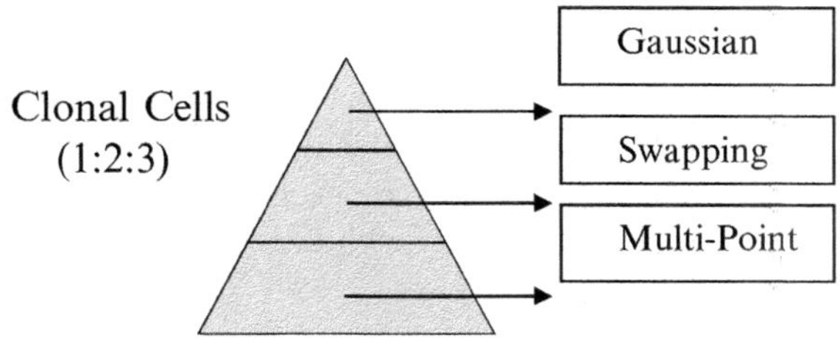

Fig. 15.1 The respective mutation operators in the pyramid framework

In MCSA, an antibody composed of a gene is in a string form. We implement self-adaption Gaussian mutation in a random gene of the mutated antibodies, which are at the top of the pyramid framework and then add Gaussian noise as in Eqs. 15.1 and 15.2.

$$gene_{new} = gene_{original} \times (1 + \alpha \cdot Gaussian(0, \sigma)) \tag{15.1}$$

$$\sigma = 1 + \frac{f - f_{\min}}{f_{\min}} \tag{15.2}$$

where the gene original is the mutated value of the antibody, α stands for a constant, f is the affinity of the respective antibody, and $f_{\min}$ is the minimum affinity of the top-level memory cells in the pyramid at the present generation.

Gaussian noise is obtained from a normally distributed random variable which has a mean value of the gene original and a standard deviation value of σ. With this scheme a new gene value can exceed the range at either end if the original gene value is sufficiently far away from the midpoint of the range. These values can be treated as truncating the value back to the endpoint if it is exceeded; that gives a value within the range.

15.3.3 Swapping Mutation

At the middle level of the pyramid framework, we carry out the swapping mutation; this method arbitrarily exchanges any two fragments in a single antibody. An example is shown below.

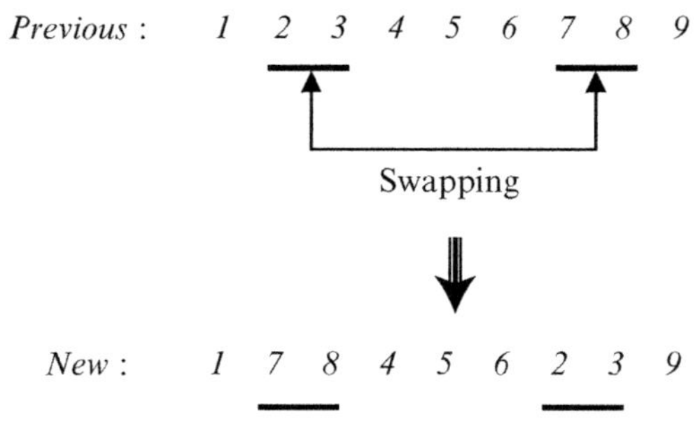

A swapping mutation operator [7] introduces a new antibody to a new population, according to the following rules. First, in the antibody chosen to be mutated, the genes of whose the swapping fragment sites should be situated, are randomly selected. Next, both fragment sites are exchanged. The swapping mutation is performed as long as a new population with the same numbers of antibodies is obtained.

15.3.4 Multipoint Mutation

At the bottom level of the pyramid framework, we execute multipoint mutation; this method arbitrarily replaces multiple random points in elected antibodies. Through multipoint mutation, the new antibodies of population could be more varied. An example is shown below.

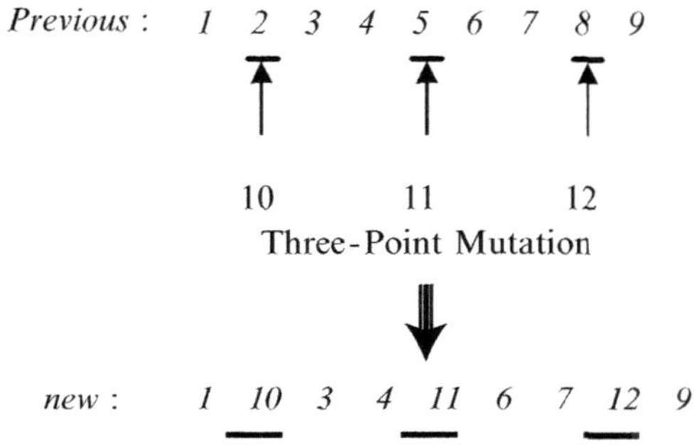

15.4 Cellular Neural Network

The cellular neural network is a brilliant alternative to conventional computers for image processing. In this chapter, the discrete-time cellular neural network (DTCNN) model is considered. Chua et al. [8, 9] have shown the dynamics of each cell described by the following equations.

$$x_{ij}(k+1) = \sum_{c(g,l)\in N_y(i,j)} A_{ij;gl}y_{gl}(k) + \sum_{c(g,l)\in N_u(i,j)} B_{ij;gl}u_{gl}(k) + I \quad (15.3)$$

$$\begin{aligned} y_{ij}(k) &= f(x_{ij}(k)) \\ &= \begin{cases} 1 & \text{if} \quad x_{ij}(k) > 0 \\ -1 & \text{if} \quad x_{ij}(k) < 0 \end{cases} \\ i &= 1,\ldots,\bar{M}\,; \quad j = 1,\ldots,\tilde{N} \end{aligned} \quad (15.4)$$

where x_{ij}, u_{ij}, and y_{ij} denote the state, input, and output of a cell, respectively. The parameters $A_{ij;gl}$ represent the feedback operators which described the interaction between the cell $C(i,j)$ and the output y_{gl} of each cell $C(g,l)$ that belongs to the neighborhood $N_y(i,j)$. Similarly, $B_{ij;gl}$ represents the control operators and the parameter I represents the bias item. These describe the interaction between the cell $C(i,j)$ and the input u_{gl} of each cell $C(g,l)$ within the neighborhood $N_u(i,j)$.

Then, Eqs. 15.3 and 15.4 can be written in vector form by renumbering the cells from 1 to n, with $n = \bar{M} \times \bar{N}$. Therefore, the model of DTCNN can be described as follows,

$$\mathbf{x}(k+1) = Ay(k) + Bu(k) + I \qquad (15.5)$$
$$\mathbf{y}(k) = \mathbf{f}(\mathbf{x}(k))$$

where $\mathbf{x}(k) = [\mathbf{x}_1(k), \ldots, \mathbf{x}_n(k)]^T$ is the state vector, $\mathbf{y}(x) = [\mathbf{y}(x_1), \ldots, \mathbf{y}(x_n)]^T$ is the output vector, $\mathbf{u} = [\mathbf{u}_1, \ldots, \mathbf{u}_n]^T$ is a constant input vector, and $\mathbf{f} = [f(x_1), \ldots, f(x_n)]^T$ is the output functions, whereas the matrices A $\in \Re^{n\times n}$ and B $\in \Re^{n\times n}$ are the known constant feedback matrix and control matrix.

15.5 MCSA-CNN Templates Optimization

We present a heuristic method for the template optimization of the modified clonal selection algorithm-cellular neural network (MCSA-CNN). The modified clonal selection algorithm was inspired by the artificial immune system (AIS) and used to define the basic features of an immune response to an antigenic stimulus, and reach optimization performance for many engineering problems.

In this section, we use MCSA for the automatic template optimization of DTCNN for solving image noise cancellation. Operations performed by an asymptotically stable CNN can be described by a triplet of signal arrays, for example, for images: the input, initial state, and settled output of the network mapped into scale values of pixels. According to the above section, the templates of DTCNN are categorized into three parameters: the feedback matrix A, the control matrix B, and the bias term I. The problem of optimization is to find the optimal template triplet, A, B, and I. These were designed as the following pattern structures.

$$A_{ij;gl} = \begin{bmatrix} a_2 \, a_1 \, a_2 \\ a_1 \, a_0 \, a_1 \\ a_2 \, a_1 \, a_2 \end{bmatrix}, \quad B_{ij,gl} = \begin{bmatrix} b_2 \, b_1 \, b_2 \\ b_1 \, b_0 \, b_1 \\ b_2 \, b_1 \, b_2 \end{bmatrix}, \quad I = i$$

$$Antibody\ type = [a_0, a_1, a_2, b_0, b_1, b_2, i],$$

where a_0, a_1, a_2 are components of matrix A; the rest may be deduced by analogy for B and I. Therefore, the solutions of the problem are represented as string forms: antibodies, constructs of A, B, and I. The training sample consists of the pair input image/desired output shown in Fig. 15.2. The input image is contaminated by uniform random noise and the desired output image is clear. Figure 15.3 shows the diagram of MCSA-CNN.

Step 1. Generating a set Ab of candidate solutions, antibodies, composed of the subset of memory cells $Ab\{m\}$ added to the remaining $Ab\{r\}$ population

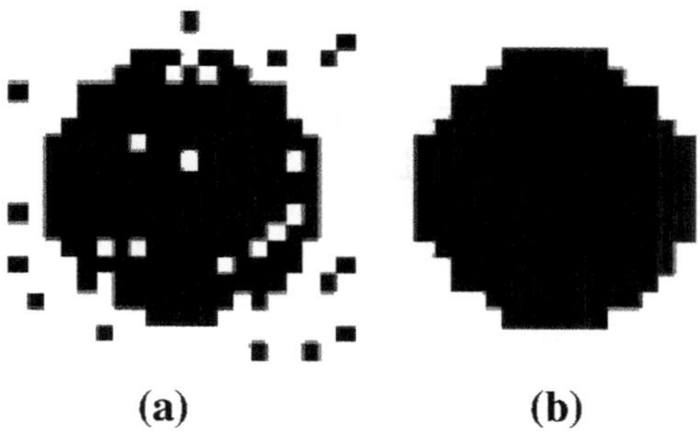

Fig. 15.2 The training samples with 8% noise: **a** input image; **b** desired output

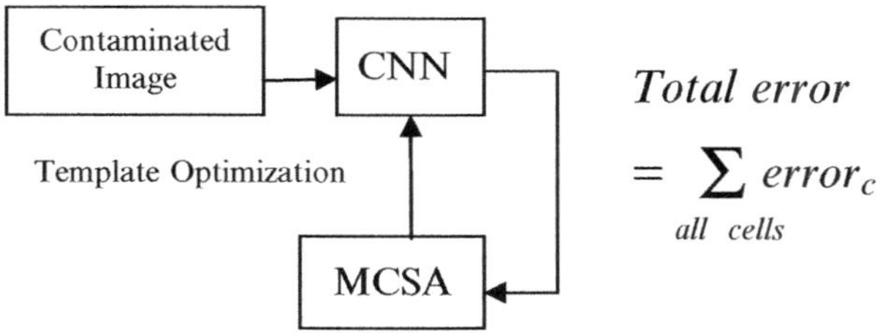

Fig. 15.3 The diagram of MCSA-CNN

$Ab = Ab\{r\} + Ab\{m\}$; and antibodies were indicated that were constituents of the templates.

Step 2. Determining (selecting) the n best individuals of the population $Ab\{p\}$, based on an affinity measure, to organize the pyramid framework; the affinity function is as the following equation presented by Lopez et al. [10].

$$error^{\,c} = (y^c(k_{end}) - y_d^c)^2,$$

where $y^c(k_{end})$ is the output of cell c which depends on the size of the templates and is reached at time interval k_{end}, and y_d^c is the desired output value. The total error is computed over all the cells of the network.

Step 3. Reproducing (cloning) the best individuals of the population in the pyramid, giving rise to a temporary population of clones (C). The clone size is an increasing function of their affinity.

Step 4. Submitting the population of clones to the respective maturation strategy operators, where the self-adaption mutation operations are proportional to their affinity. A maturated antibody population is generated ($C*$).

Step 5. Reselecting the improved individuals from $C*$, if the evolution process is stuck for N times the generation, to compose the memory set $Ab\{m\}$ through the response mechanism; else if not, to compose the memory set $Ab\{m\}$ directly. Some members of Ab can be replaced by other improved members of $C*$.

Step 6. Replace $Ab\{d\}$ antibodies by novel ones (diversity introduction). The lower affinity cells have higher probabilities of being replaced.

Step 7. Circulate from Step 1 to Step 6 until the solutions have satisfied certain conditions in the memory cells.

15.6 Simulation Results

In this section, using CNN with the MCSA approach for image noise cancellation with a 350*350 bipolar computed tomography (CT), images are contrasted with the effects of a Zeng stack smoother.

15.6.1 Example: Elbow Computed Tomography (CT) Image

Zeng [11] introduced a simple stack smoother design procedure based on a Boolean function that can preserve certain structural features of an image, such as straight lines or corners, while taking advantage of the optimal noise reduction properties of the median smoother.

Computed tomography (CT) is a familiar technology in the medical field, and it is often mainted by external interference. First, given is the elbow image in Fig. 15.4, coded by bipolar such that +1 corresponds to black pixels and –1 to white ones. This image was the network input for the DTCNN, through computer simulating; the result for the final output image obtained in Figs. 15.5–15.7, respectively.

Using our proposed MCSA-CNN algorithm, initially, we defined the parameters of our algorithm in Table 15.1.

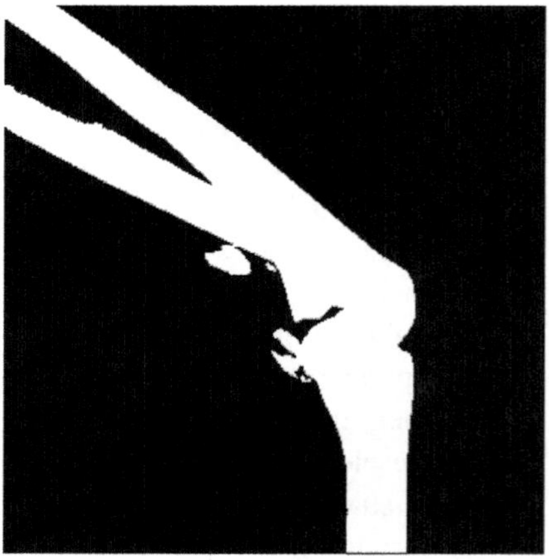

Fig. 15.4 The original elbow image

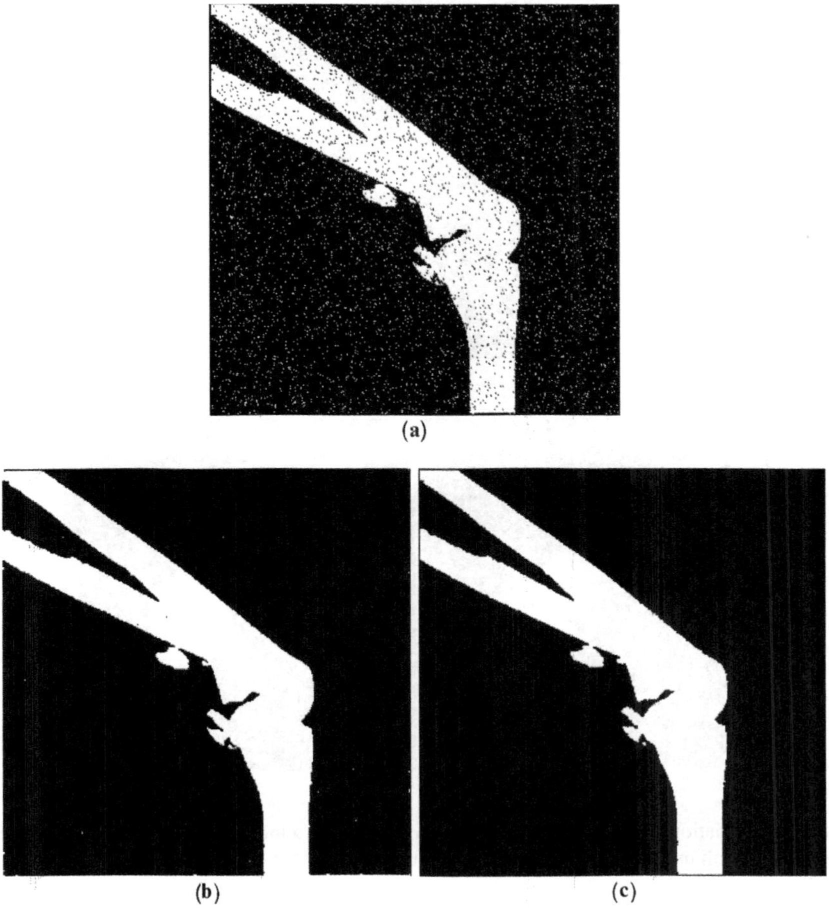

Fig. 15.5 Simulation results of elbow image with 8% noise: **a** the contaminated image with 8% noise; **b** the result using stack smoother; **c** the result using MCSA-CNN

Furthermore, in order to contrast with each state according to the above parameters, several training samples had interference by the salt and pepper noise with different noise density and were simulated by the MCSA-CNN algorithm. Therefore, the corresponding elements of the approximated optimal templates A, B, and bias I for the respective conditions were received in Table 15.2. By combining the above templates, the clarity of the results for the final output image could be obtained. In Figs. 15.5–15.7 are shown the outcomes of the experiment, and the contrasted simulations using the Zeng stack smoother. Relatively, comparing with these consequences, our proposed MCSA-CNN algorithm could effectively restrain the noise of the contaminated image.

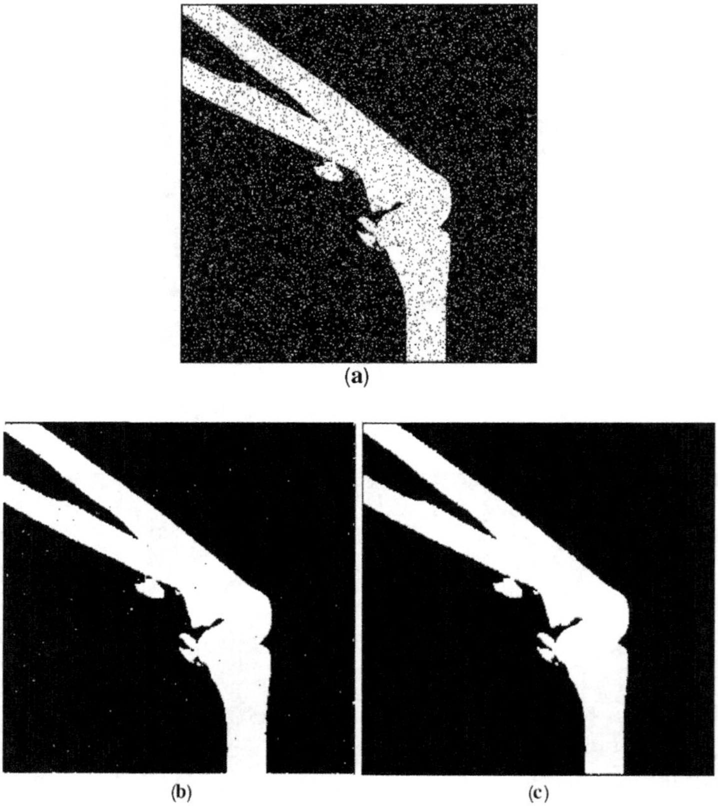

Fig. 15.6 Simulation results of elbow image with 15% noise: **a** the contaminated image with 15% noise; **b** the result using stack smoother; **c** the result using MCSA-CNN

15.7 Conclusion

We proposed a hybrid method MCSA-CNN for image noise cancellation. The optimum corresponding templates of DTCNN have been developed through consecutive generations of MCSA.

The noise of the bipolar contaminated image is effectively retrained using this method. Computer simulations show the advantage of the proposed MCSA-CNN algorithm for image noise cancellation contrasted with the Zeng stack smoother. Moreover, we will research the technique for gray or color image noise cancellation, and enhance the quality of the handled image by the modified hybrid algorithm in future.

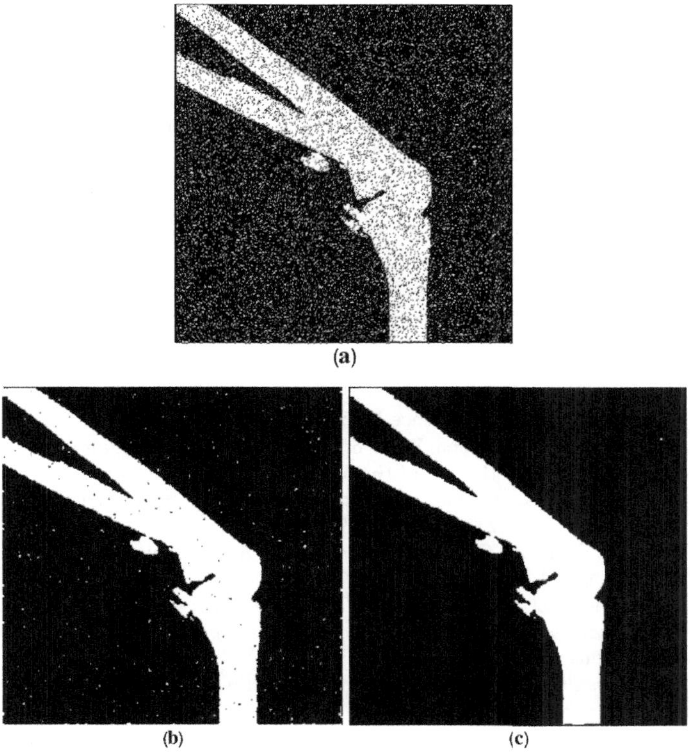

Fig. 15.7 Simulation results of elbow image with 23% noise: **a** the contaminated image with 23% noise; **b** the result using stack smoother; **c** the result using MCSA-CNN

Table 15.1 Established parameters in MCSA

Number of antibodies generated	10
Number of generations	300
Modes of mutation	Maturation Strategy Operators
Mutation probability boundary	Top Level: 0.01∼0.4
	Middle Level: 0.2∼0.6
	Bottom Level: 0.4∼0.8
Percentage of random new antibodies each generation	20%

Table 15.2 The elements of the template A, B, and I for respective conditions

Template		Noise		
		8%	15%	23%
Feedback	a_0	−3.8869	−5.3343	−9.1297
matrix A	a_1	2.5248	5.272	6.4426
	a_2	0.9415	0	2.7171
Control	b_0	5.738	7.6186	9.3077
matrix B	b_1	0	4.6706	5.6713
	b_2	0.6548	2.0759	0
Bias I	i	0	−0.2075	0

References

1. Hajela, P. and Lee, J. "*Constrained genetic search via schema adaptation: An immune network solution*," Structural Optimization, vol. 12, no. 1, pp. 11–15, 1996.
2. Fukuda, T., Mori, K., and Tsukiama, M. "*Parallel search for multi-modal function optimization with diversity and learning of immune algorithm.*" In (Ed.) D. Dasgupta, Artificial Immune Systems and Their Applications, Springer-Verlag, pp. 210–220, 1999.
3. de Castro, L.N. and Von Zuben, F.J. "*Artificial Immune System: Part I—Basic Theory and Application.*" TR-DCA 01/99, 1999.
4. de Castro, L.N. and Von Zuben, F.J. "*Artificial Immune System: Part II—A Survey of Application.*" DCA-RT 02/00, 2000.
5. de Castro, L.N. and Von Zuben, F.J. "*Learning and optimization using the clonal selection principle*," IEEE Transactions on Evolutionary Computation, vol. 6, no. 3, pp. 239–251, 2002.
6. Hinterding, R. "*Gaussian mutation and self-adaption for numeric genetic algorithms*," IEEE Evolutionary Computation Conference 1995, pp. 384–389, 1995.
7. Hong, T.P., Wang, H.S. and Chen, W.C. "*Simultaneously applying multiple mutation operators in genetic algorithms*," Journal of Heuristics, 6, pp. 439–455, 2000.
8. Chua, L.O. and Yang, L., "*Cellular neural networks: Theory.*" IEEE Transactions on Circuits and Systems, vol. 35, pp. 1257–1272, Oct.1988.
9. Chua, L.O. and Yang, L., "*Cellular neural networks: Applications.*" IEEE Transactions Circuits and Systems, vol. 35, pp. 1273–1290, Oct. 1988.
10. Lopez, P., Vilarino, D.L. and Cabello, D. "*Design of multilayer discrete time cellular neural networks for image processing tasks based on genetic algorithms.*" IEEE International Symposium on Circuits and Systems, pp. 133–136, 2000.
11. Zeng, B. "*Optimal median-type filtering under structural constraints*," IEEE Transactions on Image Processing, pp. 921–931, July 1999.

Chapter 16
An Integrated Approach Providing Exact SNP IDs from Sequences

Yu-Huei Cheng, Cheng-San Yang, Hsueh-Wei Chang, Li-Yeh Chuang, and Cheng-Hong Yang

16.1 Introduction

Most of the polymorphisms among genomes are single nucleotide polymorphisms (SNPs). An SNP is a variation of the DNA sequence caused by the change of one nucleotide by another, or insertion or deletion of one or more nucleotides. SNPs provide useful information for personalized medicine [1, 2]. Although many methodologies are reported or reviewed for genetic association studies [3–5], most of the previously reported SNPs are written in nucleotide/amino acid position formats without providing an SNP ID. For example, C1772T and G1790A SNPs in exon 12 of the HIF gene are found to be associated with the renal cell carcinoma phenotype [6], and TNF gene polymorphisms for three SNPs in the TNF gene, at positions −857, −863, and −1031, are reported to be associated with osteoporosis [7]. This anonymous SNP makes the associated SNPs hard to be analyzed or organized systemically.

Recently, NCBI SNP [8] containing a BLAST program for SNPs called SNP-BLAST [9] was developed. SNP-BLAST is designed to perform the BLAST function among various SNP databanks for many species. This BLAST program uses heuristic algorithms, which are less time-consuming and simple, to search for similar sequences across species. Even so, it cannot provide exact SNP IDs by sequences. When using the blastn function of SNP-BLAST with megablast and blastn without megablast to blast a partial sequence, results do not always show the originally entered rs#; even using megablast with IUPAC format sequences often show, "No significant similarity found," such as rs8169551 (rat), rs7288968 (human), rs2096600 (human), and so on. UCSC BLAT [10] uses the index to find regions in the genome likely to be homologous to the query sequence.

It is more accurate and faster than other existing alignment tools. It rapidly scans for relatively short matches (hits), and extends these into high-scoring pairs (HSPs). However, it usually hits so many sequences distributed in different chromosomes that sometimes the result does not show the originally entered rs# in selecting the

Oscar Castillo et al. (eds.), *Trends in Intelligent Systems and Computer Engineering.*

option of the SNPs of the title, "Variation and Repeats," such as rs8167868 (rat), rs2096600 (human), s2844864 (human), and so on. Previously, we utilized a Boyer–Moore algorithm [11] to match sequences with the SNP fasta sequence database for the human, mouse and rat genomes. However, this method does not address the problems of nucleotide change, insertion, or deletion in sequences. It will fail to obtain SNP IDs in the case described above. In other words, in-del (insertion and deletion) sequences were not acceptable.

In order to solve this problem, a dynamic programming method [12] was chosen. However, this method occupies a lot of memory and is time-consuming when applying to the huge human SNP database; therefore it is impracticable. Finally, we took notice of Uni Marker [13] and generated the following idea. We used SNP flanking markers that are extracted from the SNP fasta sequence and then combined the Boyer–Moore algorithm with search markers in the query sequences to identify possible SNPs. Then, we employed dynamic programming to validate these SNPs to obtain exact SNP IDs. The proposed method greatly reduces matched time and memory space. The experimental results show that our proposed approach is efficient, exact, and stable. Thus, it is a valuable approach when identifying SNP IDs from the literature, and could greatly improve the efficiency of systematic association studies.

16.2 Method

This integrated approach is proposed as being effective, stable, and exact. It is based on the SNP fasta database, and uses the Boyer–Moore algorithm and dynamic programming method. The following illustrates the implementation.

16.2.1 The Application of the Boyer–Moore Algorithm

The proposed approach uses a Boyer–Moore algorithm to search for SNP flanking markers in sequences. The Boyer–Moore algorithm usually matches strings from right to left, in contrast to the usual string-matching methods. It is thought to be a faster string-matching algorithm than others.

Boyer–Moore algorithms use a bad-character shift function and a good-suffix shift function. Figure 16.1 describes the process of the Boyer–Moore algorithm's bad-character shift, in which T represents a text, and P represents the pattern to be aligned. As shown in Fig. 16.1(a), P is aligned from left to right: $P(12) = T(13)$, $P(11) = T(12)$, but $P(10) \neq T(11)$, which means the position within P(10) and T(11) is mismatched. By using a bad-character shift rule, the mismatch can be shown to occur in P, in our case P(10). Then, searching from the left of P(10), the same character mismatch is shown for T(11); that is, $P(7) = T(11)$. At this stage, the bad-character shift rule will move the P window and align P(7) to T(11) as shown

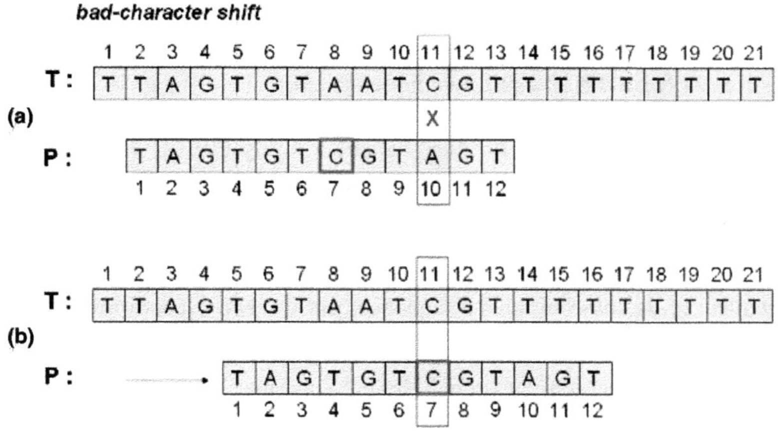

Fig. 16.1 The bad-character shift process

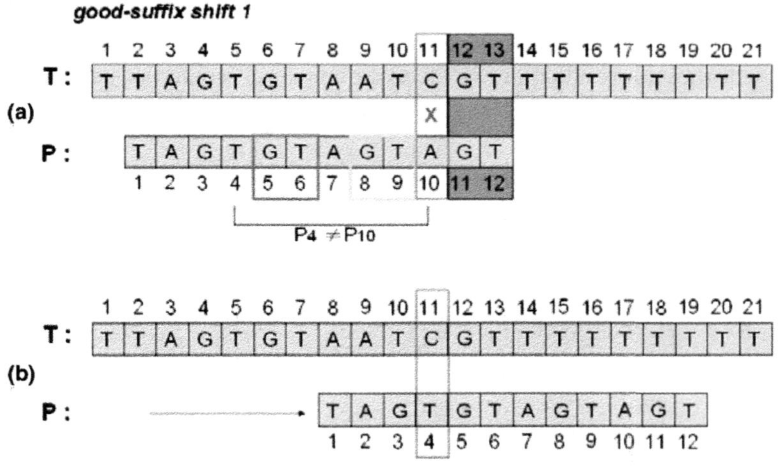

Fig. 16.2 Good-suffix shift1 process

in Fig. 16.1(b). After that, the alignment from right and left of P(12) and T(16) will start again.

The good-suffix shift rule is divided into a good-suffix shift1 and a good-suffix shift2. The process for the good-suffix shift1 is described in Fig. 16.2. In Fig. 16.2(a), P is aligned from right to left, $P(12) = T(13)$, $P(11) = T(12)$, but $P(10) \neq T(11)$. This means that a mismatch is present within P(10) and T(11). Good-suffix shift1 then searches from the right of the P mismatch position, that is, from the right of the character of P(10) and finds the match T(12,13), which is a suffix string of P, P(12,13). Also, the right character of the P suffix string cannot be the same as the mismatch P(11). As shown in Fig. 16.2(a), P(8,9) is the suffix string found, but because $P(7) = P(10)$, the search process continues from the left until $P(5,6)$ and $P(4) \neq P(11)$ are found. The good-suffix shift1 rule will then move the P window and align P(4) to T(11) as shown in Fig. 16.2(b). However, if no suffix string can be found in P, but the prefix string is the suffix substring of the suffix string in P, good-suffix shift2 is implemented.

Figure 16.3(a) shows that P(8) mismatches T(9), and P(9,12) is the suffix string of P. The suffix string P(1,3) matches the suffix string P(9,12); that is, $P(1,3) = P(10,12) = T(11,13)$. Therefore, the good-suffix shift2 rule will move the P window and align P(1) to T(11) as shown in Fig. 16.3(b). After that, alignment from right to left of P(12) and T(22) continues.

When using a Boyer–Moore algorithm to select possible SNPs from the SNP fasta sequences database by query sequence, the following conditions have to be considered.

Condition 1. Sequence-only match SNP flanking marker 3′, but SNP flanking marker 5′ is mismatched. The SNP flanking marker 5′ could possibly appear near the left side of the sequences; it resulted in SNP flanking marker 5′ could not been matched, as shown in Fig. 16.4. This condition will be a candidate of possible SNPs.
Condition 2. Sequence-only match SNP flanking marker 5′, but SNP flanking marker 3′ is mismatched. The SNP flanking marker 3′ may appear at the right side of the sequences; it resulted in SNP flanking marker 3′ could not be matched, as shown in Fig. 16.5. This condition will be a candidate of possible SNPs.

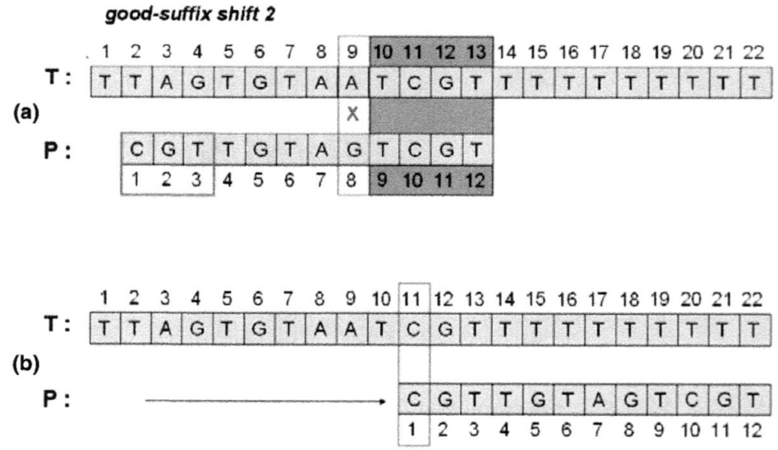

Fig. 16.3 Good-suffix shift2 process

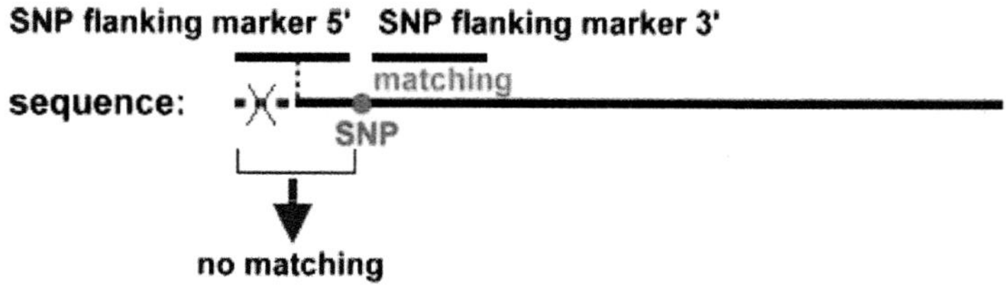

Fig. 16.4 Sequence-only matches to SNP flanking marker 3′

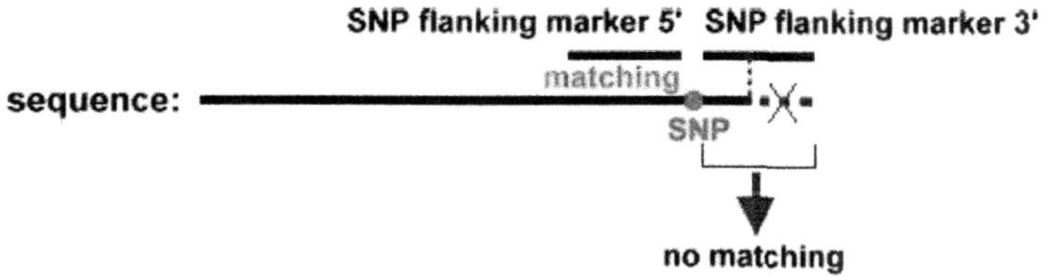

Fig. 16.5 Sequence-only matches to SNP flanking marker 5′

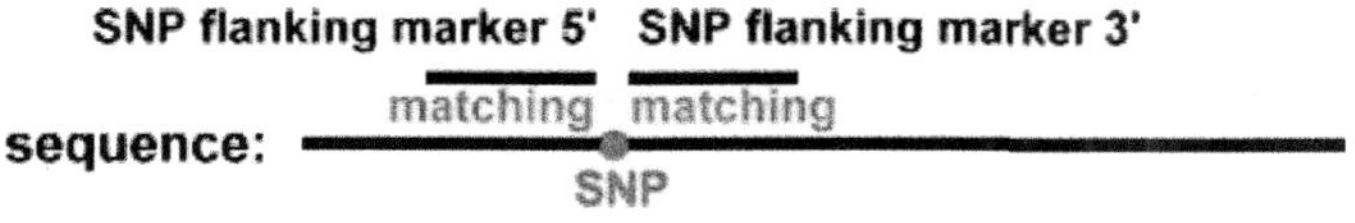

Fig. 16.6 SNP exists within sequence

Fig. 16.7 SNP does not exist within sequence, because of the distance of the matched SNP flanking markers

Fig. 16.8 SNP does not exist within sequence, because the orientation and distance of the matched SNP flanking markers

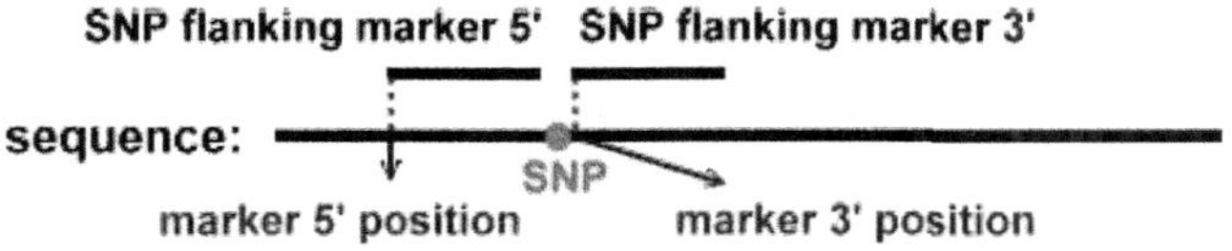

Fig. 16.9 Discriminable criterion for possible SNPs

Condition 3. Sequence matches to SNP flanking marker 5′ and SNP flanking marker 3′. In this case, two possibilities exist: (a) a SNP exists within the sequences, as shown in Fig. 16.6. It will be a candidate of possible SNPs. (b) A SNP does not exist within the sequences, but SNP flanking markers exist, as shown in Figs. 16.7 and 16.8. In Figs. 16.7 and 16.8, the SNP flanking marker 5′ and the SNP flanking marker 3′ are separated from each other, so the existence of an SNP is impossible. We eliminate it from the candidates of possible SNPs.

Possible SNPs will be selected by a criterion. The discriminable criterion is presented below and illustrated in Fig. 16.9.

$$\text{if } ((\text{marker } 5' \text{ position} + \text{marker } 5' \text{ length} + 1) = \text{marker } 3' \text{ position}) \qquad (16.1)$$

If Eq. 16.1 is confirmed, the sequence will possibly contain an SNP corresponding to one of the SNP fasta sequences database. The "+1" of this Eq. 16.1 represents the base of the SNP.

16.2.2 Revision of SNP Flanking Marker

Because of the exact character matching of a Boyer–Moore algorithm, we must consider three conditions when applying SNP flanking markers. These conditions are illustrated below.

Condition 1. SNP flanking marker 5′ has one SNP and upward in it, which will result in mismatch using Boyer–Moore algorithm. And the SNP flanking marker 3′ is at the right side of the sequence and mismatched. It is illustrated in Fig. 16.10. This condition is, "Not any SNPs found."
Condition 2. SNP flanking marker 3′ has one or more SNPs in it, which will result in a mismatch using the Boyer–Moore algorithm. And the SNP flanking marker 5′ is at the left side of the sequence and mismatched. It is illustrated in Fig. 16.11. Again, no SNPs are found in this condition.
Condition 3. Both SNP flanking marker 5′ and SNP flanking marker 3′ contain SNPs within them. This will result in no markers to match using the Boyer–Moore algorithm, but actually SNP markers exist in sequence as shown in Fig. 16.12. Still no SNP is found.

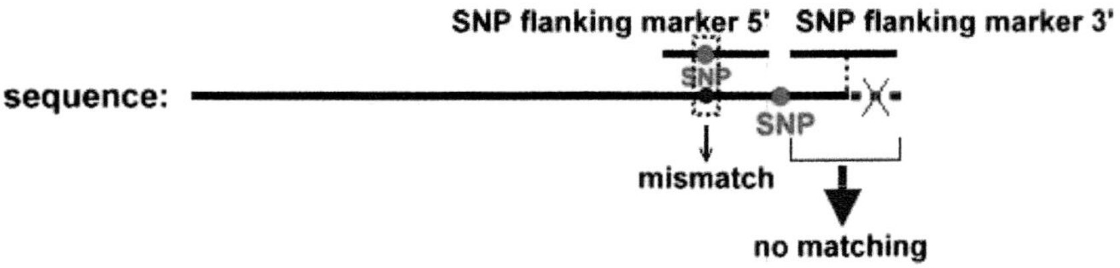

Fig. 16.10 SNP flanking marker 5′ contains SNPs in it and SNP flanking marker 3′ is not matched to the sequence; no SNPs found

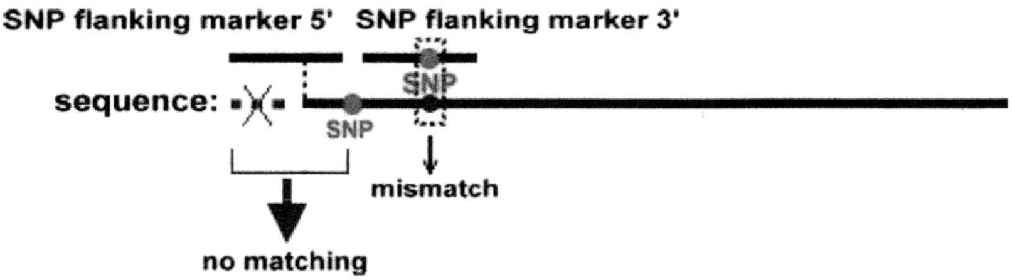

Fig. 16.11 SNP flanking marker 3′ contains SNPs in it and SNP flanking marker 5′ is not matched to the sequence; also no SNPs found

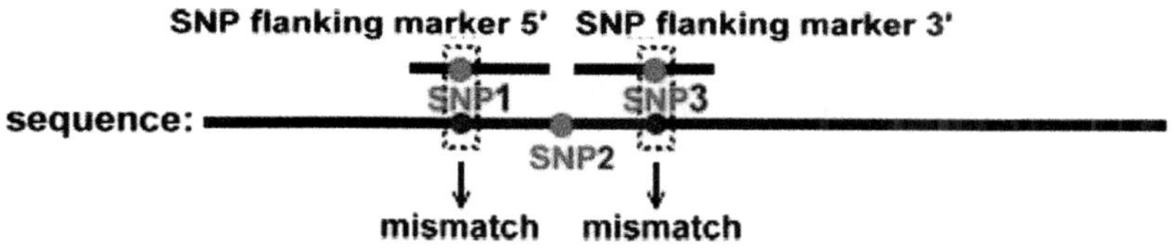

Fig. 16.12 Both SNP flanking marker 5′ and SNP flanking marker 3′ contain SNPs within them, but no SNP is found

Table 16.1 Example of the revised SNP flanking marker table

SNPs	SNP flanking marker 5′	SNP flanking marker 3′
SNP1	None	SNP2
SNP2	SNP1	SNP3
SNP3	SNP2	None

In order to improve the above faults, we constructed a revised SNP flanking marker table. It uses the SNP chromosome position from dbSNP to find existing SNPs within the SNP flanking marker 5′ and SNP flanking marker 3′. For example, under Condition 3 shown in Fig. 16.12, the flanking marker 5′ of SNP2 contains SNP1 and flanking marker 3′ of SNP2 contains SNP3, respectively. A search process for the flanking markers of SNP2 using the Boyer–Moore algorithm will result in a failure. Therefore, we revised the SNP flanking marker table to correct the condition. As shown in Table 16.1, the flanking marker 5′ of SNP2 contains SNP1 and the flanking marker 3′ of SNP2 contains SNP3. In this case, the SNP will be considered a possible SNP.

16.2.3 Alignment Using Dynamic Programming

Through the steps described above, possible SNPs within a query sequence can be retrieved. However, the query sequence must match with the fasta sequences; only matched SNP flanking markers cannot prove the existence of an SNP in sequences. If nucleotide bases outside the SNP flanking marker cannot be matched to the SNP fasta sequences, the above effort is futile. The SNP flanking marker is too short to make a complete estimate.

Consequently, we employ a dynamic programming method to match with fasta sequences of the possible SNPs in order to discover valid SNPs. The dynamic programming method contains an error-tolerant function which resolves problems associated with changes, insertions, or deletions in sequences. The corresponding SNP fasta sequences will provide the SNP ID. It works as follows. First, the SNP fasta sequences and the input sequences of the suffix edit distance $E(i, j)$ is calculated.

Suppose Tj is the SNP fasta sequences, $j = 1,2,\ldots,n$, where n is the SNP fasta sequences' length. P_i is a user's input sequences, $i = 1,2,\ldots,m$, and m is the user's input sequences length. The procedure for the suffix edit distance is given below.

```
The procedure for the suffix edit distance:
// initialization
for i ← 0 to m do
    E(i, 0) ← i
next i
for j ← 0 to n do
    E(0, j) ← 0
next j

// suffix edit distance E(i, j)
for i ← 1 to m do
    for j ← 0 to n do
        if (T(j) = P(i)) then
                E(i,j) ←(i-1, j-1)
        else
                min ← MIN[E(i-1, j), E(i, j-1)]
                E(i, j) ← min+1
        end if
    next j
next i
return E(i,j)
```

In order to obtain partially homologous sequences, the maximum tolerance error rate for the input sequences is accepted. Once the error count is equal to or smaller than the maximum tolerance error rate, the input sequences are aligned successfully to the SNP fasta sequences.

$$\text{Maximum\_tolerant\_error\_number} = (\text{input\_sequences\_length}) \times (\text{tolerant\_error\_rate}) \tag{16.2}$$

The homologous sequences can be found by using previously obtained suffix edit distances $E(i, j)$ and the maximum tolerance error number based on backward dynamic programming. Once the suffix edit distance $E(i, j)$ is smaller than or equal to the maximum tolerance error number, it is processed. The backward sequences are the homologous sequences that fit with the analogue. For example, if input sequences contain the bases (nucleotides) TAGC, the maximum tolerance error rate is 20%. When the input sequences are aligned with SNP fasta sequences of 10 bps, for example, TGGATACCAT, the maximum tolerance error number is $10 \times 0.2 = 2$. In other words, only two or fewer error alignments are allowed in this case (Fig. 16.13). The boldface arrows in Fig. 16.13 indicate the output of an agreeable homologous alignment; the homologous sequences are (1)TG, (2)TGG, (3)TGGA, and (4)TA.

		T	G	G	A	T	A	C	C	A	T
	0	0	0	0	0	0	0	0	0	0	0
T	1	0	1	1	1	0	1	1	1	1	0
A	2	1	2	2	1	1	0	1	2	1	1
G	3	2	1	2	2	2	1	2	3	2	2
G	4	3	2	1	2	3	2	3	4	3	3
			(1)	(2)	(3)		(4)				

Fig. 16.13 Homologous alignment and possible homologous sequences

16.3 Results and Discussion

This research utilizes the NCBI SNP [14] rs_fasta sequences database, which contains the human (ftp://ftp.ncbi.nih.gov/snp/organisms/human_9606/), mouse (ftp://ftp.ncbi.nih.gov/snp/organisms/mouse_10090/), and rat (ftp://ftp.ncbi.nih.gov/snp/organisms/rat_10116/) genomes. To implement the proposed method, a SNP flanking marker database must be built with data from the SNP fasta sequences database. In order to ensure that exact SNP IDs can be found, selection of the length of the SNP flanking marker is important.

When using shorter SNP flanking markers, possible SNPs are more rapidly identified by using the Boyer–Moore algorithm, but many of the select SNPs are insignificant. These insignificant SNPs will increase the load for the following process of determining exact SNP IDs. Longer SNP flanking markers will fail to obtain SNP IDs using the Boyer–Moore algorithm, because the sequence may contain changes, that is, an insertion or a deletion, or long markers may contain SNPs with high frequency.

Therefore, this research adopted a length of 10 bps of SNP flanking sequences of the fasta database as a standard for the SNP flanking marker length. Although the marker length influences the matching results, it is compensated by the revised SNP flanking marker table that we introduce in the following. The chromosome position of the table SNPContigLoc in dbSNP [8] b126 was employed to find SNPs within the SNP flanking marker, and then build the revised SNP flanking marker table.

The proposed approach using Microsoft Windows XP, a 3.4 G MHZ processor, 1 GB of RAM memory, and JRE (Java Runtime Environment) with a maximum JAVA heap size of 800 MB to discover SNP rs28909981 [Homo sapiens]. We mainly aimed at the following three sequences.

- Sequence 1.

AAGAGAAAGTTTCAAGATCTTCTGTSTGAGGAAAATGAATCCACAGCTCTA

- Sequence 2.

AAGAGAAAGTTTCAAGATCTTCTGTCTGAGGAAAATGAATCCACAGCTCTA

- Sequence 3.

AAGAGAAAGTTTCAAGATCTTCTGTGTGAGGAAAATGAATCCACAGCTCTA

1. For test sequence 1, we set the dynamic programming method with error tolerant bases = 0. rs28909981 was successfully identified and had 27 SNP flanking marker matches. Runtime was 2844 ms.
2. For test sequence 2, we set the dynamic programming method with an error tolerant bases = 1, because the C allele was mismatched with the SNP in fasta sequence. rs28909981 and rs17883172 were identified and had 36 SNP flanking marker matches. Runtime was 3313 ms. rs17883172 is similar to rs28909981. The rs17883172 sequence was as follows.

   GAGAAAGTTTCAAGATCTTCTGTCTRAGGAAAATGAATCCACAGCTCTACC

   The C allele represents SNP rs28909981. We searched rs28909981 successfully and discovered SNP rs17883172 in this sequence.
3. For test sequence 3, we set the dynamic programming method with error tolerant base = 1, because the G allele is mismatched with the SNP in fasta sequence. The result found rs28909981 successfully and had 34 SNP flanking marker matches. Runtime was 3141 ms.
4. For test sequence 1, we adjusted the dynamic programming method with error-tolerant bases = 5. rs28909981 and rs17883172 could be found, and 27 SNP flanking marker matches were identified. Runtime was 2750 ms. We also discovered that test sequences 2 and 3 with error-tolerant bases = 5 still find rs28909981 and rs17883172.

The above results show that the presented approach indeed provides exact SNP IDs from sequences. The advantages of this approach are effective, stable, and exact. It seeks through the SNP fasta database and only aims at a specific database. By this property, it can decrease unknown errors and perform more exact output. The proposed approach can be used for specialized application of SNP IDs discovery. It will help biologists to find SNP IDs in sequences and have the chance to find invalidated SNPs. It would be useful for biologists in association studies.

16.4 Conclusion

SNPs are very useful for the application of personalized medicine. In order to identify SNPs in sequences, this research proposes the use of SNP flanking markers and combines it with a Boyer–Moore algorithm with dynamic programming to provide exact SNP IDs from sequences. It is mainly built of dbSNP, SNP fasta, and SNP flanking sequences of 10 bps for the rat, mouse, and human organisms from NCBI, and improves on methods we previously proposed. After implementation, verified SNP IDs could be obtained from sequences in a fast and efficient way. This integrated approach constitutes a novel application to identify SNP IDs, and can be used for systematic association studies.

References

1. Erichsen HC, Chanock SJ: SNPs in cancer research and treatment. Br J Cancer 2004, 90(4):747–751.
2. Suh Y, Vijg J: SNP discovery in associating genetic variation with human disease phenotypes. Mutat Res 2005, 573(1–2):41–53.
3. Lunn DJ, Whittaker JC, Best N: A Bayesian toolkit for genetic association studies. Genet Epidemiol 2006, 30(3):231–247.
4. Newton-Cheh C, Hirschhorn JN: Genetic association studies of complex traits: Design and analysis issues. Mutat Res 2005, 573(1–2):54–69.
5. Su SC, Kuo CC, Chen T: Inference of missing SNPs and information quantity measurements for haplotype blocks. Bioinformatics 2005, 21(9):2001–2007.
6. Ollerenshaw M, Page T, Hammonds J, Demaine A: Polymorphisms in the hypoxia inducible factor-1alpha gene (HIF1A) are associated with the renal cell carcinoma phenotype. Cancer Genet Cytogenet 2004, 153(2):122–126.
7. Furuta I, Kobayashi N, Fujino T, Kobamatsu Y, Shirogane T, Yaegashi M, Sakuragi N, Cho K, Yamada H, Okuyama K, et al.: Bone mineral density of the lumbar spine is associated with TNF gene polymorphisms in early postmenopausal Japanese women. Calcif Tissue Int 2004, 74(6):509–515.
8. Sherry ST, Ward MH, Kholodov M, Baker J, Phan L, Smigielski EM, Sirotkin K: dbSNP: The NCBI database of genetic variation. Nucleic Acids Res 2001, 29(1):308–311. [http://www.ncbi.nlm.nih.gov/SNP/]
9. SNP BLAST. [http://www.ncbi.nlm.nih.gov/SNP/snp_blastByOrg.cgi]
10. Kent WJ: BLAT—The BLAST-like alignment tool. Genome Res. 2002, 12: 656–664.
11. Charras C, Lecroq T: Handbook of Exact String Matching Algorithms, King's College London Publications, 2004.
12. Eddy SR: What is dynamic programming? Nat Biotechnol 2004, 22(7):909–910.
13. Leslie YY, Chen, S-H, Shih ESC, Hwang M-J: Single nucleotide polymorphism mapping using genome-wide unique sequences. Genome Res. 2002 12: 1106–1111.
14. Altschul SF, Gish W, Miller W, Myers EW, Lipman DJ: Basic local alignment search tool. J. Mol. Biol. 1990, 215:403–410.

Chapter 17
Pseudo-Reverse Approach in Genetic Evolution

Sukanya Manna and Cheng-Yuan Liou

17.1 Introduction

Important insights into evolutionary processes can be determined by the rates of substitution in protein-coding sequences. Increase in the availability of coding sequence data has enabled researchers to estimate more accurately the coding sequence divergence of pairs of organisms. The use of different data sources, alignment protocols, and methods to estimate the substitution rates leads to widely varying estimates of key parameters that define the coding sequence of orthologous genes.

The rates of molecular evolution generally vary among lineages. Different studies have predicted that the source of this variation has differential effects on the synonymous and nonsynonymous substitution rates [3]. Changes in generation length or mutation rates are likely to have an impact on both the synonymous and nonsynonymous substitution rates. Hence, the number of substitutions per site between nucleotide sequences has become one of the most fundamental quantities for molecular evolution studies. It provides a valuable means for characterizing the evolutionary divergence of homologues. Thus accurate quantification of genetic evolutionary distances in terms of number of nucleotide substitutions between two homologous DNA sequences is an essential goal in evolutionary genetics. When two coding regions are analyzed, it is important to distinguish between the numbers of synonymous and nonsynonymous nucleotide substitutions per site. Estimation of calculation of these rates is not very simple; several methods have been developed to obtain these estimates from a comparison of two sequences [4, 5]. The early methods have been improved or simplified by many authors [1, 6–9]. Those methods follow almost the same strategy. The numbers of synonymous (S) and nonsynonymous (N) sites in the sequence and the numbers of synonymous (Sd) and nonsynonymous (Nd) differences between the two sequences are counted. Corrections for multiple substitutions are then applied to calculate the numbers of synonymous (ds) and nonsynonymous substitutions per site (dn) between two sequences. These methods assume an equal base and codon frequencies.

Oscar Castillo et al. (eds.), *Trends in Intelligent Systems and Computer Engineering.*

Enzymes being protein in nature, belong to the subset of existing proteins. Hence, we believe that like other proteins they too play an important role in the evolutionary process. So, we have used them here for this case study. The approach used here is pseudo-reverse in the sense that we converted the amino acid sequences of the respective genes for the enzymes back to the nucleotide sequences based on the cumulative probability of the codons of the genomes of the species taken into account. We then applied comparative genomics and the nucleotide substitution process to analyze and test this experiment. Comparative genomics is applied to align the sequences of each species' pairs: human, mouse, and rat.

17.2 Methods

17.2.1 Assumptions

This work proceeded on the basis of three major assumptions. First, mammalian species, such as human and mouse share a vast majority of their genes [10, 11]. Second, most genes are subject to much stronger selective constraints on nonsynonymous changes than on synonymous ones [12, 13]. Finally, the genes found for an enzyme for a species are closely related to one another. The first two are common assumptions with [14] about comparative genomics.

Nei and Gojobori's model is the simplest model for nucleotide substitution schemes. Hence, we have used this along with Jukes and Cantor's model to discover the nucleotide substitution rates. We implemented a much generalized model of the above-mentioned algorithms. The unweighted version of Nei and Gojobori's model is used for estimating the number of synonymous and nonsynonymous substitutions. Here, instead of using the transition matrix or codon substitution matrix, we directly calculated the aligned codon positions to compute this. Besides this, previously used models used a phylogenetic approach of codon comparison [15], but we used here simple codon-by-codon comparison of the two sequences using a sliding buffer of the length of three characters. We estimated the divergence time for the species pairs using the formula $E = K/2T$, where E is rate of evolution, T is species' divergence time, and K is base pair substitutions per site [16].

17.2.2 Approach

We basically collected a set of enzymes from the enzyme database BRENDA.[1] Then we used the Swiss-Prot knowledgebase[2] to collect the related genes' amino acid sequences for each enzyme for each of these three species. For this case study, we

[1] BRENDA: Enzyme database; electronically available at http://www.brenda.uni-koeln.de/.

[2] Swiss-Prot: Protein knowledgebase; electronically available at http://ca.expasy.org/sprot/.

considered only those enzymes for which we found valid genes in all three species. We then filtered out the data by separating the amino acid sequences having terms such as fragments, precursors, clones, and dominants and kept the mostly related sequences with respect to the enzymes considered. We assumed that the amino acid sequences obtained for each enzyme share a great similarity as do the genes belonging to the same group. Instead of finding out the conserved regions between two species, we found out the least mismatch in their amino acid sequences for respective enzymes for each species' pair. We then collected those amino acid sequences which satisfy this condition for each of the species pairs. In fact we believe that, the more similarity between the sequences, the less is the mismatch between their amino acid sequences. We used here the amino acid sequences that have multiple numbers of least mismatches. We present here the two different approaches. The first one is best matched pairs and the other one is all pairs with least mismatch.

Let $H = [h_1, h_2, \ldots, h_n]$ be the set of genes for human, $M = [m_1, m_2, \ldots, m_m]$ be the set of genes for mouse, and $R = [r_1, r_2, \ldots, r_k]$ be the set of genes for rat, and n, m, and k are the number of genes found in each species, respectively, for each enzyme. Suppose h_1m_1, h_1m_3, h_2m_2, h_1r_2, h_2r_5, h_1r_1, m_1r_1, m_1r_2, and m_2r_6 have the least mismatch in their sequences when compared among species pairs. Now, for best matched pairs, we have considered h_1m_1 for human–mouse, m_1r_2 for mouse–rat, and h_1r_2 for human–rat analysis thus forming the trio of $h_1m_1r_2$. We checked for the common ones as shown. If there is more than one best matched pair, we then choose any one randomly forming the trio. Then we generate the sequences for the ones belonging to this category (e.g., h_1, m_1, and r_2, respectively, for that particular enzyme). Then for all pairs with least mismatch, we use all these pairs for the specieswise sequence comparison for estimating the nucleotide substitution rates. So, for accomplishing this, we generated the random nucleotide sequences for the amino acids h_1, h_2, m_1, m_2, m_3, r_1, r_2, r_5, and r_6, respectively, for that particular enzyme.

The role of the pseudo-reverse mechanism comes into the picture when we convert the amino acid sequences back to the nucleotide sequences. But the conversion of all possible sequences was an absurd idea to be accomplished because of very high time as well as the space complexity. So we retrieved the total frequency of all the codons from the genomes of each species separately. Later we calculated the cumulative probability of the codons from the frequency obtained, and generated the random nucleotide sequences for all the amino acid sequences having the least mismatch for a particular enzyme. Figure 17.1 shows the frequencies of codons

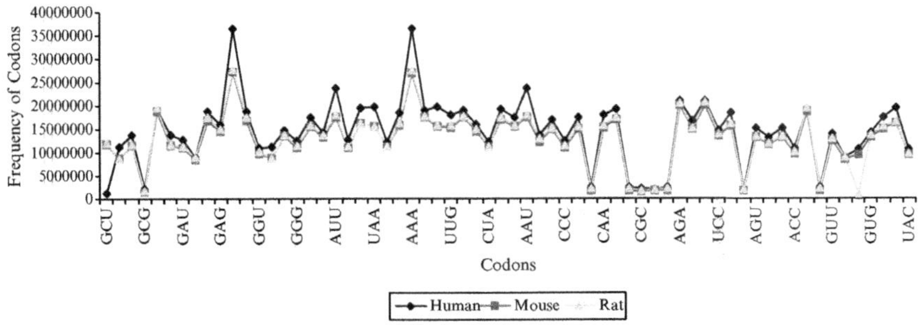

Fig. 17.1 Frequency of the codons obtained from the genome of each species

obtained. We generated 100 sequences for each of these amino acid sequences, because we were aware of the false positive and false negative outcomes. Next we compared these random sequences species' pairwise (such as human–mouse, mouse–rat, and mouse–rat, respectively) to calculate the dn/ds ratio as mentioned earlier. There were 10,000 possible comparisons for each pair per enzyme proteins (or genes). Out of these some returned valid results and others could not due to very low count of synonymous substitutions per site. We then plotted the graphs based on the valid results of the enzymes obtained.

17.3 Experimental Results

In this section, the results obtained from this work are illustrated in detail. Figures 17.2–17.4 illustrate the variation of the dn/ds ratio with different enzymes along with the species pair comparison for the experiments with best matched pairs. Here the numbers in brackets along the x-axis denote the number of codons compared for each case. Figures 17.5–17.10 are the illustrations for all pairs with least

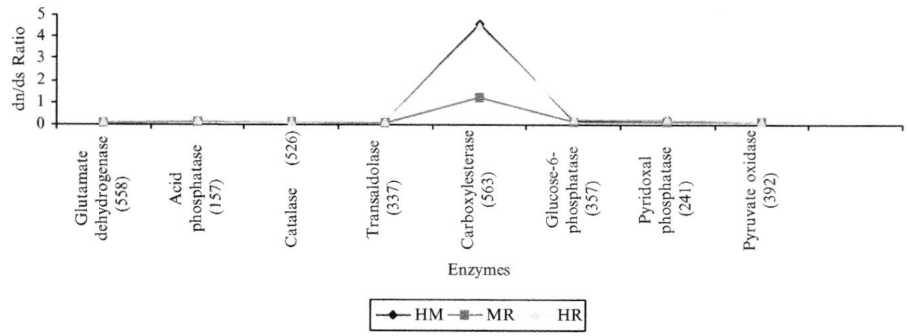

Fig. 17.2 dn/ds Ratio of the human–mouse, mouse–rat, and human–rat comparison for the enzymes common in all

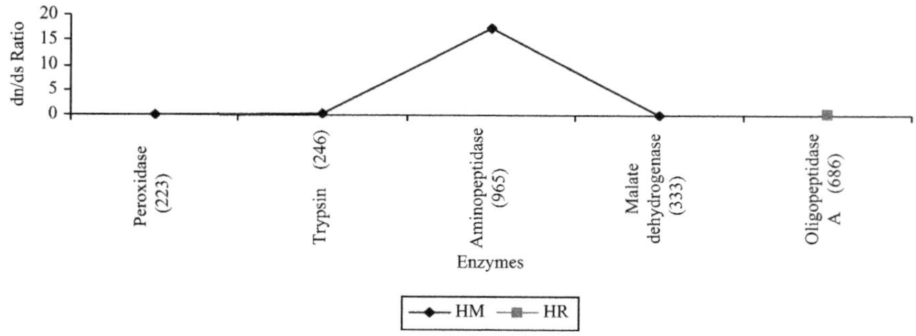

Fig. 17.3 dn/ds Ratio of human–mouse and mouse–rat comparison for the enzymes not common in them

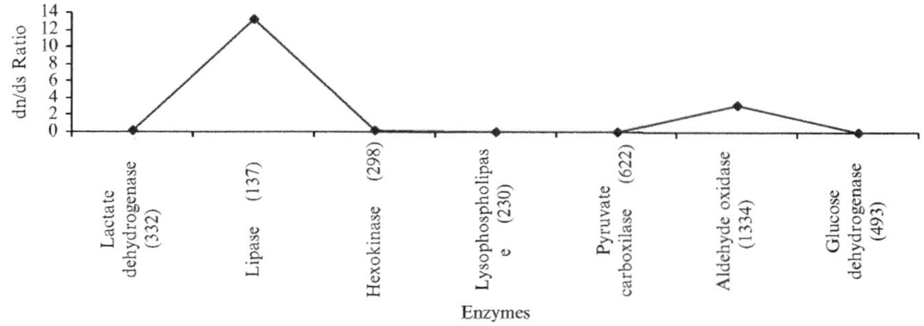

Fig. 17.4 Valid dn/ds ratio of the mouse–rat comparison for the enzymes found only in these two species but not human

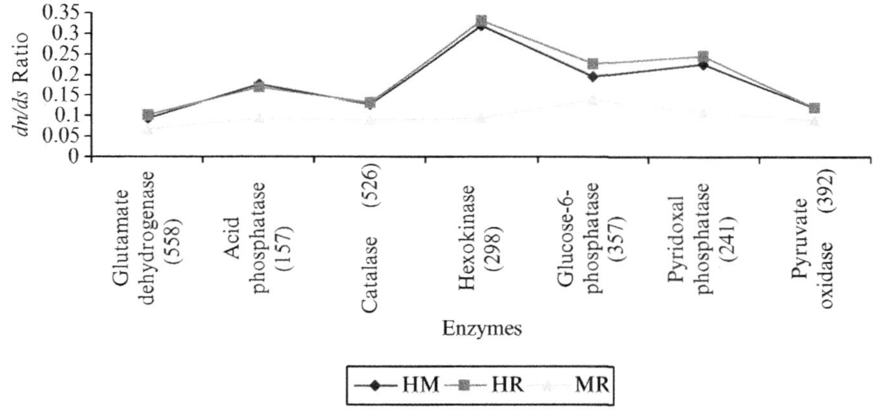

Fig. 17.5 Comparison between dn/ds ratio of the enzymes common in all

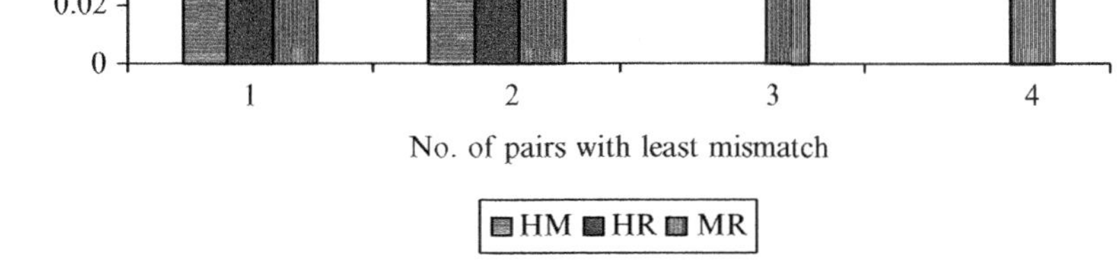

Fig. 17.6 Comparison between dn/ds ratio of the enzyme transaldolase for all pairs with least mismatches

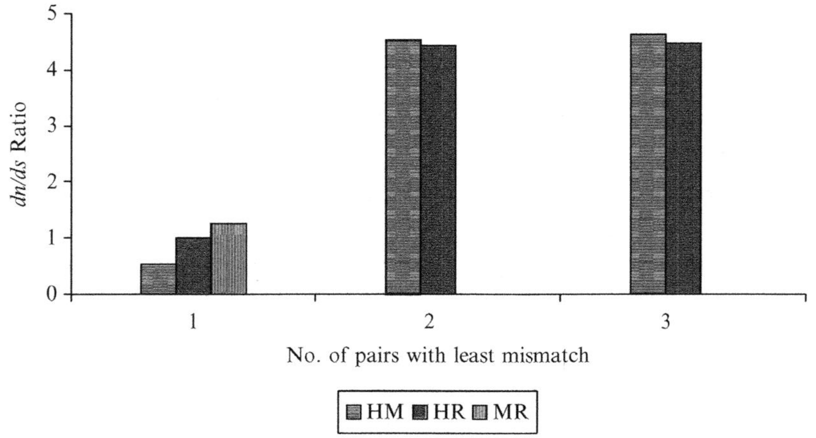

Fig. 17.7 Comparison between dn/ds ratio of the enzyme carboxylesterase for all pairs with least mismatches

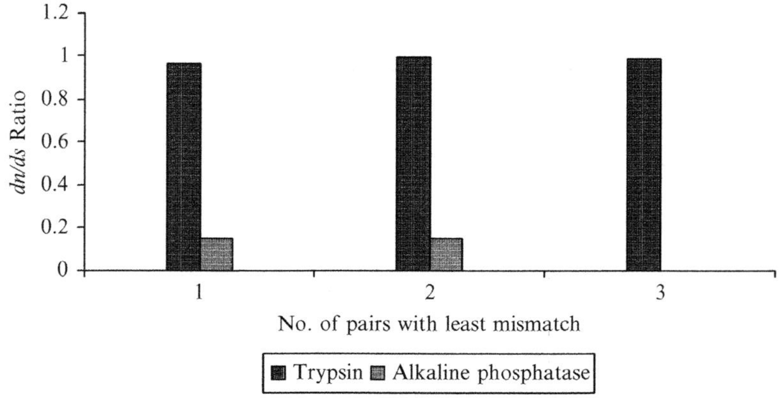

Fig. 17.8 dn/ds Ratio for the enzymes in HM having more than one least mismatch

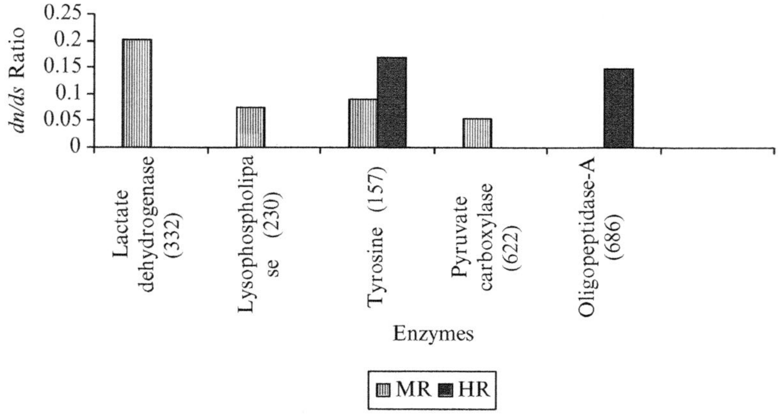

Fig. 17.9 dn/ds Ratio of species pairs with purifying result

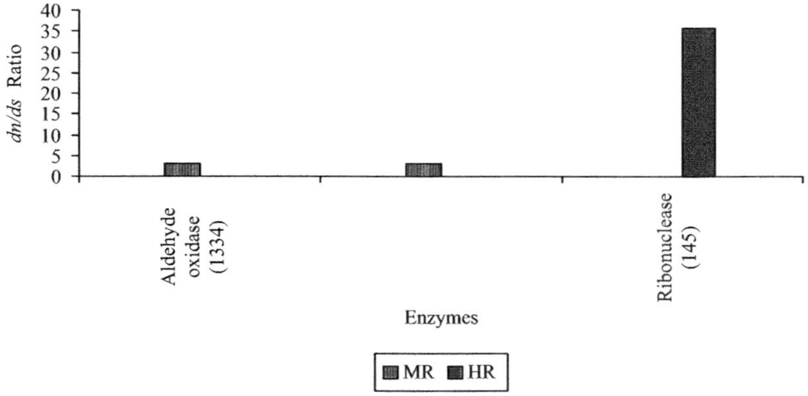

Fig. 17.10 dn/ds Ratio of species pairs with diversifying result

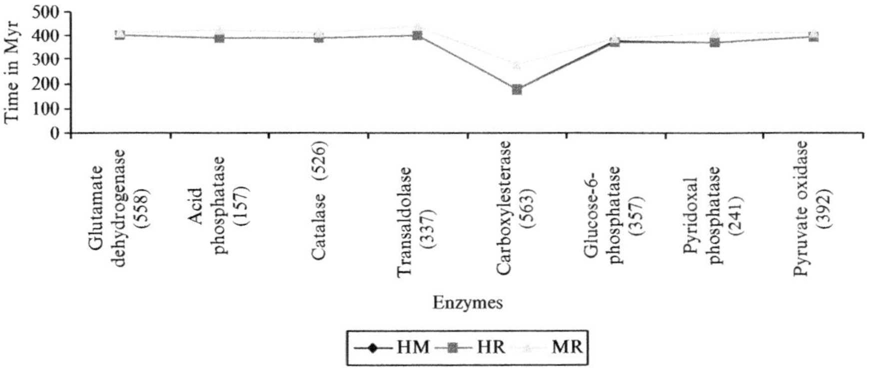

Fig. 17.11 Estimated time for amino acid substitution per site for the enzymes common in all the three species

mismatch. The last two figures, that is, Figs. 17.11 and 17.12 depict the estimated time that we have obtained from this study. Abbreviations such as HM, MR, and HR signify Human–Mouse, Mouse–Rat, and Human–Rat species' pairwise sequence comparisons respectively.

In Figure 17.2, the enzymes found in all three species are used to plot the valid dn/ds ratio for them. Here, we see that except for carboxylesterase all the other enzymes showed the same results as the protein coding exons where the dn/ds ratio is less than one, which is the normal expected behaviour. These two enzymes deviate highly from the neutral theory of evolution whose ratio exceeds 1. The mouse and rat's comparison with human shows almost similar results. But still its behaviour deviated with the already mentioned enzymes when they are compared. Figures 17.3 and 17.4 show the behaviour of the enzymes that were not common among the three species. We again notice that the enzyme aminopeptidase, in Fig. 17.3, deviates in the same manner as already shown in Fig. 17.2 for some other enzymes. We observe

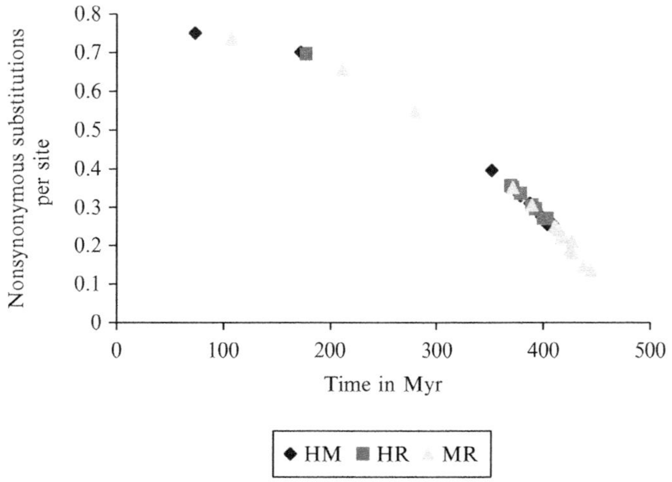

Fig. 17.12 Estimated time for amino acid substitutions per site for enzymes

a similar result for lipase and aldehyde oxidase in Fig. 17.4. For the mouse and rat comparisons, we found the similarity between the enzymes to be more.

Figures 17.5–17.10 illustrate the variation of the dn/ds ratio with different enzymes along with the species pair comparison. Now, in Fig. 17.5, the diagram clearly depicts the behaviour of the enzymes in different species pair comparisons. The ratio for the HM and HR is almost consistent here for these enzymes, but varies in the case of MR. All these show purifying selection. Figures 17.6 and 17.7, show similar kinds of results for two different enzymes, transaldolase and carboxylesterase, respectively. But the former shows purifying selection and the latter shows diversifying selection for the corresponding species pair comparisons. In both cases, we found more than one least mismatch in their corresponding amino acid sequences. Now, Fig. 17.8 shows the enzymes found only in HM comparison for which we got a valid result, but not in the other pairs.The enzymes trypsin and alkaline phosphatase belong to this category. Figure 17.9 shows the comparison between the dn/ds ratio for the enzymes found only in the MR and HR species pairs. Both of them show purifying selection as the dn/ds ratio is less than one. On the other hand, we see a drastic change in Fig. 17.10. It shows diversifying selection and the value for the enzyme ribonuclease, which shows a very high value. This means that the genes taken into account for this enzyme vary a lot in their behaviour. Here, we plotted the individual cases for MR and HR, respectively, like disjoint sets.

In Figs. 17.11 and 17.12, the estimated divergence time for human and rat/mouse in cases of enzyme proteins seems to be five times higher than ordinary proteins which is around 80 Myr [16]. The estimated range is ~400 Myr. Figure 17.8 shows the variation for the enzymes found in all three species' pair comparisons. It is clearly seen from Fig. 17.7, the amino acid replacements take longer times for all the species considered here in the cases of these enzymes.

Table 17.1 Comparison between already established work and our approach[a,b]

Enzymes	Li's approach		Our approach			
	Codons compared	dn/ds Ratio (H-M/R)	Codons compared	dn/ds Ratio (H-M)	Codons compared	dn/ds Ratio (H-R)
Ald A	363	0.03	363	0.10	NVR	NVR
Cr ki M	380	0.06	381	0.10	381	0.10
La de A	331	0.02	332	0.50	332	0.53
Gl 3 ph de	332	0.09	NVR	NVR	NVR	NVR
Gl sy	371	0.08	372	0.10	372	0.11
Ad ph	179	0.19	NVR	NVR	NVR	NVR
Ca an I	260	0.26	NVR	NVR	259	0.26

[a] NVR signifies nonvalid result.

[b] Aldolase A — *Ald, A*
Creatine kinase M — *Cr, ki, M*
Lactate dehydrogenase A — *La, de, A*
Glyceraldehyde-3-phosphate dehydrogense — *Gl, 3, ph, de*
Glutamine synthetase — *Gl, sy*
Adenine phosphoribosyltransferase — *Ad, ph*
Carbonyc anhydrase I — *Ca, an, I*

In Table 17.1, we have shown our results with the same set of enzymes as in [16]. We have calculated the dn/ds ratio from the original source, and used this in the table. Here, we also see that many enzymes do not give us valid results using our reverse approach, in spite of having data in the already established work. We represent these in the form of NVR. We illustrated the results separately for HM as well HR.

17.4 Conclusion

This work has emphasized some important facts in regard to the evolutionary trends of enzymes. Normally, the rates of nucleotide substitution vary considerably from gene to gene. But the ones closer to each other show an almost similar type of behaviour. Here, we have noticed, that many enzymes, in spite of being proteins in nature, do not provide us any valid result as shown by NVR in Table 17.1. In these cases, the rate of synonymous change was so small that a proper valid ratio could not be computed. For such a case the nonsynonymous sites were comparative higher. Thus in this approach we found the accuracy rate to be around 50%–55%. The possible reason behind this result may be the random generation of the nucleotide sequences from the amino acid sequences which might have highly deviated from the original one or the divergence between the two species may be very high for certain genes in those enzymes. We estimated here the divergence time between the species. We found that it is almost five times higher (∼400 Myr) than ordinary proteins. So, we can say that these enzymes are five times stronger than the ordinary proteins. Because enzymes are considered to be biocatalysts, it remains unchanged even after the reaction is over. Thus, these take a much longer time to mutate because during evolution, accumulation of mutation is very slow.

Table 17.1 shows a comparative study between the already established work, and our approach with the same set of enzymes. As far as the results are concerned, we can only classify them according to neutral, purifying, or diversifying. We feel that this idea can establish some new concepts in biological evolution [17–20], to trace back the relation among the genes.

Acknowledgements This work is supported by National Science Council under the project no: NSC 94-2213-E-002-034.

References

1. Nei M, Gojobori T (1986) *Molecular Biology and Evolution* 3:418–426.
2. Jukes TH, Cantor CR (1969) *Evolution of Protein Molecules. Mammalian Protein Metabolism*, Academic Press, New York.
3. Seo TK, Kishino H, Thorne JL (2004) *Molecular Biology and Evolution* 21:1201–1213.
4. Miyata T, Yasunaga T (1980) *Journal of Molecular Evolution* 16:23–36.
5. Li WH, Wu CI, Luo CC (1985) *Molecular Biology and Evolution* 2:150–174.
6. Yorozu Y, Hirano M, Oka K, Tagawa Y (1982) *IEEE Translation Journal on Magnetics in Japan* 2:740–741.
7. Li WH (1993) *Journal of Molecular Evolution* 36:96–99.
8. Pamilo P, Bianchi NO (1993) *Molecular Biology and Evolution* 10:271–281.
9. Comeron JM (1995) *Journal of Molecular Evolution* 41:1152–1159.
10. Waterston RH, et al. (2002) *Nature* 420:520–562.
11. Lamder ES, et al. (2001) *Nature* 409:860–921.
12. Li WH (1997) *Molecular Evolution*. Sinauer, Sunderland, MA.
13. Makalowski W, Boguski MS (1998) *Proceedings of the National Academy of Sciences U.S.A* 95:9407–9412.
14. Nekrutenko A, Wu WY, Li WH (2003) *Trends in Genetics* 19:306–310.
15. Yang Z (1997) *Computer Applications in the Biosciences* 13:555–556.
16. Graur D, Li WH (2000) *Fundamentals of Molecular Evolution*. 2nd edn. Sinauer, Sunderland, MA.
17. Liou CY, Wu JM (1996) *Neural Networks* 9:671–684.
18. Liou CY, Yuan SK (1999) *Biological Cybernetics* 81:331–342.
19. Liou CY, Lin SL (2006) *Natural Computing* 5:15–42.
20. Liou CY (2006) *The 16th International Conference on Artificial Neural Networks*, LNCS 4131:688–697, Springer, New York.

Chapter 18
Microarray Data Feature Selection Using Hybrid GA-IBPSO

Cheng-San Yang, Li-Yeh Chuang, Chang-Hsuan Ho, and Cheng-Hong Yang

18.1 Introduction

DNA microarray examples are generated by a hybridization of mRNA from sample tissues or blood to cDNA (in the case of a spotted array), or hybridization of oligonucleotide of DNA (in the case of Affymetrix chips, on the surface of a chip-array). DNA microarray technology allows for the simultaneous monitoring and measurement of thousands of gene expression activation levels in a single experiment. Class memberships are characterized by the production of proteins, meaning that gene expressions refer to the production level of proteins specific for a gene. Thus, microarray data can provide valuable results for a variety of gene expression profile problems, and contribute to advances in clinical medicine. The application of microarray data on cancer type classification has recently gained in popularity. Coupled with statistical techniques, gene expression patterns have been used in the screening of potential tumor markers. Differential expressions of genes are analyzed statistically and genes are assigned to various classes, which may (or may not) enhance the understanding of the underlying biological processes.

Microarray gene expression technology has opened the possibility of investigating the activity of thousands of genes simultaneously. Gene expression profiles show the measurement of the relative abundance of mRNA corresponding to the genes. Thus, discriminant analysis of microarray data has great potential as a medical diagnosis tool, because results represent the state of a cell at the molecular level. The goal of microarray data classification is to build an efficient model that identifies the differentially expressed genes and may be used to predict class membership for any unknown samples. The challenges posed in microarray classification are the availability of a relatively limited number of samples in comparison to the high-dimensionality of the sample, and experimental variations in measured gene expression levels.

The classification of microarray data samples involves feature selection and classifier design. Generally, only a small number of gene expression data show a strong

Oscar Castillo et al. (eds.), *Trends in Intelligent Systems and Computer Engineering.*

correlation with a certain phenotype compared to the total number of genes investigated. That means that of the thousands of genes investigated, only a small number show significant correlation with a certain phenotype. Thus, in order to analyze gene expression profiles correctly, feature (gene) selection is most crucial for the classification process. The goal of feature selection is to identify the subset of differentially expressed genes that are potentially relevant for distinguishing the sample classes. A good selection method for genes relevant for sample classification is needed in order to increase predictive accuracy, and to avoid incomprehensibility; it should be based on the number of genes investigated. Several methods have been used to perform feature selection, for example, genetic algorithms [1], branch and bound algorithms [2], sequential search algorithms [3], mutual information [4], tabu search [5], entropy-based methods [6], regularized least squares [7], random forests [8], instance-based methods [9], and least squares support vector machines [10].

In our study, we used a combination of a genetic algorithm (GA) and improved binary particle swarm optimization (IBPSO) to implement feature selection. IBPSO was embedded in the GA to serve as a local optimizer for each generation. The *K*-nearest neighbor method (*K*-NN) with leave-one-out cross-validation (LOOCV) based on Euclidean distance calculations served as an evaluator of the GA and IBPSO for five classification problems taken from the literature. This procedure can improve the performance of populations by having a chromosome approximate a local optimum, reducing the number of features and preventing the GA from getting trapped in a local optimum.

18.2 Methods

18.2.1 Genetic Algorithms

Genetic algorithms are stochastic search algorithms modeled on the process of natural selection underlying biological evolution. They can be applied to many search, optimization, and machine learning problems [11]. The basic concept of GAs is designed to simulate evolutionary processes in natural systems, specifically those that follow the principle of survival of the fittest first laid down by Charles Darwin. As such, they represent an intelligent exploitation of a random search within a defined search space to solve a problem. GAs proceed in an iterative manner by generating new populations of strings from old ones. Every string is the encoded binary, real, and so on, version of a candidate solution. An evaluation function associates a fitness measure with every string, indicating its fitness for the problem. Standard GAs apply genetic operators such as selection, crossover, and mutation on an initially random population in order to compute a whole generation of new strings.

GAs have been successfully applied to a variety of problems, such as scheduling problems [12], machine learning problems [13], multiple objective problems [14], feature selection problems [15], data-mining problems [16], and traveling salesman problems [17]. Further details on the mechanisms of GAs can be found in John Holland [18].

18.2.2 Improved Binary Particle Swarm Optimization (IBPSO)

Particle swarm optimization (PSO) is a population-based stochastic optimization technique, which was developed by Kennedy and Eberhart in 1995 [19]. PSO simulates the social behavior of organisms, such as birds in a flock and fish in a school. This behavior can be described as an automatically and iteratively updated system. In PSO, each single candidate solution can be considered a particle in the search space. Each particle makes use of its own memory and knowledge gained by the swarm as a whole to find the best solution. All of the particles have fitness values, which are evaluated by a fitness function to be optimized. During movement, each particle adjusts its position by changing its velocity according to its own experience and according to the experience of a neighboring particle, thus making use of the best position encountered by itself and its neighbor. Particles move through the problem space by following a current of optimum particles. The process is then iterated a fixed number of times or until a predetermined minimum error is reached [20].

PSO was originally introduced as an optimization technique for real-number spaces. It has been successfully applied in many areas: function optimization, artificial neural network training, fuzzy system control, and other application problems. A comprehensive survey of the PSO algorithms and their applications can be found in Kennedy et al. [20]. However, many optimization problems occur in a space featuring discrete qualitative distinctions between variables and between levels of variables. Kennedy and Eberhart introduced binary PSO (BPSO), which can be applied to discrete binary variables. In a binary space, a particle may move to near corners of a hypercube by flipping various numbers of bits; thus, the overall particle velocity may be described by the number of bits changed per iteration [21].

Gene expression data characteristically have a high dimension, so we expect superior classification results in different dimensional areas. Each particle adjusts its position according to two fitness values, *pbest* and *gbest*, to avoid getting trapped in a local optimum by fine-tuning the inertia weight. *pbest* is a local fitness value, whereas *gbest* constitutes a global fitness value. If the *gbest* value is itself trapped in a local optimum, a search of each particle limit in the same area will occur, thereby preventing superior classification results. Thus, we propose a method that retires *gbest* under such circumstances and uses an improved binary particle swarm optimization (IBPSO). By resetting *gbest* we can avoid IBPSO getting trapped in a local optimum, and superior classification results can be achieved with a reduced number of selected genes.

18.2.3 K-Nearest Neighbor

The K-nearest neighbor (K-NN) method was first introduced by Fix and Hodges in 1951, and is one of the most popular nonparametric methods [22, 23]. The purpose of the algorithm is to classify a new object based on attributes and training samples. The K-nearest neighbor method consists of a supervised learning algorithm where the result of a new query instance is classified based on the majority of the K-nearest neighbor category. The classifiers do not use any model for fitting and are only based on memory, which works based on a minimum distance from the query instance to the training samples to determine the K-nearest neighbors. Any tied results are solved by a random procedure.

The K-NN method has been successfully applied in various areas, such as statistical estimation, pattern recognition, artificial intelligence, categorical problems, and feature selection. The advantage of the K-NN method is that it is simple and easy to implement. K-NN is not negatively affected when the training data are large, and is indifferent to noisy training data. In this study, the feature subset was measured by the leave-one-out cross-validation of one-nearest neighbor (1-NN). Neighbors are calculated using their Euclidean distance. The 1-NN classifier does not require any user-specified parameters, and the classification results are implementation independent.

18.2.4 Hybrid GA-IBPSO (IBPSO Nested in a GA)

The hybrid GA-IBPSO procedure used in this study combines a genetic algorithm and particle swarm optimization for feature selection. It adheres to the following pattern. Initially, each chromosome is coded to a binary string $S = F_1, F_2, \ldots, F_n$, $n = 1, 2, \ldots, m$; the bit value $\{1\}$ represents a selected feature, whereas the bit value $\{0\}$ represents a nonselected feature. The initial population is generated randomly. The predictive accuracy of a 1-NN determined by the leave-one-out cross-validation (LOOCV) method is used to measure the fitness of an individual. In the LOOCV method, a single observation from the original sample is selected as the validation data, and the remaining observations as the training data. This is repeated so that each observation in the sample is used once as the validation data. Essentially, this is the same as K-fold cross-validation where K is equal to the number of observations in the original sample. The obtained classification accuracy is an adaptive functional value.

The rand-based roulette-wheel selection scheme was used in this chapter. Standard genetic operators, such as crossover and mutation, are applied without modification. A two-point crossover operator is used, which chooses two cutting points at random and alternatively copies single segments out of each parent. If a mutation is present, either one of the offspring is mutated, and its binary representation changes from 1 to 0 or from 0 to 1 after the crossover operator is applied. If the mutated chromosome is superior to both parents, it replaces the worst chromosome of the parents; otherwise, the inferior chromosome in the population is replaced. Then, the embedded PSO is implemented and serves as a local optimizer in order to improve the performance of the population in the GA with each successive generation.

Each chromosome of the GA represents a single particle of the PSO. The position of each particle is represented by $X_p = \{x_{p1}, x_{p2}, \ldots, x_{pd}\}$ and the velocity of each particle is represented by $V_p = \{v_{p1}, v_{p2}, \ldots, v_{pd}\}$. The particle is updated at each iteration by following two "best" (optimum) values, called *pbest* and *gbest*. Each particle keeps track of its coordinates in the problem space, which are associated with the best solution (fitness) the particle has achieved thus far. This fitness value is stored, and represents the position called *pbest*. When a particle takes the whole population as its topological neighborhood, the best value is a global optimum value called *gbest*.

Once the adaptive values *pbest* and *gbest* are obtained, the features of the *pbest* and *gbest* particles can be tracked with regard to their position and velocity. Each particle is updated according to the following equations.

$$v_{pd}^{new} = w \times v_{pd}^{old} + c_1 \times rand_1 \times (pbest_{pd} - x_{pd}^{old}) - c_2 \times rand_2 \times (gbest_d - x_{pd}^{old}) \tag{18.1}$$

$$\text{if } v_{pd}^{new} \notin (V_{\min}, V_{\max}) \text{ then } v_{pd}^{new} = \max(\min(V_{\max}, v_{pd}^{new}), V_{\min}) \tag{18.2}$$

$$S(v_{pd}^{new}) = \frac{1}{1 + e^{-v_{pd}^{new}}} \tag{18.3}$$

$$\text{If } (rand < S(v_{pd}^{new})) \text{ then } x_{pd}^{new} = 1;\ \text{else } x_{pd}^{new} = 0 \tag{18.4}$$

In these equations w is the inertia weight, c_1 and c_2 are acceleration (learning) factors; $rand$, $rand_1$, and $rand_2$ are random numbers between 0 and 1. Velocities v_{pd}^{new} and v_{pd}^{old} are those of the updated particle and the particle before being updated, respectively, x_{pd}^{old} is the original particle position (solution), and x_{pd}^{new} is the updated particle position (solution).

In Eq. 18.2, particle velocities of each dimension are tried to a maximum velocity $V_{\max}$. If the sum of accelerations causes the velocity of that dimension to exceed $V_{\max}$, then the velocity of that dimension is limited to $V_{\max}$. $V_{\max}$ and $V_{\min}$ are user-specified parameters (in our case $V_{\max} = 6$, $V_{\min} = -6$).

The PSO converges rapidly during the initial stages of a search, but then often slows considerably and particles can get trapped in local optima. In order to avoid particles getting trapped in a local optimum, the *gbest* value has to be evaluated before each particle position is updated. If *gbest* has the same value for a preset number of times (in our case three times), the particle could conceivably be trapped in a local optimum. In such a case, the *gbest* position is reset to zero in the fitness function (classification accuracy), meaning that zero features are selected and *pbest* is kept. In the next iteration, particles in the neighborhood of the local optimum will adjust their position by congregating towards the *gbest* position.

The features after updating are calculated by the function $S(v_{pd}^{new})$ Eq. 18.3 [24], in which v_{pd}^{new} is the updated velocity value. If $S(v_{pd}^{new})$ is larger than a randomly produced disorder number that is within $\{0.0 \sim 1.0\}$, then its position value $S_n, n = 1, 2, \ldots, m$ is represented as $\{1\}$ (meaning this feature is selected as a required feature for the next update). If $S(v_{pd}^{new})$ is smaller than a randomly produced disorder number that is within $\{0.0 \sim 1.0\}$, then its position value $F_n, n = 1, 2, \ldots, m$ is represented as $\{0\}$ (meaning this feature is not selected as a required feature for the next update) [21].

The GA was configured to contain ten chromosomes and was run for 30 generations in each trial. The crossover and mutation rates were 0.8 and 0.1, respectively. The number of particles used was ten. The two factors $rand_1$ and $rand_2$ are random numbers between (0, 1), whereas c_1 and c_2 are acceleration factors; here $c_1 = c_2 = 2$. The inertia weight w was 1.0. The maximum number of iterations used in our IBPSO was ten. The pseudo-code of the proposed method is given below.

Pseudo-code for hybrid GA-IBPSO procedure

```
1: begin
2:      Randomly initialize population
3:      while (number of generations, or the stopping criterion is not met)
4:          Evaluate fitness value of chromosome by 1-Nearest Neighbor ()
5:          Select two parents chrom1 and chrom2 from population
6:          offspring = crossover (chrom1, chrom2)
7:          mutation (offspring)
8:          replace (population, offspring)
9:          Improved Binary Particle Swarm Optimization ()
10:       next generation until stopping criterion
11: end
```

18.3 Results and Discussion

Selecting relevant genes for gene expression classification is a common challenge in bioinformatics. Classification and prediction of gene expression data is a prerequisite for current genetic research in biomedicine and molecular biology, inasmuch as a correct analysis of results can help biologists solve complex biological problems. Gene expression data can effectively be used for gene identification, cell differentiation, pharmaceutical development, cancer classification, and disease diagnosis and prediction. However, due to the fact that gene expression data is of a high dimensionality and has a small sample size, classification of gene expression data is time-consuming. Choosing feature selection as a pretreatment method prior to the actual classification of gene expression data can effectively reduce the calculation time without negatively affecting the predictive error rate.

Due to the peculiar characteristics of gene expression data (high number of genes and small sample size) many researchers are currently studying how to select genes effectively before using a classification method to decrease the predictive error rate.

Pseudo-code for Improved Binary Particle Swarm Optimization procedure

begin
  while (number of iterations, or the stopping criterion is not met)
   Evaluate fitness value of particle swarm by ***1-Nearest Neighbor*** ()
   for $p = 1$ to number of particles
    if fitness of X_p is greater than the fitness of $pbest_p$ **then**
     $pbest_p = X_p$
    end if
    if fitness of any particle of particle swarm is greater than $gbest$ **then**
     $gbest = position\ of particle$
    end if
    if fitness of $gbest$ is the same Max times **then** give up and reset $gbest$
    end if
    for $d = 1$ to number of features of each particle
     $v_{pd}^{new} = w \times v_{pd}^{old} + c_1 \times rand_1 \times \left(pbest_{pd} - x_{pd}^{old}\right) + c_2 \times rand_2 \times \left(gbest_d - x_{pd}^{old}\right)$
     if $v_{pd}^{new} \notin (V_{\min}, V_{\max})$ **then** $v_{pd}^{new} = \max(\min(V_{\max}, v_{pd}^{new}), V_{\min})$
     end if ↵
     $S\left(v_{pd}^{new}\right) = \frac{1}{1+e^{-v_{pd}^{new}}}$
     if $\left(rand < S\left(v_{pd}^{new}\right)\right)$ **then** $x_{pd}^{new} = 1$ **else** $x_{pd}^{new} = 0$
     end if
    next *d*
   next *p*
  next generation until stopping criterion
end

Pseudo-code for 1-Nearest Neighbor procedure

begin
  for $i = 1$ to sample number of classification problem
   for $j = 1$ to sample number of classification problem
    for $k = 1$ to dimension number of classification problem
     $dist_i = dist_i + (data_{ik} - data_{jk})^2$
    next *k*
    if $dist_i < nearest$ **then**
     $class_i = class_j$
     $nearest = dist_i$
    end if
   next *j*
  next *i*
  for $i = 1$ to sample number of classification problem
   if $class_i = real\ class\ of\ testing\ data$ **then** $correct = correct + 1$
   end if
  next *i*
  $Fitness\ value = correct / number\ of\ testing\ data$
end

In general, gene selection is based on two aspects: one is to obtain a set of genes that have similar functions and a close relationship, the other is to find the smallest set of genes that can provide meaningful diagnostic information for disease prediction without diminishing accuracy. Feature selection uses relatively fewer features because only selective features need to be used. This does not affect the predictive error rate in a negative way; on the contrary, the predictive error rate can even be improved.

In this study, the datasets consist of six gene expression profiles, which were downloaded from http://www.gems-system.org. They include tumor, brain tumor, leukemia, lung cancer, and prostate tumor samples. The dataset formats are shown in Table 18.1, which contains the dataset name and a detailed description.

Table 18.2 compares experimental results obtained by other methods from the literature and the proposed method. Non-SVM and MC-SVM results were taken from

Table 18.1 Format of gene expression classification data

Dataset Name	Number of					Diagnostic task
	Samples	Categories	Genes	Genes Selected	Percentage of Genes Selected	
9 Tumors	60	9	5726	2140	0.37	9 different human tumor types
Brain_Tumor1	90	5	5920	2030	0.34	5 human brain tumor types
Brain_Tumor2	50	4	10367	3773	0.36	4 malignant glioma types
Leukemia1	72	3	5327	1802	0.34	Acute myelogenous leukemia (AML), acute lymphoblastic leukemia (ALL) B-cell and ALL T-cell
SRBCT	83	4	2308	1175	0.51	Small, round blue cell tumors (SRBCT) or childhood
Average					0.38	

Table 18.2 Accuracy of classification for gene expression data

Methods	*Non-SVM*			*MC-SVM*					*NEW*[9]
Datasets	*KNN*[1]	*NN*[2]	*PNN*[3]	*OVR*[4]	*OVO*[5]	*DAG*[6]	*WW*[7]	*CS*[8]	*KNN*
9_Tumors	43.90	19.38	34.00	65.10	58.57	60.24	62.24	65.33	73.33
Brain_Tumor1	87.94	84.72	79.61	91.67	90.56	90.56	90.56	90.56	**92.22**
Brain_Tumor2	68.67	60.33	62.83	77.00	77.83	77.83	73.33	72.83	**86.00**
Leukemia1	83.57	76.61	85.00	97.50	91.32	96.07	97.50	97.50	**100.0**
SRBCT	86.90	91.03	79.50	**100.00**	**100.00**	**100.00**	**100.00**	**100.00**	**100.0**
Prostate_Tumor	85.09	79.18	79.18	**92.00**	**92.00**	**92.00**	**92.00**	**92.00**	90.20
Average	76.01	68.54	70.02	87.21	85.05	86.12	85.94	86.37	90.29

Statnikov et al. for comparison [25]. Various methods were compared to our proposed method. They include: support vector machines: (1) one-versus-rest and one-versus-one [26], (2) DAGSVM [24], (3) the method by Weston and Watkins [27], and (4) the method by Crammer and Singer [28]. The non-SVM methods include the *K*-nearest neighbor method [22, 29], backpropagation neural networks [30], and probabilistic neural networks [31].

The average highest classification accuracies of non-SVM, MC-SVM, and the proposed method are 76.01, 87.21, and 90.29, respectively. The proposed method obtained five of the highest classification accuracies for six test datasets, that is, for the 9_Tumors, Brain_Tumor1, Brain_Tumor2, Leukemia1, and SRBCT datasets. The classification accuracy of the 9_Tumors and Brain_Tumor2 datasets obtained by the proposed method are 73.33% and 86.00%, respectively, an increase of (29.43% and 8.00%) and (17.33% and 8.17%) classification accuracy compared to the Non-SVMs and MC-SVMs methods. For the Prostrate_Tumor dataset, the classification accuracy obtained by the proposed method is better than the classification accuracy of Non-SVMs and is comparable to the MC-SVM method.

GAs have been shown to outperform SFS (sequential forward search), PTA (plus and take away), and SFFS (sequential forward floating search) in Oh et al. [15]. PSO shares many similarities with evolutionary computation techniques such as GAs. PSO is based on the idea of collaborative behavior and swarming in biological populations. Both PSO and GAs are population-based search approaches that depend on information sharing among their population members to enhance the search processes by using a combination of deterministic and probabilistic rules. However, PSO does not include genetic operators such as crossover and mutation per se.

Particles update some of their inherent characteristics, that is, their velocity, according to their individual experience. This updating of information due to social interactions between particles is very similar to crossover in a GA. Furthermore, random parameters ($rand_1$ and $rand_2$ in Eq. 18.1 affect the velocity of particles in a way similar to mutation in a GA. In fact, the only difference between both is that the crossover and mutation in a GA is probabilistic (crossover rate and mutation rate), but the updated particle in PSO should be processed at each iteration without any probability.

Compared to GAs, the information-sharing mechanism in PSO is considerably different. In GAs, chromosomes share information with each other, so the whole population moves as one group towards an optimal area. In the PSO version applied in our study (improved binary PSO), the social model gives out the information to others. It is a one-way information-sharing mechanism. Evolution only looks for the best solution. In most cases all the particles tend to converge to the best solution more quickly than in a GA, even in the local version.

(1) *KNN*: *K*-nearest neighbors. (2) *NN*: backpropagation neural networks. (3) *PNN*: probabilistic neural networks. (4) *OVR*: one-versus-rest. (5) *OVO*: one-versus-one. (6) *DAG*: DAGSVM. (7) *WW*: method by Weston and Watkins. (8) *CS*: method by Crammer and Singer. (9) *NEW*: the proposed method (GA-IBPSO). Highest values are in bold type.

18.4 Conclusion

We used a hybrid of a GA and improved binary PSO (GA-IBPSO) to perform feature selection. The *K*-NN method with LOOCV served as an evaluator of the GA and IBPSO fitness functions. Experimental results show that the proposed method could simplify feature selection by reducing the total number of features needed effectively, and it obtained a higher classification accuracy compared to other feature selection methods in most cases. The classification accuracy obtained by the proposed method had the highest classification accuracy in five of the six data test problems, and is comparable to the classification accuracy of the other test problem. GA-IBPSO can serve as a preprocessing tool to help optimize the feature selection process, because it either increases the classification accuracy, reduces the number of necessary features for classification, or both. The proposed GA-IBPSO method could conceivably be applied to problems in other areas in the future.

Acknowledgements This work is partly supported by the National Science Council in Taiwan under grants NSC94-2622-E-151-025-CC3, NSC94-2311-B037-001, NS93-2213-E-214-037, NSC92-2213-E-214-036, NSC92-2320-B-242-004, NSC92-2320-B-242-013, and by the CGMH fund CMRPG1006.

References

1. Raymer, M.L., Punch, W.F., Goodman, E.D., Kuhn, L.A., and Jain, A.K. (2000). Dimensionality reduction using genetic algorithms. *IEEE Transactions on Evolutionary Computation*, 4(2): 164–171.
2. Narendra, P.M. and Fukunage, K. (1997). A branch and bound algorithm for feature subset selection. *IEEE Transactions on Computers*, 6(9): 917–922.
3. Pudil, P., Novovicova, J., and Kittler, J. (1994). Floating search methods in feature selection. *Pattern Recognition Letters*, 15: 1119–1125.
4. Roberto, B. (1994). Using mutual information for selecting features in supervised neural net learning. *IEEE Transactions on Neural Networks*, 5(4): 537–550.
5. Zhang, H. and Sun, G. (2002). Feature selection using tabu search method. *Pattern Recognition*, 35: 701–711.
6. Liu, X., Krishnan, A., and Mondry, A. (2005). An entropy-based gene selection method for cancer classification using microarray data. *BMC Bioinformatics*, 6: 76.
7. Ancona, N., Maglietta, R., D'Addabbo, A., Liuni, S., and Pesole, G. (2005). Regularized least squares cancer classifiers from DNA microarray data. *Bioinformatics*, 6(Suppl 4): S2.
8. Diaz-Uriarte, R. and Alvarez de Andres, S. (2006). Gene selection and classification of microarray data using random forest. *Bioinformatics*, 7: 3.
9. Berrar, D., Bradbury, I., and Dubitzky, W. (2006). Instance-based concept learning from multiclass DNA microarray data. *Bioinformatics*, 7: 73.
10. Tang, E.K., Suganthan, P., and Yao, X. (2006). Gene selection algorithms for microarray data based on least squares support vector machine. *Bioinformatics*, 7: 95.
11. Goldberg, D.E. (1989). *Genetic Algorithms in Search, Optimization and Machine Learning*, Reading, MA: Addison-Wesley.
12. Hou, E.S., Ansari, N., and Ren, H. (1994). A genetic algorithm for multiprocessor scheduling, *IEEE Transactions on Parallel and Distributed Systems*, 5(2): 113–120.

13. Vafaie, H. and De Jong, K. (1992). Genetic algorithms as a tool for feature selection in machine learning. In: *Proceedings of the 4th International Conference on Tools with Artificial Intelligence*, pp. 200–204.
14. Deb, K. Agrawal, S. Pratap, A., and Meyarivan, T. (2002). A fast elitist non-dominated sorting genetic algorithm for multi-objective optimization: NSGA-II, *IEEE Transactions on Evolutionary Computation*, 6, 182–197.
15. Oh et al. (2004). Hybrid genetic algorithm for feature selection, *IEEE Transactions on Pattern Analysis and Machine Intelligence*, 26(11): 1424–1437.
16. Kim, S. and Zhang, B.-T. (2001). Evolutionary learning of web-document structure for information retrieval. In: *Proceedings of the 2001 Congress on Evolutionary Computation*, vol. 2, pp. 1253–1260.
17. Pullan, W. (2003). Adapting the genetic algorithm to the traveling salesman problem, *IEEE Congress on Evolutionary Computation*, 1209–1035.
18. Holland, J. (1992). *Adaptation in Nature and Artificial Systems*, Cambridge, MA: MIT Press.
19. Kennedy, J. and Eberhart, R.C. (1995). Particle swarm optimization. In: *Proceedings of the IEEE International Conference on Neural Networks*, vol. 4, pp. 1942–1948.
20. Kennedy, J., Eberhart, R.C., and Shi, Y. (2001). San Mateo, CA: Morgan Kaufman.
21. Kennedy, J. and Eberhart, R.C. (1997). A discrete binary version of the particle swarm algorithm. In: *Systems, Man, and Cybernetics, 1997 IEEE International Conference on 'Computational Cybernetics and Simulation'*, vol. 5, Oct. 12–15, pp. 4104–4108.
22. Cover, T. and Hart, P. (1967). Nearest neighbor pattern classification. In: *Proceedings of the IEEE Transactions Information Theory*, pp. 21–27.
23. Fix, E. and Hodges, J.L. (1951). Discriminatory analysis—Nonparametric discrimination: Consistency properties. Technical Report 21-49-004, Report no. 4, US Air Force School of Aviation Medicine, Randolph Field, pp. 261–279.
24. Platt, J.C., Cristianini, N., and Shawe-Taylor, J. (2000). Large margin DAGS for multiclass classification. In: *Advances in Neural Information Processing Systems* 12, Cambridge, MA: MIT Press, pp. 547–553.
25. Statnikov, A., Aligeris, C.F., Tsamardinos, L., Hardin, D., and Levy, S. (2004). A comprehensive evaluation of multicategory classification methods for microarray gene expression cancer diagnosis. *Bioinformatics*, 21(5), Sept.: 631–643.
26. Kreßel, U. (1999). Pairwise classification and support vector machines. In: *Advances in Kernel Methods: Support Vector Learning*, Cambridge, MA: MIT Press, pp. 255–268.
27. Weston, J. and Watkins, C. (1999). Support vector machines for multi-class pattern recognition. In: *Proceedings of the Seventh European Symposium on Artificial Neural Networks (ESANN 99)*, Bruges, April 21–23.
28. Crammer, K. and Singer, Y. (2000). On the learnability and design of output codes for multiclass problems. In: *Proceedings of the Thirteen Annual Conference on Computational Learning Theory (COLT 2000)*, Stanford University, Palo Alto, CA, June 28–July 1.
29. Dasarathy, B.V. (Ed.) (1991). *NN Concepts and Techniques, Nearest Neighbor (NN) Norms: NN Pattern Classification Techniques*. Washington, DC: IEEE Computer Society Press, pp. 1–30.
30. Mitchell, T.M. (1997). *Machine Learning*. New York: McGraw-Hill.
31. Specht, D.F. (1990). Probabilistic neural network. *Neural Networks*, 3, 109–118.

Chapter 19
Discrete-Time Model Representations for Biochemical Pathways

Fei He, Lam Fat Yeung, and Martin Brown

19.1 Introduction

Based on much experimentation, traditional biochemists and molecular biologists have developed many qualitative models and hypotheses for biochemical pathway study [7, 26, 28]. However, in order to evaluate the completeness and usefulness of a hypothesis, produce predictions for further testing, and better understand the interaction and dynamic of pathway components, qualitative models are no longer adequate. There has recently been a focus on a more quantitative approach in systems biology study. In the past decade, numerous approaches for quantitative modeling of biochemical pathway dynamics have been proposed (e.g., [1, 4, 15, 29, 30, 34, 36], among others).

Among these approaches, the most prominent method is to use ordinary differential equations (ODEs) to model biochemical reactions based on mass action principles. It should be noted that using ODEs to model biochemical reactions assumes that the system is well-stirred in a homogeneous medium and that spatial effects, such as diffusion, are irrelevant, otherwise partial differential equations (PDEs) should be used [17]. In the literature, almost all publications related to pathway modeling are based on continuous-time ODE model representations. Using continuous-time ODEs facilitates researchers in analytical study and analysis, whereas it also brings difficulties for numerical computation and computer-based simulation. Therefore, constructing the corresponding discrete-time model representations is particularly important in systems biology study.

There are many reasons to formulate discrete-time model representations in pathway modeling research. First, the real biochemical kinetic reactions take place in continuous-time, whereas experimental data are measured by sampling the continuous biochemical reaction outputs, and computer-based analysis and simulation all depend on discrete-time datasets. Therefore, a discrete-time model could be an interface between real kinetic reaction, experimentation, and computer-based simulation. A delicate discrete-time model can not only assist people to better understand pathway reaction dynamics and reproduce the experimental data, but also to

Oscar Castillo et al. (eds.), *Trends in Intelligent Systems and Computer Engineering.*

generate predictions for computer-based analysis which leaves out the expensive and time-consuming experiment process. Moreover, it can be a crucial tool for further experimental design study, such as state measurement selection and sampling time design.

Second, as we know, parameter estimation is an active and important research topic in systems biology study. Estimating parameters of continuous ODEs is usually a computationally costly procedure, as even for linear ODEs it is a nonquadratic optimization problem. When considering discrete-time-based models, although we cannot change the fundamental nature of the optimization problem, an iterative polynomial discrete-time model could possibly simplify the structure of continuous ODEs, especially for some nonlinear cases. This can help researchers to develop new parameter estimation approaches based on the discrete-time models.

Furthermore, dynamic sensitivity analysis plays an important role in parameter selection and uncertainty analysis for system identification procedure [8]; in practice, local sensitivity coefficients are obtained by solving the continuous sensitivity functions and model ODEs simultaneously. As sensitivity functions are also a set of ODEs with respect to sensitivity coefficients, it would be worthwhile to calculate sensitivity coefficients in a similar discrete-time iterative way.

In practice, there are several methods that can be considered for discretization of ODEs. One type of methods is based on Taylor or Lie series expansion which is a one-step-ahead discretization stratagem. For models represented by linear ODEs which means linear in states, the discrete-time model representation is given in discrete-time control system textbooks [16] as linear ODEs can be expressed as linear state-space equations. Unfortunately, for nonlinear ODEs there is no such general direct discretization mapping. In mathematical and control theory, some discretization techniques related to nonlinear ODEs are comparatively reviewed and discussed in Sect. 19.3.

Considering real biological signaling pathway cases, we investigate a time-varying linear approach for bilinear in states ODEs situation based on Taylor expansion and Carleman linearization. However, even for this method the mathematical model expression would be complex when considering higher-order approximation. Another important discretization strategy discussed in this work is the multistep discretization approach based on the Runge–Kutta method. One advantage of this approach is it improves the discretization accuracy by utilizing multistep information for approximation of one-step-ahead model prediction. Moreover, it provides a general exact discrete-time representation for both linear and nonlinear biochemical ODE models.

19.2 Continuous-Time Pathway Modeling

19.2.1 Continuous-Time Model Representation

In the literature [31, 39], signal pathway dynamics can usually be modeled by the following ODE representation

$$\begin{cases} \dot{\mathbf{x}}(t) = \mathbf{f}(\mathbf{x}(t), \mathbf{u}(t), \theta), & \mathbf{x}(t_0) = x_0 \\ \mathbf{y}(t) = \mathbf{g}(\mathbf{x}(t)) \end{cases} \tag{19.1}$$

where $\mathbf{x} \in \mathbb{R}^m$, $\mathbf{u} \in \mathbb{R}^p$, and $\theta \in \mathbb{R}^n$ are the state, input, and parameter column vectors. x_0 is the initial states vector at t_0. From a biochemical modeling viewpoint, $\mathbf{x}$ represents molecular concentrations; $\mathbf{u}$ generally represents external cellular signals; θ stands for reaction rates. $\mathbf{f}(\cdot)$ is a set of linear or nonlinear functions which correspond to a series of biochemical reactions describing pathway dynamics. $\mathbf{g}(\cdot)$ here is the measurement function that determines which states can be measured. For the simplest case, if all the states can be measured, the measurement function $\mathbf{g}$ in Eq. 19.1 is an identity matrix. Otherwise, $\mathbf{g}$ is a rectangular zero–one matrix with corresponding rows with respect to unmeasurable states deleted from the full-rank identity matrix $\mathbf{I}_m$.

When model ODEs are linear and time invariant in states, which is also known as linear ODEs, Eq. 19.1 can be simplified as:

$$\begin{aligned} \dot{\mathbf{x}}(t) &= \mathbf{A}\mathbf{x}(t) + \mathbf{B}\mathbf{u}(t) \\ \mathbf{y}(t) &= \mathbf{g}(\mathbf{x}(t)) \end{aligned} \tag{19.2}$$

Here $\mathbf{A}_{m\times m}$ is the parameter matrix, and $\mathbf{B}_{m\times p}$ is the known input matrix. For most systems biology pathway models $\mathbf{A}$ is typical a sparse matrix; only the corresponding reaction state terms appear in the forcing function. However, this kind of linear ODE is not prevalent in representation of most biochemical reactions, only if it is an irreversible chain reaction. An illustrative example is the first-order isothermal liquid-phase chain reaction [9, 33]:

$$A \xrightarrow{\theta_1} B \xrightarrow{\theta_2} C$$

It starts from liquid reaction component A to liquid product B and then to liquid product C. This reaction process can be modeled by the following ODEs.

$$\begin{aligned} \dot{x}_1 &= -\theta_1 x_1 \\ \dot{x}_2 &= \theta_1 x_1 - \theta_2 x_2 \end{aligned} \tag{19.3}$$

where x_1, x_2 denote concentrations of components A and B, which were the only two concentrations measured. Therefore, component C does appear in the model. This ODE model Eq. 19.3 can be readily represented as linear state space form Eq. 19.2.

More generally, and more applicably, we can consider ODEs that are linear in their unknown parameters, but not necessarily states. For instance, the Michaelis–Menten enzyme kinetics [5, 13], JAK-STAT [32], ERK [6], *TNFα-NF-κB* [5], and *IκB-NF-κB* [39] pathway models are all bilinear in the states but linear in parameters. The state function of this kind of ODEs can be represented as

$$\dot{\mathbf{x}}(t) = \mathbf{F}(\mathbf{x}(t))\theta \tag{19.4}$$

where $\mathbf{F}(\cdot)$ represents a set of nonlinear functions which is also commonly presented as a sparse matrix. For example, considering a bilinear in states model, when a reaction only takes place in the presence of two molecules, most elements of the corresponding row in **F** would be zeros except for those related to these two states. Here we do not take account of model inputs $\mathbf{u}(t)$ as most of these published pathways are not considered subject to external cellular signals.

19.2.2 Parameter Estimation

Given the model structure and experiment data of measured state variables, the aim of parameter estimation is to calculate the parameter values that minimize some loss function between the model's prediction and measurement data. Considering the set of $\{\tilde{y}_i(k)\}_{i,k}$ as the measurement data and the corresponding model's predictions $\{y_i(k,\theta)\}_{i,k}$, which is simply discrete-time sampling of the continuous-time ODE model's output $\mathbf{y}(t)$, then a standard least squares loss function along the trajectory gives:

$$\hat{\theta} = \arg\min \frac{1}{2} \sum_i \sum_k \omega_i (\tilde{y}_i(k) - y_i(k,\theta))^2 \tag{19.5}$$

where the double sum can be taken simply as taking the expected value over all states (i) and over the complete trajectory (k); ω_i are the weights to normalize the contributions of different state signals and can be taken as

$$\omega_i = \left(\frac{1}{\max\limits_k (\tilde{y}_i(k))} \right)^2 \tag{19.6}$$

We assume that the model hypothesis space includes the optimal model, so that when $\hat{\theta} = \theta^*$, the model's parameters are correct and $y_i(k,\hat{\theta}) = y_i^*(k)$. Typically, we assume that the states are not directly measurable, but are subject to additive measurement noise:

$$\tilde{y}_i(k) = y_i^*(k) + \aleph(0, \sigma_i^2) \tag{19.7}$$

Here, the noise is zero mean Gaussian, where the variances depend on the state.

We have to take into account the fact that we can only measure and estimate the states rather than the first-order derivative of states to time (left-hand side of ODEs), so even for linear models, the optimization problem is not quadratic. If employing a global optimization method, such as a genetic algorithm, numerous function evaluations are required which is computationally expensive; if considering traditional local optimization methods such as the quasi-Newton or Levenberg–Marquardt method, due to the existence of local minima, nonlinear regression must be implemented from many starting points. In the literature, parameter estimation of pathway ODEs is usually reduced into solving nonlinear boundary value problems using a multiple shooting method [24, 25, 35].

19.2.3 Sensitivity Analysis

Dynamic sensitivity analysis plays an important role in parameter selection and uncertainty analysis for the system identification procedure. The first-order local sensitivity coefficient $s_{i,j}$ is defined as the partial derivative of ith state to the jth parameter

$$s_{i,j}(t) = \frac{\partial x_i(t)}{\partial \theta_j} = \frac{x_i(\theta_j + \Delta\theta_j, t) - x_i(\theta_j, t)}{\Delta\theta_j} \tag{19.8}$$

In Eq. 19.8, the sensitivity coefficient is calculated using the finite difference method (FDM), however, the numerical values obtained may vary with $\Delta\theta_j$, and repeated measurement of state is required at least once for each parameter. In practice [39], the direct differential method (DDM) is employed as an alternative by taking the partial derivative of Eq. 19.1 with respect to parameter θ_j, and the absolute parameter sensitivity equations can be written as

$$\frac{d}{dt}\frac{\partial \mathbf{x}}{\partial \theta_j} = \frac{\partial \mathbf{f}}{\partial x}\frac{\partial \mathbf{x}}{\partial \theta_j} + \frac{\partial \mathbf{f}}{\partial \theta_j} \Leftrightarrow \dot{S}_j = J \cdot S_j + P_j, \quad S_j(t_0) = S_0 \tag{19.9}$$

where J and P_j are the Jacobian matrix and parameter Jacobian matrix. By solving the m equations in Eq. 19.1 and $n \times m$ equations in Eq. 19.9 together as a set of differentical equations, both $\mathbf{x}(t)$ and $\mathbf{S}(t)_{m\times n}$ can be determined simultaneously.

For ODEs that are linear in both parameters and states, a special case described in Eq. 19.2, and when assuming biochemical reactions are autonomous which means unaffected by external inputs $\mathbf{u}$, the corresponding linearized sensitivity equations can be expressed as

$$\dot{\mathbf{S}}(t) = \mathbf{A}\mathbf{S}(t) + \mathbf{P}(t) \tag{19.10}$$

where $\mathbf{P}$ is the $m \times n$ parameter Jacobian matrix.

For linear in the parameters ODEs Eq. 19.4, the corresponding sensitivity equations can be simplified as

$$\dot{\mathbf{S}}(t) = \frac{\partial(\mathbf{F}(\mathbf{x}(t)\theta)}{\partial \mathbf{x}}\mathbf{S}(t) + \mathbf{F}(\mathbf{x}(t)) \tag{19.11}$$

Parameter sensitivity coefficients provide crucial information for parameter importance measurement and further parameter selection. A measure of the estimated parameters' quality is given by the Fisher information matrix (FIM):

$$\mathbf{F} = \sigma^2 \left(\frac{d\mathbf{x}}{d\theta}\right)^T \left(\frac{d\mathbf{x}}{d\theta}\right) = \sigma^2 \mathbf{S}^T \mathbf{S} \tag{19.12}$$

which is a lower bound on the parameter variance/covariance matrix. This is a key measure of identifiability which determines how easily the parameter values can be reliably estimated from the data, or alternatively, how many experiments would need to be performed in order to estimate the parameters to a predefined level of confidence.

In the literature, several algorithms for parameter selection are proposed based on parameter sensitivity analysis [20, 38]. Besides, many optimal experimental design methods [10, 19, 38] are developed based on maximizing the information of FIM according to commonly used optimal design criteria [2].

19.3 Discrete Time-Model Representation

Equation 19.1 describes pathway dynamics in continuous-time. However, in real experiments the measurement results are obtained by sampling continuous time series, and later on system analysis, parameter estimation, and experimental design are all based on these discrete datasets. Therefore, it is important to formulate a discrete-time model representation.

19.3.1 One-Step-Ahead System Discretization

For linear in the states ODEs, the exact discrete-time representation of system ODEs Eq. 19.2 will take the form:

$$\mathbf{x}(k+1) = \mathbf{G} \cdot \mathbf{x}(k) + \mathbf{H} \cdot \mathbf{u}(k) \tag{19.13}$$

If we denote $t = kT$ and $\lambda = T - t$
where

$$\mathbf{G} = e^{\mathbf{A}T}, \quad \mathbf{H} = \left(\int_0^T e^{\mathbf{A}\lambda} d\lambda \right) \mathbf{B} \tag{19.14}$$

If matrix $\mathbf{A}$ is nonsingular, then $\mathbf{H}$ given in Eq. 19.14 can be simplified to

$$\mathbf{H} = \left(\int_0^T e^{\mathbf{A}\lambda} d\lambda \right) \mathbf{B} = \mathbf{A}^{-1}(e^{\mathbf{A}T} - \mathbf{I})\mathbf{B} = (e^{\mathbf{A}T} - \mathbf{I})\mathbf{A}^{-1}\mathbf{B} \tag{19.15}$$

Similarly, if the biochemical reactions are autonomous, the discrete-time sensitivity equations can be written as

$$\mathbf{S}(k+1) = \mathbf{G} \cdot \mathbf{S}(k) + \mathbf{B}_d \mathbf{P}(k) \tag{19.16}$$

where

$$\mathbf{B}_d = \int_0^T e^{\mathbf{A}\lambda} d\lambda \tag{19.17}$$

The linear discrete-time representation discussed above had been mentioned in some pathway modeling literature [11, 12]. Unfortunately, there is no general exact mapping between continuous- and discrete-time systems when ODEs are nonlinear. In numerical analysis and control theory, there are several methods that have been discussed for one-step discretization of nonlinear ODEs, such as the finite difference method [37], which comprises Euler's method and finite-order Taylor series approximation, Carleman linearization [18], Jacobian linearization [14], feedback linearization [14], and Monaco and Normand-Cyrot's [22, 23] method among others.

However, as mentioned previously not all biochemical pathway systems are subject to external cellar signals, therefore Jacobian and feedback linearization approaches which aim at designing a complex input signal are not discussed in this chapter. In the next section, we propose a time-varying linear approach based on Taylor expansion and the Carleman linearization method for discretization of bilinear in the states pathway ODEs, and also briefly investigate Monaco and Normand-Cyrot's scheme.

19.3.1.1 Taylor–Carleman Method

We initially consider a generally nonlinear model of the form:

$$\dot{\mathbf{x}}(t) = \mathbf{f}(\mathbf{x}(t), \theta) \tag{19.18}$$

Then using Taylor expansion around the current time instant $t = t_k$, the state value at the next sample point $t = t_k + T$ is given by:

$$\mathbf{x}(t_k + T) = \mathbf{x}(t_k) + \sum_{l=1}^{\infty} \frac{T^l}{l!} \left. \frac{\partial^l \mathbf{x}(t)}{\partial t^l} \right|_{t=t_k} \tag{19.19}$$

which can be further simplified as

$$\mathbf{x}(k+1) = \mathbf{x}(k) + \sum_{l=1}^{\infty} \frac{T^l}{l!} \mathbf{x}^{[l]}(k) \tag{19.20}$$

As discussed in Sect. 19.2, for some bioinformatics systems, the nonlinear ODEs are simply bilinear in the states:

$$\dot{\mathbf{x}} = \mathbf{A}\mathbf{x} + \mathbf{D}\mathbf{x} \otimes \mathbf{x} \tag{19.21}$$

where $\otimes$ denotes the Kronecker product and $m \times m^2$ matrix $\mathbf{D}$ is assumed to be symmetric in the sense that the coefficients corresponding to $x_1 x_2$ and $x_2 x_1$ are the same in value. For many biochemical pathway models, $\mathbf{D}$ is generally very sparse. So let us evaluate the first few derivative terms of Eq. 19.20 to deduce the overall structure of the exact discrete-time model of Eq. 19.18:

$$\begin{aligned}
\dot{\mathbf{x}} &= \mathbf{A}\mathbf{x} + \mathbf{D}\mathbf{x} \otimes \mathbf{x} \\
\ddot{\mathbf{x}} &= \mathbf{A}\dot{\mathbf{x}} + 2\mathbf{D}\dot{\mathbf{x}} \otimes \mathbf{x} \\
&= \mathbf{A}^2\mathbf{x} + (\mathbf{A}\mathbf{D} + 2\mathbf{D}(\mathbf{A} \otimes \mathbf{I}))(\mathbf{x} \otimes \mathbf{x}) + 2\mathbf{D}(\mathbf{D} \otimes \mathbf{I})(\mathbf{x} \otimes \mathbf{x} \otimes \mathbf{x}) \\
\dddot{\mathbf{x}} &= \mathbf{A}^2\dot{\mathbf{x}} + 2(\mathbf{A}\mathbf{D} + 2\mathbf{D}(\mathbf{A} \otimes \mathbf{I}))(\dot{\mathbf{x}} \otimes \mathbf{x}) + 6\mathbf{D}(\mathbf{D} \otimes \mathbf{I})(\dot{\mathbf{x}} \otimes \mathbf{x} \otimes \mathbf{x}) \\
&= \mathbf{A}^3\mathbf{x} + (\mathbf{A}^2\mathbf{D} + 2\mathbf{A}\mathbf{D}(\mathbf{A} \otimes \mathbf{I}) + 4\mathbf{D}(\mathbf{A} \otimes \mathbf{I})^2)(\mathbf{x} \otimes \mathbf{x}) \\
&\quad + (2\mathbf{A}\mathbf{D}(\mathbf{D} \otimes \mathbf{I}) + 4\mathbf{D}(\mathbf{A} \otimes \mathbf{I})(\mathbf{D} \otimes \mathbf{I}) \\
&\quad + 6\mathbf{D}(\mathbf{D} \otimes \mathbf{I})(\mathbf{A} \otimes \mathbf{I} \otimes \mathbf{I}))(\mathbf{x} \otimes \mathbf{x} \otimes \mathbf{x}) \\
&\quad + 6\mathbf{D}(\mathbf{D} \otimes \mathbf{I})(\mathbf{D} \otimes \mathbf{I} \otimes \mathbf{I})(\mathbf{x} \otimes \mathbf{x} \otimes \mathbf{x} \otimes \mathbf{x})
\end{aligned} \tag{19.22}$$

It can be seen that it is a polynomial in $\mathbf{x}$ of degree $m+1$. Hence the infinite sum in Eq. 19.19 is an infinite polynomial. We can notice that the coefficient of $\mathbf{x}$ in nth-order derivative expansion is $\mathbf{A}^n$, and the coefficient of the second-order terms $\mathbf{x} \otimes \mathbf{x}$ can be expressed recursively:

$$\begin{aligned} q^1 &= \mathbf{D} \\ q^n &= \mathbf{A}^{n-1}\mathbf{D} + 2q^{n-1}(\mathbf{A} \otimes \mathbf{I}) \end{aligned} \tag{19.23}$$

The exact representation of Eq. 19.21 in discrete time should have the form:

$$\mathbf{x}(k+1) = \mathbf{p}(\mathbf{x}(k)) \tag{19.24}$$

where $\mathbf{p}$ is required to be vector-valued. Here, instead of treating the system as a global nonlinear discrete time representation, it would be possible to treat it as a time-varying linear system, where the time-varying components depend on $\mathbf{x}$. It is obviously an infinite-degree polynomial and some finite-length approximation must be used instead. For instance, if we only consider the second-order approximation of derivative terms in Eq. 19.22 and using jth-order Taylor expansion, the discrete-time representation of Eq.19.21 can be expressed as

$$\begin{aligned} \mathbf{x}(k+1) &= \mathbf{x}(k) + \sum_{l=1}^{j} \frac{T^l}{l!}\mathbf{x}^{[l]}(k) \\ &= \mathbf{x}(k) + \sum_{l=1}^{j} \frac{T^l}{l!}(\mathbf{A}^l\mathbf{x}(k) + q^l\mathbf{x}(k) \otimes \mathbf{x}(k)) \end{aligned} \tag{19.25}$$

The advantage of this approach is it gives a finite polynomial discrete time representation for bilinear in states models. However, as shown in Eq. 19.22 the model expression would be complex when considering exact higher-order derivative expansion, otherwise, the lower-order approximation of derivatives has to be employed as in Eq. 19.25, and corresponding discretization accuracy would decrease accordingly.

19.3.1.2 Monaco and Normand–Cyrot's Method

Instead of approximating derivatives, a recent algebraic discretization method proposed by Monaco and Normand-Cyrot is based on Lie expansion of continuous ODEs. When considering nonlinear ODEs with the form Eq. 19.18, the corresponding discretization scheme can be expressed as

$$\mathbf{x}(k+1) = \mathbf{x}(k) + \sum_{l=1}^{j} \frac{T^l}{l!}\mathcal{L}_{\mathbf{f}}^l(\mathbf{x}(k)) \tag{19.26}$$

where the Lie derivative is given by

$$\mathcal{L}_{\mathbf{f}}(\mathbf{x}(k)) = \sum_{i=1}^{m} f_i \frac{\partial \mathbf{x}}{\partial x_i} \tag{19.27}$$

and the higher-order derivatives can be calculated recursively

$$\mathcal{L}_{\mathbf{f}}^{l}(\mathbf{x}(k)) = \mathcal{L}_{\mathbf{f}}(\mathcal{L}_{\mathbf{f}}^{l-1}(\mathbf{x}(k))) \tag{19.28}$$

Thus, Eq. 19.18 can be rewritten as

$$\mathbf{x}(k+1) = \mathbf{x}(k) + T\mathbf{f} + \frac{1}{2}T^2 J(\mathbf{f}) * \mathbf{f} + \frac{1}{6}T^3 J(J(\mathbf{f}) * \mathbf{f}) * \mathbf{f} + \ldots \tag{19.29}$$

where $J(\cdot)$ is the Jacobian matrix of the argument. This truncated Taylor–Lie expansion approach has been shown with accurate discrete-time approximation and superior robust performance especially when considering a large sampling timestep in discretization [21, 23]. However, this approach could also be computationally expensive as a series of composite Jacobian matrices need to be calculated.

19.3.2 Multistep-Ahead System Discretization

Runge–Kutta methods [27] which are widely used for solving ODEs' initial value problems can be a natural choice for discrete-time system representation. Runge–Kutta methods propagate a solution over an interval by combining the information from several Euler-style steps (e.g., Eq. 19.30), and then using the information obtained to match a Taylor series expansion to some higher order.

$$\mathbf{x}(k+1) = \mathbf{x}(k) + h\mathbf{f}(\mathbf{x}(k)) \tag{19.30}$$

Here, h is the sampling interval, $t_{k+1} = t_k + h$.

The discrete time representation of the continuous-time pathway model (19.1) can be written as follows using the Runge–Kutta method.

$$\begin{aligned} \mathbf{x}(k+1) &= \mathbf{x}(k) + \mathbf{R}(\mathbf{x}(k)) \\ \mathbf{y}(k+1) &= \mathbf{g}(\mathbf{x}(k+1)) \end{aligned} \tag{19.31}$$

Here, $\mathbf{R}(\mathbf{x}(k))$ represents the Runge–Kutta formula. According to desired modeling accuracy, different order Runge–Kutta formulas can be employed, and the corresponding $\mathbf{R}(\mathbf{x}(k))$ would be different, for instance, the second-order Runge–Kutta formula with the expression:

$$\begin{aligned} d_1 &= h\mathbf{f}(\mathbf{x}(k)) \\ d_2 &= h\mathbf{f}(\mathbf{x}(k) + d_1/2) \\ \mathbf{R}(\mathbf{x}(k)) &= d_2 \end{aligned} \tag{19.32}$$

where $\mathbf{f}(\cdot)$ is the right-hand side of the model's ODEs in Eq. 19.1.

The most often used classical fourth-order Runge–Kutta formula is:

$$\begin{aligned} d_1 &= h\mathbf{f}(\mathbf{x}(k)) \\ d_2 &= h\mathbf{f}(\mathbf{x}(k) + d_1/2) \\ d_3 &= h\mathbf{f}(\mathbf{x}(k) + d_2/2) \\ d_4 &= h\mathbf{f}(\mathbf{x}(k) + d_3) \\ \mathbf{R}(\mathbf{x}(k)) &= d_1/6 + d_2/3 + d_3/3 + d_4/6 \end{aligned} \tag{19.33}$$

In practice, fourth-order Runge–Kutta is generally superior to second-order due to four evaluations of the right-hand side per step h.

Compared with one-step discretization approaches discussed in the previous subsection, the main advantage of using the Runge–Kutta method for discretization is that it utilizes multistep information for approximation of one-step-ahead predictions. This strategy enhances the discretization accuracy and reduces the complexity of the mathematical expressions compared with using higher-order derivative approximation in one-step discretization. Moreover, the Runge–Kutta method could provide a general discrete-time ODE representation for either linear or nonlinear ODEs, whereas for a one-step strategy using Taylor expansion it is difficult to formulate a close representation for nonlinear ODEs; instead some finite-order approximation has to be used.

Similarly, now the parameter sensitivity equations Eq. 19.9 can also be discretized using the Runge–Kutta method:

$$\mathbf{S}(k+1) = \mathbf{S}(k) + \mathbf{R}_S(\mathbf{S}(k)) \tag{19.34}$$

Here within the Runge–Kutta formula $\mathbf{R}_S(\mathbf{S}(k))$, $\mathbf{f}_S(\cdot)$ represents the right-hand side of sensitivity equations Eq. 19.9, for example,

$$\mathbf{f}_S(\mathbf{S}(k)) = J\mathbf{S}(t(k)) + P \tag{19.35}$$

Thus, parameter sensitivity coefficients Eq. 19.8 and the Fisher information matrix Eq. 19.12 can now be represented and calculated iteratively using the discrete-time formula Eq. 19.34.

In this section, two sorts with three kinds of discrete-time model representation methods are investigated for pathway ODEs in depth. The Runge–Kutta-based method shows superiority in mathematical model expression especially for discretization of nonlinear ODEs. In the next section, the simulation results of five discrete-time models based on these three approaches are discussed and compared numerically and graphically using the Michaelis–Menten kinetic model.

19.4 Simulation Results

In this section, a simple pathway example is employed for illustration of the different discussed discretization approaches. The accuracy and computational cost of different methods are compared as well. The example discussed here is the well-known

Michaelis–Menten enzyme kinetics. The kinetic reaction of this signal transduction pathway can be represented as

$$S+E \underset{\theta_2}{\overset{\theta_1}{\rightleftarrows}} ES \overset{\theta_3}{\rightarrow} E+P \tag{19.36}$$

Here, E is the concentration of an enzyme that combines with a substrate S to form an enzyme substrate complex ES. The complex ES holds two possible outcomes in the next step. It can be dissociated into E and S, or it can further proceed to form a product P. Here, θ_1, θ_2, and θ_3 are the corresponding reaction rates. The pathway kinetics described in Eq. 19.36 can usually be represented by the following set of ODEs.

$$\begin{aligned} \dot{x}_1 &= -\theta_1 x_1 x_2 + \theta_2 x_3 \\ \dot{x}_2 &= -\theta_1 x_1 x_2 + (\theta_2 + \theta_3) x_3 \\ \dot{x}_3 &= \theta_1 x_1 x_2 - (\theta_2 + \theta_3) x_3 \\ \dot{x}_4 &= \theta_3 x_3 \end{aligned} \tag{19.37}$$

Here four states x_1, x_2, x_3, and x_4 refer to S, E, ES, and P in Eq. 19.36, respectively. As it is a linear set in the parameter and bilinear in state ODEs, Eq. 19.37 can be written in matrix form:

$$\dot{\mathbf{x}} = \mathbf{A}\mathbf{x} + \mathbf{D}\mathbf{x} \otimes \mathbf{x} \tag{19.38}$$

First, we consider the one-step-ahead discretization methods discussed in Sect. 19.3. For the Taylor series and Carleman linearization based method, according to Eq. 19.20, the first- and second-order local Taylor series approximation can be expressed as

$$\mathbf{x}(k+1) = \mathbf{x}(k) + T\dot{\mathbf{x}}(k) \tag{19.39}$$

$$\mathbf{x}(k+1) = \mathbf{x}(k) + T\dot{\mathbf{x}}(k) + T^2/2\ddot{\mathbf{x}}(k) \tag{19.40}$$

As shown in Eq. 19.22, here

$$\begin{aligned} \dot{\mathbf{x}}(k) &= \mathbf{A}\mathbf{x}(k) + \mathbf{D}\mathbf{x}(k) \otimes \mathbf{x}(k) \\ \ddot{\mathbf{x}}(k) &= \mathbf{A}\dot{\mathbf{x}}(k) + 2\mathbf{D}\dot{\mathbf{x}}(k) \otimes \mathbf{x}(k) \end{aligned} \tag{19.41}$$

In this example,

$$\mathbf{x} = \begin{bmatrix} x_1 \\ x_2 \\ x_3 \\ x_4 \end{bmatrix}, \quad \mathbf{A} = \begin{bmatrix} 0 & 0 & \theta_2 & 0 \\ 0 & 0 & \theta_2+\theta_3 & 0 \\ 0 & 0 & -(\theta_2+\theta_3) & 0 \\ 0 & 0 & \theta_3 & 0 \end{bmatrix} \tag{19.42}$$

$$\mathbf{D} = \begin{bmatrix} 0 & -\theta_1 & 0 & 0 & 0 & 0 & 0 & 0 & 0 & 0 & 0 & 0 & 0 & 0 & 0 & 0 \\ 0 & -\theta_1 & 0 & 0 & 0 & 0 & 0 & 0 & 0 & 0 & 0 & 0 & 0 & 0 & 0 & 0 \\ 0 & \theta_1 & 0 & 0 & 0 & 0 & 0 & 0 & 0 & 0 & 0 & 0 & 0 & 0 & 0 & 0 \\ 0 & 0 & 0 & 0 & 0 & 0 & 0 & 0 & 0 & 0 & 0 & 0 & 0 & 0 & 0 & 0 \end{bmatrix}$$

We can notice that, for this simple kinetic example with only four state variables, matrix $\mathbf{D}(m \times m^2)$ is already in large scale.

For Monaco and Normand-Cyrot's method, here we employ a truncated third-order Lie expansion for simulation. As shown in Eq. 19.29, we only need to notice $\mathbf{f}$ represents the right-hand side expression of model ODE,

$$\mathbf{f}(\mathbf{x}) = \begin{bmatrix} -\theta_1 x_1 x_2 + \theta_2 x_3 \\ -\theta_1 x_1 x_2 + (\theta_2 + \theta_3) x_3 \\ \theta_1 x_1 x_2 - (\theta_2 + \theta_3) x_3 \\ \theta_3 x_3 \end{bmatrix} \tag{19.43}$$

The general Lie series expansion Eq. 19.29 seems concise in expression, however, the composite Jacobian expression would be large in scale and computationally costly.

Now we consider the multistep discretization approach based on Runge–Kutta method. It is straightforward to formulate the second and fourth order Runge–Kutta discrete-time expression using Eqs. 19.31–19.33. For this example, we just need to replace $\mathbf{f}(\cdot)$ with Eq. 19.43, and as all the states can be measured for this example, the measurement function $\mathbf{g}$ is an identity matrix $\mathbf{I}_4$, therefore

$$\mathbf{y}(k+1) = \mathbf{x}(k+1) \tag{19.44}$$

Comparing the discrete-time model expression using local Taylor–Carleman expansion, Monaco and Mormand-Cyrot's method, and the Runge–Kutta method, the expression using the Runge–Kutta method is more straightforward and compact; we only need to replace $\mathbf{f}(\mathbf{x})$ with the right-hand side expression of the specific biochemical reaction ODEs. On the contrary, the discrete representation using second-order Taylor expansion is already a bit complicated as the scale of the $\mathbf{D}$ matrix would become very large as the parameter dimension increases. As discussed in Sect. 19.3.1.1, it would be even more difficult to formulate a model expression using third- or higher-order Taylor expansion.

The time-series simulations using different discrete-time model expressions are displayed and compared in Fig. 19.1. Here, five discrete-time models, which are based on first- and second-order Taylor expansion, second- and fourth-order Runge–Kutta, and Monaco and Mormand-Cyrot, respectively, are employed for comparison. The initial values of states are: $x_1(0) = 12$, $x_2(0) = 12$, $x_3(0) = 0$, and $x_4(0) = 0$. Parameter values are set to be: $\theta_1 = 0.18$, $\theta_2 = 0.02$, and $\theta_3 = 0.23$. The simulation time period is from 0 to 10 with sampling interval 0.3. We solve the model's ODEs (19.37) using the *ode45* function in MATLAB with sampling interval 0.1, and suppose the result as an approximation of real system observation and to be a judgment of different discretization methods. The residual mean squared errors (RMSE) between different models' outputs and the observation are listed and compared in Table 19.1.

Figure 19.1 and Table 19.1 provide states' trajectory simulation results based on five different discrete-time models. It is clear to see that the discrete-time model based on first-order Taylor series gives the worst approximation results with the largest RMSE, and the one based on fourth-order Runge–Kutta gives the closest simulation result to the real observation with the smallest RMSE. Besides, the discrete-time model using Monaco and Normand-Cyrot's method and the second-order Taylor series also provide an acceptable approximation result with relatively small RMSEs to the real observation.

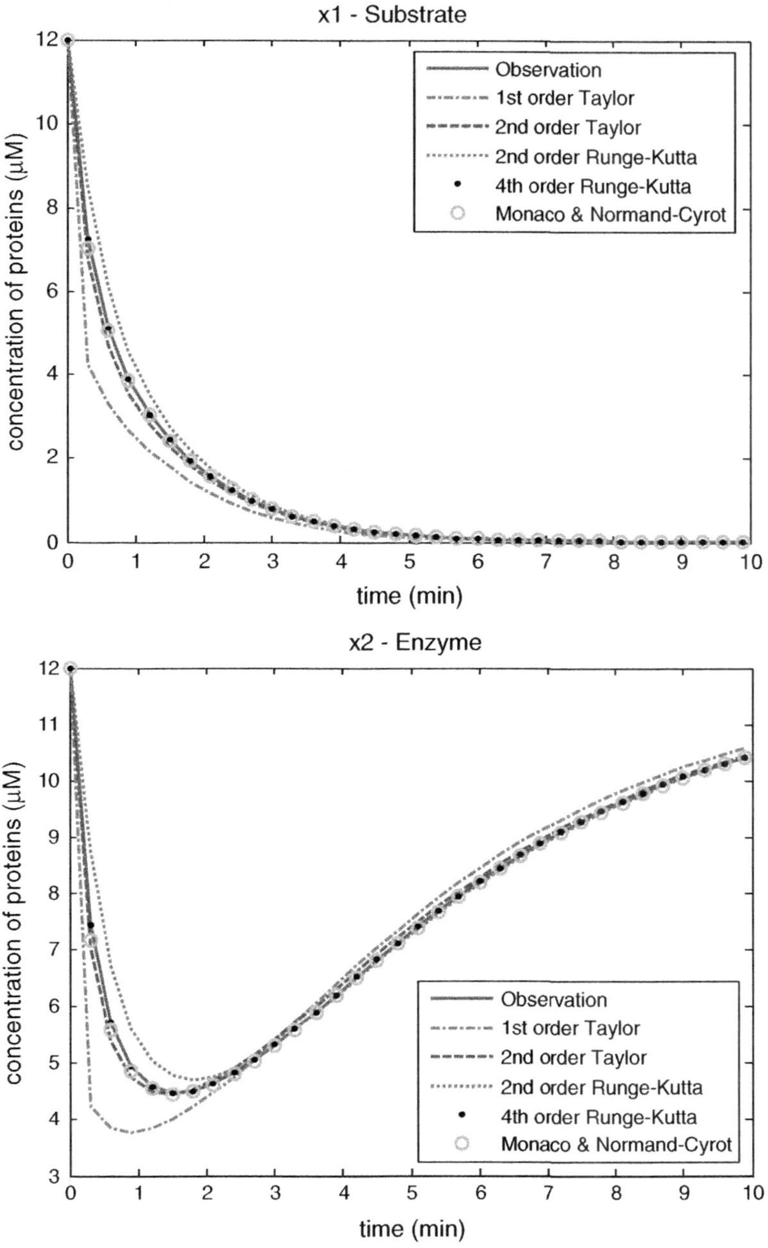

Fig. 19.1 Time series simulation results of using five different discrete-time models

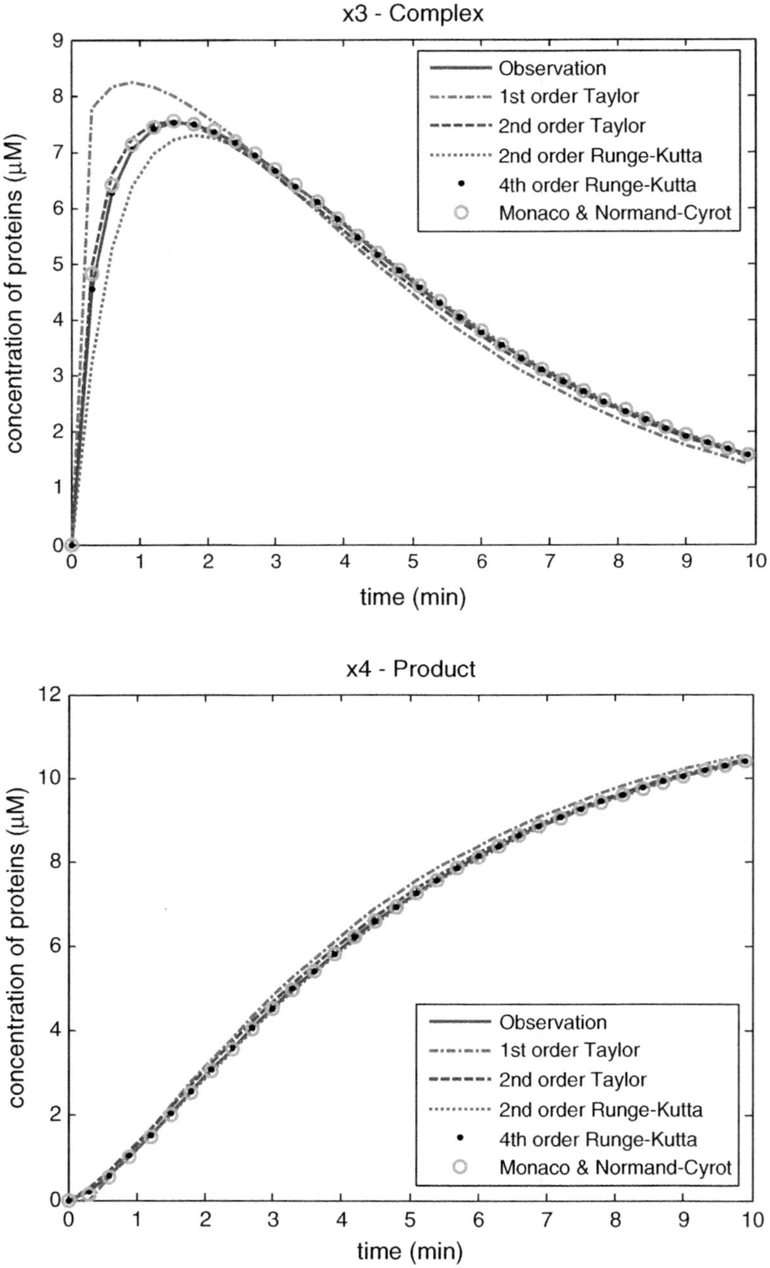

Fig. 19.1 Continued

Table 19.1 Time series simulation residual MSE of different models

RMSE	1st-Order Taylor	2nd-Order Taylor	2nd-Order Runge–Kutta	4th-Order Runge–Kutta	Monaco–Normand–Cyrot
x1	0.4771	0.0205	0.0942	0.0306e-4	0.0020
x2	0.5054	0.0144	0.1019	0.0774e-4	0.0030
x3	0.5054	0.0144	0.1019	0.0774e-4	0.0030
x4	0.0503	0.0105	0.0020	0.1770e-4	0.0002
Total	1.5381	0.0598	0.3000	3.6249e-5	0.0083

19.5 Conclusions

Quantitative discrete-time model representations are important as a link between continuous-time biochemical kinetic reactions and discrete-time experimentation. It will receive more and more attention as computer-based simulation and analyses are widely used in current biochemical pathway modeling studies. Two important types of discretization methods are mainly investigated in this work. One strategy is based on one-step-ahead Taylor or Lie series expansion. This kind of method could give an exact discrete-time representation for linear ODEs, however, for more typical bilinear or nonlinear ODEs pathway models, truncated finite-order Taylor/Lie series approximation have to be used. The mathematical discrete-time expression using higher-order Taylor/Lie expansion can be very complex and it would be computationally costly as well.

The alternative is the Runge–Kutta-based approaches, which are multistep discretization strategies. The mathematical model representation using this method is straightforward and compact, and the simulation approximation result using fourth-order Runge–Kutta is superior to others as well. Synthetically speaking, the Runge–Kutta-based discretization method can be a better choice for discrete-time model representation in a pathway modeling study, and the corresponding discrete-time model structure will be a useful and promising tool in future systems biology research. Further work can focus on dynamic analysis of discrete-time models and comparison with the corresponding continuous model; here dynamic analysis should include model zero dynamics, equilibrium property, chaotic behavior when varying the sampling step, and so on. In addition, discrete-time local and global parametric sensitivity analysis methods would also be a significant further focus for pathway modeling study.

References

1. Anand RA, Douglas AL (2000). Bioengineering models of cell signaling. *Annual Review of Biomedical. Engineering* 2:31–53.
2. Atkinson AC (1992). *Optimum Experimental Designs*. Oxford University Press, New York.

3. Bernard O (2001). Mass balance modelling of bioprocess, *Lecture Notes, Summer School on Mathematical Control Theory*, Trieste.
4. Cho K-H, Wolkenhauer O (2003). Analysis and modeling of signal transduction pathways in systems biology. *Biochemical Society Transactions*, 31(6):1503–1509.
5. Cho K-H, Shin S-Y, Lee H-W, Wolkenhauer O (2003a). Investigations in the analysis and modelling of the TNFα mediated NF-κB signaling pathway. *Genome Research*, 13:2413–2422.
6. Cho K-H, Shin S-Y, Kim H-W, Wolkenhauer O, McFerran B, Kolch W (2003b). *Mathematical Modeling of the Influence of RKIP on the ERK Signaling Pathway. Computational Methods in Systems Biology* (CMSB'03). Lecture Notes in Computer Science, 2602, Springer-Verlag, New York.
7. Eker S, Knapp M, Laderoute K, Lincoln P, Meseguer J, Sonmez K (2002). Pathway logic: Symbolic analysis of biological signaling. *Pacific Symposium on Biocomputing*, pp. 400–412.
8. Eldred MS, Giunta AA, van Bloemen Waanders BG, Wojtkiewicz SF, William WE, Alleva M (2002). DAKOTA, A multilevel parallel object-oriented framework for design optimization, parameter estimation, uncertainty quantification, and sensitivity analysis Version 3.0. Technical Report, Sandia National Labs, USA.
9. Esposito WR, Floudas CA (2002). Deterministic global optimization in isothermal reactor network synthesis. *Journal of Global Optimization*, 22:59–95.
10. Faller D, Klingmuller U, Timmer J (2003). Simulation methods for optimal experimental design in systems biology. *Simulation*, 79(12):717–725.
11. Gadkar KG, Gunawan R, Doyle FJ (2005a). Iterative approach to model identification of biological networks. *BMC Bioinformatics*, 6:155.
12. Gadkar KG, Varner J, Doyle FJ (2005b). Model identification of signal transduction networks from data using a state regulator problem. *IEE Systems Biology*, 2(1):17–30.
13. Ihekwaba AEC, Broomhead DS, Grimley RL, Benson N, Kell DB (2004). Sensitivity analysis of parameters controlling oscillatory signalling in the NF-kB pathway: The roles of IKK and IkBα. *IET Systems Biology*, 1(1):93–103.
14. Isidori A (1995). *Nonlinear Control Systems*, 3rd edn, Springer, London.
15. Jeff H, David M, Farren I, James JC (2001). Computational studies of gene regulatory networks: In *numero molecular biology. Nature Reviews Genetics*, (2):268–279.
16. Katsuhiko O (1995). *Discrete-time Control System*, 2nd edn, Prentice-Hall, Upper Saddle River, NJ, pp. 312–515.
17. Kell DB, Knowles JD (2006). The role of modeling in systems biology. In *Systems Modeling in Cellular Biology: From Concept to Nuts and Bolts*, eds. Z. Szallasi, J. Stelling and V. Periwal, MIT Press, Cambridge, MA.
18. Kowalski K, Steeb W-H (1991). *Nonlinear Dynamical Systems and Carleman Linearization*, World Scientific, Singapore.
19. Kutalik Z, Cho K-H, Wolkenhauer O (2004). Optimal sampling time selection for parameter estimation in dynamic pathway modeling, *Biosystems*, 75(1–3):43–55.
20. Li R, Henson MA, Kurtz MJ (2004). Selection of model parameters for off-line parameter estimation, *IEEE Transactions on Control Systems Technology*, 12(3):402–412.
21. Mendes E, Letellier C (2004). Displacement in the parameter space versus spurious solution of discretization with large time step, *Journal of Physics A: Mathematical and General* (37):1203–1218.
22. Monaco S, Normand-Cyrot D (1985). On the sampling of a linear control system. In *Proceedings of IEEE 24th Conference on Decision and Control*, pp. 1457–1482.
23. Monaco S, Normand-Cyrot D (1990). A combinatorial approach to the nonlinear sampling problem. *Lecture Notes in Control and Information Sciences*, (114):788–797.
24. Mueller TG, Noykova N, Gyllenberg M, Timmer J (2002). Parameter identification in dynamical models of anaerobic wastewater treatment. *Mathematical Biosciences* (177–178):147–160.
25. Peifer M, Timmer J (2007). Parameter estimation in ordinary differential equations for biochemical processes using the method of multiple shooting. *IET Systems Biology*, 1(2):78–88.
26. Peleg M, Yeh I, Altman RB (2002). Modeling biological processes using workflow and Petri net models, *Bioinformatics*, 18(6):825–837.

27. Press WH, Teukolsky SA, Vetterling WT, Flannery BP (1992). *Numerical Recipes in C: The Art of Scientific Computing*, 2nd edn, Cambridge University Press, UK.
28. Regev A, Silverman W, Shapiro E (2001). Representation and simulation of biochemical processes using π-calculus process algebra. *Pacific Symposium on Biocomputing*, pp. 459–470.
29. Robert DP (1997). Development of kinetic models in the nonlinear world of molecular cell biology. *Metabolism*, 46:1489–1495.
30. Robert DP, Tom M (2001). Kinetic modeling approaches to in vivo imaging. *Nature Reviews Molecular Cell Biology*, 2:898–907.
31. Sontag ED (2005). Molecular systems biology and control. *European Journal of Control* 11:1–40.
32. Timmer J, Muller TG, Swameye I, Sandra O, Klingmuller U (2004). Modeling the nonlinear dynamics of cellular signal transduction. *International Journal of Bifurcation and Chaos*, 14(6):2069–2079.
33. Tjoa IB, Biegler LT (1991). Simultaneous solution and optimization strategies for parameter estimation of differential-algebraic equation systems. *Industrial and Engineering Chemistry Research*, 30:376–385.
34. Tyson JJ, Kathy C, Bela N (2001). Network dynamics and cell physiology. *Nature Reviews Molecular Cell Biology*, 2:908–916.
35. van Domselaar B, Hemker PW (1975). Nonlinear parameter estimation in initial value problems. Technical Report NW 18/75, Mathematical Centre, Amsterdam.
36. Wolkenhauer O (2001). Systems biology: The reincarnation of systems theory applied in biology? *Briefings in Bioinformatics*, 2(3):258–270.
37. Wylie CR, Barrett LC (1995). *Advanced Engineering Mathematics*. 6th edn, McGraw-Hill, New York.
38. Yao KZ, Shaw BM, Kou B, McAuley KB, Bacon DW (2003). Modeling ethylene/butene copolymerization with multi-site catalysts: Parameter estimability and experimental design. *Polymer Reaction Engineering*, 11:563–588.
39. Yue H, Brown M, Kell DB, Knowles J, Wang H, Broomhead DS (2006). Insights into the behaviour of systems biology models from dynamic sensitivity and identifiability analysis: A case study of an NF-kB signalling pathway. *Molecular BioSystems*, 2(12):640–649.

Chapter 20
Performance Evaluation of Decision Tree for Intrusion Detection Using Reduced Feature Spaces

Behrouz Minaei Bidgoli, Morteza Analoui, Mohammad Hossein Rezvani, and Hadi Shahriar Shahhoseini

20.1 Introduction

Attack is a serious problem in computer networks. Computer network security is summarized in CIA concepts including confidentiality, data integrity, and availability. Confidentiality means that information is disclosed only according to policy. Data integrity means that information is not destroyed or corrupted and that the system performs correctly. Availability means that the system services are available when they are needed. Security threats have different causes, such as flood, fire, system failure, intruders, and so on.

There are two types of intruders: external intruders who are illegitimate users attacking a machine and internal intruders who have access permission to a system despite some limitations.

Traditional techniques for protection such as user authentication, data encryption, avoidance of programming errors, and firewalls are all located in the first line of defense for network security establishment. If a weak password is chosen then authentication cannot prevent illegal users from entering the system. Firewalls are not able to protect against malicious mobile codes and undefined security policies. Programming errors cannot be avoided as the complexity of system and application software grows rapidly. Therefore it is necessary to use intrusion detection techniques as a second line of defense.

An intrusion detection system (IDS) is a system that monitors events to detect intrusions. Each intrusion causes a series of anomalous behaviors in the system. Thus, the concept of IDS was proposed as a solution for network vulnerabilities [1]. Of course it is important to note that IDS cannot play the role of prevention-oriented techniques such as authentication and access control, but it can be a complementary technique to try to detect most previous suspicious accesses to the network and immunize the network from the next attacks.

Most IDSs work in a real-time fashion, but some of them do not have a real-time nature and only can operate offline, that is, collect the previous data and inject them into their built-in classifier. There is a variety of approaches for modeling IDS. We

Oscar Castillo et al. (eds.), *Trends in Intelligent Systems and Computer Engineering.*

use the data-mining approach for this purpose. Data-mining is one of the machine learning techniques. An IDS that operates based on a data-mining approach is called an expert IDS (EIDS). The stored records in an IDS database have many features with complex interrelationships that are hard for humans to interpret. To detect these relationships, it is necessary for IDSs (especially for real-time IDSs) to use reduced feature sets.

There are two methods for intrusion detection: misuse detection and anomaly detection. Misuse detection is based on knowledge about weak points of a system and known patterns of attacks. To detect each attack we must model the attack scenario. The main difference between techniques of this method is how to describe or model the behavior of an attacker.

The anomaly detection method assumes that each attack causes a deviation from the normal pattern. This method can be implemented in two ways: static and dynamic. The static anomaly detection method is based on the fact that the system under study has no variations. We usually assume that the hardware is fixed and the software is varying. For example, the operating system and the data which are contained in bootstrap never change. If the fixed part of the system deviates from its original form it means that an error or an attack has occurred. Therefore, data integrity is the main challenge in the field of static method.

Dynamic anomaly detection method studies the audit data that are stored after previous operations of the network. With data-mining techniques such as the decision tree we can recognize previous attacks that happened on the system. The main disadvantage of the misused detection method is that it can only detect attacks trained for them and cannot detect new or unknown attacks. The main disadvantage of an anomaly detection method is that if well-known attacks do not match with a user profile they may not be detected.

Another weakness of most systems is that if the attacker knows that his profile is stored he can change his profile slightly and train the system in such a way that the system will consider the attack as a normal behavior. The main advantage of the anomaly detection method is the ability to detect new or unknown attacks.

There are two types of IDS that employ one or both of the intrusion detection methods outlined above: host-based IDS and network-based IDS. Host-based IDSs make their decisions based on information obtained from a host. A generic rule-based intrusion detection model was proposed in 1987 that works based on pattern matching in which any record (object) of audit data is compared against existing profiles. If the object has a deviation from the normal pattern, it will be reported as an anomaly. Several well-known IDSs were developed based on this idea [2]. Network-based IDSs gather data by monitoring the network traffic in which several hosts are connected. The TCP/IP protocol can also be exploited by intrusions such as IP spoofing, port scanning, and so on. Therefore, network-based IDSs not only protect the network but also protect each host implicitly.

In the literature several machine learning paradigms including decision trees [3–5], neural networks [6], linear genetic programming [7], support vector machines [7], Bayesian network [8], multivariate adaptive regression splines [8], fuzzy inference systems [9], and others have been used to develop IDSs.

An IDS must reduce the amount of data to be processed. This is extremely important if real-time detection is desired. In this chapter we investigate and evaluate the performance of the decision tree for several KDDCUP99 feature sets.

The rest of this chapter is organized as follows. In Sect. 20.2, we discuss the DARPA intrusion detection dataset. Section 20.3 discusses related works about the decision tree and feature deduction. In Sect. 20.4, we explain the decision tree and C4.5 algorithm. Section 20.5 reports the results of our experiments on building an intrusion detection model using the audit data from the DARPA evaluation program and reduced datasets obtained from other research. Section 20.6 offers discussion of future work and conclusive remarks.

20.2 KDDCUP99 Data

In 1998, DARPA funded an "Intrusion Detection Evaluation Program (IDEP)" at the Lincoln Laboratory at the Massachusetts Institute of Technology [10]. DARPA intrusion detection data were re-created on the simulated military network environment along with the different attacks embedded in it. The victim machines subjected to these attacks ran Linux, SunOS™, and Solaris™ operating systems. Three kinds of data were collected: transmission control protocol (TCP) packets using the "tcp-dump" utility, basic security module (BSM) audit records using the Sun Solaris BSM utility, and system file dumps. Stolfo et al. [11], among the participants in the DARPA 1998 program, used TCP packets to build the KDD dataset, which consists of records based on individual TCP sessions. Each record has 41 features and the method used to derive these features is discussed in [12]. Data-mining techniques were utilized to generate features using the TCP packets for different connections. The KDD dataset is accessible through the UCI KDD archive [13].

The accuracy of the computational intelligent paradigms was verified by some simulations using the 1998 DARPA intrusion detection evaluation program by MIT Lincoln Labs. The LAN was operated as a real environment, but was blasted with multiple attacks. For each TCP/IP connection, 41 various quantitative and qualitative features were extracted. The 41 features are labeled in order as *A, B, C, D, E, F, G, H, I, J, K, L, M, N, O, P, Q, R, S, T, U, V, W, X, Y, Z, AA, AB, AC, AD, AF, AG, AH, AI, AJ, AK, AL, AM, AN, AO*, and the class label is named as *AP*.

The dataset contains 24 attack types that can be classified into these main categories:

1. Denial of service (DOS) is a class of attack where an attacker makes a computing or memory resource too busy or too full to handle legitimate requests, thus denying legitimate users access to a machine.
2. A remote to user (R2L) attack is a class of attack where an attacker sends packets to a machine over a network, then exploits the machine's vulnerability to illegally gain local access as a user.

3. User to root (U2R) exploits are a class of attack where an attacker starts out with access to a normal user account on the system and is able to exploit vulnerability to gain root access to the system.
4. Probing is a class of attack where an attacker scans a network to gather information or find known vulnerabilities. An attacker with a map of machines and services that are available on a network can use the information to look for exploits.

The designing of an IDS involves training and testing phases. In the training phase, the different computational intelligence paradigms are constructed using the training data to give maximum generalization accuracy on the unseen data. The test data are then passed through the saved trained model to detect intrusions in the test phase.

Some features are derived features, which are useful in distinguishing normal activities from attacks. These features are either nominal or numerical. Some features examine only the connection in the past two seconds that have the same destination host as the current connection, and calculate statistics related to protocol behavior, service, and so on. These are called "same host" features. Some features examine only the connections in the past two seconds that have same service as the current connection and are called "same service features".

Some other connection records are also stored by the destination host, and features are constructed using a window of 100 connections to the same host instead of a time window. These are called "host-based traffic features". R2L and U2R attacks do not have any sequential patterns such as DOS and Probe because the former attacks have the attacks embedded in the data packets whereas the latter have many connections in a short amount of time. Thus some features that look for suspicious behavior in the data packets such as number of failed logins are constructed and these are called "contents features".

20.3 Related Work

In 1999, the KDD conference hosted a classifier learning contest, in which the learning task was to build a predictive model to distinguish attacks from normal connections. Contestants trained and tested their classifiers on an intrusion dataset provided by MIT Lincoln Labs. Each record of this dataset has 41 features consisting of three categories: basic features of individual TCP connections, content features within a connection, and traffic features computed using a two-second time window. The results of the contest were based on the performance of the classifier over a testing dataset of 311,029 cases. Surprisingly, the top three classifiers were all decision tree classifiers [3–5]. These results show the capability of learning and classification of decision trees.

Later works retried the above task with naïve Bayes and decision tree classifiers [14]. They concluded that the naïve Bayes classifier is competitive and requires less training time than the decision tree classifier, although the latter has slightly better performance. All the above works use all 41 features of the KDDCUP99 training and testing data. But the literature survey shows that feature selection is very important in data-mining because the quality of data is an important factor that can affect the success of data-mining algorithms on a given task.

Good surveys reviewing works on feature selection can be found in [15, 16]. In the experiments of these papers, every connection record in the original dataset has 41 features, meaning that the corresponding data space is a 41-dimensional space. Without question, so many features will reduce the efficiency of detection, and some of these features have little effect on detection. In fact, feature selection can be performed without reducing the accuracy of detection remarkably [17].

Sung and Mukkamala [18] have demonstrated that a large number of features are unimportant and may be eliminated, without significantly lowering the performance of the IDS. Their algorithm reduces the 41 variables to 19 variables using SVM and neural networks. A genetic algorithm for feature deduction is given to find an optimal weight for *K* nearest neighbor (*K*NN) classifiers [15, 16].

One of the feature reduction algorithms explained in the literature is the Markov blanket (MB). The MB model algorithm helps to reduce the 41 variables to 17 variables. These 17 variables are *A, B, C, E, G, H, K, L, N, Q, V, W, X, Y, Z, AD*, and *AF* [19]. Another algorithm found in the literatures is the flexible neural tree (FNT). The FNT method helps to reduce the features as given below [17].

Normal: *C, K, U, AN*
Probe: *A, C, L, R, T, U, W, Z, AA, AE, AK, AO*
DOS: *A, H, J, K, P, Q, T, U, W, AB, AB, AC, AE*
U2R: *K, N, Q, AB, AC, AF, AJ, AL*
R2L: *A, C, K, L, M, R, T, W, Y, AL*

Another feature reduction algorithm is the classification and regression tree (CART). CART can reduce the 41 variables to 12 variables. These 12 variables are *C, E, F, L, W, X, Y, AB, AE, AF, AG*, and *AI* [17].

Another algorithm is CLIQUE, which is a density-based clustering algorithm that is specifically designed for finding clusters in subspaces of high-dimensional data. The CLIQUE method helps to reduce the features as given below [20].

Normal: *C, J, W, X, Y, Z, AC, AD, AF, AG, AH, AI, AL, AM, AN, AO*
Probe: *C, W, X, Y, AA, AC, AD, AF, AG, AH, AI, AJ, AL, AN*
DOS: *A, C, E, F, J, K, V, W, X, Y, Z, AA, AB, AC, AD, AE, AF, AG, AH, AI, AJ, AK, AL, AM, AN, AO*
U2R: *C, W, X, AG*
R2L: *B, C, W, X, AG*

The important note that all the literature has pointed to is that the performance of these approaches strongly depends to the classifier that uses them.

20.4 Decision Tree

The decision tree classifier by Quinlan [21] is one of the most well-known machine learning techniques. ID3 [22] and C4.5 [21] are algorithms introduced by him for inducing decision trees. C4.5 is an extension of ID3 that accounts for unavailable values, continuous attribute value ranges, pruning of decision trees, rule derivation, and so on. A decision tree is made of decision nodes and leaf nodes. We are given a set of records. Each record has the same structure, consisting of a number of attribute/value pairs. One of these attributes, called a target attribute, represents the label of the record. The problem is to determine a decision tree that on the basis of answers to questions about the nontarget attributes correctly predicts the value of the target attribute. Usually the target attribute takes only the values $\{true, false\}$, $\{success, failure\}$, or something equivalent. In any case, one of its values will mean failure. For example, in IDS applications the target attribute has two values: "*attack*" and "*no attack*".

The process of constructing a decision tree is basically a divide and conquer process. A set T of training data consists of k classes $(c_1, c_2, \ldots, c_k)$ denoted as C. The number of features in the dataset is n and each attribute is denoted as a_i, where $1 \leq i \leq n$. Each attribute a_i has m possible values denoted as $v1, \ldots, vm$. If T only consists of cases of one single class, T will be a leaf. If T contains no case, T is a leaf and the associated class with this leaf will be assigned with the major class of its parent node. If T contains cases of mixed classes (i.e., more than one class), a test based on some attribute a_i of the training data will be carried out and T will be split into m subsets $(T_1, T_2, \ldots, T_m)$, where m is the number of outcomes of the test over attribute a_i. The same process of constructing a decision tree is recursively performed over each T_j, where $1 \leq j \leq m$, until every subset belongs to a single class. Given n attributes, a decision tree may have a maximum height of n. The algorithm for constructing the decision tree is shown below.

Function ID3

Input: (C: a set of nontarget attributes, L: the target attribute, T: a training set) returns a decision tree;

Begin

If T is empty, return a single node with value Failure;
If T consists of records all with the same value for the target attribute, return a single leaf node with that value;
If C is empty, then return a single node with the value of the most frequent values of the target attribute that are found in records of T; [in that case there may be errors, examples that will be improperly classified];
Let a_i be the attribute with largest $Gain(a_i, T)$ among attributes in C;
Let $\{v_j | j = 1, 2, \ldots, m\}$ be the values of attribute a_i;
Let $\{T_j | j = 1, 2, \ldots, m\}$ be the subsets of T consisting respectively of records with value v_j for a_i;

Return a tree with root labeled a_i and arcs labeled $v_1, v_2, \ldots, v_m$ going respectively to the trees (ID3($C - \{a_i\}, L, T_1$), ID3($C - \{A\}$, L, T_2), ..., ID3 ($C - \{a_i\}$, L, T_m);
Recursively apply ID3 to subsets $\{T_j | j = 1, 2, \ldots, m\}$
Until they are empty

End

The choice of test condition depends on the objective measure used to determine the goodness of a split. Some of the widely used measures include entropy, the Gini index, and the statistical tests. Impurity measures such as entropy tend to favor attributes that have a large number of distinct values. The criterion that C4.5 chooses is the gain ratio criterion. According to this criterion, at each splitting step, an attribute which provides the maximum information gain is chosen. This is done with respect to reducing the bias in favor of tests with many outcomes by normalization.

After construction of the decision tree, it can be used to classify test data that have the same features as the training data. Starting from the root node of the decision tree, the test is carried out on the same attribute of the testing case that the root node represents. The decision process takes the branch whose condition is satisfied by the value of the tested attribute. This branch leads the decision process to a child of the root node. The same process is recursively executed until a leaf node is reached. The leaf node is associated with a class that is assigned to the test case. Because all KDD99 features are continuous the decision tree constructed by C4.5 is a binary tree.

20.5 Experiment Setup and Results

The data for our experiments were prepared by the 1998 DARPA intrusion detection evaluation program by MIT Lincoln Lab. We use the 10% of the KDDCUP99 for training and testing the decision tree in our experiments [23]. This dataset has five different classes, namely Normal, Probe, DOS, U2R, and R2L. The training and test are comprised of 5,092 and 6,890 records, respectively. As the dataset has five different classes we performed a five-class binary classification. The normal data belong to class 1, Probe belongs to class 2, DOS belongs to class 3, U2R belongs to class 4, and R2L belongs to class 5. We ran the experiments for all four attack categories and built a decision tree for each category. All experiments were performed using a full-cache 2.8 GHz Intel processor with 2×512 MB of RAM.

Evaluating the performance of a classification model requires counting the number of test records predicted correctly and wrongly by the model. The counts are tabulated in a table known as a confusion matrix. Table 20.1 depicts the confusion matrix for a binary classification problem. Each entry f_{ij} in this table denotes the number of records from class i predicted to be of class j. For instance, f_{01} is the number of records from class 0 wrongly predicted as class 1. Suppose attack transactions are denoted as class 1 and normal transactions are denoted as class 0. In this

Table 20.1 Confusion matrix for a two-class problem

		Predicted class	
		Class 1	Class 0
Actual class	Class 1	f_{11}	f_{10}
	Class 0	f_{01}	f_{00}

case, records that belong to class 1 are also known as positive examples whereas those from class 0 are called negative examples. The following terminology can be used to describe the counts within a confusion matrix [24].

f_{11} = True Positive (TP)
f_{10} = False Negative (FN)
f_{01} = False Positive (FP) or False Alarm
f_{00} = True Negative (TN)

Based on the entries in the confusion matrix, the total number of correct predictions made by the model is $(f_{11} + f_{00})$ and the total number of wrong predictions is $(f_{10} + f_{01})$.

Although a confusion matrix provides the complete information we need to determine how good the classification model is, it is useful to summarize this information into a single number. There are several performance metrics available for doing this. One of the most popular metrics is model accuracy, which is defined as Eq. (20.1).

$$\text{accuracy} = \frac{f_{00} + f_{11}}{f_{00} + f_{01} + f_{10} + f_{11}} \tag{20.1}$$

We construct the decision tree classifier using the training data and then pass the test data through the saved trained model. The results of classifying with each feature set are shown in Tables 20.2–20.7. Each table shows the training time, the test time, and the accuracy. Table 20.2 shows the performance of the decision tree constructed by C4.5 using the 41-variable original dataset and Tables 20.3–20.5 show the performance using the 19-variable, the 17-variable, and the 12-variable reduced datasets. The performance of the reduced datasets generated by the CLIQUE and FNT algorithms are shown in Tables 20.6 and 20.7. From the results, we can conclude that the 12-variable dataset gives the best performance to detect a normal class with 100% accuracy whereas the 19-variable and the FNT datasets give the worst performance of 95% and 83% accuracies, respectively.

Also from the results, we can conclude that the 17-variable dataset gives the best performance to detect the Probe class with 100% accuracy and the 41-variable, 19-variable, and 12-variable datasets give the worst performance. By using the C4.5 decision tree model, a DOS attack cannot be detected with 100% accuracy, however, the 41-variable, 17-variable, and CLIQUE datasets give a better performance than other datasets and the 12-variable dataset gives the worst performance with only 85% accuracy. The 19-variable dataset gives the best performance to detect the U2R

Table 20.2 Performance of 41-variable feature set

Class	Train time (s)	Test time (s)	Accuracy (%)
Normal	1.53	0.17	99.50
Probe	1.67	0.03	83.28
DOS	2.42	0.05	97.13
U2R	1.43	0.02	13.17
R2L	1.73	0.03	8.36

Table 20.3 Performance of 19-variable feature set

Class	Train time (s)	Test time (s)	Accuracy (%)
Normal	1.15	0.02	95
Probe	1.25	0.14	82
DOS	1.20	0.12	94
U2R	0.90	0.03	50
R2L	1.02	0.09	30

Table 20.4 Performance of 17-variable feature set

Class	Train time (s)	Test time (s)	Accuracy (%)
Normal	1.10	0.04	99.53
Probe	1.21	0.14	100
DOS	1.00	0.11	97.30
U2R	0.70	0.02	43
R2L	0.97	0.09	24

Table 20.5 Performance of 12-variable feature set

Class	Train time (s)	Test time (s)	Accuracy (%)
Normal	0.85	0.02	100
Probe	0.90	0.04	83.10
DOS	1.00	0.08	85
U2R	0.49	0.03	30
R2L	0.81	0.02	20

Table 20.6 Performance of CLIQUE algorithm feature set

Class	Train time (s)	Test time (s)	Accuracy (%)
Normal	1.00	0.13	96.62
Probe	0.97	0.07	95.20
DOS	1.60	0.13	98
U2R	0.40	0.01	17
R2L	0.35	0.01	23

Table 20.7 Performance of FNT algorithm feature set

Class	Train time (s)	Test time (s)	Accuracy (%)
Normal	0.23	0.01	83
Probe	0.89	0.05	89
DOS	1.00	0.09	90
U2R	0.30	0.03	40
R2L	0.50	0.02	19

class with 50% accuracy and the 41-variable datasets give the worst performance with only 13% accuracy. Finally, the 19-variable dataset gives the best performance to detect the R2L class with 30% accuracy and the 41-variable dataset gives the worst performance with only 8% accuracy. It is also found that the C4.5 can classify more accurately on smaller datasets.

As illustrated in the tables, except U2R and R2L, all the other classes were classified well by the C4.5 algorithm. A literature survey shows that the intrusion detection models proposed for R2L attacks fail to demonstrate desirable performance with high detection and low false alarm rates using the KDD dataset [25]. Heuristic rules seem to be popular to detect R2L attacks possibly due to the nature of these attacks. The intrusion detection models perform well on the KDD training dataset but fail to detect R2L attacks in the KDD testing dataset. This indicates that the attack signatures present in KDD training and testing datasets may not be correlated.

This lack of correlation can occur if there are many new attacks in the testing dataset that have signatures different than those present in the training dataset. Hence to build a successful R2L detection model using the KDD data, both training and testing datasets will need to be analyzed. Further analysis of failure for various models in the literature indicates that R2L attacks vary significantly in terms of signatures and hence the models that try to detect all R2L attacks using the same algorithm are highly likely to fail. This observation leads us to the finding that each R2L attack must be individually evaluated with its specialized detection algorithm.

Also our experiments show that the 17-variable dataset is the most successful dataset for detection of most attacks. Furthermore, with respect to attack detection accuracies obtained from each dataset we can extract the essential features for each attack. As an example, the 41-variable, 17-variable, and 12-variable datasets give a better performance compared to other datasets for detection of the normal class. Therefore, we can extract the essential features for detection of the normal class from common features among these three superior datasets. The essential features for each attack are shown in Table 20.8. Clearly, in future research, we have to pay more attention to these essential features for each class.

Our results show that if the number of features is reduced, the training time will be reduced too. This is quite reasonable because as the number of features is reduced the depth of the decision tree will probably be shorter than before and consequently there will be fewer choices in each decision node. However, with respect to the tables it reveals that reducing the number of features will not necessarily reduce the testing time. Of course, for the normal class, usually reducing the number of features

Table 20.8 The essential features derived from our experiments

Class	Essential features
Normal	C, E, L, W, X, Y, AF
Probe	A, B, C, E, G, H, K, L, N, Q, V, W, X, Y, Z, AD, AF
DOS	C, E, K, V, W, X, Y, Z, AD, AF
U2R	N, Q, AF
R2L	C, W, X

leads to a reduction in the test time but in the other classes it sometimes causes an increase. This quite depends on the existing relationship between the features in a dataset, not on the number of its features.

20.6 Conclusion

In this research we evaluated the performance of several reduced datasets on the DARPA benchmark intrusion data. We used the reduced datasets obtained by Markov blanket, flexible neural tree, support vector machine, and CLIQUE feature selection methods and analyzed the performance of a decision tree classifier. Following this, we explored some essential features for each attack category based on the results obtained from the superior datasets for each class. We concluded that in future research it is necessary to pay more attention to these essential features for each class.

Empirical results showed that by using C4.5 decision tree, Normal and Probe could be detected with 100% accuracy, DOS with close to 100% accuracy, and U2R and R2L with poor accuracies. It seems that we need more heuristic rules to detect R2L attacks. This is probably due to the nature of these types of attacks. Our experiments showed that the 17-variable dataset is the most successful dataset for detection of most attacks. We found that reducing the number of features will not necessarily reduce the test time. This quite depends on the existing relationship between dataset features, not on the number of features.

References

1. Denning D (1987). An intrusion detection model. *IEEE Transactions on Software Engineering*, SE-13(2), pp. 222–232.
2. Lunt TF, Jagannathan R, Lee R, Listgarten S, Edwards DL, Javitz HS (1988). IDES: The enhanced prototype-A real-time intrusion-detection expert system. Number SRI-CSL-88-12. Menlo Park, CA: Computer Science Laboratory, SRI International.
3. Pfahringer B (2000). Winning the KDD99 classification cup: Bagged boosting. *SIGKDD Explorations*, 1(2), pp. 65–66.
4. Levin I (2000). KDD-99 classifier learning contest LLSoft's results overview. *SIGKDD Explorations*, 1(2), pp. 67–75.

5. Vladimir M, Alexei V, Ivan S (2000). The MP13 approach to the KDD'99 classifier learning contest. *SIGKDD Explorations*, 1(2), pp. 76–77.
6. Mukkamala S, Sung AH, Abraham A (2003). Intrusion detection using ensemble of soft computing paradigms. In: *Third International Conference on Intelligent Systems Design and Applications, Intelligent Systems Design and Applications*, Advances in Soft Computing, Springer Verlag, Germany, pp. 239–248.
7. Mukkamala S, Sung AH, Abraham A (2004). Modeling intrusion detection systems using linear genetic programming approach. In: *The 17th International Conference on Industrial & Engineering Applications of Artificial Intelligence and Expert Systems, Innovations in Applied Artificial Intelligence*, Robert Orchard, Chunsheng Yang, Moonis Ali (Eds.), LNCS 3029, Springer Verlag, Germany, pp. 633–642.
8. Mukkamala S, Sung AH, Abraham A, Ramos V (2004). Intrusion detection systems using adaptive regression splines. In: *Sixth International Conference on Enterprise Information Systems, ICEIS'04*, Portugal, I. Seruca, J. Filipe, S. Hammoudi and J. Cordeiro (Eds.), Vol. 3, pp. 26–33.
9. Shah K, Dave N, Chavan S, Mukherjee S, Abraham A, Sanyal S (2004). Adaptive neuro-fuzzy intrusion detection system. In: *IEEE International Conference on Information Technology: Coding and Computing (ITCC'04)*, USA, IEEE Computer Society, Vol. 1, pp. 70–74.
10. MIT Lincoln Laboratory. URL: http://www.ll.mit.edu/IST/ideval/.
11. Lee W, Stolfo SJ, Mok KW (1999). A data mining framework for building intrusion detection models. In: *IEEE Symposium on Security and Privacy*, Oakland, CA, pp. 120–132.
12. Lee W, Stolfo SJ, Mok KW (1999). Mining in a data-flow environment: Experience in network intrusion detection. In: *Proceedings of the Fifth ACM SIGKDD International Conference on Knowledge Discovery and Data Mining*, San Diego, CA, pp. 114–124.
13. KDD99 dataset (2003). URL: http://kdd.ics.uci.edu/databases/kddcup99/kddcup99.html.
14. Amor NB, Benferhat S, Elouedi Z (2004). Naive Bayes versus decision trees in intrusion detection systems. In: *Proceedings of the 2004 ACM Symposium on Applied Computing*, pp. 420–424.
15. Punch WF, Goodman ED, Pei M, Chia-Shun L, Hovland P, Enbody R (1993). Further research on feature selection and classification using genetic algorithms. In: *Proceedings of the Fifth International Conference on Genetic Algorithms*, pp. 557–560.
16. Pei M, Goodman ED, Punch WF (1998). Feature extraction using genetic algorithms. In: *Proceedings of the International Symposium on Intelligent Data Engineering and Learning*, pp. 371–384.
17. Chebrolu S, Abraham A, Thomas J (2005). *Feature Deduction and Ensemble Design of Intrusion Detection Systems. Computers and Security*, Vol. 24/4, Elsevier Science, New York, pp. 295–307.
18. Sung AH, Mukkamala S (2003). Identifying important features for intrusion detection using support vector machines and neural networks. In: *Proceedings of International Symposium on Applications and the Internet*, pp. 209–210.
19. Tsamardinos I, Aliferis CF, Statnikov A (2003). Time and sample efficient discovery of Markov blankets and direct causal relations. In: *Ninth ACM SIGKDD International Conference on Knowledge Discovery and Data Mining, USA*: ACM Press, New York, pp. 673–678.
20. Agrawal R, Gehrke J, Gunopulos D, Raghavan P (1998). Automatic subspace clustering of high dimensional data for data mining applications. In: *Proceedings of ACMSIGMOD'98 International Conference on Management of Data*, Seattle, WA, pp. 94–105
21. Quinlan JR (1993). *C4.5, Programs for Machine Learning*. Morgan Kaufmann, San Mateo, CA.
22. Quinlan JR (1968). Introduction of decision trees. *Machine Learning*, 1, pp. 86–106
23. KDDcup99 Intrusion detection dataset http://kdd.ics.uci.edu/databases/kddcup99/kddcup.data_10_percent.gz.
24. Fawcett T (2004). *ROC Graphs: Notes and Practical considerations for Researchers*. Kluwer Academic, Dordrecht.
25. Sabhnani M, Serpen G (2003). KDD feature set complaint heuristic rules for R2L attack detection. *Journal of Security and Management*.

Chapter 21
Novel and Efficient Hybrid Strategies for Constraining the Search Space in Frequent Itemset Mining

B. Kalpana and R. Nadarajan

21.1 Introduction

Association rule mining was originally applied in market basket analysis which aims at understanding the behaviour and shopping preferences of retail customers. The knowledge is used in product placement, marketing campaigns, and sales promotions. In addition to the retail sector, the market basket analysis framework is also being extended to the health and other service sectors. The application of association rule mining now extends far beyond market basket analysis and includes detection of network intrusions, attacks from Web server logs, and prediciting user traversal patterns on the Web.

FIM algorithms could be broadly classified as candidate generation algorithms or pattern growth algorithms. Within these categories further classification can be done based on the traversal strategy and data structures used. Apart from these, several hybrid algorithms which combine desirable features of different algorithms have been proposed. A priori Hybrid, VIPER, Max Eclat, and KDCI are some of them. Our work has been motivated by the Zaki [5], which is a hybrid strategy.

We propose two hybrid strategies which make an intelligent combination of a bottom-up and top-down search to rapidly prune the search space.The intelligence gained from each phase is used to optimally exploit the upward and downward closure properties.The strategies are found to outperform the Eclat and Maxeclat as indicated in Sect. 21.7. In this chapter we also give a comparative performance of the strategies on tidset and the diffset organizations. Zaki and Gouda [6] have proved to occupy a smaller footprint in memory and hence are reported to be advantageous.

21.1.1 Problem Statement

The association mining task, introduced in [2] can be stated as follows.

Given a set of transactions, where each transaction is a set of items, an association rule is an expression $X \Rightarrow Y$ where X and Y are sets of items.

Oscar Castillo et al. (eds.), *Trends in Intelligent Systems and Computer Engineering.*

The meaning of such a rule is that transactions in the database which contain the items in X also tend to contain the items in Y. Two measures which determine the interestingness of such a rule are support and confidence. For a given rule expressed as Bread $\Rightarrow$ Cheese [support $= 5\%$, Confidence $= 90\%$], the measure "support $= 5\%$" indicates that 5% of all transactions under consideration show that bread and cheese are purchased together; "Confidence $= 90\%$" indicates that 90% of the customers who purchased bread also purchased cheese. The association rule mining task is a two-step process.

- Find all frequent itemsets. This is both computation and I/O intensive. Given m items there can be potentially 2^m frequent itemsets. It constitutes an area where significant research findings have been reported.
- Generating confident rules. Rules of the form $X/Y \Rightarrow Y$ where $Y \subset X$ are generated for all frequent itemsets obtained in Step I provided they satisfy the minimum confidence.

Our focus is on the generation of frequent itemsets. Table 21.1a shows a sample database with six transactions. The frequent itemsets generated at minimum support 50% is shown in Table 21.1b.

The number in brackets indicates the number of transactions in which the itemset occurs. We call an itemset frequent if it satisfies the minimum support. A frequent itemset is termed maximal frequent if it is not a subset of any other frequent set for a given minimum support. In our example

$\{$A, B, C, D$\}$ is a maximal frequent itemset at minimum support set to 50%. The proposed hybrid strategies aim at finding out the maximal frequent sets and generating their subsets.

Table 21.1a Sample database

Transactions	Items
1.	A, B, C, D
2.	A, B
3.	A, B, C, D, E
4.	A, B, C, D
5.	A, C, E
6.	A, B, C

Table 21.1b Frequent itemsets

Frequent itemsets	Support (Min. Supp = 50%)
A	100%(6)
B, C, AC, AB	83%(5)
ABC, BC	67%(4)
BCD, D, ACD, ABCD	50%(3)
AD, ABD	

21.2 Connecting Lattices and Hybrid Search Strategies

We review some of the definitions from lattice and representation theory [3]. We propose Lemmas 21.1 and 21.2 which form the basis of our itemset pruning strategy.

Definition 21.1. *Let P be a set. A partial order on P is a binary relation $\leq$, such that for all X, Y, $Z \in P$, the relation is:*

1. *Reflexive: $X \leq X$.*
2. *Antisymmetric: $X \leq Y$ and $Y \leq X$; implies $X = Y$.*
3. *Transitive: $X \leq Y$ and $Y \leq Z$; implies $X \leq Z$.*

The set P with relation $\leq$ is called an ordered set.

Definition 21.2. Let P be a nonempty ordered set.

1. If $X \vee Y$ and $X \wedge Y$ exist for all X, $Y \in P$, then P is called a lattice.
2. If $\vee S$ and $\wedge S$ exist for all $S \subseteq P$, then P is called a complete lattice.

For a set I, given the ordered set $P(I)$, the power set of I is a complete lattice in which join and meet are given by union and intersection, respectively.

$$\vee \{A_i / i \in I\} = \bigcup_{i \in I} A_i \tag{21.1}$$

$$\wedge \{A_i / i \in I\} = \bigcap_{i \in I} A_i \tag{21.2}$$

The top element of $P(I)$ and the bottom element of $P(I)$ are given by $T = I$ and $\perp = \{\}$, respectively. For any $L \subseteq P(I)$, L is called a lattice of sets if it is closed under finite unions and intersections; that is, $(L, \subseteq)$ is a lattice with partial order specified by the subset relation $\subseteq$, $X \vee Y = X \cup Y$ and $X \wedge Y$ [5].

The power set lattice for our sample database $I = \{\text{A, B, C, D, E}\}$ shown in Fig. 21.1 constitutes the search space. Maximal frequent sets are indicated by dark circles. Frequent itemsets are grey circles and infrequent itemsets are plain circles. It has been observed that the set of all frequent itemsets forms a meet semi-lattice. For any frequent itemset X and Y, $X \cap Y$ is also frequent. The infrequent itemsets form a join semi-lattice.

Definition 21.3. *Let P be an ordered set and $Q \subseteq P$.*

1. *Q is a down-set (decreasing set and order ideal) if, whenever, $x \in Q$, $y \in P$, and $y \leq x$, we have $y \in Q$.*
2. *Dually, Q is an up-set (increasing set and order filter) if whenever $x \in Q$, $y \in P$, and $y \geq x$, we have $y \in Q$.*

Given an arbitrary subset Q of P and $x \in P$, we define

$\downarrow Q = \{y \in P / (\exists x \in Q)\ y \leq x\}$ *and*
$\uparrow Q = \{y \in P / (\exists x \in Q) y \geq x\}$;
$\downarrow x = \{y \in P / y \leq x\}$ *and* $\uparrow x = \{y \in P / y \geq x\}$

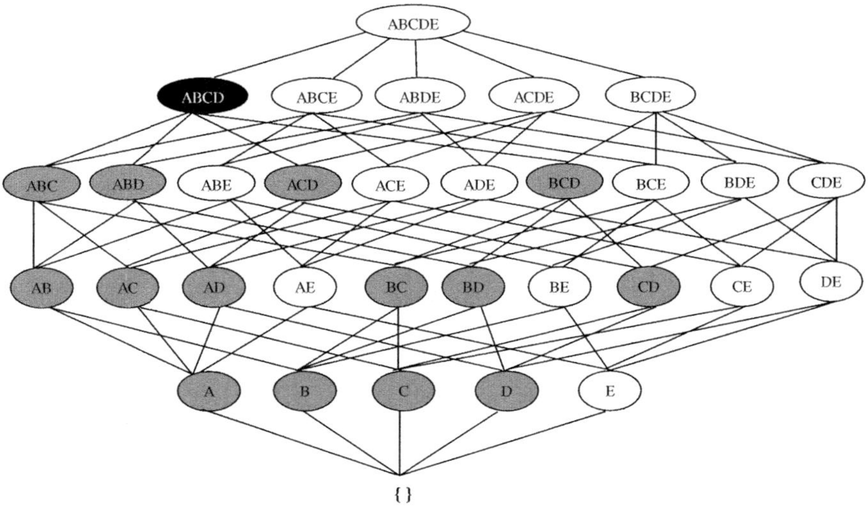

Fig. 21.1 The powerset lattice $P(I)$

Lemma 21.1. *For a maximal frequent itemset $Q \subseteq P$ all down-sets $Q1 = \downarrow Q$; $Q1 \subseteq P$ will also be frequent.*

This is a consequence of the above definition. Fast enumeration of the frequent itemsets is possible in the bottom-up phase once the first maximal frequent set is detected. Examining only the potentially frequent itemsets avoids unnecessary tidlist intersections.

Lemma 21.2. *For a minimal infrequent set $Q \subseteq P$ all up-sets $Q1 = \uparrow Q$; $Q1 \subseteq P$ will be infrequent.*

The topdown phase detects the minimal infrequent sets. In the powerset lattice shown in Fig. 21.1 AE is infrequent and it is observed that all up-sets $Q1 = \uparrow Q$ leading to the top element are also infrequent. Both algorithms alternate the phases in the search heuristically based on the detection of down-sets and up-sets.

21.3 Itemset Enumeration

The enumeration of frequent itemsets forms the computationally intensive task. For a consideration of m distinct items, we have a combination of 2^m subsets, which results in an exponential growth of the search space. Itemset enumeration research thus focuses on reducing the dataset I/O and containing the exploration. There are four applicable classes of I/O reduction suggested in [1]. They are

1. *Projection*: The projection of the database onto an equivalent condensed representation reduces the storage requirement. It may also result in computational optimization through efficient algorithmic techniques.

2. *Partitioning*: Dataset partitioning minimizes I/O costs by enabling memory resident processing of large datasets, thus reducing costly disk accesses.
3. *Pruning*: Dataset pruning techniques dynamically reduce the dataset during processing by discarding unnecessary items. This is significant in reducing the processing time.
4. *Access reduction*: Reducing the number of times that disk resident datasets need to be accessed to identify all frequent itemsets.The hybrid strategies that we propose are directed at maximal pruning of the search space by an optimal exploitation of the upward and downward closure properties.

21.4 Dataset Organization

Dataset organizations are typically horizontal or vertical. In the horizontal form each row contains an object or a transaction id and its related attributes, whereas in the vertical representation, items are represented as columns each containing the transaction where it occurs. Traditional methods used the horizontal format, whereas some of the recent methods have increasingly relied on the vertical format [4–6]. Tidsets, diffsets, and vertical bit vectors are some of the commonly used vertical data formats.

In [4] compressed vertical bitmaps or snakes were introduced to reduce the vertical representation in comparison to the equivalent horizontal representation. Here we make a comparative study of the performance of two novel hybrid strategies on tidsets and diffsets. In the diffset format we keep track of the differences of the tidlist of an itemset from its generating pattern. Diffsets are reported to reduce memory requirements and because they are shorter than the tidlists, the support computations are faster [6].

Figure 21.2 shows the tidlist format and diffset format for the sample database. It is obvious that the diffsets are smaller than the tidlists. In the tidlist format the support for an itemset ABC is computed by intersecting the tidlists of any two of its subsets, say AB and BC. The cardinality of the set obtained by this intersection gives the support. The support computation is different in diffsets. The differences in

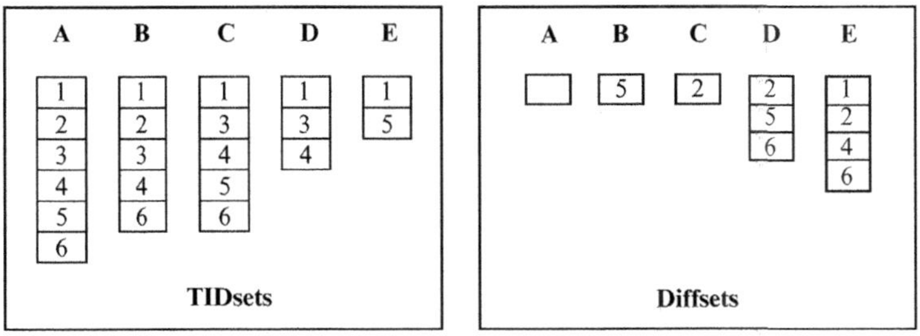

Fig. 21.2 TIDSets and DiffSets for sample database

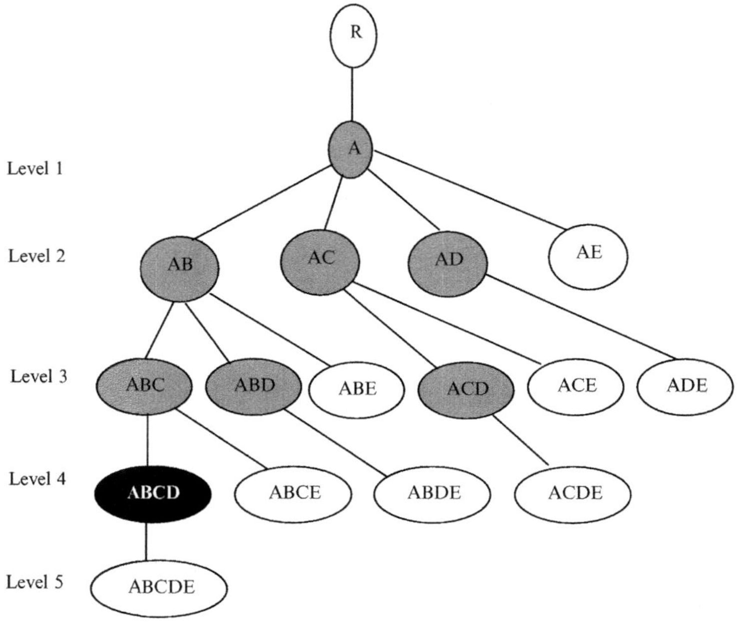

Fig. 21.3 Equivalence class for item A

the tidlists of a class member and its prefix itemset are used. The original database is maintained in tidlist format. Support of an itemset ABC is computed recursively as

$$(\text{ABC}) = (\text{AB}) - |d(\text{ABC})|$$

applying it recursively, we have,

$$(\text{ABC}) = d(\text{AC}) - d(\text{AB}).$$

For a more detailed explanation one is referred to [6]. Also see Fig. 21.3.

21.4.1 Description of Hybrid Miner I

The search starts with a bottom-up phase to identify the maximal frequent item sets. It starts at level n and performs a breadth-first search moving to the next lower level if no maximal frequent itemsets are found at the current level. Once maximal frequent itemsets are found at a given level, we determine items missing from the maximal frequent sets and start a top-down phase that lists the minimal length infrequent sets. Faster search is possible because we examine nodes which contain the missing items only. This phase starts at level 2. If no infrequent sets are found at level 2 we go to the next higher level. The top-down phase ends when minimal infrequent sets are detected.

```
Hybrid Miner I : /* Bottom up phase discovers the maximal
frequent itemsets, top down Phase discovers the minimal
infrequent itemsets */
Begin
        Set flag = false;
        for all sub lattices S induced by θ_k do
        Bottom-up(S):
                Mfreq = ϕ;
                Repeat until flag = true
                     for R ∉ Mfreq
                             L(R) = ∩{L(A_i)/A_i ∈ S};
                     if σ(R) ≥ min-supp then
                        Mfreq = Mfreq ∪ R; flag = true;
                top-down(S):
                level = 2 /* starts for 2-length itemsets */
                infreq = ϕ;
                Repeat for all atoms not in Mfreq
                      ∨A_j ∈ S with j > i do
                              L(R) = L(A_i) ∩ L(A_j);
                              if σ(R) < min_supp then
                                      infreq = infreq ∪ R; break;
                              else
                                      level = level + 1; continue;
        Repeat Bottom up for nodes not containing infrequent
        subsets; /* generate freq itemsets */
                           Max length
                freq = {       ∪       A_i ⊆ Mfreq ∧ A_i ⊄ infreq};
                              i=1
end.
```

Fig. 21.4 Pseudocode for Hybrid Miner I

The bottom-up phase then resumes listing the other maximal frequent itemsets and frequent itemsets after eliminating the nodes containing infrequent itemsets generated in the top-down phase. The computationally intensive support computation task is thus reduced by cleverly alternating the bottom-up and top-down phases every time a maximal itemset is detected. The process of generating the frequent itemsets is then a simple task of enumerating the subsets of all maximal frequent sets. We also make a check to avoid duplicates. The heuristic here is based on the assumption that the items missing from the maximal frequent itemsets are likely to lead to infrequent combinations. The top-down phase thus examines only potentially infrequent nodes. See Fig. 21.4.

21.4.2 Description of Hybrid Miner II

Hybrid Miner II starts with a top-down phase to enumerate the minimal length infrequent itemsets. This method examines the nodes in the ascending order of supports. The bottom-up phase starts when minimal length infrequent itemsets are found in

```
Hybrid Miner II /* Top down phase identifies minimal length
infrequent itemsets. Bottom up phase examines potential nodes
only */
Begin
      for all sub lattices S induced by θk do
      /* atoms sorted on ascending order of support */
       topdown(S):
      begin
            level = 2; infreq = ϕ; flag = false;
            Repeat for all nodes at level while flag = false
                ∨Aj ∈ S with j > i do
                  L(R) = L(Ai) ∩ (Aj);
                  if σ(R) < min_supp then infreq = infreq ∪ R;
                  if lastnode and flag = true then break;
                  else level = level + 1;
        end; /*Top down(S)*/
        Bottom_up(S):
        begin
             Mfreq = ϕ; level = n;
             for R ∉ Mfreq and R ⊄ Nfreq
                   L(R) = ∩ {L(Ai)/Ai ∈ S};
                   if σ(R) ≥ min_supp then MFreq = Mfreq ∪ R;
                   else level = level −1; continue;
end;
            Max length
   freq = {     ∪        Ai ⊂ Mfreq ∧ Ai ⊄ infreq};
               i=1
end.
```

Fig. 21.5 Pseudocode for Hybrid Miner II

an equivalence class. In this phase, the maximal frequent itemsets are generated by only examining nodes not containing the minimal infrequent itemsets. Generating the remaining frequent itemsets is as described for Hybrid Miner I. It is a variation of the Hybrid Miner I in that it attempts to avoid the intensive computation of supports which are encountered for the candidate nodes in the bottom-up phase in the initial stage itself. Hence efficient subset pruning is incorporated at the start of the algorithm itself. See Fig. 21.5. We now highlight some of the strengths of our algorithms.

1. Significant reduction in I/O and memory.
2. The sorting of itemsets at the second level imposes an implicit ordering on the tree. Each child is attached to the parent with the highest support. Redundancy and overlapping among classes is avoided.
3. On comparison with the approaches in [5] it is found that the number of tidlist intersections and nodes examined is reduced by optimally using heuristics to alternate between the top-down and bottom-up phases.

We further draw a theoretical comparison with the best performing Maxeclat proposed in [5]. We manually trace the Hybrid Miner I, Hybrid Miner II, and Maxeclat for the powerset lattice which is shown in Fig. 21.1. Hybrid Miner I examines only 10 nodes to generate the maximal frequent set $\{A, B, C, D\}$. Hybrid Miner II examines 12 nodes whereas Maxeclat will examine 18 nodes for generating the maximal frequent itemset. Our methods thus achieve a search space reduction of almost 50% over the Maxeclat. The savings in computation time and overhead are significant for large databases.

21.5 Experimental Results

The experiments were carried out on synthetic and real datasets. Synthetic databases were generated using the Linux version of the IBM dataset generator. The data mimic the transactions in a retailing environment. The performance of our algorithms is illustrated for synthetic and real datasets in Figs. 21.6–21.8. T, I, and D indicate the average transaction size, the size of a maximal potentially frequent itemset, and the number of transactions, respectively.

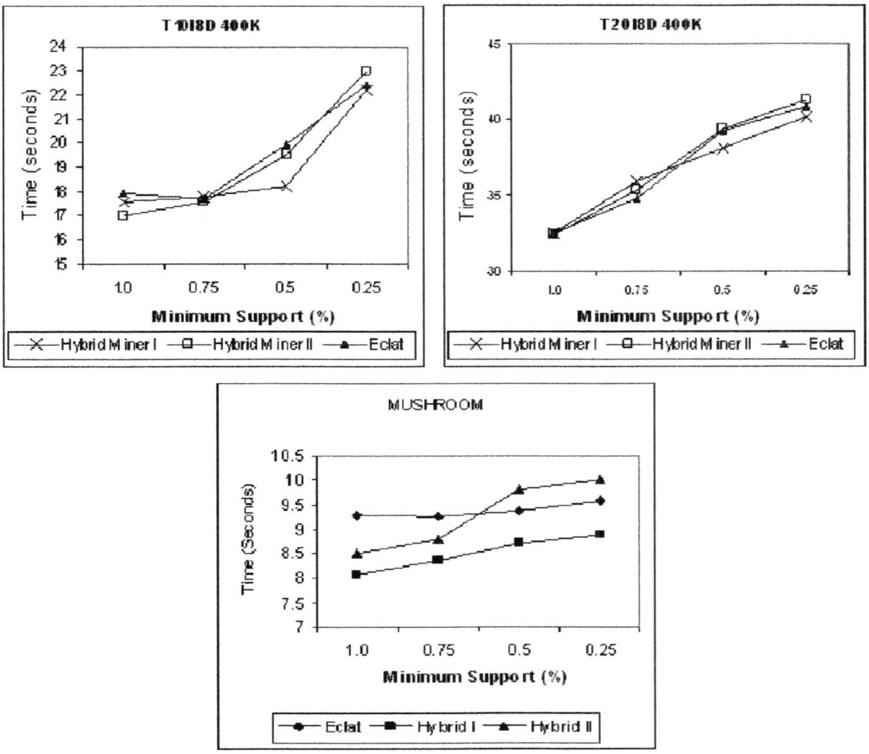

Fig. 21.6 Comparative performance of hybrid strategies with Eclat

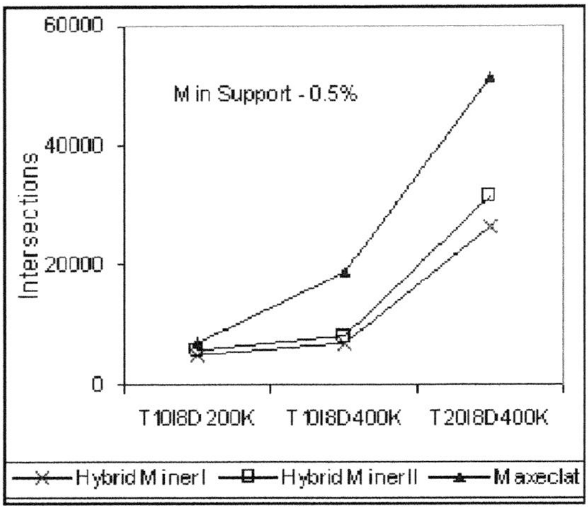

Fig. 21.7 Tidlist intersections

On the tidlist format the execution times of the proposed algorithms in comparison to Eclat are illustrated in Fig. 21.6. Hybrid Miner I performs better than Hybrid Miner II and Eclat for lower supports whereas Hybrid Miner II performs better for higher supports.

Figure 21.7 shows the tidlist intersections. Both Hybrid Miner I and Hybrid Miner II perform about half the number of intersections compared to Maxeclat [5].We give a comparison of the tidlist intersections only with Maxeclat because it is a hybrid strategy. Further reduction in time may be possible through more efficient and compressed data structures.

Figure 21.8 shows the performance of the two strategies on the tidlist and diffset organizations. The hybrid strategies on the diffset format are advantageous on the dense datasets. T10I8D100K is a relatively sparse dataset. There is no significant advantage here. However, on T10I8D400K, T20I8D400K, and the mushroom dataset, the hybrid strategies benefit from reduced execution times. The results indicate that the diffset organization may be more suitable for dense datasets. The choice of the traversal strategy may also favour a particular data format. Because the hybrid strategies use a combination of traversal mechanisms,in our experiments the diffset organization offers a moderate advantage in terms of execution times for dense datasets.

21.6 Conclusion

Our experiments have proved that Hybrid Miner I and Hybrid Miner II are efficient search strategies. Both methods benefit from reduced computations and incorporate excellent pruning that rapidly reduces the search space. From the experiments on

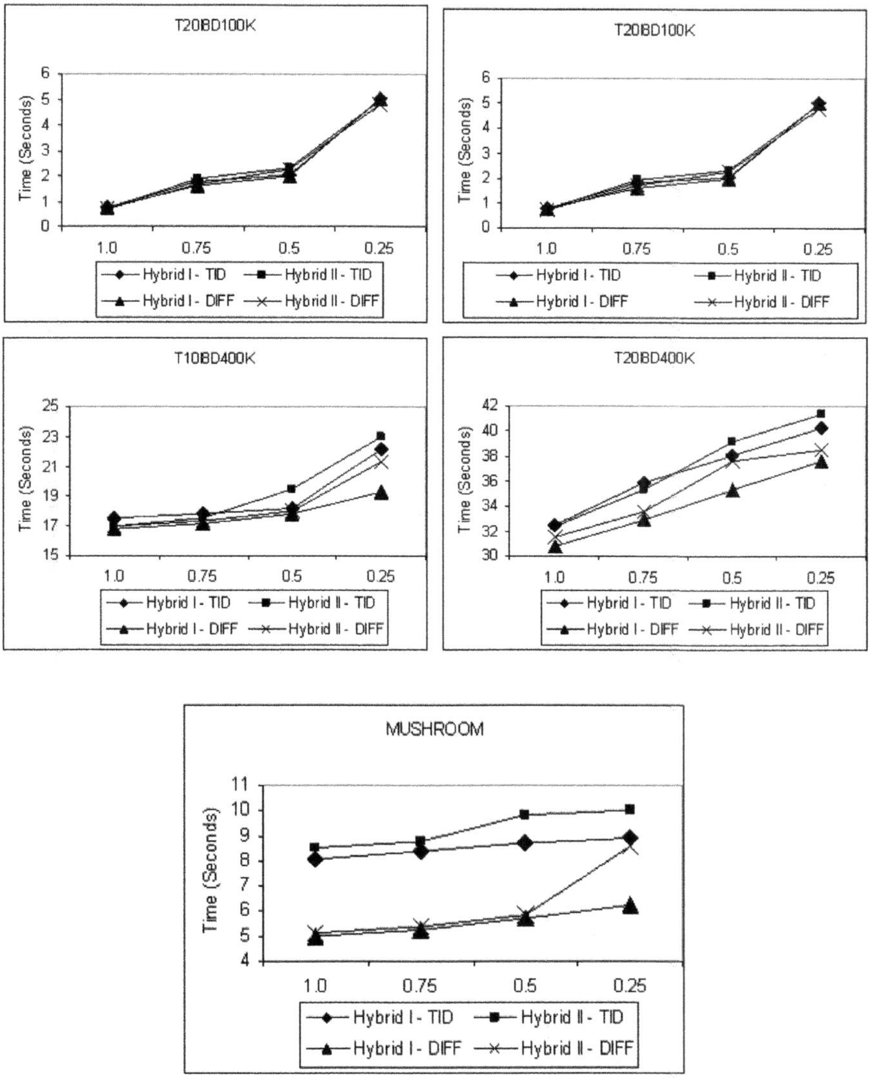

Fig. 21.8 Comparative performance of hybrid strategies on tidsets and diffsets

two different data formats, we find that the diffset format is better in the case of dense datasets Furthermore, both the upward and downward closure properties have been efficiently and optimally utilized. The objective has been to optimize the search for frequent itemsets by applying appropriate heuristics.

Acknowledgements We would like to thank Professor M. J. Zaki for providing the link to the source code of Maxeclat and Eclat.

References

1. Ceglar, A. and Roddick, J.F. (2006). Association mining, *ACM Computing Surveys*, vol. 38, no. 2.
2. Agrawal, R., Imielinski T., and Swami, A. (1993). Mining association rules between sets of items in large databases, *ACM SIGMOD Conference on Management of Data*.
3. Davey, B.A. and Priestley, H.A. (1990). *Introduction to Lattices and Order*, Cambridge University Press, UK.
4. Shenoy, P. et al. (2000). Turbo charging vertical mining of large databases, *International Conference on Management of Data*.
5. Zaki, M.J. (2000). Scalable algorithms for association mining, *IEEE Transactions on Knowledge and Data Engineering*, vol. 12, no. 3, pp. 372–390.
6. Zaki, M.J. and Gouda, K. (2003). Fast vertical mining using diffsets, SIGKDD'.

Chapter 22
Detecting Similar Negotiation Strategies

Lena Mashayekhy, Mohammad A. Nematbakhsh, and Behrouz T. Ladani

22.1 Introduction

Automated negotiation is a key form of interaction in complex systems composed of autonomous agents. Negotiation is a process of making offers and counteroffers, with the aim of finding an acceptable agreement [1]. The agents (negotiators) decide for themselves what actions they should perform, at what time, and under what terms and conditions [1, 2]. The outcome of the negotiation depends on several parameters such as the agents' strategies and the knowledge which one agent has about the opponents [2–5]. In recent years, the problem of modeling and predicting negotiator behavior has become increasingly important because this can be used to improve negotiation outcome and increase satisfaction of results [2–6].

Similarity is a fundamental notion that has to be defined before one can apply various statistical, machine learning, or data-mining methods [5]. Previous works have attempted to exploit the information gathered from opponent's offers during the negotiation to infer similarity between offers of the opponent to predict future offers. Bayesian classification [7] and similarity criteria work [2, 3] are examples of such efforts. When an agent has knowledge of the opponent's strategy, this knowledge can be used to negotiate better deals [1, 6]. However, an agent negotiates with incomplete information about the opponent and therefore using similarity between opponents' strategies makes this information for a negotiator [6]. The main problem is that there is not any measure for calculating similarity between negotiators' strategies.

Sequences of offers are a common form of data in negotiation that an agent can use to discover valuable knowledge in order to achieve its goal [2]. A session is defined as an ordered sequence of offers that an agent creates during negotiation based on its strategy [3]. To detect similarity between negotiators' strategies, we use session data. As data sequences, one method is to reduce sessions to points in a multidimensional space and use Euclidean distance in this space to measure similarity, but in negotiation, sessions do not have the same lengths. One solution

Oscar Castillo et al. (eds.), *Trends in Intelligent Systems and Computer Engineering.*

discussed in [8] for sequences, is to select n data of each sequence. The problem with this approach is: which n offers in each session represent the strategy of the negotiator. Another method is to represent sessions in k-dimensional space using k features for each session [8]. Using the feature vector representation not only needs definition of features to model the negotiator strategy, but also the problem of sessions' similarity is transformed into the problem of finding similar features in k-dimensional space.

In this chapter we consider the problem of defining strategies' similarity or distance between strategies. We start with the idea that similarity between negotiators should somehow reflect the amount of work that has to be done to convert one negotiation session to another. We formalize this notion as Levenshtein or edit distance [8, 9] between negotiations. We apply dynamic programming for computing the edit distances and show the resulting algorithm is efficient in practice.

In detail, the chapter is organized as follows. In Sect. 22.2 we present the problem in negotiations. The definition of similarity between negotiation strategies is given in Sect. 22.3. In Sect. 22.4 we review the negotiation protocol used in our experimentation. We use some negotiation strategies in our simulation discussed in Sect. 22.5. In Sect. 22.6 we present some results of computing similarity measures. Section 22.7 contains conclusions and remarks about future directions.

22.2 Statement of Problem

One way of modeling negotiations is to consider a given set $S = (o_1, \ldots, o_m)$ of offers. S shows a negotiator exchanges m offers during his negotiation session. An offer o consists of one or multiple issues.

The basic problem we consider in this chapter is how one should define a concept of similarity or distance between negotiation sessions. Such a notion is needed in any knowledge discovery application on negotiation. Exchanged offers during negotiation show negotiator session strategy [1–4, 10]. For finding a similar negotiator strategy, if one cannot say when two negotiation sessions are close to each other, the possibility for contrasting them is quite limited. For example, consider three buyers negotiating with a seller who wants to compare behavior of these buyers. The seller observation of these sessions (received offers) is shown in Fig. 22.1. Each of the buyers has its initial offer, deadline, and strategy to generate offers.

Consider the problem of clustering these three buyers. When comparing two buyers to see if they are similar, we need a similarity measure. The meaning of similarity may vary depending on the domain and the purpose of using similarity. For example, someone might group buyer 1 and 2 together, with buyer 3 as the out-group because of the number of exchanged offers. But in this chapter we want to define similarity of negotiators based on their strategy. When a seller observes that received offers from different buyers are similar during their sessions, then this seller finds these buyers have similar strategies. In the next section we discuss this similarity measure.

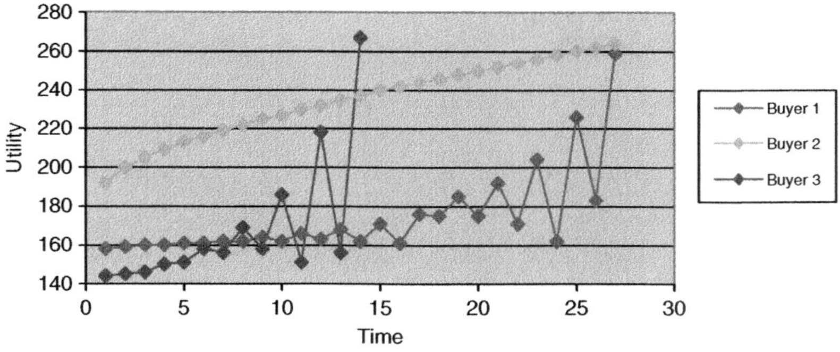

Fig. 22.1 Buyers' offers

22.3 Similarity Measure

In this section, we define two key concepts: first, distance between two sessions and second, distance between two offers.

22.3.1 Distance Between Sessions

We propose a new session similarity measure and use this measure to calculate the similarity between strategies of negotiators. Because offers are made during negotiation, we can refer to them as sequence data.

The idea behind our definition of similarity, or distance, between negotiation sessions is that it should somehow reflect the amount of work needed to transform one negotiation session to another [8, 9]. The definition of similarity is formalized as edit distance $d(S, T)$ for two sessions S and T.

Operations: For calculating the edit distance we need to define a set of transformation operations. We have chosen to use three operations:

- $ins(o)$: inserts an offer of the type o to the negotiation session.
- $del(o)$: deletes an offer of the type o from the negotiation session.
- $update(o, o')$: changes an existing offer from o to o' in the negotiation session.

Cost of operations: Instead of checking equality between two offers o_S and o_T from two sessions S and T, respectively, for each operation we associate a cost $c(op)$ based on distance of offers. The cost of an insertion operation is defined by Eq. 22.1 where o' is a previous offer of o in the negotiation session.

$$c(ins(o)) = distance(o', o), \tag{22.1}$$

With this definition the cost of adding an outlying offer into the negotiation session is higher than the cost of adding in a neighboring offer. The cost of a deletion operation is defined to be the same as the cost of an insert operation. It is proved that if the cost of insertion is equal to the cost of deletion then for each negotiation session *S*, *T* then we have [9]:

$$d(S,T) = d(T,S) \tag{22.2}$$

The cost of an update-operation is defined as Eq. 22.3 where V is a constant value.

$$c(update(o,o')) = V.distance(o,o'), \tag{22.3}$$

With this definition a low distance has a lower cost that a higher distance.

Definition of distance: If the cost of an operation op_i is $c(op_i)$, and k is the number of operations in the sequence Op_j, Eq. 22.4 calculates the cost of operation sequence $Op_j = op_1, op_2, \ldots, op_k$.

$$c(Op_j) = \sum_{i=1}^{k} c(op_i) \tag{22.4}$$

The distance $d(S,\ T)$ is defined as the sum of costs of the cheapest sequence of operations transforming *S* to the *T* as shown in Eq. 22.5.

$$\begin{aligned} d(S,T) = min\,\{&c(Op_j)|Op_j \text{ is an operation sequence} \\ &\text{transforming} \\ &\text{a session } S \text{ to a session } T\} \end{aligned} \tag{22.5}$$

That is $d(S,\ T)$ is the minimum sum of costs of operations transforming S to T.

The problem of finding the edit distance of two sessions (sequence of offers) *S* and *T* is solved using a dynamic programming approach.

22.3.2 Distance Between Offers

Distance between two offers in insert, delete, and update operations can be defined in a different way for each type of negotiation. Let o and o' be two offers.

In single-issue negotiation where each offer has numeric value such as price, $distance(o,\ o')$ is defined as Eq. 22.6.

$$distance(o,o') = |o - o'| \tag{22.6}$$

For nonnumeric issue distance can be calculated based on equality. In that case, the distance between any two offers is defined to be 0 if they are equal, and a positive number if they are not equal.

In multi-issue negotiation, distance is calculated for each issue based on the numeric or nonnumeric value as discussed above. Then Euclidean distance is used for calculating the distance of offers. For instance, if the buyer and seller negotiate on price and delivery time, for calculating distance between two offers first calculate the distance of price in each offer *d(p)* and then distance of delivery time in each offer *d(dt)*. Euclidean distance of *d(p)* and *d(dt)* are set as the distance of the two offers. If issues have different importance, importance has an influence on distance. Let $j \in \{1,\ldots,n\}$ be the issues under negotiation so offer o described as $(o_1, \ldots, o_n)$. The relative importance that an agent assigns to each issue under negotiation is modeled as a weight w_j. Equation 22.7 shows how to calculate distance between two offers.

$$distance(o,o') = \sqrt{\sum_j w_j(o_j - o'_j)^2} \tag{22.7}$$

22.4 Negotiation Protocol

In order to understand the notation used in our model, we first describe its basics. Automated negotiation is a set of agents equipped with a common protocol for bilateral negotiation. In this negotiation model, seller and buyer negotiate on a single issue such as price. We adopt an alternating offers protocol; that is, both of them can send and receive offers and decide whether to accept or reject the received offer until they reach their own deadlines [5, 11]. Each of them has incomplete information about the opponent.

Let $a \in \{b,s\}$ represent the negotiating agents and a' denote agent a's opponent. Let $[\min^a, \max^a]$ denote the range of values for price that are acceptable to agent a. In this model $\min^b$ means initial price and $\max^b$ means reservation price of the buyer, $\max^s$ means initial price and $\min^s$ means reservation price of the seller. A value for price is acceptable to both agents if it is in the zone of agreement $([\min^s, \max^b])$ [10]. This information is shown in Fig. 22.2.

The agents alternately propose offers at times in $T = \{0, 1, \ldots\}$. Each agent has a deadline. T^a denotes agent a's deadline by when the agent must complete the negotiation. Let $p^t_{b\to s}$ denote the price offered by agent b at time t. The agent who makes the first offer is chosen randomly. When an agent receives an offer from her opponent at time t, she rates the offer using its utility function U^a and responses that is defined as [4]:

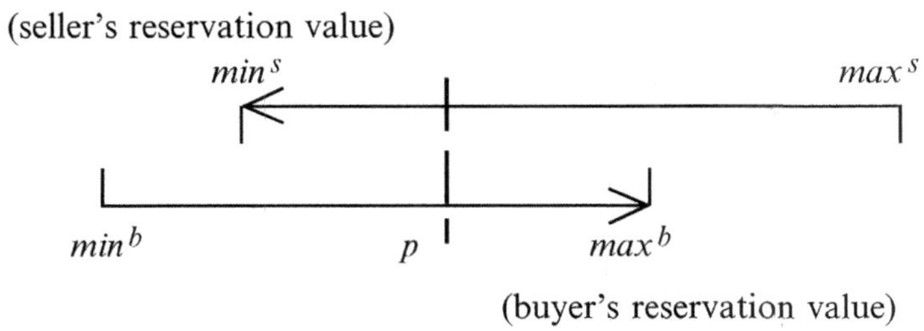

Fig. 22.2 Zone of agreement

$$Action^a(t, p^t_{a' \to a}) = \begin{cases} Quit & If\ t > T^a \\ Accept & \\ If\ U^a(p^1_{a' \to a}) \geq U^a(p^{t+1}_{a \to a'}) & \\ Offer\ p^{t+1}_{a \to a'} & Otherwise \end{cases} \tag{22.8}$$

Offers are generated by the agent's strategy which is discussed in Sect. 22.5.

If the agent's deadline passes, the agent withdraws from the negotiation. An agent accepts an offer when the value of the offered contract is higher than the offer which the agent is ready to send at that moment in time.

The agent's utility function is defined as

$$U^a(p^t) = \begin{cases} \dfrac{max^a - p^t}{max^a - min^a} & if\ a = b \\ \dfrac{p^t - min^a}{max^a - min^a} & if\ a = s \end{cases} \tag{22.9}$$

A negotiation session between b and s at time t_n is a finite sequence of offers from one agent to the other ordered over time. The last element of the sequence is $\{accept, reject\}$.

22.5 Negotiation Strategies

Offers are generated by negotiation strategy [4]. A strategy generates a value for a single negotiation issue. Two types of strategies that we used in our work are time dependent and behavior dependent.

22.5.1 *Time Dependent*

This strategy is parameterized and hence it covers a large number of distinct strategies.

As time passes, the agent will concede more rapidly trying to achieve an agreement before arriving at the deadline. The offer to be uttered by agent a for a decision variable (price) at time $t (0 < t < T^a)$ is computed as follows [1].

$$p^t_{a \to a'} = \begin{cases} min^a + \varphi^a(t)(max^a - min^a) \\ \quad if\ a = b \\ min^a + (1 - \varphi^a(t))(max^a - min^a) \\ \quad if\ a = s \end{cases} \tag{22.10}$$

where $\varphi^a(t)$ is a function depending on time $(0 \leq \varphi^a(t) \leq 1)$ and parameterized by a value β.

$$\varphi^a(t) = \left(\frac{t}{T^a}\right)^{\frac{1}{\beta}} \tag{22.11}$$

A wide range of time-dependent strategies can be defined by varying the way in which $\varphi^a(t)$ is computed [3]. However, depending on the value of β, three qualitatively different patterns of behavior can be identified:

- Boulware if $\beta < 1$
- Linear if $\beta = 1$
- Conceder if $\beta > 1$.

22.5.2 Behavior Dependent

The key feature of this strategy is that it offers based on opponent's behavior [4].

$$p^{t+1} = \begin{cases} min^a & If\ P \leq min^a \\ max^a & If\ P > max^a \\ P & Otherwise \end{cases} \tag{22.12}$$

The parameter P determines the type of imitation to be performed. We can find the following families.

Relative Tit-For-Tat: The agent reproduces, in percentage terms, the behavior that its opponent performed $\delta > 1$ steps ago.

$$P = \frac{p^{t-2\delta}}{p^{t-2\delta+2}} p^{t-1} \tag{22.13}$$

Absolute Tit-For-Tat: The same as before, but in absolute terms.

$$P = p^{t-1} + p^{t-2\delta} - p^{t-2\delta+2} \tag{22.14}$$

Averaged Tit-For-Tat: The agent applies the average of percentages of changes in a window of size $\lambda \geq 1$ of its opponent's history.

$$P = \frac{p^{t-2\lambda}}{p^t} p^{t-1} \tag{22.15}$$

We compute the values for the decision variables under negotiation according to each strategy.

22.6 Experimental Results

In this section we describe how we have evaluated the effectiveness of using this measure for detecting similar negotiator strategies under different negotiation situations. In this experiment we use 2,500 negotiation sessions. In each session a buyer

Table 22.1 Buyers' strategies

Strategy	Percent
Relative TFT	15.4
Random Absolute TFT	19.6
Average TFT	17.6
Boulware	15.8
Linear	15.4
Conceder	16.2
Total	100.0

Table 22.2 Sellers' strategies

Strategy	Percent
Relative TFT	17.2
Random Absolute TFT	12.8
Average TFT	18.2
Boulware	16.8
Linear	16.4
Conceder	18.6
Total	100.0

and a seller negotiate for price. They choose one of the implemented strategies that discussed above (Conceder, Linear, Boulware, Relative TFT, Absolute TFT, or Average TFT). This information is shown in Tables 22.1 and 22.2.

Buyers and sellers save information about their strategies, outcome, and all exchanged offers during the process of negotiation. We show how this measure finds similar strategies.

After gathering data of all sessions, we choose the data of buyers with an accepted result for detecting similarity of these agents. We use our measure for generating distance of these sessions. After calculating all distances we use the k-medoids algorithm [12] for clustering based on these distances to evaluate our measure. This method is helpful because the center of each cluster is one of the existing data in that cluster. This fact is important because we have the distance between sessions and do not need data (offers) in sessions; therefore, to find a cluster center we just need a session which has the minimum distance with other sessions in the cluster. As a result comparisons between sessions and the cluster center is simple. Furthermore, to cluster a new buyer we can compare it with the cluster center if we have data of the session cluster center. If the cluster center is not one of the existing data, we do not have real offers of the cluster center to compute the distance between the cluster center and offers of a new buyer.

After clustering, if two buyers use similar strategy and with the clustering these are in same cluster, and if two buyers use dissimilar strategy and are in different clusters, our method to measure strategy similarity is efficient. Given the strategy of a buyer in his session, this experiment shows sessions that use the same strategy for negotiation form one cluster. As we know the number of buyers strategies we choose $k = 6$ for k-medoids. Table 22.3 shows the top strategy in each session.

Table 22.3 Percent of top strategy in each cluster

Number of cluster	Strategy	Percent
1	Relative TFT	98
2	Random Absolute TFT	100
3	Average TFT	90
4	Boulware	88
5	Linear	89
6	Conceder	100

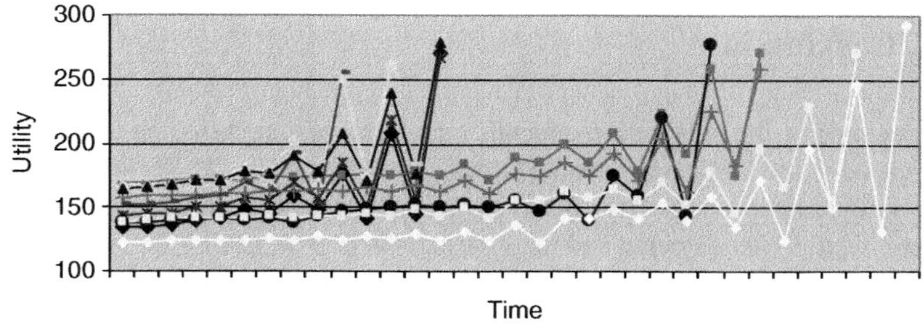

Fig. 22.3 Sessions in first cluster

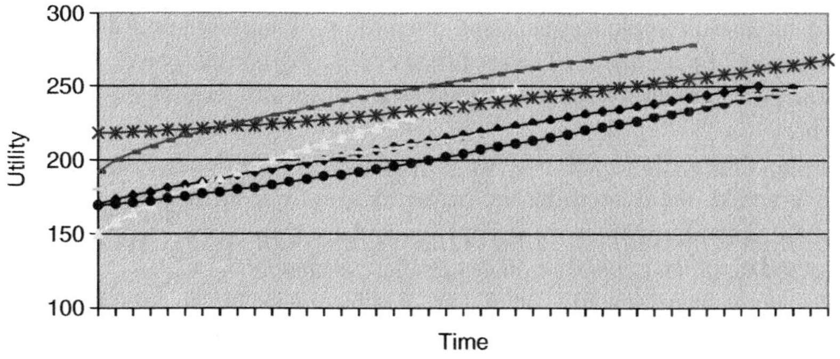

Fig. 22.4 Sessions in second cluster

These results show our method is useful for calculating similarity between buyers' strategies. But in some clusters such as 5, all strategies are not the same; this is because a buyer uses a strategy that is very close to the other strategy. Data of this cluster show that the other buyers' strategies are Boulware with $\beta \cong 1$ which is similar to the Linear strategy. Therefore results show buyers in each cluster have similar behavior.

Figure 22.3 shows changing offers of some sessions in cluster number 2.

In Fig. 22.4 some sessions of cluster number 5 are shown. This cluster contains some Boulware and Conceder strategies which are close to the Linear strategy.

The experiments are repeated with different numbers of clusters and with different negotiation strategies. All experiments show each cluster has buyers that use similar strategies.

As we mentioned above, our experiment was based on data of buyers with accepted outcome, but for other data one can do same as this experiment.

In this chapter we mainly consider a simplified model of negotiation, where each offer has only one issue. As we discussed in Sect. 22.3 the model presented above can be extended for multi-issue negotiation.

22.7 Conclusion

The outcome of the negotiation depends on several parameters such as the agents' strategies and the knowledge one agent has about the others. The problem of modeling and predicting negotiator behavior is important because this can be used to improve negotiation outcome and increase satisfaction of result. Finding similar behavior is one way to solve this problem. We have described a simple method for defining similarity between negotiation strategies. This method is based only on the sequence of offers during negotiation. This characteristic gives the method significant practical value in negotiation; for example, the result can be used in knowledge discovery.

This method is implemented using dynamic programming and it is tested in a simple model of negotiation. The results of comparing our measure for finding similar strategies to chosen strategies are illustrated. Results show that this measure is efficient.

For the future, there are two ways in which this research can be extended. First, we would like to consider the performance of our method against additional strategies. Second, in this work we only consider a single-issue negotiation model; our method could be applied to other negotiation models.

We plan to experimentally use this method for predicting an opponent's strategy during negotiation.

References

1. P Braun, J Brzostowski, G Kersten, J B Kim, R Kowalczyk, S Strecker, and R Vahidov (2006) E-negotiation systems and software agents: Methods, models and applications. In J. Gupta; G. Forgionne; M. Mora (eds.): *Intelligent Decision-Making Support System: Foundation, Applications, and Challenges*, Springer: Heidelberg, Decision Engineering Series, 503, p. 105.
2. R M Coehoorn, N R Jennings (2004) Learning an opponent's preferences to make effective multi-issue negotiation tradeoffs. In *Proceedings of the 6th International Conference on Electronic Commerce (ICEC2004)*, pp. 113–120, Delft, The Netherlands.
3. P Faratin, C Sierra, and N R Jennings (2002) Using similarity criteria to make issue trade-offs in automated negotiations. In *Artificial Intelligence*, 142, pp. 205–237.

4. C Hou (2004) Modelling agents behaviour in automated negotiation. Technical Report KMI-TR-144. Knowledge Media Institute, The Open University, Milton Keynes, UK.
5. H Lai, H-S Doong, C-C Kao, and G E Kersten (2006) Understanding behavior and perception of negotiators from their strategies. *Hawaii International Conference on System Science.*
6. L Mashayekhy, M A Nematbakhsh, and B T Ladani (2006) E-negotiation model based on data mining. In *Proceedings of the IADIS e-Commerce 2006 International Conference*, Barcelona, pp. 369–373.
7. G Tesauro (2002) Efficient search techniques for multi-attribute bilateral negotiation strategies. In *Proceedings of the 3rd International Symposium on Electronic Commerce*, Los Alamitos, CA, IEEE Computer Society, pp. 30–36.
8. M L Hetland (2001) A survey of recent methods for efficient retrieval of similar time sequences. *First NTNU CSGSC.*
9. H V Jagadish, A O Mendelzon, and T Milo (1995) Similarity-based queries. In *Proceedings of the Fourteenth ACM SIGACT-SIGMOD-SIGART Symposium on Principles of Database Systems*, pp. 36–45, ACM Press
10. S S Fatima, M Wooldridge, and N R Jennings (2004) An agenda-based framework for multi-issue negotiation. *Artificial Intelligence*, 152(1), pp. 1–45.
11. C Li, J Giampapa, and K Sycara (2003) A review of research literature on bilateral negotiations. Technical report CMU-RI-TR-03-41, Robotics Institute, Carnegie Mellon University.
12. J Han and W Kamber (2000) *Data Mining: Concepts and Techniques*, Morgan Kaufmann: San Mateo, CA.

Chapter 23
Neural Networks Applied to Medical Data for Prediction of Patient Outcome

Machi Suka, Shinichi Oeda, Takumi Ichimura, Katsumi Yoshida, and Jun Takezawa

23.1 Introduction

Prediction is vital in clinical fields, because it influences decision making for treatment and resource allocation. At present, medical records are readily accessible from hospital information systems. Based on the analysis of medical records, a number of predictive models have been developed to support the prediction of patient outcome. However, predictive models that achieve the desired predictive performance are few and far between.

Approaches to developing predictive models vary from traditional statistical methods to artificial intelligence methods. Multivariate regression models, particularly logistic regression, are the most commonly applied models, and have been for some time. As a potential alternative to multivariate regression models, interest in the use of neural networks (NNs) has recently been expressed [1, 9, 11, 14]. Because each modeling method has its own strengths and limitations [2, 8, 9, 11, 14], it is hard to determine which modeling method is most suitable for the prediction of patient outcome. Medical data are known to have their own unique characteristics, which may impede the development of a good predictive model [7]. Comparative studies using real medical data are expected to pave the way for more effective modeling methods.

In this chapter, we describe the capability of NNs applied to medical data for the prediction of patient outcome. Firstly, we applied a simple three-layer backpropagation NN to a dataset of intensive care unit (ICU) patients [12, 13] to develop a predictive model that estimates the probability of nosocomial infection. The predictive performance of the NN was compared with that of logistic regression using the cross-validation method.

Secondly, we invented a method of modeling time sequence data for prediction using multiple NNs. Based on the dataset of ICU patients, we examined whether multiple NNs outperform both logistic regression and the application of a single NN in the long-term prediction of nosocomial infection. According to the results of these studies, careful preparation of datasets improves the predictive performance of

Oscar Castillo et al. (eds.), *Trends in Intelligent Systems and Computer Engineering.*

NNs, and accordingly, NNs outperform multivariate regression models. It is certain that NNs have capabilities as good predictive models. Further studies using real medical data may be required to achieve the desired predictive performance.

23.2 Medical Data

Medical records consist of a wide variety of data items. Aside from baseline data, treatments, complications, and all that happened to the patient during the hospital stay are sequentially recorded in his or her medical record. Such medical data are characterized by their sparseness, redundancy, conflict, and time sequence.

23.2.1 Dataset of Intensive Care Unit Patients

The Japanese Ministry of Health, Labour, and Welfare established the Japanese Nosocomial Infection Surveillance (JANIS) system, for which participating hospitals routinely report their nosocomial infection surveillance data to a national database. The details of data collection and quality control in the JANIS system are described elsewhere [13]. At each hospital, trained physicians and nurses are responsible for prospective data collection using database-oriented software.

For all patients admitted to the ICU, the following data items are collected between ICU admission and hospital discharge: sex, age, ICU admission (date, time, route, and underlying diseases), APACHE II [6], operation, device use (ventilator, urinary catheter, and CV catheter), infection (pneumonia, urinary tract infection, catheter-related bloodstream infection, sepsis, wound infection, and others), ICU discharge (date, time, route, and outcome), and hospital discharge (date and outcome).

A total of 16,584 patient records were obtained from the JANIS database for the development of predictive models that estimate the probability of nosocomial infection. Part of the dataset of ICU patients is shown in Table 23.1.

23.2.2 Unique Characteristics of Medical Data

23.2.2.1 Sparseness

Predictive models are designed to estimate the probabilities of outcome events of interest. Most of the outcome events occur infrequently (less than 10% probability). Thus, a dataset contains a small number of patient records in which the outcome event occurred (positive cases), compared with a large number of patient records in which the outcome event did not occur (negative cases). In Table 23.1, the outcome event, nosocomial infection, occurs only in record No. 21. Positive cases accounted for 2.1% of the total sample.

Table 23.1 Dataset of ICU patients[a]

No.	Sex	Age	APACHE II	Operation	Ventilator	Urinary catheter	CV catheter	Nosocomial infection
1	F	70+	0–10	Elective	Yes	Yes	Yes	No
2	F	40–69	0–10	Elective	Yes	Yes	Yes	No
3	M	16–39	0–10	Elective	Yes	Yes	Yes	No
4	F	40–69	0–10	Elective	Yes	Yes	Yes	No
5	F	16–39	0–10	Elective	Yes	Yes	Yes	No
6	F	40–69	0–10	Elective	Yes	Yes	Yes	No
7	F	40–69	11–20	Elective	Yes	Yes	Yes	No
8	M	40–69	0–10	Elective	Yes	Yes	Yes	No
9	M	70+	0–10	Urgent	Yes	Yes	Yes	No
10	F	40–69	0–10	Elective	Yes	Yes	Yes	No
11	M	70+	11–20	Elective	Yes	Yes	Yes	No
12	F	40–69	11–20	Elective	Yes	Yes	Yes	No
13	M	16–39	21+	None	Yes	Yes	Yes	No
14	F	40–69	11–20	Urgent	Yes	Yes	Yes	No
15	F	70+	0–10	Elective	Yes	Yes	Yes	No
16	F	70+	21+	None	Yes	Yes	Yes	No
17	M	70+	11–20	Elective	Yes	Yes	Yes	No
18	M	70+	0–10	Elective	Yes	Yes	Yes	No
19	M	40–69	0–10	Elective	Yes	Yes	Yes	No
20	F	40–69	11–20	None	Yes	Yes	Yes	No
21	F	40–69	11–20	Elective	Yes	Yes	Yes	Yes
22	M	70+	21+	Urgent	Yes	Yes	Yes	No
23	M	70+	11–20	Elective	Yes	Yes	Yes	No
24	M	40–69	11–20	None	Yes	Yes	Yes	No

[a]Only the first 24 records are shown.

23.2.2.2 Redundancy

Predictive models incorporate a limited number of predictors. Thus, a dataset contains multiple patient records that consist of the same series of variables (predictors and an outcome). In Table 23.1, there are five groups of redundant records: Nos. 1 and 15; Nos. 2, 4, 6, and 10; Nos. 7 and 12; Nos. 8 and 19; and Nos. 11, 17, and 23. The number of groups of redundant records increases concomitantly with a decrease in the number of predictors.

23.2.2.3 Conflict

Most outcome events depend on a number of unknown factors. Patient records that consist of the same series of predictors do not always have the same outcome. In Table 23.1, there is one group of conflicting records: Nos. 12 and 21. As in the case with redundant records, the number of groups of conflicting records increases concomitantly with a decrease in the number of predictors.

23.2.2.4 Time Sequence

All that happened to a patient during the hospital stay is sequentially recorded in his or her medical record. Thus, a dataset contains a wide variety of data items that are collected at different times. In Table 23.1, "Sex," "Age," "APACHE II," and "Operation" are collected at ICU admission, whereas "Ventilator," "Urinary catheter," "CV catheter," and "Nosocomial infection" are collected at different times after ICU admission. Ideally, predictive models should represent possible causal relationships. Predictors must therefore be selected from data items that are collected prior to the occurrence of the outcome event.

23.2.3 Modeling Methods

Many different modeling methods have been applied in an effort to develop a good predictive model [2, 7, 8]. Logistic regression is the most popular method of modeling the prediction of patient outcome. NNs have recently been proposed as a potential alternative to multivariate regression models [1, 9, 11, 14].

23.2.4 Logistic Regression

Logistic regression is a multivariate regression model that is well suited to binary classification. The probability of the outcome event of interest is related to a series of predictors according to a simple equation:

$$\log[P/(1-P)] = \beta_0 + \beta_1\chi_1 + \beta_2\chi_2 + \mathrm{L} + \mathrm{L} + \beta_n\chi_n \qquad (23.1)$$

where P is the probability of the outcome event of interest, β_0 is an intercept, and β_i are regression coefficients for corresponding predictors $\chi_i (i = 1, 2, \ldots, n)$ [4].

Although the use of logistic regression makes assumptions about the linear relationships between the predictor and the outcome, logistic regression is easy to implement, and can explicitly identify possible causal relationships, as well as specify the magnitude of the relationship between the predictor and the outcome.

23.2.5 Neural Networks

NNs are a computational model that may be used for the same tasks as multivariate regression models. As shown in Fig. 23.1, a NN (three-layer perceptron) consists of a series of nodes, or "neurons," that are arranged into three layers: input, hidden, and output [1].

NNs are generally considered a "black box." Compared with multivariate regression models, NNs have poor ability to explicitly identify possible causal relation-

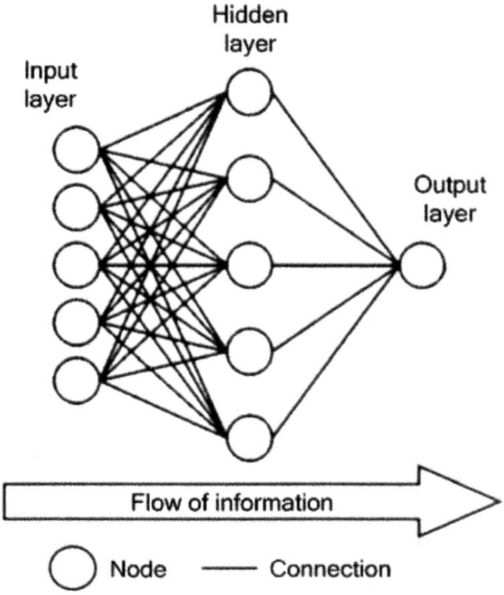

Fig. 23.1 Three-layer perceptron

ships or specify the magnitude of the relationship between the predictor and the outcome. On the other hand, NNs have the following advantages over multivariate regression models.

1. NNs automatically model complex nonlinear relationships among all of the predictors as well as the outcome.
2. NNs automatically deal with the possible interactions between the predictors.
3. The use of NNs does not make assumptions about multivariate normality and homoscedasticity.
4. NNs are relatively insusceptible to multicollinearity and singularity.

A review of the literature has suggested that NNs tend to be equivalent or to outperform multivariate regression models [11].

NNs are programmed to adjust their internal weights based on the mathematical relationships identified between the predictors and the outcome in a training dataset. Careful preparation of datasets may be the key to the predictive performance of NNs, especially in cases where real medical data are used.

23.3 Comparative Study Using Real Medical Data

We applied a simple three-layer backpropagation NN to the dataset of ICU patients for the development of a predictive model that estimates the probability of nosocomial infection. The predictive performance of the NN was compared with that of logistic regression using the cross-validation method.

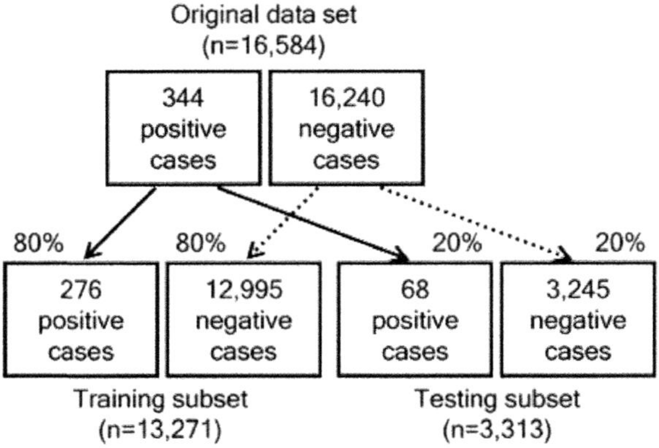

Fig. 23.2 Preparation of datasets

23.3.1 Preparation of Datasets

As mentioned in Sect. 23.2.1, the dataset of ICU patients contained 16,584 patient records. The outcome event of this study was determined by the diagnosis of nosocomial infection during the first four days of ICU stay. There were 344 patient records (2.1%) in which the outcome event occurred (positive cases). After classifying as positive or negative cases, the original dataset was randomly divided into 80% training and 20% testing subsets (Fig. 23.2).

23.3.2 Development of Predictive Models

Two predictive models, one based on logistic regression and the other based on NNs, were developed using the training subset. The predictive models were designed to estimate the probabilities of nosocomial infection based on the following seven predictors: "Sex," "Age," "APACHE II," "Operation," "Ventilator," "Urinary catheter," and "CV catheter". The distribution of predictors in the training subset is shown in Table 23.2. Six out of the seven were significantly associated with the outcome (i.e., nosocomial infection). Moreover, significant interactions were observed between the predictors; for example, old age, urgent operation, ventilator, and CV catheter were more frequently observed in patient records with high APACHE II ($p < 0.001$ with chi-square test). The relationships between the predictors and the outcome and the interactions between the predictors in the dataset of ICU patients are summarized schematically in Fig. 23.3.

Table 23.2 Distribution of predictors in the training subset

Predictors		N	Nosocomial infection, %
Sex	Men	8,447	2.3**
	Women	4,824	1.6
Age	16–39	1,194	2.4*
	40–69	6,338	2.4
	70+	5,739	1.7
APACHE II	0–10	5,692	0.9***
	11–20	5,084	2.4
	21+	2,495	4.1
Operation	None	5,644	2.2**
	Elective	4,982	1.6
	Urgent	2,645	2.7
Ventilator	No	6,139	0.9***
	Yes	7,132	3.1
Urinary catheter	No	1,484	2.0
	Yes	11,787	2.1
CV catheter	No	4,526	1.0***
	Yes	8,745	2.6

*$p < 0.05$.
**$p < 0.01$.
***$p < 0.001$ with chi-square test.

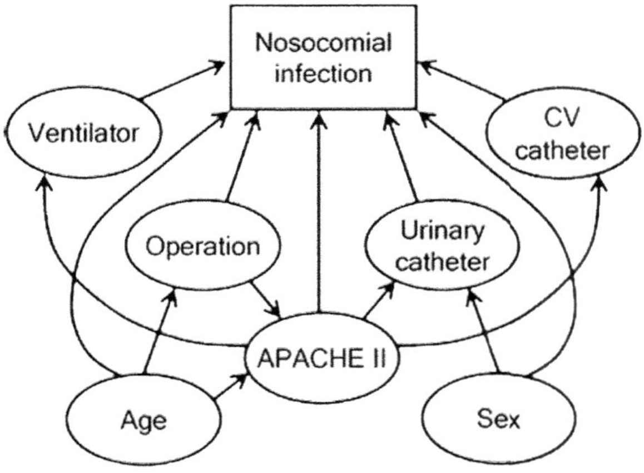

Fig. 23.3 Schematic diagram showing the relationships between the predictors and the outcome and the interactions between the predictors in the dataset of ICU patients

23.3.3 Assessment of Predictive Performance

Predictive performance was assessed using the testing subset in terms of total classification accuracy (TCA) [7] and the area under a receiver operating characteristic curve (AUC) [3]. TCA represents the proportion of correctly classified cases. The values of TCA, which range from 0 (worst) to 1 (best), indicate the discriminatory ability at a reasonable threshold level. A receiver operating characteristic curve represents the trade-off between sensitivity (classification accuracy for positive cases) and specificity (classification accuracy for negative cases). The values of AUC, which range from 0.5 (worst) to 1 (best), indicate the overall discriminatory ability independent of the setting of the threshold level.

23.3.4 Predictive Model Based on Logistic Regression

Logistic regression was performed using the LOGISTIC procedure in SAS/STAT release 8.2 (SAS Institute Inc., Cary, NC). The regression coefficients for the predictors in the training subset are shown in Table 23.3.

The probability of nosocomial infection is expressed as

$$P = \frac{1}{[1+\exp\{-(\beta_0+\beta_1\chi_1+\beta_2\chi_2+\mathrm{L}+\mathrm{L}+\beta_7\chi_7)\}]} \tag{23.2}$$

where β_0 is the intercept and β_i are the regression coefficients for the corresponding predictors $\chi_i (I = 1, 2, \ldots, 7)$.

The predictive performance of logistic regression for the testing subset is shown in Table 23.4. The TCA and AUC values are satisfactory, given that the predictive model was developed based on the statistical analysis of real medical data.

Table 23.3 Regression coefficients (β) for the predictors in the training subset

Predictors		β	(SE)
Sex	Men	0	(Reference)
	Women	−0.3733	(0.1365)
Age	16–39	0	(Reference)
	40–69	−0.1290	(0.2123)
	70+	−0.5277	(0.2213)
APACHE II	0–10	0	(Reference)
	11–20	0.7633	(0.1721)
	21+	1.0654	(0.1904)
Operation	None	0.4867	(0.1569)
	Elective	0	(Reference)
	Urgent	0.4044	(0.1697)
Ventilator	No	0	(Reference)
	Yes	0.7986	(0.1730)
Urinary catheter	No	0	(Reference)
	Yes	−0.4915	(0.2171)
CV catheter	No	0	(Reference)
	Yes	0.7621	(0.1929)

Table 23.4 Predictive performance of logistic regression[a]

TCA	0.76
Sensitivity	0.66
Specificity	0.77
AUC	0.79

[a]TCA and AUC are described in Sect. 23.3.3.

23.3.5 Predictive Model Based on Neural Networks

A three-layer NN with 13 input neurons, 10 hidden neurons, and 1 output neuron employed backpropagation learning [10] with a momentum of 0.8. The learning rate was 0.01. The outputs of the NN, which ranged from 0 to 1, were interpreted as the probabilities of nosocomial infection.

As shown in Fig. 23.2, the training subset contained a small number of patient records in which nosocomial infection occurred (positive cases), compared with a large number of patient records in which nosocomial infection did not occur (negative cases). When the NN was trained using this dataset without adjusting for the ratio of positive to negative cases, the NN could not learn the patterns for positive cases. Therefore, an equal number of patient records were randomly sampled from the 12,995 negative cases. When the NN was trained using the 552 patient records (276 positive and 276 negative cases), the classification accuracy in the testing subset increased to 0.41 (1352/3313).

Moreover, there were many redundant and conflicting records in the training subset. The use of this dataset caused inadequate biased learning. Therefore, the redundant and conflicting records were excluded from the training subset. When the NN was trained using the remaining 366 patient records (104 positive and 262 negative cases), the classification accuracy in the testing subset increased to 0.70 (2304/3313).

Improving the predictive performance required a continuous process of trial and error. The predictive performance of the NN for the testing subset is shown in Table 23.5. The AUC value was significantly larger than that for logistic regression (Table 23.4). This difference was visually confirmed by the receiver operating characteristic curves (Fig. 23.4), in which the curve for NN was nearer to upper left corner than of the curve for logistic regression.

Many different types of NNs have been developed, and each modeling method has its own strengths and limitations. We applied a simple three-layer backpropagation NN, a common method, to the dataset of ICU patients. According to the results of this study, careful preparation of datasets improves the predictive performance of NNs, and accordingly, NNs outperform multivariate regression models.

Table 23.5 Predictive performance of NN[a]

TCA	0.73
Sensitivity	0.96
Specificity	0.72
AUC	0.86

[a]TCA and AUC are described in Sect. 23.3.3.

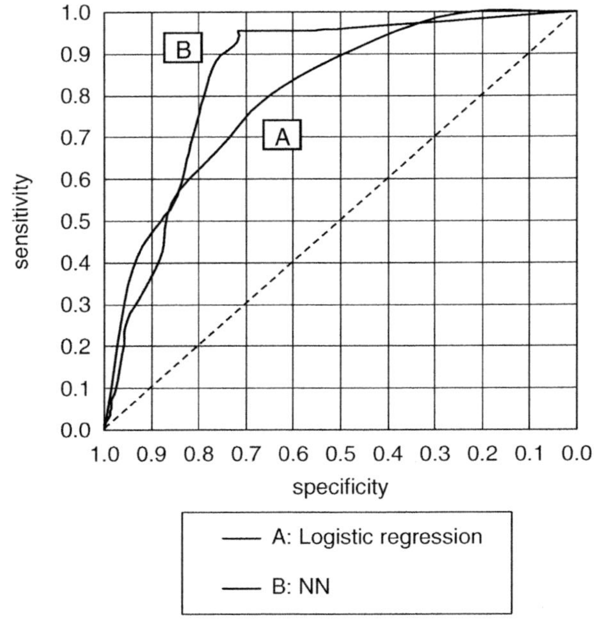

Fig. 23.4 Receiver operating characteristic curves

23.4 Challenge to Modeling Time Sequence Data

At present, medical records are readily accessible from hospital information systems. As mentioned in Section 23.2.2, medical records consist of time sequence data. Aside from baseline data, treatments, complications, and all that happened to the patient during the hospital stay are sequentially recorded in his or her medical record (Fig. 23.5).

Due to the effects of subsequent events, the probability of an outcome event is expected to change during the follow-up period. The probability of the outcome event at time t_1 should change from that estimated at time t_0 because treatment A happened at time t_1. Similarly, the probability of the outcome event at time t_2 should change from that estimated at time t_1 because treatment B and complication C happened at time t_2. Unfortunately, a method of modeling time sequence data for prediction has not yet been established. Conventional predictive models, such as multivariate regression models, are designed to estimate the probabilities of outcome events

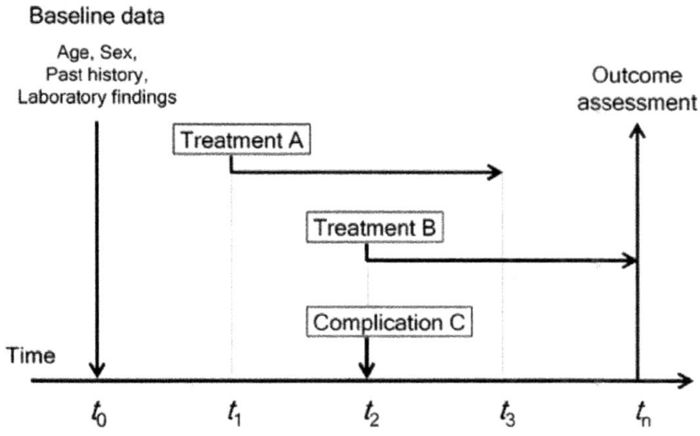

Fig. 23.5 Time sequence data of medical record

only based on baseline data; they exclude any subsequent events (using the example above: treatment A, treatment B, and complication C) that may have an effect on the patient outcome.

We invented a method of modeling time sequence data for prediction using multiple NNs. Based on the dataset of ICU patients, we examined whether multiple NNs outperform both logistic regression and the application of a single NN in the long-term prediction of nosocomial infection.

23.4.1 Multiple Neural Networks

The outcome event of this study was determined by the diagnosis of nosocomial infection during the first ten days of ICU stay. The baseline of prediction (time t_0) was set at ICU admission. Multiple NNs were designed to estimate the probabilities of nosocomial infection based on baseline data ("Sex," "Age," "APACHE II," "Operation," "Ventilator," "Urinary catheter," and "CV catheter") at ICU admission and subsequent events ("Ventilator," "Urinary catheter," and "CV catheter") at a specific period after ICU admission.

Multiple NNs consisted of four 3-layer NNs, which were responsible for predictions during the following four timeframes, respectively: Day 3–4, 5–6, 7–8, and 9–10 after ICU admission. As shown in Fig. 23.6, two neighboring NNs were connected in series to represent the dependency of subsequent periods. The estimate for period t_a, $P(t_a)$ was passed forward to a subsequent NN as an input signal, and used to enhance the estimate for the subsequent period t_b, $P(t_b)$.

As shown in Fig. 23.7, the first three-layer NN with 13 input neurons, 10 hidden neurons, and 1 output neuron was input baseline data ("Sex," "Age," "APACHE II," "Operation," "Ventilator," "Urinary catheter," and "CV catheter") at ICU admission, and used to estimate the probability of nosocomial infection at Day 3–4 after ICU

Fig. 23.6 Connection of two neighboring NNs

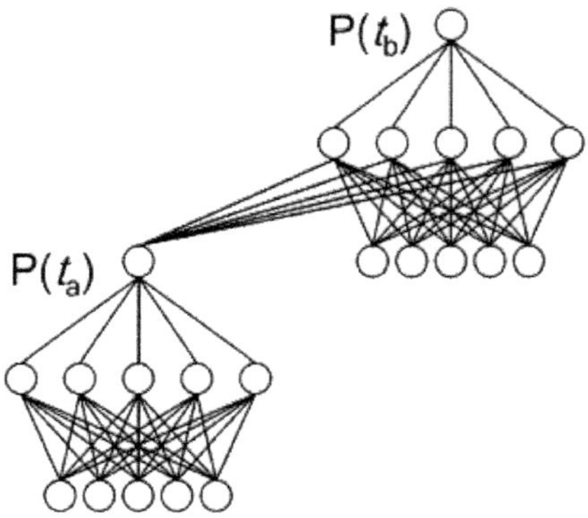

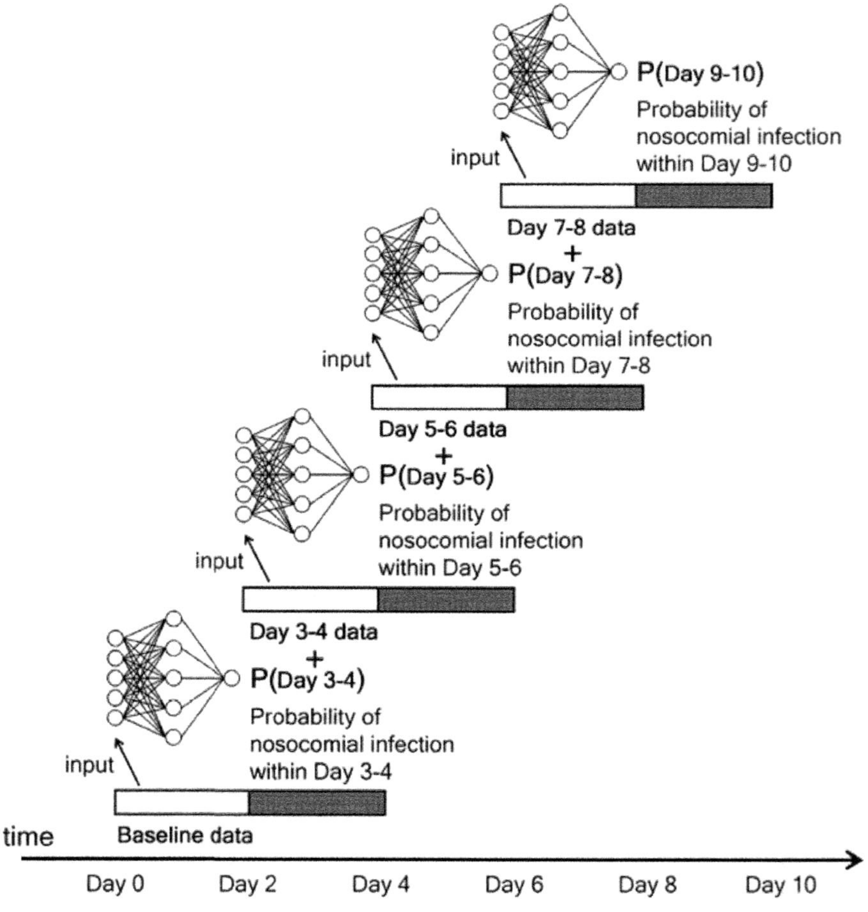

Fig. 23.7 Predictive model based on multiple NNs

admission. The second three-layer NN with 4 input neurons, 10 hidden neurons, and 1 output neuron was input the output of the first NN and the data for subsequent events ("Ventilator," "Urinary catheter," and "CV catheter") at Day 3–4 after ICU admission, and used to estimate the probability of nosocomial infection at Day 5–6 after ICU admission. The third and fourth three-layer NNs, again with 4 input neurons, 10 hidden neurons, and 1 output neuron, followed the second NN in a similar fashion. Each three-layer NN employed backpropagation learning [10] with a momentum of 0.8. The learning rate was 0.01. The outputs of the last NN, which ranged from 0 to 1, were interpreted as the probabilities of nosocomial infection.

23.4.2 Predictive Performance of Multiple Neural Networks Versus Logistic Regression and Single Neural Network

The predictive performance of multiple NNs was compared with that of logistic regression and single NN using the cross-validation method. After stratifying the dataset by the follow-up period and classifying as positive or negative cases, the original dataset was randomly divided into 80% training and 20% testing subsets (Fig. 23.8). The three predictive models were developed using the training subset.

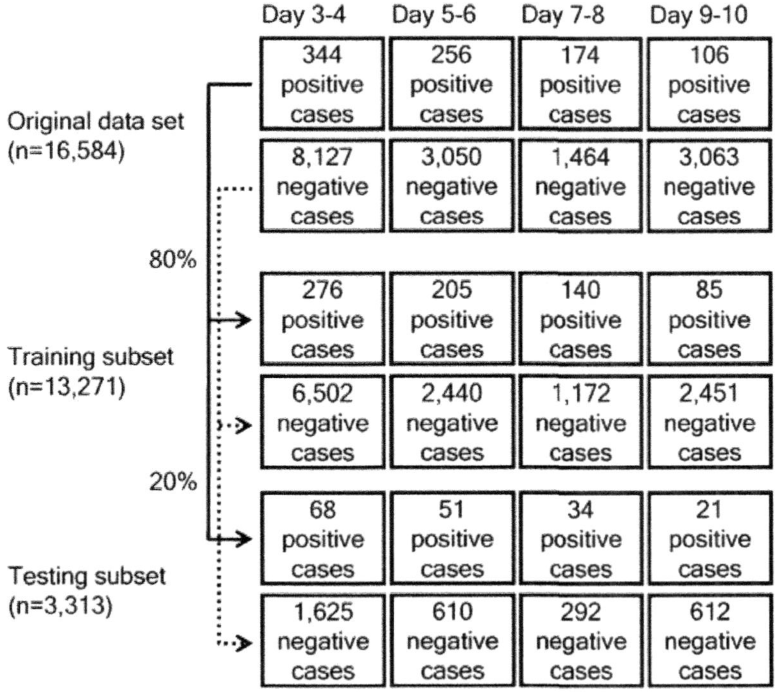

Fig. 23.8 Preparation of datasets

Multiple NNs incorporated baseline data as well as subsequent events, whereas logistic regression and single NN (a simple three-layer backpropagation NN with 13 input neurons, 10 hidden neurons, and 1 output neuron) incorporated only baseline data. Predictive performance was assessed using the testing subset in terms of TCA and AUC.

The predictive performance of each model for the testing subset is shown in Table 23.6. Multiple NNs showed higher TCA and AUC values than logistic regression and single NN. This difference was visually confirmed by the receiver operating characteristic curves (Fig. 23.9), in which the curve for multiple NNs was nearer to the upper left-hand corner than that of the curves for logistic regression and single NN.

Table 23.6 Comparison of predictive performance[a]

	Logistic regression	Single NN	Multiple NNs
TCA	0.63	0.37	0.73
Sensitivity	0.71	0.85	0.92
Specificity	0.62	0.35	0.72
AUC	0.73	0.60	0.83

[a]TCA and AUC are described in Sect. 23.3.3.

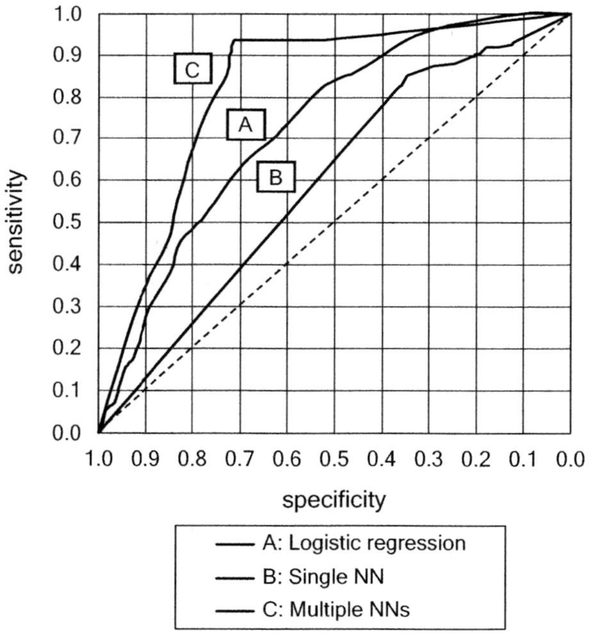

Fig. 23.9 Receiver operating characteristics curves

Table 23.7 Comparison of predictive performance by timeframe[a]

	Logistic regression		Single NN		Multiple NNs	
	TCA	AUC	TCA	AUC	TCA	AUC
Day 3–4 ($n = 1625$)	0.75	0.79	0.43	0.63	0.73	0.86
Day 5–6 ($n = 610$)	0.61	0.73	0.37	0.58	0.66	0.77
Day 7–8 ($n = 292$)	0.52	0.67	0.33	0.55	0.79	0.84
Day 9–10 ($n = 612$)	0.38	0.62	0.25	0.55	0.77	0.82

[a] TCA and AUC are described in Sect. 23.3.3

Due to the effects of subsequent events, the probabilities of nosocomial infection were expected to change during the follow-up period. Table 23.7 shows the predictive performance of each model by timeframe. Multiple NNs showed no significant change in either TCA or AUC, whereas logistic regression and single NN showed noticeable declines in both TCA and AUC.

Overall, the best predictive performance was indicated from multiple NNs, followed by logistic regression and single NN. The predictive performance of multiple NNs was maintained at a constant level, whereas that of logistic regression and single NN decreased over the follow-up period. Multiple NNs, which incorporate baseline data as well as subsequent events, can estimate the probabilities of outcome events with respect to the effects of subsequent events. Outcome events that occur at a later follow-up period may depend more on subsequent events than baseline data. The use of multiple NNs improves predictive performance, particularly at later follow-up periods, and accordingly, multiple NNs are well suited to long-term prediction.

23.5 Discussion

Medical data are characterized by their sparseness, redundancy, conflict, and time sequence, which may impede the development of a good predictive model. Careful preparation of datasets improves the predictive performance of NNs, and accordingly, NNs outperform multivariate regression models.

In this chapter, we proposed a method of modeling time sequence data for prediction using multiple NNs. Compared with conventional predictive models, multiple NNs are advantageous in that they also consider the effects of subsequent events, which results in improved predictive performance. Many different types of NNs have been developed, and each modeling method has its own strengths and limitations. Predictive performance depends entirely on the quality of the datasets used and the characteristics of the modeling methods applied. It is unlikely that one modeling method can outperform all others in every prediction task. Some select modeling methods should continue to be applied in a complementary or cooperative manner.

Comparative studies using real medical data are expected to pave the way for more effective modeling methods. The use of real medical data provides practical and convincing information. On the other hand, real medical data often contain missing values and outliers (noise) that are difficult to control using common modeling methods. The use of good predictors, selected from a wide variety of data items based on expert (background) knowledge, can reduce the effect of this problem. However, the predictive models developed exclude a number of unknown factors that may have an effect on the patient outcome.

Ichimura and his colleagues proposed the immune multiagent neural networks (IMANNs), which optimize their own network structure to adapt to a training dataset; the number of neurons in the hidden layer increases (generation) or decreases (annihilation) in the context of spatial positioning in the learning process. The IMANNs applied to real medical data have demonstrated high classification capability [5]. The use of such modeling methods may provide effective alternatives to elaborate data handling and realize the full utilization of data set.

It is certain that NNs have capabilities as good predictive models. Further studies using real medical data may be required to achieve the desired predictive performance. Extracting knowledge from NNs should also be investigated in order to deal with its "black box" nature.

Acknowledgements This study was supported by the Health and Labour Sciences Research Grant (Research on Emergent and Re-emerging Infectious Diseases) from the Japanese Ministry of Health Labour, and Welfare and the Grant-in-Aid for Scientific Research (Grant-in-Aid for Young Scientists 18790406) from the Japanese Ministry of Education, Culture, Sports, Science, and Technology.

References

1. Dayhoff JE and DeLeo JM (2001). Artificial neural networks: Opening the black box. Cancer 91(8 Suppl): 1615–1635.
2. Grobman WA and Stamilio DM (2006). Methods of clinical prediction. Am J Obstet Gynecol 194: 888–894.
3. Hanley JA and McNeil BJ (1982). The meaning and use of the area under a receiver operating characteristic (ROC) curve. Radiology 143: 29–36.
4. Harrell FE Jr, Lee KL, and Pollock BG (1988). Regression models in clinical studies: Determining relationships between predictors and response. J Natl Cancer Inst 80: 1198–1202.
5. Ichimura T, Oeda S, Suka M, Mackin KJ, and Yoshida K (2004). Adaptive learning methods of reflective neural networks and immune multi-agent neural networks. In: *Knowledge-Based Intelligent Systems for Healthcare. Advanced Knowledge International*, Ichimura T and Yoshida K (eds), Adelaide, pp. 11–49.
6. Knaus WA, Draper EA, Wagner DP, and Zimmerman JE (1985). APACHE II: A severity of disease classification system. Cril Care Med 13: 818–829.
7. Lavrac N (1999). Selected techniques for data mining in medicine. Artif Intell Med 16: 3–23.
8. Lucas PJF and Abu-Hanna A (1999). Prognostic methods in medicine. Artif Intell Med 15: 105–119.
9. Ohno-Machado L (2001). Modeling medical prognosis: Survival analysis techniques. J Biomed Inform 34: 428–439.

10. Rumelhart DE, Hinton GE, and Williams RJ (1986). Learning representations by back-propagation errors. Nature 323: 533–536.
11. Sargent DJ (2001). Comparison of artificial neural networks with other statistical approaches: Results from medical data sets. Cancer 91(8 Suppl): 1636–1642.
12. Suka M, Oeda S, Ichimura T, Yoshida K, and Takezawa J (2004). Comparison of proportional hazard model and neural network models in a real data set of intensive care unit patients. Medinfo 11(Pt 1): 741–745.
13. Suka M, Yoshida K, and Takezawa J (2006). A practical tool to assess the incidence of nosocomial infection in Japanese intensive care units: The Japanese Nosocomial Infection Surveillance System. J Hosp Inf 63: 179–184.
14. Tu JV (1996). Advantages and disadvantages of using artificial neural networks versus logistic regression for predicting medical outcomes. J Clin Epidemiol 49: 1225–1231.

Chapter 24
Prediction Method for Real Thai Stock Index Based on Neurofuzzy Approach

Monruthai Radeerom, Chonawat Srisa-an, and M.L. Kulthon Kasemsan

24.1 Introduction

The prediction of financial market indicators is a topic of considerable practical interest and, if successful, may involve substantial pecuniary rewards. People tend to invest in equity because of its high returns over time. Stock markets are affected by many highly interrelated economic, political, and even psychological factors, and these factors interact in a very complex manner. Therefore, it is, generally, very difficult to forecast the movements of stock markets. Neural networks have been used for several years in the selection of investments. Neural networks have been shown to enable decoding of nonlinear time series data to adequately describe the characteristics of the stock markets [1]. Examples using neural networks in equity market applications include forecasting the value of a stock index [2–5] recognition of patterns in trading charts [6, 7], rating of corporate bonds [8], estimation of the market price of options [9], and the indication of trading signals of selling and buying [10, 11], and so on. Feedforward backpropagation networks as discussed in Sect. 24.2 are the most commonly used networks and meant for the widest variety of applications.

Even though nearly everybody agrees on the complex and nonlinear nature of economic systems, there is skepticism as to whether new approaches to nonlinear modeling, such as neural networks, can improve economic and financial forecasts. Some researchers claim that neural networks may not offer any major improvement over conventional linear forecasting approaches [12, 13]. In addition, there is a great variety of neural computing paradigms, involving various architectures, learning rates, and the like, and hence, precise and informative comparisons may be difficult to make. In recent years, an increasing amount of research in the emerging and promising field of financial engineering is incorporating neurofuzzy approaches [14–21].

The Stock Exchange of Thailand (SET) is the stock market in Thailand whereby stocks may be bought and sold. As with every investment, raising funds in the stock exchange entails some degree of risk. There are two types of risk: systematic risk

Oscar Castillo et al. (eds.), *Trends in Intelligent Systems and Computer Engineering*.

and an erroneous one. The erroneous risk can be overcome by a sound investment strategy, called diversification. However, by using a better prediction model to forecast the future price variation of a stock, the systematic risk can be minimized if not totally eliminated.

This chapter describes a feedforward neural network and neurofuzzy system in Sect. 24.2. Subsequently, details for the methodology of stock prediction are explained in Sect. 24.3. Next, several results are presented involving neurofuzzy predictions in comparison to feedforward neural networks. Some conclusions based on the results presented in this chapter are drawn, with remarks on future directions.

24.2 Neural Network and Neurofuzzy Approaches for Time Series Stock Market Prediction

24.2.1 Neural Networks (NNs) for Modeling and Identification

The neural networks are used for two main tasks: function approximation and pattern classification. In function approximation, the neural network is trained to approximate a mapping between its inputs and outputs. Many neural network models have been proven as universal approximators; that is, the network can approximate any continuous function arbitrary well. The pattern classification problem can be regarded as a specific case of function approximation. The mapping is done from the input space to a finite number of output classes.

For function approximation, a well-known model of NNs is a feedforward multilayer neural network (MNN). It has one input layer, one output layer, and a number of hidden layers between them. For illustration purposes, consider a MNN with one hidden layer (Fig. 24.1). The input-layer neurons do not perform any computations.

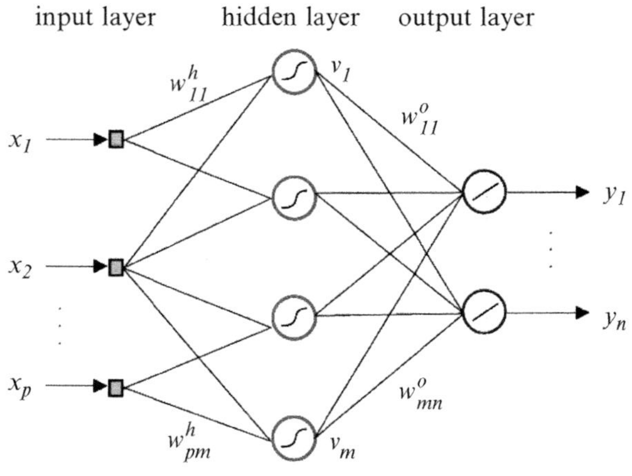

Fig. 24.1 A feedforward neural network with one hidden layer [16]

They merely distribute the inputs to the weights of the hidden layer. In the neurons of the hidden layer, first the weighted sum of the inputs is computed

$$z_j = \sum_{i=1}^{p} w_{ij}^h x_i = \left(W_j^h\right)^T X_i,\ j = 1,2,K,m \tag{24.1}$$

It is then passed through a nonlinear activation function, such as the tangent hyperbolic:

$$v_j = \frac{1-\exp(-2z_j)}{1+\exp(-2z_j)},\ j = 1,2,K,m \tag{24.2}$$

Other typical activation functions are the threshold function (hard limiter), the sigmoid function, and so on. The neurons in the output layer are linear; that is, they only compute the weighted sum of their inputs:

$$yl = \sum_{j=1}^{h} w_{jl}^o v_j = (W_l^o)^T X,\ l = 1,2,K,n \tag{24.3}$$

Training is the adaptation of weights in a multilayer network such that the error between the desired output and the network output is minimized. A network with one hidden layer is sufficient for most approximation tasks. More layers can give a better fit, but the training time takes longer. Choosing the right number of neurons in the hidden layer is essential for a good result. Too few neurons give a poor fit, whereas too many neurons result in overtraining of the net (poor generalization to unseen data). A compromise is usually sought by trial and error methods.

The backpropagation algorithm [15] has emerged as one of the most widely used learning procedures for multilayer networks. There are many variations of the backpropagation algorithm, several of which are discussed in the next section. The simplest implementation of backpropagation learning updates the network weights and biases in the direction in which the performance function decreases most rapidly.

24.2.2 Neurofuzzy System (NFs) for Modeling and Identification

Both neural networks and fuzzy systems are motivated by imitating the human reasoning process. In fuzzy systems, relationships are represented explicitly in forms of if–then rules. In neural networks, the relations are not explicitly given, but are coded in designed networks and parameters. Neurofuzzy systems combine semantic transparency of rule-based fuzzy systems with the learning capability of neural networks. Depending on the structure of if–then rules, two main types of fuzzy models are distinguished as mamdani (or linguistic) and Takagi–Sugeno models [22]. The mamdani model is typically used in knowledge-based (expert) systems, but the Takagi–Sugeno model used in data-driven systems.

In this chapter, we consider only the Takagi–Sugeno–Kang (TSK) model. Takagi, Sugeno, and Kang [22] formalized a systematic approach for generating fuzzy rules from input–output data pairs. The fuzzy if–then rules, for a pure fuzzy inference system, are of the following form.

$$if\ x_1\ is\ A_1\ and\ x_2\ is\ A_2\ and\ x_N\ is\ A_N\ then\ y = f(x) \tag{24.4}$$

where $x = [x_1, x_2, \ldots, x_N]^T$, A_1, A_2, K, A_N are fuzzy sets in the antecedent, a nd y is a crisp function in the consequent part. The function is a polynomial function of input variables x_1, x_2, x_3, K, x_N. The aggregated values of the membership function for the vector are assumed either in a form of the MIN operator or in the product form. The M fuzzy rules in Eq. 24.4 are N membership functions μ_1, μ_2, μ_3, K, μ_N. Each antecedent is followed by the consequent:

$$y_i = p_{i0} + \sum_{j=1}^{N} p_{ij} x_j \tag{24.5}$$

Where p_{ij} are the adjustable coefficients, for i=1,2,3,K, M and j=1,2,3, K, N.

The first-order TSK fuzzy model could be expressed in a similar fashion. Consider an example with two rules:

$$\begin{aligned} &if\ x_1\ is\ A_{11}\ and\ x_2\ is\ A_{21}\ and\ then\ y_1 = p_{11}x_1 + p_{12}x_2 + p_{10} \\ &if x_1\ is\ A_{12}\ and\ x_2\ is\ A_{22}\ and\ then\ y_2 = p_{21}x_1 + p_{22}x_2 + p_{20} \end{aligned} \tag{24.6}$$

Figure 24.2 shows a network representation of those two rules. The node in the first layer computes the membership degree of the inputs in the antecedent fuzzy sets. The product node $\prod$ in the second layer represents the antecedent connective

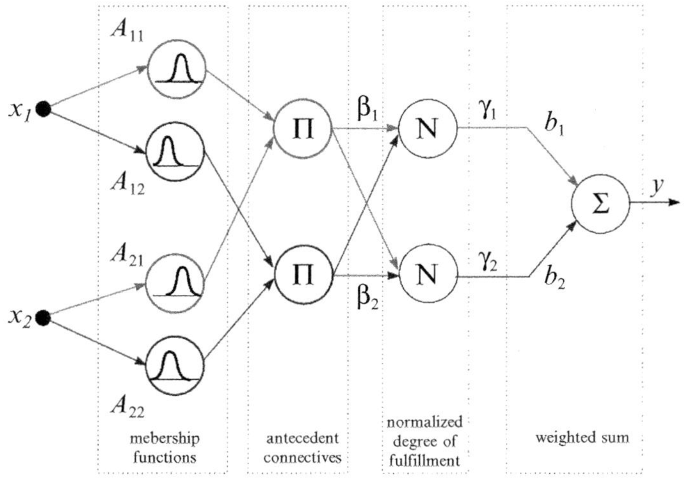

Fig. 24.2 An example of a first-order TSK fuzzy model with two rules systems [23]

(here the "and" operator). The normalization node N and the summation node $\sum$ realize the fuzzy-mean operator, for which the corresponding network is given in Fig 24.2. Applying a fuzzy singleton, a generalized bell function as membership function, and algebraic product aggregation of the input variables, at the existence of m rules the neurofuzzy TSK system output signal upon excitation by the vector, is described by

$$y(x) = \frac{1}{\sum_{r=1}^{M}\left[\prod_{j=1}^{N}\mu_r(x_j)\right]}\sum_{k=1}^{M}\left(\left[\prod_{j=1}^{N}\mu_r(x_j)\right]\left[p_{k0}+\sum_{j=1}^{N}p_{kj}x_j\right]\right) \tag{24.7}$$

The adjusted parameters of the system are nonlinear parameters of the bell function $(c_j^{(k)}, \sigma_j^{(k)}, b_j^{(k)})$, the fuzzifier functions, and linear parameters (weight) of the TSK function for every $j = 1,2,\mathbf{K}$, N and $k = 1,2,\mathrm{K}$, M. In contrary to the mamdani fuzzy inference system, the TSK model generates crisp output values instead of fuzzy ones. This network is simplified. Thus, the defuzzifier is not necessary, and the learning of the neurofuzzy network, which adapts parameters of the bell-shaped membership functions $(c_j^{(k)}, \sigma_j^{(k)}, b_j^{(k)})$ and consequent coefficients p_{ij} can be done either in supervised or self-organizing modes.

In this study, we apply a hybrid method which is a one-shot least squares estimation of consequent parameters with iterative gradient-based optimization of membership functions. The important problem in the TSK network is to determine the number of rules that should be used in modeling data. More rules mean better representation of data processing, but increase of network complexity and high cost of data processing. Therefore, the procedure for automatically determining the number of rules is required. In our solution, each rule should be associated with one cluster of data.

Fuzzy c-means is a supervised algorithm, because it is necessary to indicate how many clusters C to look for. If C is not known beforehand, it is necessary to apply an unsupervised algorithm. Subtractive clustering is based on a measure of the density of data points in the feature space [23]. The idea is to find regions in the feature space with high densities of data points. The point with the highest number of neighbors is selected as centre for a cluster. The data points within a prespecified fuzzy radius are then removed (subtracted), and the algorithm looks for a new point having the highest number of neighbors. This process continues until all data points are examined.

Consider a collection of K data points (u_k, $k = 1,2,\ldots,K$) specified by m-dimensional vectors. Without loss of generality, the data points are assumed normalized. Because each data point is a candidate for a cluster centre, a density measurement at data point u_k is defined as

$$D_k = \sum_{j=1}^{K}\exp\left(-\frac{\|u_k - u_j\|}{(r_a/2)^2}\right), \tag{24.8}$$

where r_a is a positive constant. Hence, a data point will have a high density value if it has many neighboring data points. Only the fuzzy neighborhood within the radius r_a contributes to the density measure.

After calculating the density measure for each data point, the point with the highest density is selected as the first cluster center. Let u_{c_1} and D_{c_1} be the point selected and density measure, respectively. Next, the density measure for each data point u_k is revised by the formula

$$D'_k = D_k - D_{C_1} \exp\left(-\frac{\|u_k - u_{c_1}\|}{(r_b/2)^2}\right), \tag{24.9}$$

where r_b is a positive constant. Therefore, the data points near the first cluster centre u_{c_1} will have significantly reduced density measures, thereby making the points unlikely to be selected as the next cluster centre. The constant r_b defines a neighborhood to be reduced in density measure. It is normally larger than r_a in order to prevent closely spaced cluster centres; typically $r_b = 1.5^* r_a$.

After the density measure for each point is revised, the next cluster centre is selected and all the density measures are revised again. The process is repeated until sufficient numbers of cluster centres are generated. When applying subtractive clustering to a set of input–output data, each cluster centre represents a rule. To generate rules, the cluster centres are used as the location for the premise sets in a singleton type of rule base (or the radial basis functions in a neural network).

Figure 24.3 shows an example of three clusters. The data partitioning is expressed in the fuzzy partition matrix whose elements are membership degrees of the data vector, in the fuzzy clusters with prototypes. The antecedent membership functions are then extracted by projecting the clusters onto the individual variables.

In conclusion, Fig. 24.4 summarizes the constructing neurofuzzy networks system (NFs). Process data called "training datasets" can be used to construct neurofuzzy systems. We do not need prior knowledge called "knowledge-based

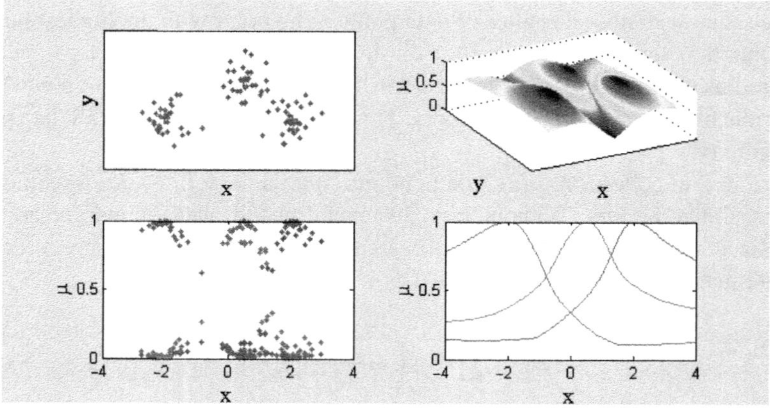

Fig. 24.3 Identifying membership functions through subtractive clustering

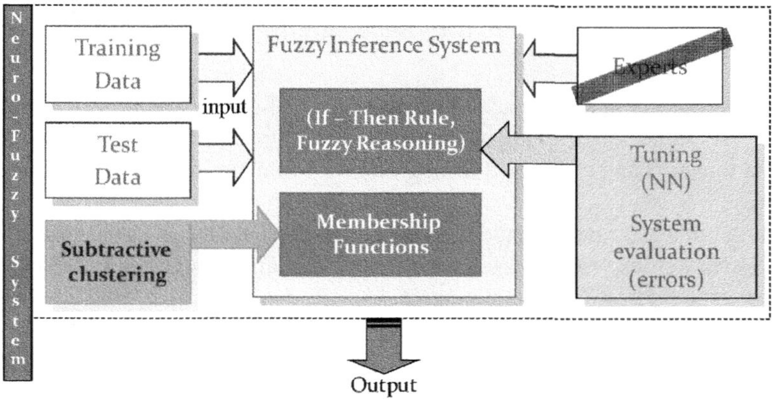

Fig. 24.4 Constructing neurofuzzy networks

(expert) systems." In this way, the membership functions of input variables are designed by the subtractive clustering method. Fuzzy rules (including the associated parameters) are constructed from scratch by using numerical data. And, the parameters of this model (the membership functions, consequent parameters) are then fine-tuned by process data.

The advantage of the TSK fuzzy system is to provide a compact system. Therefore, some classical system identification methods, such as parameter estimation and order determination algorithms, could be developed to get the fuzzy inference rules by using input–output data. Similar to neural networks, neurofuzzy systems are universal approximators. Therefore, the TSK fuzzy inference systems are general for many complex nonlinear practical problems, such as time series data.

24.3 Methodology for Stock Prediction

In this section, a methodology for stock prediction is presented. For stock prediction, much research used delay time of the old close price for predicting a future close price. Results of those methods were fairly good. However, many people who traded in the stock market always used conventional statistical techniques for decision-making in purchasing stock (buy and sell) [14, 15]. Popular techniques are fundamental analysis and technical analysis. Fundamental and technical analysis could be simulated in an intelligence method.

For fundamental methods, retail sales, gold prices, industrial production indices, foreign currency exchange rates, and so on could be used as inputs. For technical methods, the delayed time series data could be used as inputs. In this chapter, a technical method is adopted which takes not only the delayed time series data as input but also the technical indicators.

24.3.1 Time Series Forecasting with Intelligence System

Based on technical analysis, past information will affect the future. So, the threshold may be some relationship between the stock prices of today and the future. The relationship can be obtained through a group of mappings of the constant time interval.

Assume that u_i represents today's price, y_i represents the next day price. If the prediction of a stock price after ten days could be obtained using today's stock price, then there should be a functional mapping u_i to y_i, where

$$y_i = \Gamma_i(u_i) \tag{24.10}$$

Using all $(u_i,\ y_i)$ pairs of historical data, a general function Γ () which consists of Γ_i () could be obtained

$$y = \Gamma(u) \tag{24.11}$$

More generally, y_i which consists of more information in today's price could be used in function Γ (). NNs and NFS can simulate all kinds of functions, so they also can be used to simulate this Γ () function. The **u** is used as the input to the intelligence system.

24.3.2 Preprocessing the Time Series Input

Technical analysts usually use indicators to predict the future. The major types of indicators are moving average (MA), momentum (M), relative strength index (RSI) stochastic (%K), and moving average of stochastic (%D). These indicators can be derived from the real stock composite index. The target for training the neural network is the actual index.

The inputs to the neural network model are $I_{t-1}, I_t, MA_5, MA_{10}, MA_{50}, RSI$, $\%K$, and $\%D$. The output is I_{t+1}. Here I_t is the index of tth period, MA_j is the moving average after the jth period, and I_{t-1} is the delayed time series. For daily data, the indicators are calculated as mentioned above. (See Fig. 24.5) Other indicators are defined as follows.

$$RSI = 100 - \frac{100}{1 + \frac{\sum(\textit{positive change})}{\sum(\textit{negative change})}} \tag{24.12}$$

Indicators can help traders identify trends and turning points. Moving average is a popular and simple indicator for trends. The stochastic and relative strength indices are some simple indicators that help traders identify turning points.

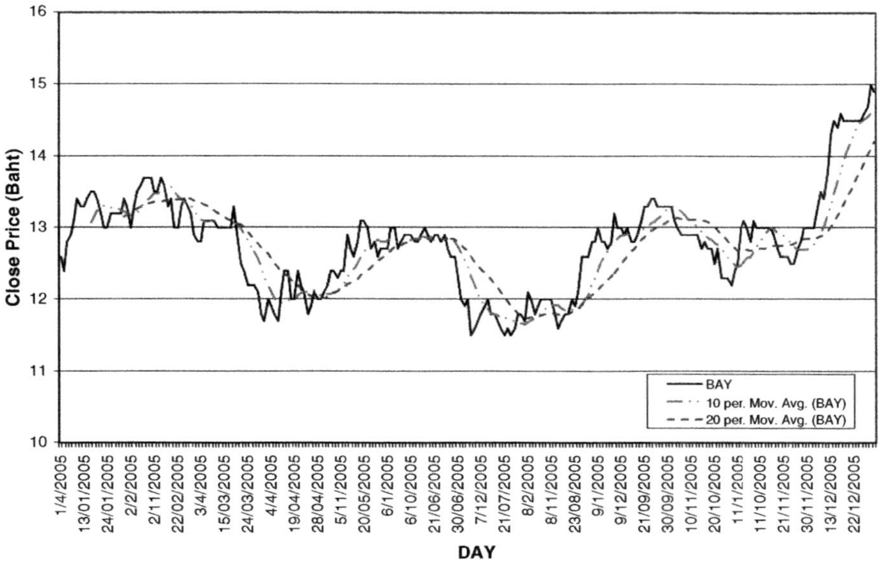

Fig. 24.5 Daily stock price (solid line), MA 10 Days (dash and dot line), and MA 20 Days (dash line) of BAY

In general, the stock price data have biases due to differences in name and spans as shown in Table 24.1. Normalization can be used to reduce the range of the dataset to values appropriate for input and output data being used. The normalization and scaling formula is

$$y = \frac{2x - (\max + \min)}{(\max - \min)} \tag{24.13}$$

where x is the data before normalizing, and y is the data after normalizing.

Because the index prices and moving averages are in the same scale, the same maximum and minimum data are used to normalize them. The max is derived from the maximum value of the linked time series, and the same applies to the minimum. The maximum and minimum values are from the training and validation datasets. The outputs of the neural network and neurofuzzy system will be rescaled back to the original value according to the same formula as shown in Table 24.5.

For the above explanation, many technical analysis indicators were found. The relation between input and output is significant for successful stock prediction. We select that relation based on the correlation coefficient. Correlation, also called the correlation coefficient, indicates the strength and direction of a linear relationship between two random variables. In general statistical usage, correlation or co-relation refers to the departure of two variables from independence, although correlation does not imply causation. In this broad sense there are several coefficients, measuring the degree of correlation, adapted to the nature of the data. A number of different

Table 24.1 Example of input variables from closing price, its technical index, and other index

Historical Quotes of Bank of Ayudhya Public Company Ltd. (BAY)													
Date	High	Low	Avg.	Close	10 Days Moving Average of Close	20 Days Moving Average of Close	10 Days Relative Strength Index	20 Days Relative Strength Index	Volume (thousand)	Value (M. Baht)	SET Index	P/E	P/BV
18/01/2005	13.6	13.4	13.5	13.5	13.16	12.81	66.7	57.89	3,511	47.44	709.55	9.2	1.2
19/01/2005	13.7	13.4	13.54	13.4	13.26	12.84	75.0	57.43	9,132	123.71	709.03	9.2	1.2
20/01/2005	13.4	13.1	13.23	13.2	13.3	12.87	58.8	57.43	9,315	123.25	706.90	9.06	1.18
21/01/2005	13.1	12.9	12.98	13	13.31	12.89	53.8	55.88	10,835	140.7	696.85	8.92	1.16
24/01/2005	13.1	12.9	12.98	13	13.3	12.9	49.3	56.52	1,871	24.31	695.92	8.92	1.16
25/01/2005	13.2	12.9	13.06	13.2	13.28	12.93	43.8	56.87	6,337	82.77	702.14	9.06	1.18
26/01/2005	13.3	13	13.15	13.2	13.27	12.98	44.4	60.91	3,095	40.71	702.66	9.06	1.18
27/01/2005	13.3	13.1	13.18	13.2	13.26	13.02	44.4	60.91	3,579	47.18	701.25	9.06	1.18
28/01/2005	13.3	13.1	13.24	13.2	13.24	13.08	40.7	66.67	3,497	46.31	701.66	9.06	1.18
31/01/2005	13.4	13.2	13.31	13.4	13.23	13.15	47.8	66.85	6,599	87.85	701.91	9.2	1.2

coefficients are used for different situations. The best known is the Pearson product–moment correlation coefficient, which is obtained by dividing the covariance of the two variables by the product of their standard deviations. Despite its name, it was first introduced by Francis Galton [24].

The correlation $\rho_{x,y}$ between two random variables x and y with expected values μ_x and μ_y and standard deviations σ_x and σ_y is defined as

$$\rho_{x,y} = \frac{\mathrm{cov}(x,y)}{\sigma_x \sigma_y} = \frac{E((x-\mu_x)(y-\mu_y))}{\sigma_x \sigma_y}. \tag{24.14}$$

where E is the expected value of the variable and cov means covariance. Because $\mu_x = E(x)$, $\sigma_x^2 = E(x^2) - E^2(x)$ and likewise for y.

The correlation is 1 in the case of an increasing linear relationship, -1 in the case of a decreasing linear relationship, and some value in between in all other cases, indicating the degree of linear dependence between the variables. The closer the coefficient is to either -1 or 1, the stronger the correlation between the variables. If the variables are independent then the correlation is 0. We choose a correlation of more than 0.8 for the relationship between input and output stock predicting data.

24.4 Results and Discussion

The dataset including the SET index and BAY stock index have been decomposed into two different sets: the training data and test data. The data for the BAY index are from December 14, 2004 to September 8, 2006 totaling 425 records, and the first 375 records will be training data and the rest of the data (i.e., 40 records) will be test data. The data for the BAY stock price include high price, low price, average price, buy–sell volume, P/BV, P/E, value, closing price, technical data, and SET index. To avoid interaction between the factors, we test each factor using correlation analysis and identify the factor that will affect the final forecasted results significantly. The final combination of the factors is finalized after the analysis. Consequently, normalization can be used to reduce the range of the dataset to values appropriate for input and output data being used as the training method.

24.4.1 Input Variables

Technical indices are calculated from the variation of stock price, trading volumes, and time following a set of formulas to reflect the current tendency of the stock price fluctuations. These indices can be applied for decision making in evaluating the phenomena of oversold or overbought in the stock market. Basically, the technical index can be classified as an index for BAY movement or particular stock price changes, such as %K%D, RSI, MA, and the like. For example, several technical indices are described as shown in Table 24.1.

24.4.2 Correlation Analysis for Input and Output

At first, we analyze a relation between input and output. The input is the delay time of the closing price. And, output is the closing price on the next day. For example, T-1 means today's close price. Table 24.2 shows the correlation between the technical index and close price on the next day. Table 24.3 shows the correlation between input and close price on the next day as output. Selecting input data should have a correlation of more than 0.8. Thus, in Table 24.4, inputs are High, Low, AVG, P/BV, MA10, MA20, RSI10, and RSI25. And, in Table 24.3, input is T-1, T-2, T-3, T-4, T-5, T-6. Thus, the number of inputs is 14 inputs.

Table 24.2 Correlation between technical index and next day closing price

	High	Low	AVG	Volume	Value	SET
Correlation	0.9986	0.9983	0.9993	0.1009	0.2638	0.5819
	P/E	P/BV	MA10	MA25	RSI10	RSI20
Correlation	0.5564	0.8724	0.901	0.81	0.91	0.85

Table 24.3 Pearson correlation between before days and next day

Historical Quotes of Bank of Ayudhya												
Date	Close T-11	Close T-10	Close T-9	Close T-8	Close T-7	Close T-6	Close T-5	Close T-4	Close T-3	Close T-2	Close T-1	Close
22/08/2006	16.8	17	17	17	16.8	16.8	17	17.2	17.6	17.4	17.4	17.6
23/08/2006	17	17	17	16.8	16.8	17	17.2	17.6	17.4	17.4	17.6	17.5
24/08/2006	17	17	16.8	16.8	17	17.2	17.6	17.4	17.4	17.6	17.5	17.4
25/08/2006	17	16.8	16.8	17	17.2	17.6	17.4	17.4	17.6	17.5	17.4	17.4
28/08/2006	16.8	16.8	17	17.2	17.6	17.4	17.4	17.6	17.5	17.4	17.4	17.3
29/08/2006	16.8	17	17.2	17.6	17.4	17.4	17.6	17.5	17.4	17.4	17.3	17.4
30/08/2006	17	17.2	17.6	17.4	17.4	17.6	17.5	17.4	17.4	17.3	17.4	17.4
31/08/2006	17.2	17.6	17.4	17.4	17.6	17.5	17.4	17.4	17.3	17.4	17.4	17.9
9/1/2006	17.6	17.4	17.4	17.6	17.5	17.4	17.4	17.3	17.4	17.4	17.9	18.1
9/4/2006	17.4	17.4	17.6	17.5	17.4	17.4	17.3	17.4	17.4	17.9	18.1	18.4
9/5/2006	17.4	17.6	17.5	17.4	17.4	17.3	17.4	17.4	17.9	18.1	18.4	18.4
9/6/2006	17.6	17.5	17.4	17.4	17.3	17.4	17.4	17.9	18.1	18.4	18.4	18.5
9/7/2006	17.5	17.4	17.4	17.3	17.4	17.4	17.9	18.1	18.4	18.4	18.5	18.2
9/8/2006	17.4	17.4	17.3	17.4	17.4	17.9	18.1	18.4	18.4	18.5	18.2	18.3
Max	19.5	19.5	19.5	19.5	19.5	19.5	19.5	19.5	19.5	19.5	19.5	19.5
Min	11.5	11.5	11.5	11.5	11.5	11.5	11.5	11.5	11.5	11.5	11.5	11.5
Correlation	0.671	0.691	0.719	0.765	0.776	0.811	0.841	0.874	0.914	0.937	0.966	1

Table 24.4 Summary of input variables for stock price prediction for BAY index (selected correlation >0.8)

Close price t -1	Close price on day
Close price t -2	Close price before one day
Close price t -3	Close price before two day
Close price t -4	Close price before three day
Close price t -5	Close price before four day
Close price t -6	Close price before five day
High	High price index on today
Low	Low price index on today
AVG	Average of stock price
P/BV	Close price/ value stock price
MA10	Moving Average on 10 days
MA20	Moving Average on 20 days
RSI10	Relative Strength Index on 10 Days
RSI20	Relative Strength Index on 20 Days

Thus, input variables are 14 inputs and output variables are the next day closing price. Before training neural network and neurofuzzy model, normalized data are required as shown in Table 24.5.

24.4.3 Comparison Between NN and Neurofuzzy System

It is very difficult to know which training algorithm will be the fastest for a given problem. It will depend on many factors, including the complexity of the problem, the number of data points in the training set, the error goal, the number of inputs or the number of outputs. In this section we perform a number of benchmark comparisons of a backpropagation neural network (BPN) within various training algorithms and our proposed neurofuzzy system. Their learning method of BPN are Fletcher–Reeves update (TRAINCGF), Powell–Beale restarts (TRAINCGB), one step secant algorithm (TRAINOSS), Broyden, Fletcher, Goldfarb, and Shanno (BFGS) algorithm (TRAINBFG), automated regularization (TRAINBR), Polak–Ribiére update (TRAINCGP), resilient backpropagation (TRAINRP), Levenberg–Marquardt (TRAINLM), and scaled conjugate gradient (TRAINSCG) methods. The BPN model has one hidden layer with 30 nodes. And, learning iteration is 10,000 epochs. Table 24.6 gives some example convergence times (elapsed times) for the various algorithms on one particular stock price prediction problem. In this problem a 1–30–1 network was trained on a training dataset with 366 input–output pairs until a mean square error performance of 0.00001 was obtained. For our neurofuzzy system, we varied a number of membership functions for several inputs from 5, 10, 15, 20, and 25, subsequently. Twenty test runs were made for each training algorithm on a Dell Inspiron E1705 (centrino Duo T2400 at 1.83 GHz with RAM 1 GB) to obtain the average numbers shown in Table 24.4.

Table 24.5 Preprocessing of the time series input and output pairs (after normalization)

	Historical Quotes of Bank of Ayudhya Public Company Ltd. (BAY)								After Nomalization						
Date	High	Low	Avg.	10 Days Moving Average of Close	20 Days Moving Average of Close	10 Days Relative Strength Index	20 Days Relative Strength Index	P/B V	Close T-6	Close T-5	Close T-4	Close T-3	Close T-2	Close T-1	Close
28/08/2006	0.450	0.526	0.478	0.578	0.540	0.493	0.097	0.061	0.475	0.475	0.525	0.500	0.475	0.475	0.450
29/08/2006	0.475	0.553	0.509	0.589	0.542	0.440	0.086	0.061	0.475	0.525	0.500	0.475	0.475	0.450	0.475
30/08/2006	0.500	0.579	0.527	0.594	0.551	0.389	0.200	0.020	0.525	0.500	0.475	0.475	0.450	0.475	0.475
31/08/2006	0.600	0.605	0.621	0.603	0.559	0.459	0.245	0.143	0.500	0.475	0.475	0.450	0.475	0.475	0.600
9/1/2006	0.650	0.711	0.664	0.622	0.571	0.659	0.275	0.224	0.475	0.475	0.450	0.475	0.475	0.600	0.650
9/4/2006	0.775	0.789	0.784	0.650	0.593	0.693	0.592	0.306	0.475	0.450	0.475	0.475	0.600	0.650	0.725
9/5/2006	0.725	0.842	0.774	0.672	0.613	0.699	0.589	0.306	0.450	0.475	0.475	0.600	0.650	0.725	0.725
9/6/2006	0.725	0.789	0.761	0.699	0.636	0.768	0.562	0.347	0.475	0.475	0.600	0.650	0.725	0.725	0.750
9/7/2006	0.725	0.789	0.735	0.721	0.653	0.461	0.354	0.265	0.475	0.600	0.650	0.725	0.725	0.750	0.675
9/8/2006	0.700	0.763	0.707	0.746	0.672	0.387	0.431	0.265	0.600	0.650	0.725	0.725	0.750	0.675	0.700
MAX	1.00	1.00	1.00	1.00	1.00	1.00	1.00	1.00	1.00	1.00	1.00	1.00	1.00	1.00	1.00
MIN	−1.00	−1.00	−1.00	−1.00	−1.00	−1.00	−1.00	−1.00	−1.00	−1.00	−1.00	−1.00	−1.00	−1.00	−1.00

Table 24.6 Comparison among various backpropagation and neurofuzzy systems

Acconym	Training Algorithm	Hidden Node	Epochs	Elap Time (Sec)	VAF Training Set (Accuracy %)	VAF Test Set (Accuracy %)
			Neural Network with various learning Methods			
CGF	TRAINCGF	30	10,000	257.76	99.08	74.65
CGB	TRAINCGB	30	10,000	276.44	98.97	75.31
OSS	TRAINOSS	30	10,000	371.69	98.98	76.16
BFG	TRAINBFG	30	10,000	402.78	98.98	76.58
AR	TRAINBR	30	10,000	1,097.80	99.13	77.44
CGP	TRAINCGP	30	10,000	264.72	99.04	77.53
RP	TRAINRP	30	10,000	190.08	99.13	77.71
LM	TRAINLM	30	10,000	758.69	99.42	78.15
SCG	TRAINSCG	30	5,000	162.31	99.17	79.17
SCG	TRAINSCG	30	10,000	335.38	99.18	79.42
SCG	TRAINSCG	30	15,000	483.39	98.98	76.87
SCG	TRAINSCG	30	50,000	1,933.90	99.21	83.56
			NeuroFuzzy with various memberships			
MFS	Membership	3		1.53	99.45	86.09
		5		1.39	99.48	84.81
		10		1.76	99.50	81.53
		15		2.77	99.44	83.68
		20		4.54	99.48	83.95
		25		6.86	99.47	85.43

We evaluate the BPN with various learning methods and neurofuzzy modeling for relations in both training datasets and test datasets with 40 input–output pairs by the percentile variance accounted for (VAF) [23]. The VAF of two equal signals is 100%. If the signals differ, the VAF is lower. When $y1$ and $y2$ are matrices, the VAF is calculated for each column. The VAF index is often used to assess the quality of a model, by comparing the true output and the output of the model. The VAF between two signals is defined as follows.

$$VAF = 100\% * \left[1 - \frac{\mathrm{var}(y1 - y2)}{\mathrm{var}(y1)}\right] \tag{24.15}$$

The comparisons of different models such as BPN and the neurofuzzy model after training the learning method are listed in Table 24.4. We found scaled conjugate gradient (TRAINSCG) methods better than other learning methods. But, the forecast results from the neurofuzzy model are much better than the neural network with scaled conjugate gradient learning which justifies the neurofuzzy model as the best. Moreover, membership functions of input variables are significant for neurofuzzy modeling. In this result, the suitable number of membership functions is three, the same as a result based on the subtractive clustering method.

Table 24.6 lists the algorithms that were tested and the acronyms used to identify them.

24.4.4 Forecast Results in BAY Index

From Table 24.4, the neurofuzzy is the best result for this benchmark of stock index prediction. The desired BAY next day closing price is greatly influenced by the close price, high price, low price, various technical terms, and so on. In addition, input data of the neurofuzzy system are 14 data items and one output data item, namely close next day. After the subtractive clustering method, numbers of memberships are three for several inputs. The initial membership function for input data, namely RSI_{10}, is shown in Fig. 24.6a. Figure 24.6b shows its membership after the learning method. Training datasets can absolutely be used to construct neurofuzzy systems. And, the parameters of this model (the membership functions, consequent parameters) are then fine-tuned by data.

Figure 24.7 shows the next day closing price based on the neurofuzzy stock predictor for the training set of the next day closing price. The neurofuzzy close price is the dash line and the training close price is the solid line. The VAF value of this model is 99.448%.

Likewise, Fig. 24.8 shows the next day closing price based on the neurofuzzy stock predictor versus the test set of the next day closing price. The VAF value of this model is 86.087%. Moreover, we use our proposed method because of preparing a number of inputs and constructing a neurofuzzy model for another stock index such as the Siam Commercial Bank (SCB). For the SCB stock index, the number of inputs is 15 and there is one output; see Table 24.7. The datasets of the SCB stock index have been decomposed into two different sets: the training data and test data. The data for the BAY index are from December 14, 2004 to September 8, 2006 totaling 425 records and the first 375 records are training data and the rest (i.e., 40 records) are test data. After training process, Fig. 24.9 shows the next day closing price based on the neurofuzzy stock predictor versus the test set of the SCB next day closing price. The VAF value of this model is 95.59% as shown in Fig. 24.9. In summary, our proposed neurofuzzy system and preparation method basically succeed and generalize for stock prediction.

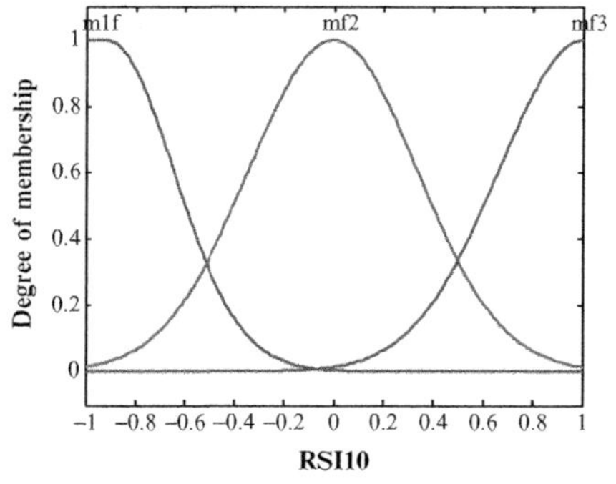

(a) before learning

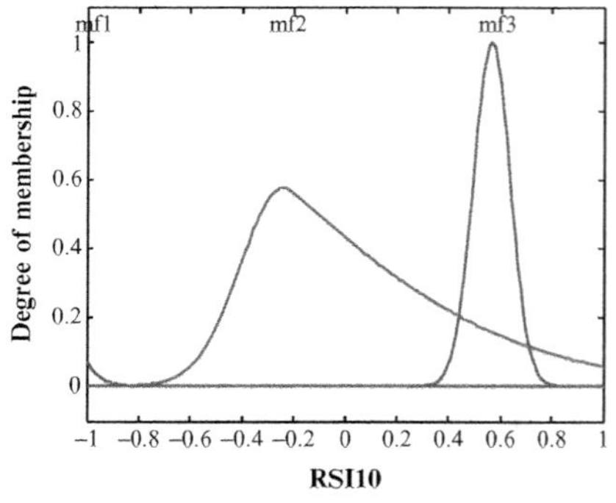

(b) after learning

Fig. 24.6 Fuzzy membership function for RSI_{10} input datasets: (**a**) before learning; (**b**) after learning

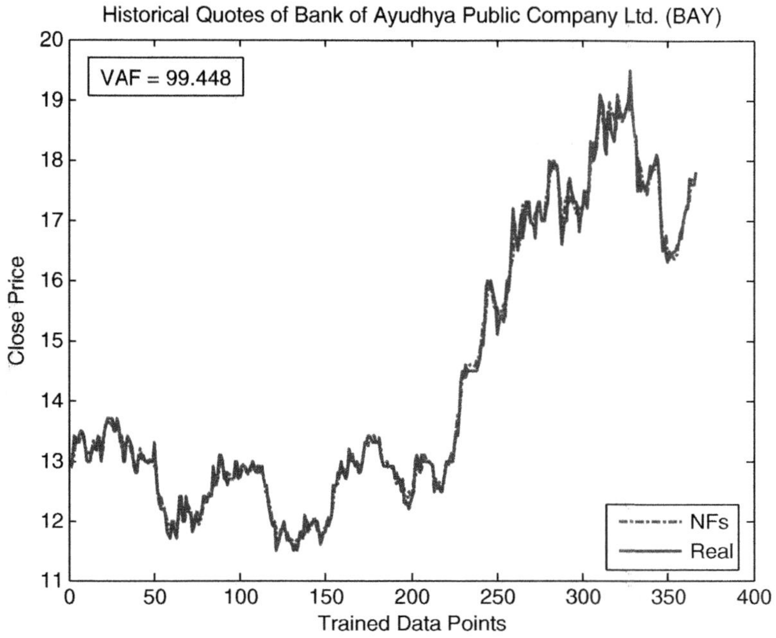

Fig. 24.7 Neurofuzzy close price (dash line) and training close price (solid line)

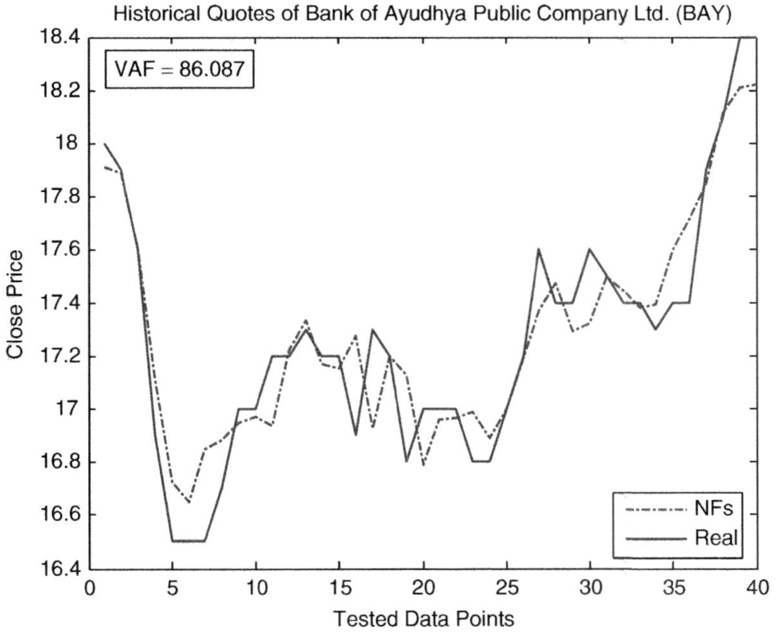

Fig. 24.8 Neurofuzzy close price (dash line) and test close price (solid line) of Bank of Ayudhya Public Company Ltd. (BAY)

Table 24.7 Preprocessing of the time series input and output pairs of (after normalization) for historical quotes of Siam Comercial Bank (SCB)

High	Low	Avg.	10 Days Moving Average of Close	20 Days Moving Average of Close	P/E	Close T-8	Close T-7	Close T-6	Close T-5	Close T-4	Close T-3	Close T-2	Close T-1	Close
−0.56	−0.64	−0.61	−0.53	−0.61	−0.89	−0.49	−0.53	−0.49	−0.46	−0.46	−0.53	−0.66	−0.58	−0.58
−0.58	−0.56	−0.58	−0.53	−0.60	−0.89	−0.53	−0.49	−0.46	−0.46	−0.53	−0.66	−0.58	−0.58	−0.56
−0.56	−0.52	−0.55	−0.53	−0.60	−0.88	−0.49	−0.46	−0.46	−0.53	−0.66	−0.58	−0.58	−0.56	−0.49
−0.53	−0.51	−0.51	−0.53	−0.60	−0.79	−0.46	−0.46	−0.53	−0.66	−0.58	−0.58	−0.56	−0.49	−0.46
−0.46	−0.49	−0.48	−0.52	−0.59	−0.75	−0.46	−0.53	−0.66	−0.58	−0.58	−0.56	−0.49	−0.46	−0.38
−0.38	−0.41	−0.40	−0.50	−0.58	−0.66	−0.53	−0.66	−0.58	−0.58	−0.56	−0.49	−0.46	−0.38	−0.35
−0.38	−0.37	−0.37	−0.49	−0.57	−0.61	−0.66	−0.58	−0.58	−0.56	−0.49	−0.46	−0.38	−0.35	−0.38
−0.38	−0.37	−0.37	−0.48	−0.56	−0.66	−0.58	−0.58	−0.56	−0.49	−0.46	−0.38	−0.35	−0.38	−0.42
−0.38	−0.41	−0.39	−0.47	−0.55	−0.70	−0.58	−0.56	−0.49	−0.46	−0.38	−0.35	−0.38	−0.42	−0.49
−0.46	−0.45	−0.46	−0.45	−0.55	−0.79	−0.56	−0.49	−0.46	−0.38	−0.35	−0.38	−0.42	−0.49	−0.46
−0.46	−0.41	−0.45	−0.43	−0.54	−0.75	−0.49	−0.46	−0.38	−0.35	−0.38	−0.42	−0.49	−0.46	−0.49
−0.53	−0.56	−0.54	−0.42	−0.54	−0.79	−0.46	−0.38	−0.35	−0.38	−0.42	−0.49	−0.46	−0.49	−0.55

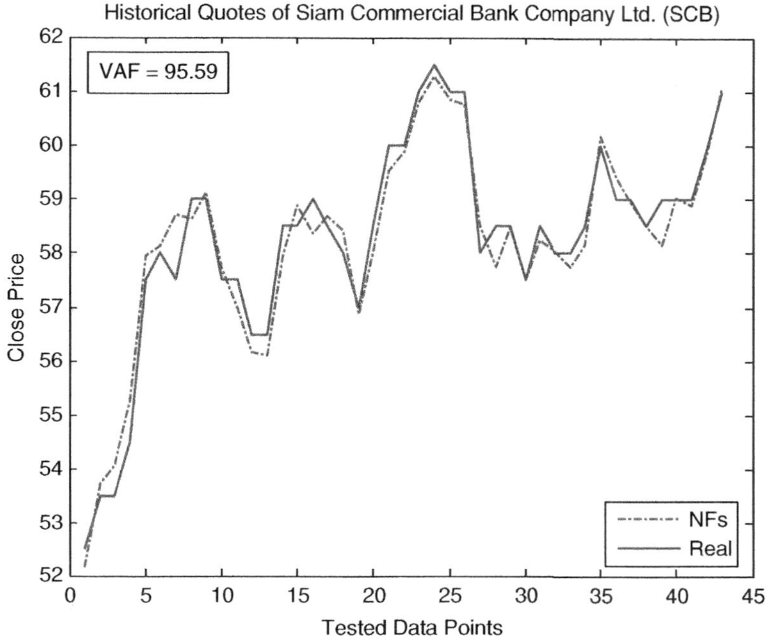

Fig. 24.9 Neurofuzzy closing price (dash line) and test closing price (solid line) of Siam Commercial Bank. (SCB)

24.5 Conclusion

A TSK fuzzy-based system is presented in the chapter by applying a linear combination of the significant technical index as a consequent to predict the stock price. Input variables are effectively selected through the correlation method from the set of technical indices. Therefore the forecasting capability of the system is greatly improved. Finally, the system is tested on the BAY and all the performance results outperform other approaches such as several BPN models. The number of memberships based on clusters of the TSK system is subjectively set from 3 to 25 and the empirical results on the two datasets show that a membership number of 3 has as good forecasting results as the same result from the subtractive clustering method. Through the intensive experimental tests, the model has successfully forecast the price variation for stocks from different sectors with an accuracy close to 99.448% in the training set and 86.087% in the testing set for BAY. Detailed experimental design can be set up for decision of the set of parameters such as different input of the technical index. Moreover, the systematic method can be further applied for the daily trading purpose.

Acknowledgments Part of this article has been presented at International MultiConference of Engineers and Computer Scientists 2007 which was organized by the International Association of Engineers (IAENG). We would like to thank the participants for their helpful comments and invaluable discussions. The authors are grateful to the anonymous referees whose insightful comments enabled us to make significant improvements. This work was partly under the Graduate Fund for Master Student, Rangsit University, Pathumthanee, Thailand. The assistance of somchai lekcharoen is gratefully acknowledged.

References

1. Lapedes, A. and Farber, R.: Nonlinear signal processing using neural networks. *IEEE Conference on Neural Information Processing System—Natural and Synthetic* (1987). 101–107.
2. Yao, J.T. and Poh, H.-L.: Equity forecasting: A case study on the KLSE index, neural networks in financial engineering. *Proceedings of 3rd International Conference on Neural Networks in the Capital Markets* (1995). 341–353.
3. White, H.: Economic prediction using neural networks: A case of IBM daily stock returns. *IEEE International Conference on Neural Networks*, Vol. 2 (1998). 451–458.
4. Chen A.S., Leuny, M.T., and Daoun, H.: Application of neural networks to an emerging financial market: Forecasting and trading the Taiwan Stock Index. *Computers and Operations Research*, Vol. 30 (2003). 901–902.
5. Conner, N.O. and Madden, M.: A neural network approach to pre-diction stock exchange movements using external factor. *Knowledge Based System*, Vol. 19 (2006). 371–378.
6. Tanigawa, T. and Kamijo, K.: Stock price pattern matching system: Dynamic programming neural network approach. *IJCNN'92*, Vol. 2, Baltimore (1992). 59–69.
7. Liu, J.N.K. and Wong, R.W.M.K.: Automatic extraction and identification of chart patterns towards financial forecast. *Applied Soft Computing*, Vol. 1 (2006). 1–12.
8. Dutta, S. and Shekhar, S.: Bond rating: A non-conservative application of neural networks. *IEEE International Conference on Neural Networks* (1990). 124–130.
9. Hutchinson, J.M., Lo, A., and Poggio, T.: A nonparametric approach to pricing and hedging derivative securities via learning networks. *International Journal of Finance*, Vol. 49 (1994). 851–889.
10. Chapman, A. J.: Stock market reading systems through neural networks: Developing a model. *International Journal of Applying Expert Systems*, Vol. 2, No. 2 (1994). 88–100.
11. Liu, J.N.K. and Wong, R.W.M.K.: Automatic extraction and identification of chart patterns towards financial forecast. *Applied Soft Computing*, Vol. 1 (2006). 1–12.
12. Farber, J.D. and Sidorowich, J.J.: Can new approaches to nonlinear modeling improve economic forecasts? In *The Economy As An Evolving Complex System*. CA, Addison-Wesley (1988). 99–115.
13. LeBaron, B. and Weigend, A. S.: Evaluating neural network predictors by bootstrapping. In *Proceedings of International Conference on Neural Information Processing (ICONIP'94)*, Seoul, Korea (1994). 1207–1212.
14. Doeksen, B., Abraham, A., Thomas, J., and Paprzycki, M.: Real stock trading using soft computing models. *IEEE International Conference on Information Technology: Coding and Computing (ITCC'05)* (2005). 123–129.
15. Refenes, P., Abu-Mustafa, Y., Moody, J.E., and Weigend, A.S. (Eds.): *Neural Networks in Financial Engineering*. Singapore: World Scientific (1996).
16. Trippi, R. and Lee, K.: *Artificial Intelligence in Finance & Investing*. Chicago: Irwin (1996).
17. Hiemstra, Y.: *Modeling Structured Nonlinear Knowledge to Predict Stock Markets: Theory. Evidena and Applications*, Chicago: Irwin (1995). 163–175.
18. Tsaih, R. Hsn, V.R., and Lai, C.C.: Forecasting S&P500 stock index future with a hybrid AI system. *Decision Support Systems*, Vol. 23 (1998). 161–174.
19. Cardon, O., Herrera, F., and Villar, P.: Analysis and guidelines to obtain a good uniform fuzzy rule based system using simulated annealing. *International Journal of Approximate Reasoning*, Vol. 25, No. 3 (2000). 187–215.

20. Li, R.-J. and Xiong, Z.-B.: Forecasting stock market with fuzzy neural network. *Proceedings of 4th International Conference on Machine Learning and Cybernetics*, Guangaho (2005). 3475–3479.
21. Yoo, P.D., Kim, M.H., and Jan, T.: Machine learning techniques and use of event information for stock market prediction; A survey and evaluation. *International Conference on Computational Intelligence for Modeling, Control and Automation, and International Conference of Intelligent Agents, Web Technologies and Internet Commerce (IMCA – IAWTIC 2005)* (2005). 1234–1240.
22. Takagi, T. and Sugeno, M.: Fuzzy identification of systems and its application to modeling and control. *IEEE Transaction on System Man and Cybernetics*, Vol. 5 (1985). 116–132.
23. Babuska, R.: Neuro-fuzzy methods for modeling and identification. In *Recent Advances in intelligent Paradigms and Application*, New York: Springer-Verlag (2002). 161–186.
24. Correlation. Available: http://en.wikipedia.org/wiki/Correlation.

Chapter 25
Innovative Technology Management System with Bibliometrics in the Context of Technology Intelligence

Hua Chang, Jürgen Gausemeier, Stephan Ihmels, and Christoph Wenzelmann

25.1 Introduction

Technology has become a decisive factor for technology-intensive companies because of its significant influence on product development and process optimization. It is important to identify advantages or barriers of technologies, to compare them as well as to analyze the probability of being substituted. Therefore, scientific researchers and decision-makers in companies address their attention to technology intelligence, which is the sum of methods, processes, best practices, and tools used to identify business-sensitive information about technological developments or trends that can influence a company's competitive position.

The technology intelligence process strides across four levels: data, information, knowledge, and decisions. Data are symbols with no meaning. Information is data that has been given meaning by way of relational connection. Knowledge is the output of scouting, processing, and analyzing information. Decisions are made on the basis of knowledge [4]. Within the framework of technology intelligence, the main task is to procure accurate information about performances and developments of technologies, that is, to identify technology indicators. Technology indicators are those indexes or statistical data that allow direct characterization and evaluation of technologies throughout their whole lifecycles, for example, technological maturity, market segment, degree of innovation, or key player (country, company, ...). Those technology indicators offer a direct view of technologies to decision-makers.

People usually read documents one by one and collect the key information manually. But, the amount of information has dramatically increased in recent years. It is no longer possible to evaluate or characterize technologies by reading documents. Therefore, there is a demand for methods that support technology intelligence by systematically analyzing documents for the purpose of extraction of information relevant to technology indicators. One of the methods, which fulfils the requirements, is bilbiometrics.

Oscar Castillo et al. (eds.), *Trends in Intelligent Systems and Computer Engineering.*

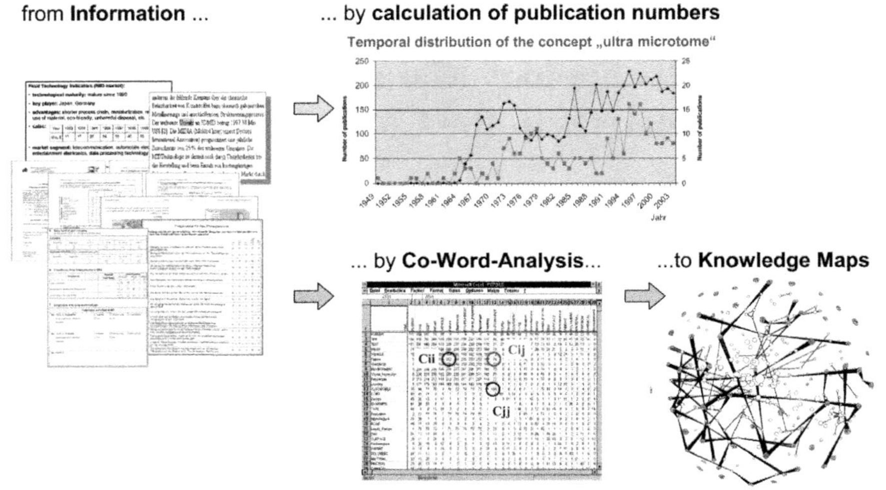

Fig. 25.1 Bibliometric methods: Publication analysis and content analysis

Bibliometrics is a type of research method used originally in library and information science. It utilizes quantitative analysis and statistics to describe patterns of publication within a given field or body of literature. Researchers may use bibliometric methods to determine the influence of a single writer, for example, or to describe the relationship between two or more writers or works. One common way of conducting bibliometric research is to use the Social Science Citation Index, the Science Citation Index, or the Arts and Humanities Citation Index to trace citations [9, 10].

Bibliometric analyses encompass traditional publication analysis and content analysis (Fig. 25.1).

Publication analysis deals with the counting of publication numbers according to time, region, or other criteria. The hypothesis is: the numbers of publications can reveal present and past activities of scientific work. Regarding content analysis, the most important method used is co-word analysis, which counts and analyzes co-occurrences of keywords in the publications on a given subject [11]. Based on the co-occurrences, the keywords can be located in a knowledge map (Fig. 25.2) by using multidimensional scaling (MDS).

The knowledge map can be read according to the following rules. Every pellet in the map stands for a keyword. The diameter means the text frequency of the keyword which is represented by the pellet. The hypothesis for co-word analysis is: the more often the keywords appear together in documents, more content-similar they are. So the keywords describing similar topics are positioned in the vicinity. For example, the word "mechatronics" is always located near the words "mechanics" and "electronics," because they always appear together in the same documents. The thickness of the lines between the keywords represents the relative co-frequency.

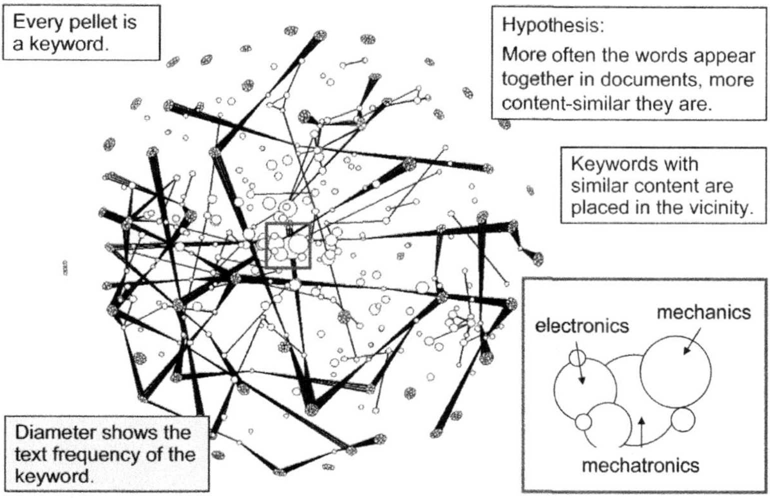

Fig. 25.2 Knowledge map based on co-word analysis

Based on bibliometric analyses, we developed a methodology for the identification of technology indicators, which fulfils the requirements within the technology intelligence process. The next part of this chapter introduces this methodology with a case study.

25.2 Methodology for the Identification of Technology Indicators

The methodology for the identification of technology indicators is developed using the three basic methods of information retrieval,[1] bibliometric analysis, as well as expert consultation.[2] The process model (Fig. 25.3) of the methodology is divided into five phases. The results for every phase are shown on the right side. The tasks that should be done and the methods used in every phase are listed in the middle.

Determination of research objective: The first step is to analyze problems and to determine research objectives, that is, to answer the question, "Who wants to know what kind of technologies in which areas?" The result is the target technology that is investigated in the following steps.

Literature search: The second step is to search for literature thematically relevant to target technology. The method used for literature search is information retrieval.

[1] Information retrieval is the art and science of searching for information in documents. It deals with the representation, storage, organization of, and access to information items [2]. For example, search engines such as Google or Yahoo are the most visible applications of information retrieval.

[2] Expert consultation is the traditional way to investigate technologies by seeking the views of experts. But normally this method is carried out only among a small number of experts because obtaining opinions from experts is always expensive and time-consuming. And if the survey is small, then its representativeness is open to question. Anyway, although there are some drawbacks of expert consultation, it is still used in most cases.

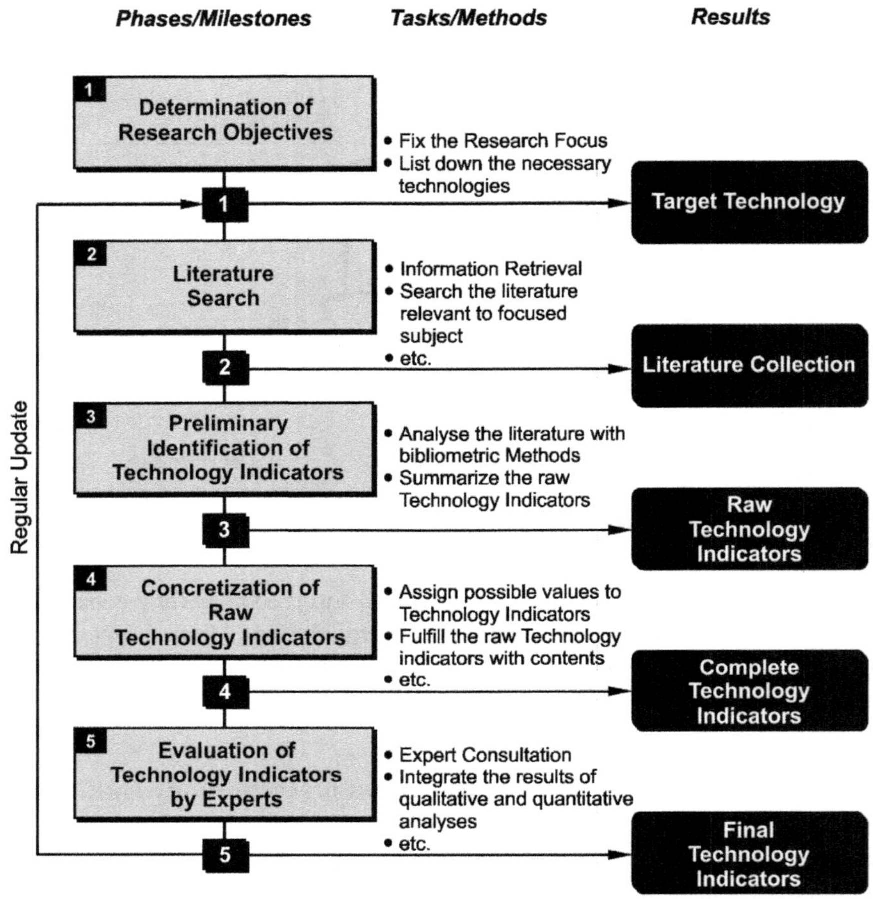

Fig. 25.3 Process model of the methodology for the identification of technology indicators

A group of phrases is defined, which can describe the target technology briefly and concisely. Those phrases are used as search queries to search the desired literature in several databases or search engines. Retrieved documents are collected and stored together. As a result, a collection of literature is available, which is analyzed in the next step.

Preliminary identification of technology indicators: In the third phase all the literature is analyzed with bibliometric methods. First, the publication numbers are counted according to time in order to reveal the historic development of technologies. Then the contents of the literature are consolidated into keywords, whose relationships are analyzed by means of co-word analysis for the purpose of characterizing technologies in detail. Based on co-occurrences, the keywords are located in a knowledge map by using MDS. Inspecting all the keywords from the knowledge map, the keywords that can directly characterize technologies or indicate the development of technologies are selected, such as technological maturity, R & D cost, and sales. Those keywords are defined as raw technology indicators, which are detailed in the next step.

Concretization of raw technology indicators: In this step, it is necessary to fulfil raw technology indicators with contents and to assign values to them. This process is supported by interpreting the knowledge map. Keywords that co-occurred most frequently with raw technology indicators are focused, especially the adjectives, numbers, units, time, and so on. The relationship between those keywords is interpreted logically. After summarizing all the interpretations of co-relationships, the contents are assigned to raw technology indicators. The result for this step is complete technology indicators with names and contents.

Evaluation of technology indicators by experts: So far, all the analyses are based on statistics. So in the fifth step, it is necessary to ask the experts' opinion from a qualitative perspective. Within the expert consultation, the definitions, values, and so on of technology indicators are evaluated and supplemented by experts. After integrating the results of qualitative and quantitative analyses, the final technology indicators are identified and documented.

Regular update: Technology is changing fast and its lifecycle is always getting shorter. Decision-makers always need firsthand information to have agile reactions to a sudden change of technologies. For those reasons, it is indispensable to update the information in the TDB regularly.

25.3 Case Study

In order to verify the proposed methodology for identification of technology indicators, it is exemplified in this section with a case study. The research objective in this case is to characterize and evaluate MID technology especially on its market development. MID technology is an emerging technology that allows the integration of injection-molded thermoplastic parts with circuit traces (Fig. 25.4). It opens a new dimension of circuit carrier design.

Aiming at searching literature relevant to MID technology, the phrases such as "moulded interconnected devices," "3D-MID technology," "MID & integration

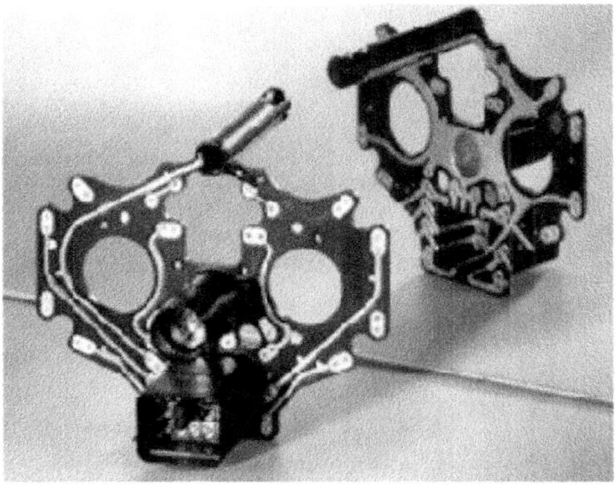

Fig. 25.4 MID technology

of mechanics and electronics" were used as search queries in the Midis Database (3D-MID e.V.), Google Scholar, and other databases. As a result, more than 700 documents thematically relevant to MID were retrieved from different information sources.

The literature was collected and analyzed by bibliometric analysis. First, publication analysis was carried out (Fig. 25.5).

As depicted in Fig. 25.5, the first publication of MID appeared in 1965 and the number of publications has dramatically increased since 1990. It is estimated that the idea of MID was first proposed in 1965 and has been widespread since 1990. So the technology indicator identified from temporal distribution of publications is "spreading speed." Similarly, from the regional distribution of publications, it is concluded that Japan is most active in the area of MID technologies, followed by Germany, and so on. The second technology indicator identified is key player.

After traditional publication analysis, content analysis was carried out. All the contents of literature about MID were consolidated into keywords. The stopwords and the keywords that seldom appear in documents were filtered out. And then the co-occurrences of keywords were calculated, based on which the knowledge map of MID was created (Fig. 25.6).

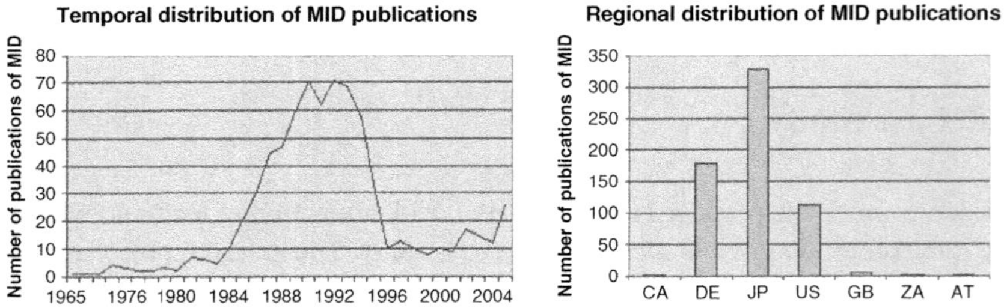

Fig. 25.5 Publication analyses of documents about MID

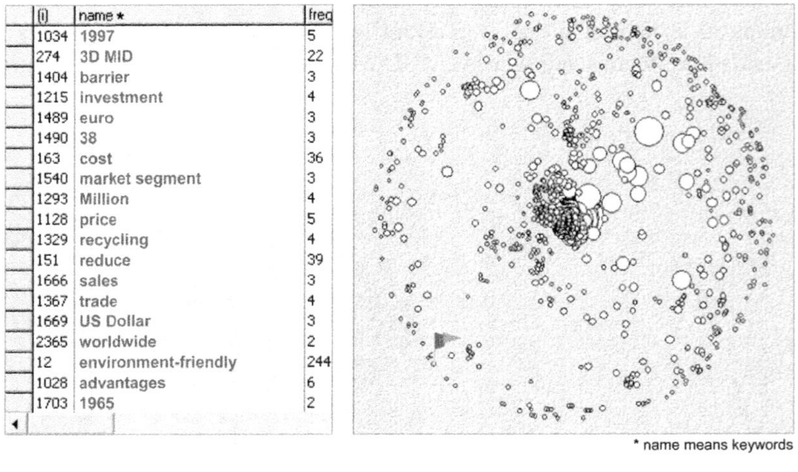

Fig. 25.6 From information by co-word analysis to knowledge map aided by the software BTM (BibTechMon)

From the knowledge map, the keywords that indicate market development of MID technology or those which have a direct influence on market development were selected out, such as sales, market share, market segment, price, and investment. Those keywords were defined as raw technology indicators for the MID market.

In the next step, the raw technology indicators were concretized with contents. There are two kinds of contents. One of them is represented qualitatively, for example, the content of raw technology indicator "advantages." As illustrated in Fig. 25.7, "advantages" and the keywords that appear most frequently with "advantages" were

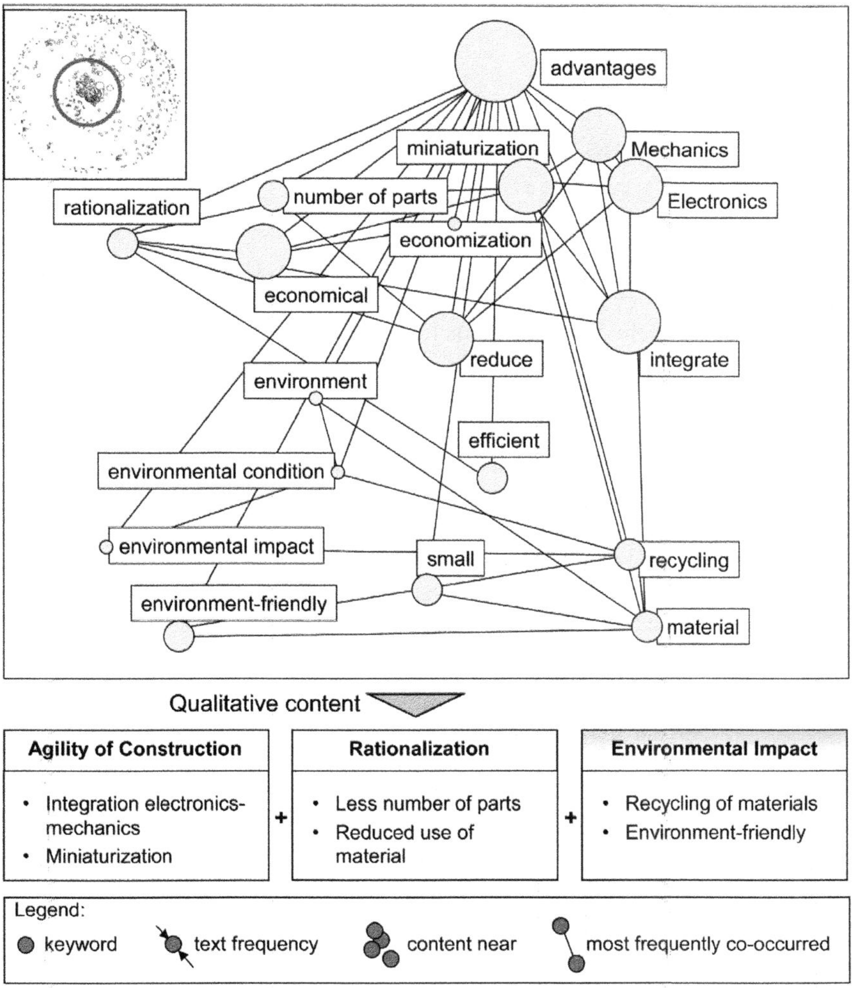

Fig. 25.7 Concretization of the raw technology indicator "advantages" to complete technology indicator

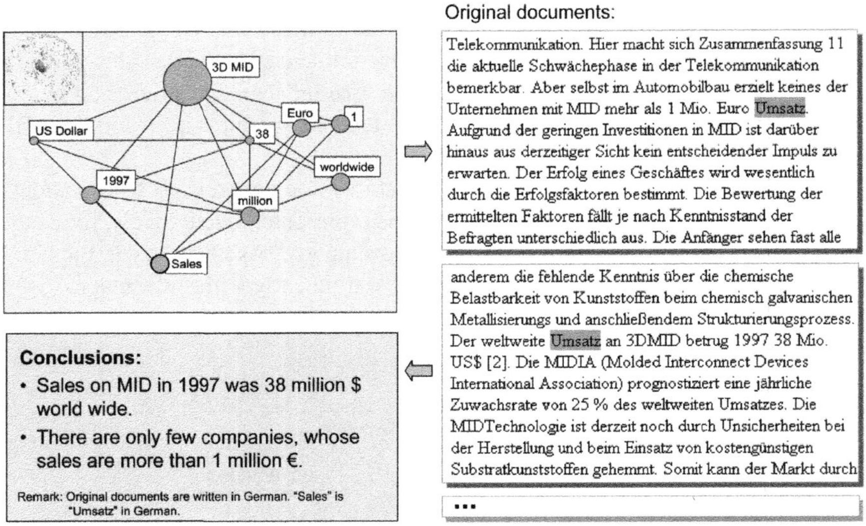

Fig. 25.8 Assignment of values to the raw technology indicators "sales"

focused. The co-relationship between those keywords was logically interpreted as follows. The technology integrates mechanics and electronics; the number of parts is reduced, so the technology is rational; the material used in this technology is recyclable, and so on. Summarizing all the interpretations, the complete technology indicator "advantages" is fulfilled with contents.

The other kind of content is represented quantitatively. As shown in Fig. 25.8, the raw technology indicator "sales" and the keywords associated most strongly with it were concentrated. Among those keywords, the numbers, units, and time were especially taken into consideration. But in this example, the logical relationships between those words were not evident. The original documents should be referred to for validation. It is not necessary to read through the whole documents, but only the segments with the focused keywords highlighted. In conclusion from the original documents, quantitative values were assigned to "sales."

Following a similar procedure, all the other raw technology indicators were fulfilled with contents.

Thus far, all the analyses are based on statistical methods. Expert consultation should be carried out in order to verify the results obtained from analyses of literature. Questionnaires (Fig. 25.9) were constructed and sent to 20 German companies and research institutes. By combining the feedback from experts and the results after quantitative analyses, the final technology indicators were identified and documented.

5 Sales trend of your company
[annual growth rate in %, average in last three years]

Company					
Reduction	Stagnation	0 – 5 %	5 – 10 %	10 – 20 %	> 20 %
☐	☐	☐	☐	☐	☐
Business field MID (Sales of unassembled MID-parts)					
Reduction	Stagnation	0 – 5 %	5 – 10 %	10 – 20 %	> 20 %
☐	☐	☐	☐	☐	☐

6 Investment of your company in 2002

Investment	**Share in Total Sales**	**Development** decrescent --	-	stagnant -/+	increasing +	++
Total Investment	%	☐	☒	☐	☐	☐
Investment in Research & Development	%	☐	☐	☒	☐	☐
Investment in the field of MID	%	☐	☐	☐	☐	☒

7 Cooperation with companies/institutes

Cooperation in the field of MID			
Yes – focus on **Companies**	☐ national	☐ Europe-wide	☐ international
Is the cooperation effective?	☐ Yes	☐ No	

Fig. 25.9 Segment from questionnaires of MID technology

25.4 Combination with Technology Database (Heinz Nixdorf Institute)

The Heinz Nixdorf Institute has developed a Technology Management System for better technology planning and product innovation (Fig. 25.10). The core of that system is a technology database, in which accumulated knowledge, emerging information of technologies, and applications, as well as the output from the bibliometrics-based methodology are stored.

On the left side of Fig. 25.10, the methods as well as information sources of information procurement are introduced, in which bibliometric methods and the methodology for the identification of technology indicators can be embedded. The identified technology indicators are stored in the relational technology database (TDB) in the middle, the core of this system.

The technology database consists of four main entities, with their relationship shown in Fig. 25.11 [6].

- *Technology*: The metadata-based information (definitions, publications, figures, etc.) relevant to technologies are stored in the TDB.

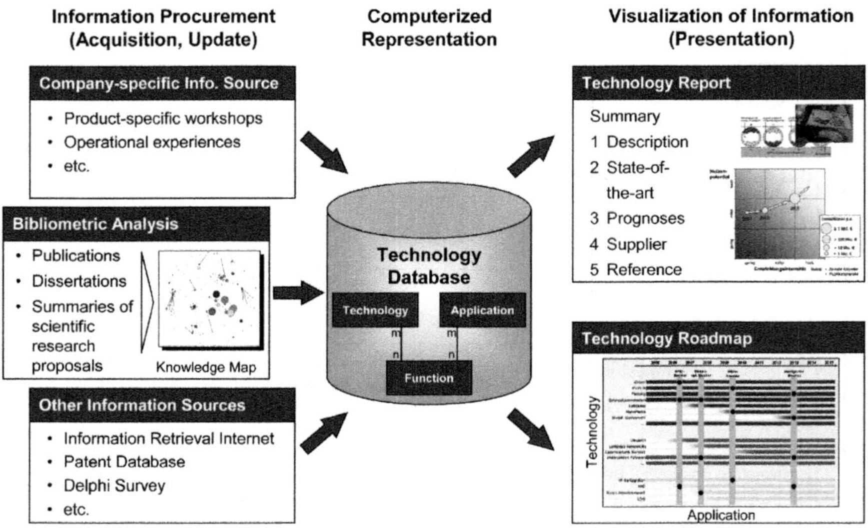

Fig. 25.10 Concept of Technology Management System (Heinz Nixdorf Institute)

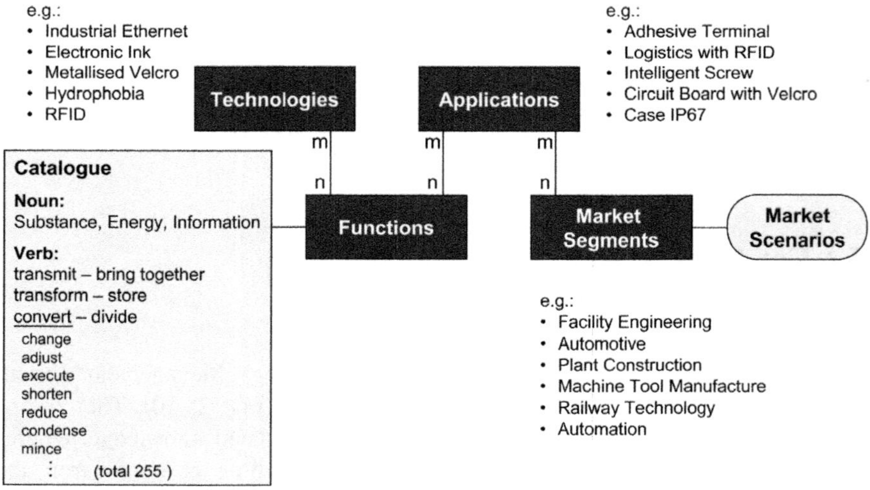

Fig. 25.11 Technology database, simplified relational data model

- *Applications*: Applications are practical solutions to problems, such as products or services, which satisfy customers' requirements. Similar to Technology above, the necessary information (description, market analysis, supplier, etc.) of applications is also available in the TDB.
- *Function*: It concerns a fixed list of general functions based on the corresponding scientific works of Birkhofer [3] and Langlotz [12]. A technology fulfils certain functions; and an application is also based on a function structure. In that way, the functions are assigned to every technology and application. Thus, technologies and applications are consequently correlated through functions.

- *Market Segments*: An application can be assigned to one or more market segments. In our database market segments are described in detail and future-orientated by market scenarios [7]. These scenarios are descriptive information, attached to the respective market segments.

Based on the four entities and their relationships, the Technology Management System allows various queries and visualizes their output automatically in two major representation forms. One of them is the Technology Report, which is detailed and is constructed in a default format. The other visualization form is the Technology Roadmap (Fig. 25.12). The Technology Roadmap is a plan that shows which technology can be used in which products at what time [5, 13]. In the horizontal row the relevant technologies for the enterprise are specified. When the respective technology is mature for employment in a series product is indicated on the time axis. Usually some technologies have to cooperate in order to realize a beneficial application. In Fig. 25.12 four applications are shown as examples. Each application is connected with its possibly usable technology through a black mark, which stands for a function that matches both the technology and application.

By visualizing all the possible connections between technologies and applications through functions in the Technology Roadmap, it offers decision makers advice on options for action. A classification of the options for action based on the Technology Roadmap, which follows the product market matrix of Ansoff [1], is also represented in Fig. 25.12. First of all, it should be figured out whether the up-to-date operated business still carries the enterprise, or business innovations are already necessary. If business innovations are necessary, the other three options for action

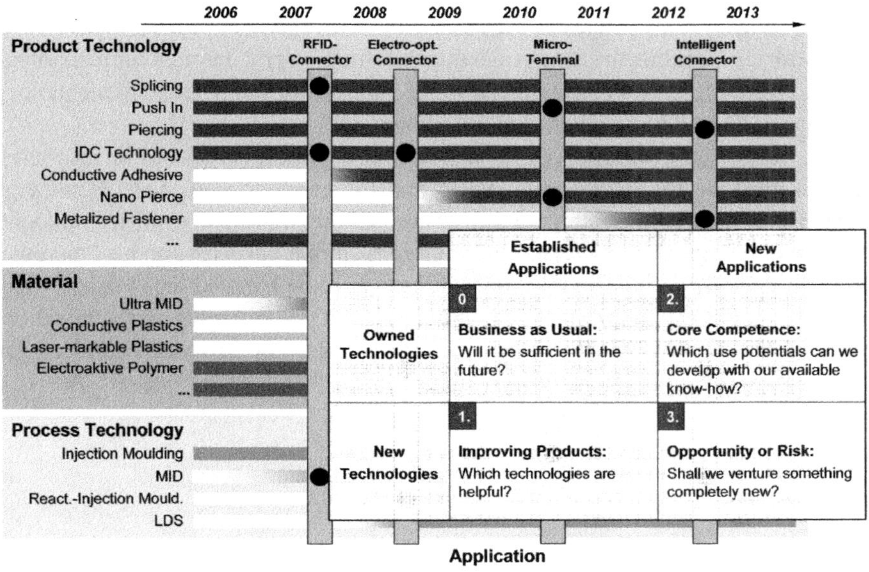

Fig. 25.12 Example of a Technology Roadmap (strongly simplified)

should be taken into consideration. Because the uncertainty of success increases accordingly, the remaining three options for action are sorted by priority as follows.

Product improvement: This option deals with the answer to the question of which external technologies can improve the cost–performance ratio of the existing products.

Core competence approach: The technologies that are mastered by the enterprise frequently represent competencies, which cannot be copied easily by competitors. Here the question arises: Which new application fields can be developed on the basis of the existing competencies in order to generate benefit for the customers and/or to satisfy them?

Departure to new shores: A completely new business has to be set up; both the technologies and the customers are new. Naturally this comes along with the highest risk and is therefore usually only considered if the two options mentioned before do not offer approaches for the advancement of the business.

Both the Technology Report and Technology Roadmap can be generated automatically from the TDB [8].

25.5 Conclusions

The proposed methodology for identification of technology indicators, which is based on bilbiometrics, has proven feasible. It combines both quantitative analysis and qualitative analysis to make the results more reliable and accurate. It standardizes the procedure of information procurement and consequently optimizes information-processing processes. Furthermore, the methodology realizes semiautomatic analysis of literature for the purpose of investigation of technologies.

Bibliometrics fulfils the requirements of information procurement for technology intelligence. Further possible applications of bilbiometrics in the field of technology intelligence should be researched and evaluated.

The Technology Management System has also proven successful in several industrial projects. Our experience shows that the generation of such Technology Roadmaps must be computer-aided—because of the high number of technologies which can be regarded, it can easily be more than one hundred—and on the other hand the often high number of applications cannot be handled any longer with a manually generated graphics. The combinations of technologies and applications based on the Technology Roadmap offer product developers numerous ideas and can lead to successful innovative products.

References

1. Ansoff HI (1965). Corporate Strategy. McGraw-Hill, New York.
2. Baeza-Yates R, Ribeiro-Neto B (1999). Modern Information Retrieval. ACM Press, New York.

3. Birkhofer H (1980). Analyse und Synthese der Funktionen technischer Produkte. Dissertation, Fakultät für Maschinenbau und Elektrotechnik. TU Braunschweig, VDI-Verlag, Düsseldorf.
4. Davis S, Botkin J (1995). The Monster Under the Bed—How Business is Mastering the Opportunity of Knowledge for Profit. Simon & Schuster, New York.
5. Eversheim W (Hrsg.) (2002). Innovationsmanagement für technische Produkte. Springer-Verlag, Berlin, Heidelberg.
6. Gausemeier J, Wenzelmann C (2005). Auf dem Weg zu den Produkten für die Märkte von morgen. In: Gausemeier, J. (Hrsg.): Vorausschau und Technologieplanung. 1. Symposium für Vorausschau und Technologieplanung Heinz Nixdorf Institut, 3–4 November 2005, Schloss Neuhardenberg, HNI-Verlagsschriftenreihe, Band 178, Paderborn.
7. Gausemeier J, Ebbesmeyer P, Kallmeyer F (2001). Produktinnovation—Strategische Planung und Entwicklung der Produkte von morgen. Carl Hanser Verlag, München, Wien.
8. Gausemeier J, Hahn A, Kespohl HD, Seifert L (2006). Vernetzte Produktentwicklung. Der erfolgreiche Weg zum Global Engineering Networking. Carl Hanser Verlag, München.
9. Gorraiz J (1992). Zitatenanalyse—Die unerträgliche Bedeutung der Zitate. In: Biblos. Jg. 41, H. 4, S.193–204.
10. Kopcsa A, Schiebel E (1995). Methodisch-theoretische Abhandlung über bibliometrische Methoden und ihre Anwendungsmöglichkeiten in der industriellen Forschung und Entwicklung. Endbericht zum Projekt Nr. 3437 im Auftrag des Bundesministeriums für Wissenschaft, Forschung und Kunst.
11. Kopcsa A, Schiebel E (1998). Ein bibliometrisches F & E-Monitoringsystem für Unternehmen. Endbericht zum Projekt S.61.3833 im Auftrag des Bundesministeriums für Wissenschaft und Verkehr GZ. 49.965/2-II/4/96.
12. Langlotz G (2000). Ein Beitrag zur Funktionsstrukturentwicklung innovativer Produkte. Dissertation, Institut für Rechneranwendung in Planung und Konstruktion RPK, Universität Karlsruhe, Shaker Verlag, Aachen.
13. Westkämper E, Balve P (2002). Technologiemanagement in produzierenden Unternehmen. In: Bullinger, H.-J.; Warnecke, H.-J.; Westkämper, E. (Hrsg.): Neue Organisationsformen im Unternehmen. Springer Verlag, Berlin.

Chapter 26
Cobweb/IDX: Mapping Cobweb to SQL

Konstantina Lepinioti and Stephen Mc Kearney

26.1 Introduction

Data-mining algorithms are used in many applications to help extract meaningful data from very large datasets. For example, the NetFlix [12] Web site uses hundreds of thousands of past movie ratings stored in an Oracle database to propose movies to returning customers.

Existing data-mining algorithms extract data from databases before processing them but this requires a lot of time and expertise from database administrators. One method of simplifying this process is to develop the algorithms as part of the database management system (DBMS) and to make them accessible using standard database querying tools. However, there are many challenges to be overcome before data mining can be performed using off-the-shelf query tools. One challenge is to make the process of asking a question and interpreting the results as simple as querying a database table. A second challenge is to develop data-mining algorithms that use the database efficiently because database access can have major performance implications.

This chapter suggests one solution to the challenge of making the data-mining process simpler. It discusses an implementation of a popular conceptual clustering algorithm, Cobweb [4], as an add-on to a DBMS. We call our implementation Cobweb/IDX. Section 26.2 is a discussion of the Cobweb algorithm. Section 26.3 discusses the motivation for choosing Cobweb as the basis for our work. Section 26.4 discusses Cobweb/IDX and how it maps the Cobweb algorithm to SQL. Section 26.5 talks about the advantages and disadvantages of the Cobweb/IDX implementation. Section 26.6 presents other work on integrating data mining with databases and finally Sect. 26.7 contains a summary and directions for future work.

Oscar Castillo et al. (eds.), *Trends in Intelligent Systems and Computer Engineering.*

26.2 Cobweb

The simplicity of Cobweb and its relevance to databases comes from (i) the category utility (CU) function that it uses to assess the similarity and differences between data records (Eq. 26.1) and (ii) the set of operations it uses to apply the CU measure to the problem of clustering data records.

26.2.1 Category Utility

Category utility is a measurement that has its roots in information theory [17]. It was introduced by Gluck and Corter [7] with the aim of predicting the basic level in human classification hierarchies. The basic level is considered to be the most natural level of categorisation, for example, dog is the basic level in the hierarchy animal–dog–poodle.

Given a partition of clusters $\{C_1, C_2, \ldots, C_n\}$, CU is the difference between the expected number of attribute values that can be guessed when the clusters are known, $P(C_k)\sum_i \sum_j P(A_i = V_{ij}|C_k)^2$, and the expected number of attribute values when there is no knowledge about the clusters, $\sum_i \sum_j P(A_i = V_{ij})^2$ [4].

$$CU(C_1, C_2, \ldots, C_n) = \frac{\sum_{k=1}^{n} P(C_k)[x - y]}{n} \tag{26.1}$$

$$x = \sum_i \sum_j P(A_i = V_{ij}|C_k)^2$$

$$y = \sum_i \sum_j P(A_i = V_{ij})^2 \tag{26.2}$$

CU is used in Fisher's algorithm to indicate cluster partitions with high intraclass similarity and interclass dissimilarity that are good for prediction purposes. An interesting observation about the measurement is that it is based on probabilities of attribute values that can be calculated using aggregate queries if the data are stored in a DBMS.

26.2.2 The Cobweb Algorithm

Cobweb uses CU to build a tree of clusters, by assessing whether each new record should be included in an existing cluster, used to create a new cluster, or used to combine/split existing clusters. Cobweb represents each cluster as a series of attribute/value counts and calculates CU from the probability distributions of these values.

The algorithm has four operators that it can apply to each level in the cluster hierarchy and it uses CU to evaluate which operator produces the best clusters at each level (adapted from [4]):

```
Function Cobweb (tuple, root)
  Incorporate tuple into the root;
  If root is a leaf node Then
    Return expanded leaf node with the tuple;
  Else
    Get the children of the root;
    Evaluate operators and select the best:
      a) Try incorporate the tuple into the best cluster;
      b) Try creating a new cluster with the tuple;
      c) Try merging the two best clusters;
      d) Try splitting the best cluster into its children;
    If (a) or (c) or (d) is best operator Then
      call Cobweb (tuple, best cluster);
```

The *incorporate* and *disjunct* operators are used to build the tree and the *merge* and *split* operators are used to correct any data ordering bias in the clusters by re-ordering the hierarchy.

- *Incorporate*: Cobweb tries the new tuple in every cluster of the assessed level. As a result, it recalculates the conditional probabilities of every cluster in the level.
- *Disjunct*: Cobweb tries the new tuple in a new cluster that covers only this tuple.
- *Split*: Cobweb tries the new tuple in every child of the best cluster as defined by the incorporate operator.
- *Merge*: Cobweb tries the new tuple in the cluster resulting from merging the best and second best clusters.

Cobweb has an additional operator used to predict missing values, the *predict* operator. The predict operator classifies a tuple down the tree using the incorporate operator but does not add the tuple to the clusters in the tree.

26.3 Motivation for Using Cobweb

Cobweb has a number of characteristics that make it suitable for a DBMS.

- It is an unsupervised learning algorithm that requires no user involvement for classifying the tuples.
- The algorithm is simple to use because it requires few parameters to be set before it produces acceptable results [19] and it produces clusters that are easy to interpret [8].
- It produces a hierarchy of clusters that helps to support progressive database queries such as, "Show more records that are similar to X."

- It is an incremental algorithm that supports incorporation of new tuples in an existing hierarchy of clusters. This is an important property considering the dynamic characteristic of operational databases.
- Although Cobweb was originally intended for clustering categorical datasets it has been extended to numeric and hybrid sets.
- It has proved to be successful when predicting missing data [1] which is a common database problem.

26.4 Implementing Cobweb/IDX

The implementation of Cobweb/IDX remains faithful to the original Cobweb algorithm. Our goal in this implementation is to improve the user's interaction with the data-mining process. In adding data mining to the database environment we have been inspired by the use of indexes in commercial database systems. Although index data structures are complex, modern relational database systems succeed in hiding much of this complexity behind a set of simple commands that create or destroy the index. The most common index structures are zero maintenance tools that can be easily integrated into any database environment. Our objective in implementing Cobweb has been to produce a data-mining tool that has many of these advantages.

26.4.1 Algorithm Design

26.4.1.1 Representing Clusters in the Relational Data Model

Cobweb/IDX stores its clusters in standard database relations. One advantage of this approach is that the algorithm can be implemented using stored procedures that are optimized for use in the database management system. A second advantage of storing the clusters in relations is that the clusters can be queried using existing SQL interface tools and so the algorithm can be used in most database environments.

The core Cobweb tree is stored in three tables. First, the tree structure itself is stored as a two-column table using a traditional parent/child hierarchical relationship, called *cw_tree*. The Oracle database management system (and others) provides efficient extensions to SQL that query this table structure. The second table is the *node values* structure that describes the attribute/value counts for each node, called *cw_values*. This table is used to calculate the probability distributions of each attribute that are needed by the category utility measure. Finally, the *node content* table, called *cw_node_content*, describes the content of each node as a $(node\_identifier, primary\_key)$ pair and links the cluster hierarchy to the original data. In addition to these three tables, there are a number of tables that improve the ease of use or performance of the algorithm.

26.4.1.2 Implementing the Operators

Implementing Cobweb in PL/SQL allows many of the algorithm's calculations to be performed within the database management system as database queries. For example, counting the number of attribute values across the dataset can be executed efficiently using an aggregate query and full table scan. Similarly, counting a subset of attribute values can be performed equally efficiently using an indexed search.

The standard Cobweb algorithm has four operators as discussed in Sect. 26.2. In Cobweb/IDX each operator is implemented as a stored procedure, for example:

```
FUNCTION EVALUATE_INCORPORATE(curr INTEGER, r &1%ROWTYPE )
   RETURN PREVIEW_RECORD IS
   o PREVIEW_RECORD;
   cu REAL;
   CURSOR c1 IS
      SELECT child FROM cw_tree_pseudo_all
      WHERE parent = curr;
   c INTEGER;
BEGIN
   o.rtype := 1;
   FOR children IN c1 LOOP
      UPDATE cw_nodes_pseudo SET node = children.child;
      UPDATE cw_values_pseudo SET node = children.child;
      cu := calcCU_pseudo( curr );
      IF cu > o.bestCU1 THEN
         o.bestCU2 := o.bestCU1;
         o.bestNode2 := o.bestNode1;
         o.bestCU1 := cu;
         o.bestNode1 := children.child;
      ELSIF cu > o.bestCU2 THEN
         o.bestCU2 := cu;
         o.bestNode2 := children.child;
      END IF;
   END LOOP;
   RETURN o;
END EVALUATE_INCORPORATE;
```

A single controlling procedure `add_instance` is responsible for recursively stepping down the cluster tree and applying a *preview* procedure to evaluate each of the four operators and to select the best operator at each level. As the algorithm evaluates each operator it changes the Cobweb data tables and assesses the quality of the resulting clusters. After each step the changes to the data tables must be reversed. Three strategies were evaluated for implementing this process:

1. Apply the changes in main memory after querying the underlying tables but without modifying them.

2. Apply the changes directly to the underlying tables and use database transaction rollback to reverse the changes.
3. Apply the changes to a set of temporary tables and merge the tables using standard SQL set operators.

Strategy 1 does not provide many benefits over existing implementation techniques as it uses the database as a repository and does not benefit from the performance improvements afforded by the database management system. Strategy 2 would support the application and reversal of each operator but transaction management is a heavyweight task and is not suitable for such fine-grained operations. Strategy 3 is the method that we have implemented as it supports full integration of SQL into the data-mining process with the database executing many of the aggregate functions rather than processing them in main memory.

Although updating the temporary tables is an overhead for the algorithm each of the working tables is small and so performance is not a major issue. Standard database views are used to combine the contents of the main Cobweb tables and the pseudo tables. For instance, the following is an example of one of the views.

```
CREATE OR REPLACE VIEW
   CW_TREE_PSEUDO_ALL (PARENT,CHILD)
AS
SELECT parent, child
   FROM cw_tree
UNION ALL
SELECT parent, child
   FROM cw_tree_pseudo
```

Storing intermediate workings in the database allows the implementation of the algorithm to make more use of SQL and stored procedures and also makes the implementation memory-independent.

This approach could be inefficient as it involves updating a number of tables when each operator is evaluated and applied. However, in Oracle the temporary table structures can be created as global temporary tables which are memory-resident and hence have little performance overhead.

26.4.1.3 Implementing Category Utility in SQL

Category utility is calculated using a series of aggregate queries. For example, the calculation of conditional probability (Eq. 26.1) of the attributes in node 10 includes the query:

```
select att,value,sum(instances)
from cw_values
where node=10
group by att, value
```

Similar queries are used to find the unconditional probabilities and to combine the results to calculate the overall category utility of a dataset.

26.4.2 User Interface Design

The interface to Cobweb/IDX supports two processes: updating the clustered dataset and predicting similar records using the cluster hierarchy. The update process monitors the indexed relation for new records and incorporates them into the cluster hierarchy. The prediction process takes a sample record and proposes similar records or missing values using the cluster hierarchy.

26.4.2.1 Updating Cobweb/IDX

From the user's perspective, updating Cobweb/IDX is intended to be similar to updating existing database index structures. Typically index structures run in the background and are updated automatically when the data are changed. For example, the B+-tree index [9] is created using the CREATE INDEX command and requires no further intervention from the user. To achieve this level of integration, Cobweb/IDX is created or dropped using a script and the update process is triggered when new records are inserted into the indexed relation. This design also allows the update process to be separated from the data insertion process and helps to improve overall performance.

The architecture of the update process is shown Fig. 26.1.

26.4.2.2 Predicting Using Cobweb/IDX

Predicting missing values using Cobweb/IDX is modeled on the query by example [21] approach to querying relational databases.

The prediction process uses two tables: input and output. The input table is empty except when the user inserts (incomplete) records into it. These records are removed from the input during the prediction process. When Cobweb/IDX reads a new input record it uses the *predict* operator to process the record and identify any missing values. At present, missing values are indicated by *null* values in the input record. The predict operator proposes values for the null attributes based on the other values in the identified cluster. The output table contains the input record but with the missing values replaced with the value predicted by the Cobweb/IDX index. At present the index predicts one value for each null but could be adapted to predict more than one value with an appropriate probability.

The architecture of the search process is shown in Fig. 26.2.

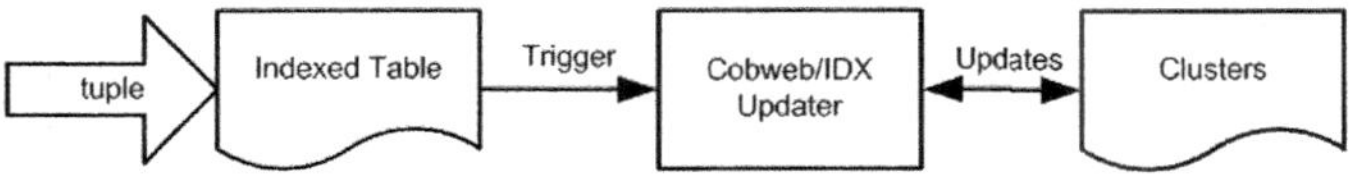

Fig. 26.1 Cobweb/IDX update process

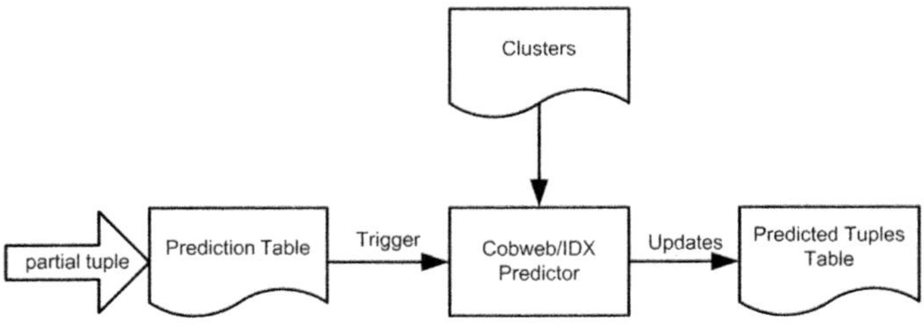

Fig. 26.2 Cobweb/IDX search process

This input/output structure provides a very convenient method of using the index and fits well with existing database query tools. For example, the input/output tables can be directly linked into a Microsoft Access database through the *linked table* facility and a convenient form-based interface built on top for nontechnical users. Predicting missing values is simply a matter of inserting records into the input table and reading the results as they appear in the output table. Our experiments with nontechnical users have produced very positive results.

26.5 Advantages and Disadvantages

26.5.1 Advantages

Our implementation of Cobweb in a DBMS has many advantages.

- It supports simple data mining. Data in the database can be directly indexed and clusters queried using standard SQL. In this way the running of the algorithm is transparent to a user.
- The DBMS maintains the data structures without user intervention.
- It provides logical and physical independence. It is possible to add or drop the Cobweb/IDX index without affecting any other DBMS structure or operation.
- It achieves memory independence. As the algorithm uses the DBMS for most of the CU computations it requires very little data in main memory.
- It is based on a good incremental algorithm that does not suffer from ordering effects in the data.

26.5.2 Disadvantages

The implementation of Cobweb/IDX is important mainly because it supports a simple application of data mining from a user's point of view. However, the algorithm has some disadvantages that must be considered.

- It uses an unsuitable cluster representation. Cobweb/IDX has to perform a large number of queries to classify a tuple. A new cluster representation is necessary to improve upon performance by reducing the number of aggregate queries required.
- It makes insufficient use of the available resources. The algorithm makes limited use of main memory even when the entire dataset can fit in main memory. A better implementation will make use of the DBMS when there is inadequate main memory to complete the task.

26.6 Relevant Work

Work in the area of integrating data mining and databases follows three general approaches: (i) loose coupling, (ii) extending the DBMS, and (iii) tight coupling.

Loose coupling. This is the simplest integration attempted in the literature. It uses an application programming interface (API) such as the open database connectivity (ODBC) standard. Systems such as Clementine [3] and algorithms such as BIRCH [22] have employed this type of connection with the DBMS. The disadvantage of this approach is that the algorithm executes completely outside the DBMS and uses the DBMS only as a data repository.

Extending the DBMS. This approach reduces the gap between data mining and databases and makes better use of the DBMS. It aims to support a range of data-mining techniques by extending the DBMS with data-mining operators that are common to a number of algorithms. Sattler and Dunemann [15] introduced primitives for supporting decision tree classifiers. These primitives are intended to serve a range of classification algorithms. The approach looked at common functions between classification algorithms and developed primitives that support these functions, for example, the computation of the gini-index measure [20]. Geist and Sattler [5] discuss the implementation of operators (such as *intersection* and *difference*) to also support building decision trees in a DBMS. Clear et al. [2] on the other hand introduced more general primitives. An example of their work is the sampling primitive developed in the commercial DBMS SQL/MX to support any type of data mining by reducing the size of the dataset.

Tight-coupling. This approach more fully integrates data mining with the DBMS [16]. MIND [20] performs classification using the gini-index measurement to find the best split in the records and grow a tree. MIND translates the data-mining problem into a DBMS problem by mapping the algorithm's classification processes into SQL. By mapping the algorithm's operations to SQL the algorithm achieves scalability through the parallel option of the DBMS but shows good performance even when applied on nonparallel systems. Similar work includes Sousa et al.'s [18] implementation of a classification algorithm that maps complex calculations of the algorithm to SQL operations. More recent work [14] translates the clustering algorithm K-means to SQL. By using improved data organisation and efficient indexing, the proposed algorithm can cluster large datasets.

Our approach is most similar to the tight-coupling of MIND although we have focused on implementing a more user-friendly implementation that works well with existing database tools.

26.7 Conclusion

This chapter discussed Cobweb/IDX which is an implementation of a popular data-mining algorithm using standard database management tools. Our implementation demonstrates a simple method of integrating data mining into a database management system and a method of accessing the results using standard database tools. We are currently working on a new version of the algorithm that will have all the advantages of Cobweb/IDX but will have better performance characteristics by making better use of memory and disk resources.

References

1. Biswas G, Weinberg JB, Fisher DH (1998) ITERATE: A conceptual clustering algorithm for data mining. *IEEE, Transactions on Systems, Man, Cybernetics - Part C: Applications and Reviews*, 28(2), 219–229.
2. Clear J, Dunn D, Harvey B, Heytens ML, Lohman P, Mehta A, Melton M, Rohrberg L, Savasere A, Wehrmeister RM, Xu M (1999) Nonstop SQL/MX primitives for knowledge discovery. *Proceedings of the Fifth International Conference on Knowledge Discovery and Data Mining (KDD-99)*, San Diego.
3. Clementine, Data Mining, Clementine, Predictive Modeling, Predictive Analytics. http://www.spss.com/clementine/, accessed on July 2006.
4. Fisher DH (1987) Knowledge acquisition via incremental conceptual clustering. *Machine Learning*, (2), 139–172.
5. Geist I, Sattler K (2002) Towards data mining operators in database systems: Algebra and implementation. *Proceedings of 2nd International Workshop on Databases, Documents, and Information Fusion (DBFusion 2002)*, Karlsruhe.
6. Gennari JH, Langley P, Fisher D (1990) Models of incremental concept formation. In J. Corbonell (ed.), *Machine Learning: Paradigms and Methods*, MIT Press/Elsevier.
7. Gluck MA, Corter JE (1985) Information, uncertainty, and the utility of categories. *Proceedings of 7th Annual Conference of the Cognitive Science Society*, 283–287.
8. Hammouda K (2002) Data mining using conceptual clustering. *International Conference on Data Mining (ICDM)*.
9. Knuth D (1997) *The Art of Computer Programming, Volume 3: Sorting and Searching*. Third Edition, Addison-Wesley.
10. Liu H, Lu H, Chen J (2002) A fast scalable classifier tightly integrated with RDBMS. *Journal of Computer Science and Technology*, 17(2), 152–159.
11. McKusick K, Thompson K (1990) COBWEB/3: A portable implementation. NASA Ames Research Center, Artificial Intelligence Research Branch, Technical Report FIA-90-6-18-2, June 20.
12. Netflix, www.netflix.com, 2006.
13. Oracle, www.oracle.com, 2006.
14. Ordonez C (2006) Integrating K-means clustering with a relational DBMS using SQL. *IEEE Transactions on Knowledge and Data Engineering*, 18(2), 188–201.

15. Sattler K, Dunemann O (2001) SQL database primitives for decision tree classifiers. *Proceedings of the 10th ACM CIKM International Conference on Information and Knowledge Management*, November 5–10, Atlanta, GA.
16. Sarawagi S, Thomas S, Agrawal R (1998) Integrating mining with relational database systems: Alternatives and implications. *SIGMOD Conference*, 343–354.
17. Shannon CE, Weaver W (1949) *The Mathematical Theory of Communication*, University of Illiniois Press.
18. Sousa MS, Mattoso MLQ, Ebecken NFF (1998) Data mining: A database perspective. *Proceedings, International Conference on Data Mining*, WIT Press, Rio de Janeiro, Brasil, September, 413–432.
19. Theodorakis M, Vlachos A, Kalamboukis TZ (2004) Using hierarchical clustering to enhance classification accuracy. *Proceedings of 3rd Hellenic Conference in Artificial Intelligence*, Samos, May.
20. Wang M, Iyer B, Vitter JS (1998) Scalable mining for classification rules in relational databases. *Proceedings, International Database Engineering & Application Symposium*, Cardiff, UK, July 8–10, 58–67.
21. Zloof M (1975) Query by Example. *AFIPS*, 44.
22. Zhang T, Ramakrishnan R, Livny M (1996) BIRCH: An efficient data clustering method for very large databases. *Proceedings — ACM — SIGMOD International Conference on Management of Data*, Montreal, 103–114.

Chapter 27
Interoperability of Performance and Functional Analysis for Electronic System Designs in Behavioural Hybrid Process Calculus (BHPC)

Ka Lok Man and Michel P. Schellekens

27.1 Introduction

Hybrid systems are systems that exhibit discrete and continuous behaviour. Such systems have proved fruitful in a great diversity of engineering application areas including air-traffic control, automated manufacturing, and chemical process control. Also, hybrid systems have proven to be useful and powerful representations for designs at various levels of abstraction (e.g., system-level designs and electronic system designs[1]).

Formal languages with a semantics formally defined in computer science increase understanding of systems, increase clarity of specifications and help in solving problems and removing errors. Over the years, several flavours of formal languages have been gaining industrial acceptance. *Process algebras* [1] are formal languages that have formal syntax and semantics for specifying and reasoning about different systems. They are also useful tools for verification of various systems. Generally speaking, process algebras describe the behaviour of processes and provide operations that allow us to compose systems in order to obtain more complex systems. Moreover, the analysis and verification of systems described using process algebras can be partially or completely carried out by mathematical proofs using equational theory. In addition, the strength of the field of process algebras lies in the ability to use algebraic reasoning (also known as equational reasoning) that allows rewriting processes using axioms (e.g., for commutativity and associativity) to a simpler form. By using axioms, we can also perform calculations with processes. These can be advantageous for many forms of analysis. Process algebras have also helped to achieve a deeper understanding of the nature of concepts such as observable behaviour in the presence of nondeterminism, system composition by interconnection of system components modelled as processes in a parallel context, and notions of behavioural equivalence (e.g., bisimulation [1]) of such systems.

[1] For the use in this chapter, we informally refer to electronic system designs as digital, analog, or mixed-signal designs.

Oscar Castillo et al. (eds.), *Trends in Intelligent Systems and Computer Engineering.*

Serious efforts have been made in the past to deal with systems (e.g., real-time systems [2] and hybrid systems [3–6]) in a process algebraic way. Over the years, process algebras have been successfully used in a wide range of problems and in practical applications in both academia and industry for analysis of many different systems.

Recently, through novel language constructs and well-defined formal semantics in a standard structured operational semantics (SOS) style [7], several process algebras/calculi (Hybrid Chi [3, 8], HyPA [4], ACP^{srt}_{hs} [5], Behavioural Hybrid Process Calculus-BHPC [6], and ϕ-Calculus [9]) have been developed for hybrid systems. Also, several attempts [10, 11] have been made over the last two years to apply hybrid process algebras in the context of the formal specification and analysis of electronic system designs. On the other hand, in order to efficiently model electronic system designs of ever-increasing complexity and size, and to effectively analyse them, powerful techniques or approaches (particularly for analysis) are needed. In industry, performance analysis of electronic system designs has so far been mainly addressed in a simulation and/or emulation context (i.e., traditional and popular techniques for performance analysis). Over the years, simulation and emulation have been shown to be well-established and successful techniques for the analysis of the dynamical behaviour of electronic system designs. For functional analysis, the most popular technique used in industry for verifying functional properties of electronic system designs is model checking. This technique has also proven to be a successful formal verification technique to algorithmically check whether a specification satisfies a desired property.

In this chapter, we propose an approach to interoperate[2] the performance and functional analysis of electronic system designs together to obtain full-blown performance as well as functional analysis. Our approach is to formally describe electronic system designs using process algebras (i.e., formal languages) that can be reasonably easily translated (even in a formal way) into models written in various input languages of existing tools for performance analysis and functional analysis. While doing these, we can also compare the input languages, the techniques used for such tools, as well as the analysis results.

For illustration purpose, in this chapter, we choose a hybrid process algebra/calculus (among Hybrid Chi, HyPA, ACP^{srt}_{hs}, BHPC, and ϕ-Calculus) as the main reference formalism for the specification of electronic system designs because of the following.

1. It comprises mathematical specifications for electronic system designs.
2. It allows for description and (syntax-based) analysis of electronic system designs in a compositional fashion.
3. It offers the possibility to apply algebraic reasoning on specifications (e.g., to refine the specifications).

[2] In the literature, there are different definitions/views for the terminology *interoperability*. For us, interoperability is the ability of components, systems, or processes to work together to accomplish a task/goal (based on the definitions given in [12]).

4. It has structured operational semantics of the specifications.
5. Its specifications can be reasonably easily translated into other formalisms even in a formal way (to guarantee that such translations preserve a large class of interesting properties).
6. Its specifications are suitable for both performance and functional analysis.

Among several hybrid process algebras/calculi, BHPC has been chosen for use in this chapter. This particular choice is immaterial and other above-mentioned hybrid process algebras/calculi may be used as well. For this chapter, we chose the tools: the OpenModelica system [13] (for performance analysis) and the model checker PHAVer [14] (for functional analysis). There are several reasons behind this.

1. These tools are well known and have been widely used in both academia and industry.
2. The two tools are free in distribution, well maintained, and well documented.
3. The two tools have many users.

The example used in this chapter is a half-wave rectifier circuit and we chose this particular example because:

1. This example is a mixed-signal circuit (i.e., a hybrid model).
2. It has been studied and analysed from many different domains (e.g., [3, 11, 15]).
3. It is suitable for both performance analysis and functional analysis.

In this chapter, we aim to show that it is reasonably easy to translate the half-wave rectifier circuit described in BHPC to the input languages of the above-mentioned tools and to analyse them in those environments. Hence, general translation schemes from BHPC to other formalisms (i.e., input languages of the OpenModelica system and the model checker PHAVer) are briefly described. However, for brevity, translations presented in this chapter between BHPC and other formalisms are not studied and discussed at the semantic level (to ensure that interesting properties can be preserved by the translations). Nevertheless, it is worth mentioning that the translation from BHPC to the input language of the OpenModelica system and from BHPC to the input language of PHAVer could already be automated (see Sect. 27.6 for details).

This chapter is set up as follows. Section 27.2 provides a brief overview of the behavioural hybrid process calculus (BHPC). A sample (modelling a half-wave rectifier circuit) of the application of BHPC is shown in Sect. 27.3. Section 27.4 first briefly presents the tool OpenModelica system and its input languages, the Modelica language [15]; and then shows how to do performance analysis on the half-wave rectifier circuit described in BHPC using the OpenModelica system. Similarly, Sect. 27.5 first gives a brief summary of the model checker PHAVer and its input language: the theory of hybrid I/O-automata [16]; and then illustrates how to perform functional analysis on the half-wave rectifier circuit described in BHPC using PHAVer. Related works are given in Sect. 27.6. Finally, concluding remarks and future works can be found in Sect. 27.7.

27.2 Behavioural Hybrid Process Calculus (BHPC)

In this section we present, just for illustration purposes, an brief overview of BHPC (that is, relevant for the use of this chapter); a more extensive treatment can be found in [6]. Note that the main concepts of BHPC presented in this section are taken from [6].

27.2.1 *Trajectories, Signal Space, and Hybrid Transition System*

In BHPC, the continuous behaviour of hybrid systems is considered as a set of continuous-time evolutions of system variables (i.e., trajectories). Such trajectories are defined over bounded time intervals $(0,t]$ (where $t \in \mathbb{R}_{>0}$) and mapped to a *signal space*, which defines the evolution of the system.

The signal space ($\mathbb{W}$) specifies the potentially observable continuous behaviour of the systems. Components of the signal space correspond to the different aspects of the continuous behaviour of the system, which are associated with trajectories qualifiers that identify them.

A hybrid transition system (HTS) is a tuple $\langle S, \mathbb{A}, \rightarrow, \mathbb{W}, \Phi, \rightarrow_c \rangle$ such that:

- S is a state space.
- $\mathbb{A}$ is a set of discrete action names.
- $\rightarrow \subseteq S \times \mathbb{A} \times S$ is a discrete transition relation.
- $\mathbb{W}$ is a signal space.
- Φ is a set of trajectories.
- $\rightarrow_c \subseteq S \times \Phi \times S$ is a continuous-time transition relation.

27.2.2 *Formal Syntax, Formal Semantics, and Congruence Property*

The formal syntax of BHPC is presented in Backus–Naur form (BNF) notation:

$$B ::= \mathbf{0} \quad | \quad a.B \quad | \quad [f \mid \Phi].B \quad | \quad \textstyle\sum_{i \in I} B_i \quad | \quad B \|_A^H B \quad | \quad \mathbf{new}\ w.B$$
$$| \ B[\sigma] \quad | \quad P$$

- $\mathbf{0}$ is a deadlock, which represents no behaviour.
- $a.B$ is an action prefix, where $a \in \mathbb{A}$ and B is a process.
- $[f \mid \Phi].B$ is a trajectory prefix, where f is a trajectory variable. It takes a trajectory or a prefix of a trajectory in Φ. In the case where a trajectory or a part of it was taken and there exists a continuation of the trajectory, then the system can continue with a trajectory from the trajectory continuations set. If a whole trajectory was taken, then the system continues as B. Furthermore, the notation $\Downarrow$ is used to separate exit conditions when it is required (see Sect. 27.3.2 for details).

- $\sum_{i \in I} B_i$ is a choice of processes over the arbitrary index sets I. The binary version of it is denoted by $B_1 + B_2$.
- $B \|_A^H B$ is a parallel composition of two processes which explicitly attaches the sets of synchronising action names A and of synchronising trajectory qualifiers H. Synchronisation on action names has an interleaving semantics; and the trajectory prefix can evolve in parallel only if the evolution of coinciding trajectory qualifiers is equal.
- **new** $w.B$ is a hiding operator, where w is a set of discrete action names and trajectory qualifiers to hide.
- $B[\sigma]$ is a renaming operator, where σ is a renaming function. $B[\sigma]$ behaves as B but with the actions and trajectory qualifiers renamed according to σ.
- P is a recursion equation, where $P \triangleq B$ is process identifier.

The formal semantics of BHPC is defined by means of SOS deduction rules that associate a hybrid transition system (as shown in Sect. 27.2.1) with each state. In the field of process algebras, a *congruence* [1] is an equivalence notion (i.e., reflexive, symmetric, and transitive) that has the substitution property. This means that equivalent systems can replace each other within a larger system without affecting the behaviour of that system. *Hybrid Strong Bisimulation* (an equivalence notion) as defined in [6] is a congruence with respect to all operations in BHPC.

27.3 Example

This section illustrates an example of a half-wave rectifier circuit and its specification described in BHPC.

27.3.1 Half-Wave Rectifier Circuit

Figure 27.1 shows the half-wave rectifier circuit. It consists of an ideal diode D, two resistors with resistance R_0 and R_1, respectively, a capacitor with capacity C_0, a voltage source with voltage v_0, and a ground voltage v_G.

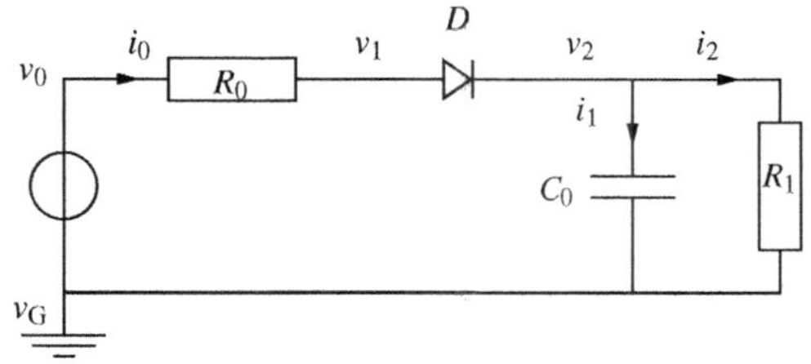

Fig. 27.1 Half-wave rectifier circuit

27.3.1.1 Ideal Diode

An ideal diode can either be in the on mode (i.e., conducts the current) or off mode. When it is in the off mode, the diode voltage must be smaller than or equal to zero and the diode current equal to zero. When it is in the on mode, the diode voltage equals zero and the diode current must be greater than or equal to zero. Thus, the two modes of the ideal diode can be described as follows.

$$\begin{aligned} \text{on}: &\quad v_1 = v_2 \wedge i_0 \geq 0 \\ \text{off}: &\quad v_2 \geq v_1 \wedge i_0 = 0 \end{aligned}$$

27.3.1.2 State Equations

The state equations of other components of the half-wave rectifier circuit are given by:

$$\begin{aligned} &v_0 = F_{\text{time}},\ v_0 - v_1 = i_0 R_0,\ C_0(\dot{v_2} - \dot{v_G}) = i_1, \\ &v_2 - v_G = i_2 R_1,\ v_G = 0,\ i_0 = i_1 + i_2 \end{aligned}$$

Note that F_{time} is an arbitrary function of time, $v_0, i_0, v_1, i_1, v_2, i_2, v_G$ are continuous variables, and R_0, R_1, C_0 are constants.

27.3.2 *Half-Wave Rectifier Circuit in BHPC*

Here, we give the BHPC specification of the half-wave rectifier circuit as described in Sect. 27.3.1. Several processes are defined for such a BHPC specification.

27.3.2.1 IdealDiode

$$\begin{aligned} \text{IdealDiode}(i_0{}^\circ, v_1{}^\circ, v_2{}^\circ) &\triangleq \text{IdealDiodeOff}(i_0{}^\circ, v_1{}^\circ, v_2{}^\circ) \\ \text{IdealDiodeOff}(i_0{}^\circ, v_1{}^\circ, v_2{}^\circ) &\triangleq [i_0, v_1, v_2 \mid \Phi_{\text{off}}(i_0{}^\circ, v_1{}^\circ, v_2{}^\circ) \Downarrow i_0 \geq 0].\mathtt{on}. \\ &\quad \text{IdealDiodeOn}(i_0, v_1, v_2) \\ \text{IdealDiodeOn}(i_0{}^\circ, v_1{}^\circ, v_2{}^\circ) &\triangleq [i_0, v_1, v_2 \mid \Phi_{\text{on}}(i_0{}^\circ, v_1{}^\circ, v_2{}^\circ) \Downarrow v_2 \geq v_1].\mathtt{off}. \\ &\quad \text{IdealDiodeOff}(i_0, v_1, v_2) \end{aligned}$$

$$\begin{aligned} \Phi_{\text{off}}(i_0{}^\circ, v_1{}^\circ, v_2{}^\circ) &= \{i_0, v_1, v_2 : (0,t] \mapsto \mathbb{R} \mid i_0(0) = i_0{}^\circ, v_1(0) = v_1{}^\circ, \\ &\quad v_2(0) = v_2{}^\circ, v_2 \geq v_1, i_0 = 0\} \\ \Phi_{\text{on}}(i_0{}^\circ, v_1{}^\circ, v_2{}^\circ) &= \{i_0, v_1, v_2 : (0,t] \mapsto \mathbb{R} \mid i_0(0) = i_0{}^\circ, v_1(0) = v_1{}^\circ, \\ &\quad v_2(0) = v_2{}^\circ, v_2 = v_1, i_0 \geq 0\} \end{aligned}$$

27.3.2.2 Others

$$\begin{aligned}\text{Others}(v_0{}^\circ, v_1{}^\circ, v_2{}^\circ, v_G{}^\circ, i_0{}^\circ, i_1{}^\circ, i_2{}^\circ) \triangleq [&v_0, v_1, v_2, v_G, i_0, i_1, i_2 \mid \\ &\Phi_{\text{others}}(v_0{}^\circ, v_1{}^\circ, v_2{}^\circ, v_G{}^\circ, i_0{}^\circ, i_1{}^\circ, i_2{}^\circ) \Downarrow \mathit{true}]. \\ &\text{others.Others}(v_0, v_1, v_2, v_G, i_0, i_1, i_2)\end{aligned}$$

$$\begin{aligned}\Phi_{\text{others}}(v_0{}^\circ, v_1{}^\circ, v_2{}^\circ, v_G{}^\circ, i_0{}^\circ, i_1{}^\circ, i_2{}^\circ) = \{&v_0, v_1, v_2, v_G, i_0, i_1, i_2 : (0,t] \mapsto \mathbb{R} \mid \\ &v_0(0) = v_0{}^\circ, v_1(0) = v_1{}^\circ, v_2(0) = v_2{}^\circ, \\ &v_G(0) = v_G{}^\circ, i_0(0) = i_0{}^\circ, i_1(0) = i_1{}^\circ, \\ &i_2(0) = i_2{}^\circ, v_0 - v_1 = i_0 R_0, \\ &C_0(\dot{v_2} - \dot{v_G}) = i_1, v_2 - v_G = i_2 R_1, \\ &v_G = 0, i_0 = i_1 + i_2\}\end{aligned}$$

27.3.2.3 Generator

$$\text{Generator}(v_0{}^\circ) \triangleq [v_0 \mid \Phi_{\text{Generator}}(v_0{}^\circ) \Downarrow \mathit{true}].\text{generator.Generator}(v_0)$$

$$\Phi_{\text{Generator}} = \{v_0 : (0,t] \mapsto \mathbb{R} \mid v_0(0) = v_0{}^\circ, v_0 = F_{\text{time}}\}$$

- Process IdealDiode models the switching-mode behaviour of the ideal diode by means of processes IdealDiodeOn and IdealDiodeOff. Initially, the ideal diode is in the off mode (described by the process IdealDiodeOff) and the trajectory prefix defines the current rise of i_0. When $i_0 \geq 0$, the process may perform action on (an unimportant action name) and switch to the process IdealDiodeOn. Analogously, IdealDiodeOn defines the period of the ideal diode being in the on mode. Notice that $i_0{}^\circ, v_1{}^\circ$, and $v_2{}^\circ$ are the initial values for i_0, v_1, and v_2, respectively; and off is an unimportant action name.
- Process Others models the behaviour of all components of the half-wave rectifier circuit excluding the ideal diode and the generator (i.e., the voltage source with voltage v_0), according to the dynamics defined by the trajectory prefix by Φ_{others}. Notice that $v_G{}^\circ, i_1{}^\circ$, and $i_2{}^\circ$ are the initial values for v_G, i_1, and i_2, respectively; *true* denotes the predicate "*true*" and others is an unimportant action name.
- Process Generator models the behaviour of the voltage source with voltage v_0, according to the dynamics defined by the trajectory prefix by $\Phi_{\text{Generator}}$. Notice that $v_0{}^\circ$ is the initial value for v_0 and Generator is an unimportant action name.

27.3.2.4 HalfWaveRectifier

The complete system is described by the process HalfWaveRectifier, which is the composition of processes IdealDiode, Others, and Generator in a parallel context and it is defined as follows.

$$\text{HalfWaveRectifier}(v_0{}^\circ, v_1{}^\circ, v_2{}^\circ, v_G{}^\circ, i_0{}^\circ, i_1{}^\circ, i_2{}^\circ) \triangleq \text{IdealDiode}(i_0{}^\circ, v_1{}^\circ, v_2{}^\circ) \|_{\emptyset}^{H} (\text{Others}(v_0{}^\circ, v_1{}^\circ, v_2{}^\circ, v_G{}^\circ, i_0{}^\circ, i_1{}^\circ, i_2{}^\circ) \|_{\emptyset}^{\{v_0\}} \text{Generator}(v_0{}^\circ))$$

Notice that $H = \{v_1, v_2, i_0\}$ is the set of trajectory qualifiers for the synchronisation of trajectories and $\emptyset$ denotes an empty set.

27.4 Performance Analysis

This section first briefly presents the tool OpenModelica system and its input languages, the Modelica language; and then shows how to do performance analysis on the half-wave rectifier circuit described in BHPC using the OpenModelica system. For a more extensive treatment of the OpenModelica system and the Modelica language, the reader is referred to [13, 15].

27.4.1 OpenModelica System and Modelica Language

The OpenModelica system is an efficient interactive computational environment for the Modelica language. The Modelica language is primarily a modelling language that allows one to specify mathematical models of complex physical systems. It is also an object-oriented equation-based programming language, oriented towards computational applications with high complexity requiring high performance. The four most important features of the Modelica language are (taken from [15]):

1. It is based on equations instead of assignment statements. This allows acausal modelling that gives better reuse of models because equations do not specify a certain dataflow direction. Thus a Modelica model can adapt to more than one dataflow context.
2. It has multidomain modelling capability, meaning that model components corresponding to physical objects from different domains including hybrid systems.
3. It is an object-oriented language with a general class concept that unifies classes, generic (known as templates in C++), and general subtyping into a single language construct. This facilitates reuse of components and evolution of models.
4. It has a strong software components model, with constructs for creating and connecting components. Thus it is ideally suited as an architectural description language for complex physical systems and to some extent for software systems.

Loosely speaking, a Modelica model (also called class) contains variable declarations (possibly with initial values) and equation sections containing equations. For illustration purposes, below is a sample of a Modelica model:

```
model Sample
 Real x(start = 1);    // variable declarations, x starts at 1
 parameter Real a = 1;
```

```
equation
 der(x) = -a*x;        // equation sections
end Sample;
```

To handle large models, in Modelica, a model can be built up from connections. Various components can be connected using the "connect" statement. Furthermore, Modelica has a electrical component library which consists of many electrical components (e.g., resistor, capacitor, inductor, and ideal diode). Such components can be freely instantiated for reuse and are also the key to effectively modelling complex systems. For illustration purposes, we provide a resistor model in Modelica as follows.

```
model Resistor
 Pin p, n;    // "positive" and "negative" pins
 parameter Real R "Resistance";
equation
 n.i = p.i; // assume both n.i and p.i to be positive
            // when currents flows from p to n
 R*p.i = p.v - n.v;
end Resistor;
```

27.4.2 Analysis and Results

This section shows how to analyse the half-wave rectifier circuit specification in BHPC using the OpenModelica system through the translation to Modelica. Note that Modelica models obtaining from the translations (presented in the next section) of BHPC processes may be slightly different from those made from scratch (using the Modelica language). We aim to have translations resemble the original BHPC processes closely in such a way that analysis results in these translations being able to be related back to the BHPC processes.

27.4.2.1 Translation

1. Process IdealDiode is translated to the corresponding ideal diode model of the electrical component library of Modelica. Below is the Modelica ideal diode model.

```
model Diode "Ideal diode"
  extends TwoPin;
  Real s;
  Boolean off;
equation
  off = s < 0;
```

```
    if off then v = s else v = 0; // conditional equations
    i = if off then 0 else s;     // conditional expression
  end Diode;
```

It is worth mentioning that the above ideal diode model is a parameterised description, where both the voltage v and the current i are functions of the parameter s which is a real number. This is another modelling style to describe the switching behaviour of the ideal diode instead of using recursion equations (as in the process IdealDiode). However, both process IdealDiode and the Modelica ideal diode model behave the same. This means that when the ideal diode is off no current flows and the voltage cannot be positive, whereas when it is on there is no voltage drop over the ideal diode and the current flows.

2. Intuitively, process Others is translated to the connection of models resistor(s), capacitor, and ground from the electrical component library of Modelica.
3. Similarly, process Generator is translated to the voltage source model from the electrical component library of Modelica.
4. Finally, the translation of the process HalfWaveRectifier is obtained by the interconnection with appropriated instantiations of the above-mentioned Modelica models from the electrical component library of Modelica (using the connect statements). Below is such a translation.

```
model HalfWaveRectifier
  Modelica.Electrical.Analog.Basic.Resistor R0(R=10);
  Modelica.Electrical.Analog.Basic.Resistor R1(R=100);
  Modelica.Electrical.Analog.Ideal.IdealDiode DD;
  Modelica.Electrical.Analog.Basic.Capacitor C0(C=0.01);
  Modelica.Electrical.Analog.Sources.SineVoltage AC(V=4);
  Modelica.Electrical.Analog.Basic.Ground G;
equation
  connect(AC.p, R0.p);
  connect(R0.n, DD.p);
  connect(DD.n, C0.p);
  connect(C0.p, R1.p);
  connect(C0.n, AC.n);
  connect(R1.n, AC.n);
  connect(AC.n, G.p);
end HalfWaveRectifier;
```

27.4.2.2 Interactive Session in OpenModelica System

Following the OpenModelica system commands *loadFile*, *simulate*, *instantiateModel*, and *plot*, the Modelica half-wave rectifier model was loaded into the system, was instantiated with appropriated parameters, and was simulated. OpenModelica system reported all these as follows.

```
>>loadFile("C:/OpenModelica1.4.2/testmodels/HalfWaveRectifier.mo")
true
>> simulate(HalfWaveRectifier,startTime=0.0,stopTime=100.0)
record
resultFile = "HalfWaveRectifier_res.plt"
end record

>> instantiateModel(HalfWaveRectifier)

"fclass HalfWaveRectifier
Real R0.v "Voltage drop between the two pins (= p.v - n.v)";
Real R0.i "Current flowing from pin p to pin n";
Real R0.p.v "Potential at the pin";
Real R0.p.i "Current flowing into the pin";
Real R0.n.v "Potential at the pin";
Real R0.n.i "Current flowing into the pin";
parameter Real R0.R = 10 "Resistance";
Real R1.v "Voltage drop between the two pins (= p.v - n.v)";
Real R1.i "Current flowing from pin p to pin n";
Real R1.p.v "Potential at the pin";
Real R1.p.i "Current flowing into the pin";
Real R1.n.v "Potential at the pin";
Real R1.n.i "Current flowing into the pin";
parameter Real R1.R = 100 "Resistance";
Real DD.v "Voltage drop between the two pins (= p.v - n.v)";
Real DD.i "Current flowing from pin p to pin n";
Real DD.p.v "Potential at the pin";
Real DD.p.i "Current flowing into the pin";
Real DD.n.v "Potential at the pin";
Real DD.n.i "Current flowing into the pin";
parameter Real DD.Ron(min = 0.0) = 1e-05 "Forward state-on differential
resistance (closed diode resistance)";
parameter Real DD.Goff(min = 0.0) = 1e-05 "Backward state-off conductance
(opened diode conductance)";
parameter Real DD.Vknee(min = 0.0) = 0 "Forward threshold voltage";
Boolean DD.off(start = true) "Switching state";
Real DD.s "Auxiliary variable: if on then current, if opened then voltage";
Real C0.v "Voltage drop between the two pins (= p.v - n.v)";
Real C0.i "Current flowing from pin p to pin n";
Real C0.p.v "Potential at the pin";
Real C0.p.i "Current flowing into the pin";
Real C0.n.v "Potential at the pin";
Real C0.n.i "Current flowing into the pin";
parameter Real C0.C = 0.01 "Capacitance";
Real AC.v "Voltage drop between the two pins (= p.v - n.v)";
Real AC.i "Current flowing from pin p to pin n";
Real AC.p.v "Potential at the pin";
Real AC.p.i "Current flowing into the pin";
Real AC.n.v "Potential at the pin";
Real AC.n.i "Current flowing into the pin";
parameter Real AC.offset = 0 "Voltage offset";
parameter Real AC.startTime = 0 "Time offset";
Real AC.signalSource.y "Connector of Real output signal";
parameter Real AC.signalSource.amplitude = AC.V "Amplitude of sine wave";
parameter Real AC.signalSource.freqHz = AC.freqHz "Frequency of sine wave";
parameter Real AC.signalSource.phase = AC.phase "Phase of sine wave";
parameter Real AC.signalSource.offset = AC.offset "Offset of output signal";
parameter Real AC.signalSource.startTime = AC.startTime "Output = offset for
time < startTime";
constant Real AC.signalSource.pi = 3.14159265358979;
parameter Real AC.V = 4 "Amplitude of sine wave";
parameter Real AC.phase = 0 "Phase of sine wave";
parameter Real AC.freqHz = 1 "Frequency of sine wave";
Real G.p.v "Potential at the pin";
Real G.p.i "Current flowing into the pin";
equation
R0.R * R0.i = R0.v; R0.v = R0.p.v - R0.n.v;
```

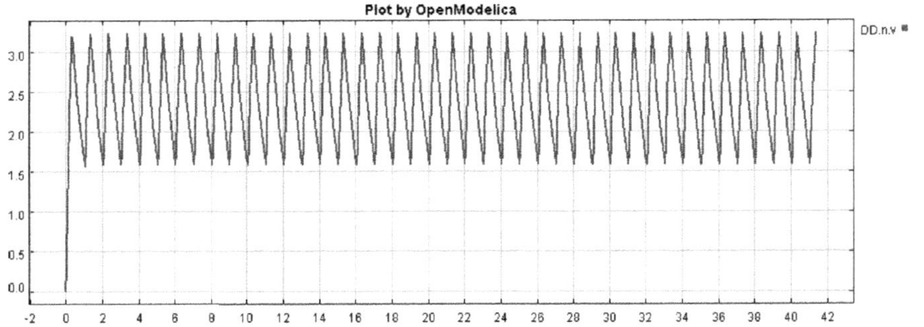

Fig. 27.2 Simulation results of the Modelica half-wave rectifier model

```
0.0 = R0.p.i + R0.n.i; R0.i = R0.p.i;
R1.R * R1.i = R1.v; R1.v = R1.p.v - R1.n.v;
0.0 = R1.p.i + R1.n.i; R1.i = R1.p.i; DD.off = DD.s < 0.0;
DD.v = DD.s * if DD.off then 1.0 else DD.Ron + DD.Vknee;
DD.i = DD.s * if DD.off then DD.Goff else 1.0 + DD.Goff * DD.Vknee;
DD.v = DD.p.v - DD.n.v; 0.0 = DD.p.i + DD.n.i;
DD.i = DD.p.i; C0.i = C0.C * der(C0.v);
C0.v = C0.p.v - C0.n.v; 0.0 = C0.p.i + C0.n.i; C0.i = C0.p.i;
AC.signalSource.y = AC.signalSource.offset + if time <
AC.signalSource.startTime then 0.0 else AC.signalSource.amplitude *
Modelica.Math.sin(6.28318530717959 * AC.signalSource.freqHz * (time -
AC.signalSource.startTime) + AC.signalSource.phase );
AC.v = AC.signalSource.y; AC.v = AC.p.v - AC.n.v;
0.0 = AC.p.i + AC.n.i; AC.i = AC.p.i;
G.p.v = 0.0; R1.n.i + C0.n.i + AC.n.i + G.p.i = 0.0;
R1.n.v = C0.n.v; C0.n.v = AC.n.v;
AC.n.v = G.p.v; DD.n.i + C0.p.i + R1.p.i = 0.0; DD.n.v = C0.p.v; C0.p.v = R1.p.v;
R0.n.i + DD.p.i = 0.0; R0.n.v = DD.p.v; AC.p.i + R0.p.i = 0.0; AC.p.v = R0.p.v;
end HalfWaveRectifier;
"
>> plot({DD.n.v})
true
>>
```

27.4.2.3 Analysis by Means of Simulation

The waveform shown in Fig. 27.2 was obtained by simulating the Modelica half-wave rectifier model with the OpenModelica system. It is not hard to see that the value of DD.n.v in the Modelica half-wave rectifier model (i.e., the voltage v_2 as shown in Fig. 27.1) is never negative and is always above 1.5 V. These are also the general properties of the half-wave rectifier circuit (see also Sect. 27.5.3 for details).

27.5 Functional Analysis

Similar to the previous section, this section first gives a brief summary of the model checker PHAVer and its input language: the theory of hybrid

I/O-automata;[3] and then illustrates how to perform functional analysis on the half-wave rectifier circuit described in BHPC using PHAVer. For a more extensive treatment of the model checker PHAVer and the theory of hybrid I/O-automata, the reader is referred to [14, 16].

27.5.1 PHAVer

PHAVer (polyhedral hybrid automaton verifyer) is a tool for analysing linear hybrid I/O-automata (i.e., a subclass of hybrid I/O-automata which only allows linear dynamics) with the following characteristics.

1. Exact and robust arithmetic based on the Parma Polyhedra Library
2. On-the-fly overapproximation of piecewise affine dynamics
3. Conservative limiting of bits and constraints in polyhedral computations
4. Supports for compositional and assume-guarantee reasoning

27.5.2 Hybrid I/O-automata

In the definition of hybrid I/O-automata, some functions and notations may be used.

- Given a set Var of variables, a *valuation* β: Var $\rightarrow \mathbb{R}$ maps a real number to each variable.
- Let $\mathrm{V}(\mathrm{Var})$ denote the set of valuations over Var.
- An *activity* is a function f: $\mathbb{R}_{\geq 0} \rightarrow V$ in C^{∞} (i.e., a function is C^{∞} if the nth derivative exists and is continuous for all n) and describes the change of valuations over time.
- Let $\mathrm{Act}(\mathrm{Var})$ denote the set of activities over Var.
- Let $f+t$ for $f \in \mathrm{Act}(\mathrm{Var})$ and $t \in \mathbb{R}_{\geq 0}$ be defined by $(f+t)(d) = f(d+t)$, $d \in \mathbb{R}_{\geq 0}$.
- A set $S \subseteq \mathrm{Act}(\mathrm{Var})$ of activities is *time invariant* if for all $f \in S, t \in \mathbb{R}_{\geq 0} : f+t \in S$.

A hybrid I/O-automaton *HIOA* $=$ (Loc, $\mathrm{Var_S}$, $\mathrm{Var_I}$, $\mathrm{Var_O}$, Lab, $\rightarrow$, Act, Inv, Init) consists of the following components.

- A finite set Loc of locations.
- A finite and disjoint set of state and input variables, $\mathrm{Var_S}$ and $\mathrm{Var_I}$, and of output variables $\mathrm{Var_O} \subseteq \mathrm{Var_S}$, and let $\mathrm{Var} = \mathrm{Var_S} \cup \mathrm{Var_I}$.
- A finite set Lab of labels.
- A finite set of discrete transitions $\rightarrow \subseteq \mathrm{Loc} \times \mathrm{Lab} \times 2^{\mathrm{V(Var)} \times \mathrm{V(Var)}} \times \mathrm{Loc}$.

[3] In the literature, many different hybrid automaton definitions already exist. Loosely speaking, the theory of hybrid I/O-automata is a dialect of the (most common) theory of hybrid automata with two additional disjoint sets of variables (in the syntax) representing input and output variables of a hybrid automaton.

- A transition $(l, a, \mu, l') \in \rightarrow$ is also written as $l \xrightarrow{a,\mu}_H l'$.
- A mapping (a labelling function) Act: Loc $\rightarrow 2^{\text{Act(Var)}}$ from locations to time invariant sets of activities.
- A mapping Inv: Loc $\rightarrow 2^{\text{V(Var)}}$ from locations to sets of valuations.
- A set Init $\subseteq$ Loc $\times$ V(Var) of initial states.

The semantics of a hybrid I/O-automaton is defined in terms of a time transition system. Let *HIOA* = (Loc, Var_S, Var_I, Var_O, Lab, $\rightarrow$, Act, Inv, Init) be a hybrid I/O-automaton. A *state* of *HIOA* is a pair $(l, v) \in$ Loc $\times$ V(Var) of a location and a valuation. The transition system interpretation of *HIOA*, written [*HIOA*], is the time transition system $(\text{Loc}, \text{Var}_\text{S}, \text{Var}_\text{I}, \text{Var}_\text{O}, \Sigma', \rightarrow_{LH}, \text{Init})$, where $\Sigma' = \text{Lab} \cup \mathbb{R}_{\geq 0} \cup \{\epsilon\}$ and $\rightarrow_{LH}$ is the union of $\xrightarrow{a}$, for $a \in \Sigma'$. The transition relations of such a time transition system are defined as follows.

1. $(l, v) \xrightarrow{a}_{LH} (l', v')$ iff $l \xrightarrow{a,\mu}_H l'$, $(v, v') \in \mu$, $v \in \text{Inv}(l)$, $v' \in \text{Inv}(l')$ (discrete transitions).
2. $(l, v) \xrightarrow{t}_{LH} (l', v')$ iff $l = l'$ and there exists $f \in \text{Act}(l)$, $f(0) = v$, $f(t) = v'$, and $\forall t', 0 \leq t' \leq t : f(t') \in \text{Inv}(l)$ (timed transitions).
3. $(l, v) \xrightarrow{\epsilon}_{LH} (l', v')$ iff $l = l'$, $v \restriction \text{Var}_\text{S} = v' \restriction \text{Var}_\text{S}$, and $v, v' \in \text{Inv}(l)$ (environment transitions).

These three kinds of transition relations are differentiated by their labels, the time elapse involved, and a special label ϵ that represents changes in the input variables by the environment.

27.5.3 Analysis and Results

This section shows how to analyse the half-wave rectifier circuit specification in BHPC using PHAVer through the translation to hybrid I/O-automata. For brevity, in what follows, we may refer to hybrid I/O-automaton/a to hybrid automaton/a.

27.5.3.1 Translation

Defining the formal translation scheme from BHPC to the theory of hybrid I/O automata and studying the correctness of the translation at the semantic level are far beyond the scope of this chapter. Nevertheless, related works in this direction can be found in [3]. For simplicity, we briefly describe the informal translation from the half-wave rectifier circuit specification in BHPC to the corresponding hybrid automaton model as follows.

- Process IdealDiode is translated to a hybrid automaton *diode* with two locations in such a way that two locations represent the behaviour of processes IdealDiodeOff and IdealDiodeOn. Also, the dynamics defined in Φ_{off} and Φ_{on}

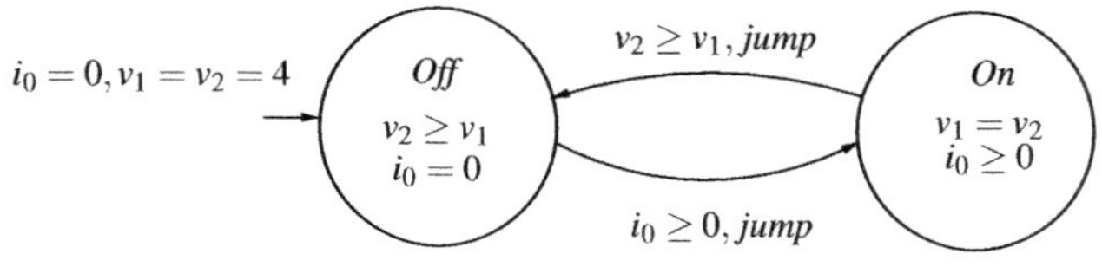

Fig. 27.3 A hybrid automaton model of the process *diode*

for such processes are translated to the activities of the corresponding locations with some appropriated initial conditions (that are assumed for the analysis). The exit conditions (e.g., $i_0 \geq 0$) of the dynamics of the processes are translated to the corresponding discrete transitions associating to the hybrid automaton *diode*. For brevity, the two unimportant action names off and on of processes IdealDiodeOff and IdealDiodeOn are translated to two unimportant synchronisation labels (e.g., on is renamed *jump*). Figure 27.3 captures the main ideas of this translation as explained previously (due to the reason of space, only the graphical representation of the hybrid automaton *diode*, i.e., the translation of the process IdealDiode, is shown).

- In a similar way, process Others is translated to the corresponding hybrid automaton *others* with two locations.
- The translation of the process Generator is discussed in the next section.

27.5.3.2 Approximation and Refinement

Because most of the formal verification tools (including PHAVer) only allow linear dynamics, we have to approximate/refine the behaviour of the half-wave rectifier circuit specification in BHPC as follows.

1. We apply the same approximation used in [17][4] for F_{time} in the process Generator as a sinusoidal voltage source. So, see [17] and the PHAVer code of the generator (as the translation of the process Generator) the Sect. 27.5.3.3 for details.
2. We refine the predicate/equation $C_0(\dot{v_2} - \dot{v_G}) = i_1$ in the process Others to $C_0(\dot{v_2}) = i_1$. This is allowed, because the predicate/equation $v_G = 0$ must always hold in the process Others.

27.5.3.3 PHAVer Codes of the Half-Wave Rectifier Circuit

The input language of PHAVer [18] is a straightforward textual representation of linear hybrid I/O-automata. Using such an input language of PHAVer to describe the

[4] It is worth mentioning that a full-wave rectifier was analysed using PHAVer in [17]. However, this analysis was done by leveraging abstraction to build a system with few variables (only the input voltage source and the output voltage of the circuit were modelled).

hybrid I/O-automata *diode*, *others*, and *generator*, we obtain PHAVer codes (i.e., the translations of the processes IdealDiode, Others, and Generator) as follows.

```
// -------------------
// Half wave rectifier
// -------------------
// ---------
// Constants
// ---------
C0:=0.01; R0:=10; R1:=100;
al:=0.01272; au:=0.01274;
bl:=4; bu:=4; cu:=1.4143;
v0min := -bu; v0max :=  bu;
x0min := -au; x0max :=  au;
// -----------
// System description
// -----------
automaton diodo
state_var: v1, v2, i0;
synclabs: jump;
loc off: while v2>=v1 & i0==0 wait {true}
    when i0>=0 sync jump do {i0'==i0} goto on;
loc on: while  v1==v2 & i0>=0 wait {true}
    when v2>=v1 sync jump do
      {v1'==v1 & v2'==v2} goto off;
initially: on & v1==4 & v2==4 & i0==0;
end

automaton others
state_var: v0, v1, v2, vG, i0, i1, i2;
synclabs: jumpp;
loc rect: while v0-v1==i0*R0 & v2-vG==i2*R1
          & vG==0 & i0==i1+i2
          wait {C0*v2'==i1} when true sync jumpp
          do {true} goto dead;
loc dead: while false  wait {false}
    when true sync jumpp do {true} goto dead;
initially: rect & v0==0 & v1==4 & v2==4
                & vG==0 & i0==0 & i1==0
                & i2==0;
end

automaton generator
state_var: x0, v0;
synclabs: B;
loc CC: while x0min <= x0 & x0 <= x0max &
          v0min <= v0 & v0 <= v0max &
          v0 >= bl-bl/al*x0&
          v0 <= cu*bu-bl/au*x0max &
           0 <= x0&v0 <= x0max &
           0 <= v0 & v0 <= v0max
          wait {x0'==v0 & v0'== -98596*x0};
          when true sync B goto DD;
          when true sync B goto FF;
loc DD: while
          x0min <= x0 & x0 <= x0max &
          v0min <= v0 & v0 <= v0max &
          v0 >= bl-bl/al*(-x0)&
          v0 <= cu*bu-bl/au*(-x0)&
          x0min <= x0 & x0 <= 0 &
          0 <= v0 & v0 <= v0max
          wait {x0'==v0 & v0'== -98596*x0};
          when true sync B goto CC;
          when true sync B goto EE;
loc EE: while
          x0min <= x0 & x0 <= x0max &
          v0min <= v0 & v0 <= v0max &
```

```
            -v0 >= bl-bl/al*(-x0)&
            -v0 <= cu*bu-bl/au*(-x0)&
            x0min <= x0 & x0 <= 0 &
            v0min <= v0 & v0 <= 0
            wait {x0'==v0 & v0'== -98596*x0};
            when true sync B goto DD;
loc FF: while
            x0min <= x0 & x0 <= x0max &
            v0min <= v0 & v0 <= v0max &
            -v0 >= bl-bl/al*x0&
            -v0 <= cu*bu-bl/au*x0&
            0 <= x0 & x0 <= x0max &
            v0min <= v0 & v0 <= 0
            wait {x0'==v0 & v0'== -98596*x0};
            when true sync B goto EE;
initially: $ & x0== -0.01273 & v0==0;
end
```

In order to increase the readability of the system description, some simplifications are introduced in the PHAVer codes and some unimportant labels are used.

27.5.3.4 Analysis by Means of Model Checking

We analyse, by means of model checking, the half-wave rectifier circuit specification in BHPC using PHAVer through the translation to hybrid I/O-automata. We want to verify two basic properties (for safety) of the half-wave rectifier circuit as follows.

1. The voltage v_2 is never negative.
2. For a given $v_0 = 4\sin(2\pi(50)t)$ V and the initial condition $v_2(0) = 4$ V (as already indicated in the PHAVer codes), $v_2(t)$ does not drop below a threshold 1.5 V for any time t.

The following PHAVer analysis commands were used to specify the two safety properties as forbidden states.

```
sys = diodo&others&generator;
reg = sys.reachable;
reg.remove(x0);
reg.print;
forbidden = sys.{$&v2<0|v2<1.5};
echo "";
reg.intersection_assign(forbidden);
echo "intersection with forbidden states:";
reg.is_empty;
```

PHAVer reported that these two safety properties held in all locations of the system (i.e., the parallel composition of hybrid automata *diode*, *others*, and *generator*).

27.6 Related Works

Because there exist many different formalisms for hybrid systems, it is not our intention to list them all in this chapter. For the use here, we only focus our attention on the following hybrid process algebras/calculi Hybrid Chi, HyPA, $\mathrm{ACP}^{\mathrm{srt}}_{\mathrm{hs}}$,

Behavioural Hybrid Process Calculus-BHPC, and ϕ-Calculus. Some comparisons and related works of the above-mentioned algebras/calculi can already be found in [3–6, 19]. HyPA was shown to be useful for the formal specification of hardware and analog circuits in [10]. Also, as reported in [11], ACP^{srt}_{hs} could reasonably and effectively be used for the formal specification and analysis of mixed-signal circuits. Below is a summary for the above-mentioned hybrid process algebras/calculi tools: *Hybrid Chi Python simulator* [3] is a symbolic simulator for Hybrid Chi specifications; and *Chi2HA translator* [3] translates (a subset of) Hybrid Chi specifications into corresponding hybrid automata that can be verified directly by the model checker PHAVer. A number of HyPA specifications can be linearised by the *HyPA linearisation tool* [20]; and *HyPA simulator* [21] is a simulator for HyPA specifications. *BHAVE* [6] is a prototype of the BHPC simulation tool. *SPHIN* [22] is a hybrid model checker for ϕ-calculus specifications.

As reported recently in [23], automatic translator BHPC2MOD accepts a BHPC specification (restricted to a subset of BHPC) as input and translates it to the corresponding Modelica specification. Clearly, Hybrid Chi and BHPC are similar. A manual translation from a BHPC specification to the corresponding Hybrid Chi specification can serve as a preprocessing step for using the Chi2HA translator (which aims to translate such a specification to the corresponding hybrid automata as the input format for PHAVer).

There are also several other specification languages and tools that can be used for the performance and functional analysis of electronic system designs. An overview of such specification languages and tools is given in [17].

Our research work presented in this chapter was inspired by [24] which intended to combine performance or functional analysis for industrial system designs through timed process algebra-based formalisms.

27.7 Conclusions and Future Work

In order to illustrate our work clearly, only a simple electronic system design was given in this chapter. Nevertheless, our proposed approach is generally applicable to all sizes and levels of electronic system designs.

As we have seen in Sects. 27.4 and 27.5, specification languages/formalisms (i.e., BHPC, Modelica language, and the theory of hybrid I/O-automata) have much in common. Therefore, translations among them are quite feasible. It is not hard to see that the (obtained) analysis results for the half-wave rectifier circuit described in BHPC (via translations) by means of the performance analysis approach and the functional analysis approach (as shown in Sects. 27.4.2 and 27.5.3, respectively) are correlated with various properties as expected. However, it is clear that we need to keep the performance and the functional analysis apart, because one cannot replace the other. Nevertheless, our proposed approach in this chapter can be generally applied to carry out performance analysis as well as functional analysis for electronic system designs.

Our future work will focus on the correctness proofs of the translations (among different specification languages/formalisms) at the semantic level and on applying our approach to complex examples in order to gain further confidence in it.

Acknowledgements K.L. Man wishes to thank Jos Baeten, Bert van Beek, Mohammad Mousavi, Koos Rooda, Ramon Schiffelers, Pieter Cuijpers, Michel Reniers, Kees Middelburg, Uzma Khadim, and Muck van Weerdenburg for many stimulating and helpful discussions (focusing on process algebras for distinct systems) in the past few years. He also would like to thank Rolf Theunissen for some helpful discussions to master the use of the model checker PHAVer.

Special thanks go to Andrea Fedeli and Menouer Boubekeur for their cooperation and comments for the research works on *Process Algebras for Electronic System Designs* (PAFESD) [25]. He is also grateful to Jong-Kug Seon for his support on the research works of PAFESD.

References

1. Baeten, J.C.M., Weijland, W.P.: *Process Algebra.* Volume 18 of Cambridge Tracts in Theoretical Computer Science. Cambridge University Press, Cambridge, UK (1990).
2. Baeten, J.C.M., Middelburg, C.A.: *Process Algebra with Timing.* EACTS Monographs in Theoretical Computer Science. Springer-Verlag, New York (2002).
3. Man, K.L., Schiffelers, R.R.H.: Formal specification and analysis of hybrid systems. PhD thesis, Eindhoven University of Technology (2006).
4. Cuijpers, P.J.L.: Hybrid process algebra. PhD thesis, Eindhoven University of Technology (2004).
5. Bergstra, J.A., Middelburg, C.A.: Process algebra for hybrid systems. *Theoretical Computer Science* **335**(2/3) (2005) 215–280.
6. Krilavičius, T.: Hybrid techniques for hybrid systems. PhD thesis, University of Twente (2006).
7. Plotkin, G.D.: A structural approach to operational semantics. Technical Report DIAMI FN-19, Computer Science Department, Aarhus University (1981).
8. van Beek, D.A., Man, K.L., Reniers, M.A., Rooda, J.E., Schiffelers, R.R.H.: Syntax and consistent equation semantics of hybrid Chi. *Journal of Logic and Algebraic Programming* **68**(1–2) (2006) 129–210.
9. Rounds, W.C., Song, H.: The ϕ-calculus—A hybrid extension of the π-calculus to embedded systems. Technical Report CSE 458-02, University of Michigan, USA (2002).
10. Man, K.L., Reniers, M.A., Cuijpers, P.J.L.: Case studies in the hybrid process algebra hypa. *International Journal of Software Engineering and Knowledge Engineering* **15**(2) (2005) 299–305.
11. Man, K.L., Schellekens, M.P.: Analysis of a mixed-signal circuit in hybrid process algebra ACP^{srt}_{hs}. In *International MultiConference of Engineers and Computer Scientists*, Hong Kong (2007).
12. Krilavičius, T.: Study of tools interoperability. Technical Report TR-CTIT-07-01, University of Twente, The Netherlands (2007).
13. OpenModelica System Web site: Openmodelica system http://www.ida.liu.se/~pelab/modelica/.
14. Frehse, G.: PHAVer: Algorithmic verification of hybrid systems past HyTech. In Morari, M., Thiele, L., eds.: *Hybrid Systems: Computation and Control*, 8th International Workshop. **3414**, LNCS. Springer-Verlag, New York (2005) 258–273.
15. Modelica Association: Modelica — A unified object-oriented language for physical systems modeling, http://www.modelica.org. (2002).
16. Lynch, N., Segala, R., Vaandrager, F.: Hybrid I/O automata. *Information and Computation* **185** (1) (2003) 105–157.

17. Carloni, L.P., Passerone, R., Pinto, A., Sangiovanni-Vincentelli, A.L.: Language and tools for hybrid systems design. *Journal of Foundation and Trends* **1** (2005) 1–177.
18. Frehse, G.: Language overview v.0.2.2.1 for PHAVer v.0.2.2, www.cs.ru.nl/~goranf. (2004).
19. Khadim, U.: A comparative study of process algebras for hybrid systems. Technical Report CS-Report 06-23, Eindhoven University of Technology, The Netherlands (2006).
20. van de Brand, P., Reniers, M.A., Cuijpers, P.J.L.: Linearization of hybrid processes. Technical Report CS-Report 04-29, Eindhoven University of Technology, Department of Computer Science, The Netherlands (2004).
21. Schouten, R.: Simulation of hybrid processes. Master's thesis, Eindhoven University of Technology (2005).
22. Rounds, W., Song, H., Compton, K.J.: Sphin: A model-checker for reconfigurable hybrid systems based on spin. Electronic Notes on Theoretical Computer Science **145** (2006) 167–183.
23. van Putten, A.: Behavioural hybrid process calculus parser and translator to Modelica. Master's thesis, University of Twente (2006).
24. Wijs, A.J., Fokkink, W.J.: From χ_t to μcrl: Combining performance and functional analysis. In Press, I.C.S., ed.: *The 10th Conference on Engineering of Complex Computer Systems*, Shanghai, China (2005).
25. PAFESD Web site: Process algebras for electronic system designs http://digilander.libero.it/systemcfl/pafesd/.

Chapter 28
Partitioning Strategy for Embedded Multiprocessor FPGA Systems

Trong-Yen Lee, Yang-Hsin Fan, Yu-Min Cheng, Chia-Chun Tsai, and Rong-Shue Hsiao

28.1 Introduction

Nanometer technology is gradually being applied after deep submicron technology due to the rapid progress of the VLSI fabrication process. Recently, system-on-a-chip (SoC) based products are gaining more advantages such as lower cost and power and high performance, but the new challenges are spending effort on hardware–software codesign, co-verification, and reusable intelligent property (IP). Moreover, the new fabrication process enhances transistor capacity greatly so that built-in multiprocessor system-on-a-chips (MPSoC) such as Xilinx [1] Virtex series FPGA platform are possible. Unfortunately, SoC's challenges are not solved completely; the MPSoC era is coming. Why is technology moving toward MPSoC? The reason is that the characteristic of MPSoC is simpler and more flexible than other architectures. For example, MPSoC-based architecture is easier for floating-point operation than redesigned hardware architecture. For computing mass and complex data, MPSoC is made more flexible by running software in parallel as opposed to hardware.

Hardware–software codesign methodology had been used in the field of SoC and MPSoC [2–6] traditional FPGA system design processes; the hardware–software partition usually depends on the engineer's experience [7, 8]. However, system integration is a major challenge in this way because high-level synthesis and compilation in hardware and software are developed separately. Also, the partitioning result may not be the best solution for low cost or high performance. As a result, system performance and functionality may be limited or insufficient in coverification because the system hardware and software are developed independently after the partitioning phase. On the other hand, hardware–software codesign begins system modeling according to system specification. Next, a hardware–software partitioning tool is able to divide hardware and software in order to meet system constraints such as cost, execution time, power consumption, chip area, and memory size. In the case where the partitioning result could not satisfy system constraints, the

Oscar Castillo et al. (eds.), *Trends in Intelligent Systems and Computer Engineering*.

refinement phase will force the designer to go back to the partition phase until the partitioning result meets system constraints. It is known that each constraint is hard to achieve successfully in embedded multiprocessor FPGA systems. How is it possible to meet all constraints by hardware–software codesign? Therefore, we introduce a hardware–software partitioning method, namely GHO, which profits from the advantages of a genetic algorithm and is hardware-oriented to solve the partitioning issues in embedded multiprocessor FPGA systems.

28.2 Technology Overview

Hardware–software partitioning is classified as software-oriented (SW-oriented) and hardware-oriented (HW-oriented) partitioning [9, 10]. The SW-oriented partitioning method starts with all functionalities in software and then moves portions into hardware if it can gain a better partition result. In contrast, the HW-oriented method starts with all functionalities in hardware and then moves portions into software implementation to obtain a valuable result. But, how can we guarantee gaining the best partitioning result in the case of a moving portion without any strategies?

The hardware–software partitioning method by genetic algorithm has been discussed by many researchers [11–17]. In 2001, Srinivasan et al. [11] categorized various specification sizes into fine-grained and coarse-grained behavioral partitions. Then, they performed their optimal algorithm in an FPGA platform. Saha et al. [12] and Zou et al. [13] also stated optimal algorithms for hardware–software partitioning in 1997 and 2004, respectively. Saha et al. used a genetic algorithm (GA) to transfer initially a hardware–software partition into a constraint satisfaction problem (CSP). Next, they solved CSP to obtain a partitioning result that incorporated cost, execution time, and concurrency constraints. On the other hand, Zou et al. claimed that executing mutation in GA should depend on the fitness function rather than mutation rate. Moreover, data communication time between hardware and software in a system was also taken into account where it carried out the hardware–software partition.

MPSoC is one of the solutions for solving complex functionalities that are required for built-in consumer electronics. In 2001, Lee et al. [18] adapted a multilevel partitioning (MLP) method to solve the hardware–software partitioning in distributed embedded multiprocessor systems (DEMS). They used a gradient metric on cost and performance for getting a better partitioning result. Pomante [19] developed a multilevel heuristic partitioning method which consisted of system- and function-level partitioning phases. First, a functional clustering phase aimed at reduction of the amount of communication among processors and checked the system time constraint. Second, the authors performed a classic hardware–software partition algorithm to decompose the system into subsystems.

Heterogeneous multiprocessor architecture design is another issue in MPSoC. Brandolese et al. [20] developed a design flow and cost function for a heterogeneous multiprocessor. Four constraints such as affinity index, load index, communication index, and physical index were taken into account in their research. In 2002, Sciuto et al. [21] also proposed a design flow and described each phase objective. Moreover, they used a metric in functional blocks for analysis to determine a general-purpose processor (GPP), digital signal processor (DSP), or application-specific integrated circuits (ASIC), and field programmable gate array (FPGA). In 2006, Lin et al. [22] used partitioning and scheduling phases corresponding to the recursive spectral bisection (RSB) algorithm and scheduling on embedded multiprocessor systems.

28.3 GHO Strategy

Powerful computing functionality and complex systems promote consumer electronics functionality such as embedded wireless communications and multimedia systems. Nowadays, powerful architectures such as embedded multiprocessor FPGA systems are implemented for developing multimedia or real-time systems. In general, the components of embedded multiprocessor FPGA systems consist of multiprocessor, memory, high-speed bus, and a lot of input–output ports result in highly complex systems. As we know, designers always trade off issues among system area, cost, power consumption, and execution time but the approach is rarely used to construct the best system architecture for developing embedded multiprocessor FPGA systems. Therefore, we design a GHO partitioning method to solve hardware–software partitioning issues in embedded multiprocessor FPGA systems.

First, we introduce the system model and constraints that are taken into account in embedded multiprocessor FPGA systems. Second, we describe a GHO partition methodology and a new fitness function that is used to assess partitioning results. Third, we utilize the advantages of GA and hardware-oriented partitioning (HOP) to obtain an efficient partitioning result. Finally, we describe a GHO partitioning algorithm that can gain a better partitioning result with reduced memory size and execution time.

28.3.1 System Model

A control and data flow graph (CDFG) is an acyclic graph composed of nodes and edges that is often used in high-level synthesis. It can easily model dataflow, control steps, and concurrent operations because of its graphical nature. Thus, we use a node of CDFG to represent a system hardware or software component for developing GHO. A CDFG example is shown in Fig. 28.1 which consists of a control flow graph (CFG) and dataflow graph (DFG). Also, nodes *a, b, c, d, e,* and *f* represent a set of function elements (*FE*s). In GHO, we use CDFG to construct a system model for the input model of a hardware–software partition.

Fig. 28.1 A simple examples of CDFG

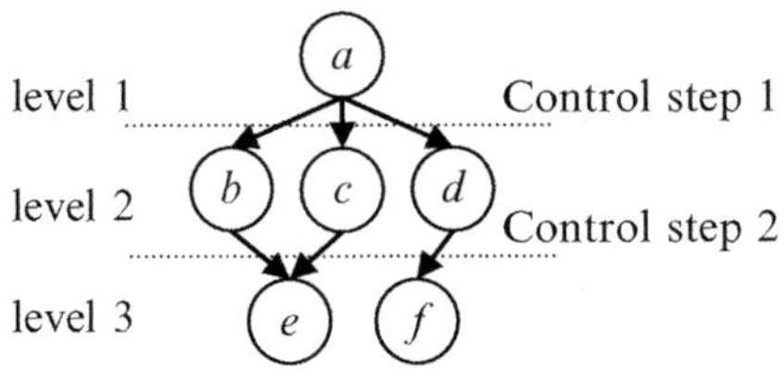

28.3.2 System Constraints

We define the limitations of a hardware–software partition as execution time, cost, power consumption, and the number of processors in a system design. Execution time represents the longest execution time of the path in CDFG such as *a* to *f* or *a* to *e* in Fig. 28.1. The system cost constraint consists of hardware and software costs which correspond to the usage of FPGA slices and memory size, respectively. The constraint of power consumption denotes the limitation of total power dissipation after the hardware–software partition. The last constraint discusses the number of processors inside a system. Under a single-processor environment, node *e* and node *f* in Fig. 28.1 cannot be assigned as software simultaneously for a processor because one job can only be executed in a single processor at a time. Similarly, nodes *b*, *c*, and *d* cannot be assigned simultaneously as software in a two-processor environment (i.e., nodes *b*, *c*, and *d* cannot be over two nodes assigned as software). But if node *b*, *c*, and *d* are assigned as software at the same time, then concurrency will happen. To avoid a concurrency partitioning result, we design a penalty value fitness function to solve the concurrency problem. Once the system constraint or concurrency do not meet system specifications, we decrease the value of the sys_pen or con_pen penalty value to obtain a partitioning result without concurrency and conflicting system constraints.

28.3.3 GHO Partition Methodology

GHO can get a system constraint partitioning result rapidly and efficiently from a huge solution space by genetic partition. In addition, GHO can also increase the utilization of FPGA slices but avoid the concurrence process by HOP. In fact, it is a big challenge to discover a partitioning result for embedded multiprocessor FPGA systems. We design a GHO to overcome it despite that many possible partitions resulting in enormous solution space may exist. GHO adapts the merits of the genetic algorithm to obtain a partitioning result which simultaneously meets all system constraints. Next, GHO improves the previous partitioning result because the utilization of slices in a field programmable gate array (FPGA) does not rely as much on running a hardware–software partition. In contrast, all slices are regarded as a cost even though the system only uses 5% slices in FPGA. In the case where more slices are used, both less memory size and fast execution time can be obtained because

Fig. 28.2 Design flow for GHO

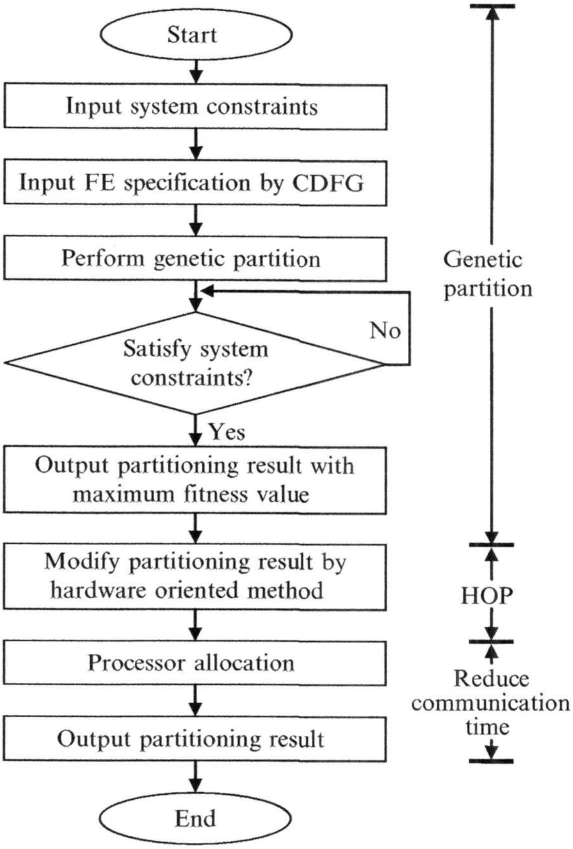

more *FE*s are assigned as hardware as it can. However, it is known that more hardware easily leads to higher power consumption. As a result, it becomes a trade-off issue among system constraints in developing an embedded multiprocessor FPGA system.

Figure 28.2 shows the design flow for GHO. Firstly, we apply a genetic partition to obtain a partitioning result *R* which meets all system constraints. Secondly, we use HOP to find the maximum execution time in FE at each level and implement by hardware under satisfaction the constraints of system power consumption and FPGA slices. Thirdly, we deploy the same path of *FE* to a processor in order to decrease the communication time among *FE*s. Finally, a better partitioning result by GHO is reported to the developer.

28.3.3.1 Fitness Function

The genetic algorithm (GA) has been proved that it can be used to solve the constraint problem in [14–17]. Why can GA gain a better result by reproduction, crossover, and mutation? The partial answer is the fitness function which is used to assess each chromosome by a system element set (*SE*) for crossover. In the case of *SE* meeting all system constraints, the value of the fitness function would be higher than others. Thus, the definition of the fitness function must distinguish a partition result from *SE*s. That is to say, a better *SE* can be obtained rapidly from a lot of *SE*s by the fitness function.

We offer a new fitness function which is shown in Eq. 28.1. It is comprised of system cost, execution time, power consumption constraints, concurrency penalty (con_pen), and system penalty (sys_pen) that are constraints in the hardware–software partition. The reality system cost, execution time, and power consumption for each *SE* is represented by cost_real, time_real, and power_real where the bias is indicated by α, β, and γ, respectively. For example, we can set a higher β to find out the minimum execution time among *SE*s. Similarly, a minimum system cost or power consumption can be obtained by setting a higher α or γ, respectively. If system cost, execution time, and power consumption constraint have the same priority, then we can set $\alpha = \beta = \gamma = 1/3$ due to the sum of α, β, and γ must be equal to 1. Two penalty values, sys_pen and con_pen, are used to judge whether *SE* meets system constraints and concurrency constraints. If not, we can set a higher value for sys_pen and con_pen to obtain a partitioning result that meets system and concurrency constraints.

$$fitness(v) = 1/(1+1/v), \tag{28.1}$$

$$\text{where } v = \left(\frac{cost - cost\_real}{cost} \times \alpha + \frac{time - time\_real}{time} \times \beta + \frac{power - power\_real}{power} \times \gamma)/(con\_pen \times sys\_pen \right)$$

and $\alpha + \beta + \gamma = 1$.

28.3.3.2 Genetic Partition

The purpose of a genetic partition in GHO is rapidly getting a partitioning result that meets all system constraints such as execution time, cost, power consumption, and the number of processors among the solution space. The steps of the genetic partition are summarized as reproduction, crossover, and mutation. In our reproduction process, we use "tournament selection" which randomly selects two or more chromosomes to form a crossover pool. According to probability theory, the higher the fitness value of chromosomes is, the more chances there are for being chosen among chromosomes. Hence, the lower fitness value of the chromosome is gradually eliminated by an evolutionary process. The next step is crossover which randomly selects a chromosome from the crossover pool. We adapt a single-point crossover method that exchanges each gene individually to form a new chromosome. The crossover probability is another factor for crossover operation that represents the crossover operation rate. We set the crossover probability equal to 1 which produces a new offspring by the crossover operation. On the contrary, if the crossover probability is 0, we only copy the chromosome from the old population.

The mutation operation is the final step in the genetic partition that uses mutation probability to determine the chromosome mutation rate. If the designer plans to mutate chromosomes every time, the mutation rate must set to 1 as opposed to 0 which sets that the mutation process will not be performed. In our case, the mutation probability is set to 0.5 for once every two times. After performing previous steps, GHO obtains a partitioning result such as $\{101000111\}$ that satisfies all system constraints as stated in Sect. 28.3.2.

28.3.3.3 Hardware-Oriented Partition

The number of FPGA slices depends completely on developing an FPGA architecture. In developing embedded multiprocessor FPGA systems, all slices within an FPGA are regarded as a hardware cost by the designer. If a design only uses 5% of slices, the cost of hardware still refers to all slices rather than 5%. As a result, up to 95% of slices are wasted in such a case. In contrast, if a partitioning result could reassign the *FE* of software to hardware as much as possible, it is very helpful for improving execution time and memory size because the final design uses more hardware. From the point of view of hardware cost, the more slices are used, the fewer costs are required. Therefore, GHO aims at increasing slice utilization for getting more efficient hardware–software partitioning results with minimum execution time and memory size. The GHO strategy is similar to the HOP method but concurrency is taken into account.

Concurrency also needs to be solved by a reassignment process. As a result, a reassignment process is a significant issue in maximizing the utilization of FPGA slices and concurrency. But, what are the best rules of a reassignment process in HOP? The rules of GHO are summarized. Firstly, we define the execution time (ET) as the maximum execution time of *FE* in each CDFG level. Secondly, we search each level for concurrence and then record it. Thirdly, we reassign ET software to hardware to improve the execution time and slice utilization. Fourthly, we check the new partitioning result as to whether it meets system constraints and is without concurrency conflict. The above procedures will be iterated until they are without concurrency. Finally, GHO decreases the communication time by allocating as a processor the same path of *FE*s.

28.3.3.4 GHO Partitioning Algorithm

Table 28.1 displays a GHO algorithm which consists of a genetic partition, Steps 1 to 10, and hardware-oriented method, Steps 11 to 30. In Step 1, an original set of *SE*, namely *parent SE* (SE_{pg}), is randomly generated and encoded randomly to correspond to *FE* as 1 or 0 where hardware and software are represented, respectively; that is, $SE = \{FE_a FE_b, \ldots, FE_v\}$, $FE_i \in \{1,0\}$. Next, each *SE* is checked

Table 28.1 GHO partitioning algorithm

GHO Partitioning Algorithm

Input : CDFG function elements (*FE*s), System constraints, Processor numbers

Output :Mapping *K* of function elements into hardware components and software components

/* **Genetic partition** */

1 Generate *N* of System Element Sets (*SE*s), $SE = \{FE_1, FE_2, \ldots, FE_K\}$ and encode randomly FE_i of *SE* into 0 as software or 1 as hardware

2 While (1) do

3 Fitness();

4 Crossover();

5 If (reach the mutation ratio) Mutation();

6 Fitness();

7 Compare();

8 If (reach the maximum number of generation)

9 Output the best *SE*, SE_{best}, which is recorded and exit

10 end While

/* **Modify partitioning by Hardware Oriented Method** */

11 If the FE_i in SE_{best} equal to 1, then mark fixed and set locked else mark unfixed and set unlocked

12 If the FE_i in SE_{best} equal to 0, find out the level *l* which FE_i located in the CDFG

13 While (1) do

14 If the FE_i had longer execution time and is unlocked and marked unfixed in SE_{best}

15 assign the FE_i to 1

16 If the total power consumption and slices of SE_{best} are satisfying system constraints

17 FE_i mark fixed

18 Else

19 assign FE_i to 0 and FE_i mark fixed

20 locked all *FE* in level *l*

21 If don't exist a unfixed and unlocked FE_i then exit

22 end While

23 While (the SE_{best} had *FE* which marked unfixed) do

24 If the FE_i had longer execution time and marked unfixed in SE_{best}

25 assign the FE_i to 1

26 If the total power consumption and slices of SE_{best} are satisfying system constraints

27 FE_i mark fixed

28 Else

29 assign FE_i to 0 and FE_i mark fixed

30 end While

31 Process or allocation

Terminate the algorithm

End of GHO partitioning algorithm

for meeting system constraints and being without concurrency conflict. Moreover, some of *SE* will be recorded if it does not meet system and concurrency constraints. In Step 3, a best *SE* of SE_{pg} will be found and recorded by calculating the fitness value for each *SE*. In Steps 4 and 5, according to the crossover and mutation rate to execute crossover and mutation which result in a set of *SE*s, namely, *child SE* (SE_{cg}), being generated. Then, GHO finds a best *SE* and record from SE_{cg} by calculating the fitness value for each *SE*. In Step 7, we compare SE_{pg} with SE_{cg} for the fitness value and record a better one. After iterating such a procedure until GHO reaches the population size, we obtain a best *SE* by genetic partition.

A partitioning result of meeting all constraints can be obtained by genetic partition. In order to use FPGA efficiently to improve the execution time and utilization of memory, we modify the result of the genetic partition by HOP. In Step 11, we mark the hardware *FE* with a fixed and locked label which stands for *FE* implementation by hardware that must not be reassigned to software. On the contrary, unfixed and unlocked are used to represent *FE* as software that can be reassigned to hardware. In Step 12, all of *FE* by software are found where located in CDFG and sort execution time by ascend. In Steps 13 to 22, the *FE* of software is reassigned for maximum execution time to hardware for obtaining a new *SE* which must meet power consumption, FPGA slices, and system constraints. If not, the *FE* reassigns processing and marks software's *FE* as fixed. In addition, the level of fixed *FE*s will be marked as locked. Alternatively, the *SE* will be marked as fixed. Although the SE_{best} has *FE* which is marked unfixed, Steps 23 to 30 will perform the reassigning software to hardware procedure. Finally, GHO deploys a partitioning result for *FE* related to paths the same as processors.

28.4 Experiment Results

We use joint photographic experts group (JPEG) to illustrate the feasibility of GHO. We also implement the GA [13], HOP, and Lin [22] partitioning methods in this example for showing the advantages of GHO. Before this example, we finished all of the function elements in Verilog and C language form. Table 28.2 lists system constraints for each *FE*. Our experiment platform is Xilinx FPGA ML310.

28.4.1 JPEG Encoding Example

Figure 28.3 shows the diagram of an encoding system for JPEG. First, the RGB of the image will be translated into the luminance (refer to Y) and chrominance (V and U) signals. The Y, U, and V signals are divided individually into 8*8 blocks. Then, each 8*8 block will be transferred to the frequency-domain by a forward discrete cosine transform (FDCT) for centralization of the low frequency of the image. After FDCT, the data of each block reserve merely the integer approximate value by quantization. The result after quantization may have a lost effect that is not recognized by humans. Finally, each block is encoded and saved.

Table 28.2 Measured data of *FE*s for system constraints

FE	Execution time		Cost		Power consumption	
	HW	SW	HW	SW	HW	SW
a(Level Offset)	155.264 ns	9.38 μs	0.00731	0.00058	4 mw	0.096 mw
b(DCT)	1844.822 ns	20 ms	0.378	0.00288	274 mw	45 mw
c(DCT)	1844.822 ns	20 ms	0.378	0.00288	274 mw	45 mw
d(DCT)	1844.822 ns	20 ms	0.378	0.00288	274 mw	45 mw
e(Quant.)	3512.32 ns	34.7 μs	0.011	0.00193	3 mw	0.26 mw
f(Quant.)	3512.32 ns	33.44 μs	0.00964	0.00193	3 mw	0.27 mw
g(Quant.)	3512.32 ns	33.44 μs	0.00964	0.00193	3 mw	0.27 mw
h(DPCM)	5.334 ns	0.94 μs	0.002191	0.000677	15 mw	0.957 mw
i(ZigZag)	399.104 ns	13.12 μs	0.035	0.000911	61 mw	0.069 mw
j(DPCM)	5.334 ns	0.94 μs	0.002191	0.000677	15 mw	0.957 mw
k(ZigZag)	399.104 ns	13.12 μs	0.035	0.000911	61 mw	0.069 mw
l(DPCM)	5.334 ns	0.94 μs	0.002191	0.000677	15 mw	0.957 mw
m(ZigZag)	399.104 ns	13.12 μs	0.035	0.000911	61 mw	0.069 mw
n(VLC)	2054.748 ns	2.8 μs	0.00774	0.0144	5 mw	0.321 mw
o(RLE)	1148.538 ns	43.12 μs	0.00256	0.006034	3 mw	0.021 mw
p(VLC)	2197.632 ns	2.8 μs	0.00862	0.0144	5 mw	0.321 mw
q(RLE)	1148.538 ns	43.12 μs	0.00256	0.006034	3 mw	0.021 mw
r(VLC)	2197.632 ns	2.8 μs	0.00862	0.0144	5 mw	0.321 mw
s(RLE)	1148.538 ns	43.12 μs	0.00256	0.006034	3 mw	0.021 mw
t(VLC)	2668.288 ns	51.26 μs	0.01921	0.0167	6 mw	0.018 mw
u(VLC)	2668.288 ns	50 μs	0.00191	0.0167	6 mw	0.018 mw
v(VLC)	2668.288 n	50 μs	0.00191	0.0167	6 mw	0.018 mw

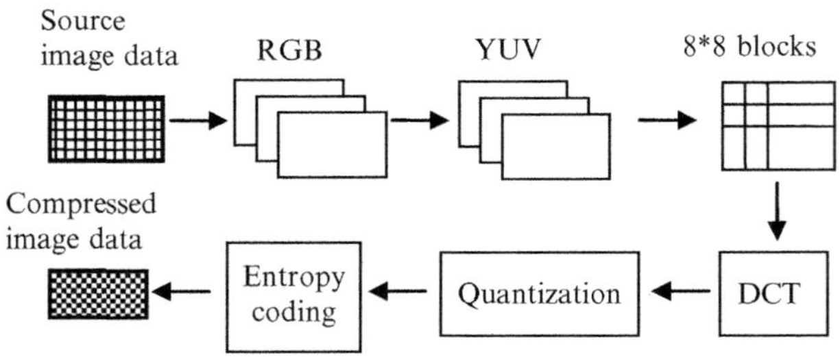

Fig. 28.3 Encoding system diagram of JPEG

The CDFG of JPEG is shown in Fig. 28.4. We develop two applications of software for transferring BMP to YUV format and YUV to 8*8 blocks. Other *FE*s will be performed by GA [13], HOP, Lin [22], and GHO. Moreover, we denote the JPEG function element in Fig. 28.4 as $FE_s = \{FE_a, FE_b, \ldots, FE_v\}, FE_a =$ Level Offset, $FE_b =$ DCT, $FE_c =$ DCT, and so on.

28.4.2 Partitioning Results Analysis

The resources of Xilinx FPGA ML310 are 13,696 slices, 2448 kb memory size, and two embedded microprocessors. In this example, we perform GHO, GA [13], HOP,

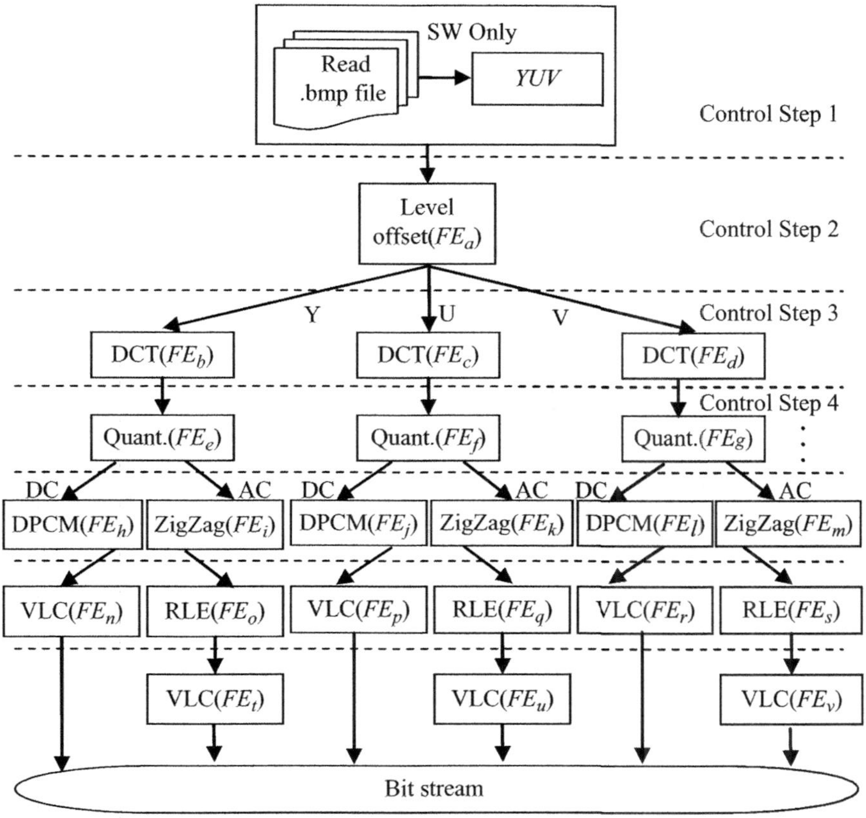

Fig. 28.4 The CDFG for JPEG encoding

Table 28.3 Partitioning results for JPEG encoding system by GHO, GA [13], HOP, and Lin [22]

Partition method	FE i = abcdefghijklmnopqrstuv	Execution time (us)	Memory (kbyte)	Satisfied slice usage	Satisfied power consumption
GHO (proposed)	1010 1111 1111 0111 1111 11	20021.66	16.507	Yes	Yes
GA [13]	0010 0101 0111 0110 1110 10	20111.26	146.509	Yes	Yes
HOP	0100 1111 1010 1110 1011 10	20066.64	129.680	Yes	Yes
Lin [22]	0000 0000 0000 0000 0000 01	20151.58	279.995	Yes	Yes

and Lin [22] to obtain a partitioning result with power consumption limited under 600 mw. The parameters of GA are set as $\alpha = 1/3$, $\beta = 1/3$, $\gamma = 1/3$, $con\_pen = 100$, $sys\_pen = 100$, $crossover\ probability = 1$ and $mutation\ probability = 0.5$.

Table 28.3 shows the partitioning result of GHO, GA [13], HOP, and Lin [22]. The memory is the sum of FE_i and multiplies 2448 which is the FPGA memory size. From execution time of view, HOP should gain the best result among the four kinds of partition methods because all *FE*s start with hardware then move a portion into software. But, according to Table 28.3, GHO is faster than GA [13] and Lin [22], especially HOP. For memory size, GHO is obviously less than GA [13], HOP, and Lin [22]. The last two columns in Table 28.3 indicate that four partition methods are satisfied constraints of FPGA slices and power consumption.

For various low-power designs, we perform a series of experiments from 600 mw to 900 mw by GHO, GA [13], HOP, and Lin [22]. Figure 28.5a displays that GHO

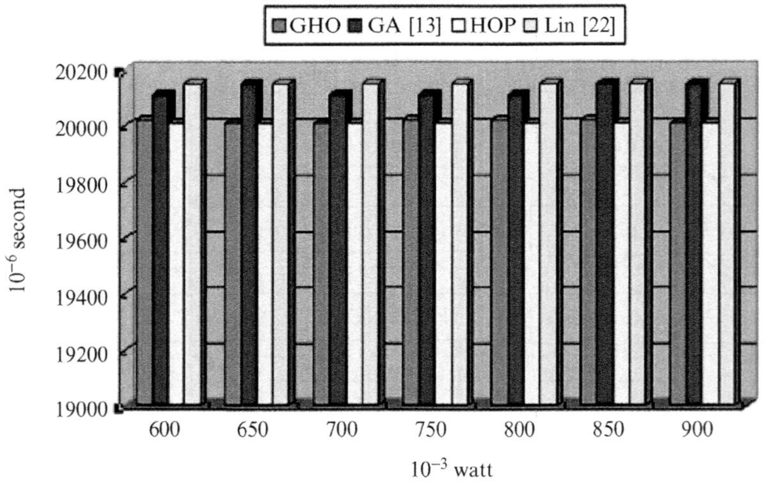

(a) Execution time

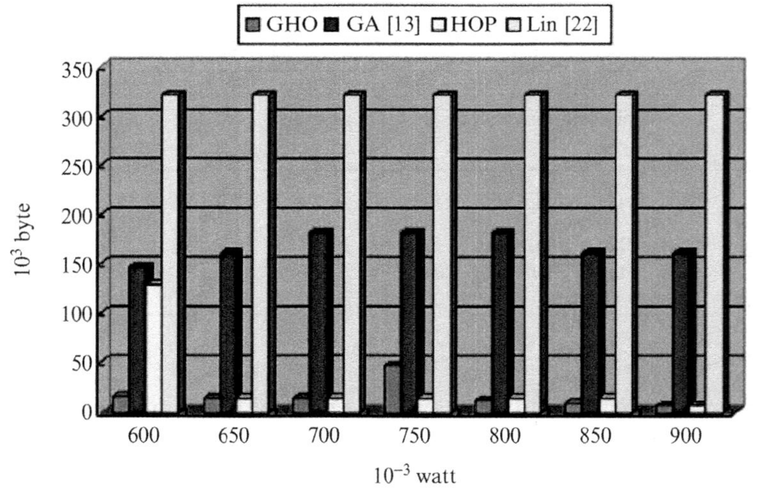

(b) Memory size

Fig. 28.5 Comparisons with GHO, GA [13], HOP, and Lin [22] using JPEG encoding system implemented: **(a)** execution time; **(b)** memory size gains more minimizing execution time than GA [13] and Lin [22]. For memory size, Fig. 28.5b shows that GHO is better than GA [13] and Lin [22] in all cases. Comparing GHO with HOP in Fig. 28.5b, GHO is also better than HOP in some cases especially where power consumption is limited under 600 mw.

28.5 Conclusion

Hardware-software partitioning method is a major issue in the development of an embedded multiprocessor FPGA system. We discuss the strategy of hardware–software partitioning by GHO to solve the partitioning problem in embedded multiprocessor FPGA systems. Experiment results illustrate that GHO can gain a better partitioning result with reduced execution time and memory size than GA [13] and Lin [22]. Moreover, concurrency conflict is also taken into account in GHO; the result in communication time can be reduced by allocating process.

References

1. Xilinx website, http://www.xilinx.com/
2. W. Wolf, A decade of hardware/software codesign, *IEEE Computer*, Vol. 36, pp. 38–43, 2003.
3. R. Ernst, Codesign of embedded systems: Status and trends, *IEEE Design and Test of Computers*, Vol. 15, No. 2, pp. 45–54, Apr.–Jun. 1998.
4. N.S. Woo, A.E. Dunlop, and W. Wolf, Codesign from cospecification, *IEEE Computer*, Vol. 27, pp. 42–47, Jan. 1994.
5. R.K. Gupta and G.D. Micheli, System synthesis via hardware-software co-design, Technical Report No. CSL-TR-92-548, Computer System Laboratory, Standford Univ., pp. 1247–1263, 1993.
6. R.K. Gupta, N.C. Claudionor, and G. De Micheli, Program implementation schemes for hardware-software systems, *IEEE Computer*, Vol. 27, No. 1. pp. 48–55, Jan. 1994.
7. G.F. Marchioro, J.M. Daveau, and A.A. Jerraya, Transformational partitioning for co-design of multiprocessor systems, *Proceeding of IEEE International Conference Computer Aided Design*, pp. 508–515, Nov. 1997.
8. M. Srivastava and R. Brodersen, SIERA: A unified framework for rapid-prototyping of system-level hardware and software, *IEEE Trans. on Computer-Aided Design of Integrated Circuits and Systems*, pp. 676–693, Jun. 1995.
9. C. Coelho Jr, D. da Silva Jr, and A. Fernandes, Hardware-software codesign of embedded systems, *Proceedings of the XI Brazilian Symposium on Integrated Circuit Design*, pp. 2–8, 1998.
10. R. Ernst, J. Henkel and T. Benner, Hardware-software cosynthesis for microcontrollers, *IEEE Design & Test of Computer*, Vol. 10, pp. 64–75, Dec. 1993.
11. V. Srinivasan, S. Govindarajan, and R. Vemuri, Fine-grained and coarse-grained behavioral partitioning with effective utilization of memory and design space exploration for multi-FPGA architectures, *IEEE Transactions on Very large Scale Integration (VLSI) Systems*, Vol. 9, pp. 140–158, Feb. 2001.
12. D. Saha, R.S. Mitra, and A. Basu, Hardware software partitioning using genetic algorithm, *Proceedings of the 10th International Conference on VLSI Design*, pp. 155–160, Jan. 1997.

13. Y. Zou, Z. Zhuang, and H. Cheng, HW-SW partitioning based on genetic algorithm, *Proceedings of Congress on Evolutionary Computation (CEC2004)*, Vol. 1, pp. 628–633, Jun. 19–23, 2004.
14. H. Kanoh, M. Matsumoto, and S. Nishihar, Genetic algorithms for constraint satisfaction problems, *IEEE International Conference on Man and Cybernetics Systems*, Vol. 1, pp. 626–631, Oct. 22–25, 1995.
15. A.L. Buczak and H. Wang, Optimization of fitness functions with non-ordered parameters by genetic algorithms, *Proceedings of the Congress on Evolutionary Computation*, Vol. 1, pp. 199–206, May 27–30, 2001.
16. K. Deb and S. Agrawal, *Understanding Interactions among Genetic Algorithm Parameters: Foundations of Genetic Algorithms* 5, Morgan Kaufmann, San Francisco, 1999.
17. S. Palaniappan, S. Zein-Sabatto, and A. Sekmen, Dynamic multi-objective optimization of war resource allocation using adaptive genetic algorithms, *IEEE Southeast Congress Proceedings*. pp. 160–165, Apr. 30, 2001.
18. T.Y. Lee, P.A. Hsiung, and S.J. Chen, Hardware-software multi-level partitioning for distributed embedded multiprocessor systems, *IEICE Transactions on Fundamentals of Electronics, Communications and Computer Sciences*, pp. 614–626, 2001.
19. L. Pomante, Co-design of multiprocessor embedded systems: An heuristic multi-level partitioning methodology, *Proceeding of the IFIP International Conference on Chip Design Automation*, pp. 421–425, 2000.
20. C. Brandolese, W. Fornaciari, L. Pomante, F. Salice, and D. Sciuto, Affinity-driven system design exploration for heteregenerous multiprocessor SoC, *IEEE Transactions on Computer*, Vol. 55, No. 5, pp. 508–519, 2006.
21. D. Sciuto, F. Salice, L. Pomante, and W. Fornaciari, Metrics for design space exploration of heterogeneous multiprocessor embedded systems, *IEEE/ACM International Workshop on Hardware Software Co-Design*, pp. 55–60, 2002.
22. T.-Y. Lin, Y.-T. Hung, and R.-G. Chang, Efficient hardware/software partitioning approach for embedded multiprocessor systems, *International Symposium on VLSI Design, Automation and Test (VLSI-DAT)*, pp. 231–234, Apr. 26–28, 2006.

Chapter 29
Interpretation of Sound Tomography Image for the Recognition of *Ganoderma* Infection Level in Oil Palm

Mohd Su'ud Mazliham, Pierre Loonis, and Abu Seman Idris

29.1 Introduction

Basal stem rot (BSR) disease in oil palm caused by a group of decaying fungi called *Ganoderma* is considered as the most serious disease faced by the oil palm plantations in SouthEast Asia [1]. Significant yield losses can be observed when the number of palms infected by the fungus increases in the plantation as the infected palms will produce less quality fruit and eventually die thus requiring an early replanting.

Flood et al. [2] and Idris and Ariffin [1] reported that the disease spread through contact of roots starting with the contact of an exiting inoculum in the soil and later spreading to the other palms through contact between the infected root with roots of other palms. The infection then moved upwards in the palm stem. The disease normally concentrated in the lower 1 meter of the trunk.

Ganoderma produces enzymes that will degrade the oil palm tissue and affect the infected oil palm xylem thus causing serious problems to the distribution of water and other nutrients to the top of the palm tree. Because oil palm stems have no vascular cambium, they are essentially devoid of secondary growth. Therefore, palms cannot repair injuries to their stems.

As the infection happens inside the palm stem, no physical symptoms can be detected at the early stage of infection. Basidiocarp is the most identifiable structure associated with the fungus. The conk originates from the fungus that grows in the infected trunk. However, most of the time, the conk does not appear at the early stage of the infection, making early detection of the disease very difficult.

The foliar symptoms of BSR are reported to be the most apparent visual sign of the infection. However, these symptoms can only be considered as symptoms of *Ganoderma* infection if they are accompanied by the development of basidiocarp. The problem is that by the time these symptoms appear, usually over half of the lower internal stem tissues have been killed by the fungus.

Oscar Castillo et al. (eds.), *Trends in Intelligent Systems and Computer Engineering.*

It was shown by Idris et al. [3] that hexaconazole dissolved in 10 liters of water applied with pressure injection gave a higher percentage of palm survival. However, as the infection is detected only after the visible symptoms appear this method is only used on palm trees that have already been severely infected by the fungus.

The ability to perform early detection of the infection will enable the palm to be treated at the early stage of the infection and thus avoid more extensive damage in the palm.

Until today, few methods based on a biochemistry process were used to detect *Ganoderma* infection which can be divided into two categories: culture-based by using *Ganoderma* selective medium (GSM) [4] or molecular DNA-based such as a polymerase chain reaction (PCR) amplification [5]. These methods require the collection of stem samples from four angles of the palm for further tests in the lab. There are no methods available that are able to give the result of the test directly on the site.

In this framework, we suggest a system capable of identifying *Ganoderma* infection inside the palm stem and localizing the infected area. The identification of the infection is based on developing expert's rules to identify the presence of *Ganoderma* fungus in the palm stem based on the recognition of the lesion in the stem using tomography images. These rules will enable us to perform automatic detection directly on the plantation.

The results of this study are presented as follows. Section 29.2 is devoted to the presentation of the expert's current knowledge of the *Ganoderma* infection pattern in the palm stem. The analysis of tomography images based on the sound propagation in the stem to detect abnormalities in the stem is presented in Sect. 29.3. A fuzzy inference system classifying each suspected lesion pattern observed in the tomography image using rules established with the help from experts and the lesion features extracted from the tomography images into three possible hypotheses: *Ganoderma* infection (*G*), non-*Ganoderma* infection (N), or intact stem tissue (I) is proposed in Sect. 29.4. The result of this classification is presented in Sect. 29.5. The result is then used to generate a basic probability assignment or mass function of lesion condition according to the above-mentioned hypotheses in Sect. 29.6. Section 29.7 discusses the method to obtain overall oil palm health condition belief function through the combination of data obtained in each lesion observed in the stem.

29.2 Identifying *Ganoderma* Infection Pattern in the Oil Palm Stem

29.2.1 Expert's Knowledge on Ganoderma *Infection Pattern*

According to Turner [6], Ariffin et al. [7], and Idris et al. [1], *Ganoderma* infection can be well defined by its lesion in the stem. The cross-section of infected palm stem shows that the lesion appears as a light brown area of rotting tissue with a distinctive irregularly shaped darker band at the borders of this area. Turner [6] and Schwarze and Ferner [8] also noted that in very old lesions, the infected tissue may become an ashen-grey powder and if the palm remains standing, the infected trunk may become hollow.

Elliott and Broschat [9] noticed that the fungus colonizes and degrades the palm trunk tissue closest to the soil line first before moving up to the center or near center of the trunk. The tissues infected by the fungus will degrade and die thus developing cavities in the trunk.

The position, number, and condition of these lesions depend on the upwards evolution of the attack and the origins of the infection. As the *Ganoderma* attack location is often limited to the roots and the lower zone of the palm tree, a detection approach should focus on the stem area close to the ground.

Figure 29.1 shows an example of the cross-section of an infected stem. Note that in Fig. 29.1a the stem was severely infected by the fungus. In this condition, the lesion appears as a light brown area of rotting tissue. In Fig. 29.1b, the infected stem is degraded around the infected area with a distinctive irregularly shaped darker band at the borders of this area. Note that in this picture, the black spot at the center

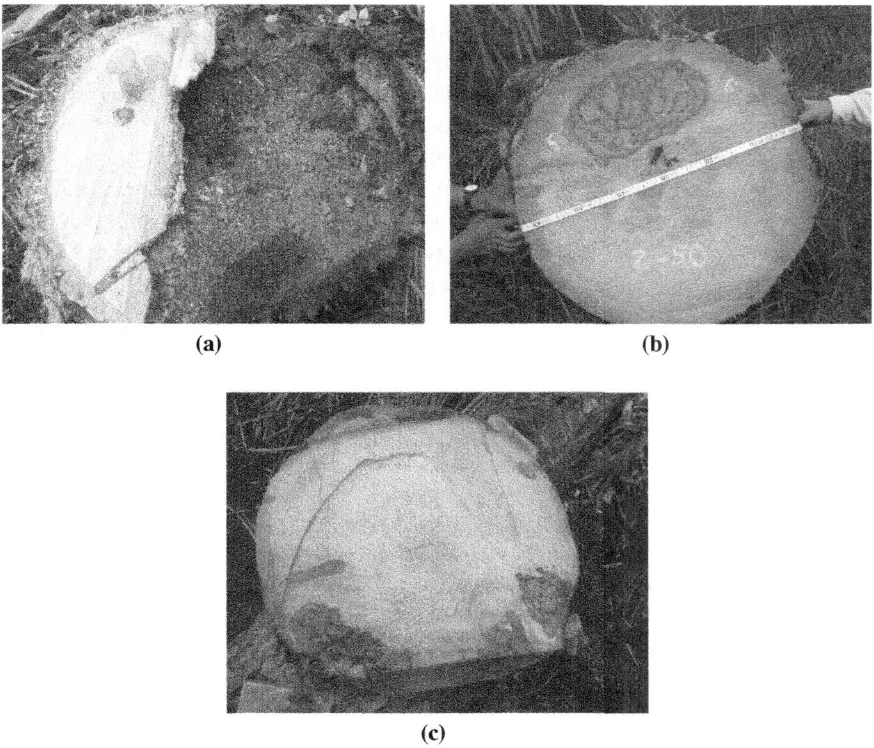

(a) (b)

(c)

Fig. 29.1 (a)–(c): *Ganoderma* infection lesion in a cross-section of an palm oil stem of the stem is the natural central hole that exists in some palm stems. This is not to be identified as a *Ganoderma* infection if it is detected in the tomography image. In Fig. 29.1c, the pencil shaped line observed on the left of the stem is caused by chemical injection for treatment against bagworm. This is not to be identified as a *Ganoderma* lesion which can be seen at the bottom of the stem in this picture. A yellow reaction zone is also clearly identified in this figure.

The experts' belief is that except for the cases mentioned earlier, most abnormalities found inside the stem are caused by *Ganoderma* infection. Other possible abnormalities can be caused by insect attack near the peripheral zone of the stem.

29.2.2 Towards an Automatic Detection

The automatic detection methods of *Ganoderma* infection in palm oil stem are still not available due to the fact that only partial knowledge of the infection process of this disease in the palm stem is available and only a very small amount of available samples for verification (the tree must be cut to verify the sensors' measurements).

The need for a nondestructive approach which is able to explain explicitly the reasoning in a readable way is then still blocked because (*i*) the characteristics of the infection symptoms can be confused with other kinds of less important infections and (*ii*) the infections are hidden inside the stem.

That is the reason why we have proposed a solution based on a tomography image integrated in a quasi nondestructive method. Our prototype is based on the extraction of reliable parameters onto specific areas of this image. The choice of these features is defined by studying with biologists the various cases of degradation in the palm stem. Then, the use of rules to model the knowledge allows us to have a convenient way to represent the complementarity between acquired physical data and experts' knowledge, in order to tune the set of rules, improving both the physical model and the knowledge.

29.3 Noninvasive Testing to Evaluate Physical Properties of Wood

Several noninvasive evaluations have been tested to analyze physical and chemical properties of wood related to wood decay or fungus infection such as [10–14]. Most of these techniques have been developed using tomography investigations that show the cross-section of the tree. Examples of tomography techniques used on wood and timber structures are ultrasonic tomography, electric tomography, sonic tomography, and georadar tomography.

Nicolotti et al. [15] compared these techniques and noticed that tomography investigation is able to detect small anomalies in the tree trunk with ultrasonic tomography giving the best resulting detection of early fungal infection or small decay.

Table 29.1 Comparison of Tomography techniques to detect decay in wood

Tomography technique	Type of measurement supplying tomography	Problem
Ultrasound	Elastic properties. Reconstruct the distribution of the velocity of ultrasonic propagation in the investigated section	Transducers coupling with the bark, Anisotropy of wood caused the low-velocity peripheral zone, Signal attenuation due to high frequency and spatial resolution
Electrical	Tree conductivity. Construction of image of resistivity distribution on a section of a tree	High resistivity of the bark preventing current flow and voltage measurement Depending on the measuring device impedance and software that consider the nonperfect body of the trunk
Georadar	Tree conductivity and permittivity	Antennas coupling with the bark and data interpretation and signal attenuation
Sonic	Elasticity and density of wood. Measure the network of sound velocities across the cross-section selected.	Wood characteristics vary among tree species, only large decay can be detected

Table 29.1 provides some detail on the tomography technique to detect decay in a tree and the problem related to each technique.

Each of the techniques has specific problems as indicated in the table. Although [15] has proven that ultrasonic tomography with a frequency of 54 kHz is able to detect anomalies of about 5 cm, there are some problems due to the wood anisotropy and signal attenuation because of high frequency. The latter problem is not observed in sonic tomography. Although electrical tomography is less sensitive than the ultrasonic tomography, it measures a different physical characteristic of wood and thus provides tomography that can give complementary information to the ultrasonic and sonic tomographies [15] also explained that white rot basidiomycete such as *Ganoderma boninense* will accumulate the cations and K-ions than lower wood resistivity even from the beginning of the decaying process thus increasing the possibility of electric tomography to perform an early detection of *Ganoderma* attack.

29.3.1 Sonic Tomography

The use of a sonic tomography sensor to detect the presence of lesions inside the stem is presented in this work. This allows us to solve the constraints, taking into account the experts' understanding of the effects of this disease on the tree and on the other hand, of performing noninvasive acquisition.

Tomography refers to the cross-sectional imaging of an object from either transmission or reflection data collected by exposing the object under study to the source wave from many directions. In the case of the sonic tomograph, the source wave is the sound. The sound velocity (V) in wood is governed by its elasticity (E) and density (D) as in the formula:

$$V = \sqrt{\frac{E}{D}}. \qquad (29.1)$$

Rotting tissues decrease the speed of sound in the palm which allows the equipment to detect the presence of a lesion in the stem.

In this work, the equipment used is PICUS Sonic manufactured by Argus Electronic. The equipment consists of a set of sensors that are strategically placed around the tree. Figure 29.2 shows PICUS Sonic equipment installed on a tree.

The PICUS Sonic Tomograph uses the calculation of the time of flight of sound waves manually induced by knocking with a small hammer on a nail placed on the tree connected to the sensors. Apparent sound velocities are calculated based on the times of flight of sound waves and the distance between sensors. This calculation is repeated on all nails around the tree as indicated in Fig. 29.3.

Finally, the times of flight are segmented into five classes from the slow velocities in decayed wood (white, blue) or degraded wood (violet), up to the high ones in solid wood (brown, black colors). The medium velocities are colored in green.

Fig. 29.2 PICUS Sonic Tomograph installed on an oil palm

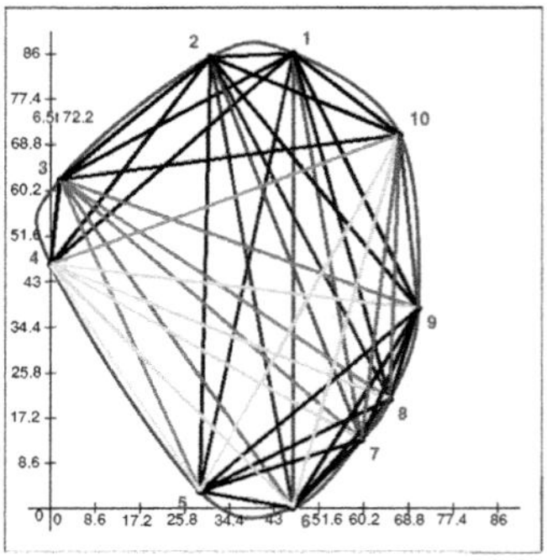

Fig. 29.3 Sound lines obtained by knocking on all nails placed on the palm tree

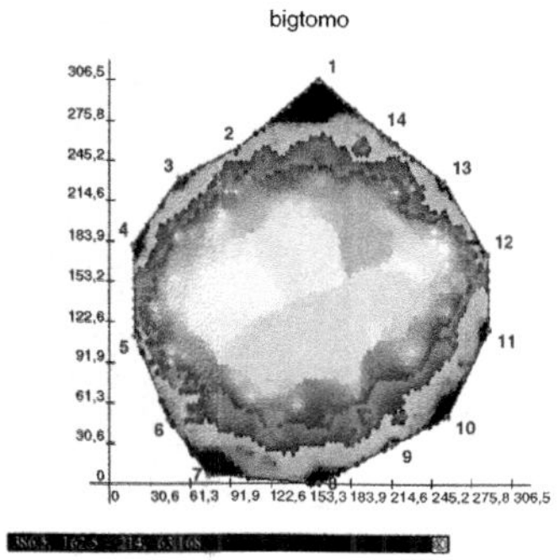

Fig. 29.4 Example of tomography of a decay using PICUS Sonic

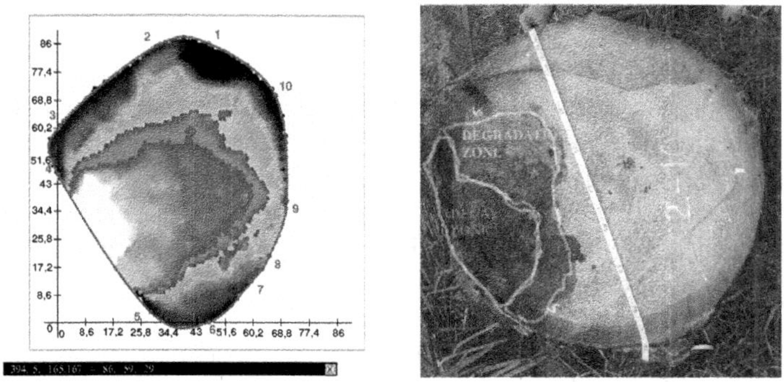

Fig. 29.5 Comparing cross-section photo with sonic tomography image

Figure 29.4 shows an example of tomography produced by the device.

Schwarze et al. [16] noted that the equipment is able to accurately determine the size of a decay zone in a tree. However, due the different stem structure between the palm and other trees, we found that some adjustment must be made for oil palms. Comparing the cross-sectional photo with the tomography image of a decayed palm, we concluded that it is best to identify the light blue area in the tomography image as decay and the blue area as the degradation area in a palm stem (Fig. 29.5). This assumption is used in this work. A new compensated tomography calculation for oil palm will be developed in cooperation with the device manufacturer.

29.3.2 Image Analysis

Because chromatic levels are used for coding the tomographic image, it is better to use a color space able to distinguish the three layers in luminosity and chromatic ones. The most used one is the CIE $L^*a^*b^*$, in which luminosity L^* (or the brightness) layer is separated from both the first chromaticity layer a^* indicating where color falls along the red–green axis, and the second chromaticity layer b^*, indicating where the color falls along the blue–yellow axis ([17]). One of the main reasons to

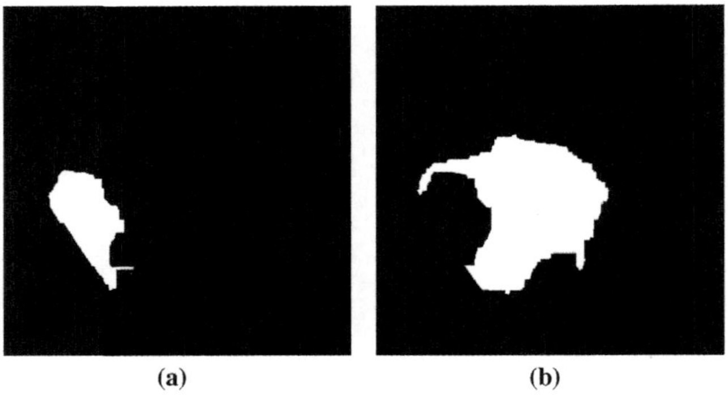

Fig. 29.6 Decay (a) and degradation (b) zones detected

use this color space instead of *RGB* is that the difference between two colors in the $L^*a^*b^*$ space is identically perceived by the human eye.

Consequently, the decay zone (blue), the degradation zone (violet), and the solid zone (brown) are easily detected. Then, for each zone i in the image, we calculate the average value $\widehat{a_i^*}$ of a_i^* and $\widehat{b_i^*}$ for b_i^*. In order to extract zone i with average value $\widehat{a_i^*}$ and $\widehat{b_i^*}$, we calculate the distance between all values a^* and b^* to the $\widehat{a_i^*}$ and $\widehat{b_i^*}$ for every pixel in the image.

Then all the pixels related with the minimum value of $(distance((a^*,\widehat{a_i^*}),(b^*,\widehat{b_i^*})))$ belong to region i. Finally, each region is converted to a binary pattern for further analysis. Figure 29.6 presents the detection of one decay pattern and one degradation pattern from the cross-section scan shown in Fig. 29.5.

29.3.3 Relevant Feature Extractions and Experts' Rules to Identify Ganoderma Lesion

An infection pattern condition rule was established with the help of biologists according to the known characteristics of the infections, which are divided into two categories: infection near the center of the stem and infection located on the external layer of the stem. The experts' belief is that most of the infection happens near the peripheral zone of the stem. However, any lesion detected in this region should be differentiated from the lesion caused by chemical injection for bagworm treatment[1] and lesions caused by insect attack. Added to the location, the shape of this infection zone (quite a regular pencil mark shape) can also be used to be differentiated with a *Ganoderma* infection (quite an irregular circular pattern). On the contrary, decay detected at the center of the stem can be confused by the natural holes that exist in certain stems. The size of the hole is much smaller than the size of *Ganoderma* infection in this area.

[1] Biologists drill a hollow to inject chemical treatment.

Then, the following features were identified by asking the experts to explain the way they perform visual *Ganoderma* recognition and classification by assigning probability values into the three classes G, N, and I (for *Ganoderma* infection, non-*Ganoderma* infection, and intact) for both decay and degradation zones detected:

1. Infection pattern's eccentricity
2. Infection pattern's orientation
3. Infection pattern's solidity
4. Infection pattern's roundness
5. Infection pattern's area and size
6. Infection pattern's position in the cross section

As the pattern observed in the image varies, there exist no exact values of eccentricity, orientation, solidity, and other features to define into which classes the pattern belongs. Experts' rules (632) were established to evaluate each lesion detected in the tomography image to enable each pattern observed in the tomography image to be classified into one of the classes mentioned in the above paragraph. Some of the rules are presented in the following.

1. **IF** lesion type is decay **AND** pattern's area is small **AND** pattern's eccentricity is circle **AND** pattern's solidity is low **AND** pattern's position is close to stem center **THEN** G is medium, NG is zero, and I is medium.
2. **IF** lesion type is decay **AND** pattern's area is big **AND** pattern's eccentricity is circular **AND** pattern's solidity is high **AND** pattern's position is in the middle **THEN** G is high, NG is zero, and I is zero.
3. **IF** pattern's type is decay **AND** pattern's area is small **AND** pattern's eccentricity is line **AND** pattern's orientation is 0° **AND** pattern's solidity is high **AND** pattern's position is close to stem center **THEN** G is middle, NG is zero, and I is middle.
4. **IF** lesion type degradation **AND** pattern's area is big **AND** pattern's eccentricity is middle **AND** pattern's solidity is low **AND** number of points in the lesion touching the border is low) **AND** pattern's position is near the peripheral zone **THEN** G is high, NG is low, and I is zero.
5. **IF** lesion type is degradation **AND** pattern's area is big **AND** pattern's eccentricity is circular **AND** pattern's solidity is high **AND** pattern's position is in the middle **THEN** G is high, NG is zero, and I is zero.
6. **IF** pattern's type is degradation **AND** pattern's area is small **AND** pattern's eccentricity is line **AND** pattern's orientation is 0° **AND** pattern's solidity is high **AND** pattern's position is close to stem center **THEN** G is middle, NG is zero, and I is middle.

Using the rules, we are able to propose the experts' degree of belief on pattern classification in three classes of *Ganoderma* degradation, non-*Ganoderma* degradation, and intact. The rules defined in the table are used to identify both decay and degradation patterns observed in the image. An extensive explanation dealing with the extraction of these rules was presented in [18].

29.4 Fuzzy Inference System for *Ganoderma* Infection Pattern Classification

In pattern recognition, a nonfuzzy classification technique assumes that a pattern X belongs to only one of the possible classes whereas fuzzy classification algorithms assign a pattern X with a distributed membership value to each class.

A fuzzy inference system (FIS) is used to formulate the mapping from a given input to an output using fuzzy logic [19, 20]. The mapping provides results in which decisions can be made or patterns can be classified. We propose to use the Mamdani-type inference system in this work. The FIS rule base is made of rules in the following form.

R_i : if S_1 is L_1^i and ... S_N is $L_{N_I}^i$ then Y_1 is 0_1^i and ... Y_N is $O_{N_O}^i$
with

R_i the ith rules of the rule base
S the input vector
L the linguistic term (fuzzy label) of input variable S in rule R with membership function μ_L
Y the output vector O the linguistic term (fuzzy label) of output variable Y in rule R with membership function μ_O

Mazliham et al. [21] proposed a FIS using the rules established in the previous paragraph. The membership function used in this work is presented in Figs. 29.7 and 29.8.

Eccentricity input uses three membership functions: circle, middle, and line to differentiate the infection patterns that are normally close to a circle from the effect of chemical injection which normally has a pencil shape.

Three membership functions are also used for the closeness inputs that are close to center, middle, and external. In this input, the position of the pattern is analyzed to reflect a more confident infection belief in the pattern found at the center of the stem from the one found at the near to the external area which can be due to chemical injection or insect attack.

The orientation has five membership functions: $-90°, -45°, 0°, 45°$, and $90°$. The orientation allows differentiation of chemical injection from the infection as the injection is made at a certain angle whereas the orientation of an infection pattern is not regular.

We used three membership functions for percentage area input to measure the severeness of the attack. The bigger the pattern area is can imply the more severe infection is. The membership functions for this input are: low, middle, and high.

For number of points at border input, two membership functions are used: not many and many. The position points at the border of each pattern observed in the tomography image are compared to the position of the stem circumference. The number of points that are positioned at the same position as the circumference indicate that there is a possibility of external infection or attack on the stem under study. If the number is small, the decay or degradation pattern observed might be caused by insect injection or the drill effect in the process of doing chemical injection.

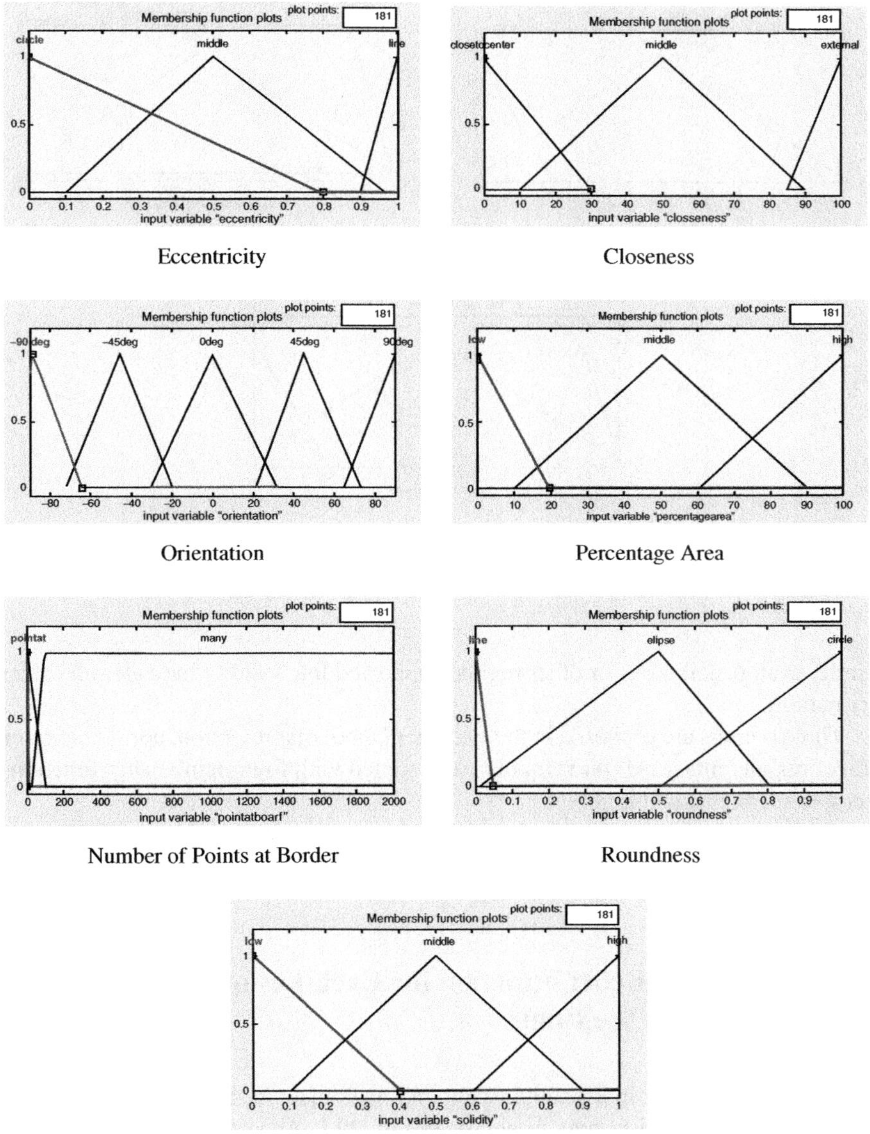

Eccentricity

Closeness

Orientation

Percentage Area

Number of Points at Border

Roundness

Solidity

Fig. 29.7 Membership function inputs

Roundness inputs have three membership functions: line, ellipse, and circle. These are also used to differentiate the irregular infection pattern from other regular patterns.

Three membership functions are used for solidity input: low, middle, and high. These inputs provide information about the pattern shape. A solid shape indicates a

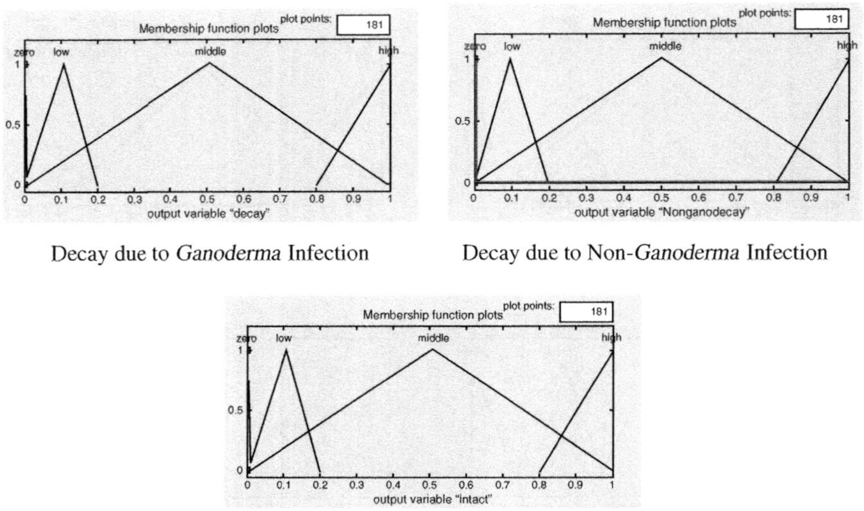

Decay due to *Ganoderma* Infection　　Decay due to Non-*Ganoderma* Infection

Intact

Fig. 29.8 Membership function outputs

circle, oval, pencil trace, or other regular shape and low solidity indicates an irregular pattern.

Three outputs are proposed in the system: *Ganoderma* infection, non-*Ganoderma* infection, and intact. All three inputs are assigned with four membership functions: zero, low, middle, and high.

Zero membership function indicates that the output is impossible, and high membership function implies that the output is very possible.

29.5 Generating Belief Function for Each Lesion Pattern Observed in the Image

In the example shown in Fig. 29.5, we found one object in the decay zone and one object in the degradation zone as shown in Fig. 29.6. As such we consider that we have two patterns, S_{D_1} and S_{d_1}, where S_{D_1} is the pattern obtained from the decay object found in the decay zone and S_{d_1} is the pattern observed in the degradation zone.

As mentioned earlier, each object found in both the decay and degradation zones will be analyzed using the FIS to provide the degrees of membership in the following hypotheses.

1. *Ganoderma* infection (G)
2. non-*Ganoderma* infection (N)
3. Intact (I)

Table 29.2 Degree of membership for the example tomography image

Pattern	*Ganoderma* infection	Non-*Ganoderma* infection	Intact
S_{D_1}	0.912	0.1	0
S_{d_1}	0.914	0	0.1

We obtained the degree of membership as shown in Table 29.2. The output of this inference system gave the degree of membership for each class.

We noticed in this example that the FIS assigns a pattern S_i with a distributed membership value to each possible class: G, N, and I. Pattern S_{D_1}, for example, has a membership value $\mu_G(S_{D_1}) = 0.912$ for G, $\mu_N(S_{D_1}) = 0.1$ and $\mu_I(S_{D_1}) = 0$.

29.6 Mass Function Initialization and Distribution

The degree of membership found in section 5 can be used to initialize mass function for each found pattern. These mass functions will be used in section 7 to establish the overall palm health condition.

The initialization of mass using the fuzzy approach is proven to improve the mass distribution assigned to compound hypotheses in the case where the value to be distributed is inaccurate [22]. In the same way, fuzzy sets were also used as an interpretation of focal elements by Straszecka in [23]. We suggest using the membership function obtained to initialize the mass function.

Suppose that we have two hypotheses θ_1 and θ_2, the definition of the nonnormalized masses, $\tilde{m}(\theta_1)$ and $\tilde{m}(\theta_2)$, from the membership function, $\mu_{\theta_1}(x)$ and $\mu_{\theta_2}(x)$ is [24]:

$$\tilde{m}(\theta_1) = \mu_{\theta_1}(x) \tag{29.2}$$

$$\tilde{m}(\theta_2) = \mu_{\theta_2}(x). \tag{29.3}$$

Thus the definition of the mass for compound hypothesis $\theta_1 \cup \theta_2$ can be defined as a function of $\mu_{\theta_1}(x)$ and $\mu_{\theta_2}(x)$,

$$\tilde{m}(\theta_1 \cup m(\theta_2)) = \Im(\mu_{\theta_1}(x), \mu_{\theta_2}(x)). \tag{29.4}$$

Once the nonnormalized mass $\tilde{m}$ is processed the final mass is computed after normalization with respect to the entire set 2^Θ,

$$m(\theta_i) = \frac{\tilde{m}(\theta_i)}{\sum_{\theta_j \in 2^\Theta} \tilde{m}(\theta_j)}. \tag{29.5}$$

Our example deals with the three classes G, N, and I, such as our frame of discernment is

$$2^\Theta = \{G, N, I, G \cup N, G \cup I, N \cup I, G \cup I \cup N, \phi)\}. \tag{29.6}$$

Table 29.3 Nonnormalized mass distribution according to DT method

Item	S_{D_1}	S_{d_1}
$\tilde{m}(G)$	0.912	0.914
$\tilde{m}(N)$	0.1	0
$\tilde{m}(I)$	0	0.1
$\tilde{m}(G \cup N)$	0	0
$\tilde{m}(G \cup I)$	0	0
$\tilde{m}(N \cup I)$	0	0
$\tilde{m}(\theta)$	1.988	1.986
$\tilde{m}(\phi)$	0	0

Table 29.4 Normalized mass distribution according to DT method

Item	S_{D_1}	S_{d_1}
$m(G)$	0.304	0.305
$m(N)$	0.033	0
$m(I)$	0	0.033
$m(G \cup N)$	0	0
$m(G \cup I)$	0	0
$m(N \cup I)$	0	0
$m(\theta)$	0.663	0.662
$m(\phi)$	0	0

Using the initialization method suggested in [25] which is based on work of Sadiq and Rodriguez [26] who propose that the ignorance mass be assigned to θ, we obtained the mass shown in Table 29.3 for each pattern found in the cross-section under study in section 2.

Applying normalization to the result we found the normalized basic probability assignments in Table 29.4.

29.7 Palm Health Belief Function

Each source of information only contains part of the total information of the area. Combining the data collected from these two sources, S_{D_1} and S_{d_1}, will enable us to determine the condition of the total area under study. We propose to combine data found in the decay and degradation zones using the Dempster–Shafer (D–S) combination to generate the overall mass function for the cross-section under study.

The Dempster combination rule calculated the aggregation of two bpas m_1 and m_2 using the following equation.

$$\begin{aligned} \forall A \neq 0 \qquad m_{1-2}(A) &= \frac{\sum_{B \cap C = A} m_1(B) m_2(C)}{1-K} \\ m_{1-2}(\phi) &= 0 \end{aligned} \tag{29.7}$$

where $K = \sum_{B \cap C = \phi} m_1(B) m_2(C)$.

The purpose of combining the data found in the previous section is to obtain an overall mass for the cross-section thus giving an indication of the *Ganoderma* infection situation in the palm.

The D–S theory assumes that the source of information is independent. In our context, the available sources of information are information on the health condition of objects found in the decay and degradation zones in the cross-section. The sources of information considered in this case are independent as data found in one object will not imply any changes to the other information in other objects. However, as each object only reflects local *Ganoderma* infection evidence, possibilities of having totally conflicting data when we combine data from several objects must be expected.

In D–S theory, the likelihood of one hypothesis is also represented with lower and upper probabilities represented by the belief (Bel) and Plausibility (Pl).

Consider a set Θ_1 a subset of Θ; Bel for Θ_1 is defined as the sum of all masses or basic probability assignment (bpa) of all subsets of Θ_1. Pl of Θ_1 is given by the sum of all bpas of subset Θ_2 that intersect with Θ_1. Belief and plausibility are define in Eqs. 29.8 and 29.9.

$$\forall\Theta_1,\Theta_2 \in \Theta \quad Bel(A) = \Sigma_{\Theta_2 \subset \Theta_1} m(\Theta_2) \tag{29.8}$$

$$\forall\Theta_1,\Theta_2 \in \Theta \quad Pl(A) = \Sigma_{\Theta_2 \cap \Theta_1 \neq \emptyset} m(\Theta_2). \tag{29.9}$$

Implementing the suggested combination algorithm as suggested in [25], we combine the data from both decay and degradation zones as both of them have maximum mass assigned to the *Ganoderma* infection hypothesis. Mazliham et al. [25] also noted that the combination rule using the conventional Dempster rule is the most suitable to this work. The following mass is obtained after combination.

Table 29.5 shows nonnormalized mass for palm under study. Table 29.6 shows the combined mass after normalization. We observe that most of the mass is given to hypothesis G.

Implementing 29.8 and 29.9 in the combined bpa found in Table 29.6 we obtained the belief and plausibility for the hypothesis in Table 29.7.

In this case, as belief for G is higher than all other plausibility of other hypotheses, we can easily decide that the palm under study is infected by *Ganoderma*.

Table 29.5 Nonnormalized mass for palm under study

Item	Mass
$\tilde{m}(G)$	0.496
$\tilde{m}(N)$	0.022
$\tilde{m}(I)$	0.022
$\tilde{m}(G \cup N)$	0
$\tilde{m}(G \cup I)$	0
$\tilde{m}(N \cup I)$	0
$\tilde{m}(\theta)$	0.439
$\tilde{m}(\phi)$	0.021

Table 29.6 Normalized mass for palm under study

Item	Mass
$m(G)$	0.507
$m(N)$	0.023
$m(I)$	0.023
$m(G \cup N)$	0
$m(G \cup I)$	0
$m(N \cup I)$	0
$m(\theta)$	0.448
$m(\phi)$	0

Table 29.7 Belief and plausibility

Item	Belief	Plausibility
$m(G)$	0.507	0.955
$m(N)$	0.023	0.471
$m(I)$	0.023	0.471

29.8 Conclusion

In this chapter a method to extract *Ganoderma* infection rules based on selected features was introduced. The features were observed and extracted from segmented tomography images. The rules were designed based on experts' knowledge of the *Ganoderma* infection pattern in the oil palm stem. A fuzzy inference system was introduced to assign membership functions to the output vectors which classify the pattern into three hypotheses: *Ganoderma* infection (G), non-*Ganoderma* infection (NG), and intact stem (I).

The assigned membership function is then used to initialize a basic probability assignment (bpa) for all members of power set $(G, NG, I, G \cup NG, G \cup I, NG \cup I, \Theta, \Phi)$. The same step will be repeated to all decay and degradation patterns recognized in the tomography image.

The data from several patterns found in the cross-section under study were combined using the D–S method. The combination generates the overall health condition of the palm under study. We have shown in this work that this will lead to the automatic recognition of *Ganoderma* infection lesions based on the classification of lesion patterns observed in the sonic tomography image.

In future, it is suggested that a study of how to interpret the severity level of the infection based on the combined mass be conducted. This can be done by considering the ratio between the area of each object believed to be infected with the total area of the cross-section under study. A threshold value can be used to separate the level of infection. A weighted operator can also be introduced to differentiate between decay and degradation objects as the severity of the infection can also derive from the condition of the lesion in the stem.

References

1. Idris AS and Ariffin D. Ganoderma penyakit reput pangkal batang dan kawalannya. *Risalah Sawit*, No. 11, 2003.
2. Flood J, Keenan L, Wayne S, and Hasan Y. Studies on oil palm trunks as sources of infection in the field. *Mycapathologia*, (19):101–107, 2005.
3. Idris AS, Ismail S, Ariffin D, and Ahmad H. Prolonging the productive life of ganoderma infected palms with hexaconazole. *MPOB Information Series*, 2004.
4. Arifin D and Idris AS. The ganoderma selective medium. *MPOB Information-Series*, 1992.
5. Idris AS, Yamaoka M, Hayakawa S, Basri MW, Noorhashimah I, and Ariffin, D. Pcr technique for detection of ganoderma, mpob information. *MPOB Information*, Series TT No. 188, 2003.
6. Turner PD. *Palm Oil Diseases and Disorers*. Oxford University Press, 1981.
7. Ariffin D, Idris AS, and Abdul Halim H. Significance of the balack line within oil palm tissue decayed by ganoderma boninense. *Elaeis*, 1(1), 1989.
8. Schwarze and Fermer. Ganoderma on tree—differentiation of species and studies of invasiveness. available online www.enspec.com/articles/.
9. Elliott ML and Broschat TK. Ganoderma butt rot of palms. http://edis.ifas.fl.edu/, 2000. University of Florida, Institute of Food and Agricultural Science.
10. Steffen R. A new tomographic device for the non destructive testing of trees. 2000.
11. Sandoz JL. Ultrasonic solid wood evaluation in industrial application. *10th International Symposium on Nondestructive Testing of Wood*, 1996.
12. Veres IA and Sayir MB. Wave propogation in a wooden bar. *Ultrasonic Elsevier*, 2004.
13. L. Iancu et al. Quantification of defects in wood by use of ultrasonic in association with imagistic method. *15t WCNDT Roma*, 2000.
14. T. Tanaka. Wood inspection thermography. *Wood NDT*, 2000.
15. Nicolotti G, Socco LV, Martinis R, Godio A, and Sambuelli L. Application and comparison of tree tomographic techniques for detection of decay in trees. *Journal of Arboriculture*, 29(2):66–78, March 2003.
16. Schwarze F.W.M.R, Rabe C, Ferner D, and Fink S. Detection of decay in trees with stress waves and interpretation of acoustic tomograms. *Arboricultural Journal*, 28(1/2):3–19, 2004.
17. Gonzalez RC and Woods RE. *Digital Image Processing* (2nd Edition). Addison-Wesley, Reading, MA, 2002.
18. Mazliham MS, Loonis P, and Idris AS. Extraction of information based on experts knowledge rules to recognize ganoderma infection in tomography image. In *Proceedings of IAENG International MultiConference of Engineers and Computer Scientists, Hong Kong*. IAENG, 2007.
19. Nelson BN. Automatic vehicle detection in infrared imagery using a fuzzy inference-based classification system. *IEEE Transaction on Fuzzy Systems*, 9(1), February 2001.
20. Aydin. Fuzzy set approaches to classification of rock masses. *Engineering Geology*, (74):227–245, 2004.
21. Mazliham MS, Loonis P, and Idris AS. Towards automatic recognition and grading of ganoderma infection pattern using fuzzy systems. In *ENFORMATIKA Transaction on Engineering, Computing and Technology Advances in Computer, Infromation and Systems Science and Engineering,* Vol 19, Bangkok, 2007.
22. Germain M, Voorons M, Boucher JM, and Benie GB. Fuzzy statistical classification method for multiband image fusion. In *ISIF*, 2002.
23. Straszecka E. An interpretation of focal elements as fuzzy sets. *International Journal of Intelligent Systems*, 18:821–835, 2003.
24. Bentabet L, Zhu YM, Dupuis O, Kaftandjian V, Babot D, and Rombaut M. Use of clustering or determining mass function Dempster Shafer theory. In *5th International Conference on ICSP*, 2000.

25. Mazliham MS, Loonis P, and Idris AS. Mass function initialization rules for ganoderma infection detection by tomography sensor. In *Proceedings of Second IASTED International Conference on Computer Intelligence, San Francisco*. IASTED, 2006.
26. Sadiq R and Rodriguez MJ. Interpreting drinking water quality in the distribution system using dempster shafer theory of evidence. *Chemosphere*, 59(2):177–188, 2005.

Chapter 30
A Secure Multiagent Intelligent Conceptual Framework for Modeling Enterprise Resource Planning

Kaveh Pashaei, Farzad Peyravi, and Fattaneh Taghyareh

30.1 Introduction

Facing the challenge of responding more rapidly to changing markets, manufacturing industries have been motivated to explore many exciting innovations in their business practices and procedures such as MRP, ERP, EIS, CIM, and the like [1–3]. However, economic globalization is forcing companies to look further for world-class manufacturing levels. The limitation of resources, such as technology, people, or money, makes enterprises focus on their own competence as well as try to get involved with others. The core idea here is that enterprises should share their resources and abilities without owning them, in order to exploit the market opportunity that is beyond the ability of a single enterprise [4–6].

In the late 1970s and early 1980s, the need for enterprisewide integrated systems intensified as global competition became inevitable, and product customization and innovation became important factors to retain customers and subsequently to gain market share [7]. Systems thinking-based management philosophies such as total quality management and just-in-time systems were introduced, which necessitated the management of relationships among functional areas and cross-organizational processes. The development of such systems slowly evolved from standalone systems (e.g., a standard inventory control system) to material requirement planning/manufacturing resource planning (MRP I and MRP II) systems, and subsequently, to enterprisewide systems to include other functional areas such as sales and marketing, financial accounting, and human resource management. However, attempts to provide a complete enterprisewide software solution were not successful until the mid-1990s due to technical complexity, lack of resource availability, and unclear vision [8].

In the mid-1990s, the Gartner Group coined the term "ERP" to refer to next-generation systems which differ from earlier ones in the areas of relational database management, graphical user interface, fourth generation languages, client–server architecture, and open system capabilities [9]. The integration is normally

Oscar Castillo et al. (eds.), *Trends in Intelligent Systems and Computer Engineering*.

implemented through the use of a common database among subsystems. The information is updated as changes occur, and the new status is available for everyone to use for decision making or for managing their part of business. The decisions made in different functional areas are based on the same current data to prevent nonoptimal decisions from obsolete or outdated data. Expected benefits from ERP implementation include lower inventory, fewer personnel, lower operating costs, improved order management, on-time delivery, better product quality, higher productivity, and faster customer responsiveness [10].

Multiagent enterprise resource planning is an area that has indeed received a lot of attention amongst researchers in the past two decades. The MetaMorph II agent-based architecture, a distributed intelligent environment that integrates manufacturing enterprise was proposed in Shen and Norrie [11]. The AARIA (Autonomous Agents for Rock Island Arsenal) architecture [12] describes the capabilities of a distributed manufacturing complex to configure itself in order to satisfy an individual customer's desires. Agent-based techniques were developed in Turoski [13] for coordinating activities of e-commerce and an Internet-based supply chain system for mass customization markets. Agent-based architectures were proposed in Li and Fong [14] to facilitate the formation and organization of virtual enterprises for order management. An approach is presented in Zhang et al. [15] that would enable manufacturing organizations to dynamically and cost-effectively integrate their own manufacturing systems in a coordinated manner to cope with the dynamic changes occurring in a global market.

Furthermore in Wong and Sycara [16], Soshi and Maekawa [17], and Thirunavukkarasu et al. [18] the general analysis and classifications of attacks and possible countermeasures for securing agent technology as part of published agent systems is described. The security requirement and design for mobile agents is addressed in Corradi et al. [19] and Korba [20]. Several challenges remain and there has not been work on developing a security model for multiagent ERP systems. Furthermore, there is not an evaluating factor for measuring the data migration time in secure and nonsecure modes. In this context, security mechanisms are used to capture the privileges and part of security policies required in distributed applications which can then be used as a dynamic capability in providing distributed authorization and confidentiality.

The purpose of this chapter is to investigate the use of software agents to achieve the secure system integration of ERP software packages. A secure multiagent-based intelligent ERP (SMAIERP) architecture is proposed to take advantage of the existing information systems and security techniques to simulate the secure ERP system using capabilities and characteristics of the software agent-based computer systems. The rest of the chapter is organized as follows. In the next section, we discuss the MetaMorph I as a model that describes two architectures of the multiagent system, namely, the brokering and recruiting mechanisms for selecting our model. Next architecture of SMAIERP system is provided to highlight its properties, various types, and applications in order to establish its viability for developing an ERP type system. In Sect. 30.4, security issues are discussed. Section 30.5 shows how agents communicate and synchronize with each other. Experimental environment and simulation results in a secure and nonsecure manner is described in Sect. 30.6. Finally, the implications of the SMAIERP system and future research directions are provided.

30.2 MetaMorph I

MetaMorph (now referred to as MetaMorph I) [21] is a multiagent architecture for intelligent manufacturing developed at The University of Calgary. The architecture has been named MetaMorphic, because a primary characteristic is its changing form, structure, and activity as it dynamically adapts to emerging tasks and changing environment. Additionally, mediator agents assume the role of system coordinators by promoting cooperation among intelligent agents and learning from the agents' behavior. Mediator agents provide system associations without interfering with lower-level decisions unless critical situations occur. Mediator agents are able to expand their coordination capabilities to include mediation behaviors, which may be focused upon high-level policies to break decision deadlocks. Mediation actions are performance-directed behaviors. Mediator agents can use brokering and recruiting communication mechanisms [22] to find related agents for establishing collaborative subsystems. The brokering mechanism consists of receiving a request message from an intelligent agent, understanding the request, finding suitable receptors for the message, and broadcasting the message to the selected group of agents. This mechanism is shown in Fig. 30.1.

The recruiting mechanism is a superset of the brokering mechanism, because it uses the brokering mechanism to match agents. However, once appropriate agents have been found, these agents can be directly linked. The mediator agent then can step out of the scene to let the agents proceed with the communication themselves. This mechanism is shown in Fig. 30.2. Both mechanisms have been used in MetaMorph I. To efficiently use these mechanisms, mediator agents need to have sufficient organizational knowledge to match agent requests with needed resources. Organizational knowledge at the mediator level is basically a list of agent-to-agent relationships that is dynamically enlarged.

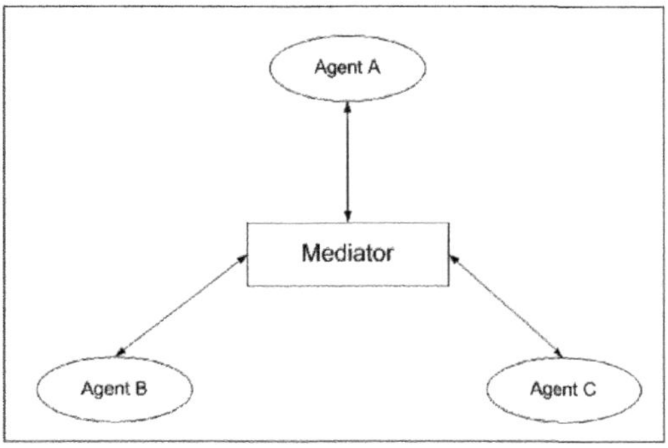

Fig. 30.1 Brokering mechanism

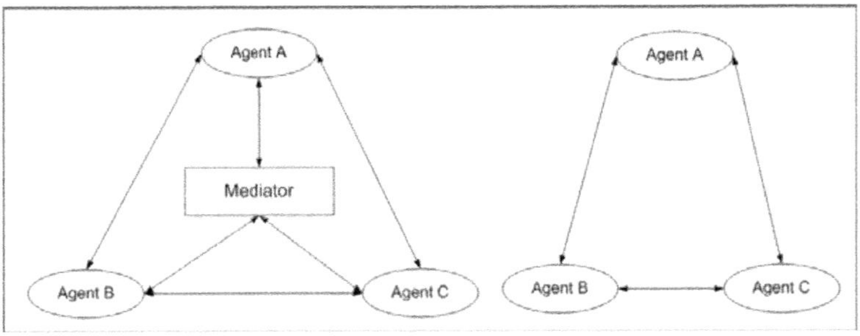

Fig. 30.2 Recruiting mechanism

The brokering and recruiting mechanisms generate two relevant types of collaboration subsystems. The first corresponds to an indirect collaboration subgroup, because the requester agent does not need to know about the existence of other agents that temporarily match the queries. The second type is a direct collaboration subgroup, because the requester agent is informed about the presence and physical location of matching agents to continue with direct communication. One common activity for mediator agents involved in either type of collaboration is interpreting messages, decomposing tasks, and providing processing times for every new subtask. These capabilities make mediator agents very important elements in achieving the integration of dissimilar intelligent agents. Federation multiagent architectures require a substantial commitment to supporting intelligent agent interoperability through mediator agents.

In MetaMorph I [21], mediators were used in a distributed decision-making support system for coordinating the activities of a multiagent system. This coordination involves three main phases: subtasking, creation of virtual communities of agents (coordination clusters), and execution of the processes imposed by the tasks. These phases are developed within the coordination clusters by distributed mediator agents together with other agents representing the physical devices. The coordination clusters are initialized through mediator agents, which can dynamically find and incorporate those other agents that can contribute to the task.

30.3 Architecture of SMAIERP System

The proposed SMAIERP architecture uses the recruiting mechanism from MetaMorph I and is composed of a set of five software agents: a coordinating agent, a planning agent, an interface agent, a data collecting agent, and a set of task agents. We propose that this set of software agents within each functional area (say, department A and department B) interacts through a coordinating agent. Figure 30.3 illustrates the abstract level of the SMAIERP system architecture with coordination agents serving as the representatives of each department and communicating with each other over the company's network. There is a user interface agent who serves

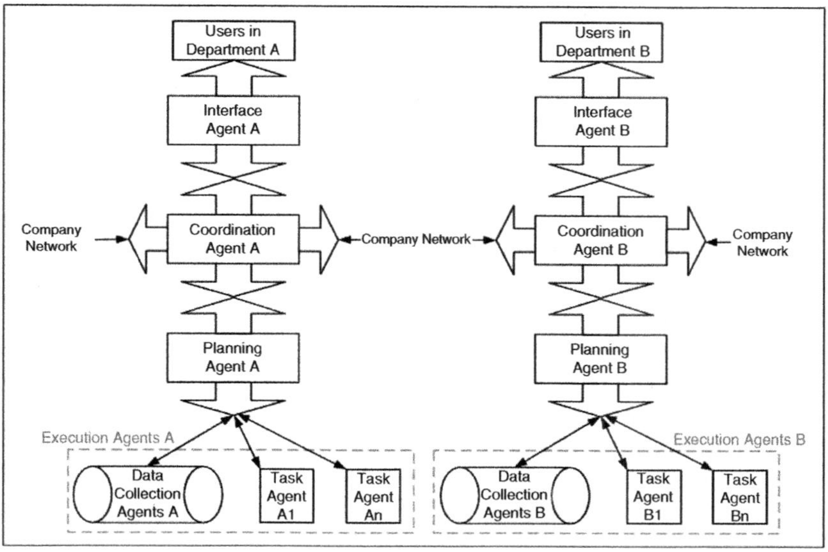

Fig. 30.3 Architecture of SMAIERP system

as a communication tool between the user and the SMAIERP system, and there is a collection of execution agents which is composed of several task agents and data collecting agents that perform specific tasks within the department. Furthermore, there is a planning agent who establishes its plan and broadcasts this to execution agents. Next, various functions/responsibilities undertaken by each type of software agent are discussed.

30.3.1 The Coordination Agent

The coordination agent is the heart of this multiagent intelligent ERP architecture, is the representative of the department when communicating with other coordination agents, and is the controller of the other agents within the department. A department can have one or many coordination agents depending on the nature of task complexity. The major responsibilities for the coordination agent include:

- Receiving instructions and reporting to the human user through an interface agent
- Communicating with execution agents and exchanging data with them through planning agent
- Communicating with and providing requested data to other coordination agents

With their domain knowledge, the coordination agents have the ability to monitor, communicate, and collaborate with other agents, as well as react to various requests.

30.3.2 The Planning Agent

The planning agent is in charge of department planning functions. To keep the agents relatively simple, it is desirable to limit the number of functions encapsulated to a small number. In pinpointing the planning agents in the department, we need to first identify the main planning intensive functions and processes in the planning level of the organization. These are selected from a complete set of planning processes which are tasked to the various resources. As highlighted by Huin et al. [23], there are usually only a limited number of planning agents in a department. However, if one looks at the interface between any two operational departments, it can be seen that the information transmitted between the agents is in the form of "orders" (e.g., purchase order, material issue vouchers, firmed planned orders, etc.), and the "commitment" to the orders (e.g., confirmed sales orders). These are the "Demands" and "Production Orders" which the departments must react to quickly to fulfill the needs of their customers. These should be matched as quickly as possible to ensure the competitiveness of the organization. Therefore the responsibilities of planning agent include:

- Assigning data collection to and receiving data from a data collection agent
- Relaying the dataset, assigning tasks to, and receiving feedback from task agents
- Assign tasks to proper task agents and data collection agents

30.3.3 The Data Collection Agent

The objective of the data collection agents is to query specific database(s) within the department and obtain the information requested by its own coordination agent. It possesses specific domain knowledge needed to carry out its tasks. The "intelligence" in the data collection agents identifies invalid data and missing values so that the data are complete and applicable when being returned to the coordination agent. However, the structures or abilities of data collection agents need not be the same in different departments because each department may have a different database management system (DBMS) or data warehouse. The responsibility of a data collection agent is to:

- Retrieve information requested by its own planning agent.
- Query specific database(s) within the department.
- Perform data warehousing and prepare dataset upon request from planning and coordination agent.

30.3.4 The Task Agent

The task agents usually possess mobility and can act autonomously within their own domain knowledge without the intervention of coordination agents. For example, a task agent that is assigned to monitor price change would go to and stay in a supplier's site to monitor the supplier's price and report any price change that crosses given threshold values with or without the instruction from its coordination agent. The number of task agents varies by the number and complexity of tasks within a department. The functions of a task agent may also vary from department to department depending on what needs to be accomplished. In general, the responsibilities of a task agent include:

- Receiving data from the planning agent
- Performing data analysis by running specific program and/or algorithm

30.3.5 The Interface

The interface agent possesses the ability to learn and store preferences of users and the ability to monitor and inform users when tasks have been completed without the inquiry of users. With enhancement, the interface agent may observe and record the user's disposition to follow the recommendations of coordination agents and invoke machine learning to determine many of the user's preferences. The primary responsibilities of an interface agent include:

- Communicating between human users and coordination agents
- Interpreting results
- Preparing reports for human users

30.4 Security Implications

In this architecture, security mechanisms should be considered for agent communications, for example, when the planning agent of the accounting department requests the planning agent of the inventory department to send its prediction about the costs of goods that were provided for the whole organization for the next year, it is necessary to make sure that other agents could not access this report or guarantee authentication of this message. Thus to ensure the security issues for agent communications we use one of the cryptography techniques, namely, RSA. This algorithm provides these following security properties [23].

- *Confidentiality*: Assurance that communicated information is not accessible to unauthorized parties
- *Data integrity*: Assurance that communicated information cannot be manipulated by unauthorized parties without being detected
- *Authentication of origin*: Assurance that communication originates from its claimant
- *Availability*: Assurance that communication reaches its intended recipient in a timely fashion

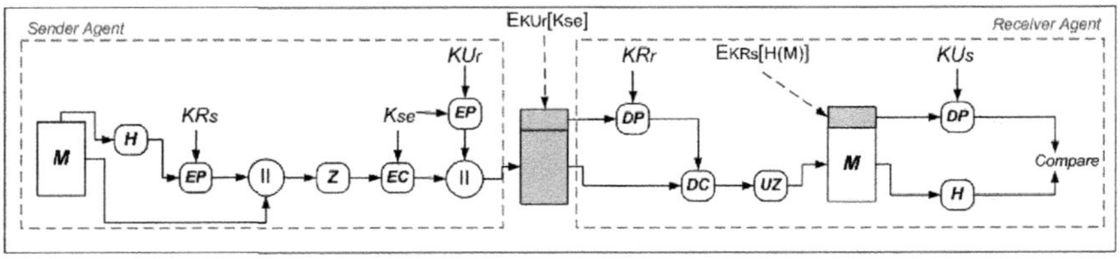

Fig. 30.4 Encryption/decryption of messages using PGP toolkit

- *Nonrepudiation*: Assurance that the originating entity can be held responsible for its communications

We use open the PGP tool for encrypt and decrypt messages in the planning agent of each department. This process is done as shown in Fig. 30.4 [23].

The encryption/decryption process of messages between data communicating agents has the following steps

1. Sender agent creates a message M.
2. SHA-1 is used to generate 160 bit hash code of message.
3. Hash code of message is encrypted using sender agent's private key.
4. This encrypted message concatenates with original message M and then zips them.
5. The zipped message is encrypted using session key that produced by sender agent.
6. Session key is encrypted with receiver's public key and concatenates with the message produced in step 5. The message produced in this step contain two parts; the first part is the session key which is encrypted with receiver public key and the second part contains the encrypted message using the sender agent private key for determining authentication of sender agent and the original message that was zipped and encrypted with session key.
7. Receiver agent decrypts the first part using its private key to obtain session key.
8. Decrypt the second part using session key.
9. Unzip the message.
10. This message contains the original message M and a header for authentication check.
11. The header is decrypted with the sender public key and compared to H(M) which denotes the hashed initial message for authentication check.

We use this process for important messages such as critical reports for each department and current inventory status to guarantee the authentication and confidentiality of important messages.

30.5 Communication and Synchronization of Agents

The four types of agents have the same logical structure. They exchange information with internal entities and the outside world by receiving and dispatching messages through independent interfaces. Action is taken upon the reception of the

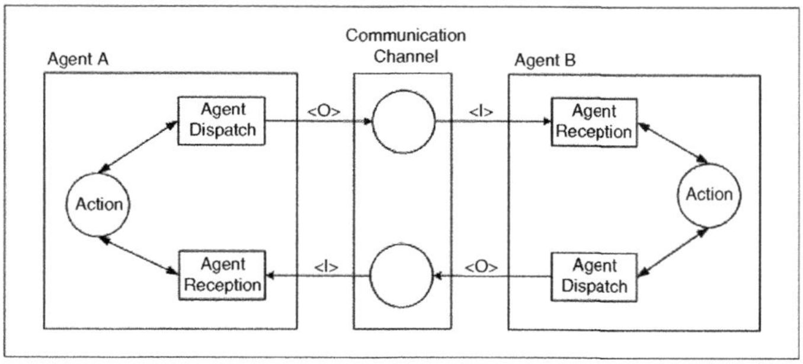

Fig. 30.5 Communication and interface of the various resources agents within department. <O> indicates outflow of information. <I> indicates the inflow of information

corresponding message. The changed state, particularly for the execution agents, is transmitted to other agents. Also the planning results from each planning agent may affect the planning process of other agents.

Agents need to exchange information in order to adapt to each other processes (see Fig. 30.5). This helps ensure that essential project information is collated and communicated at the various levels. There are two possibilities for exchanging messages. The agents can broadcast information about their intended actions to each other. This incurs a large amount of system overhead because more information must be exchanged and because agents may duplicate each others' rational decisions. Such weakness can bog down any ERP project management in the organization. In this four-tier architecture, the agents transfer their plans to the coordinating agents only before it converges.

This arrangement transforms the traditional organizational dynamics of a multilateral affair between the agents involved into a set of bilateral negotiations. In this case, the planning agents do not only interact directly with each other but each agent must communicate his final plans with the coordination agent. In Fig. 30.4, the messages exchanged between the planning agents are simply the outcome of the plans as and when they are processed, such as changes to the orders. No other internal information is normally exchanged except that from the execution agents which are submitted and kept with the planning agents. The information that a planning agent submits to the coordination agent is the final project plan information provided to the adjacent planning agent. In this manner, the information confidentiality and transfer problems are properly handled within the system. The coordination agent is in full control and has knowledge of the project plans and actions.

There are two approaches in representing events and their effects on the execution phase: state-based representation [24] and event-based representation [25]. Actions taken by the execution agents are triggered by the messages from the department planning agents. After execution, the systems states are dispatched back to the planning agents for further actions. The various agents act as controller whose responsibility is to ensure that the plans are executed properly and the systems states are well maintained. The agents transfer their plans to the coordinating agents only before it converges.

30.6 Experimental Environment and Simulation Results

In order to illustrate the proposed SMAIERP system in a manufacturing firm, we demonstrate our model and implement it in a secure manner, and then we compare our result in nonsecure mode, secure mode, and a combination secure and nonsecure mode. We use time factor versus data size to exemplify our results. We assume that one coordination agent is assigned to the marketing (i.e., Coordination Agent M), accounting (i.e., Coordination Agent A), inventory (i.e., Coordination Agent I), and logistics/distribution (i.e., Coordination Agent D) departments. Due to the numbers and complexity of tasks, two coordination agents are assigned to the production department for the product mix optimization system (i.e., Coordination Agent PO) and for the master production scheduling system (i.e., Coordination Agent PS). Furthermore, for each coordination agent, there is one interface agent, and one execution agent which consists of the data collection agent, and several tasks agents. Figure 30.6 provides an overview of SMAIERP system.

In our model the encryption/decryption process is done in the coordination agent of each department. Each coordination agent uses the PGP toolkit for decryption of messages that must be sent in secure mode. For implementation we assume that these agents are distributed nodes and we simulate them according to Fig. 30.7

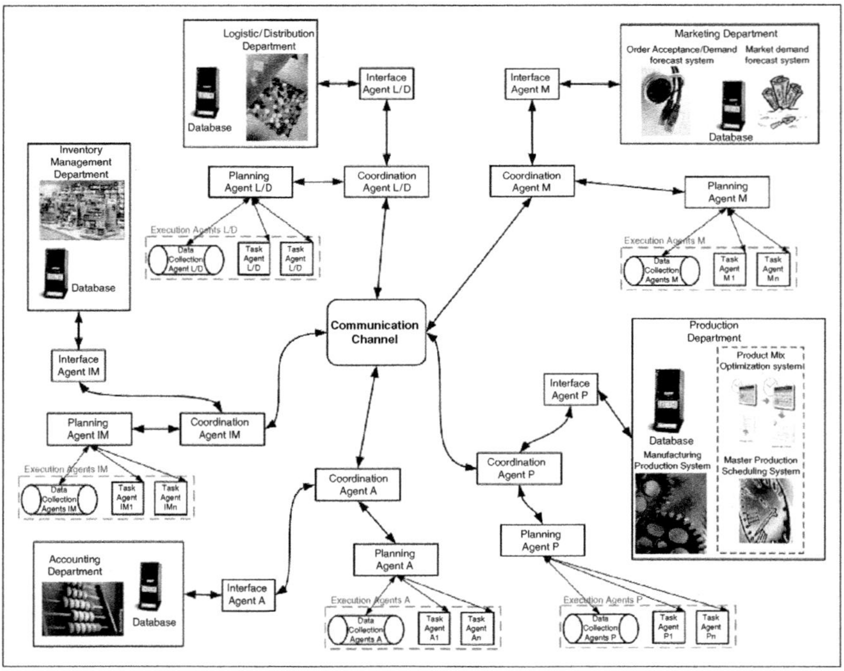

Fig. 30.6 An SMAIERP system application in a manufacturing firm

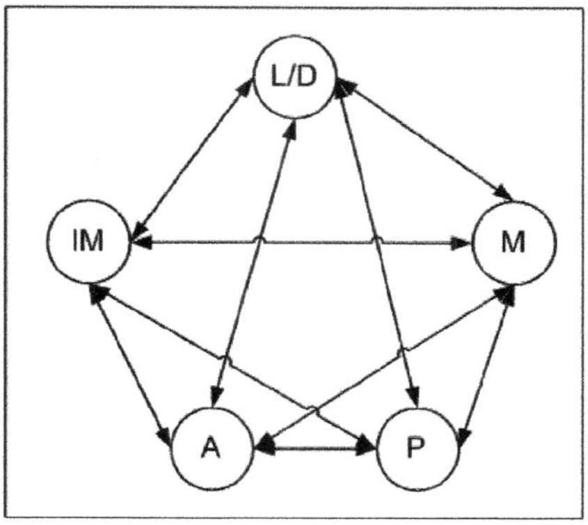

Fig. 30.7 Coordination agents in our topology

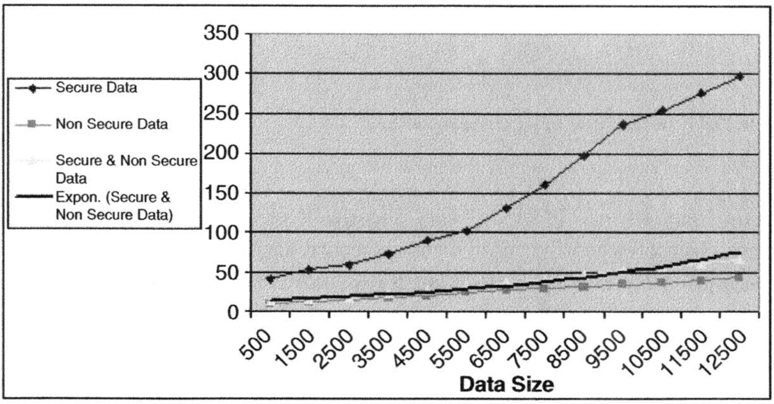

Fig. 30.8 Average time spent for data migration in secure, nonsecure, and secure and nonsecure manner

topology. Each of these nodes accesses other nodes through the communication channel, so in the topology we connect these coordination agents directly.

We developed software to simulate secure and nonsecure environments and compare them. This simulator is configurable for testing data transmission in a security manner and nonsecurity manner and different data request conditions. Our simulator marks each data send and receive time. In our experiments, we consider two factors: average time spent for moving data from one coordination agent to another (data migration time) and response time which is the sum of waiting for processing request, processing request, waiting for communication channel, and data migration time. We investigate the effect of different data sizes on these factors. First we sent data from one coordination agent to another in a secure manner and then we sent data in a nonsecure manner. We assume that from 110 messages interchanged between agents, 10 messages must send in secure manner and 100 messages must send in nonsecure manner. Figure 30.8 shows the effect of data size versus time factor.

According to Fig. 30.8 for all data sizes the average time spent for data migration in the secure manner is much larger than the nonsecure manner. The reason is that

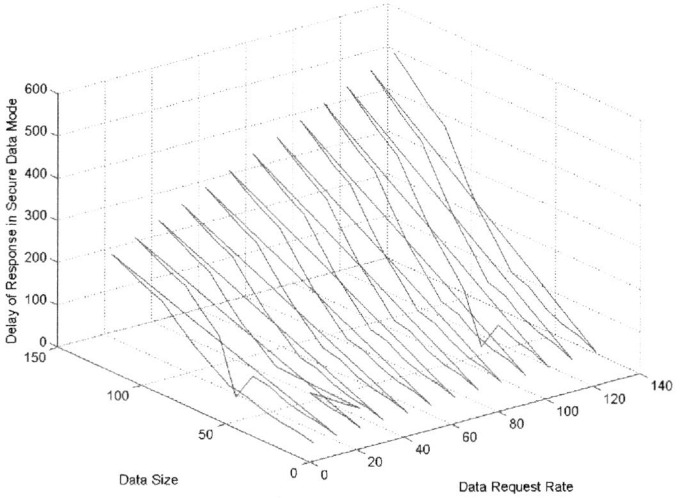

Fig. 30.9 Delay of response in secure manner

for secure data we use PGP software and the RSA algorithm to encrypt/decrypt messages. In the case of combination of secure and nonsecure data according to our assumption the average time spent for data migration is close to the nonsecure condition. As shown in Fig. 30.8, the equation of exponential regression of data migration to data request in secure and nonsecure manner is:

$$y = 12.749e^{01367x}$$

For calculating the delay of response to a data request from the coordinator agent we consider two factors: data size and data request rate. As the data request rate increased, the delay of response time for fixed data size increased. This effect is demonstrated in Fig. 30.9. Additionally the delay of response for fixed data request rate is increased when the data size is enlarged. We calculate this delay for evaluation of response time to a data request from the coordination agent of each department.

Furthermore our experiments show that the data request rate from the coordinator agent of each department has a Poisson distribution which has an average $\lambda = 23$ in the 100 min intervals. Figure 30.10 illustrates the response time for nonsecure, secure, and secure and nonsecure manner versus data size according the Poisson distribution of data requests.

Figure 30.10 depicts that for data size between 500 and 3,500 the response time in secure manner is relatively close to nonsecure manner. For data size larger than 3,500 response time for secure manner is larger than nonsecure manner. The reason is that for secure data we use processing time to encrypt/decrypt messages. Figure 30.11 represents the processing time for encrypt/decrypt message in secure manner. In the case of combination of secure and nonsecure data according to our

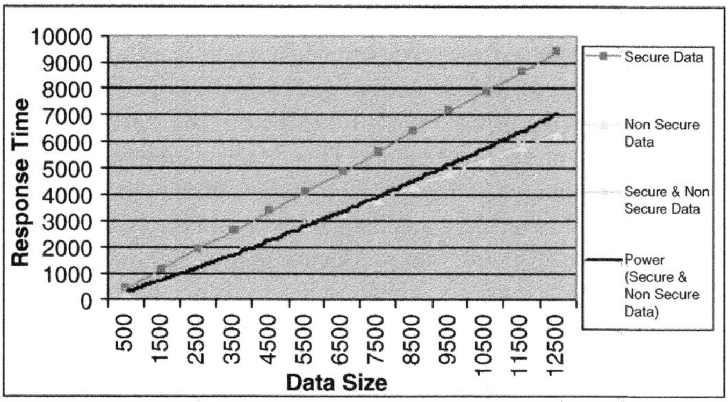

Fig. 30.10 Response time in secure, nonsecure, and secure and nonsecure manner

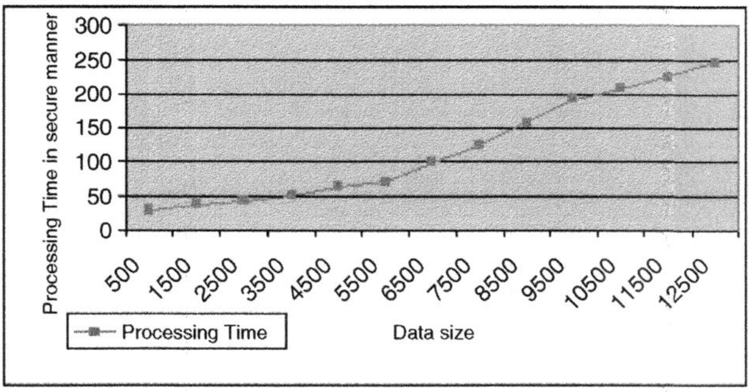

Fig. 30.11 Processing time in secure manner

assumption the response time is close to nonsecure condition. As shown in Fig. 30.9, the equation of power regression of response time to data request in secure and nonsecure manner is:

$$y = 326.16x^{1.1981}$$

The results show that there is little difference between nonsecure manner and the combination of secure and nonsecure manner in both data migration time and response time, so we can use this model to transfer agents' important messages and ensure the security of our model.

30.7 Conclusion

In this chapter, we present a secure multiagent system that is capable of providing enterprisewide integration. With this approach, we demonstrate that a set of software agents with specialized expertise can be quickly assembled to gather relevant information and knowledge, and more important, to cooperate with each other in a secure manner in order to arrive at timely decisions in dealing with various enterprise scenarios. We introduce four types of agents, namely, interface agent, coordination agent, planning agent, and execution agent which cooperate with one another to handle the process of the organization. Furthermore, we introduce a mechanism that ensures the security of our architecture. This mechanism uses public key, private key, and session key for encrypting and decrypting important messages interchanged between agents. Using this security mechanism in our topology demonstrates that if we use the combination of secure and nonsecure mode, the average time spent for data migration and response time to a data request is close to the nonsecure mode. Thus we can send our important messages through the secure mode. In this agent-based model each department has its own set of agents. This secure four-level model can thus be applied to solve organizational dynamics issues in enterprise resources planning projects in organizations. Further research is needed to extend the current work and to address its limitations.

References

1. E. Teicholz and J. Orr, *Computer Integrated Manufacturing Handbook*. New York: McGraw-Hill, 1987.
2. AMICE and CIMOSA, *Open System Architecture for CIM*. New York: Springer Verlag, 1993.
3. P. Bernus and L. Nemes, *Modeling and Methodologies for Enterprise Integration: IFIP*. New York: Chapman and Hall, 1996.
4. W.H. Davidow and M.S. Malone, *The Virtual Corporation: Structuring and Revitalizing the Corporation for the 21st Century*. New York: Harper Collins, 1992.
5. J. Browne, "The extended enterprise—Manufacturing and the value chain," presented at *Proceedings of BASYS95*, 1995.
6. L.M. Camarinha-Matos and H. Afsarnianesh, *Handbook of Life Cycle Engineering: Concepts, Tools and Techniques, Chapter Virtual Enterprise: Life Cycle Supporting Tools and Technologies*. New York: Chapman and Hall, 1997.
7. R. Kalakota and M. Robinson, *e-Business 2.0: Roadmap for Success*. Boston: Addison-Wesley, 2000.
8. K. Kumar and J.V. Hillegersberg, "ERP experiences and evolution," *Communication of the ACM*, 43(4), pp. 22–26, 2000.
9. C. Dahlen and J. Elfsson, "An analysis of the current and future ERP market with focus on Sweden," Stockholm: available at http://whitepapers.techrepublic.com.com/whitepaper.aspx?&docid = 3805&promo = 100511, 1999.
10. B. Robinson and F. Wilson, "Planning for the market? Enterprise resource planning systems and the contradictions of capital," *Database for Advances in Information Systems*, 32(4), pp. 21–33, 2001.
11. W. Shen and D.H. Norrie, "An agent-based approach for manufacturing enterprise integration and supply chain management," presented at *Globalization of Manufacturing in the Digital Communications Era of the 21st Century: Innovation, Agility, and the Virtual Enterprise*, 1998.
12. A.D. Baker, H.V.D. Parunak, and K. Erol, "Manufacturing over the Internet and into your living room: Perspectives from the AARIA project," ECECS Department, Technical Report 1997.
13. K. Turoski, "Agent-based e-commerce in case of mass customization," *International Journal of Production Economics*, 75(1–2), pp. 69–81, 2002.

14. T. Li and Z. Fong, "A system architecture for agent based supply chain management platform," In *Proceedings of the 2003 Canadian Conference on Electrical And Computer Engineering (CCECE): Toward a Caring and Humane Technology*, 2003.
15. D.Z. Zhang, A.I. Anosike, M.K. Lim, and O.M. Akanle, "An agent-based approach for e-manufacturing and supply chain integration," *Computers & Industrial Engineering*, 51, pp. 343–360, October 2006.
16. H.C. Wong and K. Sycara, "Adding security and trust to multi-agent systems," presented at *Autonomous Agents '99 Workshop on Deception, Fraud and Trust in Agent Societies*, 1999.
17. M. Soshi and M. Maekawa, "The Saga security system: A security architecture for open distributed systems," presented at *6th IEEE Computer Society Workshop on Future Trends of Distributed Computing Systems*, 1997.
18. C. Thirunavukkarasu, T. Finin, and J. Mayfield, "Secret agents a security architecture for the KQML Agent communication language," presented at *CIKM '95 Intelligent Information Agents Workshop*, Baltimore, December 1995.
19. A. Corradi, R. Montanari, and C. Stefanelli, "Security issues in mobile agent technology. Distributed Computing Systems," presented at *7th IEEE Workshop on Future Trends of Distributed Computing Systems*, 1999.
20. L. Korba, "Towards secure agent distribution and communication," presented at *32nd Hawaii International Conference on System Sciences*, 1999.
21. F. Maturana and D. Norrie, "Multi-agent mediator architecture for distributed manufacturing," *Journal of Intelligent Manufacturing*, 7, pp. 257–270, 1996.
22. K. Decker, "Environment centered analysis and design of coordination mechanisms," *Ph.D. Thesis*. Amherst: Massachusetts in Department of Computer Science, 1995.
23. S. Huin, L. Luong, and K. Abhary, "Managing enterprise resources planning systems deployment in SMEs," presented at *Proceedings of the 3rd International Conference on Project Management Society of Project Management, and College of Engineering*, Japan, 2002.
24. W. Stalling, *Cryptography and Network Security Principles and Practice*. Upper Saddle River, NJ: Prentice-Hall, 2003.
25. A. Lansky, "Localized event-based reasoning for multi-agent domains," presented at *SRI International*, Stanford University, 1998.

Chapter 31
On Generating Algebraic Equations for A5-Type Key Stream Generator

Mehreen Afzal and Ashraf Masood

31.1 Introduction

Algebraic cryptanalysis is a newer and quite successful technique for the security evaluation of stream ciphers as well as a threat to the structures which are resistant to other types of attacks. Originally algebraic attacks were proved to be successful against ciphers having combining or filtering Boolean function along with the linear part. Very successful attacks have been mounted on ciphers with nonlinear components with or without memory [1–5]. So far the most successful attacks are on a particular design having two components: a nonlinear filter or combining function and a linear component of one or many LFSRs. However, algebraic attack on somewhat different structures of stream ciphers has not been much studied yet. Clock control is also one of the mechanisms employed to introduce nonlinearity into a key stream generator built from linear feedback shift registers. Algebraic attacks against clock-controlled stream ciphers have recently been studied by Sultan AH, Lynn B, Bernard C, Kenneth W [6], which to our knowledge is the first work in this direction. Although LILI 128 is also a clock-controlled cipher and successful algebraic attacks can be found against it [4], for its clock-controlled part, guessing is being used. The work in [6] basically involves stream ciphers such as the stop-and-go generator, alternating step generator, self-decimated, and step1/step2 generator in which one or more LFSRs are irregularly clocked and their clocking depends on some regularly clocked LFSR. This attack is based on the general assumption that the output bit of one shift register controls the clocking of other registers in the system and produces a family of equations relating the output bits to the internal state bits. This chapter aims at finding how practical can it be to mount an algebraic attack on A5/1 in which none of the LFSRs are regularly clocked. Because the feasibility of the algebraic attack has not yet been explored against it, the objective of this research is to mount an algebraic attack against an A5/1-type clock-controlled generator.

The A5 key stream generator has a long history of cryptanalytic attacks against it. Most of the attacks are divide and conquer, time–memory trade-off attacks which exploit some of the weaknesses in the algorithm such as the relatively small number

Oscar Castillo et al. (eds.), *Trends in Intelligent Systems and Computer Engineering.*

of internal states, frequent initialization, and so on. First cryptanalytic attacks on A5/1 were based on the time–memory trade-off and divide and conquer by [8]; the design used in this work was slightly different from the actual design of A5/1, and also this attack required a huge amount of memory and time. Another attack by Biryukov et al. [9] is an improved one which required only a few seconds to a few minutes computation on a PC; this attack is also based on the time–memory trade-off and requires huge memory for precomputation storage and also a precomputation phase. An attack against A5/1 based on clock-control guessing is also found to be quite efficient [11]. A type of correlation on A5/1 by [10] is independent of shift register lengths and also does not require much precomputation but this attack is based on poor initialization of A5/1. Yet another correlation attack on the A5-type generator based on edit distance and edit probabilities can recover any two of the three stop/go clocked shift registers [7].

Finding a system of algebraic equations that relate the bits of the initial state K and bits of the key stream Z differs for each cipher and is a first step of algebraic attack. However, it is a precomputation step and once equations are formed, these can be used for attacking multiple key streams. In the case of clock-controlled stream ciphers the development of algebraic equations is not as simple as for generators with a combining or filtering function. In this chapter we present the method for developing algebraic equations which can relate the initial states of the LFSRs and key stream bits of the A5 generator depending upon its particular majority function-based irregular clocking.

31.2 Difference in Algebraic Attacks on Stream Ciphers with Clock-Controlling and Those with Combining or Filtering Function

The main concept behind algebraic attacks is to represent a cipher by a system of equations. Solving that system of equations gives the secret key which is the initial state of LFSRs in this case. The complexity of the attack is related to the degree of equations: the lower the degree is, the more efficient is the attack. Thus algebraic attacks can be viewed as a process consisting of the following two steps.

- The first step is to find a system of algebraic equations that relate the bits of the initial state K and bits of the key stream Z; this step differs for each cipher. However, it is a precomputation step and once equations are formed, these can be used for attacking multiple key streams.
- The second step is performed when some key stream bits are obtained. Observed key stream bits are substituted into the algebraic equations already formed and then these equations are solved using different methods such as linearization, XL algorithm, and Grobner bases. This solution is efficient if the degree of equations is low, and a sufficient number of equations can be formed from the observed key stream. Hence to find low-degree equations is a desirable goal in algebraic attacks.

In the case of stream ciphers with a linear update function and a Boolean output function such as nonlinear filters and combiners, equations that describe the stream cipher are:

$$z_t = f(s_t)$$
$$s_{t+1} = L^t(s_o) = L^t(k_0, \ldots, k_{n-1}),$$

where f denotes the output Boolean function and L denotes the linear update function (LFSR usually). Thus equations in key bits $k_0, k_1, \ldots, k_{n-1}$ can be derived straightforwardly as

$$f(k_0, \ldots, k_{n-1}) = z_0$$
$$f(L(k_0, \ldots, k_{n-1})) = z_1$$
$$f(L^2(k_0, \ldots, k_{n-1})) = z_2$$
$$\ldots\ldots\ldots\ldots\ldots$$

This system of equations has n unknowns corresponding to the n bits of the initial state. Because L is linear, the degree of all equations is equal to the degree of f. So in order to make this attack feasible, it is important to transform this system of equations to the one which is simpler to solve; for that it is tried that the degree of equations be reduced.

Algebraic attacks on stream ciphers with linear feedback are quite practical but so far the most successful attacks are on a particular design having two components: a nonlinear filter or combining function and a linear component of one or many LFSRs. The use of clock-controlling to introduce nonlinearity into the sequence generated from linear component LFSR dates back to some very early approaches. A thorough survey of clock-control techniques is given by Chambers and Gollman [12]. In clock-controlled ciphers some arrangement is devised so that clocking of one register is in some way dependent on another register. Sequences generated by these techniques tend to be more complex than any of the constituent sequences. Still many methods of their analysis have been devised [13, 14]. In the case of algebraic attack on clock-controlled stream ciphers equations are generated differently. Because although the linear part is the same, that is, linear feedback, nonlinearity is not due to a nonlinear higher-degree filtering or combining function; rather, nonlinearity is introduced in the output sequence of one or more LFSRs by some function that controls their clock. Therefore to obtain a relation between initial state bits and output bits is not as straightforward as in the case of the combining and filtering functions. Even if we develop such a relationship, the degree of the equations involved may change at each clock; this makes the algebraic attack on clock-controlled ciphers different.

In [6] an algebraic attack is mounted on clock-controlled stream ciphers such as the stop and go register, alternating step generator, and self-shrinking generator. In these ciphers one regularly clocked LFSR controls output bits of some other one or more LFSR thus making their output bits irregular. We aim at establishing the fact that if all the registers are irregularly clocked, then the equations developed will involve variables with large degree, and consequently lead to the infeasibility of this attack.

31.3 System of Algebraic Equations for A5 Key-Stream Generator

The structure of the A5 algorithm is given in Fig. 31.1. It has three primitive linear feedback shift registers of lengths p, m, and n, so they produce maximum length sequences when clocked regularly. Their initial states are represented as $A_1,\ldots,A_p$, $B_1,\ldots,B_m$, and $C_1,\ldots,C_n$. Each register has a single clocking bit represented as A_{clk}, B_{clk}, and C_{clk}. A majority function is calculated from the clocking bit of each register which are input bits of the function. This function outputs 1 or 0 if two or more of the input bits are 1 or 0, respectively. Based on the output of the majority function, a register is clocked if its clock bit contains a value equal to this majority value. The binary clock-controlled bits are derived from each register by using the stop/go rule described above and the key stream is then produced as the XOR of the output bits of the three registers. In the case of A5/1 algorithm of GSM, the lengths are, respectively, 19, 22, and 23.

The output bit of each LFSR depends upon the output of the majority function for which A_{clk}, B_{clk}, and C_{clk} are input, therefore at time t the output bit of each LFSR can be represented in terms of A_{clk}, B_{clk}, and C_{clk}.

In the case of a clock-controlled generator, there is no simple and direct relationship between the input bits (the initial state of LFSR in this case) and key stream

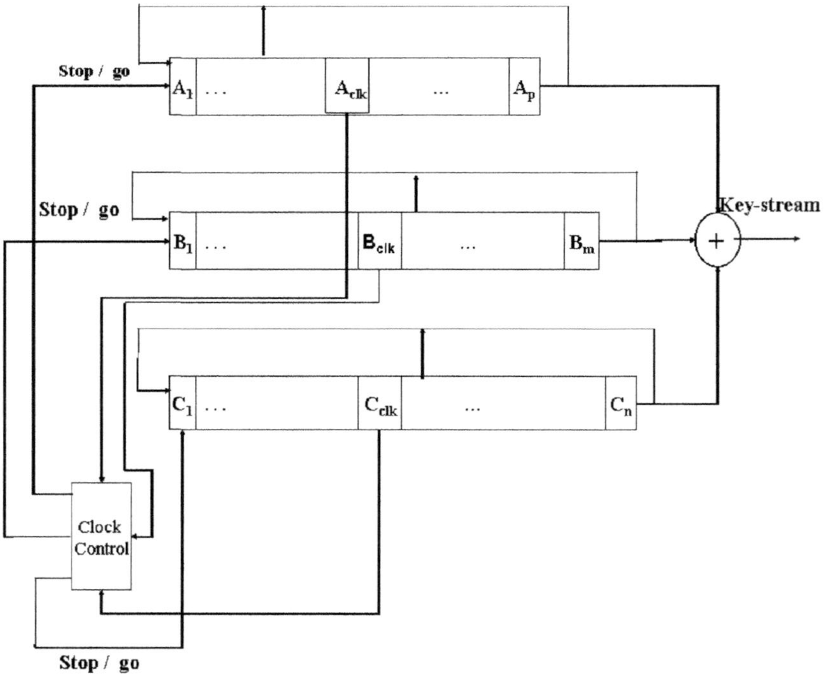

Fig. 31.1 The structure of A5 key stream generator

Table 31.1 Relation between clock bits and output bits

A_{clk}	B_{clk}	C_{clk}	$f(A)$	$f(B)$	$f(C)$
0	0	0	0	0	0
0	0	1	0	0	1
0	1	0	0	1	0
0	1	1	1	0	0
1	0	0	1	0	0
1	0	1	0	1	0
1	1	0	0	0	1
1	1	1	0	0	0

bits. In this case clocking bits and their function which affects output bits play an important role in developing the relationship that could lead to algebraic equations.

Algebraic normal form (ANF) for the majority function is developed based on Table 31.1, in which $f(A), f(B)$, and $f(C)$ are given value 0 if A, B, and C are clocked, respectively, and 1 otherwise. Using this ANF, the relation of output bit of LFSR A with its state bits can be written as

$$\begin{aligned} A_p^t = {} & A_p^{t-1}(A_{clk}^{t-1} \oplus A_{clk}^{t-1}B_{clk}^{t-1} \oplus A_{clk}^{t-1}C_{clk}^{t-1} \oplus B_{clk}^{t-1}C_{clk}^{t-1}) \\ & \oplus A_{p-1}^{t-1}(1 \oplus A_{clk}^{t-1} \oplus A_{clk}^{t-1}B_{clk}^{t-1} \oplus A_{clk}^{t-1}C_{clk}^{t-1} \oplus B_{clk}^{t-1}C_{clk}^{t-1}). \end{aligned} \tag{31.1}$$

If we represent $f_{t-1} = A_{clk}^{t-1}B_{clk}^{t-1} \oplus A_{clk}^{t-1}C_{clk}^{t-1} \oplus B_{clk}^{t-1}B_{clk}^{t-1}$ then Eq. 31.1 can be written as

$$A_p^t = A_p^{t-1}(A_{clk}^{t-1} \oplus f_{t-1}) \oplus A_{p-1}^{t-1}(1 \oplus A_{clk}^{t-1} \oplus f_{t-1}), \tag{31.2}$$

or,

$$A_p^t = A_{p-1}^{t-1} \oplus (A_{p-1}^{t-1} \oplus A_p^{t-1})(A_{clk}^{t-1} \oplus f_{t-1}).$$

Similarly:

$$B_m^t = B_{m-1}^{t-1} \oplus (B_{m-1}^{t-1} \oplus B_m^{t-1})(B_{clk}^{t-1} \oplus f_{t-1})$$

and

$$C_n^t = C_{n-1}^{t-1} \oplus (C_{n-1}^{t-1} \oplus C_n^{t-1})(C_{clk}^{t-1} \oplus f_{t-1}).$$

Because the key stream bit at time t is found by taking addition modulo 2 of the output bits at time t of the three LFSRs, therefore the key stream bit at time t can be represented as

$$\begin{aligned} & z^t = A_p^t \oplus B_m^t \oplus C_n^t \\ & \Rightarrow z^t = (A_{p-1}^{t-1} \oplus B_{m-1}^{t-1} \oplus C_{n-1}^{t-1}) \oplus (A_{p-1}^{t-1} \oplus A_p^{t-1})(A_{clk}^{t-1} \oplus f_{t-1}) \\ & \quad \oplus (B_{m-1}^{t-1} \oplus B_m^{t-1})(m_{clk}^{t-1} \oplus f_{t-1}) \oplus C_{n-1}^{t-1} \oplus C_n^{t-1})(C_{clk}^{t-1} \oplus f_{t-1}). \end{aligned} \tag{31.3}$$

This system of equations represents the A5 cipher, but it can be seen that the degree of this system of equations will increase at each step. The maximum degree an equation can have in this system is $(p+m+n)$ so not only will solving this system be infeasible, but also finding all these equations will be quite difficult.

31.4 Reducing the Degree of Systems of Algebraic Equations for A5 Key Stream Generator

Reducing the degree of the equations is primary in the algebraic attack. In this section we present some attempts made to reduce the degree of the algebraic equations representing the A5 cipher.

First we apply the method given in [6], which is taking the binary derivative to simplify the degree of equations. From Eq. 31.1, we can obtain

$$A_p^{t+1} = A_p^t(A_{clk}^t \oplus f_t) \oplus A_{p-1}^t(1 \oplus A_{clk}^t \oplus f_t). \tag{31.4}$$

From Eqs. 31.1 and 31.3, we have:

$$\begin{aligned} \Rightarrow A_p^{t+1} \oplus A_p^t &= A_p^t(1 \oplus A_{clk}^t \oplus f_t) \oplus A_{p-1}^t(1 \oplus A_{clk}^t \oplus f_t) \\ &= (1 \oplus A_{clk}^t \oplus f_t)(A_p^t \oplus A_{p-1}^t). \end{aligned}$$

Thus we obtain:

$$\begin{aligned} \Rightarrow z^t \oplus z^{t+1} = {} & (A_p^t \oplus A_{p-1}^t \oplus B_m^t \oplus B_{m-1}^t \oplus C_n^t \oplus C_{n-1}^t)(1 \oplus f_t) \\ & \oplus A_{clk}^t(A_p^t \oplus A_{p-1}^t) \oplus B_{clk}^t(B_m^t \oplus B_{m-1}^t) \oplus C_{clk}^t(C_n^t \oplus C_{n-1}^t) \end{aligned} \tag{31.5}$$

$$\Rightarrow (z^t \oplus z^{t+1}).f_t = (A_{clk}^t(A_p^t \oplus A_{p-1}^t) \oplus B_{clk}^t(B_m^t \oplus B_{m-1}^t) \oplus C_{clk}^t(C_n^t \oplus C_{n-1}^t)).f_t. \tag{31.6}$$

This system of equations represents the A5 cipher, but it can be seen that the degree of this system of equations will increase at each step. The maximum degree an equation can have in this system is $(p+m+n)$ so not only will solving this system be infeasible, but also finding all these equations will be quite difficult. But as it is known that f_t is of degree 2, therefore after the first iteration the degree of the equations will increase at each step.

Yet another approach to reduce the degree of equations can be considered as follows. From Eq. 31.4,

$$\begin{aligned} & (z^t \oplus z^{t+1})(1 \oplus A_p^t \oplus A_{p-1}^t \oplus B_m^t \oplus B_{m-1}^t \oplus C_n^t \oplus C_{n-1}^t) \\ & = (A_{clk}^t(A_p^t \oplus A_{p-1}^t) \oplus B_{clk}^t(B_m^t \oplus B_{m-1}^t) \oplus ACt_{clk}(C_n^t \oplus C_{n-1}^t)) \\ & \quad (1 \oplus A_p^t \oplus A_{p-1}^t \oplus B_m^t \oplus B_{m-1}^t \oplus C_n^t \oplus C_{n-1}^t). \end{aligned}$$

Here if we consider the case of consecutive same bits then the equations will become:

$$
\begin{aligned}
&(A^t_{clk}(A^t_p \oplus A^t_{p-1}) \oplus B^t_{clk}(B^t_m \oplus B^t_{m-1}) \oplus ACt_{clk}(C^t_n \oplus C^t_{n-1}))\\
&(1 \oplus A^t_p \oplus A^t_{p-1} \oplus B^t_m \oplus B^t_{m-1} \oplus C^t_n \oplus C^t_{n-1}) = 0\\
\Rightarrow &(A^t_p \oplus A^t_{p-1})(B^t_m \oplus B^t_{m-1})(A^t_{clk} \oplus B^t_{clk}) \oplus (A^t_p \oplus A^t_{p-1})(C^t_n \oplus C^t_{n-1})(A^t_{clk} \oplus C^t_{clk})\\
&\oplus(B^t_m \oplus B^t_{m-1})(C^t_n \oplus C^t_{n-1})(B^t_{clk} \oplus C^t_{clk}) = 0
\end{aligned}
$$

or

$$
\begin{aligned}
&A^t_{clk}(A^t_p \oplus A^t_{p-1})(B^t_m \oplus B^t_{m-1} \oplus C^t_n \oplus C^t_{n-1}) \oplus B^t_{clk}(B^t_m \oplus B^t_{m-1})(A^t_p \oplus A^t_{p-1}\\
&\oplus C^t_n \oplus C^t_{n-1}) \oplus C^t_{clk}(C^t_n \oplus C^t_{n-1})(A^t_p \oplus A^t_{p-1} \oplus B^t_m \oplus B^t_{m-1}) = 0. \qquad (31.7)
\end{aligned}
$$

Equation 31.6, also gives the product of three variables as do previous equations. But in this because we are looking for consecutive same bits, some equations obtained from the initial clocking will be discarded, and this will lead to an even more rapid increase in the degree of equations. In this case, however, different increases in the degrees are obtained for different key bits.

We have applied three forms of equations given in 31.3, 31.6 and 31.7 on a small example of an A5-type generator. For illustration we used a generator with three LFSRs of lengths 3, 4, and 5. Although it is clear that equations will take their maximum degree (which is 12 here), in all cases but the main, an attempt is made to obtain some low-degree equations in the beginning which can help solve the equations efficiently. The increase in the degree of equations for three of these cases is shown in Fig. 31.2.

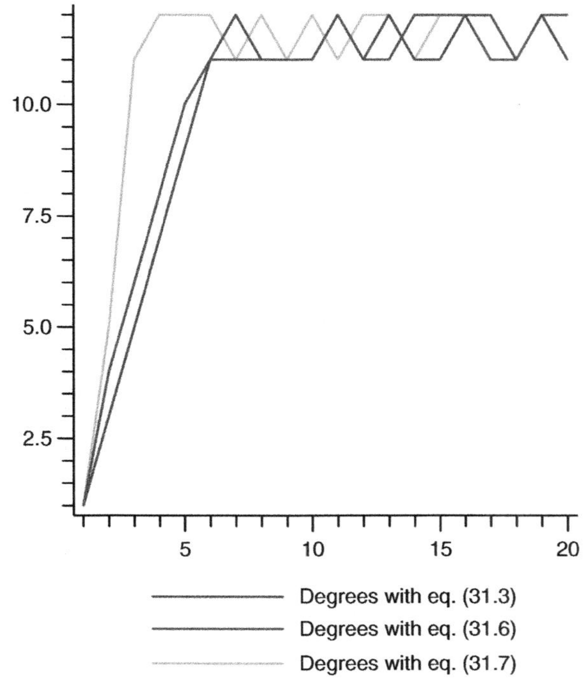

Fig. 31.2 Increase in the degree for different forms of algebraic equations

It can be seen that all three forms of the equations behave almost identically whereas Eq. 31.3 shows somewhat better performance in terms of the increase in the degree of equations. Thus it can also be concluded that in this case taking the binary derivative or changing the form of the equations will not help in decreasing the degree of the equations so as to simplify the problem of solving the equations.

31.5 Analysis of Algebraic Attack on A5 Cipher

As compared to other clock-controlled generators, an algebraic attack on A5 is not as successful because of its clocking mechanism in which all the LFSRs are irregular. Clock-controlled stream ciphers such as the stop-and-go generator, alternating step generator, self-decimated, and step1/step2 generator all have one regularly clocked cipher that controls the clocking of one or more other LFSRs; while developing equations for algebraic attack for such stream cipher an attempt is made to eliminate the effect of irregularly clocked LFSRs, whereas in the case of A5, all three LFSRs are irregularly clocked and even if the equation could be reduced to one variable, its degree cannot be less than the length of that LFSR. In this case, with each clock degree the equation increases and for LFSRs of reasonable length it becomes unreasonably difficult to generate equations, thus making this attack not less in complexity than exhaustive search.

31.6 Conclusion

In clock-controlled generators, developing equations to relate output bits with initial state is not a simple and straightforward task. In this chapter algebraic equations of an A5 key stream generator are developed using ANF of its majorityfunction. It is found that this type of clocking shows more resistance to algebraic attack than the irregular clocking used in the stop-and-go generator; alternating step generator, self-decimated, and step1/step2 generator. All of the mentioned generators have one regularly clocked cipher that controls the clocking of one or more other LFSRs. That is why while developing equations for algebraic attack for such a stream cipher an attempt is made to eliminate the effect of irregularly clocked LFSRs. This is not possible in the case of the A5 cipher and very high-degree equations are formed.

There is a long history of attacks on the A5/1 cipher, the GSM standard. Most of these attacks exploit the weaknesses in the whole encryption system, which includes its resynchronization, reinitialization, and so on. Also some of the attacks use framing information. Our attack, however, does not include such weaknesses. We have considered the generic structure of the A5/1 cipher for our attack because our objective is to analyze this type of irregular clocking for algebraic cryptanalysis rather than showing weaknesses in the GSM security system. Therefore it remains an open question whether our attack can be improved by exploiting the inherent weaknesses found in some existing attacks on A5/1.

References

1. Nicolas C (2002) Higher order correlation attacks, XL algorithm and cryptanalysis of Toyocrypt. In: *ICISC*, LNCS 2587, Springer, Berlin.
2. Nicolas C (2003) Fast algebraic attacks on stream ciphers with linear feedback. In: *Proceedings of Crypto*, LNCS 2729, Springer, New York.
3. Frederik A (2004) Improving fast algebraic attacks. In: *FSE*, LNCS 3017, Springer, New York.
4. Frederik A, Matthias K (2003) Algebraic attacks on combiners with memory. In: *Proceedings of Crypto*, LNCS 2729, Springer, New York.
5. Nicolas C (2004) Algebraic attacks on combiners with memory and several outputs. In: *ICISC*, LNCS 3506, Springer, New York.
6. Sultan AH, Lynn B, Bernard C, Kenneth W (2006) Algebraic attacks on clock-controlled stream ciphers. In: *ACISP*, LNCS 4058, Springer, New York.
7. Jovan DG, Menicocci R (2002) Computation of edit probabilities and edit distances for the A5-type keystream generator. *Journal of Complexity*, Vol. 18, 356–374.
8. Jovan DG (1997) Cryptanalysis of alleged A5 stream cipher. In: *Advances in Cryptography, Eurocrypt*, LNCS 1233, Springer, New York.
9. Biryukov A, Shamir A, Wagner D (2000) Real cryptanalysis of A5/1 on a PC. In: *FSE*, LNCS 1978, Springer, New York.
10. Patrik E, Thomas J (2003) Another attack on A5/1. *IEEE Transactions on Information Theory*, Vol. 49, No. 1.
11. Erik Z (2002) On the efficiency of the clock control guessing attack. In: *ICISC*, LNCS 2587, Springer, New York.
12. Gollmann G, Chambers WG (1989) Clock-controlled shift registers: A review. *IEEE Journal on Selected Area in Communications*, Vol. 7, No. 4.
13. Baum U, Blackburn S (1994) Clock-controlled pseudorandom generators on finite groups. In: *Proceedings of Leuven Algorithms Workshop*, Springer, New York.
14. Jovan DG, O'Conner L (1994) Embedding and probabilistic correlation attacks on clock-controlled shift registers. In: *Advances in Cryptology-Eurocrypt*, LNCS, Springer, New York.

Chapter 32
A Simulation-Based Study on Memory Design Issues for Embedded Systems

Mohsen Sharifi, Mohsen Soryani, and Mohammad Hossein Rezvani

32.1 Introduction

Due to the increasing gap between the speed of CPU and memory, cache designs have become an increasingly critical performance factor in microprocessor systems. Recent improvements in microprocessor technology have provided significant gains in processor speed. This dramatic rise has increased further the gap between the speed of the processor and main memory. Thus, it is necessary to design faster memory systems. In order to decrease the processor–memory speed gap, one of the main concerns has to be in the design of an effective memory hierarchy including multilevel cache and TLB (Translation Lookaside Buffer).

On the other hand, a notable part of the computer industry nowadays is involved in embedded systems. Embedded systems play a significant role in almost all domains of human activities including military campaigns, aeronautics, mobile communications, sensor networks, and industrial local communications. Timeliness of reactions is necessary in these systems and offline guarantees have to be derived using safe methods. Hardware architectures used in such systems now feature caches, deep pipelines, and all kinds of speculations to improve average-case performance. The speed and size are two important concerns of embedded systems in the area of memory architecture design. In these systems, it is necessary to reduce the size of memory to obtain better performance.

Real-time embedded systems often have a hard deadline to complete some instructions. In these cases, the speed of memory plays an important role in system performance. Cache hits usually take one or two processor cycles, whereas cache misses take tens of cycles as a penalty of mishandling and so the speed of memory hierarchy is a key factor in the system. Almost all embedded processors have on-chip instructions and data caches. From a size point of view, it is critical for battery-operated embedded systems to reduce the amount of consumed power.

Oscar Castillo et al. (eds.), *Trends in Intelligent Systems and Computer Engineering.*

Another factor that affects cache performance is the degree of associativity. Nowadays, modern processors include multilevel caches and TLBs and their associativity is increasing. Therefore, it is critical to revisit the effectiveness of common cache replacement policies. When all the lines in a cache memory set become full and a new block of memory needs to be replaced into the cache memory, the cache controller must replace it with one of the old blocks in the cache. If the extracted cache memory line is needed in the near future, the performance of the system will be degraded. Therefore, the cache controller should extract a proper line from the cache. However, it can only guess which cache memory line should be discarded. The state-of-the-art processors employ various policies such as LRU (Least Recently Used) [1], Random [2], FIFO (First-In First-Out) [3], and PLRU (Pseudo-LRU) [4]. This shows that the selection of a proper replacement policy is still an important challenge in the field of computer architecture. All these policies, except Random, determine which cache memory line to replace by looking only at the previous references.

LRU policy increases the cost and implementation complexity. To reduce these two factors, Random policy can be used, but potentially at the expense of desirable performance. Researchers have proposed various PLRU heuristics to reduce the cost by approximating the LRU mechanism. Recent studies have considered only pure LRU policy [5–7] and have used the compiler optimization [8] or integrated approaches including hardware/software techniques [9].

One of the goals of our study is to explore common cache replacement policies and compare them with an optimal (OPT) replacement algorithm. An OPT algorithm would replace a cache line whose next reference is the farthest away in the future among all the cache lines [7]. This policy requires knowledge of the future, and hence its real implementation is impossible. Instead, heuristics have to be used to estimate it. We study OPT, LRU, a type of PLRU, Random, and FIFO policies on a wide range of cache organizations, varying cache sizes, degree of associativity, cache hierarchy, and multilevel TLB hierarchy.

Another goal in this study is to investigate the performance of the two-level TLB against the single-level TLB. This analysis is done in conjunction with cache analysis. Virtual to physical address translation is a significant operation because this is invoked on every instruction fetch and data reference. To speed up the address translation, systems provide a cache of recent translations called TLB, which is usually a small structure indexed by the virtual page number that can be quickly looked up. Several studies have investigated the importance of TLB performance [10]. The idea of multilevel TLB has been investigated in [11–13]. We compare the performance of a two-level TLB with traditional single TLB.

The performance analysis is based on *SimpleScalar*'s [14] cache simulators executing selected SPEC CPU2000 benchmarks [15]. Using a two-level TLB, we study the degree of cache associativity that is enough to offer a low miss rate with respect to SPEC CPU2000. Some prior research has investigated this subject, but only for traditional replacement policies such as LRU [5, 6]. We offer a competitive simulation-based study to reveal the relationship between cache miss rates produced by selected benchmarks and cache and TLB configurations. Additionally, we measure the gap between miss rates of various replacement policies, especially OPT and the LRU family (such as PLRU which is less expensive than LRU). Similar measurements have been done in the literature for other modifications of LRU policy. Wong and Baer [9] demonstrated the effectiveness of a modification of the standard LRU replacement algorithm for large set associative L2 (level two) caches.

The aim of this chapter is to offer a comprehensive and simulation-based performance evaluation of the cache and TLB design issues in embedded processors such as two-level versus single TLB, split versus unified cache, cache size, cache associativity, and replacement policy.

The rest of chapter is organized as follows. Section 32.2 elaborates the problem under our study, related works on hierarchical TLB, specifications of SPEC CPU2000 benchmarks, and the reasons for selecting the benchmarks used in our study. Section 32.3 describes the setup of our experiments. Section 32.4 reports the results of our experiments, and Sect. 32.5 concludes the chapter.

32.2 Related Work

Three categories of related works have guided our study: prior research on cache replacement policies, prior research on hierarchical TLB mechanisms, and prior research on specifications of the SPEC CPU2000 benchmarks.

Several properties of the processor caches influence performance: associativity, replacement, and write policy, and whether there are separated data and instruction caches. The predictability of different cache replacement policies is investigated in [9]. The following are widely used replacement policies in commercial processors.

- LRU used in Intel Pentium and MIPS 24K/34K
- FIFO (or Round-Robin), used in Intel XScale, ARM9, ARM11
- PLRU used in Power PC 75X and Intel Pentium II, III, and IV

Cache implementations vary from direct-mapped to fully associative. With direct-mapped caches, also called one-way caches, each memory block is mapped onto a distinct cache line, whereas with fully associative cache memory, each memory block can be mapped to any of the empty cache lines. Generally, with k-way set associative cache memory, a memory block can be mapped to any of the empty lines among k cache lines within the set to which the block belongs. If all cache lines within the set are full, one of the cache lines is extracted according to a replacement policy.

As the degree of cache associativity increases, selecting an efficient replacement policy becomes more important [3]. Traditionally, most processors have chosen the LRU policy as the replacement policy. LRU replacement maintains a queue of length k, where k is the associativity of the cache. If an element is accessed that is not yet in the cache (a miss), it is queued at the front and the last element of the queue is removed. This is the least recently used element of those in the queue. At a cache hit, the element is moved from its position in the queue to the front, effectively treating hits and misses equally. The contents of LRU caches are very easy to predict. Having only distinct accesses and a strict least recently used replacement, directly yields the tight bound k, that is, the number of distinct accesses (hits or misses) needed to know the exact contents of a k-way cache set using LRU replacement is k [16]. The weak point of this policy is its requirement of time and power.

To reduce the cost of LRU policy, Random policy [2] with lower performance is used. In this policy, the victim line is chosen randomly from all the cache lines in the set. Another candidate policy is FIFO, which can also be seen as a queue: new elements are inserted at the front evicting elements at the end of the queue. In contrast to LRU, hits do not change the queue. Real implementations use a Round-Robin replacement counter for each set pointing to the cache line that will be replaced in the future. This counter is increased if an element is inserted into a set, and a hit does not change this counter. In the case of misses only, FIFO behaves as does LRU. Thus, it has the tight bound k [16].

An approximation to the LRU mechanism is PLRU [17, 18]. This policy speeds up operations and reduces the complexity of the implementation. One of the PLRU implementations is based on using the most recently used (MRU) bits. In this policy, each line is assigned an MRU-bit, stored in a tag table. Every access to a line sets its MRU-bit to 1, indicating that the line has been recently used. PLRU is deadlock-prone. Deadlock occurs if the MRU-bits for all blocks are set to 1 and therefore none of them is ready to be replaced.

In order to prevent this situation, whenever the last 0 bit of a set is set to 1, all other bits are reset to 0. At each cache miss, the line with lowest index (in our representation the leftmost line) whose MRU-bit is 0 is replaced. As an example, we represent the state of an MRU cache set as $[A, B, C, D]_{0101}$, where 0101 are the MRU-bits and $A, \ldots, D$ are the contents of the set. On this state, an access to E yields a cache miss and new state $[E, B, C, D]_{1101}$. Accessing D leaves the state unchanged. A hit on C forces a reset of the MRU-bits: $[E, B, C, D]_{0010}$.

Figure 32.1 illustrates the MRU policy with respect to the above scenario. For the MRU replacement policy, it is impossible to give a bound on the number of

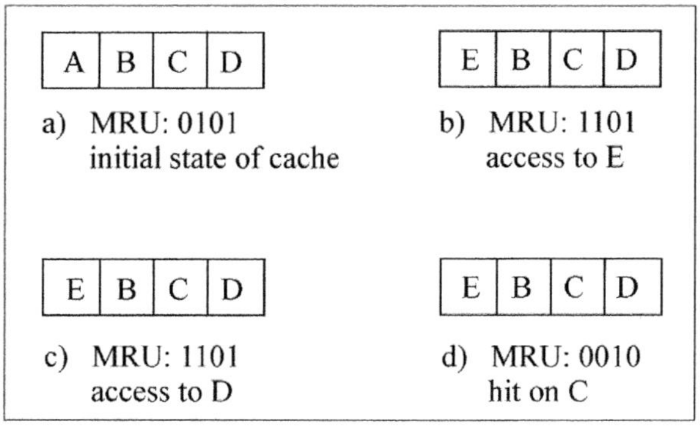

Fig. 32.1 An illustrative example of MRU replacement policy

Table 32.1 Bits used for each replacement policy

Replacement policy	Used bits
Random	Log2ways
FIFO	Nsets. Log2ways
LRU	Nsets.ways. Log2ways
MRU	Nsets.ways

accesses needed to reach a completely known cache state [16]. It seems that this method looks like LRU and FIFO; hence, we expect its performance to be better than FIFO and worse than LRU.

The LRU policy requires a number of bits to track when each cache line is accessed, whereas the MRU does not require as many bits. Indeed the MRU has less complexity than LRU. Table 32.1 shows the amount of bits used in each replacement policy [17].

Another contribution of our work is in studying two-level TLB in embedded systems and comparing its performance against single TLB. Many studies have pointed out the importance of the TLB [19–23]. Hardware TLB hierarchy and its impact on system performance are investigated in [11, 12, 20]. The advantages of multilevel TLBs over single TLBs are studied in [13, 24]. There are also real implementations of multilevel TLBs in commercial processors such as Hal's SPARC 64, IBM AS/400 Power PC, Intel Itanium, and AMD. They use either hardware or software mechanisms to update the TLB on the misses.

From the principal component analysis of raw data in [25] it is concluded that several SPEC CPU2000 benchmark programs such as *bzip, gzip, mcf, vortex, vpr, gcc, crafty, applu, mgrid, wupwise*, and *apsi* exhibit a temporal locality that is significantly worse than other benchmarks. Concerning spatial locality, most of these benchmarks exhibit a spatial locality that is relatively higher than that of the remaining benchmarks. The only exceptions are *gzip* and *bzip2* which exhibit poor spatial locality. As pointed out in [25], there is lots of redundancy in the benchmark suite. Simulating benchmarks with similar behavioral characteristics will add to the overall simulation time without providing any additional insight, so we have selected our benchmarks based on clustering results presented in [25].

We have also noticed the results presented in [26] to select our benchmarks. Some of SPEC CPU2000 benchmarks are eccentric, that is, have a behavior that differs significantly from the behavior of other benchmarks. Eccentric benchmarks are excellent candidates for case studies and it is important to include them when subsetting a benchmark suite (e.g., to limit simulation time). These benchmarks differ from the average SPEC CPU 2000 benchmark in different ways, for example, requiring high associativity or suffering from repetitive conflict misses, having low spatial locality, or benefiting from extremely large caches. For example, *crafty* has a lower cache miss rate when the block size is small. It is also somewhat more dependent on high associativity than other benchmarks. *equake* depends strongly on the degree of associativity and suffers from repetitive conflict misses. *vpr* is highly sensitive to changes in data cache associativity, having a large number of misses.

32.3 Experimental Setup

The simulator used in our study is *SimpleScalar*. Performance of two-level TLB with different cache replacement policies were evaluated using *Sim-Cache* and *Sim-Cheetah* simulators from the Alpha version of this toolset [14].

The *Sim-Cache* engine simulates associative caches with FIFO, Random, and LRU policies. The *Sim-Cheetah* engine simulates fully associative caches efficiently, as well as simulating a sometimes-optimal replacement policy. Belady [27] calls the latter policy MIN, whereas the simulator calls it OPT. Because OPT uses future knowledge to select a replacement, it cannot be really implemented. It is, however, useful as a yardstick for evaluation of other replacement policies. We modified the original simulator to support hierarchical two-level TLB and MRU replacement policy as well as their native replacement policies, that is, FIFO, Random, LRU, and OPT and its original single TLB.

As mentioned above, we have used selected benchmarks from the SPEC CPU2000 suite as a simulation workload. Given the eccentricity of some benchmarks, we filtered out benchmarks insensitive to increases in cache associativity. We selected *vpr, gcc, crafty, eon*, and *twolf* as five integer benchmarks and *mgrid, apsi, fma3d*, and *equake* as four floating-point benchmarks. Selected benchmarks were used with reference data inputs. Instructions related to initializations were skipped and the main instructions were simulated.

For each benchmark, we performed the simulation with various L1 data cache organizations with 2-, 4-, and 8-way associativity, replacement policies as FIFO, LRU, and MRU and various instruction and data cache sizes. In our experiments all TLBs are assumed to be fully associative with the LRU replacement policy. We changed the cache sizes with 4 KB, 8 KB and 16 KB. To study the impact of 2-level d-TLB (data TLB), we compared its miss rates with that of a single TLB of the same size ($32+96=128$ entries). Also in all experiments a cache line size was 32 bytes. Relating to the cache memory, we considered two scenarios: in the first scenario, the system consists of two separate data and instruction caches in the first level (L1D and L1I) and in the second scenario, it only has a unified cache in the first level (L1U) which serves both data and instruction references. Figure 32.2 shows the above two scenarios.

The defaults used in *Sim-Cache* were as follows: 8 KB L1 instruction cache and data cache, 256 KB L2 unified cache, i-TLB (instruction TLB) with 64 entry and d-TLB with 128 entries. In all memories, the default replacement policy was LRU.

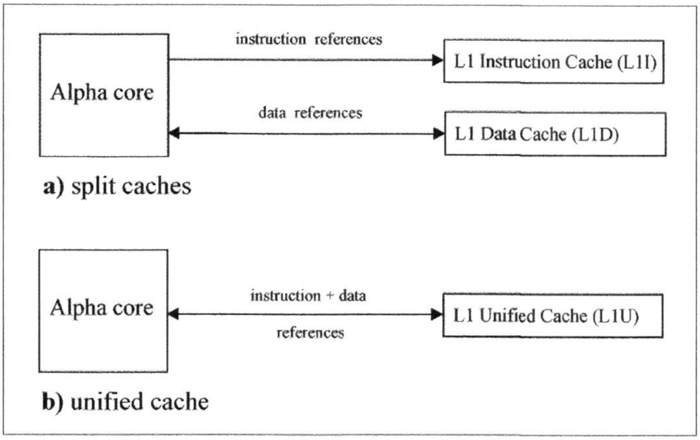

Fig. 32.2 Two scenarios for memory hierarchy

32.4 Results of Experiments

32.4.1 The Effect of Two-Level TLB on Overall Performance

Table 32.2 shows the results of using hierarchical two-level d-TLB and single level d-TLB. It shows the miss rates for each of the two levels as a percentage of the number of references to that level, as well as the overall miss rate. The overall miss rate is the percentage of references that do not find a translation in either of the two levels. The results show higher overall TLB miss rates when using two-level TLB, especially for *twolf, gcc, crafty*, and *apsi*. The average TLB miss rates for selected integer benchmarks for two-level and single TLBs are 1.91% and 1.48%, respectively, resulting in a degradation of about 0.43%. The miss rate degradation for selected floating-point benchmarks is on average 0.46%.

Despite the higher miss rates, the benefit of a two-level TLB is in reducing the access time of first level TLB and in avoiding accessing the second level. Figure 32.3 shows the normalized program execution times for the selected benchmarks. Here, the normalized execution time is the ratio of program execution time with a two-level TLB to its native execution time with single TLB. Except the two-level TLB, the other parts of simulation are the same as defaults used in the *SimpleScalar*. The two-level TLB is fully associative with LRU replacement, as is common in most commercial processors.

The average reductions in execution time, when using two-level TLB for integer and floating-point benchmarks, are about 0.32% and 0.60%, respectively. According to the results, using a 2-level TLB cannot produce a conspicuous reduction in execution time, and it degrades the overall miss rate.

Table 32.2 Miss rates of two-level TLB and single TLB

Benchmark	Miss rate of first level	Miss rate of second level	Overall miss rate of two-level TLB	Miss rate of single TLB
Twolf	3.32	44.6	1.48	1.27
Vpr	4.21	41.1	1.73	1.68
Gcc	2.19	64.84	1.42	1.13
Crafty	1.27	96.06	1.22	0.09
Eon	3.85	96.36	3.71	3.24
Mgrid	21.78	97.56	21.25	20.23
Apsi	2.60	88.77	2.30	1.98
fma3d	1.93	78.23	1.51	1.21
Equake	20.34	94.93	19.31	19.12

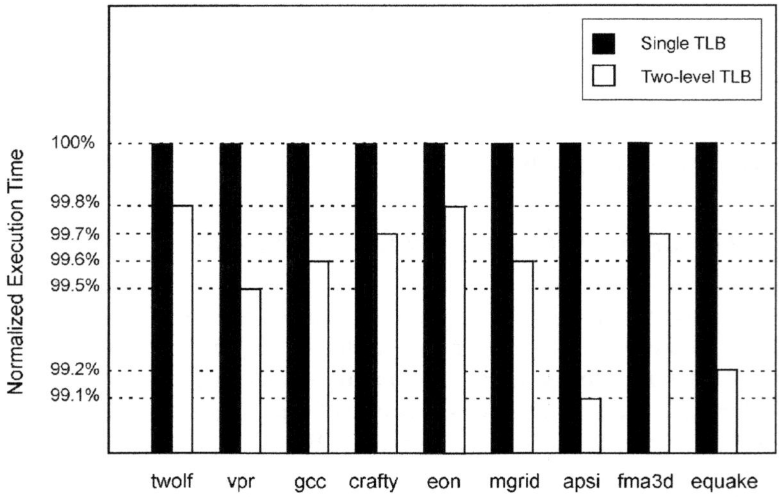

Fig. 32.3 Normalized execution time for each benchmark with two-level TLB and Single TLB

32.4.2 The Effect of Cache Associativity, Size, and Replacement Policy on Performance

Tables 32.3 and 32.4 show the average cache miss rates of integer and floating-point benchmarks for the L1 data cache. The miss rates for floating-point applications are lower than integer applications when the L1 data cache is 4 KB. For larger size L1 data caches of 8 KB and 16 KB miss rates of floating-point applications become higher. In the L1 data cache, it is hard to select a replacement policy between FIFO and Random policies, and the difference between them decreases as the cache size increases. There are applications where Random policy outperforms FIFO. The *crafty, fma3d*, and *equake* are three examples of such applications in our selected benchmarks. In other applications, FIFO has fewer misses against Random.

Table 32.3 Average L1 data cache miss rates for five SPEC CPU2000 integer applications (*vpr, gcc, crafty, eon*, and *twolf*)

		1W	2W	4W	8W	16W	32W
4 KB	FIFO	5.31	4.28	3.99	3.84	3.82	3.81
	LRU	5.31	3.99	3.57	3.40	3.29	3.29
	Random	5.31	4.36	4.09	3.98	3.92	3.92
	MRU	5.31	3.99	3.60	3.47	3.38	3.35
	OPT	5.31	3.62	3.57	2.36	2.23	2.18
8 KB	FIFO	4.12	2.94	2.65	2.53	2.49	2.45
	LRU	4.12	2.76	2.38	2.24	2.16	2.15
	Random	4.12	2.94	2.65	2.55	2.49	2.46
	MRU	4.12	2.76	2.38	2.25	2.18	2.16
	OPT	4.12	2.12	1.76	1.61	1.53	1.49
16 KB	FIFO	3.23	2.00	1.76	1.64	1.59	1.58
	LRU	3.23	1.89	1.60	1.46	1.40	1.38
	Random	3.23	2.00	1.78	1.75	1.61	1.60
	MRU	3.23	1.89	1.61	1.49	1.46	1.38
	OPT	3.23	1.89	1.28	1.17	1.12	1.10

Table 32.4 Average L1 data cache miss rates of four SPEC CPU2000 floating-point applications (*mgrid, apsi, fma3d*, and *equake*)

		1W	2W	4W	8W	16W	32W
4 KB	FIFO	5.73	4.63	4.18	3.29	3.28	3.28
	LRU	5.73	4.47	3.98	3.29	3.03	3.01
	Random	5.73	4.69	4.53	3.89	3.54	3.52
	MRU	5.73	4.53	3.99	3.29	3.03	3.03
	OPT	5.73	3.30	2.82	2.45	2.20	2.14
8 KB	FIFO	3.59	2.97	2.70	2.56	2.46	2.46
	LRU	3.59	2.80	2.45	2.30	2.26	2.25
	Random	3.59	3.13	2.92	2.80	2.55	2.46
	MRU	3.59	2.88	2.53	2.41	2.33	2.29
	OPT	3.59	2.39	2.02	1.78	1.64	1.61
16 KB	FIFO	2.20	1.93	1.84	1.83	1.80	1.74
	LRU	2.20	1.87	1.75	1.68	1.65	1.65
	Random	2.20	2.09	2.04	2.04	2.03	1.97
	MRU	2.20	1.92	1.76	1.75	1.73	1.72
	OPT	2.20	1.71	1.59	1.55	1.54	1.54

In general, we can conclude that for the rest of the benchmarks, for larger cache sizes, Random policy dominates, whereas for smaller cache sizes, FIFO dominates.

Experiments show that LRU policy is almost better than FIFO and Random, but there are some exceptions. For example, Random policy is sometimes better than LRU for *equake* and *apsi*. Compared to LRU policy, FIFO is on average about 17% worse, whereas Random is about 18% worse.

Compared to LRU policy, the performance degradation of MRU is relatively small; however, because of low complexity of MRU, we can neglect its degradation.

The gap between MRU, the best realistic replacement used in this chapter, and OPT is larger for smaller caches due to more conflict misses. The miss rate reduction is more distinct, as the size of the L1 data cache decreases. The results show that the largest reduction in miss rate is for transition from a direct-mapped to a two-way set associative L1 data cache.

For both integer and floating-point benchmarks, increased associativity has a large miss reduction with small caches. In floating-point applications, for large caches (16 KB), associativity higher than two does not efficiently reduce the miss rate, but for small caches (4 KB), the amount of miss reduction is noticeable. For L1 data cache sizes of 8 KB and 16 KB, the effect of increased associativity on miss reduction is more obvious for integer applications, than for floating-point applications.

As mentioned earlier, OPT replacement policy can be used as a yardstick to evaluate the replacement policies. From Tables 32.3 and 32.4 it can be deduced that the OPT miss rate of a certain cache size is roughly close to the MRU miss rate of a cache twice as big, with the same number of ways. For example, in Table 32.4, the miss rate of an eight-way set associative cache with size 16 KB and MRU replacement policy is 1.75%. This is approximately equal to the (1.78%) miss rate of an eight-way set associative, 8 KB, optimal replacement policy. This shows that there is still a large gap between OPT and realistic policies such as MRU. Therefore, if near optimal policy can be found in practice, the size of caches can be reduced to one-half.

32.5 The Effect of Split Cache Versus Unified Cache on Performance

Tables 32.5 and 32.6 show the results of using split data and instruction caches as the first-level caches against the common unified cache as the first-level cache. For any of the k-way set associative unified caches in the table, the miss rate is considerably higher compared to the aggregated miss rate of corresponding split instruction and data caches of equivalent size. The difference becomes smaller as the size of instruction and data caches increases.

Tables 32.5 and 32.6 show that for integer applications, a four-way set associative cache, on average across various cache sizes, reduces the miss rate about 12% for L1U, 12% for L1D, and 8% for L1I when compared to a two-way set associative cache. The reduction of four-way over two-way for floating-point applications for L1U, L1D, and L1I is 10%, 13%, and 9%, respectively. For integer applications the benefit of an eight-way organization, compared to two-way set associative, is 13% for L1U, 16% for L1D, and 12% for L1I, whereas for floating-point applications the benefit of eight-way over two-way is 15%, 20%, and 12% for L1U, L1D, and L1I, respectively. The results show that the gain of increasing associativity in the case of data and unified caches is more than instruction caches.

As expected, LRU and MRU replacement policies perform better than Random for data caches, although surprisingly, Random policy performs almost better than

Table 32.5 Average cache miss rates for L1I, L1D, and L1U caches for selected SPEC CPU2000 integer applications

		L1I of Size *i* KB			L1D of Size *i* KB			L1U of Size 2*i* KB		
		2W	4W	8W	2W	4W	8W	2W	4W	8W
$i=4$	FIFO	5.52	5.55	5.61	4.28	3.99	3.84	15.30	13.98	13.82
	LRU	5.48	5.50	5.57	3.99	3.57	3.40	14.99	13.55	13.37
	Random	5.38	5.34	5.36	4.36	4.09	3.98	15.35	14.00	13.94
	MRU	5.48	5.31	5.32	3.99	3.60	3.47	14.99	13.58	13.42
	OPT	4.36	3.90	3.70	3.62	3.57	2.36	14.22	13.17	12.20
$i=8$	FIFO	4.29	3.97	4.01	2.94	2.65	2.53	10.91	9.59	9.29
	LRU	4.25	3.90	3.91	2.76	2.38	2.24	10.74	9.31	9.23
	Random	4.01	3.61	3.47	2.94	2.65	2.55	10.91	9.62	9.50
	MRU	4.24	3.71	3.54	2.76	2.38	2.25	10.74	9.32	9.27
	OPT	3.11	2.45	2.21	2.12	1.76	1.61	9.10	7.71	7.63
$i=16$	FIFO	2.75	2.75	2.60	2.00	1.76	1.64	7.02	5.62	5.61
	LRU	2.70	2.71	2.57	1.89	1.60	1.46	6.85	5.51	5.43
	Random	2.45	2.16	1.91	2.00	1.78	1.75	7.05	5.76	5.75
	MRU	2.71	2.24	2.21	1.89	1.61	1.49	6.87	5.60	5.48
	OPT	1.78	1.36	1.20	1.89	1.28	1.17	6.40	5.27	5.19

Table 32.6 Average cache miss rates for L1I, L1D, and L1U caches for selected SPEC CPU2000 floating-point applications

		L1I of size *i* KB			L1D of size *i* KB			L1U of size 2*i* KB		
		2W	4W	8W	2W	4W	8W	2W	4W	8W
$i=4$	FIFO	6.82	6.86	6.92	4.63	4.18	3.29	17.07	16.11	15.65
	LRU	6.78	6.80	6.87	4.47	3.98	3.29	17.01	15.83	15.22
	Random	6.67	6.65	6.68	4.69	4.53	3.89	17.91	16.20	15.73
	MRU	6.78	6.61	6.62	4.53	3.99	3.29	17.05	15.91	15.30
	OPT	5.69	5.20	5.03	3.30	2.82	2.45	15.00	13.11	12.51
$i=8$	FIFO	5.58	5.27	5.30	2.97	2.70	2.56	14.85	13.55	13.27
	LRU	5.53	5.24	5.26	2.80	2.45	2.30	14.11	12.80	12.50
	Random	5.31	4.89	4.76	3.13	2.92	2.80	14.97	13.61	13.30
	MRU	5.58	5.00	4.81	2.88	2.53	2.41	14.20	12.90	12.60
	OPT	4.39	3.69	3.55	2.39	2.02	1.78	12.91	9.03	8.91
$i=16$	FIFO	3.74	3.74	3.66	1.93	1.84	1.83	10.80	8.97	8.91
	LRU	3.69	3.74	3.55	1.87	1.75	1.68	10.21	8.91	8.80
	Random	3.34	3.04	2.87	2.09	2.04	2.04	10.98	8.99	8.94
	MRU	3.69	3.26	3.25	1.92	1.76	1.75	10.29	8.92	8.85
	OPT	2.75	2.34	2.14	1.71	1.59	1.55	9.23	7.50	7.46

LRU and MRU policies for instruction caches. The temporal locality of the instruction cache is low, compared to that of the data cache. Thus, a rich replacement policy such as LRU has approximately equal performance compared to Random policy, which is a poor replacement policy.

32.6 Conclusions

The organization of cache and TLB memory is a critical issue in general-purpose embedded systems. This chapter presented a simulation-based study of the performance evaluation to find the main cache design issues such as hierarchical TLB and cache, cache size and associativity, and replacement policy in embedded processors. We selected some applications from the SPEC CPU2000 benchmark suite based on eccentricity and clustering concepts and found that the two-level TLB would not produce a significant reduction in execution time, while degrading the overall miss rate.

The experimental results showed that the gain of increasing associativity in the case of data and unified caches is more than instruction caches. In the L1 data cache and L1 unified cache, the largest miss rate reduction occurs for transition from a direct-mapped to a two-way set associative. In floating-point applications, for large caches, associativity higher than two, does not effectively reduce the miss rate, but for small caches, the amount of reduction in miss rate is noticeable.

As expected, LRU and MRU replacement policies perform better than Random for data caches, and surprisingly, for instruction caches, Random policy performs almost better than LRU and MRU. Comparing FIFO and random policies, for larger data cache sizes, Random policy dominates, whereas for smaller cache sizes, FIFO dominates. Nevertheless, in general for the L1 data cache, it is hard to select a winner replacement policy between FIFO and Random policies, and the difference between them decreases as the cache size increases.

For large caches, the MRU policy is a good approximation of LRU policy. Compared to LRU policy, MRU has less complexity and according to our results has negligible miss rate degradation.

The results of experiments also illustrated that the performance of OPT policy is nearly the same as the performance of the lower-cost MRU policy using a cache twice as big. This shows that there is still a large gap between optimal replacement policy and realistic replacement policies such as MRU. Eliminating this gap will reduce the size of caches even to one-half of their current sizes.

With respect to the attempts of memory designers to reduce the amount of power, our results offer valuable insights into the design of memory for embedded systems.

References

1. Kalavade A, Knoblock J, Micca E, Moturi M, Nicol CJ, O'Neill JH, Othmer J, Sackinger E, Singh KJ, Sweet J, Terman CJ, Williams J (2000). A single-chip, 1.6 billion, 16-b MAC/s multiprocessor DSP. *IEEE Journal of Solid-State Circuits*, 35(3), pp. 412–423.
2. Hennessy JL, Patterson D (2003). *Computer Architecture: A Quantitative Approach*. Third Edition, San Mateo. CA: Morgan Kaufmann.
3. Intel Xscale^{TM} (2000). *Core: Developer's Manual*, December (2000). URL: http://developer.intel.com.

4. Intel Pentium 4 and Intel Xeon Processor Optimization: Reference Manual^{TM}. *Reference Manual*. URL: http://developer.intel.com.
5. Cantin JF, Hill MD (2000). Cache performance of the SPEC CPU2000 benchmarks. URL: http://www.cs.wisc.edu/multifacet/misc/spec2000cachedata/.
6. Sair S, Chamey M (2000). Memory behavior of the SPEC2000 benchmark suite. IBM Thomas J. Waston Research Center, Technical Report RC-21852.
7. Thomock NC, Flangan JK (2000). Using the BACH trace collection mechanism to characterize the SPEC 2000 integer benchmarks. *Workshop on Workload Characterization*.
8. Wang Z, McKinley K, Rosenberg A, Weems C (2002). Using the compiler to improve cache replacement decisions. *The International Conference on Parallel Architectures and Compilation Techniques*, Charlottesville, Virginia.
9. Wong W, Baer JL (2000). Modified LRU policies for improving second-level cache behavior. *The 6th International Symposium on High-Performance Computer Architecture*, Toulouse, France.
10. Jacob BL, Mudge TN (1998). A look at several memory management units: TLB-refill mechanisms, and page table organizations. *Proceedings of the Eight International Conference on Architectural Support for Programming Languages and Operating Systems*, pp 295–306.
11. Talluri M (1995). Use of superpages and subblocking in the address translation hierarchy. PhD thesis, Deptartment of CS, Univiversity of Wisconsin at Madison.
12. Nagle D, Uhlig R, Stanley T, Sechrest S, Mudge T, Brown R (1993). Design tradeoffs for software managed TLBs. *Proceedings of the 20th Annual International Symposium on Computer Architecture*, pp 27–38.
13. Chen JB, Borg A, Jouppi NP (1992). A simulation based study of TLB performance. *Proceedings of the 19th Annual International Symposium on Computer Architecture*, pp 114–123.
14. Burger D, Austin T (1997). The SimpleScalar tool set, version 2.0. Technical Report #1342, Computer Sciences Department, University of Wisconsin, Madison, WI.
15. Henning JL (2000). SPEC CPU2000: Measuring CPU performance in the new millennium. *IEEE Computer*, 33(7), pp 28–35.
16. Reineke J, Grund D, Berg C, Wilhelm R (2006). Predictability of cache replacement policies. AVACS Technical Report No. 9, SFB/TR 14 AVACS.
17. Malamy A, Patel R, Hayes N (1994). Methods and Apparatus for Implementing a Pseudo-LRU Cache Memory Replacement Scheme with a Locking Feature. United States Patent 5353425.
18. So K, Rechtshaffen RN (1988). Cache operations by MRU change. *IEEE Transaction on Computers*, 37(6), pp 700–707.
19. Anderson TE, Levy HM, Bershad BN, Lazowska ED (1991). The interaction of architecture and operating system design. *Proceedings of the Fourth International Conference on Architectural Support for Programming Languages and Operating Systems*, Santa Clara, CA, pp 108–120.
20. Jacob B, Mudge T (1998). Virtual memory in contemporary microprocessors. *IEEE Micro*, 18(4), pp 60–75.
21. Clark DW, Emer JS (1985). Performance of the VAX- 1/780 translation buffers: Simulation and measurement. *ACM Transactions on Computer Systems*, 3(1).
22. Huck J, Hays J (1993). Architectural support for translation table management in large address space machines. *Proceedings of the 20th Annual International Symposium on Computer Architecture*, pp 39–50.
23. Rosenblum M, Bugnion E, Devine S, Herrod S (1997). Using the SimOS machine simulator to study complex computer systems. *ACM Transactions on Modeling and Computer Simulation*, 7(1), pp 78–103.
24. Austin TM, Sohi GS (1996). High bandwidth address translation for multiple issue processors. *The 23rd Annual International Symposium on Computer Architecture*.
25. Phansalkar A, Joshi A, Eeckhout L, John K (2004). Four generations of SPEC CPU benchmarks: What has changed and what has not. Technical Report TR-041026-01-1.
26. Vandierendonck H, De Bosschere K (2004). Eccentric and fragile benchmarks. *2004 IEEE International Symposium on Performance Analysis of Systems and Software*, pp 2–11.
27. Belady LA (1966). A study of replacement algorithms for a virtual-storage computer. *IBM Systems Journal*, 5(2), pp 78–101.

Chapter 33
SimDiv: A New Solution for Protein Comparison

Hassan Sayyadi, Sara Salehi, and Mohammad Ghodsi

33.1 Introduction

The number of known proteins is increasing every day; tens of thousands have been studied and categorized by now.

To understand the functions and behaviors of a newly found protein, one should find well-studied proteins with similar structure. In fact, the behavior of a protein is related to its sequence of amino acids and its 3D structure. So the comparison of proteins is a key technique not only in finding similarities in the structures of proteins but also to categorize them and define families and superfamilies among the proteins. As with many comparison problems, this problem is hard because there is neither an exact definition of the likelihood of proteins' structures nor an efficient algorithm that exists for it. Although there exist optimal dynamic programming algorithms for comparing the sequences of amino acids, the result is highly related to the definition of the relations of the sequences, which have not been uniquely defined [1]. The problem of comparing the 3D structures of proteins becomes even harder. There is no efficient algorithm that guarantees the optimality of the answer. In fact, this problem is NP-hard. When the proteins become more complicated, the relationship models are more varied than the models of sequence relatedness.

In this chapter, we propose a model for protein matching or extracting similar parts of two given proteins. We focus on the computational geometric approach and the graph matching method that are used to model and compare the sequence and 3D structure of proteins.

The remainder of this chapter is organized as follows. We first have a glance at the related works. There are two major methods used in the literature: Delaunay tetrahedralization and similarity flooding. We explain the required information in the next section as background knowledge, and then propose a new idea in Sect. 33.4 which can improve the current methods. We then present experimental results of the implemented method which show its effectiveness.

Oscar Castillo et al. (eds.), *Trends in Intelligent Systems and Computer Engineering.*

33.2 Related Works

Delaunay triangulation and Delaunay tessellation are common computational geometric methods used in bioinformatics. For example, in [2] the Delaunay tessellation of the α-carbons of the protein molecule is used to study the HIV-1 protease. This is because this model provides an objective and robust definition of four nearest-neighbor amino acid residues as well as a four-body statistical potential function. The other usages are studied in the fields of packing analysis [2, 3], fold recognition [4], virtual mutagenesis [5], and structure comparison.

Authors of [6] consider the Delaunay tetrahedralization determined by the alpha carbon positions of some particular protein. Starting at the amino-terminal residue, the edges of the tetrahedralization that connect to a residue that has already been encountered are recorded as a relative residue difference. For example, if there is an edge between the fifth alpha carbon and the third one, this edge is represented as 2. When the edge of a particular residue is exhausted a 0 is recorded to indicate a new residue. This linear representation will contain each edge in the tetrahedralization exactly once. Furthermore, secondary structural components will be indicated by particular subsequences. Two one-dimensional representations are then compared by a dynamic programming scheme adapted from protein sequence analysis, thus reducing protein structural similarity to sequence similarity of the appropriate structure strings. In Bostick et al. [7] the Euclidean metric for identifying natural nearest neighboring residues via the Delaunay tessellation in Cartesian space and the distance between residues in sequence space is given. In addition, the authors of [8] find recurring amino acid residue packing patterns, or spatial motifs, that are characteristic of protein structural families, by applying a novel frequent subgraph mining algorithm to graph representations of protein three-dimensional structure. Graph nodes represent amino acids, and edges are chosen in one of three ways: first, using a threshold for contact distance between residues; second, using Delaunay tessellation; and third, using the recently developed almost-Delaunay edges. Furthermore, Dafas et al. [9] present a solution that reduces the computation of PSIMAP, a protein interaction map derived from protein structures in the protein databank PDB and the structural classification of proteins SCOP [10] to know about interaction of two proteins. The original PSIMAP algorithm computes all pairwise atom/residue distances for each domain pair of each multidomain PDB entry. But they developed an effective new algorithm, which substantially pruned the search space. The basic idea of their novel algorithm is to prune the search space by applying a bounding shape to the domains. Interacting atoms of two domains can only be found in the intersection of the bounding shapes of the two domains. Generally, the proposed algorithm steps are:

1. A convex hull for each of the two domains is computed.
2. Both convex hulls are swelled by the required contact distance threshold.
3. The intersection of the two transformed convex hulls is computed.

4. All residue/atom pairs outside the intersection are discarded and for the remaining residues/atoms the number of residue/atom pairs within the distance threshold is computed. If this number exceeds the number threshold the two domains are said to interact.

33.3 Background Knowledge

33.3.1 Delaunay Tetrahedralization

Delaunay tetrahedralization is a special type of tetrahedralization which is defined based on the Voronoi diagram through the principle of duality [11]. A Voronoi box is formed through the intersection of planes and is therefore a general irregular polyhedron (Fig. 33.1). The facets of the Voronoi boxes correspond in the dual graph to the Delaunay edges which connect the points of P.

- *Voronoi*: Let $P = \{p_1, \dots, p_k\}$ be a finite set of points in the n-dimensional space R^n and their location vectors $\mathbf{x}_i \neq \mathbf{x}_j \ \forall\, i \neq j$. The region given by

$$V(p_i) = \{\mathbf{x} \mid |\mathbf{x} - \mathbf{x}_i| \leq |\mathbf{x} - \mathbf{x}_j| \ \forall\, j \neq i\}$$

  is called the Voronoi region (Voronoi box) associated with p_i and $\mathcal{V}(P) = \bigcup_{i=1}^{k} V(p_i)$ is said to be the Voronoi diagram of P.

A Voronoi box is formed through the intersection of planes and is therefore a general irregular polyhedron. The facets of the Voronoi boxes correspond in the dual graph to the Delaunay edges which connect the points of P.

- *Delaunay Edge*: Let P be a finite set of points in a subdomain Ω^n of the n-dimensional space R^n. Two points p_i and p_j are connected by a Delaunay edge e if and only if there exists a location $x \in \Omega^n$ which is equally close to p_i and p_j

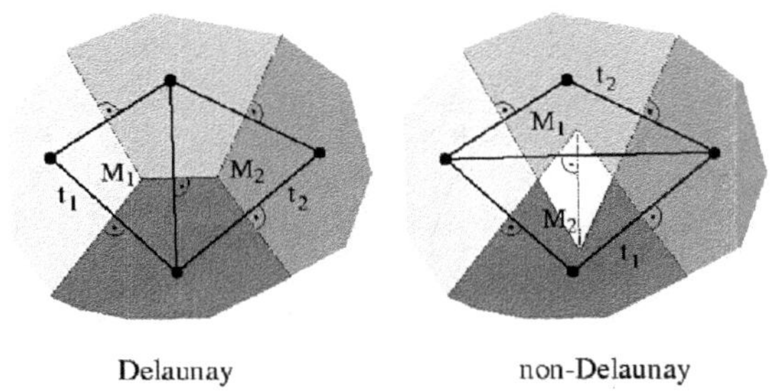

Fig. 33.1 Each Voronoi box associated with a point is differently shaded. Two triangles t_1 and t_2 with their circumcenters M_1 and M_2 which are the vertices of the Voronoi boxes are depicted for the correct Delaunay case and for the non-Delaunay case. Incorrect Voronoi boxes derived from non-Delaunay triangles overlap [11]

and closer to p_i, p_j than to any other $p_k \in P$. The location x is the center of an n-dimensional sphere which passes through the points p_i, p_j and which contains no other points p_k of P.

$$\begin{aligned} e_{Delaunay}(p_i, p_j) \Leftrightarrow\ & \exists x:\ x \in \Omega^n \\ & \wedge\ |x - p_i| = |x - p_j| \\ & \wedge\ \forall k \neq i, j:\ |x - p_i| < |x - p_k| \end{aligned}$$

Combining this criterion for the three edges of a triangle (Fig. 33.2) and furthermore for the four triangles of a tetrahedron leads to the following criterion for Delaunay tetrahedron.

- *Delaunay Triangle*: Let P be a finite set of points in a subdomain Ω^n of the n-dimensional space R^n. Three noncollinear points p_i, p_j, and p_k form a Delaunay triangle t if and only if there exists a location $x \in \Omega^n$ which is equally close to p_i, p_j, and p_k and closer to p_i, p_j, p_k than to any other $p_m \in P$. The location x is the center of an n-dimensional sphere which passes through the points p_i, p_j, p_k and which contains no other points p_m of P. For $n = 2$ only one such sphere exists which is the circumcircle of t.

$$\begin{aligned} t_{Delaunay}(p_i, p_j, p_k) \Leftrightarrow\ & \exists x:\ x \in \Omega^n \\ & \wedge\ |x - p_i| = |x - p_j| = |x - p_k| \\ & \wedge\ \forall m \neq i, j, k:\ |x - p_i| < |x - p_m| \end{aligned}$$

  implies that an empty circumcircle is necessary but not sufficient for Delaunay surface triangles in three dimensions. This is the reason why a two-dimensional Delaunay Triangulation code is of limited use to construct a three-dimensional Delaunay surface triangulation. The Delaunay edge and Delaunay triangle criteria are depicted in Fig. 33.2. A Delaunay tetrahedron corresponds to a point in the Voronoi diagram, which is the vertex of four incident Voronoi boxes.
- *Delaunay Tetrahedron*: Let P be a finite set of points in a subdomain Ω^n of the n-dimensional space R^n, where $n \geq 3$. Four noncoplanar points p_i, p_j, p_k, and p_l form a Delaunay tetrahedron T if and only if there exists a location $x \in \Omega^n$ which

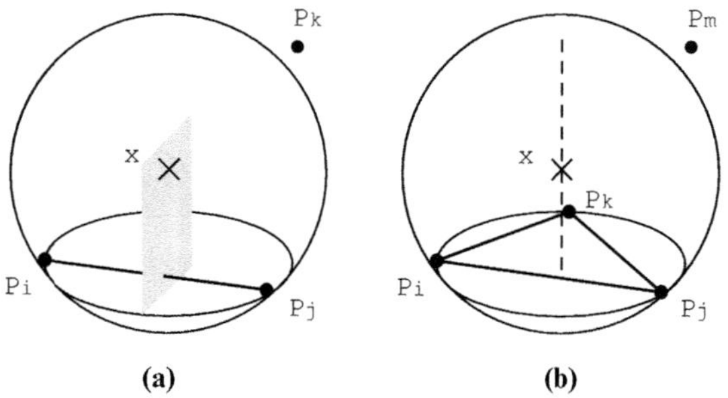

Fig. 33.2 **a** Delaunay edge; **b** Delaunay triangle criteria [11]

is equally close to p_i, p_j, p_k, and p_l and closer to p_i, p_j, p_k, p_l than to any other $p_m \in P$. The location x is the center of an n-dimensional sphere which passes through the points p_i, p_j, p_k, p_l and which contains no other points p_m of P. For $n = 3$ only one such sphere exists which is the circumsphere of T.

$$
\begin{aligned}
T_{Delaunay}(p_i, p_j, p_k, p_l) \Leftrightarrow \exists x : \; & x \in \Omega^n \\
& \wedge \; |x - p_i| = |x - p_j| = |x - p_k| = |x - p_l| \\
& \wedge \; \forall m \neq i, j, k, l \; : \; |x - p_i| < |x - p_m|
\end{aligned}
$$

A Delaunay tetrahedron must consist of Delaunay edges and Delaunay triangles. The edge and triangle criteria are implicit, because the existence of the n-dimensional sphere in the Delaunay edge criterion and in the Delaunay triangle criterion is guaranteed by the sphere in the Delaunay tetrahedron criterion.

33.3.2 Similarity Flooding

Matching or finding similar elements of two data schemas or two data instances play a key role in data warehousing, e-business, or even biochemical applications. The authors of [12] present a matching algorithm named "similarity flooding" based on a fixpoint computation that is usable across different scenarios. As in the example illustrating the similarity flooding algorithm shown in Fig. 33.3, the algorithm takes two graphs (schemas, catalogs, or other data structures) as input, and produces as output a mapping between corresponding nodes of the graphs.

As a first step, the schemas should be translated from their native format into graphs G_1 and G_2. Next, the pairwise connectivity graph (PCG) should be made that is an auxiliary data structure derived from G_1 and G_2. If N_1 represents the set of all nodes in G_1 and, respectively, N_2, each node in the connectivity graph is an element from $N_1 \times N_2$ and is called a "map-pair". Furthermore, edges in connectivity graph are defined as:

$$
\begin{gathered}
((x_1, y_1), P, (x_2, y_2)) \; \in \; PCG(G_2, G_2) \\
\Leftrightarrow \\
(x_1, P, x_2) \in G_1 \; \text{ and } \; (y_1, P, y_2) \in G_2
\end{gathered}
$$

Each map-pair contains one node from each graph and the similarity score between them. The initial similarity for each map-pair is obtained using a simple string matcher that compares common prefixes and suffixes of literals in each node. Finally, computing the similarities relies on the intuition that elements of two distinct models are similar when their adjacent elements are similar. In other words, a part of the similarity of two elements propagates to their respective neighbors as follows.

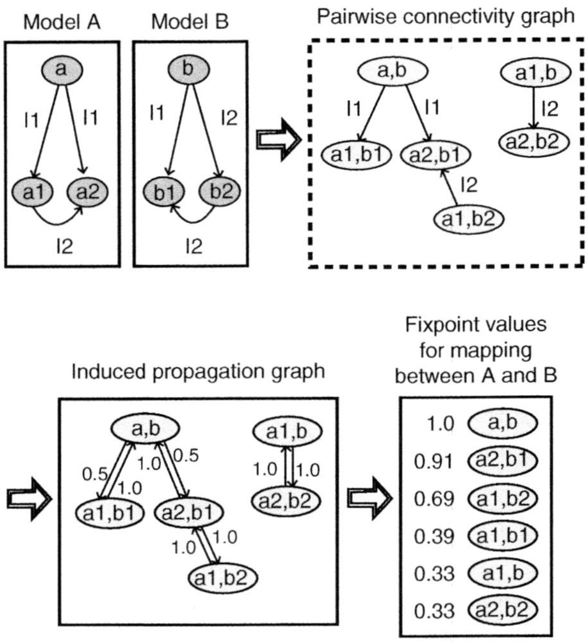

Fig. 33.3 Example illustrating the similarity flooding algorithm [12]

$$\sigma^{k+1}(x,y) = \sigma^k(x,y) + \sum_{(a_i,x)\in G_1,(b_i,y)\in G_2} \sigma^k(a_i,b_i).W((a_i,b_i),(x,y))$$
$$+ \sum_{(x,a_i)\in G_1,(y,b_i)\in G_2} \sigma^k(a_i,b_i).W((x,y),(a_i,b_i))$$

where $\sigma^k(x,y)$ shows the similarity between x and y after iteration k and $W((a_i,b_i),(x,y))$ is the propagation weight of the similarity between a_i and b_i to the similarity between x and y. The above computation is performed iteratively until the Euclidean length of the residual vector $\triangle(\sigma^n,\sigma^{n-1})$ becomes less than ϵ for some $n > 0$. If the computation does not converge, it will be terminated after some maximal number of iterations.

33.4 Proposed Method

In this section we present the proposed approach. Here we combine sequence similarity which is a simple extension of amino acid or nucleotide similarity and structural similarity which is the residue's position similarity. Using both techniques leads to an efficient method for extracting similar parts of proteins.

The different phases of the proposed method may be represented as follows.

1. Protein tetrahedralization
2. Creating a pairwise graph
3. Similarity propagation
4. Extracting similar components

We discuss each phase in the following sections.

33.4.1 Protein Tetrahedralization

Each protein is a sequence of residues in the 3D space in which each two consequent residues are connected by one edge called a "chain-edge". Firstly, for each protein, we use the Delaunay tetrahedralization algorithm to convert the protein sequence to a tetrahedralized shape (Fig. 33.4b). Because all proteins have 3D shape, using the Delaunay algorithm leads to create edges called a "tetrahedralization-edge" between atoms which are close to each other in space, regardless of their distance in the protein sequence. This closeness has an extremely high influence on structural similarity which is discussed in the proposed method for extracting similar parts of proteins.

Inasmuch as the tetrahedralization algorithm creates convex shapes, in order to have a much more similar shape to the real protein shape, we need to eliminate edges whose lengths are more than α for tetrahedralization edges and more than β for chain edges. Obviously, the value of β is more than α, because of the importance of chain edges in protein comparison (Fig. 33.4c).

We now construct a graph from the tetrahedralized shape. Each node in this graph contains one number which is the amino acid number of the corresponding atom in the protein and coordinates (x, y, z) expressing the coordination of that atom. This graph has two different types of edges:

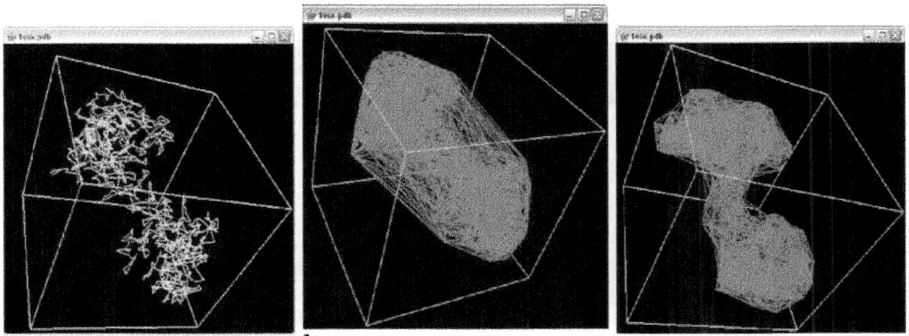

a b c

Fig. 33.4 **a** Protein chain; **b** tetrahedralized protein; **c** tetrahedralized protein after removing worthless edges

1. Edges which belong to the protein chain and also can be the part of the tetrahedrons named chain edges
2. Edges obtained from tetrahedralization which do not belong to the protein chain named tetrahedralization edges

Consequently, each protein is converted to one graph which not only contains the protein chain but also contains edges connecting atoms near each other in 3D space. These graphs are data structures for similarity flooding algorithm used in protein matching.

33.4.2 Creating Pairwise Graph

A pairwise connectivity graph (PCG) arises from two proteins' graphs P_1 and P_2 which were created through tetrahedralization in the first phase. If N_1 and N_2 show the sets of all nodes in P_1 and P_2, each node in the pairwise graph is an element from $N_1 \times N_2$. We call such nodes map-pairs. The edges of a pairwise graph are categorized in three parts depending on their map-pairs (see Fig. 33.5).

1. If a chain-edge exists between the first nodes of two map-pairs and there is a chain-edge between the second nodes of those map-pairs in their proteins, then we will connect these two map-pairs with an edge of type chain.

$$((x,y), CH, (x',y')) \in PCG(P_1,P_2) \Leftrightarrow$$
$$(x, CH, x') \in P_1 \text{ and } (y, CH, y') \in P_2$$

where CH represents the edge of type chain.

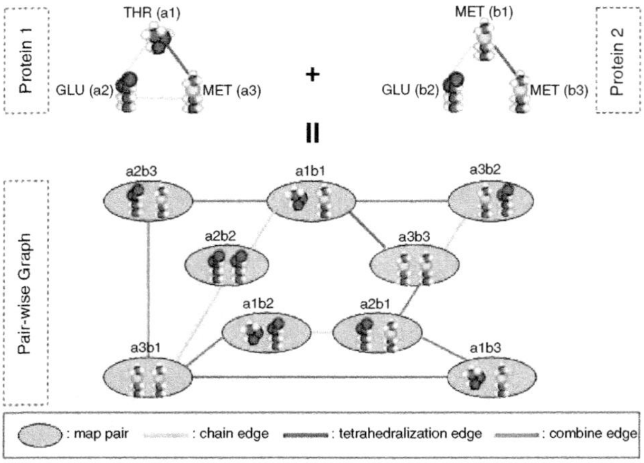

Fig. 33.5 Pairwise connectivity graph for proteins

2. If a tetrahedralization-edge exists between the first nodes of two map-pairs and there is a tetrahedralization-edge between the second nodes of those map-pairs in their proteins, then we will connect these two map-pairs tetrahedralization-edge.

$$((x,y),T,(x',y')) \in PCG(P_1,P_2) \Leftrightarrow$$
$$(x,T,x') \in P_1 \text{ and } (y,T,y') \in P_2$$

where T represents the edge of type tetrahedralization.
3. If a tetrahedralization-edge exists between the first nodes of two map-pairs and there is a chain-edge between the second nodes of those map-pairs in their proteins or vice versa, then we will connect these two map-pairs with a combined type edge.

$$((x,y),C,(x',y')) \in PCG(P_1,P_2) \Leftrightarrow$$
$$(x,T,x') \in P_1 \text{ and } (y,CH,y') \in P_2$$
$$\text{or}$$
$$(x,CH,x') \in P_1 \text{ and } (y,T,y') \in P_2$$

where C represents the combind type edge.

We categorized these edges to three types, because the influence of their nodes in similarity propagation weights in the proposed method will differ from each other.

33.4.3 Similarity Propagation

In the created pairwise graph, the primary similarity of each map-pair depends on the similarity between two nodes of that map-pair which are the atoms of the proteins. This similarity derives from the amino acids scoring matrix. The two-dimensional matrix contains all possible pairwise amino acid scores. Scoring matrices are also called substitution matrices because the scores represent relative rates of evolutionary substitutions. According to the similarity flooding algorithm the similarity of a map-pair increment is based on the similarities of its neighbors in the pairwise graph. Hence, the similarities of neighbors are effective in calculating the final similarity between two atoms in each map-pair. It seems more rational for neighbors related to chain-edges to be much more affective in protein matching. Similarly, the weight of neighbors related to combine edges is more than those of tetrahedralization-edges. Thus, over a number of iterations, the initial similarity of any two nodes propagates through the graphs. Similarity propagation in each iteration is computed as follows.

$$
\begin{aligned}
Sim^{i+1}(x,y) &= a * Sim^{i}(x,y) \\
&\quad + (1-a) * NeighborAffect(x,y) \\
NeighborAffect(x,y) &= CHF * CHAffect^{i}/NF \\
&\quad + CF * CAffect^{i}/NF \\
&\quad + TF * TAffect^{i}/NF
\end{aligned}
$$

In the above equation, $Sim^{i}(x,y)$ is defined as the similarity between x and y in each map-pair after i number of iteration(s), and a is the learning rate from the neighbors. Moreover,

- *CHAffect* is the average similarity of neighbors connecting with the chain-edge to the respective map-pair and is calculated as

$$
CHAffect^{i}(x,y) = \sum_{\substack{((x,y),CH,(x_i,y_i)) \\ \in PCG(P_1,P_2)}} \frac{Sim^{i}(x_i,y_i)}{CHSize}
$$

  and *CHF* is the propagation weight of *CHAffect*.
- *CAffect* is the average similarity of neighbors connecting with the combined edge to the respective map-pair and is defined as

$$
CAffect^{i}(x,y) = \sum_{\substack{((x,y),C,(x_i,y_i)) \\ \in PCG(P_1,P_2)}} \frac{Sim^{i}(x_i,y_i)}{CSize}
$$

  and *CF* is the propagation weight of *CAffect*.
- *TAffect* is the average similarity of neighbors connecting with the tetrahedralization edge to the respective map-pair and is computed as

$$
TAffect^{i}(x,y) = \sum_{\substack{((x,y),T,(x_i,y_i)) \\ \in PCG(P_1,P_2)}} \frac{Sim^{i}(x_i,y_i)}{TSize}
$$

  and *TF* is the propagation weight of *TAffect*.

In the above formula, *CHSize* (*CSize* and *TSize*) is the number of edges of the type chain (combine and tetrahedralization) connected to the map-pair. The sum of the *CHF*, *CF*, and *TF* must be equal to 1 to restrict *NeighborAffect* between valid ranges which are discussed in the experimental results section. Furthermore, *NF* is a normal factor applying to cases in which there are no neighbors related to edges of one type. For example, assume that there isn't any neighbor related to the tetrahedralization edge, but there are edges of chain and combine types. Hence, we should set $NF = CHF + CF$ to normalize *NeighborAffect* into a valid range.

33.4.4 Extracting Similar Components

Due to the fact that the similarity degree of each map-pair in a pairwise graph expresses the matching degree of its atoms, we should extract similar components of two proteins by eliminating map-pairs and their related edges in a pairwise graph that have similarity degree less than γ. Consequently, the pairwise graph transforms to a forest in which we have several connected components. Each connected component declares one similar part of two proteins, and each map-pair in the connected components expresses matched atoms between two proteins. Similarly, each edge in the connected components shows conforming edges between two proteins. Hence, by extracting each protein's nodes and edges from the connected component, we obtain two connected subgraphs, each of which belongs to one of two given proteins. Connected components with the number of nodes less than η are not valuable for the result of matching, therefore they should be removed. For example, assume that you have two pictures, and you want to match them. Obviously if one pixel of them is analogous, you cannot assert that these two pictures are the same or you find valuable matching. Hence, the number of nodes of each connected component should be noticed and connected components containing less than η number of atoms should be eliminated.

33.5 Experimental Results

In this section we explain the SimDive engine which extracts similar parts of proteins by using the proposed method. We read protein information from PDB files. Then we use the Visad[1] package to do Delaunay tetrahedralization on the input protein chain. This tetrahedralization contributes our sequence chain of protein to convert to a tetrahedralized shape which is needed in counting structural similarity in the proposed protein matching algorithm. We can construct filters that apply $\alpha = 2$ for creating a boundary for tetrahedralization edge length and $\beta = 5$ for removing worthless chain-edges. Hence, edges that are longer than these thresholds are removed. Figure 33.6 shows the accuracy obtained after applying the above filters in two protein shapes.

Two tetrahedralized shapes in the form of a graph data structure were used to create the pairwise graph. After this creation, we used amino acid scoring matrices presented in the Blast book[2] to assign primary similarity to each map-pair. Table 33.1 shows the amino acid scoring matrix. Inasmuch as a part of the similarity of two nodes in each map-pair propagates to their respective neighbors, we propagated the similarities of the map-pair via the proposed equation in which CHAffect, CAffect, and TAffect were defined as the affect of neighbors of the map-pairs. The final similarities were obtained after applying filters shown below.

---

[1] http://www.ssec.wisc.edu/ billh/visad.html.

[2] http://safari.oreilly.com/0596002998/blast-CHP-4-SECT-3.

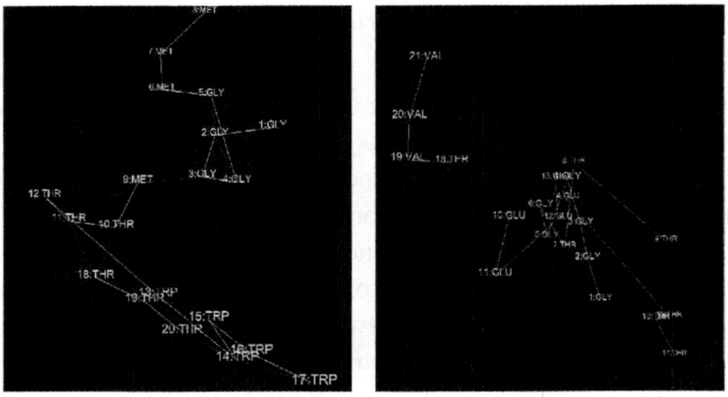

Fig. 33.6 **a** Protein number 1 after tetrahedralization; **b** protein number 2 after tetrahedralization

Table 33.1 Amino acid similarity matrix

		1	2	3	4	5	6	7	8	9	10	11	12	13	14	15	16	17	18	19	20
ALA	1	4	−1	−2	−2	0	−1	−1	0	−2	−1	−1	−1	−1	−2	−1	1	0	−3	−2	0
ARG	2	−1	5	0	−2	−3	1	0	−2	0	−3	−2	2	−1	−3	−2	−1	−1	−3	−2	−3
ASN	3	−2	0	6	1	−3	0	0	0	1	−3	−3	0	−2	−3	−2	1	0	−4	−2	−3
ASP	4	−2	−2	1	6	−3	0	2	−1	−1	−3	−4	−1	−3	−3	−1	0	−1	−4	−3	−3
CYS	5	0	−3	−3	−3	9	−3	−4	−3	−3	−1	−1	−3	−1	−2	−3	−1	−1	−2	−2	−1
GLN	6	−1	1	0	0	−3	5	2	−2	0	−3	−2	1	0	−3	−1	0	−1	−2	−1	−2
GLU	7	−1	0	0	2	−4	2	5	−2	0	−3	−3	1	−2	−3	−1	0	−1	−3	−2	−2
GLY	8	0	−2	0	−1	−3	−2	−2	6	−2	−4	−4	−2	−3	−3	−2	0	−2	−2	−3	−3
HIS	9	−2	0	1	−1	−3	0	0	−2	8	−3	−3	−1	−2	−1	−2	−1	−2	−2	2	−3
ILE	10	−1	−3	−3	−3	−1	−3	−3	−4	−3	4	2	−3	1	0	−3	−2	−1	−3	−1	3
LEU	11	−1	−2	−3	−4	−1	−2	−3	−4	−3	2	4	−2	2	0	−3	−2	−1	−2	−1	1
LYS	12	−1	2	0	−1	−3	1	1	−2	−1	−3	−2	5	−1	−3	−1	0	−1	−3	−2	−2
MET	13	−1	−1	−2	−3	−1	0	−2	−3	−2	1	2	−1	5	0	−2	−1	−1	−1	−1	1
PHE	14	−2	−3	−3	−3	−2	−3	−3	−3	−1	0	0	−3	0	6	−4	−2	−2	1	3	−1
PRQ	15	−1	−2	−2	−1	−3	−1	−1	−2	−2	−3	−3	−1	−2	−4	7	−1	−1	−4	−3	−2
SER	16	1	−1	1	0	−1	0	0	0	−1	−2	−2	0	−1	−2	−1	4	1	−3	−2	−2
THR	17	0	−1	0	−1	−1	−1	−1	−2	−2	−1	−1	−1	−1	−2	−1	1	5	−2	−2	0
TRP	18	−3	−3	−4	−4	−2	−2	−3	−2	−2	−3	−2	−3	−1	1	−4	−3	−2	11	2	−3
TYR	19	−2	−2	−2	−3	−2	−1	−2	−3	2	−1	−1	−2	−1	3	−3	−2	−2	2	7	−1
VAL	20	0	−3	−3	−3	−1	−2	−2	−3	−3	3	1	−2	1	−1	−2	−2	0	−3	−1	4

Parameter	a	CHF	CF	TF
Value	0.8	0.5	0.3	02

As you can see in the above table, owing to the importance of neighbors in the chain, we set the CHF factor as more than twice the TF factor, and correspondingly the CF factor more than the TF factor.

The similarity values between two amino acids in the scoring matrix are between −4 and +11, and these similarities between one amino acid with itself are between +4 and +11.

After propagating, the similarities still remained between −4 and +11. Hence, to avoid removing map-pairs that contain two identical nodes or two different nodes

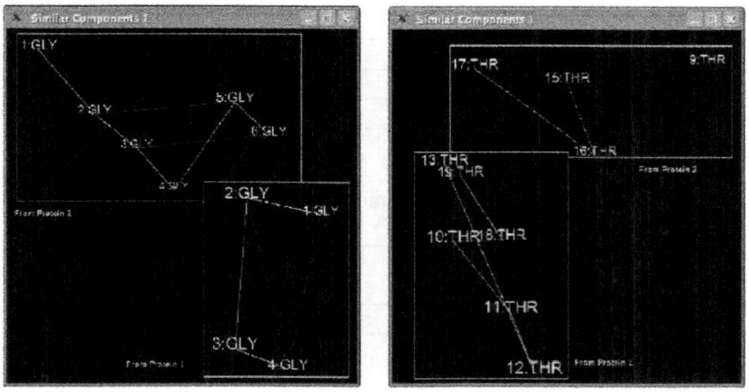

Fig. 33.7 Extracted similar components

with reasonable similitude degree, we set the threshold $\gamma = +3$. Therefore, map-pairs whose similarities are less than this threshold were eliminated. After this removal, components that included more than $\eta = 3$ atoms from each protein were extracted. Extracted similar components from small parts of two proteins shown in Fig. 33.6 are represented in Fig. 33.7. (In order to have clear figures, we choose very small parts of proteins for these figures).

Because of the large number of residues in each protein, the pairwise graph created from two proteins will contain a large number of map-pairs. Consequently, dealing with these large numbers of map-pairs needs a large amount of memory. Moreover, the process of similarity propagation and similar components extraction will become very time-consuming jobs. Hence, for managing this problem, we do protein fragmentation for larger proteins into accelerate the whole process of comparison. In this case, we divided each of our proteins into many segments and then compared each segment of protein with all segments of the other one. After that we extracted similar parts between each two fragments. Using this method reduces the model time and space complexity. If $f(n)$ presents the running time of our proposed model for comparing two proteins of size n without using fragmentation, the time complexity of the model will be exponential, However, by using fragmentation it will be linear. In other words, if m is the size of each fragment the time complexity function $f'(n)$ will be calculated as follows.

$$f'(n) = \left(\frac{n}{m}\right)^2 * f(m)$$

Nevertheless, using fragmentation leads the similar parts of two proteins which are located on fragmentation points to be broken. As you can see in Fig. 33.8, (A,B) is a similar component of the two given proteins, and you can see that after applying fragmentation the component (A,B) converts to $(A1,B1)$ and $(A2,B2)$ which is an undesirable fracture.

To avoid these undesirable fractures, we let our segments overlap with each other over their boundaries (Fig. 33.9). Hence, if m be the size of each fragment and l be the size of segment overlap, the time complexity function $f''(n)$ will be calculated as follows.

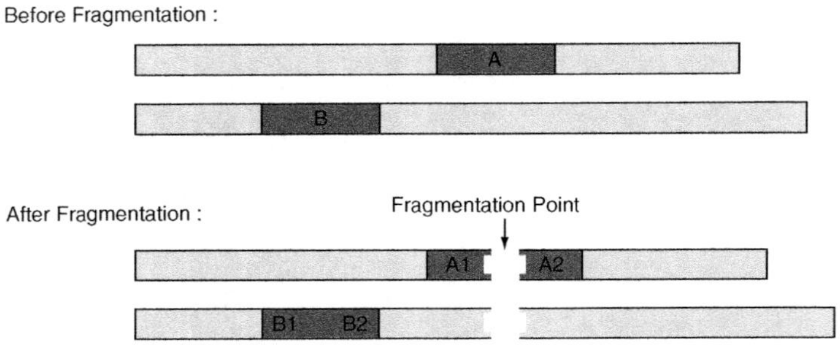

Fig. 33.8 Protein fragmentation

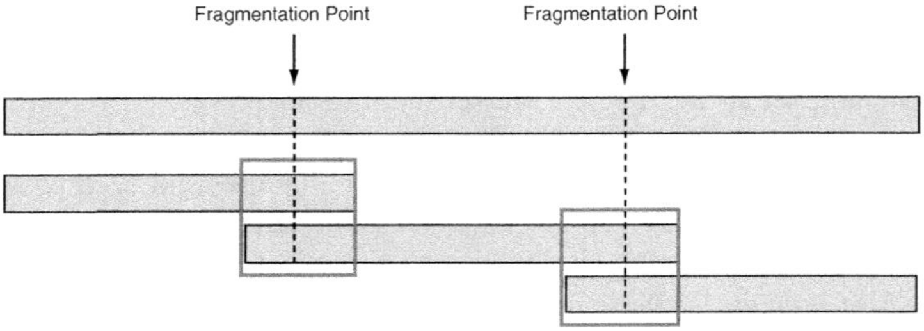

Fig. 33.9 Overlapping

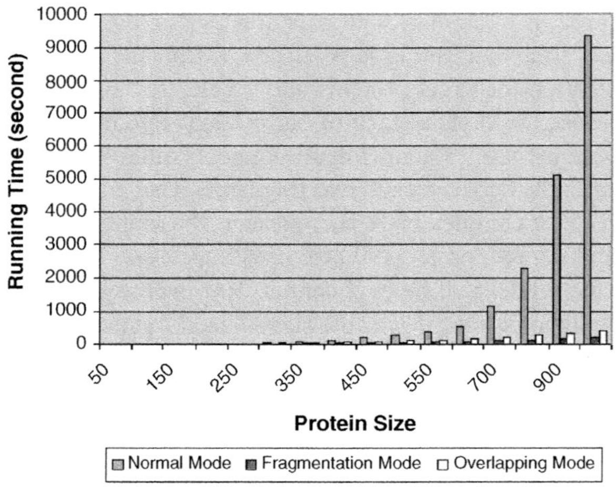

Fig. 33.10 Running time

$$f''(n) = \left(\frac{n}{m}\right)^2 * f(m + 2 * l)$$

Figure 33.10 shows our experimental results for running the whole algorithm. It is obvious that by using fragmentation we hasten the whole process. This chart related to segments of size 200 and overlapping parameter is equal to 50.

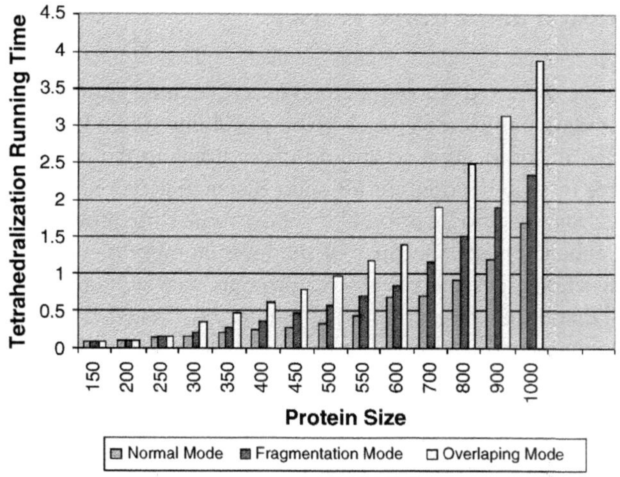

Fig. 33.11 Triangulation time

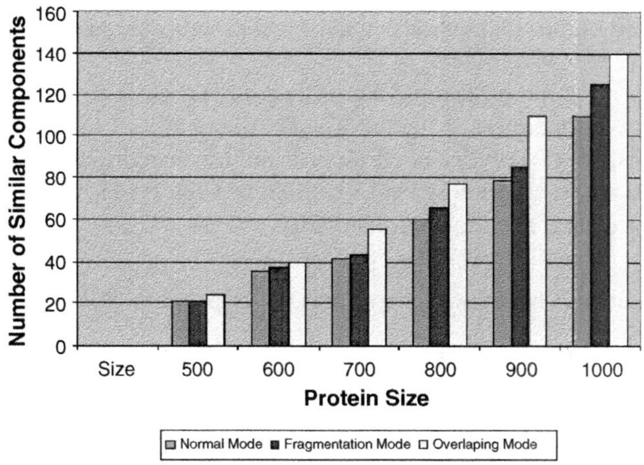

Fig. 33.12 Number of similar components

We also calculate the tetrahedralization time with and without using fragmentation. You can see the results in Fig. 33.11.

Although we mentioned that using fragmentation leads to unwanted fractures, Fig. 33.12 expresses that the number of these unwanted fractures is not worthy. On the other hand, these figures show that the number of similar components extracted in the overlapping mode is more than the basic fragmentation mode because in the overlapping mode the similar component in the overlapped section may be calculated twice. The important thing is that by overlapping we find both the fractured and unfractured components and the duplicated components should be ignored.

33.6 Conclusion and Future Works

In this chapter we proposed a novel method used to extract similar parts of proteins based on computational geometry and graph matching methods. Our method used the Delaunay tetrahedralization of the α-carbon atoms in the protein molecules to add some edges in protein structure for short distance nodes in counting structural similarity. The basic method can build a robust model for protein matching but it needed some enhancements. Because of the large number of residue in each protein, the pairwise graph created from two proteins will contain a large number of map-pairs and therefore a large amount of memory was needed. Hence, we applied protein fragmentation and overlapping to reduce the time complexity of our algorithm. We implement the SimDiv engine for testing the proposed method and the experimental results show the final verifications and effectiveness of our proposed method.

Furthermore, we need some test collections for optimizing the proposed model parameters such as the similarity propagation weight of each type in the propagation graph and thresholds used in the model. Moreover, this model will be tested on more real proteins with real size. In our experiments we use some parts of real proteins instead of the complete structure of real proteins because of the mentioned problems.

Moreover, our proposed method can extract similar parts of the two given proteins precisely. Furthermore, it includes both structural and sequential similarity. Our model has great flexibility in all aspects and by changing different model parameters such as propagation weight of different edge types, we can change the influence of structural and sequence similarity.

Acknowledgment M. Ghodsi's work has been partly supported by IPM School of Computer Science (contract: CS1385-2-01).

References

1. Eidhammer I, Jonasses I, Taylor W (2000) Structure comparison and structure patterns. Journal of Computational Biology. Volume 7. 685–716
2. Finney J (1970) Random packing and the structure of simple liquids, the geometry of the random close packing. Proceedings of the Royal Society. Volume 319. 479–493
3. Tropsha A, Carter C, Cammer S, Vaisman I (2003) Simplicial neighborhood analysis of protein packing (SNAPP): A computational geometry approach to studying proteins. Methods in Enzymology. Volume 374. 509–544
4. Cho W Z S, Vaisman I, Tropsha A (1997) A new approach to protein fold recognition based on delaunay tessellation of protein structure. In: Pacific Symposium on Biocomputing, Singapore 487–496
5. Carter C, LeFebvre B, Cammer S, Trosha A, Edgell M (2001) Four-body potentials reveal protein-specific correlations to stability changes caused by hydrophobic core mutations. Journal of Molecular Biology. Volume 311. No. 4. 625–638

6. Roach J, Sharma S, Kapustina M, Carter C (2005) Structure alignment via delaunay tetrahedralization. Proteins: Structure, Function, and Bioinformatics. Volume 60. 66–81
7. Bostick D, Shen M, Vaisman I (2004) A simple topological representation of protein structure: Implications for new, fast, and robust structural classification. Proteins: Structure, Function, and Bioinformatics. Volume 56. 487–501
8. Hun J, Bandyopadhyay D, Wang W, Snoeyink J, Prins J, Trosha A (2005) Comparing graph representations of protein structure for mining family-specific residue-based packing motifs. Journal of Computational Biology. Volume 12. 657–671
9. Dafas P, Gomoluch A K, Schroeder M (2004) Structural protein interactions: From months to minutes. In: Elsevier B.V, Parallel Computing: Software Technology, Algorithms, Architectures & Applications. 677–684
10. Park J, Lappe M, Teichmann S A (2001) Mapping protein family interactions: Intramolecular and intermolecular protein family interaction repertoires in the PDB and yeast. Journal of Molecular Biology. Volume 307. No. 3. 929–938
11. (http://www.iue.tuwien.ac.at/phd/fleischmann/node43.html)
12. Garcia-Molina S M H, Rahm E (2002) Similarity flooding: A versatile graph matching algorithm and its application to schema matching. In: Proceedings of the 18th International Conference on Data Engineering (ICDE), San Jose, CA

Chapter 34
Using Filtering Algorithm for Partial Similarity Search on 3D Shape Retrieval System

Yingliang Lu, Kunihiko Kaneko, and Akifumi Makinouchi

34.1 Introduction

Because 3D models are increasingly created and designed using computer graphics, computer vision, CAD medical imaging, and a variety of other applications, a large number of 3D models are being shared and offered on the Web. Large databases of 3D models, such as the Princeton Shape Benchmark Database [1], the 3D Cafe repository [2], and Aim@Shape network [3], are now publicly available. These datasets are made up of contributions from the CAD community, computer graphics artists, and the scientific visualization community. The problem of searching for a specific shape in a large database of 3D models is an important area of research. Text descriptors associated with 3D shapes can be used to drive the search process [4], as is the case for 2D images [5]. However, text descriptions may not be available, and furthermore may not apply for part-matching or similarity-based matching. Several content-based 3D shape retrieval algorithms have been proposed [6–8].

For the purpose of content-based 3D shape retrieval, various features of 3D shapes have been proposed [6–9]. However, these features are global features. In addition, it is difficult to effectively implement these features on relational databases because they include topological information. An efficient feature is proposed in Lu et al. [10] that can also be used in the partial similarity matching of shapes. However, they do not describe an efficient method to retrieve complex shapes by their partial similarity in Lu et al. [10] for a 3D shape database. In addition, the shock graph comparison-based retrieval method described in a previous paper [11] is based only on the topological information of the shape. A geometrical, partial similarity-based, and efficient method is needed to retrieve 3D shapes from a 3D shape database.

In this chapter, we propose a novel filtering method to filter out shapes. The proposed method is based on geometrical information rather than on topological information alone. Shapes are removed from the candidate pool if the processing part of the key shape is not similar to any part of the potential candidate shape. We

Oscar Castillo et al. (eds.), *Trends in Intelligent Systems and Computer Engineering.*

select as partly similar shapes, those shapes that have the greatest number of similar parts. In addition, the method is herein implemented on a curve-skeleton thickness histogram (CSTH) [10] based 3D shape search. Therefore, the method can also be easily implemented on other multibranch complex graph matching applications if there are heavy values on the curves.

The remainder of the chapter is organized as follows. Section 34.2 provides an overview of related work in skeleton generation and content-based retrieval. In Sect. 34.3 we describe the CSTH. In addition, we describe the segment thickness histogram (STH) and the postprocession of CSTH to produce STH. In Sect. 34.4, we describe the novel filtering algorithm and the partial similarity shape retrieval method based on the shape's STHs mentioned in Sect. 34.3. A discussion of an empirical study and the results thereof are presented in Sect. 34.5. Finally, in Sect. 34.6, we conclude the chapter and present ideas for future study.

34.2 Related Work

Work related to this chapter includes skeleton detection and 3D shape similarity searching.

A number of different approaches have been proposed for the similarity searching problem. Using a simplified description of a 3D model, usually in lower dimensions (also known as a shape signature), reduces the 3D shape similarity searching problem to comparing these different signatures. The dimensional reduction and the simple nature of these shape descriptors make them ideal for applications involving searching in large databases of 3D models. Osada et al. [8] propose the use of a shape distribution, sampled from one of many shape functions, as the shape signature. Among the shape functions, the distance between two random points on the surface proved to be the most effective at retrieving similar shapes. In Chen et al. [12], a shape descriptor based on 2D views (images rendered from uniformly sampled positions on the viewing sphere), called the Light Field Descriptor, performed better than descriptors that used the 3D properties of the object. Kazhdan et al. [13] propose a shape description based on a spherical harmonic representation.

Unfortunately, these previous methods cannot be implemented on partial matching. Another popular approach to shape analysis and matching is based on comparing graph representations of shape. Nicu et al. [9] develop a many-to-many matching algorithm to compute shape similarity on a curve-skeleton's topological information. Sundar et al. [6] develop a shape retrieval system based on the shape's skeleton graph. These previous methods only focus on the shape's topological information. Unfortunately, that the most important information of shape is that shape's geometric information is neglected. Moreover, it is highly costly to use graphs to match shapes. We proposed a novel shape feature of a 3D model [10]. That feature is called CSTH (mentioned in Sect. 34.1). It is based on the shape's geometric information.

Curve-skeletons are a 1D subset of the medial surface of a 3D object and recently are used in shape similarity matching [6, 9]. In the last decade many algorithms and applications have been developed based on them. Topological thinning methods [14] can directly produce a curve-skeleton that keeps the topological information of objects. Unfortunately, those algorithms are resolution-dependent and lose the geometric information of objects. Distance transform methods [15] use distance field of volume data to extract the skeleton. Unfortunately, they do not produce a 1D representation directly. Using those methods requires some significant postprocessing. However, it holds some geometric information on the extracted voxel.

Various types of fields generated by functions are used to extract curve-skeletons. They can produce nice curves on medial sheets. The potential at a point interior to the object is determined as a sum of potentials generated by point charges on the boundary of the object using the potential field function. Such functions include the electrostatic field function [16] and visible repulsive force function [17]. The skeleton points are found by determining the "sinks" of the field and connecting them using a force following algorithm [18] or minimizing the energy of an active contour [19], which are used to generate the initial skeleton in this chapter.

34.3 Segment Thickness Histogram

In this section, we briefly describe the methods used to build the thickness of a curve-skeleton from 3D polygonal models. For details, please refer to Lu et al. [10]. We also introduce a novel method by which to break a curve-skeleton into independent parts (called segments) by its topology. In addition, we describe in detail the normalization that normalizes the curve-skeleton thickness histogram of a single segment.

34.3.1 Skeleton Extraction

A number of methods of skeleton extraction have been reported [15, 18] (see Figs. 34.1 and 34.2). The electrostatic field function [18] can extract well-behaved curves on medial sheets. Even though the result is connected, extracted curves are divided into a number of segments based on electrostatic concentration. However, we need to split the skeleton into parts based on topology rather than on electrostatic concentration. In Lu et al. [10], the initial curve-skeleton based on the method described in a previous study [18] is first extracted. The distance transform (DT) algorithm [15] was then used to compute the DT of all voxels on the extracted curve-skeleton. Finally, in Lu et al. [10], they assumed that all of the curve-skeletons of the shape were connected and had no branches, and then introduced a similarity computation method of 3D shape models based on the curve-skeleton thickness distribution of the entire shape model.

Therefore, there must be several branches on the curve-skeleton of a complex shape (Fig. 34.1). We must first merge all of the parts that are separated from the

Fig. 34.1 A 3D model used to extract the skeleton

Fig. 34.2 The curve-skeleton with thickness of the 3D model in Fig. 34.1

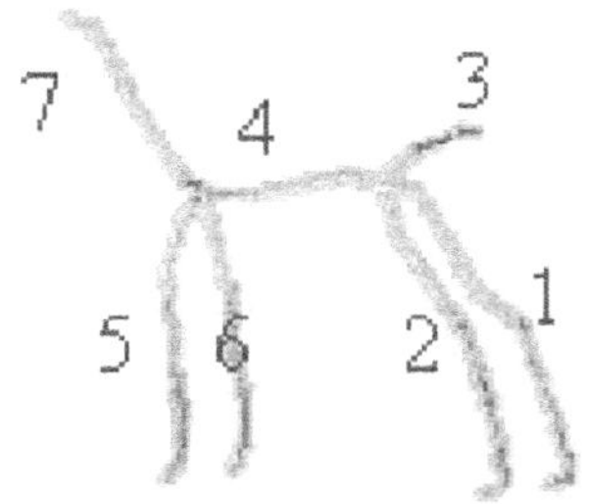

Fig. 34.3 The segments of curve-skeleton after splitting the curve-skeleton in Fig. 34.2

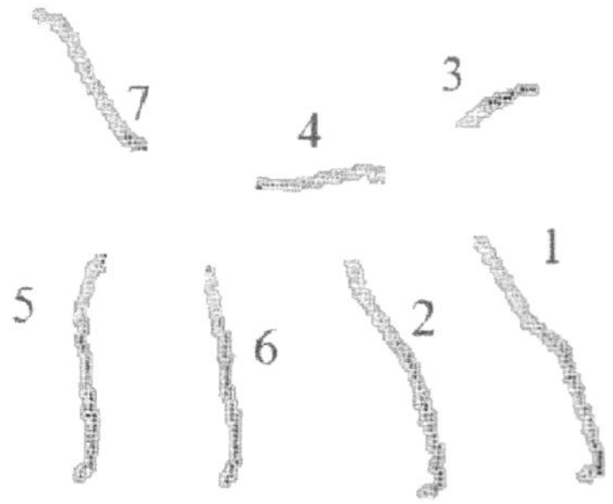

curve-skeleton by the electrostatic concentration into a connected curve. Then, we break the connected curve into parts according to its topology (Fig. 34.3).

34.3.2 Segment Thickness Histogram

We compute all voxel distance transforms that are on the segments mentioned in Sect. 34.3.1. We generate a thickness distribution histogram (Fig. 34.4) from all segments of the curve-skeleton produced in Sect. 34.3.1 that are joined together based on topological and curvature information, and use it as the shapes' partial feature to apply for partial matching.

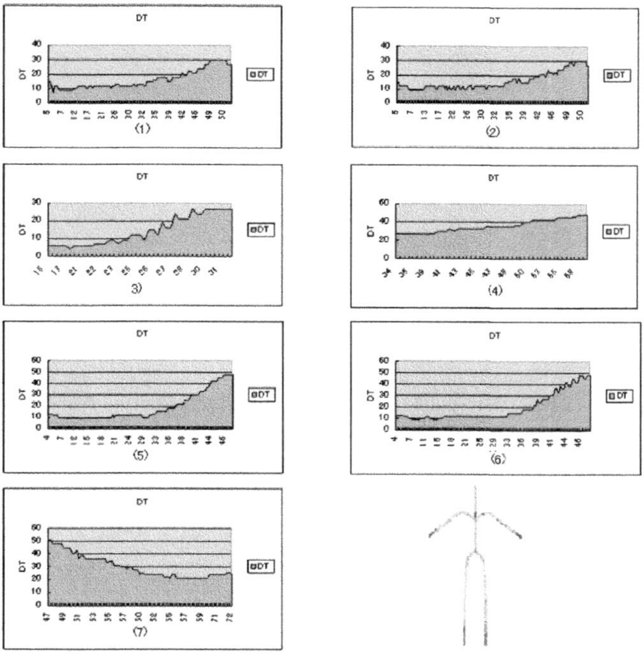

Fig. 34.4 Thickness distribution graph on the segments of the skeleton of the model in Fig. 34.1

34.3.3 Normalize the Segment Thickness

In order to obtain a segment thickness histogram (STH) representation that is invariant with the scale of a 3D model for similarity measuring, a normalization step is needed. The horizontal axis of the distribution should be normalized with a fixed value. Moreover, the vertical axis should be zoomed by a ratio that is equal to the zoom ratio of horizontal normalization. Using the normalization strategy, we use the variation of each STH of the shape as a feature of the shape. Furthermore, we treat the proportion of the length of a segment and the thickness distribution along with the segment as a component of the feature by this method.

34.4 Retrieval by Filtering

In this section, we describe the filtering algorithm used to retrieve the partial similarity shapes by their STH. Using this algorithm, we can retrieve shapes that have partial similarity, but only use some parts but not all parts of the key shape.

34.4.1 Comparison of Two Different Segments

Having constructed the segment thickness histograms of parts of two 3D models, we are left with the task of comparing them in order to produce a dissimilarity measure. In our implementation, we have experimented with a simple dissimilarity measure based on the LN norms function with $n = 2$. We use the formula as follows.

$$\text{Dissimilarity} = \sum_i (X_i - Y_i)^2, \tag{34.1}$$

where X and Y represent two STHs and X_i represent the thickness of the ith voxel on the X STHs.

In addition, because there are two different alignment ways to compare two segment thickness histograms, the different alignments will produce different dissimilarity results. For convenience, we use the minimum dissimilarity value in the experiment.

34.4.2 Filtering Algorithm

Two shapes are similar if all of their parts correspond and the relative parts are similar (Fig. 34.5). This means that their thickness histograms are similar on each segment of their curve-skeletons.

In addition, in the case of partial similarity, the two shapes are partially similar, but the part that exists in shape A must not necessarily be similar to a part that exists in shape B. We want to retrieve those 3D models that have the most similar parts.

To retrieve mostly similar shapes by their parts, we first sort the segments of their curve-skeletons by their STH volume size. Secondly, we retrieve the most n segments from the shape database. Each of the n retrieved segments belongs to different shapes. In order to improve the retrieval performance, we only compare the segments that are not in the candidate pool when we process the similarity searching

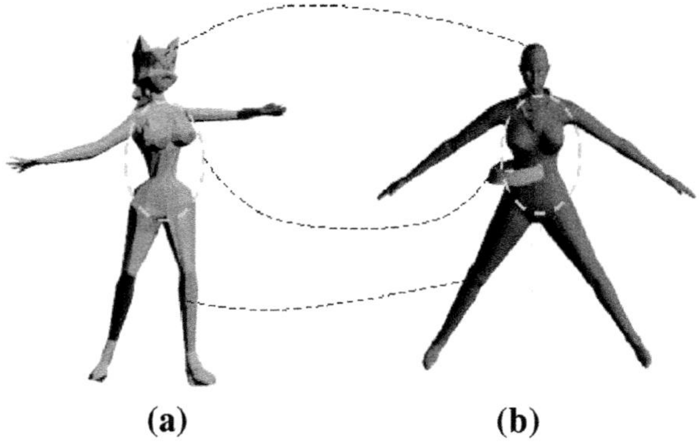

Fig. 34.5 The filtering algorithm

of the ith key segment. The result is like that mentioned in Table 34.1. The KS means the key shape with m segments of the curve-skeleton. The $KS.SG_1$ mentions the segment of the key shape that has the largest STH volume size. The $CS_{21}.SG_x$ mentions the xth segment that is on the curve-skeleton of the shape CS_{21}. The xth segment mentions that this segment has the xth largest STH volume on its shape. The candidate shapes $(CS_{i1}, CS_{i2}, \ldots, CS_{in})$ that belong to $KS.SG_i$ are different from each other. And, the candidate segments that belong to different key segments must be not be same as each other in Table 34.1. Finally, SQL is used to find the most similar shapes, based on the largest number of similar segments, when we need to retrieve some similar shapes from a key shape. In addition, the last step is to find the shapes that have the most amounts in the candidate pool of Table 34.1.

Therefore, we implement the proposed algorithm as shown in Fig. 34.6. The proposed algorithm finds the most similar n segments that belong to n different

Table 34.1 The candidate pool of the key shape

Key	Candidate pool		
$KS.SG_1$	$CS_{11}.SG_x$	...	$CS_{1n}.SG_x$
$KS.SG_2$	$CS_{21}.SG_x$	...	$CS_{2n}.SG_x$
⋮	⋮	⋮	⋮
$KS.SG_m$	$CS_{m1}.SG_x$	...	$CS_{mn}.SG_x$

```
Algorithm
  Filtering out different shapes
  Set n as the threshold value of the number of result
  Set MIN as the minimum similarity in the S_n
  For X_i := Each parts of the key shape
    numshape := 0
    For Y_j :=Each shape in database
       count := 0
       For Y_jk := Each parts of Y_j
         If similarity(X_i,Y_jk) > MIN and similarity(X_i,Y_jk) < TH_dis then
           Put Y_j into similar shapes set
           If numshape > n then
            remove the mostly dissimilar shape from the S_n.
           else
            numshape:= numshape +1
           end if
           Break;
         end if
         count :=count+1
        end for
        if count = the number of the parts of Y_j then
          remove the Y_j from the candidate data set
       end if
    end for
    insert the result set into relational database
  end for
```

Fig. 34.6 The filtering algorithm

shapes for each segment on the curve-skeleton of the key shape. The algorithm then saves the result and the similarity of the two segments on two shapes into a database.

We define S_n as the set of n shapes related to the most similar m segments. The *THdis* is the threshold of dissimilarity between the two segment thickness histograms.

There are a head, a trunk of a body, and four limbs that belong to the both shapes in Fig. 34.5. Using the proposed algorithm, we can retrieve the shape B (Fig. 34.5b) only by some parts of the shape A (Fig. 34.5a). For example, we can get the searching result that includes the shape B, only use the STHs that belong to a head, an upper extremity, a lower extremity, and a trunk of a body of the shape A (Fig. 34.5a).

34.5 Experiment and Discussion

In order to test the feasibility of the segment thickness histogram-based similarity, we conducted an empirical study with different parts of a sample model. We found that the proposed method accurately classified the different parts having different STHs.

Figure 34.7 shows the result of the comparison between the different parts on the sample model mentioned above (Fig. 34.1). The gray values in the comparison matrix indicate the quality of the match. Black indicates a perfect match, whereas white represents no match. The matrix illustrates the performance of the proposed algorithm, which is very high for the overall parts of the experimental model.

In order to test the feasibility of the similarity shape retrieval strategy proposed herein, we implement our algorithms on a Linux system by C++ and PostgreSQL. We use the Princeton shape database as the test data in the present study. We found that the proposed method works well for partial similarity shape retrieval.

The key shape in Fig. 34.8 has six segments on its curve-skeleton. These segments belong to a head, a trunk of a body, and four limbs. Because each segment has its own thickness histogram, the key shape has six independent thickness histograms. In order to find no more than 30 shapes using a segment of a key shape, we set n of S_n as 30 for the experiments. Our filtering program retrieves 30 shapes by each STH of the key shape, and then inserts them into the database. We want to

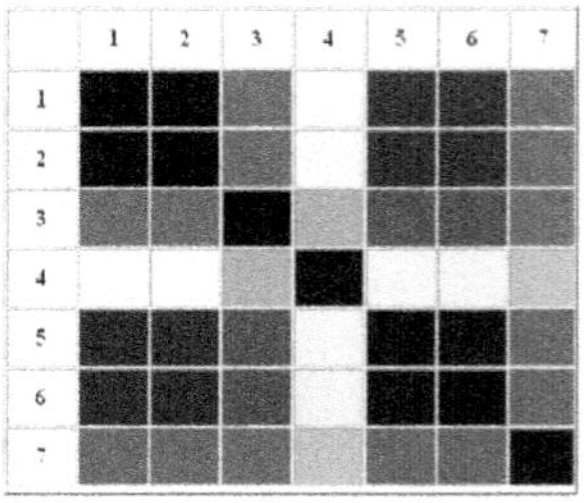

Fig. 34.7 Comparison matrix for the different parts of the example shape in Fig. 34.1

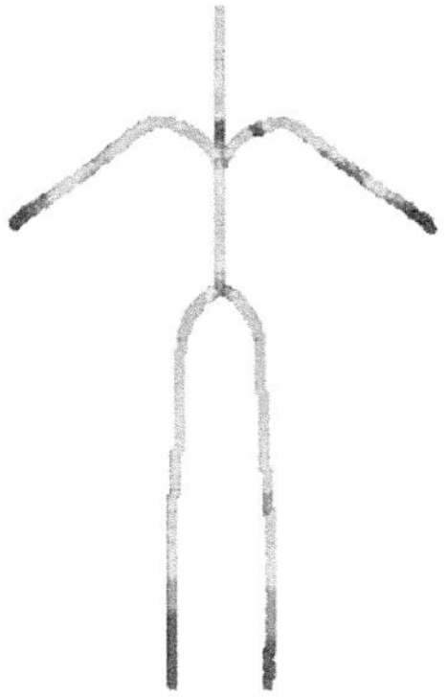

Fig. 34.8 The curve-skeleton with thickness of the key shape in Fig. 34.9

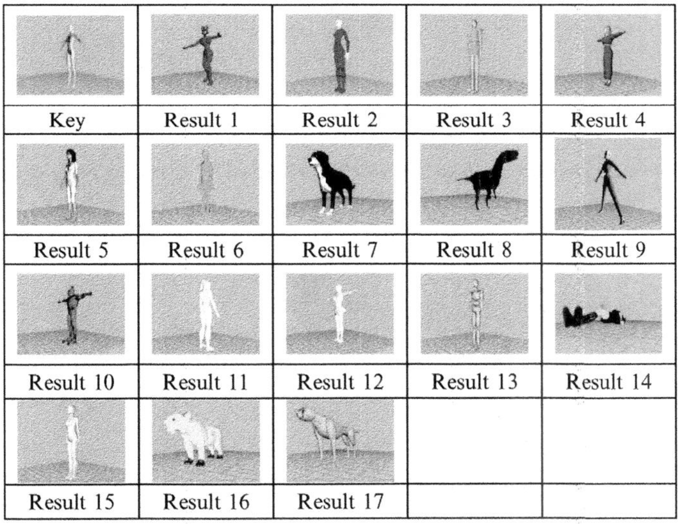

Fig. 34.9 Results of retrieval using the partial similarity based filtering method

find the shapes that match the key shape for the head, the trunk of the body, and the four limbs. Therefore, we want to find the shapes that are in each result set of the six parts. We obtain 17 shapes on which each of the six key parts has a matching part (Fig. 34.9). Figures 34.10–34.16 show some results of the shape retrieval test.

Figure 34.9 shows some results of the shape retrieval test. The results reveal that the proposed method can find the most similar shape from a 3D shape database. Furthermore, the proposed method also retrieves the models that are dissimilar with respect to the global feature but that have parts that are similar to the parts of the key shape, such as Result 7 in Fig. 34.9. The part of the tail of Result 7 does not have a part that is similar to the key shape.

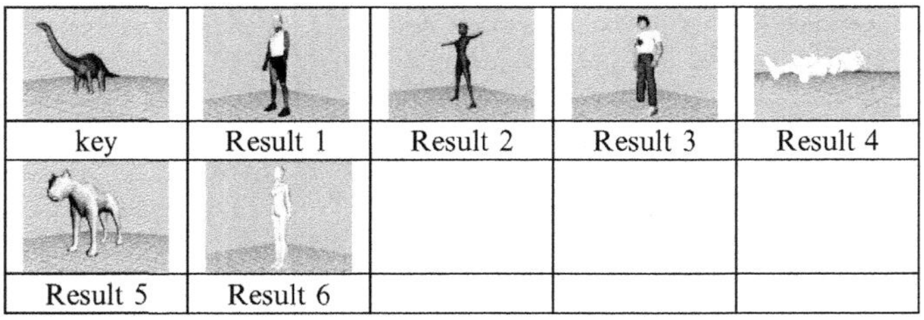

Fig. 34.10 The most similar objects to dinosaur

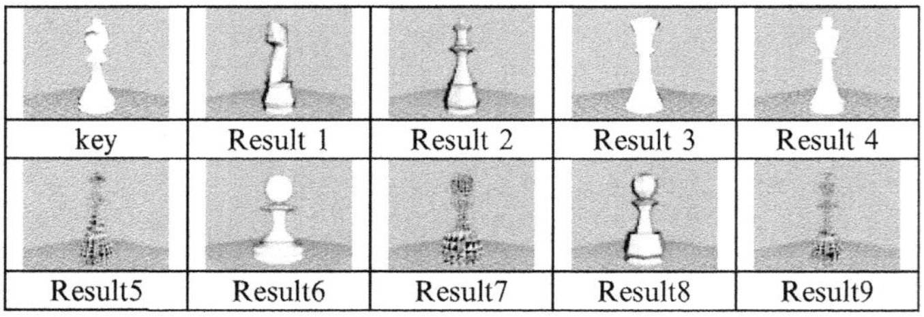

Fig. 34.11 The most similar objects to the key chess retrieved by ascending order of the similarity

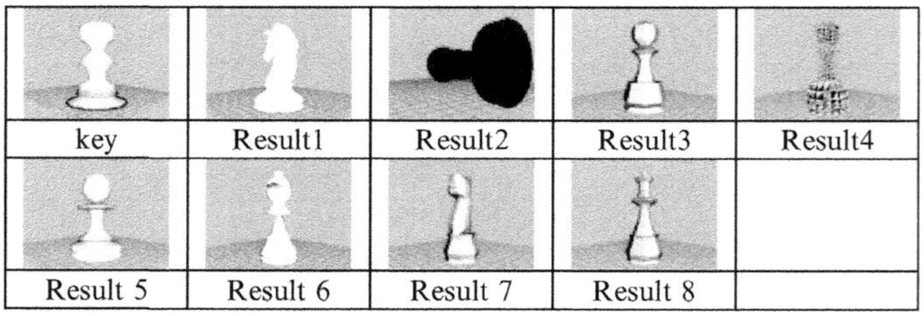

Fig. 34.12 The most similar objects to the key chess retrieved by ascending order of the similarity

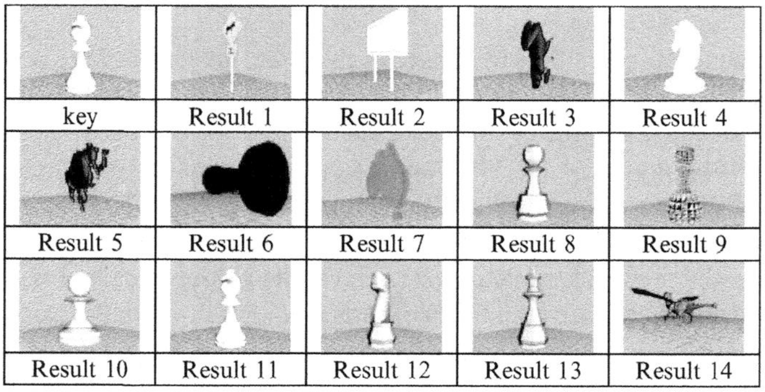

Fig. 34.13 The most similar objects to the key chess retrieved by ascending order of the similarity

Fig. 34.14 The curve-skeleton with thickness of the key chess in Fig. 34.13

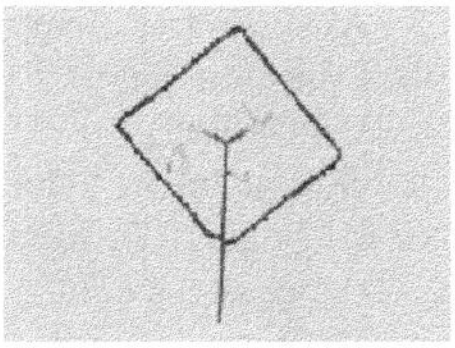

Fig. 34.15 The curve-skeleton with thickness of Result 1 in Fig. 34.13

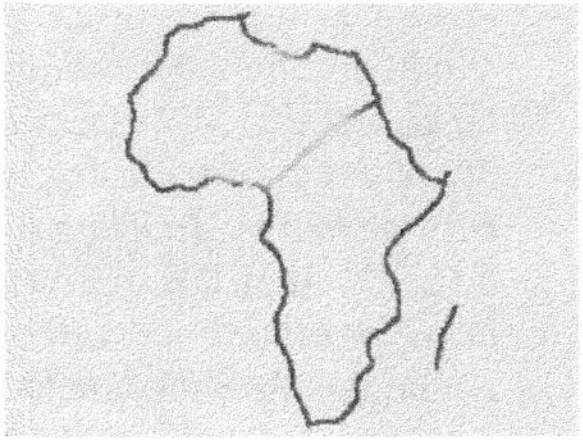

Fig. 34.16 The curve-skeleton with thickness of Result 3 in Fig. 34.13

34.6 Conclusions and Future Studies

The shape retrieval method proposed in this chapter is based on partial similarity between shapes. The proposed method first extracts a curve-skeleton. Second, we compute the dissimilarity of the segment thickness histograms (STHs) of each part with respect to the shapes. At last, we propose a novel shape retrieval strategy. It is possible to retrieve 3D models by partial similarity based on the dissimilarity of the STHs of curve-skeleton graphs.

Because these STHs are extracted from 3D shapes using their geometrical information, the shapes can be compared based on geometrical information rather than on topological information alone. Because the STH is a partial feature of a shape, the STH also can compare two shapes based on their partial features, rather than on their global features alone. It exhibits good efficiency and good results were obtained in our experiments using it.

In the future, we intend to add the thickness ratio on the connected parts as a feature of shape to filter out models, as shown by Results 7, 8, 16, and 17 in Fig. 34.9. In addition, we intend to develop an algorithm that efficiently searches 3D models from 2D drawings.

Acknowledgment Special thanks to Dr. Nicu D.Cornea for the voxelization code. This research is partially supported by the Special Coordination Fund for Promoting Science and Technology, and Grant-in-Aid for Fundamental Scientific Research 16200005, 17650031 and 17700117 from Ministry of Education, Culture, Sports, Science and Technology Japan, and by 21st century COE project of Japan Society for the Promotion of Science.

References

1. P. Shilane, M. K. P. Min, and T. Funkhouser: The Princeton shape benchmark, *Shape Modeling International*, Genoa, June 2004.
2. 3D Cafe: http://www.3dcafe.com/asp/freestuff.asp.
3. AIM@SHAPE Network of Excellence: http://www.aimatshape.net/.
4. Princeton Shape Retrieval and Analysis: 3D Model Search: http://shape.cs.princeton.edu/search.html.
5. Google Image Search: http://www.google.com/.
6. H. Sundar, D. Silver, N. Gagvani, and S. Dickenson: Skeleton based shape matching and retrieval, *SMI 2003*, pp. 130–139, 2003.
7. M. Hilaga, Y. Shinagawa, T. Kohmura, and T. Kunii: Topology matching for fully automatic similarity estimation of 3D shapes, *ACM SIGGRAPH 2001 Proceedings*, 2001.
8. R. Osada, T. Funkhouser, B. Chazelle, and D. Dobkin: *Matching 3D Models with Shape Distributions Shape Modeling International*, Genova, Italy, May 2001.
9. N. D. Cornea, M. F. Demirci, D. Silver, A. Shokoufandeh, S. J. Dickinson, and P. B. Kantor: 3D object retrieval using many-to-many matching of curve skeletons, *International Conference on Shape Modeling and Applications SMI 2005*, MIT, Boston, June 15–17, 2005.
10. Y. Lu, K. Kaneko, and A. Makinouchi: 3D shape matching using curve-skeletons with thickness, *1st International Workshop on Shapes and Semantics*, June 2006.
11. K. Siddiqi, A. Shokoufandeh, S. Dickinson, and S. Zucker: Shock graphs and shape matching, *International Journal of Computer Vision*, 30: 1–24, 1999.
12. D. Y. Chen, M. Ouhyoung, X. P. Tian, Y. T. Shen, and M. Ouhyoung: On visual similarity based 3D model retrieval, *Proceedings of Eurographics*, Granada, Spain, 2003.
13. M. Kazhdan, T. Funkhouser, and S. Rusinkiewicz: Rotation invariant spherical harmonic representation of 3d shape descriptors, *Symposium on Geometry Processing*, pp. 167–175, 2003.
14. T. Saito, S. Banjo, and J. Toriwaki: An improvement of three dimensional thinning method using a skeleton based on the Euclidean distance transformation—A method to control spurious branches, *IEICE Transaction on Information and Systems*, Japan, J84-D-II, 8, pp. 1628–1635, August 2001.
15. N. Gagvani and D. Silver: Parameter controlled volume thinning, *Graphical Models and Image Processing*, 61(3): 149–164, 1999.
16. T. Grigorishin and Y.H. Yang: Skeletonization: An electrostatic field-based approach, *Pattern Analysis and Applications*, 1: 163–177, 1998.
17. F. Wu, W. Ma, P. Liou, R. Liang, and M. Ouhyoung: Skeleton extraction of 3D objects with visible repulsive force, *Eurographics Symposium on Geometry Processing*, 2003.
18. N. Cornea, D. Silver, X. Yuan, and R. Balasubramanian: Computing hierarchical curve-skeletons of 3D objects, *The Visual Computer*, 21(11): 945–955, 2005.
19. W. Ma, F. Wu, and M. Ouhyoung: Skeleton extraction of 3D objects with radial basis functions, *IEEE SMA*, 2003.

Chapter 35
Topic-Specific Language Model Based on Graph Spectral Approach for Speech Recognition

Shinya Takahashi

35.1 Introduction

Large vocabulary continuous speech recognition techniques have greatly advanced in recent years due to the remarkable advances of computers. Even personal computers today have extraordinary computation powers so that we can perform automatic speech recognition with high performance in a small computer. This is due to not only the evolution of the computers but also the development of some efficient recognition algorithms and the utilization of statistical acoustic and language models with a large speech database.

In addition, rapid development of the WWW makes it possible to utilize enormous textual data resources for creating excellent language models. Especially, the topic-specific language model can give high performance for speech recognition if the large amount of appropriate topic-related documents can be collected.

Under these circumstances, we have been developing the broadcast news search system with the language model adaptation using the information on the WWW. The basic idea is that broadcast news has similar Web documents on the Internet news site, so the performance of news speech recognition can be improved with the adapted language model by collecting a similar article via Web crawling [1, 2].

Several methods collecting the topic-related documents automatically, which are similar to our approach, have been proposed [3–7]. For example, the method proposed in Berger and Miller [4] collects similar documents from a Web site with previous user utterances as queries to the search engine and adapts unknown words to a general language model. Almost all of the other methods, such as Sethy et al. [7], require us to select an appropriate keyword as a query term manually or using some other methods. On the other hand, a unsupervised language model adaptation method based on "query-based sampling" has been proposed [8]. This method selects the keywords from recognition results directly and collects topic-related documents with these keywords from the Web for topic adaptation.

Oscar Castillo et al. (eds.), *Trends in Intelligent Systems and Computer Engineering.*

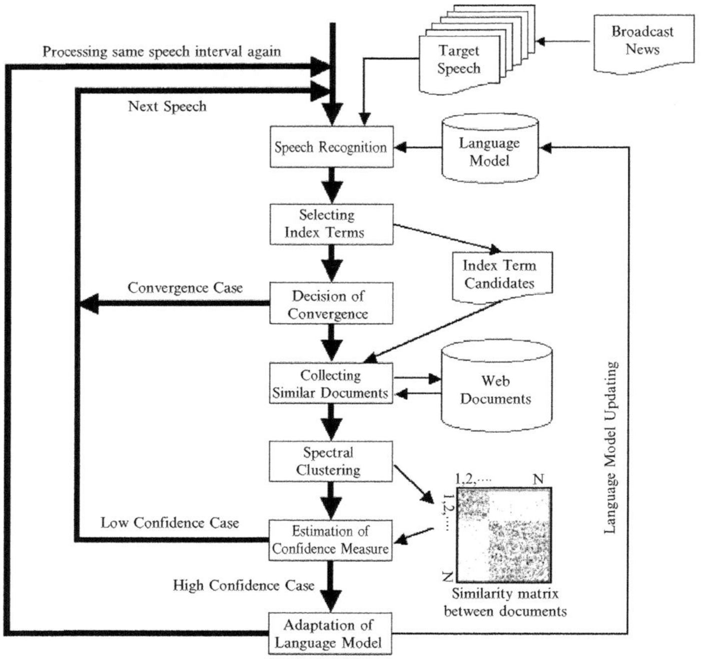

Fig. 35.1 Index term automatic extraction system

However, if the expected documents cannot be collected, the adaptation does not work well. In this chapter a document filtering method for language model adaptation using a spectral clustering approach is proposed in order to select the appropriate topic-related documents (see Fig. 35.1). To show the effectiveness of this approach, speech recognition experiments are demonstrated for broadcast news speech. Experimental results show the proposed method can give a suitable cluster that consists of the topic-related documents and can improve speech recognition performance.

35.2 System Overview

35.2.1 Automatic Extraction of Index Terms

Figure 35.1 shows an overview of the total system we have been developing now. This system processes an input news speech document as follows.

1. A news speech document is recognized using a large vocabulary speech recognizer Julius [9] with a general n-gram prepared in advance.
2. Candidates of keywords are selected from high-frequency words in a recognition result.

3. Web documents similar to the recognition result are retrieved using a search engine with the keywords.
4. The text corpus is extracted from the collected Web documents using spectral clustering.
5. A topic-specific n-gram language model is trained from the extracted corpus.
6. The input news speech document is recognized with the topic-specific n-gram.

The above procedure is executed iteratively until the index terms converge.

35.2.2 Collecting Topic-Related Documents

As mentioned before, the most critical issue in this approach is the reliability or confidence of the recognition result in the first stage. If there are a lot of misrecognition words, the system may collect a lot of dissimilar documents from the Web site. To cope with this issue, we investigate to evaluate the confidence of the recognition result using the collected documents.

The basic idea is that higher confidence results can collect more amounts of similar documents because there would be a lot of news articles similar to the input news speech document. To the contrary, lower confidence results are expected to collect dissimilar articles, so there would be several dissimilar clusters in the collected documents. So we consider applying cluster analysis to the collected documents.

35.2.3 Selecting Topic-Related Corpus

In this chapter, the keyword candidates are selected with the constraints regarding a part of speech. In concrete, the words are restricted in generic nouns, proper nouns, and verbal nouns. In the similar document retrieval stage, the nouns from the recognition result as the keywords and retrieve the documents including these keywords chosen by using Google Japan[1] and Yahoo! Japan[2] as the search engines. Here, the number of the documents including all keywords might be small, so the retrieval process is performed by deleting the lowest frequency word from the keywords iteratively until a sufficient amount of the documents is obtained.

After collecting the documents, the similarity matrix between the documents is calculated by using cosine similarity with TF weights in a vector space model [10].[3] If there is no recognition error in the index terms, we can easily select the expected documents from the collected documents in the most similar order. However, all documents similar to a recognition result are not documents similar to a target article due to misrecognition words. Therefore, it is necessary to find the actual distribution of the documents similar to the target article. So, we used spectral clustering (explained later) for cluster analysis.

[1] http://www.google.co.jp/.

[2] http://search.yahoo.co.jp.

[3] The reason why IDF isn't used is that IDF tends to depend on the target domain.

35.2.4 Language Model Updating

There are two methods for merging a topic-dependent or topic-specific language model and a general language model. One is a method that simply concatenates a training corpus and a general corpus [8]. The other is a method that merges and assigns a new n-gram constructed from the training corpus to the general n-gram [11]. However, both methods require a reliable and a large amount of resources, so it is difficult to apply to a small task such that is dealt with in this research.

So, in order to deal with unknown words, we merged a topic-specific dictionary with a general dictionary and we used the topic-specific n-gram constructed from the collected documents with the merged dictionary.

35.3 Document Filtering Based on Spectral Clustering

35.3.1 Spectral Clustering

Spectral clustering is one of the clustering methods that uses eigenvectors of the Laplacian of the symmetric matrix $\mathbf{W} = (w_{ij})$ containing the pairwise similarity between data objects i, j [12]. Spectral clustering has been proposed from the graph theory field to solve a minimum cut-set problem of a nondirectional graph. In recent years this method has been applied to many applications, such as document clustering, image recognition, and so on [13, 14].

Given n objects and the similarities between them $W = (w_{ij})$, consider a two-way clustering problem that decides a membership indicator as

$$q_i = \begin{cases} 1 & \text{if } i \in A \\ -1 & \text{if } i \in B. \end{cases} \tag{35.1}$$

Here, if nodes i and j are connected, $w_{ij} = 1$; otherwise $w_{ij} = 0$.

This problem is defined as the minimization problem of the following equation,

$$\begin{aligned} J(\mathbf{q}) &= \frac{1}{4}\sum_{i,j} w_{i,j}(q_i - q_j)^2 \\ &= \frac{1}{2}\sum_{i,j} q_i(d_i\delta_{ij} - w_{ij})q_j \\ &= \frac{1}{2}\mathbf{q}^T(D - W)\mathbf{q}, \end{aligned} \tag{35.2}$$

where δ_{ij} is the Kronecker delta, D is a diagonal matrix with each diagonal element being the sum of the corresponding row ($d_i = \sum_j w_{ij}$).

Using a Lagrangian multiplier under the following constraint

$$\mathbf{q}^T D\mathbf{q} = 1, \tag{35.3}$$

and relaxing the restriction of q_i from discrete values to continuous values in $(-1,1)$, the minimization of $J(\mathbf{q})$ becomes the following eigenvalue problem,

$$(D-W)\mathbf{q} = \lambda D\mathbf{q}. \tag{35.4}$$

Because elements of $\mathbf{q}$ are already relaxed to continuous value, each q_i takes continuous value that satisfies $-1 \leq q_i \leq 1$. Therefore, each node is classified into two classes that satisfy $\{i|q_i < 0\}$ and $\{j|q_j \geq 0\}$. Here, the eigenvector for the minimum eigenvalue is the optimal solution, because we are going to minimize the cut-size J. The first eigenvalue, however, is zero, and its eigenvector is a direct current component so that meaningful solution is the second eigenvector or later than second.

In this chapter we use the following eigenvalue equation [15] substituting $\mathbf{q} = D^{-1/2}\mathbf{z}$ into (35.4),

$$D^{-1/2}WD^{-1/2}\mathbf{z} = \lambda'\mathbf{z}, \quad \lambda' = 1-\lambda. \tag{35.5}$$

It should be noted that here the solution is the eigenvectors corresponding to the second or later largest eigenvalue because $\lambda' \leq 1$ for (35.5).

The solution to solve the graph partition problem described above can be applied to a general clustering problem by replacing weights between each node into similarities between each datum. Basically, an eigenvector gives an optimal partition, so $\log_2 k$ eigenvectors give 2^k cluster boundaries.

35.3.2 Reordering the Similarity Matrix

By sorting each element of the eigenvectors by its value and arranging the target data in the corresponding element order, more similar data can be replaced closer to the topographic relations. In addition, a segmentation with appropriate boundaries enables an approximate clustering. In Ding and He [15], the cluster boundaries were obtained by detecting the edge from a connectivity matrix emphasizing boundaries of the similarity matrix. In our case, however, there might not be a clearly independent topic in the collected documents so that cluster boundaries are calculated from the eigenvector directly. So, in this chapter based on the principle of graph partition, we decide the candidates of the cluster boundaries as the points that have small d_i as shown in Sect. 35.3.1. The point with small d_i means that a connective weight of the point between the graphs is small when a target graph is partitioned into two graphs. Selecting the boundary candidate points in the order of smaller d_i, we can obtain cluster regions sandwiched between them. Figure 35.2 shows an example of spectral clustering for a cosine similarity matrix among documents including three topics. Figure 35.2c is a reordered matrix obtained by sorting the elements of the second eigenvector.

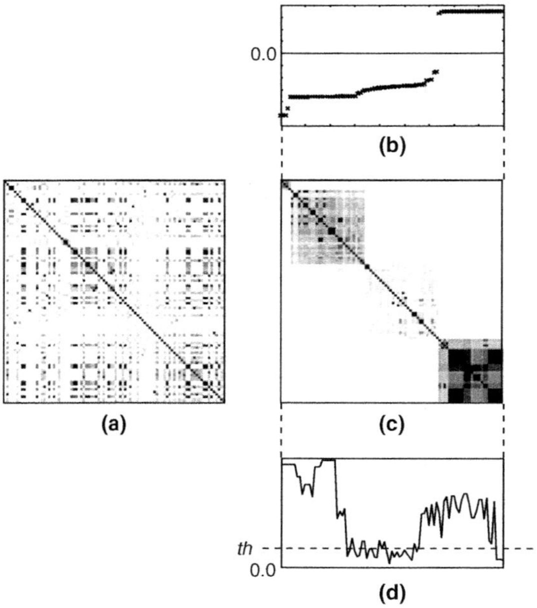

Fig. 35.2 Example of spectral clustering: **a** original cosine similarity matrix among documents including two topics and noises; **b** elements of the second eigenvector; **c** result of reordered matrix; **d** a graph of the difference of adjacent d_i, which are diagonal elements of D

35.3.3 Document Filtering Algorithm

The purposes of the proposed approach are (1) to select a cluster that consists of a certain amount of documents similar to each other and (2) to filter out unsuitable documents caused by misrecognition. It can be supposed that the document collected with low confidence recognition results tends to consist of some different words from the document collected with high confidence recognition results in the same topic.

The complete algorithm is as follows.

1. Collect the documents using the recognition results for utterances in each topic.
2. Calculate the cosine similarity matrix containing the pairwise similarity between all the documents that consist of recognized sentences and the collected documents.
3. Arrange the similarity matrix via spectral clustering.
4. Calculate the boundary points using diagonal elements of D in the reordered matrix.
5. Extract the target region that is the widest range and includes the target articles or the most number of utterances from each region sandwiched between the boundary points.

35.4 Experiments

35.4.1 Speech Data and Speech Recognition Tools

In the experiments we used two datasets from NHK-TV news programs. The specifications and the topics included in each TV program are shown in Tables 35.1 and 35.2. Web documents used in the experiment were collected a week after each news broadcast.

A bigram LM trained from newspaper documents with 20,000 words, which is included in the Julius dictation kit Ver.3.1, was used as a general LM to obtain the first recognition results. A news topic-specific LM, which is a bigram model, was created from the topic-related documents after spectral clustering using the CMU-Cambridge SLM Tool kit [16]. To deal with unknown words, the topic-specific dictionary was merged with the general dictionary. For an acoustic model, a gender-independent HMM was used. For evaluation, a word correct rate of the speech recognition result, recall and precision of the target nouns were used.

35.4.2 Experiments for Dataset A

35.4.2.1 Result of Spectral Clustering for Each Topic

At first we conducted spectral clustering for each topic independently. The similar Web documents were collected by using the recognition result of whole topic sentences of each topic for dataset A. Table 35.3 shows index terms used in the similar document retrieval stage. Note that the English words and phrases shown in this table are not correct translations. The index terms used actually are Japanese words. In this table, italic words stand for misrecognition words. Figure 35.3 shows results

Table 35.1 Dataset A (broadcast at 3 PM on April 12th, 2006)

Topic		Spoken time (s)	# Words	# Target noun
A-1	DNA testing result of South Korean abductee[a]	77	196	63
A-2	Heavy rain influenced by a wide-area low pressure[b]	78	268	94
A-3	Data falsification of Nuclear reactor flow-meter	73	251	72
A-4	Exchange and Stock	24	71	44
A-5	Inhabitant refuge due to a landslide[c]	90	287	90

[a]Includes an interview by telephone in Korean
[b]Rain sounds overlapped
[c]Includes an interview on street

Table 35.2 Dataset B (broadcast at 1 PM on January 8th, 2007)

Topic		Spoken time (s)	# Words	# Target noun
B-1	Threat of rainstorm in north Japan	86	269	41
B-2	Washington subway derailment	56	186	36
B-3	Resignation of Polish archbishop[a]	60	174	35
B-4	Thalidomide as cancer drug[b]	86	275	53

[a]With background noise by crowd
[b]With a press conference interview

Table 35.3 Index terms used in collecting Web documents for dataset A

Topic	Index terms
A-1	Korea, *Ai/Megumi*, utilize, Japan, government, *self*, *Kin/Kane(money)*, cooperation, daughter, North Korea
A-2	Range, West Japan, air pressure, rain, Wakayama, utilize, *surface*, atmosphere, branch, *bath*
A-3	Falsification, data, Toshiba, *transmission*, examination, industrial, economical, *requirement*, power generation, East
A-4	Tokyo, market, *activity*, *telephone*, quotation, *teacher*, *Library*, current stock price, *drain*, stock
A-5	Telephone, mobile, company, foundation, head, Nagasaki, *Suzuka*, quantity, prevention, collapse

of spectral clustering for five topics.[4] An arrow with a vertical dotted line at the bottom means the location of the target document, which is the recognized sentences of the input speech. The other vertical (solid) lines at the bottom stand for the boundary points selected with the 10 d_i from the smallest to the largest. The longer lines indicate smaller 5 d_i corresponding to numerals in the upper line and the shorter lines indicate the others corresponding to numerals at the bottom. Table 35.4 shows the specification of the selected clusters from five or ten clusters for each topic.

The result for topic A-1 indicates that the large cluster that consists of documents similar to the target article could be selected. On the other hand, as can be seen from Fig. 35.3e, although there is the large cluster including the recognition result of topic A-5, the recognition result is located at the leftmost side. This means that the recognition result of topic A-5 could not collect similar documents. The results for the other topics, A-2, A-3, and A-4, indicate that collecting similar documents is successful but the selected cluster is very small.

35.4.2.2 Speech Recognition Experiments with Topic-Specific Language Model for Dataset A

Next, we conducted speech recognition experiments for dataset A with the topic-specific language model constructed from the selected documents shown

[4] Note that this experiment was conducted under different conditions from the previous report.

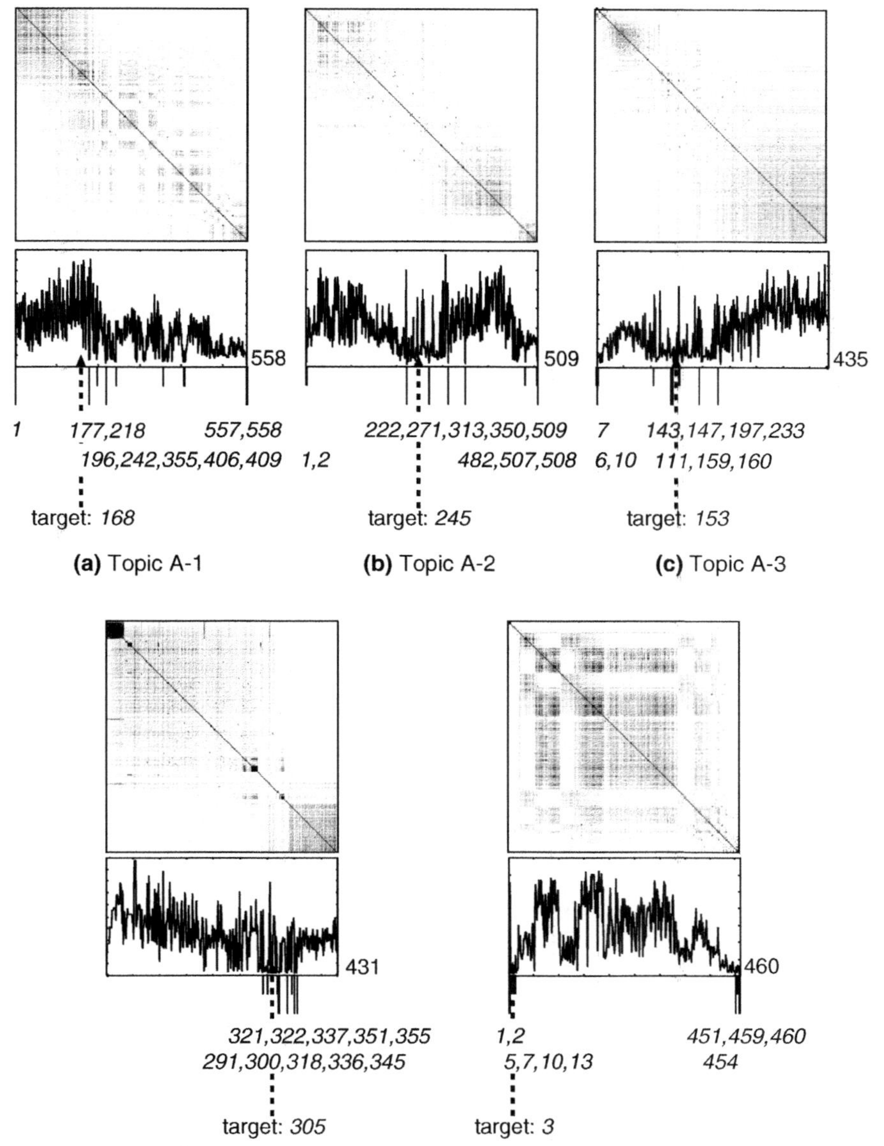

Fig. 35.3 Results of spectral clustering for dataset A

in Table 35.4. The results for each topic are shown in Table 35.5. For comparison the results with the general language model and with the language model constructed from all collected documents are also shown in this table. Table 35.6 shows the specification of the language model used in the experiments. Here, OOV stands for out of vocabulary, that is, words that are not defined in dictionary. It should

Table 35.4 Selected clusters for dataset A

Topic	# of All documents	Location of target	Selected cluster (from 5 clusters)			Selected cluster (from 10 clusters)		
			Begin	End	# of Documents	Begin	End	# of Documents
A-1	558	168	1	177	177	Same as left		
A-2	509	245	222	271	50	Same as left		
A-3	435	153	147	197	51	147	159	13
A-4	431	305	1	321	321	300	318	19
A-5	460	3	2	451	450	2	5	4

Table 35.5 Results of speech recognition with news topic-specific LM for each topic of dataset A

Topic	General LM			Topic-specified LM with all documents		
	WCR	P	R	WCR	P	R
A-1	40.9	53.7	47.9	75.0	74.4	76.3
A-2	19.6	31.8	33.3	40.2	33.3	43.2
A-3	57.8	55.3	60.0	73.8	59.5	73.5
A-4	23.9	58.3	26.9	54.8	55.6	66.7
A-5	50.9	53.6	51.7	52.4	37.7	39.0
Topic	**Topic-specified LM with selected documents from 5 clusters**			**Topic-specified LM with selected documents from 10 clusters**		
	WCR	P	R	WCR	P	R
A-1	75.0	75.7	73.7	Same as left		
A-2	41.9	29.4	40.5	Same as left		
A-3	73.8	55.6	73.5	47.7	24.6	47.1
A-4	56.2	62.5	66.7	27.4	18.8	20.0
A-5	53.1	38.3	39.0	15.3	14.7	22.7

WCR Word Correct Rate, *P* Precision, *R* Recall

be noted that a small size corpus such as the case in Table 35.6 cannot obtain the statistical information so the language model constructed from this small corpus is not actually useful.

As shown in Table 35.5, the word correct rates in all topics for the selected documents from five clusters were improved significantly as compared with that of the general language model. However, the precision and recall for topics A-2 and A-5 were not improved. Compared with the results for using all documents collected from the WWW, the precision and recall were partially improved.

The selected documents from ten clusters for A-3 through A-5 failed to construct a topic-specific language model completely. The reason for this failure is that the size of the selected region was too small; that is, similar documents weren't collected sufficiently. Needless to say, such a small corpus should not actually be used for constructing a language model.

Table 35.6 Specification of topic-specific language models

Topic A-1				Topic A-2			
LM	# of Sentences	# of Word	# of Added OOV	LM	# of Sentences	# of Word	# of Added OOV
ALL	2071	6819	15316	ALL	1605	5618	16443
S-5cl	940	4437	17309	S-5cl	172	1300	20132
S-10cl	Same as above			S-10cl	Same as above		
Topic A-3				**Topic A-4**			
LM	# of Sentences	# of Word	# of Added OOV	LM	# of Sentences	# of Word	# of Added OOV
ALL	1387	5259	16786	ALL	1036	5235	16691
S-5cl	115	1121	20302	S-5cl	632	3654	18026
S-10cl	21	315	21045	S-10cl	59	745	20624
Topic A-5							
LM	# of Sentences	# of Word	# of Added OOV				
ALL	1626	5681	16584				
S-5cl	1614	5669	16591				
S-10cl	1	12	21335				

ALL: language model constructed from all documents
S-5cl: language model constructed from a selected cluster in five clusters
S-10cl: language model constructed from a selected cluster in ten clusters

These results mean that the documents collected by using the recognition result for the whole target speech are not sufficient to construct the topic-specific language model due to the misrecognition words included in the recognition result.

35.4.2.3 Spectral Clustering with Each Utterance

Based on the above results, we next conducted the spectral clustering with each utterance for the same topics. Here, we selected the target clusters as the region including each utterance. Each utterance was separated based on the zero-crossing rate and level threshold for the speech signal by using the modified program of the silence detection included in Julius. Threshold parameters used in the experiments are same as Julius's.

Figure 35.4 shows the location of each utterance in the spectral clustering results. Each number in parentheses in the bottom lines shows an utterance number in each topic. Table 35.7 shows the specification of the selected clusters. As can be seen from these results, the large cluster that consists of documents similar to the target utterance could be selected for all topics except for topic A-5.

Table 35.8 shows the results of the speech recognition experiments with the topic-specific language model constructed for the selected documents shown in Table 35.7 for each topic. Figure 35.5 shows F-measures (harmonic means of recall and

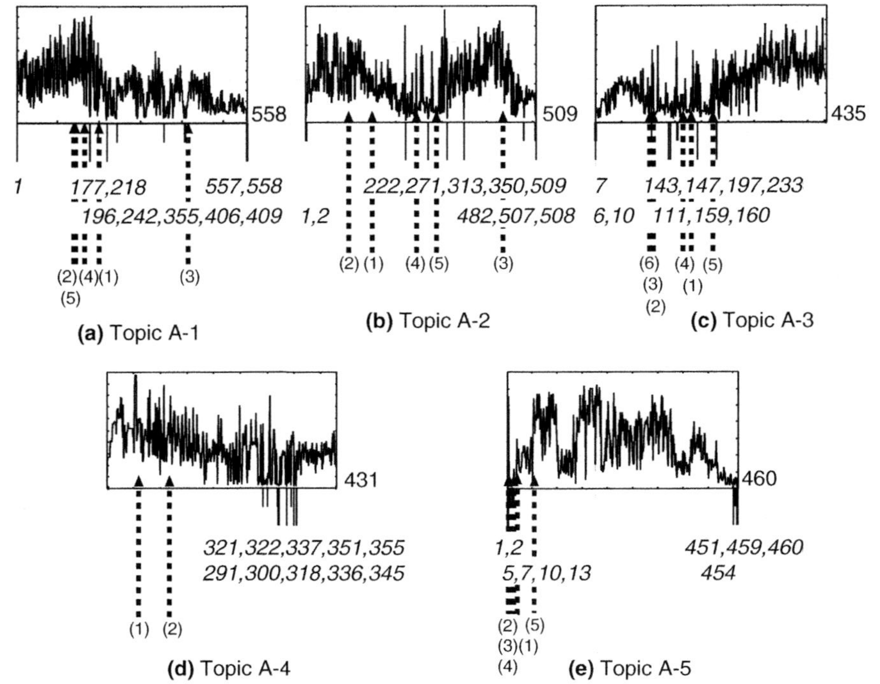

Fig. 35.4 Results of spectral clustering for each topic

Table 35.7 Selected clusters with each utterance for dataset A

Topic	Location of each utterance	Selected clusters with each utterance	# of Selected documents
A-1	136, 140, 197, 163, 412	1 to 177, 196 to 218, 409 to 557	349
A-2	97, 159, 246, 289, 435	1 to 313, 350 to 509	473
A-3	103, 107, 109, 166, 184, 222	10 to 111, 160 to 233	176
A-4	56, 94	1 to 291	291
A-5	4, 5, 6, 14, 55	2 to 7, 13 to 451	445

Table 35.8 Speech recognition result using the clusters including each utterance for dataset A

Topic	Topic-specified LM with selected documents		
	WCR	P	R
A-1	77.9	76.3	76.3
A-2	39.3	32.7	43.2
A-3	74.6	58.5	70.6
A-4	57.5	64.7	73.3
A-5	53.1	38.3	39.9

WCR word correct rate, *P* precision, *R* recall

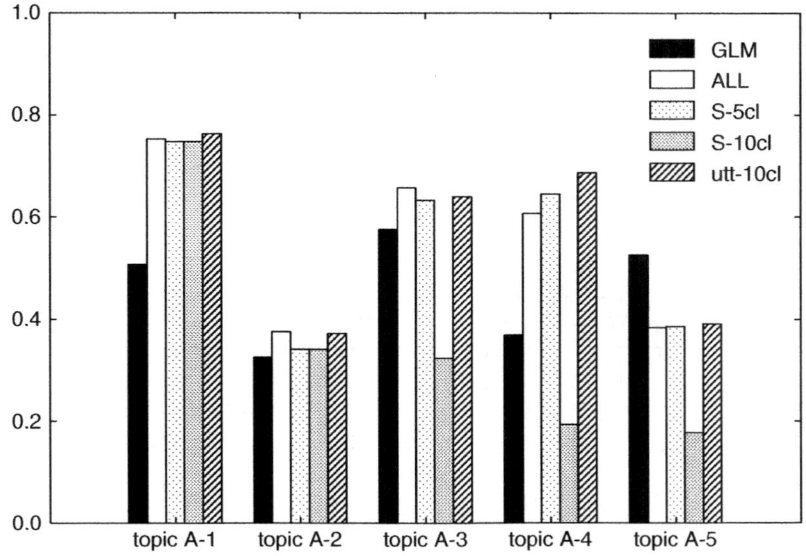

Fig. 35.5 F-measures for speech recognition results. GLM: generic language model. ALL: language model constructed from all documents. S-5cl: language model constructed from a selected cluster in five clusters. S-10cl: language model constructed from a selected cluster in ten clusters. utt-10cl: language model constructed from selected clusters including utterances in ten clusters

precision) for each method discussed above for comparison . As shown in Table 35.8 and Fig. 35.5, the results except for topic A-5 with the language model constructed from the clusters including each utterance are improved as compared with the general language model and the language model constructed from all documents.

35.4.3 Experiments for Dataset B

In the experiment described above, Web documents similar to a recognition result for a news article spoken by a newscaster were collected from the WWW. However, the influence of the misrecognition words is not small so that it resulted in collecting a lot of dissimilar documents.

So, we collected similar Web documents using the recognition result of each utterance of each topic for dataset B independently. Figure 35.6 shows results of spectral clustering for dataset B. Each number in parentheses in the bottom lines shows an utterance number in each topic. We obtained the boundaries of document clusters using the 5 d_i from the smallest to the largest as the same as the previous experiment and selected clusters that include more than two utterances. Table 35.9 shows the result of the selected clusters.

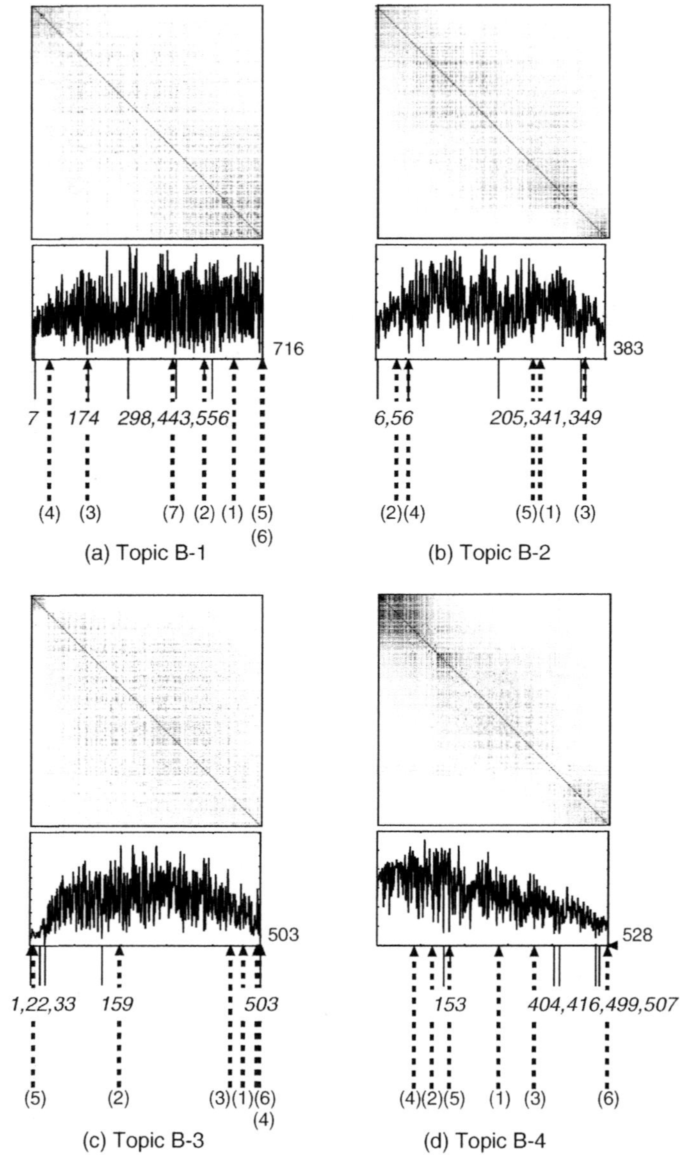

Fig. 35.6 Results of spectral clustering for each topic

35.4.3.1 Performance of Topic-Specific Language Model

Next, we conducted speech recognition experiments using the topic-specific language model constructed from the selected clusters shown in Table 35.9. Table 35.10 shows the specification of the topic-specific language models. From Table 35.10

Table 35.9 Selected clusters with each utterance for dataset B

Topic	# of All documents	Location of each utterance	Selected clusters with each utterance	# of Selected documents
B-1	716	623, (530), 171, 51, 709, 713, (435)	7 to 174, 556 to 716	329
B-2	383	274, 37, (347), 56, 262	6 to 56, 205 to 341	188
B-3	503	461, 193, 433, 496, (6), 491	159 to 503	345
B-4	528	279, 126, 360, 86, 166, (526)	1 to 404	404

Table 35.10 Specification of topic-specific language models

Topic B-1				Topic B-2			
LM	# of Sentences	# of Words	# of Added OOV	LM	# of Sentences	# of Words	# of Added OOV
utt-5cl	7991	13910	11832	utt-5cl	4466	14610	11137
Topic B-3				**Topic B-4**			
LM	# of Sentences	# of Words	# of Added OOV	LM	# of Sentences	# of Words	# of Added OOV
utt-5cl	15512	20001	8097	utt-5cl	22986	20001	8261

utt-5cl: language model constructed from clusters including more than two utterances in five clusters

it can be seen that more sentences and words could be selected as compared with Table 35.6.

The results of speech recognition for each topic are shown in Table 35.11. The results with the general language model and with the language model constructed from all of the collected documents are shown for comparison. In this table, a number in parentheses means results for an utterance that is filtered out, which is not used in the calculation of the average. As shown in Table 35.11, the recognition performance for each utterance in all topics is improved significantly as compared with that of the general language model. As compared with the language model constructed from all documents, there is no remarkable improvement for topics B-1 and B-2. However, it is shown that the performance for the utterances filtered out, shown in parentheses, are improved. This means that the document filtering method we proposed, which is a method for removing the documents collected with the low confidence recognition results from the training corpus, is useful for decreasing noise in the training corpus.

Table 35.12 shows the candidate of the index terms extracted from the recognition result for topic B-1. The candidates are selected from high-frequency words. As can be seen from this table, the topic-specific language model can extract the correct candidates of the index terms.

Table 35.11 Results of speech recognition for each utterance of dataset B

Topic B-1									
	General LM (MNP-20k)			Topic-specific LM (selected clusters)			Topic-specific LM (all clusters)		
No.	WCR	P	R	WCR	P	R	WCR	P	R
B-1-1	58.1	41.7	35.7	72.1	75.0	64.3	67.4	75.0	64.3
B-1-2	67.4	63.2	57.1	(89.1)	(94.1)	(84.2)	82.6	93.3	73.7
B-1-3	78.6	80.0	57.1	92.9	80.0	80.0	92.9	80.0	80.0
B-1-4	72.2	66.7	28.6	94.1	66.7	66.7	100.0	100.0	100.0
B-1-5	75.0	50.0	40.0	100.0	100.0	100.0	80.0	66.7	80.0
B-1-6	64.7	100.0	60.0	82.4	75.0	60.0	82.4	75.0	60.0
B-1-7	74.6	52.6	52.6	(74.2)	(72.2)	(72.2)	78.8	66.7	77.8
Ave.	70.9	64.9	47.3	88.3	79.3	74.2	83.4	79.5	76.5
Topic B-2									
	General LM (MNP-20k)			Topic-specific LM (selected clusters)			Topic-specific LM (all clusters)		
No.	WCR	P	R	WCR	P	R	WCR	P	R
B-2-1	65.0	71.4	71.4	71.4	71.4	71.4	71.4	71.4	71.4
B-2-2	55.1	57.1	50.0	75.5	66.7	62.5	79.6	76.5	81.3
B-2-3	66.7	54.5	75.0	(73.3)	(54.5)	(75.0)	62.2	46.2	75.0
B-2-4	77.4	62.5	62.5	73.3	71.4	62.5	70.0	85.7	75.0
B-2-5	65.9	81.8	90.0	73.2	88.9	80.0	73.2	88.9	80.0
Ave.	66.0	65.5	69.8	73.4	74.6	69.1	71.3	73.7	76.5
Topic B-3									
	General LM (MNP-20k)			Topic-specific LM (selected clusters)			Topic-specific LM (all clusters)		
No.	WCR	P	R	WCR	P	R	WCR	P	R
B-3-1	93.9	100.0	100.0	97.9	100.0	100.0	97.0	100.0	100.0
B-3-2	59.1	36.4	50.0	72.7	60.0	66.7	77.3	66.7	66.7
B-3-3	78.4	75.0	69.2	75.7	69.2	69.2	75.7	69.2	69.2
B-3-4	73.8	70.0	58.3	62.8	64.3	75.0	65.1	69.2	75.0
B-3-5	44.4	0.0	0.0	(40.0)	(50.0)	(25.0)	40.0	33.3	25.0
B-3-6	79.0	100.0	100.0	79.0	100.0	100.0	79.0	100.0	100.0
Ave.	71.4	63.6	62.9	77.6	78.7	82.2	72.4	73.1	72.7
Topic B-4									
	General LM (MNP-20k)			Topic-specific LM (selected clusters)			Topic-specific LM (all clusters)		
No.	WCR	P	R	WCR	P	R	WCR	P	R
B-4-1	72.1	63.2	63.2	86.9	88.9	84.2	86.9	88.9	84.2
B-4-2	83.0	86.7	81.3	90.6	87.5	87.5	90.6	87.5	87.5
B-4-3	54.2	30.8	28.6	72.9	78.6	78.6	72.9	78.6	78.6
B-4-4	77.5	75.0	69.2	82.1	91.7	84.6	82.1	91.7	84.6
B-4-5	90.0	83.3	76.9	83.3	80.0	61.5	83.3	80.0	61.5
B-4-6	11.9	12.5	11.1	(19.1)	(29.6)	(22.2)	14.3	14.3	11.1
Ave.	64.8	58.6	55.1	83.2	85.3	79.3	71.7	73.5	67.9

WCR word correct rate, *P* precision, *R* recall

Table 35.12 Extraction result for topic 1

Correct		General LM		Topic specific LM	
Hokkaido	6	Hokkaido	7	Hokkaido	7
Atomospheric pressure	5	North Japan	3	Atomospheric pressure	4
Expectation	4	site	3	Hokuriku	3
North Japan	3	forcast	2	North Japan	3
Rainstorm	3	Hokuriku	2	Wind	3
Wind	3	wind	2	Japan sea	3
(Pressure) pattern	3	Japan sea	2	Winter-type	3
			2		
Japan sea	3	East Japan	2	High wave	3
Winter-type	3	mirror	2	Forcast	2
High wave	3			Windstorm	2
Marine	3			Forming	2
Stomy weather	3			(Pressure) pattern	2
Forcast	2			Tohoku	2
Hokuriku	2			East Japan	2
Forming	2			Place	2
Tohoku	2			Peak	2
East Japan	2				
Snow	2				
Sea	2				
Peak	2				

35.5 Conclusions

In this chapter we introduced the framework of the automatic index term extraction system for broadband news and proposed the document filtering method for the language model adaptation using a spectral clustering approach in order to select the appropriate topic-related documents. The basic idea is that broadcast news has similar Web documents on the Internet news site, so the performance of news speech recognition can be improved with the adapted language model by collecting the similar article via Web crawling. To show the effectiveness of this approach, spectral clustering and speech recognition experiments were demonstrated for nine topic broadcast news speeches. From the experimental results of spectral clustering, we confirmed that the proposed method can extract the suitable cluster that consists of the topic-related documents and filter out the unsuitable clusters. In addition, we showed that the speech recognition performance is improved significantly as a whole using the topic-specific language model constructed from the selected clusters that consist of similar articles.

This approach is characterized by filtering out the useless documents dis-similar to speech input with the distribution of Web resources as external information without a so-called confidence score as internal information. In the other words, this approach estimates the confidence of the speech recognition result utilizing knowledge of word co-occurrence relations extracted automatically from the Web document.

In the future, we intend to develop a total search system for broadcast news programs using the automatic index term extraction system with the language model adaptation method we have proposed in this chapter.

References

1. Takai, D., Morimoto, T., Takahashi, S., "Extraction of index terms for retrieving multimedia news documents from World Wide Web (in Japanese)," *Proceedings of the 56th JCEEE Kyushu*, Kumamoto, Japan, August 2003
2. Takhashi, S., Morimoto, T., Irie, Y., "Adaptation of language model with iterative web crawling for speech recognition of broadcast news (in Japanese)," *Proceedings of FIT2006*, Fukuoka, Japan, pp. 381–384, September 2006
3. Zhu, X., Rosenfield, R., "Improving trigram language modeling with the World Wide Web," *Proceedings of ICASSP'01*, Salt Lake City, UUT, May 2001
4. Berger, A., Miller, R., "Just-in-time language modeling," *Proceedings of ICASSP'98*, Seattle, pp. 705–708, December 1998
5. Bulyko, I., Ostendorf, M., Stolcke, A., "Getting more mileage from Web text sources for conversational speech language modeling using class-dependent mixtures," *Proceedings of HLT-ACL*, Edmonton Canada, pp. 7–9, May 2003
6. Nishimura, R., et al., "Automatic N-gram language model creation from Web resources," *Proceedings of EUROSPEECH-2001*, Aolbarg, Denmark, pp. 2127–2130, September 2001
7. Sethy, A., Georgiou, P.G., Narayanan, S., "Building topic specific language models from webdata using competitive models," *Proceedings of INTERSPEECH'06*, Lisboa, Portugal, pp. 1293–1296, September 2005
8. Suzuki, M., Kajiura, Y., Ito, A., Makino, S., "Unsupervised language model adaptation based on automatic text collection from WWW," *Proceedings of INTERSPEECH'06*, Pittsburgh, pp. 2202–2205, September 2006
9. http://julius.sourceforge.jp/
10. Salton, G., et al., "A vector space model for automatic indexing," *Communications of the ACM*, V18, N11, pp. 613-620, 1975. Reprinted in *Readings in Information Retrieval*, Jones, K.S. and Willett, P. (Eds.), Morgan Kaufmann, San Mateo, CA, pp. 273–280, 1997
11. Nagatomo, K., et al., "Complemental back-off algorithm for merging language models (in Japanese)," *IPSJ Journal*, V43, N9, pp. 2884–2893, September 2002
12. Pothen, A., Simon, H., Liou, K., "Partitioning sparse matrices with eigenvecotors of graphs," *SIAM Journal on Matrix Analysis*, V11, N3, pp. 430–452, 1990
13. Dhillon, I.S., "Co-clustering documents and words using bipartite spectral graph partitioning," *ACM SIGKDD International Conference on Knowledge Discovery and Data Mining*, San Francisco, pp. 269–274, August 2001
14. Tsuruta, N., Aly, S.K.H., Maeda, S., Takahashi, S., Morimoto, T., "Self-organizing map vs. spectral clustering on visual feature extraction for human interface," *Proceedings of International Forum on Strategic Technology (IFOST) 2006*, Ulsan, Korea, pp. 55–58, October 2006
15. Ding, C., He, X., "Linearized cluster assignment via spectral ordering," *Proceedings of ACM International Conference on Machine Learning*, Banff, Canada, pp. 30–37, July 2004
16. Clarkson, P.R., Rosenfeld, R., "Statistical language modeling using the CMU-Cambridge toolkit," *Proceedings of ESCA Eurospeech*, Rhodes, Greek, pp. 2707–2710, September 1997

Chapter 36
Automatic Construction of FSA Language Model for Speech Recognition by FSA DP-Matching

Tsuyoshi Morimoto and Shin-ya Takahashi

36.1 Introduction

For accurate speech recognition, a well-defined language model is necessary. When a large amount of learning texts (corpus) is available, a statistical language model such as bi-gram or tri-gram generated from a corpus is quite powerful. For example, most of current dictation systems employ bi-gram or tri-gram language models. However, if the size of the corpus is not sufficiently large, reliability of statistical information calculated from the corpus decreases and then so does the effectiveness of the generated statistical language model (sparseness problem). Furthermore, preparing a sufficient amount of texts for spoken language is generally very expensive. Therefore, a finite state automaton (FSA) language model is generally used for small- or middle-size (around 1000 words) vocabulary speech recognition. However, defining a FSA model by hand requires much human effort. In some systems, a FSA model is automatically generated by converting from regular-grammar type grammar rules (e.g., see Hparse [1]), but it is still a very time- and effort-consuming task to prepare a grammar with sufficient coverage and consistency.

On generating a FSA language model from learning data, several methods have been proposed. Note that constructing an acyclic FSA from given data is a fairly simple problem; one can construct a TRIE tree in which common prefix is shared, and then minimize the tree by merging equivalent states. This method, however, is computationally quite expensive. Several methods have been proposed to improve the efficiency [2, 3], but they are still at the stage of basic research and have not yet been applied to practical applications. Meanwhile, other kinds of approaches have been proposed [4–6] aiming at applications to an actual speech recognition language model. However, because common key techniques adopted by them for improving efficiency is to use stochastic features, a sparseness problem mentioned above arises again when the corpus size is not large enough.

In this chapter, we propose a new method to construct a FSA language model, by using a FSA DP (dynamic programming) matching method. We also report some

Oscar Castillo et al. (eds.), *Trends in Intelligent Systems and Computer Engineering.*

experimental results; the method was applied to a travel conversation corpus (with about a thousand sentences) to generate a FSA model, and then, speech recognition experiments using the language model were conducted. The result shows that the recognition correct rate is high enough for closed data, but is not so satisfactory for open data. To cope with this problem, we also propose an additional mechanism that decides whether a recognized result can be accepted or should be rejected by evaluating a distance of the result from the learning corpus. After the mechanism is applied, the recognition correct rate for accepted results is considerably improved.

36.2 Overview of the Algorithm

Before executing DP matching, sentences in a corpus are grouped into several clusters based on distances between them. This is to avoid unnecessary execution of DP matching between dissimilar sentences. There are several ways[1] to calculate a distance between two sentences, but we adopted Eq. 36.1 for simplicity.

$$d(S_x, S_y) = 1 - \frac{Num(w|w \in S_x I\ S_y)}{Num(w|w \in S_x Y\ S_y)} \tag{36.1}$$

where

$d(S_x, S_y)$: distance between sentence S_x and S_y
$Num(w)$: number of words such as w

For each cluster, one randomly chosen sentence (a target) is converted to a simple one-path FSA.[2] Next, another sentence (a reference) is picked up, converted to a one-path FSA in the same way, and two FSA are DP matched (hereafter, we call this DP matching method FSA-DP matching: FDP). The detail of FDP is described in the next section. As a result of FDP, relationships between target and reference nodes (words) such as *equal, substitution, deletion*, or *insertion* are obtained. Then, according to these relationships, appropriate nodes or arcs are added to the target FSA (this process is called *merging)*. In this way, the target FSA is incrementally extended until the whole sentences in a cluster are processed.

The above procedure is repeated for all clusters. When finished, many FSA are eventually obtained; each of them corresponds to each cluster. Finally, these FSA are combined as one big FSA so that they share a common start (ST) node and a common end (ED) node.

---

[1] For instance, a distance based on a vector space model is widely used in an IR field.

[2] A FSA is defined as a DAG (directed acyclic graph). A word is represented as a node, and a connection between words is represented as a directed arc. A FSA has a start node and an end node.

36.3 FDP Matching and FSA Construction

36.3.1 FDP Matching

Nodes of a target FSA and a reference FSA are placed on a y-axis and x-axis, respectively. Here, the target FSA's nodes are linearly placed in topological order and arcs between them are maintained. In FDP matching, the basic matching algorithm is as same as normal DP matching [7], but is different in a way that matching paths are decided according to arcs in a target FSA. When a node C and a node X are DP matched as shown in Fig. 36.1, a global distance is decided as minimum one among distances calculated along five paths.

More formally, a global distance at a point $(x,\ y)$ is calculated according to Eq. 36.2.

$$gd(x,y) = \min \begin{pmatrix} gd(x-1,\ y) \\ gd(x-1,\ \Delta y) \\ gd(x,\ \Delta y) \end{pmatrix} + ld(x,y) \tag{36.2}$$

where

$gd(x,\ y)$: a global distance between word w_x and word w_y
$ld(x,\ y)$: a local distance between word w_x and word w_y, and calculated as follows.

$$ld(x,\ y) = \begin{pmatrix} 0.0\ (w_x = w_y) \\ 1.0\ (W_x \neq w_y) \end{pmatrix} \tag{36.3}$$

Δy: locations on a y-axis of the previous words

When calculating a global distance at a point $(x,\ y)$, the best incoming path that has the minimum global distance is internally kept, and after having reached the final point, the totally best matched path (having the smallest global score) is chosen by backtracing these paths.

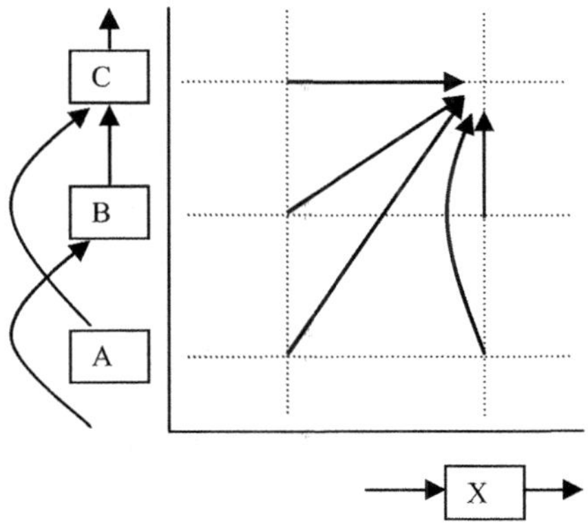

Fig. 36.1 FDP matching

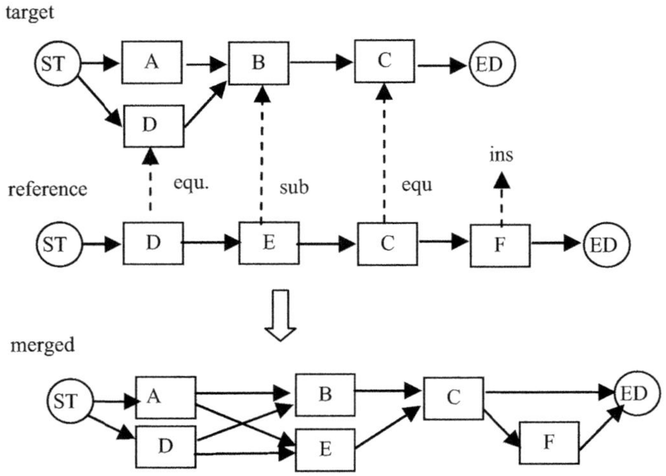

Fig. 36.2 Merging

36.3.2 Construction of a FSA

A relationship of the reference node and a target node at a position (x, y) is decided according to where a path comes from and a local distance at that point.

- equal (equ): incoming path is from $(x-1, \Delta y)$ and $ld = 0.0$
- substitution (sub): incoming path is from $(x-1, \Delta y)$ but $ld \neq 0.0$
- deletion (del): incoming path is from $(x-1, y)$
- insertion (ins): incoming path is from $(x, \Delta y)$

Next, according to these relationships, reference nodes are merged to the target as shown in Fig. 36.2.

When finished for all clusters, all FSA are combined into one FSA so that they share a common ST and ED node as mentioned before.

36.4 Speech Recognition Experiment and Discussion

36.4.1 Corpus

From four Japanese–English travel conversation textbooks, Japanese travel conversation sentences have been collected as a corpus. Here, sentences being too colloquial or fragmental were removed, and several new sentences modified slightly from the collected sentences were added to supplement insufficiency of the collected data. The number of vocabulary and sentences included in the corpus are shown in Table 36.1, and some sentence examples are shown in Fig. 36.3.

Table 36.1 Corpus

Vocabulary	1254 words
Sentences	1000 sentences
Average words per sentence	8.87 words

Fig. 36.3 Sentence examples

Sukaato-o sagashi-te i-masu.
(I'm looking for a skirt.)
Otearai-wa doko-desu-ka?
(Where is a restroom?)
Shiyakusyo-e-wa dou ittara ii-desu-ka?
(How can I get to the city-hall?)
Kuriimu-iri koohii-o ippai morae-masu-ka?
(Can I have a cup of coffee with cream?)
Konnya-juu-ni suutsu-ni puresu-o shite morae-masu-ka?
(Can I get my suite pressed by tonight?)

Table 36.2 Features of the speech recognition system and the test set

Recognizer	HVite [1]
HMM	Context-independent tri-phone model with four mixtures [8]
Test set	60 utterances (three male speakers spoke different utterances)

Table 36.3 Experiment result (closed data)

No. of clusters	No. of nodes	No. of links	Branching factor	Word correct rate (%)	Sentence correct rate (%)
30	3519	5098	1.45	98.7	90.0
50	3738	5299	1.42	98.7	90.0
70	3881	5447	1.40	98.9	91.7
90	3939	5519	1.40	98.9	93.3
Bi-gram	—	—	5.88 (Perplexity)	93.0	70.0

36.4.2 Experiment for Closed Data

All 1000 sentences were used as learning data, and several FSA were constructed for different numbers of clusters. As a test set, 60 sentences were selected randomly from the learning sentences, and were used as test data for a speech recognition experiment. Features of the speech recognition system and the test set used for the experiment are shown in Table 36.2.

In Table 36.3, static features, such as number of nodes, links, and a branching factor (i.e., an average number of branches per node, $\Sigma\, arcs/\Sigma\, nodes$) of the generated

FSA, and results of speech recognition experiments are shown. We see that one word is defined roughly as three nodes, and FSAs constructed by our method have a very small branching factor compared to the perplexity of the bi-gram. Consequently, they can attain a high speech recognition correct rate. Especially, the sentence recognition correct rate is 20 points higher than that of the bi-gram. As the number of clusters increases, a branching factor decreases and then the speech recognition correct rate slightly increases.

36.4.3 Experiment for Open Data

To evaluate for open data, we took out 60 sentences as test data from the corpus, and constructed a FSA from the remaining 940 sentences. Here, as the learning sentences are absolutely insufficient, simple elimination of test texts will cause disappearance of several words from the learning corpus. To avoid this, nouns appearing in 940 sentences were replaced by appropriate semantic classes, and then a FSA was constructed by FDP. In FDP, the equation of calculating local distance is modified as the next equation.

$$ld(x, y) = \begin{pmatrix} 0.0(w_x = w_y) \\ 0.5(w_x \neq w_y, sem_x = sem_y) \\ 1.0(w_x \neq w_y, (sem_x \neq sem_y \; or \; w_x \neq noun)) \end{pmatrix} \quad (36.4)$$

Here, two nodes are regarded as *quasi-equal* if $ld(x, y) = 0.5$, and a node representing the semantic-class is generated. In the next step, these nodes are expanded to all nouns which belong to that semantic-class. For conversion from a noun to a semantic-class, we used *Bunrui-Goi-Hyo* (BGH) [6] as a thesaurus. In BGH, each word is categorized in five levels. For example, a word "airport" is assigned a semantic code as follows. We use two decimal places as a semantic code.

Experiment results are shown in Table 36.4. Comparing with Table 36.3, we see that the speech recognition correct rate, especially the sentence recognition correct rate, drops very much. This would be because some words except nouns disappeared

Table 36.4 Experiment result (open data)

No. of clusters	No. of nodes	No. of links	Branching factor	Word correct rate (%)	Sentence correct rate (%)
30	7252	12325	1.70	78.8	26.7
50	7410	12402	1.67	79.9	21.7
70	7720	12932	1.68	79.4	28.3
90	7367	12170	1.65	69.2	21.7
Bi-gram	—	—	6.36 (Perplexity)	58.4	3.3

in the final FSA and/or even some paths did so because the size of the training data was fairly small. Although the results are not so satisfactory, they are still better than that of the bi-gram.[3]

36.5 Introducing Accept/Reject Mechanism Based on Distance of Speech Recognition Result from Learning Corpus

As mentioned above, some words or paths necessary for correct recognition do not appear in a FSA for open data. This fact implies that the speech recognition correct rate depends on the distance of a test text from the learning corpus; as a distance becomes large, the recognition correct rate drops. Note here that we do not know a correct word string of a test text, and hence we cannot calculate a distance between a test text and a learning corpus itself. However, it can be expected that a recognition result would not be quite different from a test text because the word recognition correct rate is considerably high as seen in Table 36.4, and therefore the same tendency still remains for distance between a recognition result and a learning corpus. From these considerations, we examined the relation between a distance described above and the speech recognition correct rate. The result is shown in Fig. 36.4 for the case of 70 clusters. Here, the x-axis is a distance of a test text from a learning corpus calculated by Eq. 36.5. From this figure, we can see that when a distance is less than 0.5, the speech recognition correct rate is comparatively high.

$$D(S_x, C) = \min d(S_x,\ S_y | S_y \in C) \tag{36.5}$$

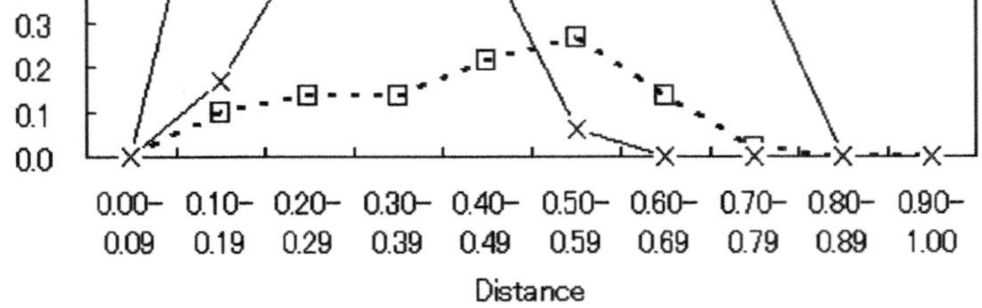

Fig. 36.4 Distribution of distance and speech recognition accuracy

[3] When constructing a bi-gram language model, 60 sentences were not eliminated but back-off (Witten–Bell) smoothing was employed.

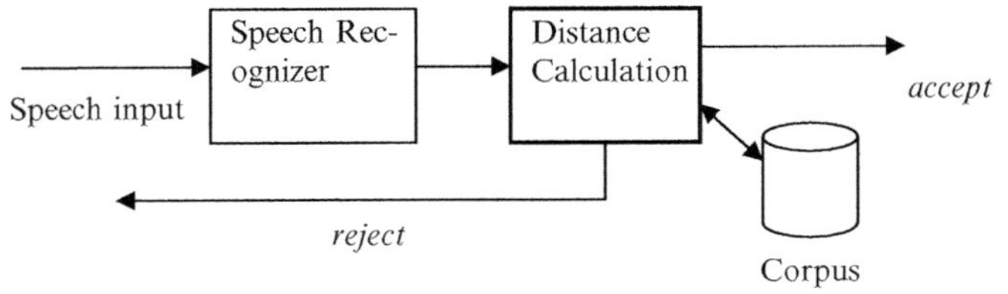

Fig. 36.5 Accept/reject mechanism

Table 36.5 Effectiveness of accept/reject mechanism

Decision	The Decision is	
	Right (%)	Wrong (%)
Accept (58.3%)	42.9	57.1
Reject (41.7%)	96.0	4.0

where

$d(S_x, S_y)$: the same within Eq. 37.1
$D(S_x, C)$: distance between sentence S_x and a corpus C

This fact encouraged us to introduce an additional mechanism which decides whether a recognized result can be accepted or should be rejected according to a distance; a result is accepted if a distance is higher than a certain threshold (in the above case, 0.5), but is rejected if a distance is less than it (see Fig. 36.5).

When the result is rejected, a user would be solicited to speak again with another wording. We evaluated the effectiveness of this mechanism for the case of 70 clusters. The ratio of accept versus reject is 58.3% to 41.7%. The result is shown in Table 36.5.

In the table, *accept-right* means that the recognition result is accepted and the decision is right (the recognition result is correct); *reject-right* means, to the contrary, that the result is rejected and the recognition result is actually erroneous, and so on. From this result, we see that, for the *accepted* results, the sentence recognition correct rate rises up to 42.9%, that is, 14.6 points up from the result in Table 36.3.

36.6 Sentence Acceptability

A generated FSA is able to accept more sentences than learning texts. We evaluated how many sentences could be accepted by a generated FSA. First, 500 sentences were randomly generated on the FSA, and they were checked by a human as to whether they were syntactically and semantically correct. For correct sentences, we counted separately whether a generated sentence was the same as one in the learning data, or a new one. The result is shown in Table 36.6.[4]

[4] Only unique sentences are counted.

Table 36.6 Correctness of generated texts

Category	No. of sentences
Same with one in learning data	75
Correct	97
Slightly wrong	18
Wrong	123

This result can be interpreted as follows. If 75 sentences are used as learning texts, new 97 unseen texts can be accepted by the FSA, and therefore, the FSA can accept $(75+97)/75 = 2.29$ times sentences as many as learning sentences.

36.7 Conclusion

A method to automatically construct a FSA language model from a learning corpus is proposed. This method is quite effective for middle-size speech recognition, because it requires neither a huge size learning corpus nor much human elaboration. We applied the method to travel conversation sentences, and conducted a speech recognition experiment with generated FSA.

It is shown that the speech recognition correct rate for closed data is quite high. On the other hand, it is not so satisfactory for open data, although still better than that of a statistical language model. To cope with this problem, we also propose an additional mechanism that decides whether a speech recognition result can be accepted or should be rejected by evaluating the distance of the result from the learning corpus. After the mechanism is applied, recognition accuracy for accepted results is considerably improved.

In the next step, we are planning to improve coverage, especially for open data, of a generated language model by introducing another kind of word category such as part-of-speech, and/or to apply to other kinds of conversational sentences.

References

1. S. Young et al.: *The HTK Book (for Ver. 3.0)*, 1999 (http://htk.eng.cam.ac.uk/)
2. K. J. Lang, B. A. Pearlmutter, and R. Price: Results of the Abbadingo one DFA learning competition and a new evidence driven state merging algorithm, *Proceedings of International Colloquium Grammatical Inference*, pp. 1–12, 1998
3. S. M. Lucas and T. J. Reynolds: Learning deterministic finite automata with a smart state labeling evolutionary algorithm, *IEEE Transactions on Pattern Analysis and Machine Intelligence*, Vol. 27, No. 7, July 2006
4. C. Kermorvant, C. de la Hinguera, and P. Dupont: Learning typed automata from automatically labeled data, *Journal Électronique d'Intelligence Artificielle*, Vol. 6, No. 45, 2004
5. J. Hu, W. Turin, and M. K. Brown: Language modeling with stochastic automata, *Proceedings of International Conference on Spoken Language Processing (ICSLP)-1996*, 1996

6. G. Riccardi, R. Pieraccini, and E. Boccieri: Stochastic automata for language modeling, *Computer Speech and Language*, Vol. 10, No. 4, pp. 265–293, 1996
7. A. V. Aho, J. D. Ullman, and J. E. Hopcroft: *Data Structures and Algorithms*, Addison-Wesley, 1983
8. T. Kawahara, A. Lee, K. Takeda, K. Itou, and K. Shikano: Recent progress of open-source lvcsr engine julius and japanese model repository—Software of continuous speech recognition consortium, *Proceedings of International Conference on Spoken Language Processing (ICSLP)-2004*, 2004 (http://julius.sourceforge.jp/en/julius.html)

Chapter 37
Density: A Context Parameter of Ad Hoc Networks

Muhammad Hassan Raza, Larry Hughes, and Imran Raza

37.1 Introduction

A mobile ad hoc network or MANET is an autonomous collection of mobile nodes that communicate over wireless links. This mobility means that the network topology may change rapidly and unpredictably over time as the nodes move or adjust their transmission and reception parameters. The network is also decentralized, meaning that the nodes must execute message delivery independently of any centralized control [1].

The number of nodes in the area of an ad hoc network is often referred to as density and has been defined as the number of neighbours within a node's transmission range, or the total number of nodes within a given area [2].

Density is one of the context parameters for ad hoc networks such as node speed, pause-time, network size, and number of traffic sources [3]. Density can influence the behaviour of an ad hoc network: increasing the number of nodes in an area can result in congestion and collisions, and when the number of nodes in an area is low, the coverage tends to be poor [4]. The power consumption also increases with the increase in the density and the efficiency of the network decreases [3].

Density is referred to when performance metrics are defined [5], when researchers define the context of an ad hoc network for experiments [6], and when different kinds of protocols are compared. There are many protocols (e.g., [7]) that select design parameters on the basis of density but do not describe the mechanism for determining density. Despite the importance of density, little research appears to have been done to determine its value in an ad hoc network environment. Although some literature such as Durresi et al. [8] mentions the importance of local knowledge of an ad hoc node, in terms of 1-hop or 2-hop knowledge, for optimizing broadcasting, a coordinated scheme for determining density is required. This is the reason that not much related work is available to compare with this work.

This chapter describes two approaches to determining density in an ad hoc network: a census of nodes and traffic analysis. The remainder of this chapter is as

Oscar Castillo et al. (eds.), *Trends in Intelligent Systems and Computer Engineering.*

follows. Section 37.2 describes two anthropogenic counting techniques (census and traffic analysis) and explains how these counting techniques may be applied to determine density. Section 37.3 describes the design of the proposed algorithms. Section 37.4 presents an application of the density-determining algorithm to CARP. Section 37.5 presents simulation results and Sect. 37.6 consists of conclusions.

37.2 Anthropogenic Counting Techniques for Determining Density

Density of an ad hoc network can be determined by applying two basic anthropogenic counting concepts: population census and traffic analysis. Ad hoc networks consist of nodes and these nodes generate traffic by following some ad hoc routing protocol. The presence of a number of nodes in an area can be exploited by the concept of population census to determine density and analysis of the traffic among nodes can also be a method of determining density. A population census is a survey of an entire population conducted on a scientific basis after a specific period of time. The census data prove to be beneficial from an individual to groups such as governments as a planning and resource allocation tool [9].

Traffic monitoring on a transport link is a fundamental concern of transportation engineering. Traffic analysis is also widely understood as the process of intercepting and examining messages in order to deduce information from patterns in communication. In general, the greater the number of messages observed or even intercepted and stored, the more can be inferred from the traffic [10].

Traffic monitoring can be performed in the context of military intelligence or counterintelligence, and is a concern in computer security. In a military context, traffic analysis is usually performed by a signals intelligence agency, and can be a source of information about the intentions and actions of the enemy. Examples of the patterns include: frequent communications (can denote planning), rapid and short communications (can denote negotiations), and a lack of communication (can indicate a lack of activity, or completion of a finalized plan [10]). The usefulness of traffic monitoring can be reduced if traffic is faked or if traffic cannot be intercepted. Both occurred in the period before the attack on Pearl Harbor [10].

Traffic analysis is also a concern in computer security; an attacker can gain important information by monitoring, for example, the frequency and timing of network packets [11].

Studies such as McIntosh et al. [12] offer insight into the effects of modem traffic on the overall call-hold time distribution. The authors state that the use of the exponential distribution for call-holding time (service time) seriously underestimates the actual numbers of very long calls (e.g., analog modem data calls that last for many hours).

37.2.1 Applying Census and Traffic Analysis for Determining Density

When determining density of an ad hoc network means to obtain the number of nodes present in the area of that ad hoc network, it becomes similar to a population census that enumerates a group within a specific geographic jurisdiction. Generally, the enumeration of population is repeated after a fixed time but occasionally this time period is varied. For example, national censuses were conducted in Canada every ten years beginning in 1851. Since 1956, they have been conducted every five years [13]. This change in time period is derived by more frequent requirement to track relatively rapid changes in demographics since the last population census.

In ad hoc networks also, there is a need to enumerate nodes and to repeat the process of enumeration after a specific period of time to track the changes in the node population. Although implementation of the population census seems to be the same in ad hoc networks, it has certain distinctions from a population census. The scale of operation is very small in ad hoc networks and population changes are more dynamic and unpredictable because of the factors affecting density such as mobility. The density determined at one moment cannot be generalized for a future census; it reveals that time period between any two enumerations may need to have different values.

The communication in an ad hoc network is of broadcast and multihop nature, and a message is relayed hop by hop until it reaches its destination. This type of communication can produce a considerable amount of network traffic. The analysis of this network traffic can be a source to assess the density of the ad hoc network in the same way as by doing traffic analysis: the number of vehicles and related transportation resources are assessed and used for planning [14]. A node with a dense neighbourhood may need to process more traffic in terms of the number of packets as compared to that of a node with sparse neighbourhood.

37.3 Design

This section describes the design of two density-determining algorithms for ad hoc networks: a census of nodes and traffic analysis. The design description of an algorithm that reacts to changes in density is also included as a part of the density-determining algorithms.

37.3.1 Determining Density from a Node Census

The objective of this algorithm is to determine the density of an ad hoc network by conducting the census of location aware ad hoc nodes for enumerating nodes

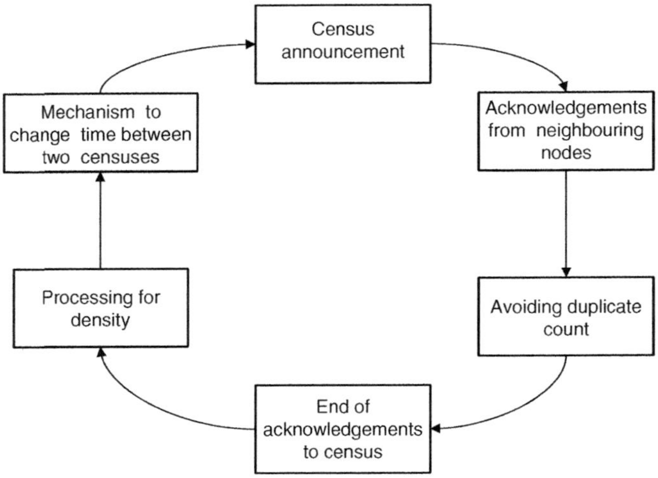

Fig. 37.1 Steps of the census algorithm

and determining related demographic information, notably the lowest and the highest Cartesian addresses of the enumerated nodes. The steps to be followed for implementing a census of ad hoc nodes are shown in Fig. 37.1 and explained in the following sections.

37.3.1.1 Census Announcement

The enumerating node announces a census by sending a message (*census_announcement*) under the controlled broadcast mechanism [4] to all other nodes asking for their respective Cartesian addresses. The controlled broadcast, in conjunction with an unique sequence number for each broadcast, and the enumerating node's identity, eliminates duplicate retransmission of the same message and reduces unnecessary traffic in the network.

37.3.1.2 Acknowledgement from Other Nodes

This part of the algorithm describes the way acknowledgements to the *census_announcement* are sent by enumerated nodes in the network.

Every node that receives a *census_announcement* sends its Cartesian address by a unicast acknowledgement to the enumerating node by using some geographical position-based protocol such as the Cartesian ad hoc routing protocol (CARP) [6], as now all the nodes know the enumerating node's Cartesian address, from the *census_announcement*.

When many neighbouring nodes send acknowledgements to a single enumerating node, it may cause an implosion. The IEEE 802.11 provides generic support for retransmission of the dropped packets [15]; however, in addition to IEEE 802.11, a delay-based protocol (DBP) addresses the implosion problem by having recipient nodes wait a random amount of time before sending an acknowledgement [16].

37.3.1.3 Avoiding Counting an Enumerated Node More Than Once

It is important to ensure that each enumerated node's Cartesian address is not counted more than once. If every incoming packet is checked for the duplicate, it may result in greater computing overhead, and the power of a node depletes at a faster rate. An approach based on statistical sampling is proposed to handle duplicate counts and economize computing resources. The enumerating node works in one of the two states: sampling or counting.

Sampling consists of collecting a sample of nonduplicate Cartesian addresses, and counting duplicate addresses detected during the collection of the sample. Whenever an address is received, it is first compared with the existing entries of the sample; if it exists, it is not included in the sample and the duplicate counter is incremented; otherwise (does not exist in the sample) it is included in the sample. When the sample has been collected, the ratio of duplicates to the sample size is known, and from this, the fraction can be obtained. Thereafter, the regular counting begins and continues until the end of the census period (determined by the enumerating node from a timeout), at which point the size of the population can be estimated. The regular counting counts every packet without consideration of the duplicates because the fraction of duplicate count to sample size is applied to filter out duplicate count from the total count.

The choice of sample size can reduce the probability of errors, while maximizing the accuracy of population estimates, and increasing the generality of the results. There are few sample size guidelines for researchers using exploratory factor analysis (EFA) and principal components analysis (PCA) [17]. Some statisticians who tried to set a cut-off point of large sample size have adopted a sample of 30 as the cut-off [18, 19]; the same sample size is to be used in the proposed density-determining algorithms.

The complete algorithm for handling sampling and counting of enumerated nodes is explained below.

1. Upon receiving an acknowledgement to the census announcement from the other (to be enumerated) nodes, the enumerating node checks each received Cartesian address for duplicate values in a *sample_table*; if no duplicate is detected then this address is recorded in the *sample_table*. If a duplicate is found, a counter dup_count is incremented and not recorded in the *sample_table*.
2. Once the number of entries in the *sample_table* is equal to the *sample_size*, regular counting begins, and every received packet is processed further without being checked for the duplicate, meaning that after collection, of the sample, the algorithm assumes that every incoming address is nonduplicate.

3. The number of duplicate Cartesian addresses (*dup_count*), determined during the initial sampling, is used to estimate the duplicate addresses in the *node_count*. Estimated number of duplicate Cartesian addresses:

$$N_{DA} = (dup\_count \div sample\_size) \times node\_count$$

Estimated number of nonduplicate Cartesian addresses:

$$N_{NDA} = node\_count - N_{DA}$$

The value of the *sample_size* is also added to N_{NDA} to get the total number of estimated nonduplicate addresses because the sample is also a part of the received nonduplicate addresses.

There may be a situation in which the total number of acknowledgements may be less than the sample; in this case, at the end of counting, the value of the *node_count* is 0, and the algorithm takes the current value of the sample as the total number of nonduplicate enumerated addresses.

37.3.1.4 End of Acknowledgement to a Census Announcement

The census of ad hoc nodes, once started, needs a mechanism to indicate that all the nodes have participated in polling and the census has ended. A population census carried out by humans has preset times for starting and ending polling, although in some circumstances, this time period can also be extended. If a preset time technique is applied to the census in ad hoc networks, it may leave some nodes unpolled or may unnecessarily wait for a response even after all nodes have responded.

An adaptive approach based on a timeout is proposed to determine the end of acknowledgement to a census announcement. During the process of receiving response from other nodes, if the time to wait for the next packet exceeds the timeout, the enumerating node assumes that there are no more acknowledgements to receive.

The value of the timeout is derived from the mean time that is calculated by dividing the sum of the duplicate addresses (*dup_count*) and the sample, by the time taken in receiving them. The value of the timeout should be greater than the mean time, and should not be large to keep the enumerating node waiting unnecessarily for more acknowledgements that are not there. The timeout is proposed to equal 3 × *meantime*, as that is considered to be sufficient time for a response.

37.3.1.5 Processing for Density

This segment of the algorithm determines density from the number and addresses of the enumerated nodes. The following abbreviations are used to describe the algorithm.

1. (x_e, y_e): is the Cartesian address of the enumerating node.
2. (x_r, y_r): is the Cartesian address of any enumerated node.

Fig. 37.2 Calculation of area

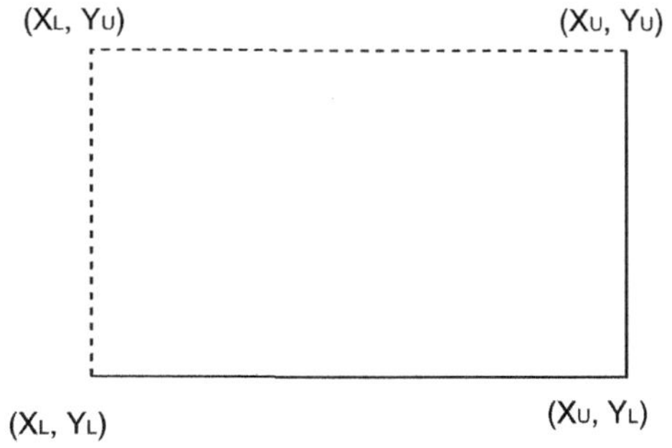

3. (x_L, y_L): is the lower Cartesian address of the enumerated addresses and is initialized to (x_e, y_e). This address also represents the lower-left corner of Fig. 37.2.
4. (x_U, y_U): is the upper Cartesian address of the enumerated addresses and is initialized to (x_e, y_e). This address also represents the upper-right corner of Fig. 37.2.

On receiving each acknowledgement, the enumerating node works as per the following algorithm to get the area of the ad hoc network.

For lower and upper x:

```
if (xr < xL)
xL = xr
else if
(xr > xU)
xU = xr
```

For lower and upper y:

```
if (yr < yL)
yL = yr
else if
(yr > yU)
yU = yr
```

From (x_L, y_L) and (x_U, y_U), the area of the network is determined that is equal to the multiple of the two sides of Fig. 37.2:

$$\text{area} = ((x_U - x_L) \times (y_U - y_L)).$$

The ratio of the total number of nonduplicate addresses (determined in Sect. 37.3.1) to the area, corresponds to the density of the ad hoc network.

37.3.1.6 A Mechanism to Change Time Between Two Censuses

Ad hoc networks are dynamic in nature, meaning that the density determined at one instant may not be the same as another due to the potential of node mobility.

This section describes an algorithm that reacts to these changes by taking corrective action by varying the frequency of the density calculation and based on feedback consisting of previously calculated values of density. A relatively dynamic network may need a more frequent determination of density and this part of the algorithm manages the changing time between two density calculations.

The spread in density is monitored by using variance that is the most commonly used measure of spread of a datum [20]. Variance (S^2) is the average squared deviation of values from the mean; for example, if the values of density are represented by ($d_1, d_2, d_3, \ldots, d_N$), the variance is calculated as

$$\text{Mean density (md)} = (d_1 + d_2 + d_3 + \cdots + d_N) \div N$$
$$\text{Variance } (S^2) = [(d_1 - md)^2 + (d_2 - md)^2 + \cdots + (d_N - md)^2] \div N.$$

If the values of the density are observed to have a wide spread, it requires a more frequent determination of density, thus needing a shorter time interval (*time_interval*). The new value of the *time_interval* is determined by applying the following formula.

$$\text{New } time\_interval = time\_interval - (time\_interval \times \text{Variance})$$

If the time to determine density is represented by *dentime*, the upper threshold for the *time_interval* between two consecutive density calculations is equal to three times *dentime* and the lower threshold for the *time_interval* equals *dentime*. If the algorithm examines *time_interval* to be 0, it replaces it with the upper threshold for the next round of tracking changes in density.

37.3.2 Determining Density from Traffic Analysis

When nodes are not location-aware and the protocols are independent of node location, some mechanism for determining density without reliance on location is needed. This section describes a mechanism that determines density by collecting statistics from the ongoing traffic in an ad hoc network. A node examines the traffic passing through it and collects IDs of the forwarding nodes to find the density. This algorithm enables all nodes in an ad hoc network to determine density independently, in parallel with the routing operation.

In contrast to the algorithm in Sect. 37.3.1, this algorithm measures density in terms of the nearest neigbouring nodes collected from traffic (passing through the node at which density is measured) over the time taken for the collection. A sample of the IDs (of forwarding nodes) is collected, and the time to collect the sample of forwarding node IDs is determined. In the description of this algorithm, the *dup_count* counts the number of duplicate packets from similar nodes, and N is the sample size. The algorithm is described as follows.

1. On receiving a packet, the ID of the forwarding node is extracted and checked against entries of the sample table for a duplicate. If no duplicate is found, the ID of the forwarding node is entered in the sample table of unique IDs.

2. If the new packet is found from any of the forwarding nodes having entry in the sample table, the *dup_count* increments.
3. The timer stops when the value of the entries in the table equals the sample size.
4. At the end of the process, the following equations are applied to calculate density:

$$\text{Total Count} = \text{N} + dup\_count$$

The total count is used to determine the mean time for receiving a packet because both the sample and *dupcount* are received in the time consumed in populating the sample table, so both are added to get the total count.

$$\text{The mean time taken by a packet } (T_{pkt}) = \text{Total Count } \div \text{Elapsed Time}$$

$$\text{Time taken by non\_duplicate packets } (T_{ND}) = \text{N} \times T_{pkt}$$

The sample of forwarding node IDs and the time taken for collecting this sample can provide an estimate about the density around the observing node. A relatively shorter value of time elapsed in collecting the sample means that the observing node is in a dense scenario and a greater value of time elapsed means that the scenario is less dense. An application-dependent and customized classification of the observed traffic from sparse to dense can also be defined.

To address the changes in density over time, the mechanism to change time between two density-determining cycles is applied as described in Sect. 37.1. We consider an application of the density-determining algorithms to Cartesian ad hoc routing protocols (CARP). The next section gives a brief description of CARP for a better understanding of these protocols.

37.4 Cartesian Ad Hoc Routing Protocols (CARP)

The Cartesian ad hoc routing protocols [7] are a set of three adaptive and connectionless protocols that address the problem of routing and power consumption in MANET. Each protocol operates at the physical layer and the network layer; all nodes are location aware. The authors claim that the design of CARP has three objectives: restrict flooding, reduce power consumption, and save bandwidth. The authors of CARP have also proved its better performance against the other leading geographical ad hoc protocols. All Cartesian ad hoc routing protocols attempt to restrict transmission to those nodes that lie between the source and the destination. These protocols are used to limit the number of forwarding nodes in a logical transmission area.

As nodes in the network are location aware, when a source node transmits a packet, a logical rectangular transmission area is formed by comparing the coordinates of the source and destination nodes. An approach of rectangular transmission area (RTA) in CARP reduces the size of the transmission area, and limits traffic within the transmission area. Although the RTA tries to reduces traffic, the volume of the flooding traffic may still be problematic in a very dense network. The trimmed transmission area (TTA) algorithm and transmission area with limiting angle (TALA) attempt to modify the shape of the transmission area through a simple optimization.

The CARP attempt to optimize the transmission area by using some simple calculations based on the location information. The CARP allow a transmitting node to vary the size of the transmission area associated with a packet, adapting it to the density of the nodes. The CARP consist of the following subsystems.

1. *The direction and location determination subsystem*: It provides location information (using GPS) and direction information (using an electronic compass) to the other subsystems. It works on the physical layer.
2. *The location verification subsystem*: It determines whether the node is inside or outside of the transmission area using the location information. It works on the network layer.
3. *The transmission area creation subsystem*: It creates a new transmission area for the next hop using the location information. It works on the network layer.
4. *The antenna selection subsystem*: It selects the right antenna(s) facing the direction of the destination. It works on the physical layer. In the next section, we examine the need of applying the density-determining for the proper operation of CARP.

One of the contributing factors in reducing the size of the transmission area is the density of a network, but no precise mechanism has been presented in CARP.

37.4.1 Application of Density Determination to CARP

In CARP [7], when a source node tries to transmit a packet in the direction of a destination node, a logical transmission area is formed. To reduce the flooding traffic, only those nodes that are located inside this area will forward the packet.

Density of an ad hoc network is considered a network parameter in CARP, and is used in calculations to form a logical transmission area. The CARP assume that the ad hoc nodes have some knowledge about the density from previous activity of the network, but no specific density-determining mechanism has been described. The density of the network in CARP is claimed to be guessed by the number of responses a node has received in the previous transmissions. Each node is supposed to maintain the statistical record of the transmission to represent the density of the network. The value of the transmission area is determined on the basis of this statistical record of transmissions.

We consider that this argument about the assessment of density may be unrealistic due to the following reasons.

1. Density assessed on the basis of the number of transmissions received by a node may be incorrect, because of the broadcast nature of the MANET. A previ-

ously received message may be received repeatedly and the density value may be incorrect.
2. It is not explained how the recording and analysis of the statistics are governed.
3. It is not clear what factors make a node start and stop collecting statistics, and at what stage density is determined.
4. It is also not clear what attributes fall under the heading of statistics.

The above-mentioned issues are addressed by the proposed density-determined algorithms and Sect. 37.6 presents the simulation results based on the application of density-determining algorithms to CARP, and facilitate CARP with the updated value of density.

37.5 Simulation Results

The density-determining algorithms are implemented in OPNET Modeler 10.5. Different network scenarios are defined to imitate an ad hoc network such as a conference, and the transmission range of each node is 50 m. A sample size of 30 is applied in all tests. The mobility model used in the network model is Random Waypoint.

37.5.1 Simulation Results for the Census of Nodes

The purpose of these simulations is to verify the node population that is defined in a test scenario, and to determine the unknown area of the network in which these nodes are present. For the first set of tests, in order to verify the performance of the census algorithm, three network scenarios of 90, 240, and 840 nodes are simulated. The nodes are aware of their respective geographical positions.

The test results are shown in Table 37.1. The first test has a node population of 90. At the end of acknowledgements, using the values of the sample, duplicates in sampling, and the number obtained during counting, the total number of nonduplicate addresses is calculated as 90 that verifies the node population of the test scenario. The area of the network is shown as 12,100 m^2, calculated from the overall lower bound (4945, 3695) and upper bound (5055, 3805). The value of density is 0.0074 nodes m^2 that is determined from the area (12,100 m^2) and the number of nonduplicate nodes in this area.

Table 37.1 Results for the census of nodes

Test	Overall area m^2	Counting	Dup in sampling	Nondup nodes	Density nodes/m^2
1	12,100	78	9	90	0.0074
2	27,225	350	20	240	0.0880
3	81,000	1350	20	840	0.0103

Table 37.2 Results for the double population in the same area

Test	Overall area m^2	Counting	Dup in sampling	Nondup nodes	Density nodes/m^2
1	12,100	156	18	180	0.0148
2	27,225	700	40	480	0.176
3	81,000	2,700	40	1,680	0.0206

The node populations for the second and the third tests were 240 and 840, respectively. The number of nonduplicate addresses that is 240 in the second test, determined as the result of the census, verifies the corresponding node population. Similarly, the number of nonduplicate addresses for the third test is 840 that equals the number of nodes in this test scenario.

For the second set of tests, the respective areas of the three test scenarios remain the same as that of the areas in the first set of tests, but the respective node population is doubled in each of three network scenarios. It is clear from the test statistics in Table 37.2 that this change results in doubling the value of the density for each of the three scenarios.

37.6 Simulations for Traffic Analysis

The simulations in this section sense density in terms of a fixed sample of the forwarding node IDs over the time elapsed in collecting the sample. The expected results should show that the same sample of nonduplicate forwarding node IDs is collected in different times depending on the respective density of nodes around an analysing node.

For simulating a traffic analysis algorithm, a simple protocol is implemented (on top of CARP) to generate the traffic to be analysed, as per which each node that receives a broadcast replaces the forwarding node ID with its node ID in the packet. The forwarding node ID changes at every hop. The monitoring node gets the forwarding node ID from each received packet and follows the traffic analysis algorithm.

The test environment for the first set of tests consists of 90 nodes that are generating traffic as per the algorithm defined in the previous paragraph. The nodes are not required to know their geographical position. The simulation data from three tests are shown in Table 37.3. These tests are carried out for different monitoring nodes with different density situations.

The results in Table 37.3 show that a fixed sample of nonduplicate forwarding node IDs is collected during three different tests in different times. For example, the analysing node in the first test collects the sample in 1.09 s and the analysing node in the third test collects the same sample in 0.90 s, meaning that the analysing node in the third test is in a dense population of nodes and is busier than the other two analysing nodes, and collects the sample earlier than the other nodes. In contrast, the

Table 37.3 Results for density from traffic monitoring

Test	Time of analysis (s)	Duplicate IDs	Time for duplicates (s)	Time for sample (s)
1	4.02	80	3.93	1.09
2	3.5	112	2.77	0.73
3	3.89	99	2.99	0.90

Table 37.4 Results with monitoring of increased traffic

Test	Time of analysis (s)	Duplicate IDs	Time for duplicates (s)	Time for sample (s)
1	1.02	320	0.93	0.0029
2	0.90	448	0.84	0.0016
3	0.95	394	0.88	0.0021

analysing node in the first test has taken more time in collecting the same sample because it was in a relatively less dense situation than the other two analysing nodes, so it has to wait longer to collect the sample. The time for the collection of the sample is calculated by applying the following formulae.

If the mean time taken by a packet is represented by (T_{pkt}),

$$(T_{pkt}) = Total\ Count \div Elapsed\ Time. \tag{37.1}$$

If time by nonduplicate packets is represented by (T_{ND}),

$$(T_{ND}) = N \times T_{pkt}. \tag{37.2}$$

For the second set of tests, the number of nodes is changed to 360 and by keeping rest of the conditions the same, we obtain another set of results as shown in Table 37.4. The results clearly indicate that when increased traffic will be produced, the time parameters of the traffic anlysis algorithm reflect this change with decrease in their respective values.

Different timings for the collection of a fixed sample of nonduplicate forwarding node IDs show that in different tests the analysing node is in different density situations and the scale of traffic around the analyzing nodes is also different in different tests.

37.6.1 Simulations for Tracking Changes in the Values of Density

The density determined at one moment may not be the same at another. The simulations in this section verify a segment of the algorithm that reacts to these changes in density. The test scenario consists of a network of 90 mobile nodes under the Random Waypoint mobility model.

The data in Table 37.5 explain the changes in time between two density calculations with respect to the variance. For each iteration in Table 37.5, the variance is determined and the next value of *time_interval* is changed with respect to the value of the variance; for example, in the first iteration, the initial time interval was set to the upper threshold of *time_interval* as 4.8 s, and with a variance 0.078, the new value of the time interval between two densities becomes 4.425 s. In the following iterations, the value of the time interval changes with respect to the variance. The data in Table 37.5 are represented in the form of line graphs in Fig. 37.3 to show the reaction of the density-determining algorithm over the changes in the network environment. The changes in density measured as the variance in density values and the difference between the existing and new time intervals are shown on the vertical axis of the graph. It is obvious that the variance in the density value triggers the next density calculation sooner or later. The corresponding points that represent variance in density and the difference in time interval show that when variance decreases the time interval between two density calculations also decreases and vice versa.

Table 37.5 Tracking changes in density

Index	Initial time_inter (s)	Variance	New time_inter (s)	Difference time_inter (s)
1	4.8	0.078	4.425	0.375
2	4.425	0.073	4.102	0.323
3	4.102	0.065	3.857	0.245
4	3.857	0.074	3.335	0.522

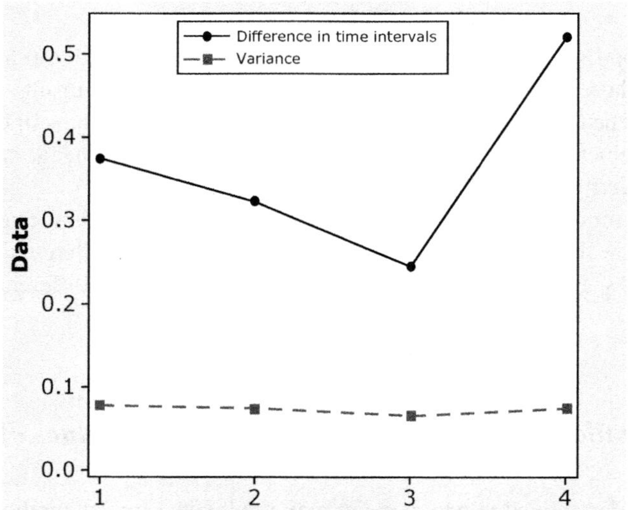

Fig. 37.3 Changes in time interval due to variance in density

The simulation results in this section have verified the ability of the density algorithms to track density, and if required, revise the time interval between two consecutive density calculations.

37.7 Conclusions

It was established that density was an important parameter for ad hoc networks, and a variation in density had an impact on the performance of an ad hoc network. Two novel density-determining algorithms, based on two anthropogenic techniques, census of nodes and traffic analysis, were presented. The census algorithm is based on the enumeration of the ad hoc nodes and finding the area in which these nodes are present. The traffic analysis gets an idea about the density by collecting a sample of nonduplicate IDs of the neighbouring nodes and the time to collect this sample. We have economized the CPU cycles and power use by adopting a counting technique based on statistical sampling.

We selected CARP as the set of protocols to authenticate the density-determining techniques. CARP need density to tune some design parameters but lack a mechanism to determine the density.

Through simulations, we verified the working of both the density-determining algorithms for CARP to facilitate CARP to get the value of the density. The density determined by these algorithms can be used as a key parameter to define a logical transmission area under CARP. A feature of these algorithms for taking care of the changing values of density and dynamically resetting the time between two consecutive density calculations was also verified by simulations.

In the future, we want to work for the three-dimensional version of these algorithms, so that the three-dimensional ad hoc networks could be enabled to get a density awareness.

References

1. S. Murthy and J. J. Garcia-Luna-Aceves (1996) An efficient routing protocol for wireless networks. In: *ACM Mobile Networks and Applications Journal*, Special Issue on Routing in Mobile Communication Networks. Volume 1. No. 2.
2. Yunjung Yi and Mario Gerla (2002) Efficient flooding in ad hoc networks using on-demand (passive) cluster formation. In: *Proceedings of MOBIHOC 2002*.
3. D. D. Perkins, H. D. Hughes, and C. B. Owen (2002) Factors affecting the performance of ad hoc networks. *Communications*, 2002. ICC 2002. IEEE International Conference.
4. Y. Tseng et al. (1999) The broadcast storm problem in a mobile ad-hoc network. In: *Proceedings of ACM International Conference on Mobile Computing and Networking (MOBICOM)*.
5. C. K. Toh (2002) Ad hoc wireless networks, protocols and systems. In: *Proceeings of IEEE Conference*. Prentice Hall, Upper Saddle River, NJ.
6. L. Hughes and Y. Zhang (2004) Self-limiting adaptive protocols for controlled flooding in ad hoc networks. In: *Proceedings of Ad-hoc, Mobile, and Wireless Networks*. IEEE, Canada.

7. L. Hughes, Y. Zhang, and K. Shumon (2003) Cartesian ad hoc routing protocol. In: *Proceedings of Second International Conference, ADHOC-NOW*. Springer, Montreal, Canada.
8. A. Durresi, V. Paruchuri, L. Barolli, and Jain Raj (2005) QoS-energy aware broadcast for sensor networks. In: *Proceedings of Parallel Architectures, Algorithms and Networks. ISPAN 2005. 8th International Symposium.*
9. J. C. Boettcher and L. M. Gaines (2005) *Industry Research Using the Economic Census: How to Find It, How to Use It*, Greenwood Press, Westport, CT.
10. Ferguson Niels et al. (2003) Practical cryptography. In: *Proceedings of IEEE Conference*, IEE.
11. Dawn Song, David Wagner, and Xuqing Tian (2001) Timing analysis of keystrokes and timing attacks on SSH. In: *Proceedings of 10th USENIX Security Symposium.*
12. A. McIntosh et al. (1994) Statistical analysis of CCSN/SS7 traffic data from working CCS subnetworks. *IEEE Journal on Selected Areas in Communications*, 12(3).
13. George G. Morgan (2004) *How to Do Everything with Your Genealogy*. McGraw-Hill Professional, New York.
14. Jon D. Fricker and Robert K. Whitford (2004) *Fundamentals of Transportation Engineering: A Multimodal Approach*. Prentice Hall, Upper Saddle River, NJ.
15. Bob O'Hara and Al Petrick (1999) *IEEE 802.11 Handbook, A Designer's Companion*. IEEE Press, Washington, DC.
16. Ranjith S. Jayaram and Injong Rhee (2003) A case for delay-based congestion control for CDMA 2.5G networks. In: *Proceedings of International Conference on Ubiquitous Computing*. Springer, Seattle.
17. E. Guadagnoli and W. F. Velicer (1988) Relation of sample size to the stability of component patterns. *Psychological Bulletin*, Volume 103. No. 2:265–275.
18. C. H. Yu and J. Behrens (1994) Misconceptions in statistical power and dynamic graphics as a remediation. Poster session presented at the *Annual Meeting of American Statistical Association*. Toronto, Canada. Volume 18. No. 2.
19. I. G. Dambolena (1984) Teaching the central limit theorem through computer simulation. *Mathematics and Computer Education*: 128–132.
20. Olive J. Dunn and Virginia A. Clark (1987) *Applied Statistics: Analysis of Variance and Regression*, 2nd Edition. John Wiley and Sons, New York.

Chapter 38
Integrating Design by Contract Focusing Maximum Benefit

Jörg Preißinger

38.1 Introduction

As mechanisms and tools that support the construction of bug-free software evolve, the complexity of software systems rises and the challenge to build dependable software, especially in the case of distributed systems, remains.

Mechanisms to increase the dependability of distributed systems have to be designed in different phases of system construction and on different layers of abstraction. Starting in the field of software engineering with mechanisms that support readable and well-defined software specification, testing, and validation, down to the design of fault-tolerant protocols such as error-correcting codes on the bit-level, we need approaches and mechanisms to avoid or tolerate faults.

One known paradigm that affects several of these aspects is design by contract (DbC), first introduced by Bertrand Meyer for the Eiffel language [1]. DbC is a software engineering technique to specify the axioms of abstract data types in the form of contracts. Adapted to object-oriented languages this means that the effects of an object's method, called postconditions, are specified. The implementation of the object's method must fulfill these postconditions, if some requirements, called preconditions, are fulfilled by the caller. This approach of an abstract specification has several advantages in the software engineering process, such as documentation and a formal semantic description. The main purpose of DbC is to enable a system to detect and locate faults and thus enable the use of fault tolerance mechanisms. For that purpose the specified contracts have to be checked at runtime.

We used the experimental distributed system MoDiS (model-oriented distributed systems) [2, 3] as a basis to adapt DbC to components of the object-based Ada-like programming language INSEL (integration and separation language). MoDiS is based on a top-down approach that enables us to extend the programming language and modify the compiler and runtime system, thus we could lay our objective on an implementation with maximum benefits due to no restrictions arising from usability or compatibility aspects.

Oscar Castillo et al. (eds.), *Trends in Intelligent Systems and Computer Engineering.*

DbC is a paradigm that enables different advantages for software construction: the implementation-independent specification of axioms, a well-defined but readable documentation, the prerequisites for static analyses and verification mechanisms, the prerequisites for test-suite generation and finally the runtime checking of the axioms to detect and locate programming errors. Furthermore, DbC opens new fields of research for automated usage of the specified contracts, which are an abstract view on the functionality of the method implementations.

After DbC was introduced with the language Eiffel, parts of the paradigm were implemented in different object-oriented languages in the last years (e.g., Java [4–9], C++ [10], Smalltalk [11], and Python [12]). In this chapter we propose our approach to an integration of DbC into a programming language in comparison to some of the existing approaches. The purpose of our work is to show up the important design and implementation issues necessary to benefit all advantages of DbC, not only runtime-checking of assertions, and to enable a flexible and efficient usage of these advantages.

The remainder of the chapter is structured as follows. In the next section we compare related work and motivate our approach. In the third section we give a brief introduction to the programming language INSEL and the concepts of the distributed system MoDiS. The fourth section contains a short description of the DbC paradigm. In the fifth section we describe our integration of the contracts into the language INSEL and explain advantages and conclusions for other implementations. At the end we summarize the chapter and give a perspective on future work.

38.2 Realisation Trend

The design by contract paradigm was first introduced by Bertrand Meyer for the Eiffel language [1], but the concepts are based on the work of Charles A. R. Hoare [13]. Since 1992 the paradigm has been adapted to several object-oriented languages, such as Java, C++, Smalltalk, and Python [4, 10–12]. For some of these languages, like Java, there even exist several implementations of different approaches for the realisation of the contract specifications and runtime checks with different focuses. We give a short summary of the existing implementations for Java to show the trend of the last years and legitimate our approach, that focuses on maximum benefit of DbC. We show that the realisation approach to integrate DbC has a huge impact on the benefits of the concept.

At least seven implementations of contract support for the Java language exist: JMSAssert [7], iContract [4], jContractor [5], HandShake [6], Jass [9], Kopi [8], and a nameless DbC toolset containing the tools DocGen, StaticAnalyzer, and ContractChecker [14]. The different implementations were implemented in the last 15 years and give an impression of the trend of how to implement this concept.

Generally, three different approaches can be distinguished.

1. *Built-in:* DbC is fully integrated into the language and its compiler. The contracts become a language feature; the code for the runtime checks can be generated in the compiler.

2. *Preprocessing:* The contracts are specified as comments or in separate files, not necessarily in the programming language. A precompiler reads the comments and generates code in the programming language for the runtime checks. The main compiler of the language cannot distinguish application code from contract code. Language and compiler need not be altered for this approach.
3. *Library-based:* This approach is also referenced as metaprogramming. The runtime environment of the language is changed, such that at runtime the code for the checks of specified contracts are integrated and executed. For this approach the contracts are often specified external, in separate files.

We briefly describe the Java approaches and classify them. For a detailed insight the mentioned reference of each realisation should be read.

JMSAssert uses comments to specify the contracts within the source code. This approach is based on a preprocessor that maps the embedded contracts to triggers written in JMScript, a Java-based scripting language. These triggers are automatically executed at runtime by an extension dynamic link library that includes the JMScript interpreter.

iContract also uses a preprocessor to generate the code for the contract-checking from JavaDoc-style comment tags. The generated code is – contrary to JMSAssert – standard Java and can be executed on standard JVMs without additional libraries. Due to the comment-based declaration the source code is in both approaches also compilable without the use of the preprocessor, of course not including the contract checks then.

jContractor uses a "pure-Java library-based system which supports contracts in Java using a design-pattern approach" [5]. The contracts are specified as Java methods with predefined naming pattern. The jContractor approach puts emphasis on the absence of special tools such as preprocessors, or modifications of compiler, runtime system, or JVM. Its class loader identifies contracts by the design patterns and rewrites the code to reflect the contracts. jContractor implements the library-based approach.

The HandShake approach is also library-based. It allows the programmer to specify contracts for Java classes and interfaces without access to their source code. The contracts are specified in separate files; a dynamic link library between the virtual machine and the operating system is used to generate classes including the contracts at load time, that can be executed on standard JVMs. This approach focuses on the use of contracts if the source code is not accessible.

The toolset described by Wang et al. [14] uses, analogous to iContract, external tools including a preprocessor to generate code on the basis of comments in the source code, thus assuring backward compatibility to standard Java compiler versions and JVMs. Additionally several tools are provided. One is for the generation of documentation based on the contracts, further a static analyser is to check for null pointer and array out of bounds exceptions. This approach focuses not only on the runtime checks, but provides tools to take advantage of several DbC aspects.

Jass is in its integration approach also comparable to iContract, it uses a preprocessor to generate the code for the runtime checks. The syntax is different and Jass allows a specification of the dynamic behaviour of objects.

Kopi is a project in which a new Java compiler was built, based on the standard Java compiler. This is the only project in which the contracts were integrated fully into the Java language according to approach 1. The compiler generates code from the specified contracts to implement the runtime checks.

All the different approaches add DbC to the Java language, but focus on different objectives in their implementations. Most of the described tools – except Kopi – reduce the DbC paradigm to a debug mechanism via runtime checks, which is only a part of DbC's benefits, as mentioned in Sect. 38.1. At the implementation of these tools, more importance was attached to the compatibility with existing environments than on maximum benefit of the paradigm.

In our view it is important to consider all different advantages of DbC (see Sect. 38.4). A full integration into a programming language is necessary to support the software engineering-oriented aspects, meaning abstract specification and documentation. The integration into the compiler of the language is in our view necessary to enable a flexible and efficient use of the mechanisms on the one hand, meaning compilation with and without contract checks and dynamic switching. Furthermore, a compiler integration enables full use of the information given by the contracts, for example, for other verification methods, test-suite generation, or fault tolerance mechanisms. Furthermore, we think that the compiler can use the information in the contracts for further analysis, if the distinction between application implementation and contract code is available in the compiler. The trend for the realisation of DbC is the preprocessing-based approach, because compatibility with the existing compiler and runtime-environment is focused. In the long-term maximum benefit of the concept for bug-free software should be focused, as we show in detail in Sect. 38.5.2. The following section gives a brief introduction to the experimental distributed system and the programming language which we used as the basis for our integration.

38.3 MoDiS and INSEL

The concepts of MoDiS are characterised best as a top-down driven approach to develop distributed systems. The term "top-down" means that the instrument programmers need to specify applications for a distributed environment efficiently, was and is the point of origin for the development of the experimental system MoDiS. The language, compiler, and runtime system were developed based on an analysis of this instrument. The language INSEL was designed to offer mechanisms to support the specification of concurrent activities, synchronisation mechanisms, and shared data objects on a high abstraction level to hide the complexity of distributed resource management from the programmer. On a high abstraction layer, the concurrent application can be programmed with INSEL objects. The communication between objects via message passing and via shared memory can be used transparently, without knowledge about the actual distribution or task placement.

The mechanisms needed to transform an application specified in INSEL to an executable system (consisting of application and management functionality), as well as the mapping to the distributed hardware configuration, are integrated in the MoDiS management. This consists mainly of the INSEL compiler *gic* (gnu INSEL compiler) [15] and the runtime management. The compiler *gic* is based on the *gcc* (gnu compiler collection) [16], which was extended by a new language front-end. The MoDiS management hides all mapping decisions necessary for the execution of an application on the distributed environment from the programmer. It offers resources such as a message-based communication system and distributed shared memory, as well as it comes to necessary runtime decisions such as the placement of components or distributed shared memory management. The necessary information to reach good management decisions, meaning nonstandard but application-dependent decisions for distributed resource management, are won statically by the compiler and dynamically during runtime. The advantages of the top-down, language-based approach pay off in this management. For a comprehensive report on the concepts of MoDiS and INSEL we refer to Spies et al. [2, 3, 15, 17].

INSEL is an imperative, object-based, and static type-safe high-level programming language. All INSEL objects, which represent the components of the distributed system, are created as instances of classes describing components, so-called generators. Generators contain other generator definitions as well as local object instance declarations. The nesting of generators leads to a hierarchy of generators that predetermines dependencies between object instances, as visibility dependencies or lifetime dependencies. The defined structures of the language enable the system management to handle the complex decisions for the distribution and resource management during runtime in an application-oriented way.

INSEL objects can either be passive or active. Active objects are called actors and are comparable to normal processes. By creating an actor, a new flow of control is established that executes the statement part of the new actor in parallel to the flow of control of its creator. An actor terminates if it has reached the end of its statement part and all its dependent objects, meaning created actors, have terminated. By creating a passive object, the flow of control of the creator switches to the newly created passive object in order to execute its statement part. Procedures and depots are examples of passive object types; the terms procedure and method are used equally in this chapter. In the case of depots, the statement part implements the initialisation of the depot; it is accessible after the flow of control switches back to its creator. There exist two possibilities for the interaction of active INSEL objects. First, they can directly interact in a client–server style and synchronise using operation-oriented rendezvous semantics (see [3]). Second, they can co-operate indirectly using shared objects (depots).

The experimental system MoDiS with the language INSEL is a suitable basis for a DbC integration that benefits all aspects of the paradigm; the integration can be used as reference for other languages. The top-down approach leads to no restrictions by existing compiler, runtime-system, or external tools. Furthermore, the integrated approach enables us to make any adjustments we consider necessary for maximum benefit. We focus on most improvement of the software-engineering with INSEL and additionally to use the information of the contracts for further system analysis and automated management.

38.4 Design by Contract

DbC is a systematic approach for the specification and implementation of object-oriented software systems. DbC describes a software methodology for designing software systems; on the other hand it introduces a feature for programming languages. An important point of this approach is the combination of the design phase of a software project with the implementation phase. The system model of DbC is based on a set of components interacting via well-defined interfaces. In the design phase each component is described by an abstract specification containing an enumeration of all exported methods. In this specification, the interfaces of the component as well as the requirements and semantics of the exported methods are stated.

In the implementation phase of a software project, the specifications of the components have to be transformed into source code. Basically, the correctness of the implementation could be verified using a formal verification tool such as the Isabelle theorem prover [18]. Practically, formal verification is still a complex and time-consuming approach. DbC supports the building of bug-free software without proving correctness. Instead of verification, DbC utilises the programming language for error detection at runtime. The programmer has the possibility to express the semantic part of the specification using assertions in the form of pre- and postconditions as well as invariants in the source code. These assertions express the correctness conditions for components and methods. The compiler creates checking code which ensures the fulfilment of the conditions at runtime. The syntax part of the specification is verified by the type system of the compiler.

The assertions of a method build the contract between the method and any potential caller. Essentially, assertions of an object exported method also describe the interaction of the object with its environment.

Preconditions are input conditions and express the requirements of a method regarding parameters and accessible variables. Postconditions ensure the correct implementation of the method as specified.

The contract between the components can be seen as follows. If the preconditions of a called method are fulfilled, the method guarantees the correct behaviour according to the postconditions. An object's invariants describe the consistent state of the object. This state is checked before and after the execution of methods that require a consistent state (see Sect. 38.5.3). The square-root function is a good example to demonstrate the principle:

$$\begin{aligned} &sqrt : in : float \mapsto out : float \\ &\mathbf{pre} \quad in \geq 0 \\ &\mathbf{post} \quad out \cdot out - in \leq 0.001 \end{aligned}$$

The function is not defined for negative input values, as specified in the precondition. In the postcondition, the correct result can be checked by multiplication of the result with itself. Rounding errors must be considered, so the difference of the square to the input value need not be zero, but below a barrier. The square-root function demonstrates the possibilities to check the correctness of the implementation. In this case, the inverse function is efficiently computable, so if the check passes, the implementation is – for the given parameters – correct without doubt. Next to the contract checks at runtime there exist several other advantages of *DbC*.

B. Meyer enumerates the benefits of *DbC* as follows [19].

- A better understanding of the object-oriented method
- A systematic approach to building bug-free object-oriented systems
- An effective framework for debugging, testing, and, more generally, quality assurance
- A method for documenting software components
- Better understanding and control of the inheritance mechanism
- A technique for dealing with abnormal cases, leading to a safe and effective language construct for exception handling

If the first approach for the realisation of DbC, a full integration to language and compiler is used, there arise even more benefits, as we describe in Sect. 38.5.2.

38.5 Integration in INSEL

From the conceptional point of view, the results of this work are valid for the task of integrating DbC in programming languages generally. Because the programming language INSEL has the properties of an object-based and not of an object-oriented programming language, INSEL does not support inheritance, thus problems regarding inheritance and DbC could be avoided in this integration and are not discussed.

38.5.1 Preconditions and Postconditions

In INSEL passive objects and active objects can be decorated with pre- and postconditions, formulated as Boolean INSEL expressions. If an expression returns `false`, the distributed system would run into an invalid state. Fault-tolerance mechanisms such as exception-handling can be used to react on the failure in this case.

Regardless of object types, the preconditions are checked after the creation of the object's variables that hold parameter values, but before the evaluation of the declaration part of the object. Therefore, unwanted side effects during evaluation of initialisation expressions are avoided. In many Java implementations of DbC, one cannot check preconditions of constructor methods before the creation and initialisation of an object's instance variables, which may lead to a change in the system state. This is an important issue in object-oriented or object-based systems. If one or more preconditions fail, the object must not leave any changes in its environment; for example, it must not alter the state of other objects, create files, or send network messages. Any side effects before evaluation of the precondition expressions would lead to the necessity for a rollback mechanism. For the same reason, the statements in pre- and postconditions must not cause side effects. If DbC is integrated in language and compiler as we recommend, we can ensure that the evaluation of contract conditions is side effect free in the compiler analysis. Application functions can be called in pre- or postconditions, which can potentially cause side effects. If the compiler has information about contracts and application implementation, and can distinguish one from the other (in contrast to preprocessing-based approaches), the side effects in contracts can be detected as faults during compilation and be reported to the programmer.

Postconditions are checked at the end of the object statement part and after all actors that were created by the object have terminated. In INSEL every object exists as long as there is at least one dependent actor alive. Because every parallel flow of control contributes to the functionality of the calling object, DbC has to wait for all created actors before checking postconditions. The postconditions can also be specified and checked for actors, because an actor can return a result as any object method in the language INSEL. Thus an actor can be seen as an object with only one method, which is executed implicitly at its creation. In the specified postconditions for an actor, its correct execution and termination can be checked.

For the specification of postconditions, *old*-expressions are provided. The *old*-value of an object is defined as the state of the object at creation time of the accessing object, for example, method or actor. The *old*-expression can be used to check the changes of an object during execution of a method or actor. In order to support *old*-expressions the DbC extension of INSEL creates for every used *old*-value an identical copy of the corresponding object. The compiler includes the copy objects directly in the abstract syntax tree, thus they are part of the INSEL specification and can be managed by the runtime system equally to the original objects. Thus it is not necessary to alter the runtime system to handle these additional objects. In the compiler, the objects get identifiers which are not allowed as identifiers for application objects, therefore compiler and runtime system can differentiate the copy objects from application objects if necessary; on the other hand the application code can never access a copy object by accident.

The *old*-values used in an assertion expression can be determined by traversing the abstract syntax tree of the expression, which allows an efficient determination due to the direct mapping of the generator structure to the abstract syntax tree. The generator structure determines the visibility of an object, thus if an object is not accessible by another object, this is also reflected in the copy objects. External implementations need to find the expressions without the tree structure, which reflects an object's visibility and thus possible expressions for *old*-values. This can be a time-consuming task in external implementations.

In INSEL every object generator defines its own lexical scope. In order to access the parameters, the preconditions are checked inside the lexical scope of the object.

Because postconditions have to ensure a correct implementation, access to local attributes of the object is necessary. Therefore postconditions are also checked inside the lexical scope of the object. As INSEL supports hiding of variables, hidden objects cannot be verified with postconditions. We believe that this limitation does not affect programmers because hidden variables can be renamed easily. Contrarywise, this leads to better readable source code.

38.5.2 Extending the Language

In order to support the concepts of DbC in a programming language, there were different approaches developed in related work, as was described in Sect. 38.2. Basically three main concepts can be distinguished: either full integration in the syntax of the programming language, first introduced in Eiffel, or metaprogramming using special libraries, or comment-based approaches with preprocessing. We enhanced the language INSEL with the concepts of DbC following the first approach. A full integration in the language leads to many different advantages, which cannot be met completely by other implementations (see Sect. 38.2).

- The syntactic and semantic analysis of the compiler is able to detect incorrect formulated assertion expressions directly. An example is infinite recursions, if the evaluation of a precondition leads indirectly to the call of the method, in which the precondition is specified.
- The binding analysis of the compiler is able to allocate for identifiers the corresponding declarations and object type clearly. This is necessary for the implementation of *old*-values. The naming scheme of the *old*-values allows the runtime environment to distinguish between copy objects and application objects.
- Preconditions can be checked before evaluation of the declaration part and postconditions can be checked after termination of all created actors.
- Detection of side effects in assertions can be easily implemented using the information of the semantic analysis of the compiler.
- Dynamic on and off switching of the assertion checking at runtime is possible without rerunning the compiler.
- Because contracts form an important part of the specification of a component, the management of the distributed system can further use this semantic information besides the runtime checks, for application-dependent management decisions.
- Contract information and static analysis information together can be used to support fault-tolerance mechanisms based on the contracts.
- Documentation output can easily be generated during compilation, without additional tools.

The assertions of an object have to be declared in INSEL in an extra block called `CONTRACT`, that is placed between the declaration and statement part of an object generator. As this block is found in the lexical scope of the object generator, the binding analysis of the compiler does not have to be modified. Furthermore, we believe that a single block for all assertions makes the source code easy to read and is helpful for the documentation of the application. The contract block can be specified as follows.

```
CONTRACT BEGIN
  PRE expression;
  POST expression;
  INV expression;
END;
```

Invariants are only specified for depots. For actors and orders, only pre- and postconditions can be specified due to their lack of different access methods. The term `expression` is any INSEL expression. The conjunction of the conditions are checked at runtime:

$$pre = \bigwedge_i pre_i$$

$$post = \bigwedge_i post_i$$

$$inv = \bigwedge_i inv_i$$

The following example shows preconditions (`PRE`) and postconditions (`POST`) of the square root function. The variable `result` is generated by the compiler and always provides access to the return value of a function in postconditions. Following to the `BEGIN`-`END`-block of the contract specification, the implementation of the function is also encapsulated by a `BEGIN`-`END`-block.

```
FUNCTION sqrt(x: real) RETURN real IS
  foo, bar: real := 0.0;
  CONTRACT BEGIN
    PRE  x >= 0.0;
    POST result*result - x <= 0.001;
  END;
BEGIN ... END;
```

38.5.3 Consistency

In the following passages, we describe some considerations about object consistency, necessary for a useful and correct DbC integration. Unfortunately, as far as we know, the problem itself and its solution possibilities are not addressed or not clearly explained in other implementations.

The passive object type `depot` encapsulates data and gives the possibility to define exported methods in order to access and manipulate data. The data are stored in local objects declared in the declaration part. Thus depots are comparable to objects of common object-oriented programming languages and are usable in distributed systems as data containers.

An important aspect of depots, and generally of objects, is the consistency of data. Several instruments of DbC can help provide consistency. First, the input given to a called exported method has to be valid and checked using preconditions. Second, the data have to be consistent before manipulating the local objects. Third, the method has to ensure that all local objects are left in a consistent state after execution of the method body. Depot invariants specify consistency conditions that must be satisfied by any instance of a depot generator. They are checked at method invocation conjunct with the preconditions and on leaving a method conjunct with the postconditions.

Depot invariants are not checked on calling or returning from private methods. We believe that the distinction of methods assuming and not assuming consistency is a crucial concept of object-based and object-orientied languages. This is based on the work of Broy and Siedersleben [20] in the field of software engineering. This distinction must be considered for a proper integration of DbC in any language in order to preserve a flexible use of object methods together with ensured consistency.

Definition 38.1: *Let x and y be objects; let a and b be methods. $x.a \leadsto y.b$ means method a of object x causes the call of method b of object y. The call of method b could be direct or indirect via another called method.*

Most implementations of DbC do not check invariants if an exported method $x.a$ calls an exported method $x.b$ ($x = y, x.a \leadsto x.b$) of the same object. This can be dangerous because the implementation of an exported method assumes object consistency, and thus could not work correctly if the object's state is inconsistent. During the execution of a method body (e.g., $x.a$), the object could be inconsistent. As a consequence, INSEL checks invariants on every call of an exported method. Thus even private methods have to ensure consistency before calling an exported method. If an object is in an inconsistent state, only a private method may be called. This strict behaviour simplifies the implementation and leads to exact specified source code, because the programmer is forced to clearly distinguish between methods ensuring or not ensuring consistency.

The following example illustrates this distinction between private and exported methods. The actor generator `System` contains the declaration of the depot generator `D`. Depot `D` is accessible by the exported method `E`. Furthermore, the depot contains the declaration of the private method `P`. The invariant `x > 0.0`, that reflects a consistent state in this example, has to be true before and after execution of the method body of `E` only. During execution of `P` in `E` the value of the object `x` is unspecified.

As the example shows, it is important to constitute the kind of methods that assume consistency, and thus in which methods the invariants are to be checked. Before and after the execution of the exported method `E`, this is the case. Thus even private methods calling `E` have to ensure that the object state is consistent. If a private method is called (e.g., P in the example), the object state may be inconsistent; invariants are not checked.

```
MACTOR System IS
  DEPOT SPEC D IS                     -- interface decl.
    PROCEDURE E(i: real);             -- access method
  END;
  DEPOT D IS                          -- implementation decl.
    x: real := 1.0;
    PROCEDURE P IS                    -- private method
    BEGIN ... END;
    PROCEDURE E(i: real) IS           -- exported method
    BEGIN
      x := 0.0;
      P;                              -- depot inconsistent
      x := 1.0;
    END;
    CONTRACT BEGIN
      INV x > 0.0;
    END;
  BEGIN ... END;
  do: D;
BEGIN
  do.E(23.0);
END;
```

38.5.4 Fault Detection and Tolerance

The benefits of the DbC paradigm, as described and referenced in Sect. 38.4, are to a large extent incapable of measurement. The improvements in the software engineering process due to formal specification, code documentation, and better code readability, for example, could at most be evaluated statistically over several projects. The types of bugs that are detectable with DbC and the time to fix them based on DbC, in relation to the types that can be found by conventional debugging techniques and the time needed to fix them, are also not measurable. But we give an example for the existence of bugs which are not easy to debug with conventional techniques, but were on the other hand easily detected by DbC in our applications. In one of our applications was an errorprone specification of mutual exclusion for the access of a depot. The fault (more than one actor accessing in parallel) only occurred seldom due to racing conditions, thus it was hard to debug. After our DbC integration we detected the fault the first time it occurred again. This happened because the second access occurred while the depot was in an inconsistent state (due to the first access). The advantage of DbC is that the fault was not only recognised but also easy to locate then.

Furthermore, DbC's contribution in the sense of fault detection is more than just for debugging, but DbC enables the tolerating of faults. Exception mechanisms can be used to trigger fault-handling routines. Recovery and retry mechanisms are state of the art to address this topic. One advantage of fault detection with DbC is that the problems of fault location and fault propagation are addressed implicitly by the concept. The fault-tolerance mechanism is provided with information about the method implementation or the calling object that caused the contract violation.

The contracts form a specification of the component's semantics. We believe that the information provided by the static analysis of the compiler together with the contracts enable further, automated fault-tolerance mechanisms. In future research, we will focus on such mechanisms, for example, forward-recovery based on the postcondition specification. Based on the defined consistent state of objects and the results demanded in postconditions, code to set the objects in a consistent or initial state can be generated. At present, there are still open questions for this approach that have to be dealt with, such as those concerning side effects and system consistency. We think that the use of the contract information generally opens a lot more fields for enquiry than only the runtime checks for error detection.

38.5.5 Adaptive Management

The specification of a method functionality in the form of contracts is a second information source for system management. If DbC is fully integrated into the programming language and compiler, the compiler can use this information for adaptive system management, as in the case of MoDiS. Application-oriented decisions in the runtime system can only be reached if enough information about the applications components is available. The specified contracts are a compact second information source next to the implementation itself, which reflect the semantics and functionality of the software. We hope to use the contract specifications in future research in this sense, to improve the automated, application-oriented management of our systems.

38.6 Summarisation

We presented an adaptation of the DbC paradigm to the object-based language INSEL. We explained why, in our view, the best way to integrate the paradigm is a full integration in language and compiler. The availability of implementation information and contract information side by side depends on the integration approach. The automated use of this information enables maximum benefit of the concept, and opens further fields for enquiry.

The benefits, of which several are not met by other implementations, include documentation, static analysis, fault detection, support for fault tolerance, and providing information for automated system management. Further efficient realisation of contract checks, as well as an error detection in the contract specification itself is possible with a full integration.

This implementation and its conclusions can be used as reference for future implementations in other programming languages. We described our language extensions and the compiler integration. We discussed the general problem of object consistency for objects with invariants combined with private and exported methods. In future investigations we will examine extended fault-tolerance mechanisms and the use of the contract information for automated system management based on our DbC implementation.

Acknowledgement We thank Prof. Dr. Peter Paul Spies, Dr. Christian Rehn, and Ulrich Dümichen for their suggestions and valuable comments in discussions on our work. Furthermore, we thank Alexander Mayer for the practical work he did within the scope of his diploma thesis [21].

References

1. B. Meyer. Design by contract. In B. Meyer and M. D., editors, *Advances in Object-Oriented Software Engineering*. Prentice–Hall, Engewood Cliffs, NJ, 1992.
2. P. Spies, C. Eckert, M. Lange, D. Marek, R. Radermacher, F. Weimer, and H.-M. Windisch. Sprachkonzepte zur Konstruktion verteilter Systeme. Technical Report TUM-I9618, SFB 342/09/96 A, Technische Universitaet Muenchen, Germany, 1996.
3. C. Eckert and M. Pizka. Improving resource management in distributed systems using language-level structuring concepts. *The Journal of Supercomputing,* 13:35–55, 1999.
4. R. Kramer. Icontract — The java design by contract tool. In *Technology of Object-Oriented Languages, TOOLS 26,* pages 295–307. IEEE Press, August 1998.
5. M. Karaorman, U. Holzle, and J. Bruno. jcontractor: A reflective Java library to support design by contract. Technical Report, Santa Barbara, CA, 1999.
6. A. Duncan and U. Hoelzle. Adding contracts to Java with handshake. Technical Report TRCS98-32, 9, 1998.
7. *Design by Contract for Java Using JMSAssert. Man Machine Systems, 2000.* http://www.mmsindia.com/DBCForJava.html.
8. M. Lackner, A. Krall, and F. Puntigam. Supporting design by contract in Java, 2002.
9. D. Bartetzko, C. Fischer, M. Moller, and H. Wehrheim. Jass - java with assertions. *Electronic Notes in Theoretical Computer Science*, 55(15):1–15, January 2004.
10. R. Ploesch and J. Pichler. Contracts: From analysis to C++ implementation. In *TOOLS '99: Proceedings of the Technology of Object-Oriented Languages and Systems*, page 248, Washington, DC, 1999. IEEE Computer Society.
11. M. Carrillo-Castellon, J. Garcia-Molina, E. Pimentel, and I. Repiso. Design by contract in Smalltalk. *Journal of Object-Oriented Programming*, 9(7):23–28, November/December 1996.
12. R. Ploesch. Design by contract for Python. In *Fourth Asia-Pacific Software Engineering and International Computer Science Conference*, page 213, Washington, DC, 1997. IEEE Computer Society.
13. C. A. R. Hoare. An axiomatic basis for computer programming. *Communications of the ACM*, 12(10), 1969.
14. J. Wang, L. Qin, N. Vemuri, and X. Jia. *A Toolset for Design By Contract For Java.* http://se.cs.depaul.edu/ise/zoom/papers/other/DBC.pdf.

15. M. Pizka. Design and implementation of the gnu insel compiler (gic). Technical Report TUM–I9713, SFB–Bericht 342/09/97 A, Technische Universitaet Muenchen, Germany, October 1997.
16. *GNU Compiler Collection*. http://gcc.gnu.org/.
17. M. Pizka and C. Rehn. Heaps and stacks in distributed shared memory. In *16th International Parallel and Distributed Processing Symposium (IPDPS '02 (IPPS, SPDP))*, page 107, Washington - Brussels - Tokyo, April 2002. IEEE.
18. T. Nipkow, L. Paulson, and M. Wenzel. Isabelle/hol — a proof assistant for higher-order logic. 2283, Springer, New York, 2002.
19. B. Meyer. Building bug-free OO software: An introduction to design by contract. *Object Currents, SIGS Publication*, 1(3), 1996.
20. M. Broy and J. Siedersleben. Objektorientierte Programmierung und Softwareentwicklung. *Informatik-Spektrum*, 25(1):3–11, 2002.
21. A. Mayer. Integration von design by contract in das sprachbasierte, verteilte system modis. Master's thesis, Technische Universität München, August 2006. German only.

Chapter 39
Performance Engineering for Enterprise Applications

Marcel Seelig, Jan Schaffner, and Gero Decker

39.1 Introduction

Performance is a key aspect for software systems. Bad responsiveness can result in a decreasing number of users. In contrast, exceptional performance can lead to a decisive competitive advantage, as the example of Google's search engine shows.

However, performance issues of enterprise applications are often poorly addressed during the software development process. In industry practice, performance issues are tackled with both "tune for performance" and "design for performance" approaches. When following a "tune for performance" strategy, important architectural design choices are normally made without having performance in mind. Later, after having fully implemented the system, hot-spots are identified and fixed. Problematically, the most critical performance problems are often due to wrong architectural decisions in early stages of the development process. Therefore, a "tune for performance" development strategy will often result in major reimplementation work and is thus ineffective. Performance concerns, as stated by Jain [1] influence every step of the software system lifecycle: requirements specification, design, development, implementation and manufacturing, sales and purchase, use and upgrade. Still it is a highly argued question when to think and act to get a reasonable performance. No one would deny that performance is important for every software project. Yet it is often considered less important than other factors and moved to a later part of the software's lifecycle as documented by Dugan [2]. This decision is often justified with Knuth's famous statement, "We should forget about small efficiencies, about 97% of the time. Premature optimization is the root of all evil [3]." In order to avoid unnecessary reimplementation work, "design for performance" tries to incorporate performance trade-offs into early architecture definition stages. Architectural patterns help to avoid common pitfalls.

Using performance prediction techniques facilitates more accurate design for performance than simple rules of thumb. Here, we can distinguish between two approaches based on different types of performance models: analytical performance models and simulation models. Analytical models such as queuing systems

Oscar Castillo et al. (eds.), *Trends in Intelligent Systems and Computer Engineering.*

explained by Kleinrock [4] are based on a set of formulas that can be used to generate performance metrics for given parameters. Simulation models are software systems that simulate the performance behavior of the real system.

We assume that the biggest part of a modern enterprise system is already available at very early development stages. Existing components (or services) are incorporated into the system and a lot of the actual work is done by the sophisticated middleware platforms. Therefore, the middleware must not be neglected when predicting a software system's performance. We argue that analytical performance models are not applicable in the industry, which is due to the fact that the respective mathematical models are very hard to create. In contrast, the complexity of modern middleware platforms is huge and can hardly be captured using mathematical models.

Based on these assumptions our simulation models go one step further than what is described, for example, in the workings of Denaro et al. [5] and Liu et al. [6]. We propose a method for analysis of a software system's performance based on its conceived architecture. In our approach, we try to use as much of the real environment as possible. We argue that much better performance predictions can be done that way while having little modeling effort for the engineers. We have implemented our concepts in a simulation framework. Our results show that a software system's performance can be predicted at very early stages in the design phase of the software development process, if architecture-based performance models are used to carry out simulations within real middleware infrastructure.

We validate our approach using a case study. We have taken an existing ABAP system, measured the performance of the system, and compared the measurements to simulation results obtained with our framework.

The remainder of this chapter is structured as follows. First, we discuss related work in Sect. 39.2. In Sect. 39.3 we introduce our methodology for simulation-based performance engineering. Section 39.4 discusses the implementation of our performance simulation framework. Afterwards, Sect. 39.5 presents the case study where our simulation-based approach has been applied. The results demonstrating the applicability of our approach are discussed in Sect. 39.6, also providing concluding remarks and an overview of our further research prospects.

39.2 Related Work

The assessment of software performance through simulation is certainly a very interesting and important topic, but very difficult to solve in a generic way, which is perhaps why not too much work in this area exists. A survey by Balsamo et al. [7] presents performance models to characterize the quantitative behavior of software systems. It also concludes that, despite successful application of the different models, performance prediction has not been integrated into normal software development. A comprehensive review of recent research in the field of model-based performance prediction at software development time can be found in the aforementioned survey, whereas some representatives are also briefly discussed here. With all of these model-based approaches it is possible to predict an application's performance, each approach offering some specific advantages or disadvantages. They are applicable at different stages of the software lifecycle and have more or less maturity to be used for meaningful performance prediction. However, all of them require a lot of expertise in the area of performance, which is, according to an expert working for 28 years in the field of optimizing business applications, acquired only after many years of experience.

Microsoft's magpie project [8], investigating performance of loosely coupled systems, develops a performance monitoring tool for Microsoft Windows which has the ability to count the resources consumed by a particular transaction throughout its lifetime. The performance application programming interface (PAPI) project [9] gives users a graphical representation of performance information that has been gathered by accessing hardware performance counters available on most modern microprocessors. Thus, it is allowing users to quickly see where performance bottlenecks are in their application. Although both projects are very useful for the refinement of workload models and localizing hot-spots, a fully implemented system is needed and therefore not really applicable during the early stages of software development.

Weyuker and Vokolos [10] discuss an approach to software performance testing using a case study. In their approach, a representative workload using history data is created and applied to an earlier version of the system to be tested, if available. They also found that those projects that are in the most trouble almost never considered performance issues during the architecture phase of development. They recognized that, with regard to performance, it was irrelevant what the software under test was actually doing, provided the resources were used in a similar way to the intended new software. This perception is very similar to our approach.

Early performance testing is investigated by Denaro et al. [5]. They state the need for early evaluation of software performance and find an existing focus on analytic models rather than testing techniques. Therefore they concentrate on performance testing applied to existing middleware products. For them, as for the authors of this work, early performance evaluation is desirable and existing software infrastructure, such as middleware, can be used for performance prediction before any other component is developed. There are two major reasons for that assumption. First, a great deal of resources are consumed on the middleware level and second, many (business) applications rely on the input of other applications or databases which is already available at the very beginning of a software project. Based upon those assumptions Denaro et al. are generating stubs for missing components and are executing the test while applying workloads derived from use cases. ForeSight [6] constructs models based on empirical testing that act as predictors of the performance effects of architectural trade-offs. They use the results of an empirical benchmarking engine to build mathematical models to predict performance. In our approach a different path is chosen: architecturally relevant components are represented by a generic business module, which can be configured to behave, with regard to performance, as the desired component.

Software performance engineering (SPE) as presented by Smith [11], is an approach to design software systems for performance. With SPE it is also possible to evaluate complex distributed enterprise applications. The performance models are basically analytical models that can be solved using formulas and input parameters to determine the system's performance. That approach distinguishes SPE massively from our technique, although both are addressing performance at the same very early stage of software development with the emphasis on software architecture. With SPE a design for performance is possible albeit a complex infrastructure will easily make this task a difficult and time-consuming one. Additionally, this task has to be done by a performance expert.

Woodside, Petriu, and Amyot, main investigators in the performance from unified model analysis (PUMA) [12] project, are developing a unified approach to building performance models from design models that specify scenarios for use cases. The performance modeling is done using the layered queuing (LQN) formalism. Their scenario input is based on the UML:SPT [13] profile. Layered queuing networks are used for component-based performance prediction by Wu, McMullan, and Woodside [14]. The LQN notation was also proposed for future enhancement of simulation-based performance testing by [5].

When applying these different approaches to the development of enterprise applications there is one central aspect that immediately attracts a software engineer's attention: it costs a lot of time to come to an appropriate performance prediction. One reason for this dilemma is the fact that enterprise applications utilize technically very complex environments. Enterprise integration platforms belong to the biggest software systems in terms of lines of code. The performance behavior of this environment is hard to capture in analytical performance models. To our knowledge no research project has ever managed to do so with satisfying precision. Another major disadvantage of model-based performance prediction and analysis is that the infrastructure has to be modeled explicitly which takes a lot of time.

Simulation-based approaches are far more promising in this case. The next section describes our approach that aims to obtain performance predictions even when it comes to the use of complex infrastructure.

39.3 A Methodology for Simulation-Based Performance Engineering

Throughout this section we describe our method for simulating software systems prior to their implementation, based on high-level architectural models. The five steps of the methodology are depicted in Fig. 39.1. Before we provide details for each step, we introduce the modeling notation used for the performance models.

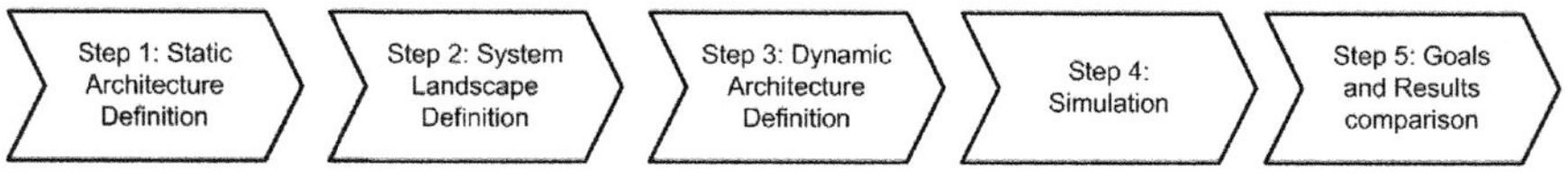

Fig. 39.1 A methodology for simulation-based performance engineering

39.3.1 FMC Modeling Notation

The modeling approach is based on the fundamental modeling concepts (FMC), an approach for describing architectural structures of computer-based systems, using a semiformal graphical notation [15, 16]. In order to support a wide variety of systems, FMC distinguishes three basic types of system structures which are fundamental aspects of any computer-based system:

- Compositional structure (i.e., the static structure consisting of the interacting components of the system)
- Dynamic structure (i.e., the behavior of the components)
- Value structure (i.e., the data structures found in the system)

Only the compositional structures are relevant in the context of this chapter and in consequence the corresponding conceptual and notational elements are discussed below. Any system can be seen as a composition of collaborating components called agents. Each agent serves a well-defined purpose and communicates via channels with other agents. If an agent needs to keep information over time, he has access to at least one storage where information can be stored. Channels and storages are (virtual) locations where information can be observed. The agents are drawn as rectangular nodes, whereas locations are symbolized as rounded nodes. In particular, channels are depicted as small circles and storages are illustrated as larger circles or rounded nodes. Any agents and locations drawn with a shadow represent more than one exemplar of that type. The possibility to read information from or write information to a location is indicated by arrows. Types of agents and locations are identified by descriptive textual labels. Arbitrary complex structures can be described because agents can be connected to multiple locations and locations can be shared by multiple agents. For example, it is possible to describe unidirectional or bidirectional channels (connecting only two agents) as well as broadcast channels (connecting more than two agents) and channels for sending requests (bidirectional, with an "R"-arrow indicating the request direction). Shared storages can be used for buffered communication. In general, agents and locations are not necessarily related to the system's physical structure. The compositional structure facilitates the understanding of a system, because one can imagine it as a physical structure (e.g., as a team of cooperating persons).

39.3.2 Iterative Methodology

Our approach focuses on the iterative process of simulating enterprise applications (i.e., the stepwise refinement of the software architecture based on the simulation results). A precondition for the iterative engineering cycles is that the following information is already available.

- The system's most important use cases have to be known.
- The workflow intensity describes the work placed on the system by the clients. Metrics such as tasks performed per user per hour, or number of users can be applied for each use case.

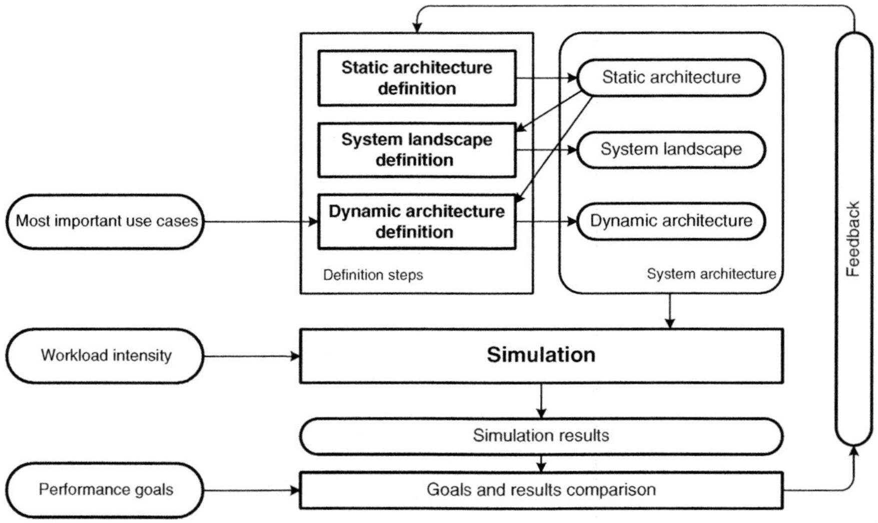

Fig. 39.2 Iterative engineering process for performance simulations

- The performance goals define acceptable performance behavior per task; for example, it could be demanded that searching for a certain item should not take longer than 0.01 sec.

The whole engineering process is depicted in Fig. 39.2. The rounded rectangles represent artifacts/information and the other rectangles represent engineering activities. The arrows show which information is consumed by a certain activity and which information is produced.

As stated earlier in this section, the following five steps make up an iteration cycle.

1. *Static Architecture Definition.* The software engineer defines which building blocks will appear in the application and how they are connected.

   The system components might have already been developed (e.g., existing services that are being reused in the new system), whereas other system components have to be developed.

   The static structure definition phase leads to the application's static architecture. This is a very important artefact, because it is consumed by the two downstream activities "system landscape definition" and "dynamic architecture definition".
2. *System Landscape Definition.* The static structure only gives a logical view of the system. During the system landscape definition the software engineer enriches the static architecture with system setup information. To do so, she describes how the system components are physically distributed (e.g., in terms of IP addresses) and defines on which technology platform the components will be realized (e.g., J2EE or ABAP). Channels are annotated with information about the protocol used by two agents for communication (e.g., HTTP or RMI). Configurations for

storages indicate parameters such as the desired fill level of the database and empirical distribution of the data.

In order to keep the process of creating these models as simple as possible, the engineer has to be restricted in the choice of parameters. The final landscape definition has to correspond to the real system setup.

3. *Dynamic Architecture Definition.* The dynamic architecture describes the interactions between agents for a set of use cases. Request/response messages and the activities performed by the agents are characterized. As depicted in Fig. 39.2, the static architecture is necessary for describing the system's dynamic architecture. This is due to the fact that only those agents that are connected by a channel in the static architecture can legally communicate via request/response messages.

   More details on the definition of the agents' activities can be found in Sect. 39.4.
4. *Simulation.* Before a simulation can start the engineer has to select a set of use cases that are to participate in the simulation. The workload intensity will apply accordingly. Additionally, workloads for other applications that are deployed in the target environment can be defined. That way the influence that different applications have on each other can be captured.

   Simulations are divided into three phases: configuration, execution, and collection of results. During the configuration phase dummies are deployed to the machines and configured according to the system architecture. Storages are set up by the accessing dummies. One generic simulation component (GSC) represents one agent from the static architecture. Because individual views on the dynamic architecture are propagated to the dummies, they know which calls they have to perform when they are triggered. After the configuration has finished the execution of the simulation can start. The system components are triggered according to the workload intensity and performance measurements are traced. As soon as the execution is over the trace data can be collected and aggregated into the simulation results.
5. *Goals and Results Comparison.* The results produced in the simulation give an insight into the conceived system's performance behavior. As performance goals are set in the requirements, we can now compare them to the simulation results. This feedback can be used in two ways: the architecture can be modified until the performance goals are met (e.g., by introducing caching mechanisms or by changing the physical setup). On the other hand, the information on potential hot-spots can be used as input for the implementation phase in order to know where special care has to be taken.

39.4 Architecture of the Performance Simulation Framework

The purpose of this section is to describe the architecture and concepts of our performance simulation framework. This framework is an implementation of our methodology together with an editor for the models. The architecture of this framework is shown in Fig. 39.3.

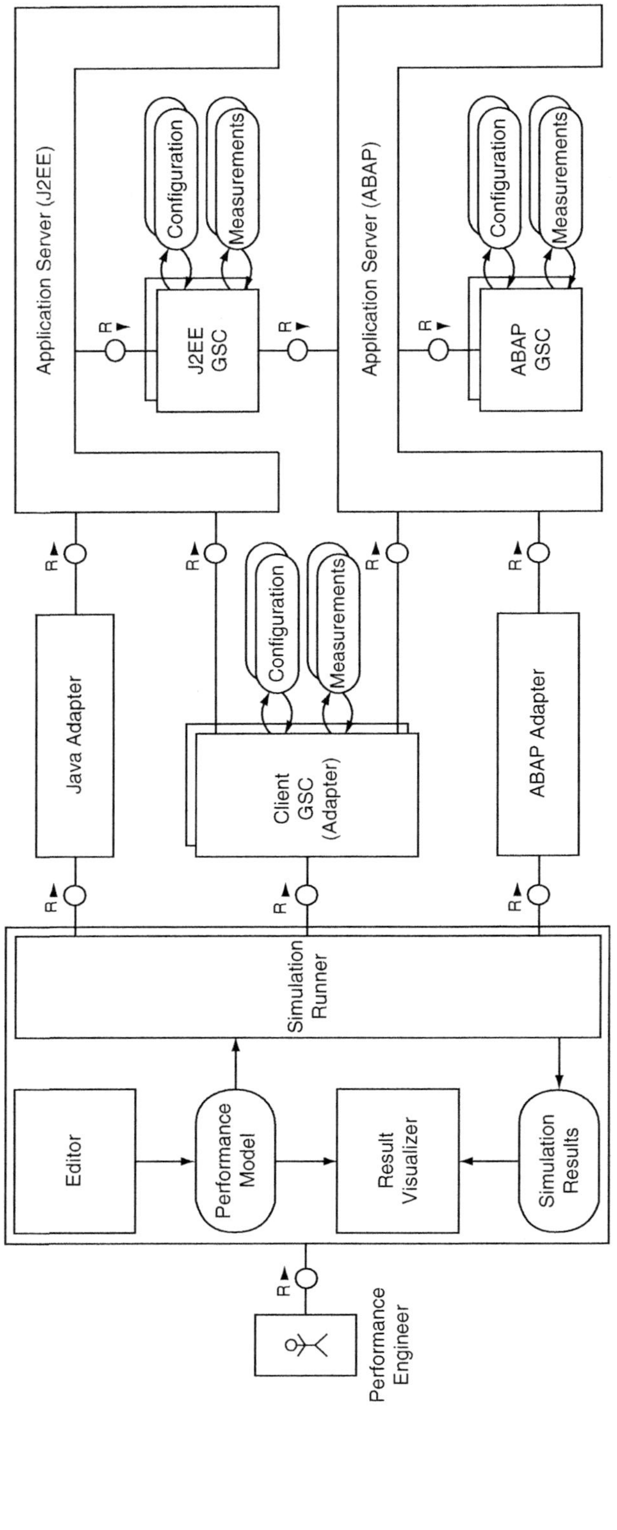

Fig. 39.3 Reference architecture (FMC block diagram)

The performance architect has access to the editor for creating the performance model. This model consists of the static architecture combined with the system landscape information and the dynamic architecture. Both the static and the dynamic architecture are defined using graphical representations of their models. The system landscape information is defined by parameters that can be added to the different parts of the models.

When the performance model is defined the simulation can be started. The simulation runner is responsible for the execution of the whole simulation. Using different adapters, GSCs (configurable generic business components) can be accessed. This enables the simulation to be independent of the technologies with which the GSCs are realized. The current implementation of the framework supports J2EE and ABAP technology.

An adapter supports the three phases of simulation: configuration, execution, and result collection. At first the simulation runner distributes the different configurations to all GSCs through the adapters. Then the actual simulation will be started, during which all GSCs trace measurements. These measurements can be collected after the simulation has finished. The simulation runner aggregates the measurements and provides them as the simulation results to the result visualizer. This component uses the graphical representations of the performance model to visualize the results. The visualization uses textual annotations as well as colors to present the results to the performance engineer. Examples of both the performance model as well as the visualization are shown in Sect. 39.5.

39.4.1 Types of GSCs

The current implementation of the framework supports three types of GSCs: J2EE dummies, ABAP dummies, and client dummies. The client GSC provides to the simulation runner the same interface as the adapters do. This way there is no difference for the simulation runner between an adapter and a client GSC. On the other hand, the client GSC does not need to forward any information to a GSC, as it itself is the GSC.

In general, each GSC has an internal structure as depicted in Fig. 39.4. The GSC controller receives a call and executes the behavior as described in its configuration. Regarding the behavior, the GSC executes a certain scenario that consists of performing computational and/or memory consuming behavior. In addition, it can access a database executing a standard statement. Besides performing the behavior defined by the configuration, each GSC has the ability to call other GSCs. In order to encapsulate the implementation of calling another GSC, a set of callers with a unique interface is used. For each possible type of GSC there exists one caller.

During all the activities the GSC performs it traces measurements. The measurements are stored asynchronously in order to keep the overhead as small as possible. After the simulation has finished the measurements can be collected via the adapter.

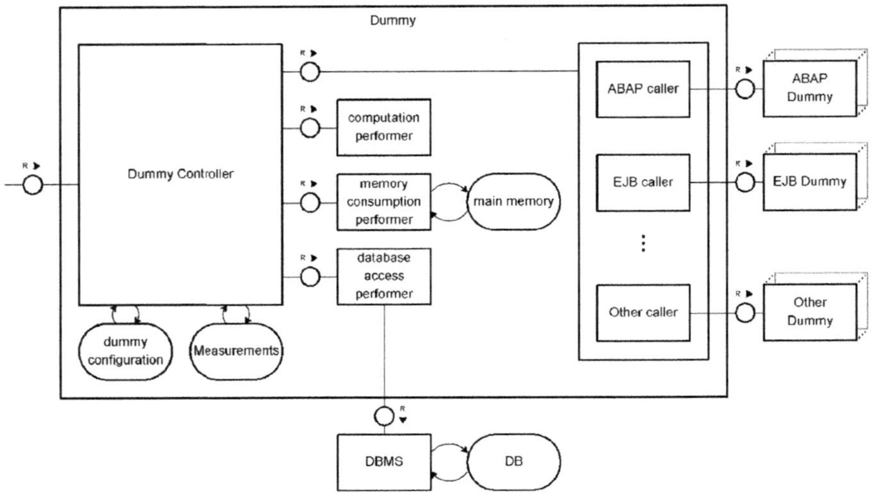

Fig. 39.4 Internal structure of a GSC (FMC block diagram)

Although in ABAP technology the GSC is a function module, for the J2EE GSC there exist two different types: the enterprise beans and the Web components. The enterprise beans are implemented as described above. They can be called via the adapter or immediately by calling their remote interfaces. Web components are servelets and Java-server-pages (JSP). Usually a servelet receives the request and prepares the data for the JSP, which provides the response data. Thus, a servelet and a JSP appear as a pair and are represented as one component type in our framework.

The communication between a client GSC and the servelet-JSP-pair is done by requesting a URL on which the servelet is listening. The servelet then performs the action and returns the rendered JSP to the client GSC. By calling this URL out of a simple browser the performance engineer can experience the performance behavior, mainly the response time, as if she were interacting with a real system.

39.4.2 Simulations with Implemented Components

Today it is not always the case that a project starts from scratch. Often an existing system needs to be modified or extended with some new components. The question regarding performance is then how the new components will affect the whole system. Thus, it is obvious to include the existing parts in the simulation in order to get the best possible results.

In the pure GSC-based simulation, there was only one call configuration: a GSC calls another GSC. Including existing components in the simulation brings up two new scenarios: a GSC calls an existing component, and a real component calls a GSC.

In the first case, the challenge is to call the existing component in such a way that it behaves as it would in the real system. This is not trivial, because the GSC has no application logic inside that produces these calls. It rather must choose a predefined call.

In the second case, when an existing component calls a GSC, the problem is more complex. The GSC must be adapted for each call in order to implement the interface the real component normally calls. But in order to continue in its usual behavior the real component needs a return value corresponding to the specific call, which is the more difficult part. Thus, the GSC needs to know something about the application logic as it otherwise would not be able to return the corresponding values. The question is how much of the application logic the GSC needs to know, if not all.

The next step after integrating existing components in the simulation is incremental performance prediction. The idea behind it is to stepwise replace the GSCs by their corresponding real components. Always when a component is implemented it will be included in the simulation in order to narrow the simulation result to the exact performance behavior of the whole system. Although the simulation results based on the GSCs provide an important input for the development of the system, the results will never match exactly the later behavior. Thus, it is important to include as many real components as possible in the simulation.

39.5 Case Study

During the realization of the simulation framework, two case studies were used to gain the information needed for validation of the GSCs. The first one was an application we got from SAP, that was developed using ABAP technology. The second one was the Java Pet Store from the J2EE BluePrints Program of Sun Microsystems [17] ported to the NetWeaver platform [18]. With both applications we did a reengineering in order to get the according performance model. Then we added traces to the applications in order to measure their performance. Finally, these measurements were compared with the simulation results.

39.5.1 Performance Model

The ABAP application is a simple batch application that analyzes a set of documents for duplicates. Its static architecture is depicted in Fig. 39.5. The boxes each represent one agent, which is realized by one GSC during the simulation. The lines between the boxes (channels) show possible communication between the associated agents. Communication is possible in both directions if two agents are connected via a channel. The cylinders represent storages on which the agents can operate. If two agents operate on two different storages, it is guaranteed by the framework that

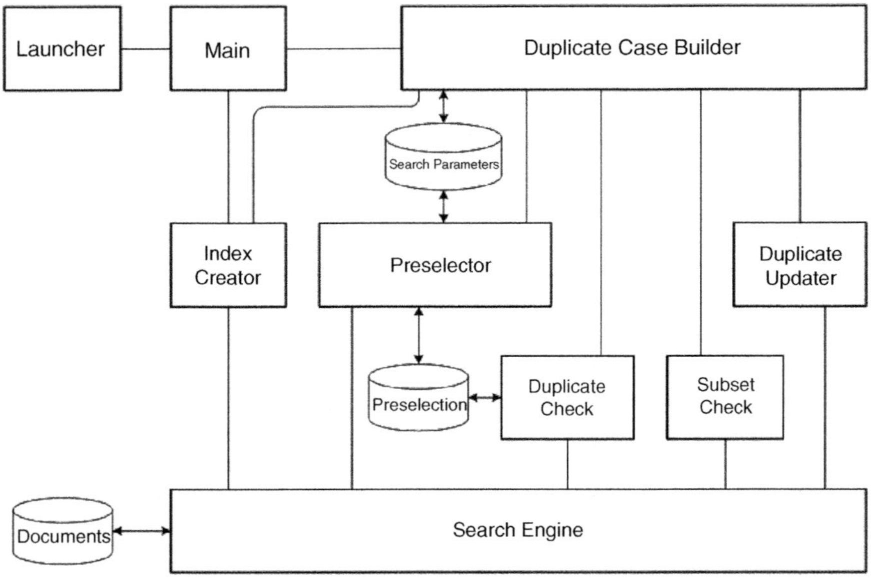

Fig. 39.5 Case study: static architecture model

no locking situation can occur. Otherwise, with a certain probability the two agents can lock each other.

The landscape information is not visualized within the diagram; instead a properties viewer is used to add the different properties to the model. Regarding the agents, this is their location, the address of the application server, and the technology type in which the agent should be realized. A storage is realized using a data schema consisting of different entities and relationships. As landscape information the number of instances for each entity can be specified.

Figure 39.6 shows the dynamic architecture of the ABAP application. The model is based on the UML 2.0 Sequence Charts [19]. The boxes on top of the diagram represent the same agents as in the static architecture. Currently agents can only do synchronous calls to other agents, but asynchronous calls are planned. For each activity the performance-relevant behavior can be defined by choosing a scenario from a predefined set. We identified work on storages, algorithmic computation, and main memory-consuming processes as the core parts of performance-relevant behavior. The predefined scenarios consist of a combination of these parts.

The iteration of a set of calls to other agents can be managed using loops. In order to keep this model type simple, we decided not to support branches. On the abstraction level of architecture definitions, the assumption can be made, that only few branches are needed. Thus, the complete dynamic architecture can be described using some few call sequences with redundant parts.

The restrictions we made lead to simple process models that are easy to execute and allow for simple aggregation of the trace data. Our framework has implemented a lot of different aggregations that can be visualized within the diagrams. Figure 39.7

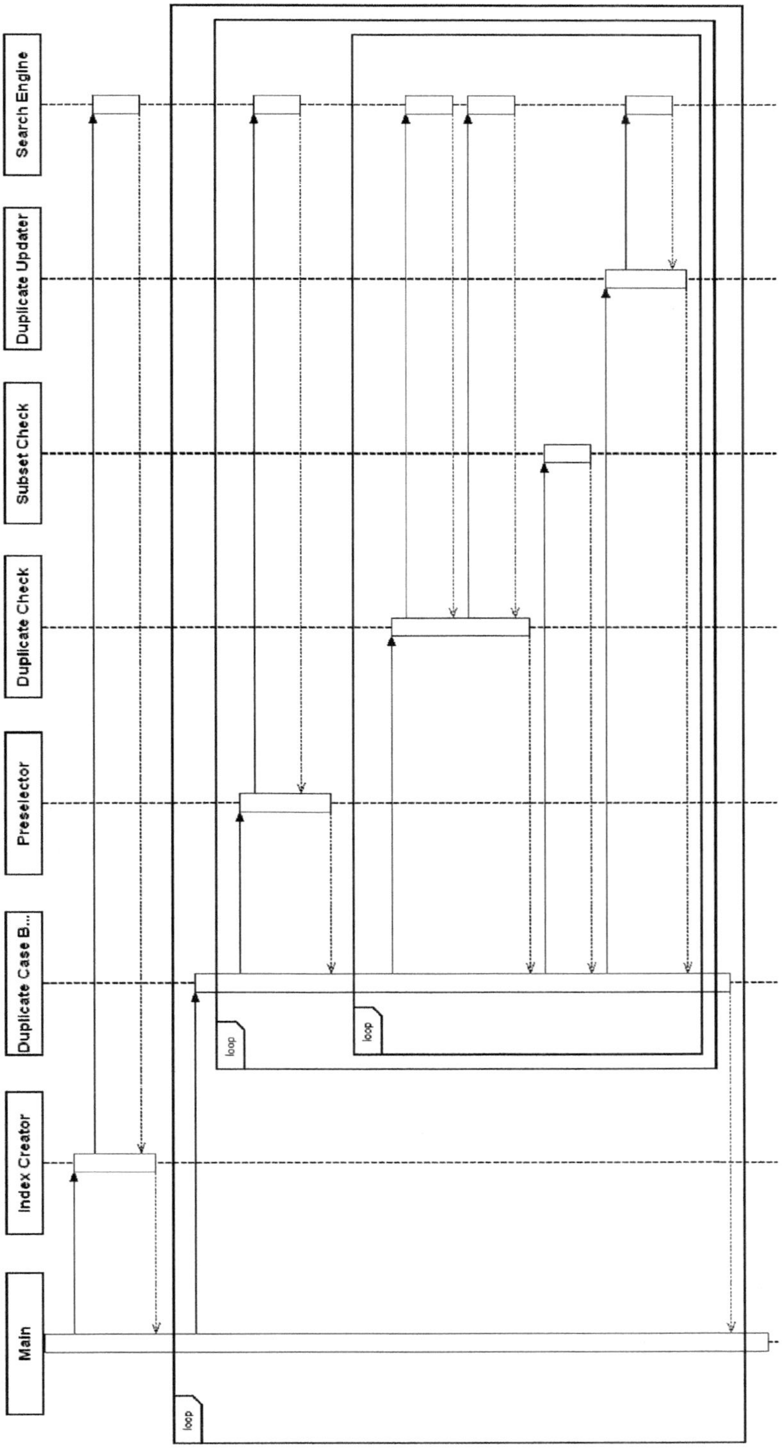

Fig. 39.6 Case study: dynamic architecture model

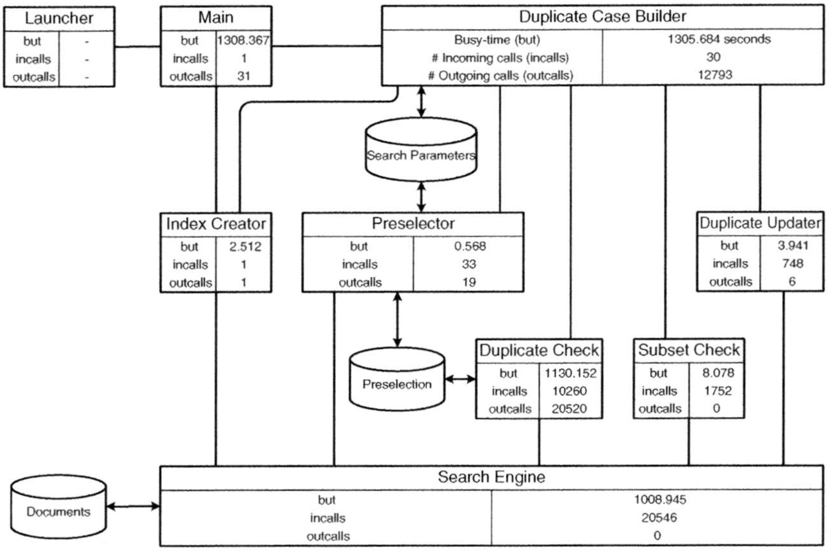

Fig. 39.7 Case study: simulation result view

shows an example where the results of a simulation of the ABAP application are visualized in the static architecture diagram.

Into each box three different types of performance figures can be projected. A fourth figure can be visualized using colors. The example shows the time the agent spent between incoming request and outgoing response aggregated for all calls, called busy time. Because this is a batch application, execution time is more relevant than response time. In addition, the number of incoming and outgoing calls is depicted. As the storages are passive components, no measurements are taken and thus, nothing is visualized. Regarding the channels, it is planned to measure and visualize the amount of data that is passed through the channel during the different calls.

Visualization of results is possible in the sequence charts in a way similar to the block diagram by adding numbers and using colors.

39.5.2 Analysis

The simulation results shown in Fig. 39.7 are a factor 2.6 lower than the measurements we took from the ABAP application. Although this seems to be a big deviation, one must consider that all components have the same relative difference.[1] Therefore, it is still possible to identify hot-spots in the architecture using our framework. If the overall response time is not the subject of concern but the distribution of resource usage, our framework can even at this early stage provide

[1] The aforementioned factor was constant for all parts.

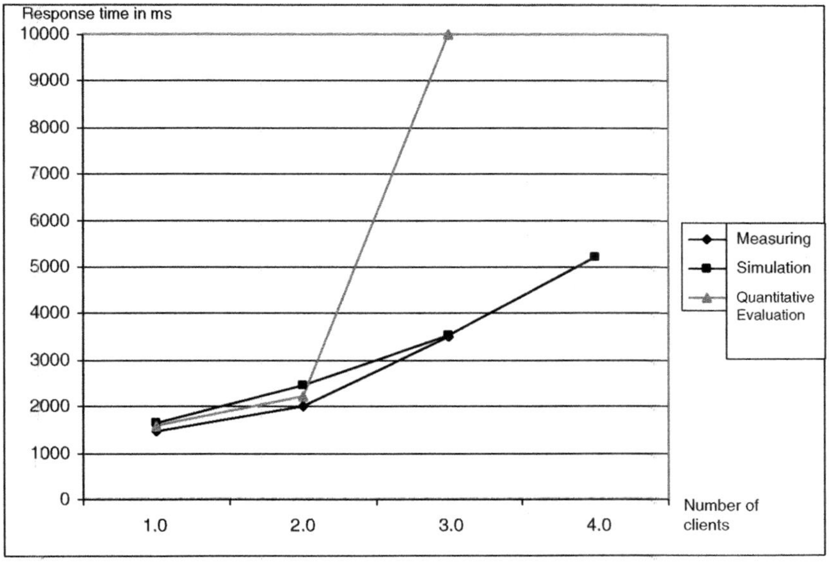

Fig. 39.8 Case study: comparison of simulation, measurement, and quantitative analysis

useful results. Also, even if the simulation runs faster than the real application, it is very helpful for the architect to have an estimate of the final performance, which is relatively close considering that none of the application's components has to be existing. Bad response time in the simulation would mean even worse response times in the application. In order to get simulation results that are closer to the measured values, the chosen performance model needs to be adjusted. The activities the GSCs execute especially need further attention.

Further validation of the simulation framework has been achieved by analysing a case study with mathematical models using techniques of quantitative analysis. The results show, as can be seen in Fig. 39.8, that the simulation as well as the mathematical model are well aligned with the actual measurements. At the upper limit of the load of the application server, the different methodologies differ slightly in their results: although the mathematical model predicts a very slow response time, the simulation is not at all affected by the high load. In reality the system is not working according to its specification any more, which declares any prediction void. That means that the actual real-world application shows a very bad and also varying response time when under high stress, which is also depicted in Fig. 39.8.

39.6 Conclusion

In this chapter we have presented a methodology for simulation-based performance engineering. With our approach, it is possible to get a feeling of a software system's performance even in a very early stage of its development: when the high-level software architecture is being conceived. The software architect therefore creates high-level architectural diagrams as well as sequence diagrams that depict the main call scenarios between the software components. This architecture is then simulated by deploying GSCs representing the modeled software components into a real, up and running middleware platform. The modeled call scenarios are then executed, and performance metrics are being recorded. The results enable the software architect to identify potential hot-spots prior to actually implementing the software components. He can then resolve these hot-spots by refining the architecture, altering the models and rerunning the simulation to evaluate his refinements.

We validated the applicability of our methodology by developing a framework that consists of a model editor with included visualizer and configurable GSCs for JAVA and ABAP technology.

We tested our frameworks with a case study, a batch processing software in ABAP. Our results so far show that our framework—in its current implementation—can be used to obtain first estimations of a software system's performance. All measured performance metrics of the simulations deviate from the real systems by the same constant factor. Thus, bottlenecks in the system's architecture can be identified by the relative distribution of computation time. Further fine-tuning of our GSCs to imitate more accurately the performance of the existing system will therefore be necessary, although our GSC concept in general has proven to be reasonable.

Apart from the accuracy of the simulations conducted within our framework, our methodology presents further advantages over analytical performance prediction. As far as industrial applicability is concerned, performance prediction should be as simple as possible. High-level system architectures and the main call scenarios (i.e., the use cases) are artifacts that are produced anyhow when conducting a software development project, as opposed to maintaining formulae used to represent a specific performance behavior. Also, the infrastructure that the software components are embedded into is captured by the simulation. Furthermore, the infrastructure does not have to be modeled as real infrastructure platforms are used. This is especially valuable as modern middleware consumes a large share of the total execution time of a software system. The rising number of integration projects underlines the importance of middleware. In contrast, complex infrastructure platforms can hardly be described using analytical models, due to their complexity. Analytical models also have to capture different scenarios regarding the availability of resources such as CPU, main memory, or network bandwidth. Our simulations in contrast can simply be run on different hardware setups. Therefore, our approach can also be used to test different deployment setups.

The use of real hardware and infrastructure is not free of drawbacks: first, the infrastructure (e.g., servers running the middleware) has to be in place before the implementation and is required to reflect the system landscape intended for the later deployment of the software under development. This can be problematic especially when our approach is to be applied to very large software development projects, due to the fact that hardware is an expensive asset. Second, our approach builds on simulations that ideally run exactly as long as the software being under development. This is appropriate for end-user applications, however, the simulation of batch processing applications may be time consuming.

Our future work will not only include further refinement of our GSCs, but also the conducting of more case studies to confirm and further demonstrate the applicability of our approach. Moreover, research on how the GSCs could be replaced by real components throughout a software project in an incremental fashion has to be undertaken. The possibility to do so would enable the software architect to gradually verify the predictions produced with our methodology during the software development process. Also, the framework could be extended to support other middleware platforms and to provide dummies for component technologies other than ABAP and J2EE.

Acknowledgments The authors would like to thank SAP for their support and valuable feedback during the development phases of this project.

References

1. Raj Jain. *The Art of Computer Systems Performance Analysis*. Wiley, New York, 1991.
2. Robert F. Dugan Jr. Performance lies my professor told me: The case for teaching software performance engineering to undergraduates. In *WOSP'04*, Redwood City, CA, January 2004. ACM.
3. Donald E. Knuth. Structured programming with *go to* statements. *Computing Surveys*, 6(4): 261–301, December 1974.
4. L. Kleinrock. *Queuing Systems, Theory*, volume 1. Wiley, New York, 1975.
5. Giovanni Denaro, Andrea Polini, and Wolfgang Emmerich. Early performance testing of distributed software applications. In *WOSP' 04*, pages 94–103, Redwood City, CA, January 14–16, 2004. ACM.
6. Yan Liu, Ian Gorton, Anna Liu, Ning Jiang, and Shiping Chen. Designing a test suite for empirically-based middleware performance prediction. In James Noble and John Potter, editors, *40th International Conference on Technology of Object-Oriented Languages and Systems (TOOLS Pacific 2002)*, volume 10. Australian Computer Society Inc., 2002.
7. Simonetta Balsamo, Antinisca Di Marco, Paola Inverardi, and Marta Simeoni. Model-based performance prediction in software development: A survey. *IEEE Transactions on Software Engineering*, 30(5):295–310, May 2004.
8. Rebecca Isaacs and Paul Barham. Performance analysis in loosely-coupled distributed systems. Technical Report, Microsoft, 2003.
9. K. London, J. Dongarra, S. Moore, P. Mucc, K. Seymour, and T. Spencer. End-user tools for application performance analysis using hardware counters. In *International Conference on Parallel and Distributed Computing Systems*, August 2001.
10. Elaine J. Weyuker and Filippos I. Vokolos. Experience with performance testing of software systems: Issues, an approach, and case study. *IEEE Transactions on Software Engineering*, 26(12):1147–1156, 2000.
11. Connie U. Smith. Software performance engineering. In *Proceedings of Computer Measurement Group International Conference XIII*, pages 5–14. Computer Measurement Group, December 1981.
12. M. Woodside, D. Petriu, and D. Amyot. PUMA project. http://www.sce.carleton.ca/rads/puma. URL retrieved 20.11.2005.
13. Object Management Group. UML Profile for Schedulability, Performance and Time. OMG Full Specification, formal/03-09-01, 2003.
14. Xiuping Wu, David McMullan, and Murray Woodside. Component-based performance prediction. In *Proceedings of Sixth ICSE Workshop on Component-Based Software*, May 2003.

15. Andreas Knoepfel, Bernhard Groene, and Peter Tabeling. *Fundamental Modeling Concepts: Effective Communication of IT Systems*. Wiley, UK, 2006.
16. Peter Tabeling and Bernhard Grone. Integrative architecture elicitation for large computer based systems. In *ECBS'05: Proceedings of the 12th IEEE International Conference and Workshops on the Engineering of Computer-Based Systems (ECBS'05)*, pages 51–61, Washington, DC, 2005. IEEE Computer Society.
17. Sun Microsystems, Inc. Java pet store demo. http://java.sun.com/blueprints/code/jps132/docs/index.html.
18. Daniel Brinkmann. Analyse des Java Pet Stores und Portierung auf den SAP Web Application-Server. Master's thesis, Hasso-Plattner-Institut für Softwaresystemtechnik an der Universität Potsdam, 2005.
19. Object Management Group. Unified Modeling Language Specification v. 2.0. OMG UML 2.0 Superstructure Specification, formal/05-07-04, 2005.

Chapter 40
A Framework for UML-Based Software Component Testing

Weiqun Zheng and Gary Bundell

40.1 Introduction

Component-based software engineering (CBSE) is becoming a widely used software engineering approach to reduce software development costs and sustain rapid software production. *Software component testing* (SCT) denotes a set of testing activities that analyse software artefacts, uncover software faults, and evaluate software correctness and quality of software components under test (CUT) and component-based software or systems (CBS) [1, 2]. SCT focuses on producing component test cases (CTCs) that are the central part of all SCT tasks. Although component functionality and reusability are always needed in software component design/development (SCD), SCT considerably influences component reliability and quality [3], which, in a certain sense, could determine whether CBS succeeds in the CBSE practice. SCT becomes an integral part of the SCD lifecycle in CBSE.

Our previous SCT work with a software component laboratory project has proposed a XML-based component test specification (CTS), and developed an accompanying testing tool to support dynamic testing with executable component programs in the runtime environment [4, 5]. The CTS test case specifications have several unique characteristics different from traditional test case representations [1], such as a well-defined and well-structured format for specifying CTCs, portability and platform neutrality for compatibility and reusability, and executability and verifiability for dynamic testing.

To further our previous SCT work, this research focuses on investigating a new paradigm of SCT methodology on how to develop and construct effective CTCs. We analyse software specification and design artefacts that develop software components, identify what test artefacts are needed, and construct effective test artefacts as a core testing foundation for test design and generation. We examine and study current software testing methods to develop a new SCT process that aims to guide the undertaking of SCT activities in a systematic way. We also utilise our XML-based CTS to represent and specify the verifiable CTCs that are designed

Oscar Castillo et al. (eds.), *Trends in Intelligent Systems and Computer Engineering.*

and generated with our SCT methodology. This chapter focuses on component integration testing (CIT) that bridges component unit testing and component system testing.

Because CTCs are usually derived for testing software components under development, test artefacts used to design and generate test cases should be based on SCD information, such as analysis and design specifications, program implementation, and so on. Accordingly, there are three basic categories of test design and generation approaches: implementation-based, specification-based, and model-based [1]. The traditional implementation-based software testing usually relies on the source code to derive tests, and is more or less limited to white-box structural testing, which is ineffective for CIT that is concerned with testing of component interfaces and interactions. By contrast, specification-based software testing is more effective for CIT, especially for more important black-box functional testing of integration interfaces that are captured with component specifications. By incorporating software testing with model-based specifications, model-based software testing effectively bases testing tasks on certain models of the software under test, designs and generates test cases, and evaluates test results from the software models and model artefacts [6]. This research incorporates widely accepted software practices based on the utilisation of UML modeling to both develop and test object-oriented software components and CBS.

Advancing from UML-based SCD to UML-based SCT poses certain challenges for new UML-based SCT approaches to effective production of functional and reliable software components. This research particularly addresses some important SCT issues about building test models and developing SCT processes with UML modeling. We argue that the importance of test models for SCT should be considered equally with development models for SCD. One of our main research goals is to bridge the gap between SCT and SCD with UML modeling and to base SCT on UML-based test models, in order to gain benefits from using a consistent specification approach. We also argue that incremental development approaches (e.g., the Unified Process (UP) [7]) should be applied to SCT practice to develop an effective iterative SCT process so that SCT activities are carried out effectively in parallel to SCD activities, which is another of our research goals.

In this chapter, we present a new UML-based SCT methodology, model-based software component testing (MBSCT), which aims to leverage SCT with software component modeling (SCM) for SCD, and achieve an effective integration of SCT with UML modeling towards SCT practice. The main contribution of this chapter is that the MBSCT methodology introduces a new SCT framework that is supported with a set of useful SCT techniques. The MBSCT framework provides a major two-phase workflow process to develop CTCs with UML modeling: the first phase builds a set of UML-based test models, which is supported with the scenario-based CIT and test by contract techniques; the second phase designs and generates CTCs from the constructed test models, which is supported with a component test mapping technique. The derived CTCs are represented and specified with our XML-based CTS to become CTS test case specifications. The MBSCT methodology was initially proposed in [8], and this chapter describes and discusses it in more detail.

The remainder of the chapter is structured as follows. Section 40.2 reviews and discusses the related research work on software testing with UML modeling. Section 40.3 presents a methodology overview. Section 40.4 discusses how UML-based test models are built from parallel SCM models. Section 40.5 describes improving testability for effective test design along with test model construction. Section 40.6 discusses how to derive CTCs from the constructed test models. A case study is illustrated throughout the chapter. Section 40.7 concludes this chapter and outlines future work.

40.2 Related Work

This research follows a small but growing body of work on UML-based software testing. Binder's work [1] provides a comprehensive literature review of object-oriented software testing with models, patterns, and tools. Binder discusses some generic test strategies and requirements for UML models and provides a general guide to software testing with UML diagrams.

Offutt and Abdurazik [9] develop a testing technique that adapts state-based specification test data generation criteria to generate test cases from a restricted form of UML state diagrams, but with the limitation that may result in some states not being entered or reached. This technique also is limited to class-level testing, and does not directly support integration testing.

Their subsequent work [10] analyses behaviour of interacting objects specified by UML collaboration diagrams at the software design level, and adapts some traditional data-flow coverage criteria (such as all definition-uses) in the context of UML collaboration diagrams to assist test generation. But their work does not discuss how to generate actual test cases and does not make use of UML sequence diagrams. Also the empirical evaluation of the test criteria has not been yet carried out in their work. Both their works also do not specifically address testing issues associated with the SCT domain.

Briand and Labiche [11] present a system test methodology that supports the derivation of functional system test requirements from UML models produced at the end of the analysis development stage. Their methodology involves UML analysis artefacts for use-cases, sequence and collaboration diagrams, class diagrams, and possible OCL constraints [12] on UML artefacts. But this work does not currently address how to generate actual test cases by using the derived test requirements.

Wu et al. [13] present a UML-based integration testing technique and analyse four key test elements: interfaces, events, context-dependence relationships, and content-dependence relationships. These elements describe the characteristics of component interactions to be considered for testing CBS. These test elements can be derived from UML models, such as collaboration/sequence diagrams and state diagrams. However, the work in their paper does not address or give practical ways on how to generate actual test cases for SCT by using their proposed test model and test criteria.

Most of the testing techniques only provide certain general test requirements or criteria, and lack detailed and operational descriptions on how to apply them to generate actual test cases. Little work has been conducted on CIT by the effective use of UML models and artefacts (such as UML sequence diagrams and interacting messages for use-case scenarios). In addition, the test cases derived are more or less deficient in some uniform representation for the special SCT needs, and they are often neither executable nor verifiable in the dynamic testing environment for testing component implementation. Lastly, little earlier work discusses the important SCT issues of building test models and developing testing processes, especially in conjunction with UML modeling, which is one of the central focuses of this research.

40.3 Model-Based Software Component Testing: A Methodology Overview

This section presents an overview of the MBSCT methodology. We first give an overall methodology summary, and then we describe main concepts and technical aspects of each MBSCT process/technique. More technical details are further discussed in the following sections and illustrated with a case study.

40.3.1 Methodology Summary

The MBSCT methodology is a new UML-based SCT approach that has been developed with a set of supporting SCT process and techniques. An iterative SCT process is introduced to build UML-based test models from parallel SCM models along with the incremental SCD process, which abides by the UP principles. A scenario-based *CIT* technique is developed to emphasise testing priority on test scenarios that examine crucial component functions with operational scenarios in their integration contexts. A contract-based SCT technique, test by contract, is developed to design and construct useful test contracts that are applied to component artefacts for consolidating the test models, which are used to establish a solid SCT foundation for test design and generation. A test mapping technique for SCT, component test mapping, is developed to map and transform testable UML artefacts and test contracts into target test data for constructing test sequences and test elements, and then to derive and generate the CTS test case specifications. Finally, executable component programs are dynamically tested by executing and verifying the CTS test case specifications in the runtime environment of the CUT/CBS.

40.3.2 Iterative SCT Process

An iterative SCT process is introduced to be an extension of general incremental development approaches (such as UP) to the SCT domain. One major objective of

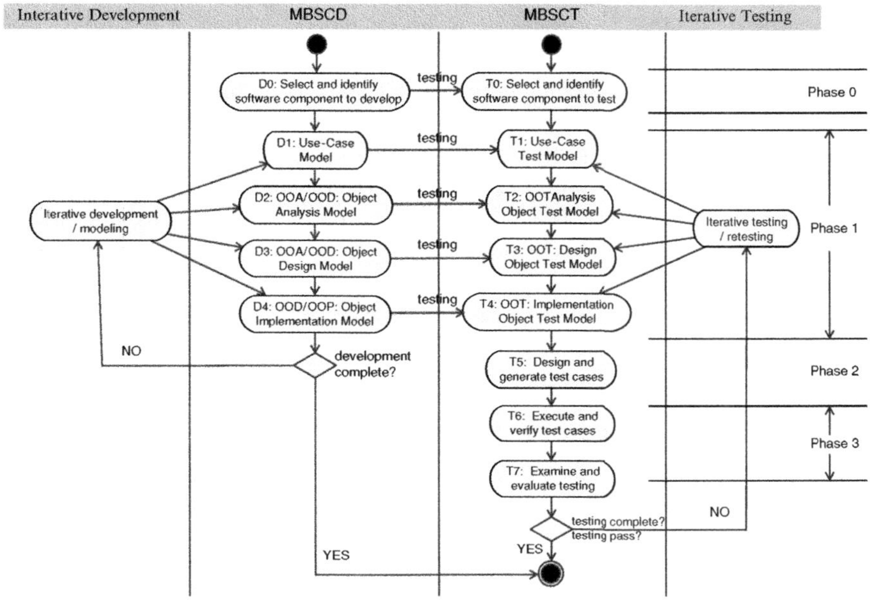

Fig. 40.1 MBSCT methodology: an iterative SCT process

the iterative SCT process is to guide the incremental construction of UML-based test models as a SCT foundation. The entire process consists of two main parallel workflow streams: model-based SCD (MBSCD) and model-based SCT (MBSCT), whose relationships are summarised and illustrated in Fig. 40.1.

Both workflows follow the proven incremental approach of the UP principles, and apply UML modeling to produce software components and CBS. In the object-oriented SCD context, the MBSCD process is composed of a number of development (with modeling) steps (marked D0, D1, D2,...), which produce a set of SCM models at different modeling levels, including the Use-Case Model, Object Analysis Model, Object Design Model, Object Implementation Model, and so on.

In parallel, the MBSCT process is linked and works closely with the MBSCD process. Working in the context of object-oriented testing, the MBSCT process covers a number of main test steps (marked T0, T1, T2,...) and builds a group of SCT models at different testing levels, typically including the Use-Case Test Model, Analysis Object Test Model, Design Object Test Model, Implementation Object Test Model, and so on. Note that a "T" (Testing) step in the MBSCT stream corresponds to a parallel "D" (Design/Development) step in the MBSCD stream at the same modeling level. Whereas a later "D" step is mainly based on its preceding "D" steps, a later "T" step is based on its parallel "D" step as well as its preceding "T" step. For example, we build the Design Object Test Model (Step T3) based on its parallel Object Design Model (Step D3) and Analysis Object Test Model (Step T2).

Consistent with the UP principles, the entire MBSCT process utilises the two parallel workflow streams to jointly establish an incremental and systematic framework with a series of SCD/SCT steps covering almost all main SCT tasks with UML modeling. Technically, for this MBSCT framework, we can group the related steps into four main phases: (1) Phase 0: including D0/T0, about component selection, and not further discussed here but referred to elsewhere [1, 2]. (2) Phase 1: including D1/T1 to D4/T4, discussed in Sects. 40.4 and 40.5. (3) Phase 2: including T5, discussed in Sect. 40.6. (4) Phase 3: including T6 and T7, about dynamic testing, and referred to in our previous work [4, 5]. This chapter focuses on the important methodological aspects in Phases 1 and 2 for developing CTCs in the MBSCT framework; that is, the first phase builds UML-based test models based on SCD models, and the second phase derives CTCs from the test models and other test information.

40.3.3 Scenario-Based CIT Technique

The MBSCT methodology described in this chapter particularly focuses SCT on CIT, which is based on the completion of its underlying test levels and bridges component unit testing and component system testing. The iterative MBSCT process has a key focus on use-case driven development, which explores certain relationships between testing and use-cases as well as scenarios. A use-case scenario illustrates a specific functional behaviour and forms a typical integration context covering interaction dynamics. Accordingly, the central CIT focus is on examining functional scenarios that specify and realise software component integration (SCI) with object interactions among integrated components and their composite objects in the specific SCI context. Using UML modeling, we can model object interactions with use-cases, interaction diagrams and class diagrams to capture scenarios, sequences, messages/operations, classes, elements (states/events), and the like, which all are important testable model artefacts.

We apply scenario-based software testing [14] to CIT, and conduct what we call scenario-based CIT. One key feature of this technique is that it clearly focuses testing priority on test scenarios to exercise and examine critical deliverable software functions with operational use-case scenarios in the related SCI contexts. Consistent with use-case driven development, the main testing task is to identify and construct test scenarios that examine the associated scenarios and test multiple components and composite objects along its scenario execution path by using test sequences and test operations.

Test scenarios naturally form typical SCI contexts to examine related software artefacts for CIT. When applying this technique to testing practice, the tester can use a single test scenario to exercise and verify the CUT's multiple objects and operations participating in the associated scenario under test. In addition, the tester can test a single CUT with multiple test scenarios for diverse test objectives and requirements, typically when the CUT is involved in multiple SCI contexts. Such multiple testings are especially necessary when software components are integrated in any new component-based applications under development.

One objective of the scenario-based CIT technique aims to gain a trade-off between testing coverage and testing costs. Usually in testing practice, full coverage testing is known to be impractical, and high-level coverage testing is also too expensive. Compared to other testing techniques, this technique prioritises test coverage by focusing on key test scenarios covering and verifying the crucial software functions that the CUT/CBS must deliver. Applying the scenario-based CIT technique to identify and construct test scenarios is a major task of building UML-based test models for undertaking CIT.

40.3.4 Test by Contract Technique

SCI must comply with certain rules or contracts between composite component modules in the SCI context. Component contracts govern and regulate operations and interactions of integrated components and component objects that are plugged into component-based applications. This means that any violation of component contracts may result in certain component faults. From the viewpoint of component contracts, one of the main CIT tasks is to verify component operations to conform to the specified contracts, and to detect possible component faults that violate the component contracts. Contracts are very important testable information to support effective testing. However, in SCD practice, usual SCM models may not actually contain sufficient testable information, which could cause some component artefacts to be nontestable although they may be appropriate simply for the purposes of component design and implementation. For those nontestable software artefacts that need to be tested according to certain testing objectives and requirements, the testing becomes very difficult to undertake test design and generation effectively. To resolve this problem in SCT practice, the MBSCT methodology applies the design by contract [15] principle to both SCD and SCT activities, and extends design by contract to develop a new contract-based SCT technique, called test by contract (TbC).

With the TbC technique, we identify and devise special test contracts based on contract artefacts that are used for test conditions/constraints. Test contracts are exerted upon the crucial component artefacts at different modeling levels, so as to effectively examine these software artefacts included in the final implementation of the CUT/CBS. Test contracts are testable software artefacts, and are commonly represented and realised with preconditions, postconditions, and invariants in the form of assertions (conditions or constraints) and associated concepts, especially when they are applied to test the component implementation. The TbC technique aims to achieve a clear testing goal: improving component testability and testing the related software artefacts that are crucial for conforming to component correctness and reliability.

In SCT practice, the tester can apply the TbC technique to identify and construct test contracts to consolidate test model construction for effective test derivation. In particular, for some nontestable software artefacts that need to be tested according to certain test objectives, the tester can apply the TbC technique to improve their testability by designing and adding appropriate test contracts that are additional to the software under test. In this way, effectively applying the TbC technique is able to make it possible to transform nontestable software artefacts to testable software artefacts that serve as the basis of test cases under design and generation.

To deal with test contracts effectively, we introduce an important test contract concept: *effectual contract scope*, which is a software context (e.g., component context or modeling context) of a test contract where the test construct can take effect. Based on this useful concept, we can classify test contracts into two main categories: (1) an internal test contract (ITC) is defined and applied to, and is also verified within the same effectual contract scope and the same software context; and (2) an external test contract (ETC) is defined and applied to a software context, but is verified outside this software context, which means that the effectual contract scope of the ETC is not the same as its software context. It is important to recognise the property of these types of test contracts when the tester is to design and apply them effectively in testing practice. Usually, ITCs are often used in component unit testing, but they are required to be re-examined in the SCI context where they are involved. By contrast, ETCs are often used in CIT, where an ETC is verified in one integration module whereas the ETC is defined and applied to another integration module.

40.3.5 Component Test Mapping Technique

The MBSCT methodology emphasises test derivation based on SCD artefacts, especially SCM artefacts. This testing strategy indicates a strong requirement for exploring the fundamental relationship between SCT artefacts and SCD/SCM artefacts so as to undertake test derivation effectively. A component test mapping (CTM) technique is introduced to develop a test mapping relationship between SCT artefacts and SCD/SCM artefacts with the aim to achieve a clear testing goal: assisting actual test derivation in a more feasible way and thus further bridging the gap between the SCT and SCD/SCM domains.

Conceptually, the CTM technique establishes a typical $(1-\mathsf{n})$ mapping relationship between two sets, which can be defined as follows.

$$(1-\mathsf{n})\ \mathsf{mapping\ CTM\colon\ SCDS} \rightarrow \mathsf{SCTS}$$

where (1) set $\mathsf{SCDS} = \{$elements of SCD specifications, e.g., UML model artefacts for SCD$\}$; (2) set $\mathsf{SCTS} = \{$elements of SCT specifications, e.g., test case artefacts represented and specified by the XML-based CTS$\}$. This CTM definition means that an element in set SCDS may be mapped and thus corresponds to one or more elements in set SCTS for constructing and specifying a test for a specific testing objective.

The CTM technique is developed to refine the process of test design and generation, and focuses on how to map and transform testable UML model artefacts and test contracts into test case data, which can be used to construct test sequences and

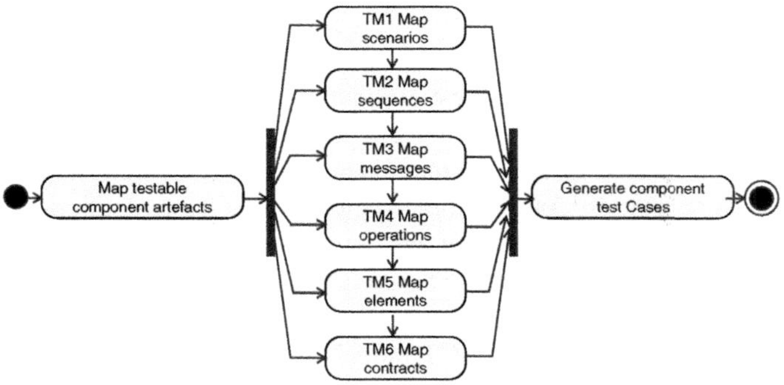

Fig. 40.2 MBSCT methodology: component test mapping

elements for CTCs. The CTM process takes a series of steps of test transformations and constructions with respect to relevant UML models and elements at different modeling levels towards the intended CTCs. Figure 40.2 illustrates the CTM process and steps as well as their relationship.

40.4 Building UML-Based Test Models

40.4.1 Main Tasks of Constructing Test Models

Following the two-phase MBSCT framework for developing CTCs, the main tasks in the first phase are to build a set of test models based on UML models for SCD. The MBSCT methodology provides the iterative SCT process, the scenario-based CIT, and TbC techniques to support constructing SCT models as the SCT foundation.

The iterative MBSCT process indicates what test models need to be constructed, and also indicates what related SCD/SCT models are needed as the basis of constructing a specific SCT model. The main tasks in building a SCT model (e.g., design object test model or DOTM) are to identify, extract, and construct testable model artefacts based on its corresponding SCD model (e.g., object design model), and then to map and transform them to test artefacts for constructing the SCT model (e.g., DOTM).

We apply the scenario-based CIT technique to undertake the construction of a particular test model. Our testing priority is to identify and construct test scenarios with test sequences and test operations to examine object interactions in SCI contexts. We model test scenarios based on UML models illustrating use-case scenarios under test (e.g., UML sequence diagrams and class diagrams), and capture test scenarios with test sequence diagrams and test class diagrams. Furthermore, to enhance testability, we apply the TbC technique to design special test contracts to enhance test models for effective test derivation.

Typical test artefacts of test models mainly include test scenarios, test sequences, test messages/operations, test classes, test elements (e.g. test states, test events), and special test contracts. We classify test artefacts constructed in the test model into two main categories: (1) *Basic test artefacts* (e.g., basic test operations) are testable model artefacts extracted from SCD models and are transformed into test artefacts in SCT models to exercise and examine typical component functions with operational scenarios. (2) *Special test artefacts* (e.g., special test operations) are special test contracts constructed for particular testing objectives in the related testing contexts (e.g., testing a class/component operation for object interaction in an integration context), or for consolidating SCT models with the enhanced testability (e.g., transforming nontestable model artefacts to testable model artefacts).

40.4.2 Case Study

We develop a case study to illustrate the MBSCT process of building UML-based test models and designing and generating CTCs for CIT, which also serves as a preliminary demonstration of the effectiveness of the MBSCT methodology. We develop a software controller simulation for a car parking system (CPS) that simulates a typical public access control system, where a flow of vehicles and parking control devices are monitored, coordinated, and regulated against certain public access requirements and safety rules (Fig. 40.3). Based on a prototype in [16], the CPS system is redeveloped and componentised into a typical pattern-based CBS, with the base component EventCommunication that is a pattern-based software component built based on the Observer pattern [17], and three application components including a device control component, a car control component, and a GUI simulation component. The following sections describe how to construct UML-based test models for CIT using an excerpt of the CPS case study [18].

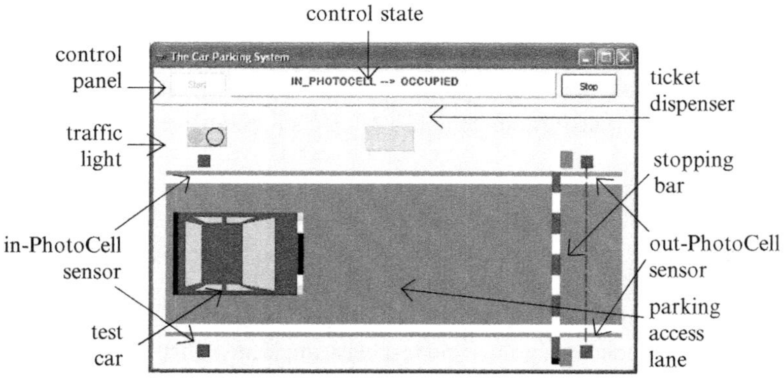

Fig. 40.3 Car parking system

40.4.3 Use-Case Test Model

The iterative MBSCT process indicates that there are two main levels of test models under development: use-case test model (UCTM) and object test model (OTM). This section discusses the first step D1 → T1 to construct the UCTM based on the use-case model (UCM). Our main task focuses on identifying and extracting testable model artefacts in the UCM, and mapping and transforming them to test artefacts for constructing the UCTM (Fig. 40.4).

With UML modeling, the UCM mainly describes the system's functions and requirements in terms of a set of actors, use-cases and their relationships. We apply the scenario-based CIT technique to identify and construct test use-cases and test scenarios as a key task in the UCTM construction. Usually, test scenarios are constructed based on the associated use-case scenarios that fulfil software functions in the integration context. Above all other CPS use-cases in the UCM for testing purposes, the three core test use-cases for testing typical CPS operations are identified as: (1) TUC1: test car enters the parking access lane (PAL); (2) TUC2: test driver withdraws ticket; (3) TUC3: test car exits the PAL. Altogether they constitute a test sequence of the three parking phases for one full parking process cycle for any car. All cars can access the PAL as many times as needed, and reiterate the same parking sequence indefinitely during their different accesses. TUCs form typical SCI contexts to examine related test scenarios for CIT.

Figure 40.5 shows a partial UCTM of the CPS system, with a test use-case diagram for three test use-cases (Fig. 40.5a) and a system test sequence diagram for the

D1: Use-Case Model		T1: Use-Case Test Model
1. Functions and requirements.		1. Testing objectives and requirements.
2. Use-case diagrams.		2. Test use case diagrams with test actors, test events.
3. Actors and descriptions.		3. Test actors and descriptions.
4. Use cases and scenario descriptions.	⇨	4. Test use cases and test scenario descriptions.
5. System sequence diagrams for system scenarios with system events.		5. System test sequence diagrams for system test scenarios with test actors, test events.
6. Contracts for system events and scenarios.		6. Test contracts for system test events and scenarios.

Fig. 40.4 Constructing the use-case test model

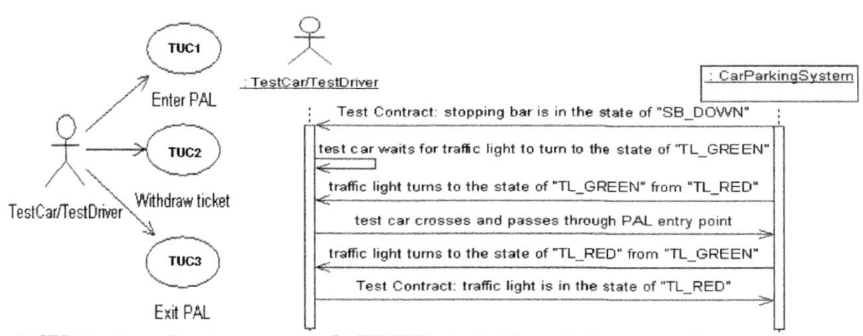

(a) CPS test use case diagram (b) CPS TUC1 test scenario: system test sequence diagram

Fig. 40.5 Use-case test model

TUC1 test scenario (Fig. 40.5b). A test scenario is a typical test use-case instance (e.g., of TUC1 in Fig. 40.5b) that exercises and examines a sequence of system test events that interact between the test actor (e.g., test car/driver) and the black-box CBS under test (e.g., our CPS), and thus tests the associated system operational use-case scenarios for the required behaviour (e.g., test car enters PAL) in the use-case under test (e.g., TUC1). The test scenario also reflects the corresponding changes of related system states (e.g., traffic light turns to the state of "TL_GREEN" from "TL_RED" or vice versa in Fig. 40.5b), which are triggered by system test events (e.g., car parking activities) and are key indicators for scenario-based testing. A clear testing objective is that certain functional requirements (e.g., the test car enters the PAL correctly in TUC1) are correctly fulfilled as expected through examination of the associated test scenario.

40.4.4 Object Test Model

Working with object-oriented testing techniques, we build a series of object test models based on parallel object models (see Fig. 40.1). In particular, the main task is to identify, extract, and construct testable model information in UML-based object models, and map and transform them to test artefacts for constructing corresponding object test models. Due to space constraints, this chapter only presents the DOTM construction and related testing activities with the CPS case study. More details on constructing test models can be found in [18].

As a SCT model for testing design objects, the DOTM is constructed with test scenarios, test sequences, test operations and test classes at the object design level (Fig. 40.6). Figure 40.7 shows a design test sequence diagram for the CPS TUC1 test scenario, which is part of CPS's DOTM. The CPS TUC1 is the first of three core test use-case scenarios, and performs CIT on how a car enters the PAL correctly in the TUC1 integration context, where the PAL entry point is jointly controlled by the traffic light and in-PhotoCell sensor devices.

We apply the scenario-based CIT technique and construct the corresponding test scenario to exercise and examine integration-participating operations of six test objects from device classes (e.g., class TrafficLight) and test classes (e.g., object testCarController) in the integration context. In the DOTM, a basic test

D3: Object Design Model		T3: Design Object Test Model
1. Design classes in software solution domain.		1. Design test classes, e.g. design classes and related test helper classes.
2. Design class diagrams with design classes.		2. Design test class diagrams with test classes.
3. Design sequence diagrams for use case realisations with objects of design classes.	⇨	3. Design test sequence diagrams for test scenarios with test classes.
4. Interaction messages/operations and sequences with objects of design classes.		4. Test scenarios, test sequences and test operations.
5. Contracts for main operations of design classes.		5. Test contracts for main operations of test classes, test states, test events.

Fig. 40.6 Constructing the design object test model

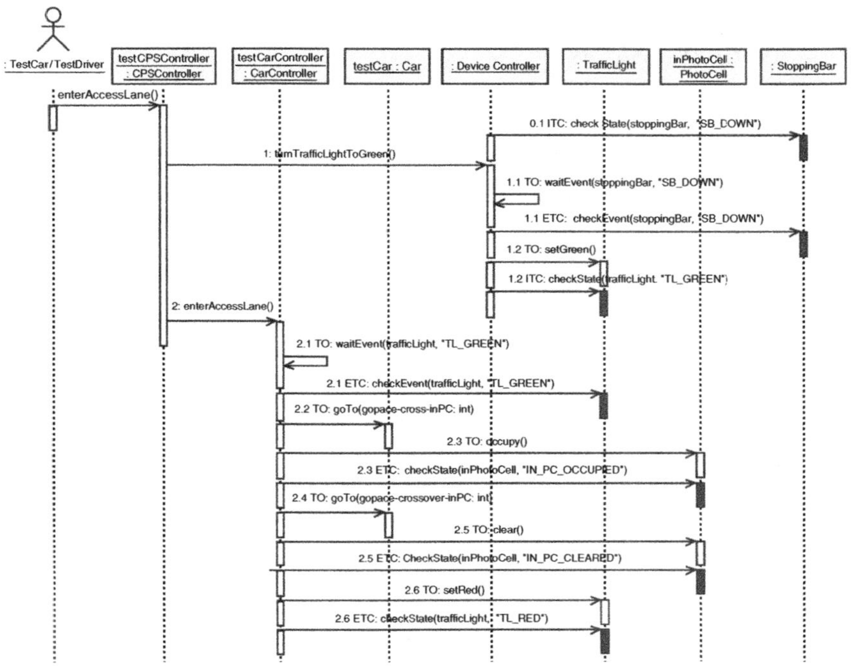

Fig. 40.7 CPS TUC1 test scenario: design test sequence diagram

operation (e.g., setGreen()) from its test object (e.g., object trafficLight) exercises and examines what the CPS does with the basic operation under test. A special test contract is constructed to verify whether an operation performs correctly to realise its related object interaction, and is represented and realised with a special test operation (e.g., checkState()), which is designed and added to its test class (e.g., class TrafficLight).

40.5 Improving Testability for Effective Test Design

It is not sufficient to simply extract certain basic testable model artefacts from UML SCD models as test artefacts for building test models. A reason is that the usual SCD models primarily aim for component design and then implementation may not actually contain adequate testing information for effective test design. The MBSCT methodology develops the TbC technique that aims to improve component testability and to facilitate test model construction. We show how the TbC technique is applied to construct test contracts to support CIT and test design for uncovering faults along with test model construction, which is enhanced with relevant test contracts.

40.5.1 Designing Test Contracts to Verify Object Interactions with Test States

An important focus of CIT is on object interactions and object state changes with object interactions, because SCI takes place with the interactions through the interfaces of component objects in the SCI context. We apply the TbC technique and design test contracts (including ITCs and ETCs) to track down dynamic changes of object states against certain expected test states, and to examine whether a particular interacting operation is performed correctly for the corresponding object interaction. Test contracts are constructed to add into related test models (e.g., DOTM in Fig. 40.7), which are consolidated for undertaking CIT. In the CPS TUC1 test scenario, we design test contracts as special test operations (illustrated with the shaded narrow rectangles in Fig. 40.7), and apply them to all controlling operations for parking control services in order to check changes of control states, which are typical test states of the CPS system. For example, test contract 2.3 ETC checkState(inPhotoCell, "IN_PC_OCCUPIED") is designed and works for CIT as follows (see Fig. 40.7): (1) It checks whether the in-PhotoCell device is in the correct state of "IN_PC_OCCUPIED" as expected, after test operation 2.3 TO occupy() is performed. (2) It is applied to operation occupy() in object inPhotoCell in the device control component, but is verified in object testCarController in the car control component. So test contract 2.3 is designed as an ETC.

40.5.2 Designing Test Contracts to Verify Object Interactions with Test Events

We now explore another important aspect of CIT: verifying a particular object interaction by checking certain communication messages that realise object interactions between collaborating objects. We illustrate this test task with retesting the pattern-based component EventCommunication in the CPS TUC1 integration context. Test contracts are constructed with the TbC technique to examine and verify event communications by checking particular test events, in order to ensure that the event communication is correctly conducted in the integration context. This testing is especially necessary when system control shifts from the device control component to the car control component, and vice versa. We show how such testing is carried out with a group of test contracts and test operations constructed as follows.

1. In the TUC1 test scenario (Fig. 40.7), the system control starts with the device control component: (1) Test operation 1.2 TO setGreen() runs on object trafficLight to set the traffic light to the new state of "TL_GREEN". (2) Execution of this test operation causes the object's state change, which results in a new event being generated. Then by conducting an event communication, the event notifier object trafficLight needs to notify the new event to all its waiting event listener objects testCarController and deviceController. (3) Test contract

1.2 ITC checkState(trafficLight, "TL_GREEN") is constructed to check whether the traffic light device is now in the correct state of "TL_GREEN" as expected, before the system control is switched over.

2. Then, system control shifts to the car control component: (1) The car waits for an incoming event notification as a parking instruction to access the PAL. This is conducted by test operation 2.1 TO waitEvent(trafficLight, "TL_GREEN") on object testCarController. (2) When the event communication is completed with the base component EventCommunication, the car needs to take some action according to the received event information. However, before the car enters the PAL, it is necessary to recheck whether the event reception is correct on the event listener object testCarController. Test contract 2.1 ETC checkEvent(trafficLight, "TL_GREEN") is constructed to check whether the event listener object testCarController receives the correct event notification (i.e., traffic light is in the correct state of "TL_GREEN"; the car is allowed to enter the PAL) from the correct event notifier object trafficLight.

40.5.3 Conducting Test Design to Detect Faults

Based on test models constructed for CIT, test design can combine certain related test contracts and test operations to detect possible component faults in the SCI context. Specifically, if a test contract returns false, a software fault has probably occurred, which is mostly likely related to the associated operation under test. The detection process is undertaken in conjunction with fault case analysis and fault-based testing techniques [1]. For example, a test design combining test operation 2.3 TO and test contract 2.3 ETC (Fig. 40.7) can examine and detect a fault case scenario of the in-PhotoCell sensor device if this test contract returns false: while the car has occupied the in-PhotoCell device, this device is not in the correct state of "IN_PC_OCCUPIED".

Fault consequences may include: the PAL entry point is not occupied as expected, and some subsequent operation (e.g., test operation 2.5 TO clear()) may not be conducted as needed in the expected sequence of control operations, which may further lead to the entire CPS operation being halted at this point. Fault analysis indicates that test operation 2.3 TO fails in the TUC1 integration context with two main fault causes:

(1) The incorrect invocation/usage of operation occupy() by integration class CarController in the car control component (as the current SCI context). This fault occurrence clearly relates to intercomponent integration testing, because class CarController in the car control component invokes operation occupy() of device class PhotoCell in the device control component, and the invocation is an object interaction that realises a component interaction between the two CPS components.

(2) The incorrect initial definition of operation occupy() by device unit class PhotoCell in the device control component, which means that there may be a physical

hardware fault of the in-PhotoCell device. This fault occurrence clearly relates to component class unit testing, because operation occupy() is defined in unit class PhotoCell.

This typical testing example demonstrates that our model-based CIT approach effectively achieves two testing benefits: examination and detection of possible component faults not only related to certain integration contexts as the central focus of CIT, but also to certain unit contexts as a secondary focus of CIT. The uncovered faults need to be corrected, and regression testing needs to repeat the integration/unit testing activities after the software change.

40.6 Deriving Component Test Cases

40.6.1 Main Tasks of Test Design and Generation

Following the MBSCT framework, the main tasks in the second phase of developing CTCs are: (1) analysing and identifying what software artefacts need to be tested for target test objectives and requirements; (2) designing test sets with test scenarios and test sequences; (3) identifying and constructing composite test case artefacts for generating test cases; and (4) generating test cases to uncover target software faults.

Our method of test design and generation is model-based, and some test design tasks are conducted in conjunction with building test models, which provides adequate testing information for designing and generating CTCs. Our XML-based CTS provides a well-structured uniform representation to specify CTCs, which become CTS test case specifications for SCT. Further describing our MBSCT framework, we incorporate the CTM technique to provide more technical details for our method and the process of CTC design and generation in practice.

40.6.2 Component Test Mapping

Test models are constructed to capture adequate test information to provide the SCT foundation for test design and generation. The CTM technique refines to detail the method and process of test design and generation, and to support deriving CTCs from test models. The CTM technique maps and transforms testable UML model artefacts and test contracts into target test data for constructing test sequences and test elements, and then to derive and generate CTCs, which are then represented and specified with the XML-based CTS to produce the CTS test case specifications.

Technically, the entire CTM process is carried out in two major phases as shown in Fig. 40.8. Firstly the test mapping maps out to produce adequate test artefacts and test data for constructing CTCs, and secondly test data are further mapped to appropriate CTS elements to generate the target CTS test case specifications. The

Component test mapping :
{testable model artefacts with UML models for CUT/CBS}
→ {test artefacts for target test data of CTCs}
→ {CTS test case specifications for CUT/CBS}

Fig. 40.8 Component test mapping phases

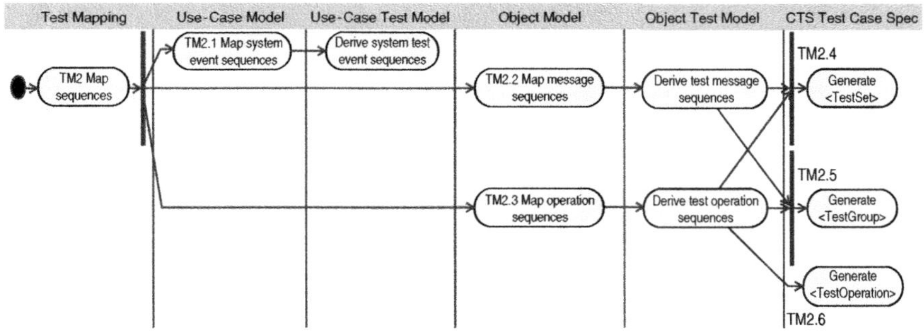

Fig. 40.9 TM2: mapping sequences

two phases apply in each individual CTM step. Note that the first CTM phase is aided by test model construction from other MBSCT techniques. Due to space limitations, this chapter only describes some of the most useful CTM Steps TM2, TM4, and TM6. For more CTM descriptions, refer to [18].

40.6.2.1 Mapping Sequences

The sequence mapping in Step TM2 carries out mapping and transforming sequences of interactions into sequences of logically ordered composite tests, which are called *test sequences*. Test sequences represent and realise test scenarios for undertaking CIT, using a sequence of test operations to examine whether object interactions correctly fulfil the required functions by integrated objects in the integration context. This test mapping may take place to derive test sequences at different mapping levels as shown in Fig. 40.9.

Our XML-based CTS provides several structural elements to represent test sequences at different levels of test granularity to streamline the structure of CTS test case specifications. After sequences are mapped out, a test sequence needs to be further mapped to one of the CTS structural elements. This test mapping phase is required to generate the hierarchical structure of the target CTS test case specification. For example, Step TM2.5 maps test sequences to test groups represented with XML element <TestGroup>. A typical test group (Fig. 40.10) is mapped from a pair of a test operation (e.g., test operation 2.3 TO) and its associated test contract (e.g., test contract 2.3 ETC) to exercise and verify an object interaction in CIT. Several test operations and their associated test contracts may be mapped to one test group if they work closely together for the same testing objective; for example, they

```
... ... ... ...
..<TestSet Name=TUC1_tests_carEnterPAL>
....<Desc>Test Set 2: tests examine car entering PAL</Desc>
... ... ... ...
....<TestGroup Name=occupy_tests>
......<Desc>grouped tests examine setting in-PhotoCell device to the state of IN_PC_OCCUPIED
      </Desc>
......<TestOperation Name=goTo_tests>
........<Desc>2.2 TO: examine the test car crossing PAL entry point</Desc>
........<TestMethod Name=goTo Target=testCar>
..........<Desc>2.2 TO: the test car crosses PAL entry point controlled by in-PhotoCell device
          </Desc>
..........<Arg Name=gopace Source=gopace-cross-inPC DataType=int />
........</TestMethod>
......</TestOperation>
......<TestOperation Name=occupy_tests>
........<Desc>2.3 TO: examine setting in-PhotoCell device to the state of IN_PC_OCCUPIED
        </Desc>
........<TestMethod Name=occupy Target=inPhotoCell>
..........<Desc>2.3 TO: set in-PhotoCell device in the state of IN_PC_OCCUPIED</Desc>
........</TestMethod>
........<TestMethod Name=checkState Target=inPhotoCell>
..........<Desc>2.3 ETC: check inPhotoCell device in the resulted correct state of
                IN_PC_OCCUPIED </Desc>
..........<Arg Name=aObservable Source=inPhotoCell DataType=java.util.Observable />
..........<Arg Name=aState Source=IN_PC_OCCUPIED DataType=java.lang.Object />
..........<Result DataType=boolean Save=y>
............<Desc>2.3 ETC result: checkState must return true</Desc>
............<Exp>true</Exp>
..........</Result>
........</TestMethod>
......</TestOperation>
....</TestGroup>
... ... ... ...
..</TestSet>
... ... ... ...
```

Fig. 40.10 CTS test sequences/groups, operations, and contracts mapped for CPS TUC1 test scenario

jointly verify the same complex component interaction. The details of specific tests included in a test group are provided with composite test operations and test elements as described below.

40.6.2.2 Mapping Operations

Step TM4 carries out the mapping and transformation of functional operations to test operations to exercise and verify whether a particular operation correctly fulfils its target function. Operation mapping may take place to derive test operations at different mapping levels as shown in Fig. 40.11. In particular, Step TM4 results in

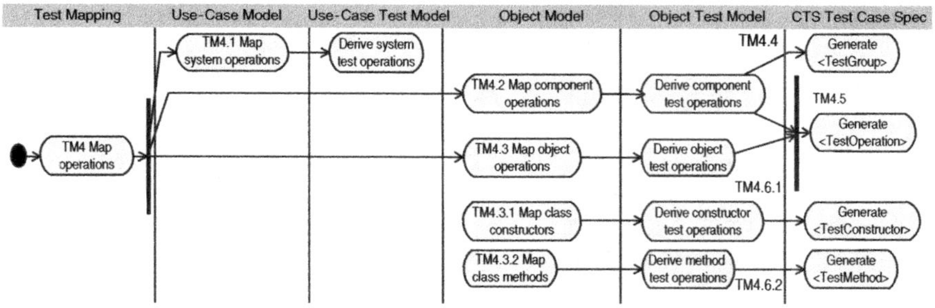

Fig. 40.11 TM4: mapping operations

system test operations mapped from system operations, component test operations mapped from component operations, object test operations mapped from object operations, or specific constructor test operations mapped from class constructors or method test operations mapped from class methods.

In practice, how Step TM4 works to produce which type of test operation depends on the complexity of the operations under test. After operations are mapped out, a test operation needs to be further mapped to one or more CTS atomic test operations and elements. Here we describe a $(1-n)$ general case of the defined CTM relationship. One operation in the set SCDS is mapped and corresponds to several atomic test operations (represented with XML element <TestMethod> or <TestContructor>) in the set SCTS. In this case, one operation is examined with several tests specified with several CTS test elements. These generated tests are then structured and organised into certain test sequences made up of related structural elements <TestOperation> or <TestGroup> as necessary. For example in the CPS TUC1, with Steps TM4.2 and TM4.4 carried out (Fig. 40.11), <TestGroup> in Fig. 40.10 is generated and composed of two atomic test operations to examine the composite operation that the in-PhotoCell device is occupied and set to the state of "IN_PC_OCCUPIED".

40.6.2.3 Mapping Contracts

Step TM6 maps and transforms contract artefacts to test contracts and then to test operations. With the TbC technique (see Sect. 40.3.4), test contracts are identified and constructed as necessary test constraints to examine and verify the related software artefacts for component correctness. As test contracts are typically represented and implemented with special test operations, Step TM6 results in system test contracts, component test contracts, object test contracts, or operation test contracts (Fig. 40.12). And also after contracts are mapped out, a test contract, depending on its complexity, needs to be further mapped to one or more test operations. We here describe a (1–1) simple case of the defined CTM relationship. A test contract is mapped and corresponds to an atomic test operation (represented

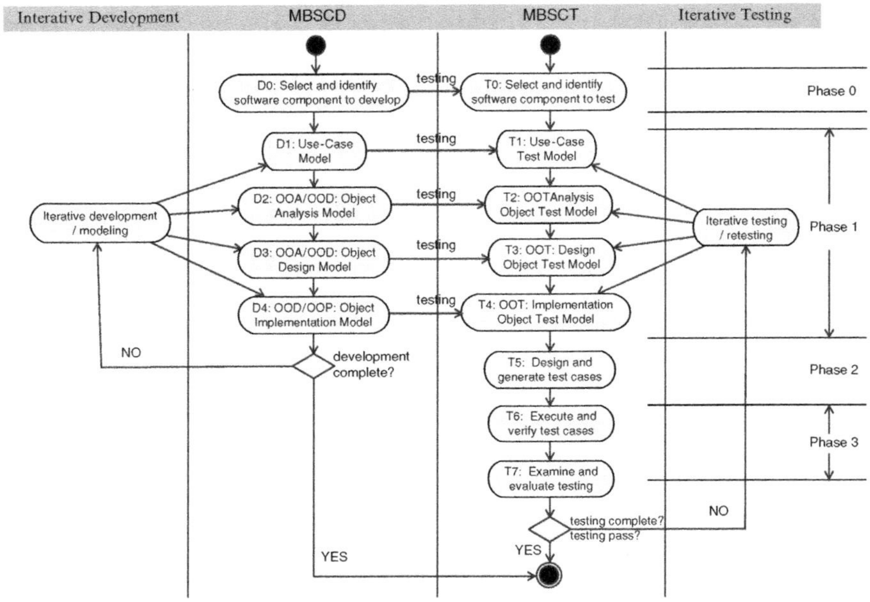

Fig. 40.12 TM6: mapping contracts

with XML element <TestMethod> or <TestContructor>) in the set SCTS. This case often occurs when the test contract is represented with a simple test operation. For example, after we conduct Steps TM6.4 and TM6.7 (Fig. 40.12), atomic test operation <TestMethod Name = "checkState" ...> (Fig. 40.10) represents test contract 2.3 ETC to check the in-PhotoCell sensor device in the expected state of "IN_PC_OCCUPIED". The test contract also requires checking the associated <Result> of the <TestMethod> to detect where the related state is correct as expected (see Fig. 40.10).

40.6.3 Generating CTS Test Case Specifications

Based on the constructed test models, we are able to apply the CTM technique to test design and generation, and derive CTS test case specifications for testing of the CPS system. Taking the CPS TUC1 test scenario as an example, the test model (Fig. 40.7) shows that test operations 2.2 TO and 2.3 TO exercise and verify setting the in-PhotoCell sensor device to the state of "IN_PC_OCCUPIED". We can map out and construct these tests into a test group that is included in the second test set (represented with XML element <TestSet>) of the CTS test case specification shown in Fig. 40.10. The test group contains two test operations and associated test contracts. Test operation 2.3 TO contains an atomic test operation occupy() and its associated test contract 2.3 ETC, which is a special test operation designed to check that the inPhotoCell device should be in the state of "IN_PC_OCCUPIED" if test operation 2.3 occupy() executes correctly.

40.6.4 Setting and Applying CTM Criteria

The MBSCT methodology develops certain CTM criteria to ensure CTM correctness and quality and to enhance the CTM technique for effective test derivation. The methodology identifies and sets two main types of CTM criteria: CTM correctness criteria and CTM optimising criteria.

40.6.4.1 CTM Correction Criteria

This type of CTM criteria focuses on dynamic testing rules, and aims to ensure that CTCs are correctly derived with the CTM technique. To ensure the sequence mapping correctness, we introduce and define a CTM correctness criterion:

Sequence consistency matching criterion. The sequence of test messages/operations that contain test elements for constructing CTCs should consistently match (in the same sequential logical order) the sequence of corresponding interacting messages/operations that derive the test messages/operations.

Based on this CTM criterion, although some individual (testable) software artefacts are mapped and transformed to become test artefacts, the sequencing logic or pattern should remain unchanged. Test contracts that are constructed and added into test sequences are also required to follow a certain consistent sequence pattern. Any mismatch may change the sequence logic and lead to incorrect test derivations, which may contradict test requirements and/or the functional logic of the CUT/CBS. For example, the test group shown in Fig. 40.10 is a test sequence composed of (1) test operation 2.2 TO, and (2) test operation 2.3 TO. This test sequence matches with the sequential working logic as illustrated with the sequence diagrams and the implementation. The test contract 2.3 ETC is constructed and added into this test sequence, whose consistent sequence pattern remains unchanged.

40.6.4.2 CTM Optimising Criteria

This type of CTM criteria focuses on static testing and structural rules, and aims to improve the test mapping and derivation to achieve better test effectiveness. We introduce and define a CTM optimising criterion as follows.

Sequence formatting/structuring criterion. Test messages/operations and underlying test elements can be structured and optimised to maintain a consistent hierarchical structure and format (e.g., recursive, nested indention rules at the same logical level) of interacting messages/operations that occur over time in sequence diagrams and/or programs for CUT/CBS.

A consistent and uniform structure between test artefacts and model artefacts can produce a well-formed CTC structure and format. The consistent structure also indicates that related test operations work closely together for a specific common testing objective and so can be organised in the same structured test sequence at the same level. For example, a collection of consecutive test operations in Fig. 40.10 jointly works to achieve a common testing objective: testing that the in-PhotoCell sensor device is occupied and set to the state of "IN_PC_OCCUPIED". So these test operations and associated test contracts are organised into the same structured test group.

40.7 Conclusions and Future Work

This research has presented a new UML-based SCT methodology and developed a set of supporting SCT techniques. The iterative MBSCT process establishes an incremental and systematic SCT framework, which includes the two main phases for developing CTCs: first constructing SCT models and second deriving CTCs from the test models. The scenario-based CIT technique emphasises testing priority on test scenarios that test component functions with operational scenarios in their integration contexts. The TbC technique constructs special test contracts and applies them to crucial testing-concerned component artefacts for verifying component correctness, which consolidates test models for effective SCT. The CTM technique refines the process of CTC design and generation, which enables test mapping and transformation to produce the target CTCs. Finally, a case study has been carried out to demonstrate the process and the effectiveness of the MBSCT methodology in SCT practice.

Our ongoing and future work with the MBSCT methodology is dedicated to the following main areas.

(1) *Methodology evaluation.* We are currently undertaking further evaluations with empirical and comparative studies about the effectiveness and applicability of our methodology with other SCT methods. More in-depth experimentation with additional complex CBS and industrial case studies is also under investigation.
(2) *Test criteria.* Furthering the test mapping criteria developed by this research, we are working on additional test criteria for improving testing coverage and the adequacy of test models and test case elements for effective test derivation.
(3) *Test automation.* Furthering the development of the CTS verification tool, a testing toolset to support the MBSCT methodology is under investigation, mainly including: a test contract generator for designing and generating appropriate test contracts applied to crucial component artefacts under test, a test case mapper for mapping and transforming test artefacts into test models for producing particular test case elements, and a test specification generator for compiling and integrating certain related test case data to generate the target CTS test case specifications.

References

1. R. V. Binder. *Testing Object-Oriented Systems: Models, Patterns, and Tools*. Addison-Wesley, 2000.
2. J. Z. Gao, H.-S. J. Tsao, and Y. Wu. *Testing and Quality Assurance for Component-Based Software*. Artech House, Sept. 2003.
3. A. Cechich, M. Piattini and A. Vallecillo (Eds.), *Component-Based Software Quality: Methods and Techniques*. LNCS2693, Springer, June 2003.
4. J. Morris, G. Lee, K. Parker, G. Bundell and C. P. Lam. Software component certification. *IEEE Computer*, vol. 34, no. 9, pp. 30–36, Sept. 2001.
5. J. Morris, C. P. Lam, G. Bundell, G. Lee and K. Parker. Setting a framework for trusted component trading. In A. Cechich, M. Piattini and A. Vallecillo (Eds.), *Component-Based Software Quality: Methods and Techniques*, LNCS 2693, pp. 101–131, Springer, June 2003.
6. A. Pretschner and J. Philipps. Methodological issues in model-based testing. In M. Broy, B. Jonsson, J. P. Katoen, M. Leucker, A. Pretschner (Eds.), *Model-Based Testing of Reactive Systems*, LNCS 3472, pp. 281–291, Springer, June 2005.
7. I. Jacobson, G. Booch and J. Rumbaugh. *The Unified Software Development Process*. Addison-Wesley, 1999.
8. W. Zheng and G. Bundell. A UML-based methodology for software component testing. *Proceedings of The IAENG International Conference on Software Engineering (ICSE 2007)*, Hong Kong, pp. 1177–1182, March 2007.
9. J. Offutt and A. Abdurazik. Generating tests from UML specifications. *Proceedings of 2nd International Conference on the Unified Modeling Language: Beyond the Standard (UML'99)*, Fort Collins, CO, Oct. 1999. LNCS 1723, pp. 416–429, Springer, 1999.
10. A. Abdurazik and J. Offutt. Using UML collaboration diagrams for static checking and test generation. *Proceedings of 3rd International Conference on the Unified Modeling Language: Advancing the Standard (UML'00)*, York, UK, Oct 2000. LNCS 1939, pp. 383–395, Springer, 2000.
11. L. C. Briand and Y. Labiche. A UML-based approach to system testing. *Journal of Software and Systems Modeling*, vol. 1, no. 1, pp. 10–42, Sept. 2002.
12. J. Warmer and A. Kleppe. *The Object Constraint Language: Getting Your Models Ready for MDA*. 2nd Edition, Addison-Wesley, 2003.
13. Y. Wu, M.-H. Chen and J. Offutt. UML-based integration testing for component-based software. *Proceedings of 2nd International Conference on COTS-Based Software Systems (ICCBSS 2003)*, Ottawa, Canada, 10–12 Feb. 2003. LNCS 2580, pp. 251–260, Springer, 2003.
14. W.-T. Tsai, R. Paul, L. Yu and X. Wei. Rapid pattern-oriented scenario-based testing for embedded systems. In Hongji Yang (Eds.), *Software Evolution with UML and XML*, pp. 222–262, Idea Group, London, 2005.
15. B. Meyer. *Object-Oriented Software Construction*. 2nd Edition, Prentice-Hall, 1997.
16. D. Brugali and M. Torchiano. *Software Development: Case Studies in Java*. Addison-Wesley, 2005.
17. E. Gamma, R. Helm, R. Johnson and J. Vlissides. *Design Patterns: Elements of Reusable Object-Oriented Software*. Addison-Wesley, 1995.
18. W. Zheng and G. Bundell. Model-based software component testing: A methodology in practice. Technical Report CIIPS_ISERG_TR-2006-01, School of Electrical, Electronic and Computer Engineering, University of Western Australia, 2006.

Chapter 41
Extending the Service Domain of an Interactive Bounded Queue

Walter Dosch and Annette Stümpel

41.1 Introduction

Modern computer systems are composed of software components which store information and provide services through interfaces. A component-based distributed system evolves by an ongoing interaction [1] between the components and the environment.

The input/output behaviour of an interactive component [2] forms a function from input streams to output streams [3]. A communication history, for short, a stream, records the sequence of messages passing on a transmission channel.

In general, an interactive component provides the contracted service [4] only for a subset of all input histories, called the service domain. For example, an interactive stack cannot serve a pop command in a regular way whenever the internal stack is empty [5].

The designer of an interactive component has no control over the environment where the component is used in different applications. So for practical purposes, the service domain of an interactive component must be well documented. Moreover, the component should also react in a predictable way to unexpected input outside the service domain. The irregular behaviour of a (sub)component widely influences the behaviour of the overall system in response to critical inputs.

Against this general background, we study possible extensions of the service domain of an interactive bounded queue. Our approach partitions the set of all input histories into the class of regular and erroneous streams; the subset of regular streams forms the service domain. As the starting point of the development, we specify the input/output behaviour of an interactive bounded queue for regular input histories (Sect. 41.4). Then we extend the regular behaviour in a systematic way to input streams outside the service domain (Sect. 41.5).

We present different types of irregular behaviour for erroneous input streams outside the service domain distinguishing between buffer underflow and buffer overflow. We separate three major types of irregular queue behaviours for buffer

Oscar Castillo et al. (eds.), *Trends in Intelligent Systems and Computer Engineering*.

underflow. An *underflow-sensitive queue* breaks upon the first buffer underflow and provides no further output whatever future input arrives (Sect. 41.6). An *underflow-tolerant queue* ignores an offending de-queue command which would lead to a buffer underflow, and continues its service afterwards (Sect. 41.7). An *underflow-correcting queue* suspends erroneous de-queue commands until further data become available (Sect. 41.8).

We characterize four major types of irregular queue behaviours for buffer overflow. An *overflow-sensitive queue* (Sect. 41.9) and an *overflow-tolerant queue* (Sect. 41.10) behave analogously to the underflow case. An *overwriting queue* enters the datum of an offending enter-queue command to the full internal buffer overwriting the datum entered most recently (Sect. 41.11). A *shifting queue* drops the first element of a full internal buffer while entering the datum of an erroneous enter-queue command at the rear of the buffer (Sect. 41.12). The different variants of an interactive bounded queue can be equipped with possible error notifications signalling the offending input command in the output stream.

The introductory sections survey streams (Sect. 41.2) and state transition machines (Sect. 41.3).

The different variants of an interactive bounded queue are discussed in greater detail employing a black-box view on the behavioural level and a glass-box view on the implementation level. On the black-box level, we characterize the input/output behaviour by a recursively defined stream function. The behavioural view reveals many similarities between the variants, but also significant differences with respect to the reaction to unexpected input.

On the glass-box level, we describe the implementations of the variants by state transition machines. The state transition machine for the variants uses a common data state for buffering the elements, but possibly different error states for handling the irregular behaviour. The state transition tables differ in the transition rules for handling erroneous inputs.

The chapter contains various scientific contributions. We present unifying functional descriptions for different variants of interactive bounded queues in the setting of stream functions. We refine the functional behaviour of the variants in a systematic way to state-based implementations. The design separates the different aspects for regular and erroneous input streams in a modular way both on the specification and implementation level.

Our contribution goes beyond a case study for interactive queues. The specification techniques, the description methods, the underlying transformations, and the overall methodology contribute to a general "engineering theory of services."

41.2 Communication Histories

In this section, we briefly survey the basic notions about streams as communication histories and interactive components as stream functions [6].

41.2.1 Streams as Communication Histories

Streams model the succession of messages in a network of components with asynchronous communication. Given an alphabet $\mathcal{A}$, the set $\mathcal{A}^\star$ comprises all (finite) streams $A = \langle a_1, a_2, \ldots, a_k \rangle$ of length $|A| = k$ with elements $a_i \in \mathcal{A}$ ($i \in [1,k], k \geq 0$).

The set $\mathcal{A}^{\leq k}$ ($\mathcal{A}^k$) comprises all streams over $\mathcal{A}$ with at most (exactly) $k \geq 0$ elements.

On communication histories, we use the following basic operations.

The concatenation $\& : \mathcal{A}^\star \times \mathcal{A}^\star \to \mathcal{A}^\star$ of two streams $A = \langle a_1, \ldots, a_k \rangle$ and $B = \langle b_1, \ldots, b_l \rangle$ yields the stream $A \,\&\, B = \langle a_1, \ldots, a_k, b_1, \ldots, b_l \rangle$.

For a natural number $k \in \mathbb{N}$ and a stream $A \in \mathcal{A}^\star$, the expression A^k denotes the stream $\underbrace{A \,\& \cdots \&\, A}_{k}$ concatenating k copies of A.

The subtraction $A \ominus B = R$ of the initial segment B from its extension $A = B \,\&\, R$ yields the final segment R.

The filter operation $.||. : \mathcal{A}^\star \times \mathcal{P}(\mathcal{A}) \to \mathcal{A}^\star$ filters a stream with respect to a subset $S \subseteq \mathcal{A}$ of elements. We have $\langle\rangle || S = \langle\rangle$, and $(\langle a \rangle \,\&\, A) || S = \langle a \rangle \,\&\, (A || S)$ for $a \in S$ and $(\langle a \rangle \,\&\, A) || S = A || S$ for $a \notin S$, respectively.

The set $\mathcal{A}^\star$ of streams over $\mathcal{A}$ forms a partial order under the prefix relation. Here a stream A approximates a stream B, denoted by $A \sqsubseteq B$, iff $A \,\&\, R = B$ holds for some stream $R \in \mathcal{A}^\star$. The prefix relation models operational progress; the shorter stream forms an initial part of the communication history.

41.2.2 Components as Stream Functions

The input/output behaviour of a component with one input and one output channel is modelled by a stream function $f : \mathcal{A}^\star \to \mathcal{B}^\star$ mapping an input stream of type $\mathcal{A}$ to an output stream of type $\mathcal{B}$.

In the sequel, we concentrate on monotonic functions where future input cannot cancel past output. A function $f : \mathcal{A}^\star \to \mathcal{B}^\star$ is called *(prefix) monotonic*, if $A \sqsubseteq B$ implies $f(A) \sqsubseteq f(B)$ for all $A, B \in \mathcal{A}^\star$.

41.3 State Transition Machines

State transition machines with input and output [7] model state-based components on an abstract level. For more refined state transition systems compare [8] and for labelled transition systems [9].

41.3.1 Structure of the Machine

A state transition machine with input and output

$$M = (\mathit{State}, \mathit{Input}, \mathit{Output}, \mathit{next}, \mathit{out}, \mathit{init}),$$

Fig. 41.1 Representing a state transition machine M by a state transition table

$M = (State, Input, Output, next, out, init)$			
State	*Input*	*State'*	*Output**
$\vdots$	$\vdots$	$\vdots$	$\vdots$
q_i	a_j	$next(q_i, a_j)$	$out(q_i, a_j)$
$\vdots$	$\vdots$	$\vdots$	$\vdots$

for short, a *state transition machine*, consists of a nonempty set *State* of *states*, a nonempty set *Input* of *input messages*, a nonempty set *Output* of *output messages*, a one-step *state transition function* $next : State \times Input \longrightarrow State$, a one-step *output function* $out : State \times Input \longrightarrow Output^\star$, and an *initial state* $init \in State$.

State transition machines are often represented by *state transition tables*; compare Fig. 41.1. The rows represent transition rules relating the current state q_i and the current input a_j to the successor state $next(q_i, a_j)$ and the output $out(q_i, a_j)$ generated by the transition.

In practical applications, software engineers work with state transition diagrams [10], in particular the widely used UML state diagrams [11, 12].

41.3.2 Implementing Stream Functions by State Transition Machines

A state transition machine processes an input stream message by message.

The *multistep state transition function* $next^\star : State \longrightarrow [Input^\star \longrightarrow State]$ yields the state reached after processing a finite input stream.

The *multistep output function* $out^\star : State \longrightarrow [Input^\star \longrightarrow Output^\star]$ yields the output stream generated by a finite input history.

A state transition machine $M = (State,\ Input,\ Output,\ next,\ out,\ init)$ is said to correctly implement a stream function $f : Input^\star \longrightarrow Output^\star$ iff the equation $f = out^\star(init)$ holds.

A comprehensive treatment of the state refinement process can be found in [13], condensed in [14, 15].

41.4 Regular Behaviour

An *interactive bounded queue* is a communicating component with one input channel and one output channel. The component provides the service of a queue with capacity $n \geq 1$ storing elements in a first-in/first-out way. We parameterize the design with the capacity $n \in \mathbb{P}$ of the bounded queue where $\mathbb{P} = \{1, 2, \ldots\}$ denotes the set of nonzero natural numbers.

An interactive bounded queue receives a stream of enter-queue and de-queue commands. An enter-queue command stores a datum and produces no output. A de-queue command requests the datum stored earliest which has not been requested so far. When time progresses, the queue component consumes an input stream of commands and produces an output stream of data.

41.4.1 Interface

The (syntactic) interface of the component is determined by the types of messages received on the input channel and sent to the output channel. The type $Data \neq \emptyset$ of data to be stored need not be specified further.

The type of output messages simply is *Data*. Input messages are either de-queue commands (deq) or enter-queue commands (enq) together with the datum to be stored:

$$Input = \{deq\} \cup enq(Data) \tag{41.1}$$

The notation $enq(Data)$ denotes the set $\{enq(d) \mid d \in Data\}$.

41.4.2 Service Domain

The component's service domain comprises the set of all regular input histories whose processing leads to no underflow and no overflow. An underflow occurs when a de-queue command meets an empty internal buffer. An overflow occurs when an enter-queue command meets a full internal buffer.

An input stream represents a regular input history from the service domain

$$ServDom_n \subseteq Input^\star \tag{41.2}$$

of an interactive bounded queue with capacity $n \in \mathbb{P}$ iff

- For each prefix the number of de-queue commands does not exceed the number of enter-queue commands.
- The difference between the number of enter-queue commands and the number of de-queue commands does not exceed the capacity n:

$$X \in ServDom_n \text{ iff } \forall Y \sqsubseteq X : 0 \leq |(Y||enq(Data))| - |(Y||\{deq\})| \leq n \tag{41.3}$$

The service domain enjoys closure properties which are characteristic for safety conditions. Initial segments of regular input histories are regular as well:

$$X \,\&\, Y \in ServDom_n \implies X \in ServDom_n \tag{41.4}$$

A bounded queue with a larger capacity provides a larger service domain:

$$m < n \iff ServDom_m \subset ServDom_n \tag{41.5}$$

We give an inductive generation to ease function definitions by pattern matching. The service domain $ServDom_n$ of an interactive bounded queue with capacity $n \in \mathbb{P}$ is the (with respect to set inclusion) least set with the following properties ($Enq_k \in (enq(Data))^{\leq k}$):

(i) $Enq_n \in ServDom_n$.

(ii) If $Enq_{n-1} \,\&\, X \in ServDom_n$,
then $\langle enq(d)\rangle \,\&\, Enq_{n-1} \,\&\, \langle deq\rangle \,\&\, X \in ServDom_n$.

Any sequence of enter-queue commands forms a regular input history if its length does not exceed the capacity n of the internal buffer (i). A regular input history can be prolonged by inserting a pair of an enter-queue and a de-queue command around an initial segment of enter-queue commands if the internal buffer is not already filled up with the initial segment of enter-queue commands (ii). Every regular input history can be constructed by a finite number of steps in a unique way using the rules (i) and (ii).

41.4.3 Regular Input/Output Behaviour

We specify the *regular behaviour* of an interactive bounded queue with capacity $n \in \mathbb{P}$ as a stream function

$$rqueue_n : ServDom_n \rightarrow Data^{\star}$$

mapping regular input histories to output histories. The input/output behaviour describes a black-box view not revealing the internal state.

The input/output function is defined by structural induction on the service domain ($Enq_k \in (enq(Data))^{\leq k}$):

$$rqueue_n(Enq_n) = \langle\rangle \tag{41.6}$$

$$rqueue_n(\langle enq(d)\rangle \,\&\, Enq_{n-1} \,\&\, \langle deq\rangle \,\&\, X) = \langle d\rangle \,\&\, rqueue_n(Enq_{n-1} \,\&\, X) \tag{41.7}$$

A sequence of at most n enter-queue commands generates no output (41.6). A de-queue command after a nonempty sequence of at most n enter-queue commands outputs the datum entered first. Afterwards the interactive bounded queue continues its service with the simplified input stream (41.7). For a regular input stream, the length of the output stream corresponds to the number of enter-queue commands in the input.

In a loose approach to system modelling, the specification of an interactive component concentrates on regular input histories in the first step. With such an

Fig. 41.2 State transition table for the regular behaviour of an interactive bounded queue with capacity $n \in \mathbb{P}$ ($D_k \in Data^{\leq k}$)

$State_n$	$Input$	$State'_n$	$Output^\star$
D_{n-1}	$enq(d)$	$D_{n-1} \,\&\, \langle d \rangle$	$\langle\rangle$
$\langle d \rangle \,\&\, D_{n-1}$	deq	D_{n-1}	$\langle d \rangle$

underspecification the designer expresses the willingness to accept every behaviour on erroneous input histories. In subsequent refinement steps, the underspecification will be resolved by adding further design decisions.

41.4.4 Implementation

The regular behaviour is implemented by a state transition machine. The major design decision amounts to introducing an appropriate state space.

The state of a bounded queue with capacity $n \in \mathbb{P}$ buffers at most n data elements:

$$State_n = Data^{\leq n} \tag{41.8}$$

The transition functions can systematically be derived from the regular behaviour using history abstractions. For the foundations of this transformation we refer to [14]. The resulting state transition table for the regular behaviour is displayed in Fig. 41.2.

The two rules form a complete case analysis for all possible combinations of states and inputs which can occur when processing regular input histories. Initially the state transition machine starts with the empty internal buffer $\langle\rangle$.

41.5 Irregular Behaviour

The irregular behaviour of an interactive component captures its reaction to erroneous input histories. The different irregular behaviours surveyed in this section are evaluated in greater detail in subsequent sections.

41.5.1 Erroneous Input Histories

The set of erroneous input histories comprises all input streams outside the service domain ($n \in \mathbb{P}$):

$$Error_n = Input^\star \setminus ServDom_n \tag{41.9}$$

Every erroneous input history can uniquely be decomposed into a maximal regular prefix and a remaining input stream. Erroneous input histories enjoy closure properties dual to input histories from the service domain; compare Eq. (41.4). Prolongations of erroneous input histories are erroneous as well:

$$X \in Error_n \implies X \,\&\, Y \in Error_n \tag{41.10}$$

41.5.2 Extensions

The irregular behaviour must be “compatible” with the regular behaviour specified so far. The extended behaviour of an interactive bounded queue depends on

- A set *ErrorMes* of *error messages*
- A set *Underflow* of *strategies for buffer underflow*
- A set *Overflow* of *strategies for buffer overflow*

We enlarge the range of the stream function by the set *ErrorMes* of error messages:

$$Output = Data \cup ErrorMes \tag{41.11}$$

The overall behaviour of the interactive bounded queue with capacity $n \in \mathbb{P}$

$$queue_n : [Input \to ErrorMes^\star] \times Underflow \times Overflow \to [Input^\star \to Output^\star]$$

depends on the possibly empty sequence of error messages in reaction to an erroneous input command, a strategy for buffer underflow, and a strategy for buffer overflow. In the sequel, we formulate three important constraints.

The overall queue behaviour must be a monotonic stream function:

$$queue_n(E,U,O)(X) \sqsubseteq queue_n(E,U,O)(X \,\&\, Y) \tag{41.12}$$

Furthermore, the overall queue behaviour must agree with the regular behaviour on the service domain. For all regular input streams $R \in ServDom_n$ we have:

$$queue_n(E,U,O)(R) = rqueue_n(R) \tag{41.13}$$

Error messages should signal an error in the input stream as soon as possible. For the prolongation $R \,\&\, \langle e \rangle \,\&\, X \in Error_n$ of a regular input stream $R \in ServDom_n$ we expect

$$queue_n(E,U,O)(R) \,\&\, E(e) \sqsubseteq queue_n(E,U,O)(R \,\&\, \langle e \rangle \,\&\, X) \tag{41.14}$$

Altogether, an erroneous input stream can be processed as a regular input stream as long as neither a buffer underflow nor a buffer overflow shows up.

The component's behaviour for a buffer underflow and for a buffer overflow, respectively, can be defined by complementing the regular behaviour with equations such as ($ENQ_k \in (enq(Data))^k$),

$$queue_n(E,U,O)(\langle deq\rangle \& X) = \dots$$
$$queue_n(E,U,O)(ENQ_{n+1} \& X) = \dots$$

The component's implementation is extended to a buffer underflow and to a buffer overflow, respectively, by complementing Fig. 41.2 with the erroneous transitions ($F_n \in Data^n$),

$$next(\langle\rangle, deq) = \dots \qquad out(\langle\rangle, deq) = \dots$$
$$next(F_n, enq(d)) = \dots \qquad out(F_n, enq(d)) = \dots$$

Some error handling strategies require extending the set $State_n$ of regular states to a set

$$ExtState_n \supseteq State_n \tag{41.15}$$

which captures information about previous errors.

41.5.3 Classification

We classify possible irregular behaviours by their envisaged properties.

A fault-sensitive queue breaks upon the first unexpected command and provides no further output whatever future input arrives. The output stems from the longest regular prefix of the erroneous input stream; compare Figs. 41.3 and 41.4. A bounded queue can be fault-sensitive with respect to buffer underflow or buffer overflow.

A fault-tolerant queue ignores an offending command in the input stream and continues its service with the remaining commands. The output stream concatenates the streams generated by the regular segments of the input stream interleaved with error notifications; compare Figs. 41.3 and 41.4. A bounded queue can be fault-tolerant with respect to buffer underflow or buffer overflow.

An underflow-correcting queue suspends erroneous de-queue commands until further data are available. After all postponed de-queue commands are served, the underflow-correcting queue continues with its original service. The underflow-correcting queue rearranges the input stream by postponing erroneous de-queue commands; compare Fig. 41.3.

buffer underflow								
input stream	$enq(1)$	deq	deq	deq	$enq(2)$	$enq(3)$	$enq(4)$	deq
underflow sensitive	$\langle\rangle$	$\langle 1\rangle$	$\langle underflow\rangle$	$\langle\rangle$	$\langle\rangle$	$\langle\rangle$	$\langle\rangle$	$\langle\rangle$
underflow tolerant	$\langle\rangle$	$\langle 1\rangle$	$\langle underflow\rangle$	$\langle underflow\rangle$	$\langle\rangle$	$\langle\rangle$	$\langle\rangle$	$\langle 2\rangle$
underflow correcting	$\langle\rangle$	$\langle 1\rangle$	$\langle underflow\rangle$	$\langle underflow\rangle$	$\langle 2\rangle$	$\langle 3\rangle$	$\langle\rangle$	$\langle 4\rangle$

Fig. 41.3 Output of different variants of an interactive bounded queue with capacity $n \geq 3$ in response to an erroneous input stream causing a buffer underflow with the error notification $E(deq) = \langle underflow\rangle$

buffer overflow								
input stream	$enq(1)$	deq	$enq(2)$	$enq(3)$	$enq(4)$	$enq(5)$	deq	deq
overflow sensitive	$\langle\rangle$	$\langle 1\rangle$	$\langle\rangle$	$\langle\rangle$	$\langle \mathit{overflow}(4)\rangle$	$\langle\rangle$	$\langle\rangle$	$\langle\rangle$
overflow tolerant	$\langle\rangle$	$\langle 1\rangle$	$\langle\rangle$	$\langle\rangle$	$\langle \mathit{overflow}(4)\rangle$	$\langle \mathit{overflow}(5)\rangle$	$\langle 2\rangle$	$\langle 3\rangle$
overwriting	$\langle\rangle$	$\langle 1\rangle$	$\langle\rangle$	$\langle\rangle$	$\langle \mathit{overflow}(4)\rangle$	$\langle \mathit{overflow}(5)\rangle$	$\langle 2\rangle$	$\langle 5\rangle$
shifting	$\langle\rangle$	$\langle 1\rangle$	$\langle\rangle$	$\langle\rangle$	$\langle \mathit{overflow}(4)\rangle$	$\langle \mathit{overflow}(5)\rangle$	$\langle 4\rangle$	$\langle 5\rangle$

Fig. 41.4 Output of different variants of an interactive bounded queue with capacity $n = 2$ in response to an erroneous input stream causing a buffer overflow with the error notification $E(enq(d)) = \langle \mathit{overflow}(d)\rangle$

Postponing erroneous enter-queue commands makes no sense for a bounded queue because the data of postponed enter-queue commands would require additional unbounded storage. Therefore we explore different strategies for handling errors in case of buffer overflow.

An overwriting queue stores the datum of an offending enter-queue command at the rear of its internal buffer. Thereby, it overwrites the datum stored most recently; compare Fig. 41.4.

A shifting queue enters the datum of an offending enter-queue command at the rear while shifting the contents of the internal buffer one position ahead. On this occasion, the first buffer element is lost; compare Fig. 41.4.

This classification identifies three major classes of possible queue behaviours for buffer underflow, viz. the underflow-sensitive (*us*), the underflow-tolerant (*ut*), and the underflow-correcting (*uc*) behaviour:

$$\mathit{Underflow} = \{us, ut, uc\} \tag{41.16}$$

For buffer overflow, we identified the overflow-sensitive (*os*) behaviour, the overflow-tolerant (*ot*) behaviour, overwriting (*ow*) the last buffer element, and shifting (*sh*) the internal buffer:

$$\mathit{Overflow} = \{os, ot, ow, sh\} \tag{41.17}$$

Furthermore, our modelling allows for different error messages depending on the offending input command.

In the sequel, we specify the input/output behaviours of the parameterized queue component by adding suitable equations for erroneous input histories. Moreover, we design state-based implementations for each variant of the queue component. We discuss the different variants of the queue component both from a black-box and a glass-box point of view.

41.6 Underflow-Sensitive Queue

An underflow-sensitive interactive queue processes the maximal regular prefix of the input stream in a regular way and breaks upon receiving the first erroneous de-queue command. The entire input stream after the first offending de-queue command is skipped.

41.6.1 Input/Output Behaviour

The input/output behaviour of an underflow sensitive queue

$$queue_n(E, us, O) : Input^\star \to Output^\star$$

is specified by one additional equation:

$$queue_n(E, us, O)(\langle deq \rangle \& X) = E(deq) \tag{41.18}$$

For all input streams starting with a de-queue command, an underflow-sensitive queue produces the (possibly empty) error notification and no further output afterwards (41.18). In operational terms, the underflow-sensitive queue enters a failure state with the first erroneous de-queue command which cannot be left any more with subsequent inputs.

The output of the underflow-sensitive queue stems from the maximal regular prefix of the input stream:

$$R \in ServDom_n \wedge R \& \langle deq \rangle \in Error_n \implies$$
$$queue_n(E, us, O)(R \& \langle deq \rangle \& X) = rqueue_n(R) \& E(deq) \tag{41.19}$$

41.6.2 Implementation

The underflow-sensitive queue is implemented by a state transition machine based on the state transition machine for the regular behaviour; cf., Fig. 41.2. The major design decision amounts to extending the state space appropriately.

The state of an underflow-sensitive queue is either the state of the state transition machine for the regular behaviour or a control state recording a failure:

$$ExtState_n \supseteq \{fail\_us\} \tag{41.20}$$

The transition functions can systematically be derived from the functional behaviour using history abstractions. The resulting extension of the state transition table is displayed in Fig. 41.5.

The first rule describes the transition to the failure state which cannot be left any more by the second rule. The two rules for the regular behaviour from Fig. 41.2 and

Fig. 41.5 Extension of the state transition table for an underflow-sensitive queue with capacity $n \in \mathbb{P}$

$ExtState_n$	$Input$	$ExtState'_n$	$Output^\star$
$\langle\rangle$	deq	$fail\_us$	$E(deq)$
$fail\_us$	x	$fail\_us$	$\langle\rangle$

the two additional rules from Fig. 41.5 form a complete case analysis for regular input histories possibly followed by an offending de-queue command.

41.7 Underflow-Tolerant Queue

An underflow-tolerant interactive queue ignores an erroneous de-queue command in the input stream and continues processing the remaining input in a regular way.

41.7.1 Input/Output Behaviour

The input/output behaviour of an underflow-tolerant queue

$$queue_n(E, ut, O) : Input^\star \to Output^\star$$

is specified by one additional equation:

$$queue_n(E, ut, O)(\langle deq \rangle \& X) = E(deq) \& queue_n(E, ut, O)(X) \quad (41.21)$$

Opposite to Eq. (41.18), the underflow-tolerant queue continues its service after an erroneous de-queue command (41.21).

An underflow-tolerant queue prolongs the output stream which is generated by an underflow-sensitive queue in response to the same input stream:

$$queue_n(E, us, O)(X) \sqsubseteq queue_n(E, ut, O)(X) \quad (41.22)$$

41.7.2 Implementation

The underflow-tolerant queue can again be implemented by extending the state transition machine for the regular behaviour. The set $ExtState_n$ of states need not be extended further because the state need not record information about previous underflows.

The resulting extension of the state transition table is shown in Fig. 41.6.

An underflow-tolerant queue possesses a simpler implementation than an underflow-sensitive queue, because it requires no control information about a buffer underflow in the past history.

Fig. 41.6 Extension of the state transition table for an underflow-tolerant queue with capacity $n \in \mathbb{P}$

$ExtState_n$	$Input$	$ExtState'_n$	$Output^\star$
$\langle\rangle$	deq	$\langle\rangle$	$E(deq)$

41.8 Underflow-Correcting Queue

An underflow-correcting interactive queue serves regular de-queue commands in the input history as expected, and postpones erroneous de-queue commands until further data become available. When all pending de-queue commands are served, the underflow-correcting queue returns to regular operation.

41.8.1 Input/Output Behaviour

The input/output behaviour of an underflow-correcting queue

$$queue_n(E, uc, O) : Input^\star \rightarrow Output^\star$$

is specified by two additional equations. The equations discriminate whether the first offending de-queue command in the input stream is eventually followed by an enter-queue command ($m \geq 0$):

$$queue_n(E, uc, O)(\langle deq \rangle^{m+1}) = (E(deq))^{m+1} \tag{41.23}$$

$$queue_n(E, uc, O)(\langle deq \rangle^{m+1} \& \langle enq(d) \rangle \& X) =$$

$$(E(deq))^{m+1} \; \& \; \langle d \rangle \; \& \; (queue_n(E, uc, O)(\langle deq \rangle^m \& X) \ominus (E(deq))^m) \tag{41.24}$$

For a sequence of de-queue commands, an underflow-correcting queue generates a sequence of error notifications (41.23). Otherwise, the datum of the next enter-queue command is emitted after error notifications are generated for the initial segment of de-queue commands. Then the remaining input is processed without producing again the error messages for the initial segment of de-queue commands (41.24).

An underflow-correcting queue serves all regular and erroneous de-queue commands as far as data are available.

41.8.2 Implementation

We extend the state transition machine for the regular behaviour in order to implement the underflow-correcting behaviour.

The extended state space incorporates the regular part and an error part reflecting the two operational modes of the component. If the queue operates in regular mode, the state buffers the collection of data entered but not yet requested. If the queue operates in error mode, the state stores the number of postponed de-queue commands:

Fig. 41.7 Extension of the state transition table for an underflow-correcting queue with capacity $n \in \mathbb{P}$ ($p \in \mathbb{P}$)

$ExtState_n$	$Input$	$ExtState'_n$	$Output^\star$
$\langle\rangle$	deq	1	$E(deq)$
p	deq	$p+1$	$E(deq)$
1	$enq(d)$	$\langle\rangle$	$\langle d\rangle$
$p+1$	$enq(d)$	p	$\langle d\rangle$

$$ExtState_n \supseteq \mathbb{P} \tag{41.25}$$

The resulting extension of the state transition table for an underflow-correcting queue is shown in Figure 41.7.

The first transition rule switches from regular mode to error mode, and the third transition rule back from error mode to regular mode.

41.9 Overflow-Sensitive Queue

Analogous to the underflow-sensitive queue, an overflow-sensitive interactive queue processes the maximal regular prefix of the input stream in a regular way and breaks upon receiving the first erroneous enter-queue command.

41.9.1 Input/Output Behaviour

The input/output behaviour of an overflow-sensitive queue

$$queue_n(E,U,os) : Input^\star \to Output^\star$$

is specified by one additional equation ($ENQ_k \in (enq(Data))^k$):

$$queue_n(E,U,os)(ENQ_n \,\&\, \langle enq(d)\rangle \,\&\, X) = E(enq(d)) \tag{41.26}$$

41.9.2 Implementation

In analogy to the underflow-sensitive case, the state space is extended by a trap state recording a previous buffer overflow:

$$ExtState_n \supseteq \{fail\_os\} \tag{41.27}$$

The resulting extension of the state transition table is shown in Fig. 41.8.

Fig. 41.8 Extension of the state transition table for an overflow-sensitive queue with capacity $n \in \mathbb{P}$ ($F_k \in Data^k$)

$ExtState_n$	$Input$	$ExtState'_n$	$Output^\star$
F_n	$enq(d)$	$fail\_os$	$E(enq(d))$
$fail\_os$	x	$fail\_os$	$\langle\rangle$

Fig. 41.9 Extension of the state transition table for an overflow-tolerant queue with capacity $n \in \mathbb{P}$ ($F_k \in Data^k$)

$ExtState_n$	$Input$	$ExtState'_n$	$Output^\star$
F_n	$enq(d)$	F_n	$E(enq(d))$

41.10 Overflow-Tolerant Queue

Analogous to the underflow-tolerant queue, an overflow-tolerant interactive queue ignores an erroneous enter-queue command in the input stream.

41.10.1 Input/Output Behaviour

The input/output behaviour of an overflow-tolerant queue

$$queue_n(E,U,ot) : Input^\star \rightarrow Output^\star$$

is specified by one additional equation ($ENQ_k \in (enq(Data))^k$):

$$\begin{aligned} &queue_n(E,U,ot)(ENQ_n \,\&\, \langle enq(d)\rangle \,\&\, X) \\ &\qquad = E(enq(d)) \,\&\, queue_n(E,U,ot)(ENQ_n \,\&\, X) \end{aligned} \tag{41.28}$$

41.10.2 Implementation

In analogy to the underflow-tolerant case, the implementation of the overflow tolerant queue does not enlarge the state space; compare Figure 41.9.

41.11 Overwriting Queue

An overwriting interactive queue stores the datum of an enter-queue command at the rear of the internal buffer. In case of a full internal buffer, it overwrites the datum entered most recently.

Fig. 41.10 Extension of the state transition table for an overwriting queue with capacity $n \in \mathbb{P}$ ($F_k \in Data^k$)

$ExtState_n$	$Input$	$ExtState'_n$	$Output^\star$
$F_{n-1}\,\&\,\langle c\rangle$	$enq(d)$	$F_{n-1}\,\&\,\langle d\rangle$	$E(enq(d))$

41.11.1 Input/Output Behaviour

The input/output behaviour of an overwriting interactive queue

$$queue_n(E,U,ow) : Input^\star \to Output^\star$$

is specified by one additional equation ($ENQ_k \in (enq(Data))^k$):

$$\begin{aligned} &queue_n(E,U,ow)(ENQ_{n-1}\,\&\,\langle enq(c)\rangle\,\&\,\langle enq(d)\rangle\,\&X) \\ &\quad = E(enq(d))\,\&\,queue_n(E,U,ow)(ENQ_{n-1}\,\&\,\langle enq(d)\rangle\,\&X) \qquad (41.29)\end{aligned}$$

The datum of an offending enter-queue command $enq(d)$ eliminates the most recently entered datum which has not been requested yet (41.29).

41.11.2 Implementation

For the implementation of the overwriting queue no new states need to be introduced because it does not administrate a previous overflow when processing the future input. The extension of the transition functions for an offending enter-queue command is shown in Fig. 41.10.

Opposite to an overflow-tolerant queue, the overwriting queue modifies the internal buffer in reaction to an offending enter-queue command.

41.12 Shifting Queue

A shifting interactive queue behaves as an input-driven shift register. When processing an offending enter-queue command, the internal buffer is advanced, dropping the first element at the front and storing the entered datum at the rear.

41.12.1 Input/Output Behaviour

The input/output behaviour of a shifting queue

$$queue_n(E,U,sh) : Input^\star \to Output^\star$$

Fig. 41.11 Extension of the state transition table for a shifting queue with capacity $n \in \mathbb{P}$ ($F_k \in Data^k$)

$ExtState_n$	$Input$	$ExtState'_n$	$Output^\star$
$\langle c\rangle \,\&\, F_{n-1}$	$enq(d)$	$F_{n-1} \,\&\, \langle d\rangle$	$E(enq(d))$

is specified by one additional equation ($ENQ_k \in (enq(Data))^k$):

$$queue_n(E,U,sh)(\langle enq(c)\rangle \,\&\, ENQ_{n-1} \,\&\, \langle enq(d)\rangle \,\&\, X)$$
$$= E(enq(d)) \,\&\, queue_n(E,U,sh)(ENQ_{n-1} \,\&\, \langle enq(d)\rangle \,\&\, X) \qquad (41.30)$$

An offending enter-queue command $enq(d)$ drops the first entered datum which has not been requested yet (41.30).

41.12.2 Implementation

A shifting queue does not record a past overflow for the future behaviour. Hence the state space need not be extended. The extension of the transition functions for offending enter-queue commands is shown in Fig. 41.11.

Upon a buffer overflow, the overwriting queue replaces the datum entered most recently, whereas a shifting queue removes the oldest datum entered.

41.13 Summary

We specified an interactive bounded queue in a modular way separating the concerns for regular and erroneous input histories. An interactive bounded queue is determined by selecting a strategy for buffer underflow and a strategy for buffer overflow. This way we provided three times four different versions of queues, without regarding the error notification. In this section, we summarize the overall input/output behaviour and the corresponding implementation.

The overall input/output behaviour of an interactive bounded queue in Fig. 41.12 combines the equations describing the regular behaviour, the erroneous behaviour for buffer underflow, and the erroneous behaviour for buffer overflow.

Figure 41.13 summarizes the implementation combining the transitions of the regular behaviour from Fig. 41.2 and of the erroneous behaviours from Figs. 41.5 to 41.11. The failure states for buffer underflow and buffer overflow can be unified to a single failure state *fail* because the transition functions agree on these states.

$$queue_n(E,U,O)(Enq_n) = \langle\rangle \quad (41.6)$$

$$queue_n(E,U,O)(\langle enq(d)\rangle \,\&\, Enq_{n-1} \,\&\, \langle deq\rangle \,\&\, X) = \langle d\rangle \,\&\, queue_n(E,U,O)(Enq_{n-1} \,\&\, X) \quad (41.7)$$

$$queue_n(E,us,O)(\langle deq\rangle \,\&\, X) = E(deq) \quad (41.18)$$

$$queue_n(E,ut,O)(\langle deq\rangle \,\&\, X) = E(deq) \,\&\, queue_n(E,ut,O)(X) \quad (41.21)$$

$$queue_n(E,uc,O)(\langle deq\rangle^{m+1}) = (E(deq))^{m+1} \quad (41.23)$$

$$queue_n(E,uc,O)(\langle deq\rangle^{m+1} \,\&\, \langle enq(d)\rangle \,\&\, X) = (E(deq))^{m+1} \,\&\, \langle d\rangle \,\&\, (queue_n(E,uc,O)(\langle deq\rangle^{m} \,\&\, X) \ominus (E(deq))^{m}) \quad (41.24)$$

$$queue_n(E,U,os)(ENQ_n \,\&\, \langle enq(d)\rangle \,\&\, X) = E(enq(d)) \quad (41.26)$$

$$queue_n(E,U,ot)(ENQ_n \,\&\, \langle enq(d)\rangle \,\&\, X) = E(enq(d)) \,\&\, queue_n(E,U,ot)(ENQ_n \,\&\, X) \quad (41.28)$$

$$queue_n(E,U,ow)(ENQ_{n-1} \,\&\, \langle enq(c)\rangle \,\&\, \langle enq(d)\rangle \,\&\, X) = E(enq(d)) \,\&\, queue_n(E,U,ow)(ENQ_{n-1} \,\&\, \langle enq(d)\rangle \,\&\, X) \quad (41.29)$$

$$queue_n(E,U,sh)(\langle enq(c)\rangle \,\&\, ENQ_{n-1} \,\&\, \langle enq(d)\rangle \,\&\, X) = E(enq(d)) \,\&\, queue_n(E,U,sh)(ENQ_{n-1} \,\&\, \langle enq(d)\rangle \,\&\, X) \quad (41.30)$$

$$Enq_k \in (enq(Data))^{\leq k} \qquad ENQ_k \in (enq(Data))^{k} \qquad m \geq 0$$

Fig. 41.12 Input/output behaviour of an interactive bounded queue with capacity $n \in \mathbb{P}$

$ExtState_n$	$Input$	$Underflow$	$Overflow$	$ExtState'_n$	$Output^\star$	Fig.
D_{n-1}	$enq(d)$			$D_{n-1} \,\&\, \langle d\rangle$	$\langle\rangle$	41.2
$\langle d\rangle \,\&\, D_{n-1}$	deq			D_{n-1}	$\langle d\rangle$	41.2
$\langle\rangle$	deq	us		$fail$	$E(deq)$	41.5
$fail$	x	us		$fail$	$\langle\rangle$	41.5
$\langle\rangle$	deq	ut		$\langle\rangle$	$E(deq)$	41.6
$\langle\rangle$	deq	uc		1	$E(deq)$	41.7
p	deq	uc		$p+1$	$E(deq)$	41.7
1	$enq(d)$	uc		$\langle\rangle$	$\langle d\rangle$	41.7
$p+1$	$enq(d)$	uc		p	$\langle d\rangle$	41.7
F_n	$enq(d)$		os	$fail$	$E(enq(d))$	41.8
$fail$	x		os	$fail$	$\langle\rangle$	41.8
F_n	$enq(d)$		ot	F_n	$E(enq(d))$	41.9
$F_{n-1} \,\&\, \langle c\rangle$	$enq(d)$		ow	$F_{n-1} \,\&\, \langle d\rangle$	$E(enq(d))$	41.10
$\langle c\rangle \,\&\, F_{n-1}$	$enq(d)$		sh	$F_{n-1} \,\&\, \langle d\rangle$	$E(enq(d))$	41.11

Fig. 41.13 State transition table for an interactive bounded queue with capacity $n \in \mathbb{P}$ ($D_k \in Data^{\leq k}$, $F_k \in Data^k$, $p \in \mathbb{P}$)

41.14 Conclusion

An interactive bounded queue allows various design decisions on how to handle an unexpected underflow or overflow of the internal buffer. The irregular behaviour influences the overall behaviour of a complex system [16] where the bounded queue is embedded as a subcomponent. Software engineering should provide sound methods and transparent guidelines how to document, specify, and implement interactive components with a restricted service domain.

The different extensions of the interactive bound queue structure erroneous input streams exist in different ways. A fault-sensitive queue splits an erroneous input stream into the maximal regular prefix and the remainder input stream which does not contribute an output. A fault-tolerant, an overwriting, and a shifting queue group an erroneous input stream into maximal regular segments separated by sequences of erroneous commands. An underflow-correcting queue splits an erroneous input history into an alternating sequence of regular segments processed in regular mode, and erroneous segments processed in error mode.

The reader should compare the high-level descriptions of the different variants as recursively defined stream functions with their implementations by state transition machines. The high-level specifications refer to entire communication histories, whereas the state-based implementations describe the component's reaction to single inputs.

The specifications of the variants document the design decisions for regular and erroneous input streams whereas the implementations tend to obscure them, because particular "situations" are encoded in control states. The local transitions have to be integrated into an overall behaviour [17] in order to understand a component-based system in a compositional way [18].

We advocate a clear separation of concerns as a specification and design methodology. In the first step, the designer should concentrate on capturing the required service of an interactive component. In the second step, the designer should consider the irregular behaviour for unexpected input streams. The effects of the regular and erroneous behaviour should be traceable during the software development. These aspects help the software engineer to modify an implementation in a disciplined way when the component must be adjusted to changing requirements.

The presented variants form a coarse classification of how to extend the service domain which allows further refinements and various other combinations. For page restrictions, we confined the presentation to three basic types of regular and four basic types of irregular behaviours to demonstrate the specification and design methodology. A more comprehensive classification would explore further dimensions such as the number of faults or the type of faults.

Over the years, error-handling has been investigated in different areas of computer science, for example, on the chip level, for operating systems, for communication systems, and for compilers. With the changing paradigm from mainframes to distributed systems, software science needs a well-established "engineering theory of services" which, among others, copes with the behaviour of interactive components in response to input inside and outside their service domain.

References

1. Peter Wegner. Why interaction is more powerful than algorithms. *Communications of the ACM*, 40(5):80–91, May 1997.
2. Manfred Broy and Ketil Stølen. *Specification and Development of Interactive Systems: Focus on Streams, Interfaces, and Refinement*. Monographs in Computer Science. Springer, New York, 2001.

3. Gilles Kahn. The semantics of a simple language for parallel programming. In J.L. Rosenfeld, editor, *Information Processing 74*, pages 471–475, 1974.
4. Manfred Broy, Ingolf H. Krüger, and Michael Meisinger. A formal model of services. *ACM Transactions on Software Engineering and Methodology*, 16(1), February 2007.
5. Walter Dosch and Gongzhu Hu. On irregular behaviours of interactive stacks. In S. Latifi, editor, *Proceedings of the Fourth International Conference on Information Technology: New Generations (ITNG 2007)*, pages 693–700. IEEE Computer Society Press, Los Alamitos, CA, 2007.
6. Robert Stephens. A survey of stream processing. *Acta Informatica*, 34(7):491–541, 1997.
7. Walter Dosch and Annette Stümpel. From stream transformers to state transition machines with input and output. In N. Ishii, T. Mizuno, and R. Lee, editors, *Proceedings of the 2nd International Conference on Software Engineering, Artificial Intelligence, Networking and Parallel/Distributed Computing (SNPD'01)*, pages 231–238. International Association for Computer and Information Science (ACIS), 2001.
8. Manfred Broy. From states to histories: Relating state and history views onto systems. In T. Hoare, M. Broy, and R. Steinbrüggen, editors, *Engineering Theories of Software Construction*, volume 180 of *Series III: Computer and System Sciences*, pages 149–186. IOS Press, Amsterdam, 2001.
9. Glynn Winskel and Mogens Nielsen. Models for concurrency. In S. Abramsky, D.M. Gabbay, and T.S.E. Maibaum, editors, *Semantic Modelling*, volume 4 of *Handbook of Logic in Computer Science*, pages 1–148. Oxford University Press, UK, 1995.
10. David Harel. Statecharts: A visual formalism for complex systems. *Science of Computer Programming*, 8:231–274, 1987.
11. James Rumbaugh, Ivar Jacobson, and Grady Booch. *The Unified Modeling Language Reference Manual*. Addison-Wesley Object Technology Series. Addison-Wesley, Reading, MA. 1998.
12. Object Management Group (OMG). *Unified Modeling Language: Superstructure*, 2.1.1 edition, 2007.
13. Annette Stümpel. *Stream Based Design of Distributed Systems through Refinement*. Logos Verlag, Berlin, 2003.
14. Walter Dosch and Annette Stümpel. Transforming stream processing functions into state transition machines. In W. Dosch, R.Y. Lee, and C. Wu, editors, *Software Engineering Research and Applications (SERA 2004)*, LNCS 3647, pages 1–18. Springer, New York, 2005.
15. Walter Dosch and Annette Stümpel. Deriving state–based implementations of interactive components with history abstractions. In I. Virbitskaite and A. Voronkov, editors, *Perspectives of Systems Informatics (PSI 2006)*, LNCS 4378, pages 180–194. Springer, New York, 2007.
16. Leo Motus, Merik Meriste, and Walter Dosch. Time-awareness and proactivity in models of interactive computation. ETAPS-Workshop on the Foundations of Interactive Computation (FInCo 2005). *Electronic Notes in Theoretical Computer Science*, 141(5):69–95, 2005.
17. Max Breitling and Jan Philipps. Step by step to histories. In T. Rus, editor, *Algebraic Methodology and Software Technology (AMAST'2000)*, LNCS 1816, pages 11–25. Springer, New York, 2000.
18. Willem-Paul de Roever, Hans Langmaack, and Amir Pnueli, editors. *Compositionality: The Significant Difference*. LNCS 1536. Springer, New York, 1998.

Chapter 42
A Hybrid Evolutionary Approach to Cluster Detection

Junping Sun, William Sverdlik, and Samir Tout

42.1 Introduction

The modern world has witnessed a surge in technological advancements that span various industries. In some sectors, such as search engines, bioinformatics, and pattern recognition, software applications typically deal with having to interpret shear amounts of data in an attempt to discover patterns that may provide great value for business analysis, development, and planning. This emphasized the importance of fields of study such as clustering, a descendant discipline of data mining, which gained momentum in recent decades. Clustering addresses this very problem of analyzing large datasets and attempting to unravel data distributions and patterns by means of a mostly unsupervised data classification [9]. Example clustering applications include multimedia analysis and retrieval [10], pattern recognition [15], and bioinformatics [5].

Research continues in the field of clustering, involving numerous disciplines. Several clustering approaches were introduced in recent decades, and brought along new challenges, such as outlier handling, detection of arbitrary shaped clusters, processing speed, and dependence on user-supplied parameters. PYRAMID, or parallel hybrid clustering using genetic programming and multi-objective fitness with density, was introduced in Tout et al. [23]. In an effort to resolve most of the above challenges, it employed a combination of data parallelism, genetic programming (GP), special genetic operators, and multiobjective density-based fitness function in the context of clustering.

PYRAMID divided the data space into cells that became the target of clustering thus eliminating dependence on the order of data input. It also used data parallelism to achieve speedup by dividing the dataset onto multiple processors each of which executed a genetic program that used a flexible individual representation that can represent arbitrary shaped clusters. The genetic program also utilized a density-based fitness function that helped avoid outliers. The experiments in Tout et al. [23] have shown positive results. The datasets used therein were characterized by various

Oscar Castillo et al. (eds.), *Trends in Intelligent Systems and Computer Engineering.*

sizes and irregular cluster shapes. They were used to compare cluster and outlier detection between PYRAMID and existing renowned algorithms such as BIRCH [25], CURE [8], DBSCAN [7], and NOCEA [20].

This chapter starts by providing an overview of existing clustering approaches. Then, it defines key concepts that are utilized by the PYRAMID algorithm. It also presents the experiments that were conducted in Tout et al. [23] as well as other experiments using various datasets that were employed in Sheikholeslami et al. [21] featuring different challenges. Finally, it explores the independence of PYRAMID on user-supplied parameters and outlines future research directions.

42.2 Background and Related Work

Clustering is the process of partitioning a dataset into meaningful groupings, called clusters [4]. In other words, it helps give a good understanding of the natural partitioning of data. Clustering is sometimes called unsupervised learning because, unlike classification, there are no predefined classes that guide the categorization of data points [9]. Clustering has been studied heavily within the realms of machine learning [18] and statistical analysis [2].

Data mining (DM), the art of discovering hidden gems within typically large datasets, is one of the umbrellas under which clustering falls. Data mining adds more stipulations to clustering algorithms such as the ability to accommodate massive amounts of data, noise that mostly comes with larger data mining datasets, finding clusters of arbitrary shapes, and insensitivity to the order of the data [9]. Other applications of clustering include information retrieval for library systems [17], image segmentation [22], and object recognition to group 3D object views [6].

Berkhin [1] indicates that some existing clustering techniques have slower processing, which is normally the result of computing distances between all data points within the clustered dataset. One of the renowned distance measures is called the Euclidean distance. Given d dimensions, Eq. 42.1 shows the squared Euclidean distance between two multidimensional points, $\mathbf{p}$ and $\mathbf{q}$, where p_i and q_i are the respective coordinates of these two points in the ith dimension.

$$\sum_{i=1}^{d} (p_i - q_i)^2 \tag{42.1}$$

In order to handle this processing issue, several approximation techniques have evolved. One of these techniques uses sampling, which operates on a representative subset of the data [13]. Another technique uses bins to generalize the data space as in Wang et al. [24] and Sarafis et al. [20]. Although Jain et al. [11] provide a taxonomy of known clustering algorithms, their hierarchy is not as comprehensive as the one found in Kolatch [14]. This is demonstrated in Fig. 42.1, which lists the traditional clustering approaches. As shown in Fig. 42.1, Kolatch [14] divides clustering into three main categories: Partitional, Hierarchical, and Locality-based. Each of these categories is further divided into multiple subcategories such as k-means, top-down,

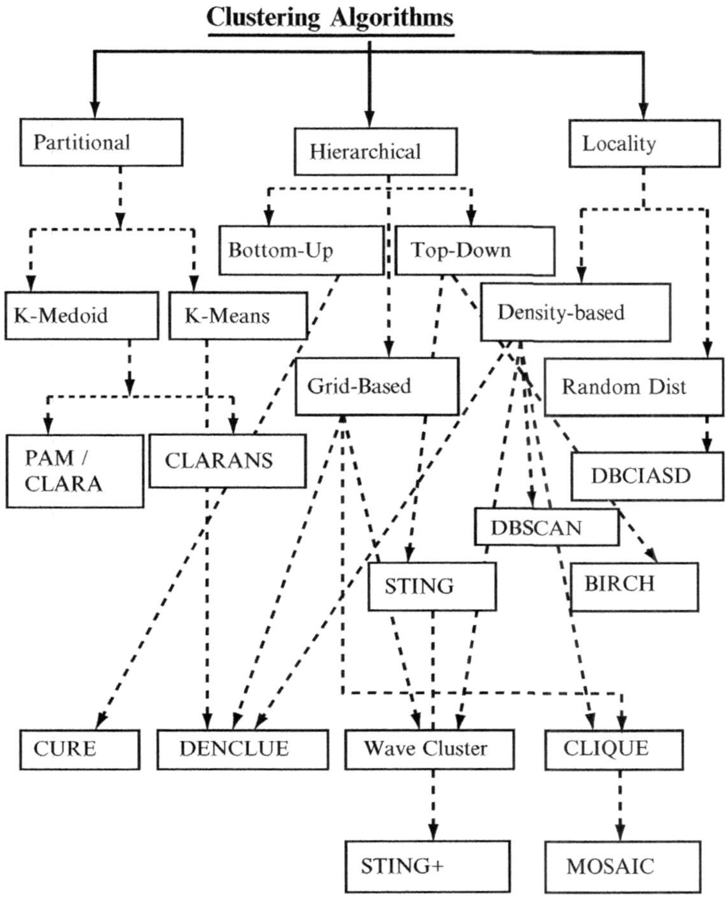

Fig. 42.1 The traditional clustering algorithms [14]

density-based, grid-based, and others. It is also noticeable that some approaches actually belong to more than one category such as wavelet transform-based clustering (Wave Cluster; [21]) and density-based clustering (DENCLUE; [10]). Some of these algorithms from the literature are briefly listed below [1, 9, 11].

BIRCH [25] focused on speed by means of data summarization but favored circular clusters [14]. CURE [8] concentrated on sampling and outlier handling. It depended on two main parameters, namely, a shrink factor and the number of representatives. DBSCAN [7] yielded good detection by favoring dense neighborhoods. As reported in Karypis et al. [12], CURE and DBSCAN did not always detect outliers and depended on user-supplied parameters. RBCGA [19] used a genetic algorithm that utilized rectangular shapes, called rules, each representing a cluster, thus only accommodating rectangular cluster shapes. NOCEA [20] improved over RBCGA by providing multiple rules per cluster, thus offering better detection than RBCGA but, as stated in Sarafis et al. [20], its crossover operator did not always detect sparse areas within those rules, thus resulting in coarse detections. The PYRAMID clustering algorithm was introduced in Tout et al. [23] and has proven to provide value in cluster and outlier detection on renowned datasets. The ensuing description of PYRAMID borrows directly from Tout et al. [23].

42.3 The PYRAMID Approach

42.3.1 Definitions

This section briefly introduces terms and concepts that are pertinent to the PYRAMID algorithm. The reader is encouraged to refer to Tout et al. [23] for further details. In all definitions, n symbolizes the number of points in a dataset and d is the number of dimensions.

Definition 42.1: *A Minimum Bounding Hyper-Rectangle (MBHR) is the smallest hyperrectangular area in the data space that contains all the d-dimensional data points in a given dataset [16].*

Definition 42.2: *Binning of dimension m within the MBHR where* $m = 1, \ldots, d$*, is the division of the m-axis into* t_m *nonoverlapping segments, called bins. All bins within a dimension m have the same bin width.*

Definition 42.3: *The intersections of the bin lines construct a d-dimensional grid that divides the MBHR into contiguous nonoverlapping d-dimensional cells, denoted quantization. Furthermore, cells have the following property:* $\forall$ *cell c, width of c with respect to dimension* $m = \mathrm{w}_m$.

Definition 42.4: *Rule r is the hyperrectangular subregion of the MBHR that contains one or more cells, denoted constituent cells of r. A rule r is said to overlap with another rule* r' *if they share at least one common constituent cell. This study does not allow overlapping rules within the same solution.*

Definition 42.5: *Individual I is the region in the MBHR that is a union of rules, called I's constituent rules. A list of the constituent rules of an individual I and their constituent cells, comma-separated, is called the individual profile, or profile(I). The size of an individual size(I) is the number of rules in I. The cardinality, volume, and density of cells, rules, and individuals are the same as outlined in Tout et al. [23].*

Definition 42.6: *Geometric division is an algorithm that divides the data space into quadrants. A quadrant encompasses a data subset formed by the data points that belong to its constituent cells. The details of this algorithm are outlined in [23] and exemplified in Fig. 42.2 in a three-dimensional data space.*

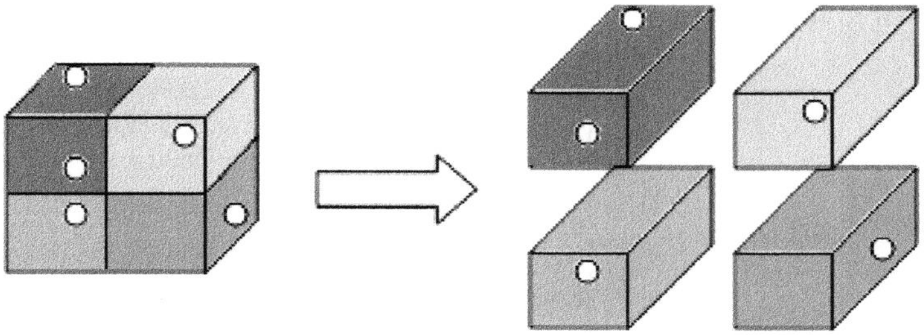

Fig. 42.2 Sample geometric division

42.3.2 The PYRAMID Algorithm

PYRAMID is a multistep hybrid algorithm that is based on several components that utilize the above concepts. For the sake of simplicity, the rest of this study focused on two-dimensional datasets as in Tout et al. [23] and leaves higher dimensions for future research. The PYRAMID algorithm is summarized in Fig. 42.3.

42.3.2.1 Data Transfer from Master to Slaves

The geometric division algorithm forms quadrants as groups of cells. The master processor sends each quadrant's data subset to a separate slave processor where a genetic program is executed.

42.3.2.2 Genetic Program

A genetic program typically represents a solution as a tree-based individual [15]. In this study, each individual is encoded as a combination of blocks (rules) to form a genetic programming tree with leaf nodes symbolizing these constituent rules. This representation offers more flexibility than genetic algorithm-based bit-strings [15]. Figure 42.4 shows an individual I_1 and its tree representation. As in standard genetic programming, the internal nodes represent the functions that apply to the leaf nodes [15]. This study uses the union function because individuals are formed as a combination of rules, represented by leaf nodes.

Genetic Operators

The main genetic operators used by PYRAMID are crossover, smart mutation, architecture altering (also called structural), and repair. The reader is encouraged

Fig. 42.3 Master and slave roles in PYRAMID

```
Master Processor
1. Conduct binning.
2. Perform geometric division.
3. Send each subset to a different slave.
4. Receive p subsets of discovered data
   points from p slaves. Determine cells that
   contain returned points.
5. Merge returned cells into global solution
   that labels every cell with a cluster.
Slave Processor
1. Receive a data subset P from master
   processor. Perform quantization on local
   data.
2. Run genetic program on local data points
   in P.
3. Send points in discovered cells to master
   processor.
```

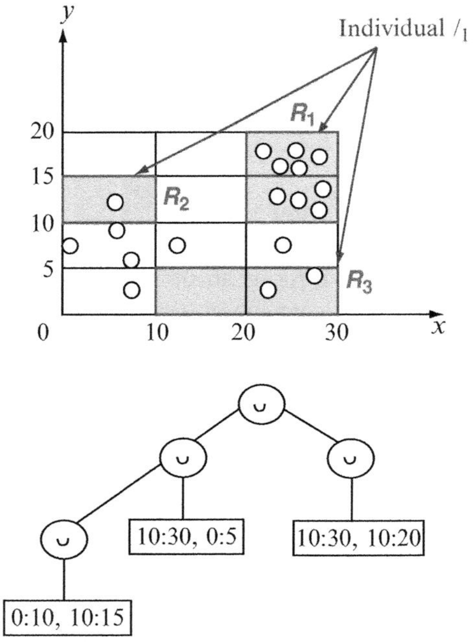

Fig. 42.4 Individual I_1 rules and tree representation

to refer to Tout et al. [23] for further details about these operators. In summary, crossover occurs at the rule level by swapping rules between individuals thus producing two new individuals. Smart mutation has two flavors: enlarge mutation, which moves towards denser neighboring cells and shrinking mutation diminishes a rule by one bin with respect to a certain dimension m. Mutation always produces one new individual. Architecture altering adds a new rule or deletes an existing one.

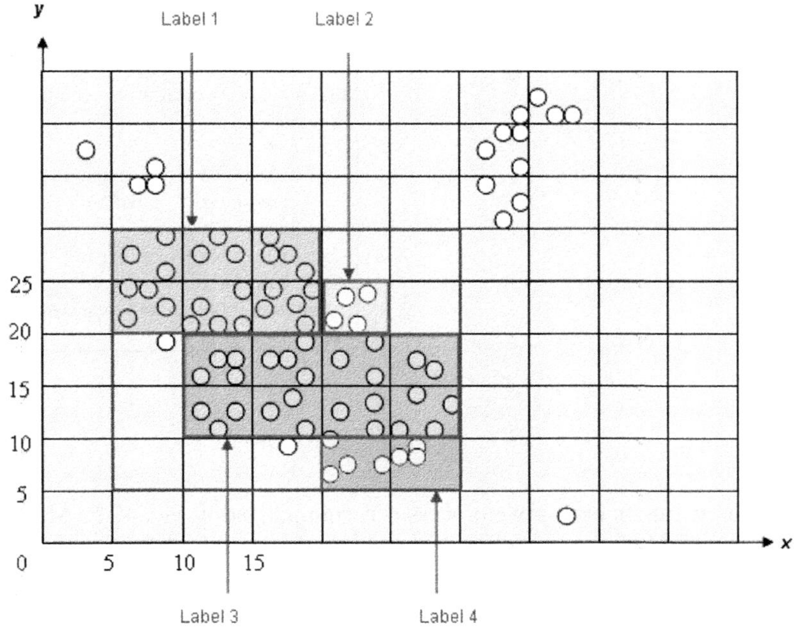

Fig. 42.5 PYRAMID repair operator

Finally, the repair operator reforms overlapping rules into new ones that align with the distribution of data points. This is shown in Fig. 42.5 where the frame depicts the area covered by the original rule.

Fitness Function

PYRAMID focuses on three main factors to achieve good solutions. It attempts to find a solution that achieves high coverage. It also tries to identify gatherings of dense areas in the MBHR by looking for solutions in the form of dense individuals composed of dense rules that contain dense cells. Finally, it attempts to avoid complex individuals by having a bias in favor of those having a smaller number of member rules, thus exercising parsimony pressure. Therefore, PYRAMID attempts to identify better solutions, or individuals, by incorporating, in its fitness function, the following three main objectives [23] shown in Eq. 42.2.

$$Fitness(I) = \frac{F_{coverage}(I) \times F_{density}(I)}{F_{size}(I)} \tag{42.2}$$

Selection Operator and Elitism

The selection operator is based on tournament selection with a tour size of three [3]. We adopt one-individual elitism, whereby in every iteration, the best performer is preserved for the next generation [1].

Fig. 42.6 Serial GP algorithm

```
t=0
Initialize population t
Evaluate population t
While (not termination condition)
 Begin
  t=t+1
  s=selection from population t-1
  c=crossover 2 individuals in t
  m=smart mutation
  a=architecture-altering
  e=elitism
  Evaluate (fitness) population t
 End
```

Main Algorithm

The GP that is run on each slave processor is summarized in Fig. 42.6. After each operator is applied, the fitness of resulting individuals is evaluated.

42.3.2.3 The Merge Phase

After the discovered points are sent back to the master, it traverses these discovered cells; assigning them cluster labels based on their neighborhoods. The merge algorithm was discussed in details in Tout et al. [23].

42.4 Experiments

42.4.1 Setup

This section outlines several experiments that were identified and performed in Tout et al. [23] to validate capability of PYRAMID in tackling some of the clustering challenges. This chapter expands on these experiments, presents their results, and provides a comprehensive discussion of their various aspects.

The experiments and their associated results were divided into two main categories: qualitative, which is more concerned with the goodness of detection, and quantitative, which concentrates on metrics and performance measurements. Additional details are provided about the experiments, including the datasets, the experimentation tool, and the experimental environment settings.

The following sections provide details about the datasets, elaborate on the data distribution and transmission between the master and slave processors followed by the experimental environment setup. Finally, it shows the experiments and provides a discussion of their results. It is worth noting that Tout et al. [23] developed an experimental tool called PYRAMID-CT (PYRAMID-clustering tool) that implemented the PYRAMID clustering algorithm. This tool is discussed in detail in Tout et al. [23].

42.4.2 Datasets

This section provides some information about the datasets used in the experiments. The reason for choosing these specific datasets is that they were employed in several other research efforts to evaluate existing state-of-the-art approaches such as CURE, DBSCAN, and NOCEA [12]. Therefore, these datasets allow comparisons to be performed between these approaches and PYRAMID. The three datasets (DS1, DS2, and DS3) were obtained from Guha et al. [8] and Karypis et al. [12].

The first dataset, DS1, has 8000 data points, with six clusters of various sizes, orientations, and shapes along with noise data points and special effects such as the streaks that run across clusters, which are considered as outliers. The second dataset, DS2, has 10,000 data points and contains nine clusters with irregular shapes. It also has noise data points in the form of vertical streaks. The third dataset, DS3, has 100,000 data points, and contains six clusters of different sizes, densities, and shapes. These datasets are tabulated in Table 42.1.

The datasets used in this study are two-dimensional. As Karypis et al. [12] stated; one reason for the use of these datasets is mostly that they have been employed extensively by other clustering research efforts. Another reason is that it is easier to evaluate the quality of clustering using two-dimensional datasets. Furthermore, most papers researched in this study used two-dimensional datasets and this study follows the same approach. Figure 42.7 shows these datasets as displayed by PYRAMID-CT.

Table 42.1 Datasets used in experiments [12, 19]

Dataset	No of data points
DS1	8,000
DS2	10,000
DS3	100,000
DS4	1,120

Fig. 42.7 Datasets DS1, DS2, and DS3 as displayed by the PYRAMID-CT tool

42.4.3 Data Transmission and Experimental Environment Setup

One integral part of the experiments, specifically for parallel runs, is the data distribution and transmission between the master and slave processors. Section 42.4.3.1 explains the way PYRAMID-CT implemented these aspects as well as the relationship between the master and slaves with respect to their cells and associated data points. Section 42.4.3.2 elaborates on the experimental environment configuration.

42.4.3.1 Data Distribution and Transmission

This section explains the method that PYRAMID-CT employs to distribute the data from the master processor to the slave processors and back. After PYRAMID finishes the geometric division, the data points that fall within the *i*th quadrant are placed into a file and sent to the *i*th slave processor using Windows sockets, as depicted by the black arrows in Fig. 42.8.

42.4.3.2 PYRAMID Environment Configuration

The genetic programming utilized in the PYRAMID experiments used 15 individuals per population and, as in NOCEA, the number of rules per individual was set to

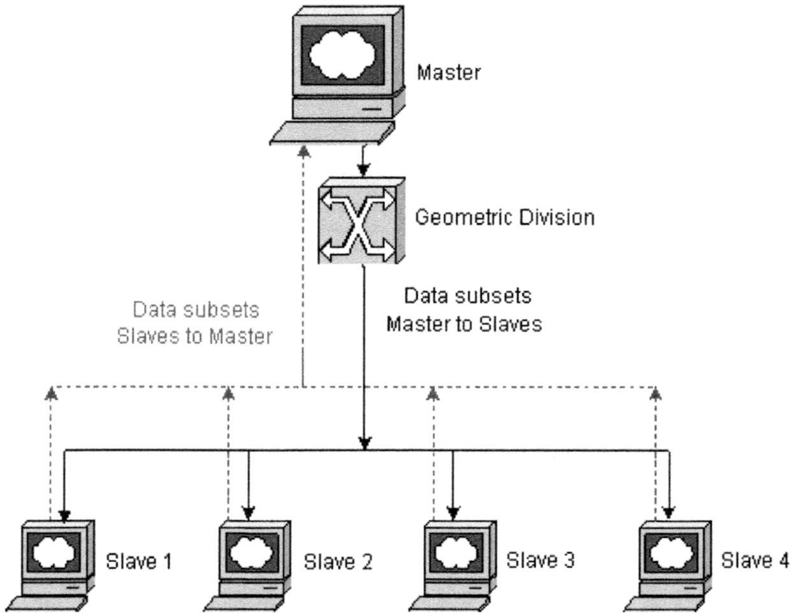

Fig. 42.8 Master–slave communication in PYRAMID

10. The crossover percentage was set to 20%, the mutation percentage to 35%, and the structural percentage to 45%.

Tout et al. [23] used 100 iterations as the termination condition, similar to existing approaches such as DPC and RBCGA [19]. This differs from NOCEA, where the termination condition was 200 iterations.

The hardware used includes five x86 single-processor machines with similar configurations, such as 1-Ghz Celeron with 512 MB of memory running on Windows 2000 SP4. The computers were networked using a 10 megabits per second Ethernet local area network.

42.4.4 Experimental Results and Discussions

This section presents the details of the PYRAMID experiments, which address the clustering challenges mentioned earlier. These include the identification of clusters with arbitrary shapes, the irrelevance of the order of input data, outlier handling, and performance evaluation using speedup measurements. The objectives are clarified and the results are explained in terms of comparisons with other approaches or within the PYRAMID approach. Each experiment falls under one of two categories: qualitative, which concentrates on the goodness of detection, or quantitative, which addresses performance measurements and other metrics.

The authors could not obtain any data for the results produced by Sarafis et al. [19] or Sarafis et al. (2003) for RBCGA and NOCEA, respectively. Hence, the only source of such data is through their published papers. Therefore, as it was conducted in Karypis et al. [12], this study has adopted an approach of comparison, which appeals to the reader to visually inspect the differences in detection that are clearly pointed out in the figures below. The figures that depict the detections of all approaches other than PYRAMID were obtained from papers discussed earlier, such as Karypis et al. [12] and Sarafis et al. [19] and (2003).

42.4.4.1 Qualitative Experiments

This category includes the experiments that address the identification of clusters with arbitrary shapes, the irrelevance of the order of input data, and outlier handling. In each of the figures that depict the PYRAMID detection, identified clusters are labeled with numbers, enclosed in parentheses, which match the numbers utilized by the other compared approach.

Experiment A: Identification of 2-D Clusters with Arbitrary Shapes

The objective of this experiment is to conduct a comparison in accuracy of detection among PYRAMID, NOCEA, and RBCGA. The hypothesis is that PYRAMID

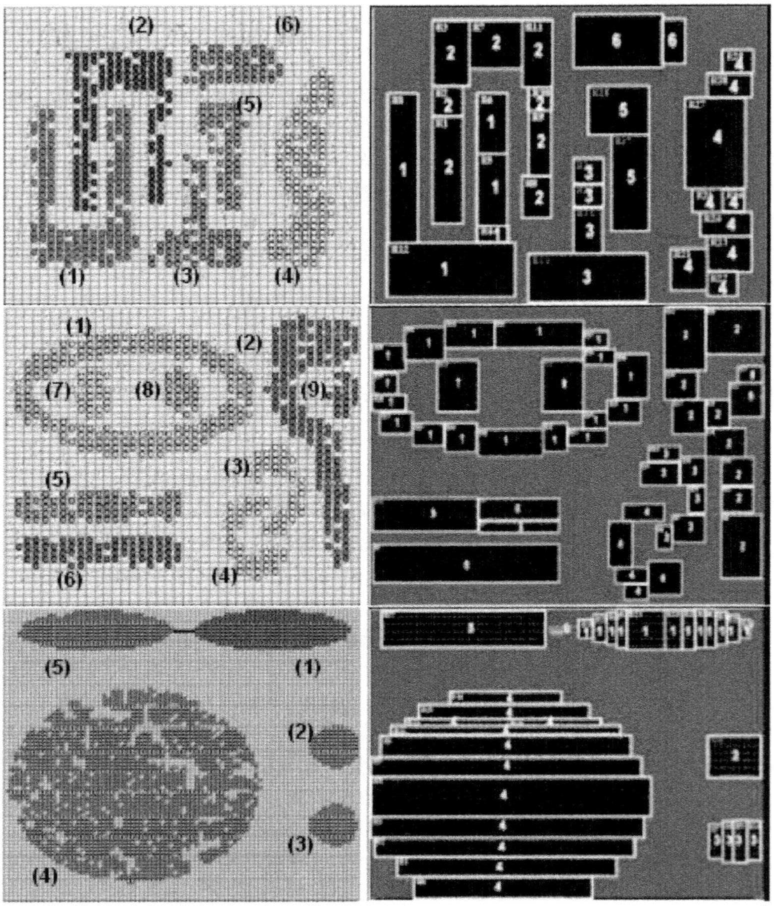

Fig. 42.9 Detection by PYRAMID and NOCEA [20]

produces smoother detections because its repair operator results in finer detections than those generated by NOCEA or RBCGA.

The comparisons between PYRAMID and NOCEA in regard to datasets DS1, DS2, and DS3 are depicted in Fig. 42.9. They show how PYRAMID detected smooth nonrectangular curves whereas the same locations were detected with a coarser representation by NOCEA.

Experiment B: Irrelevance of the Order of Input Data

The objective of this experiment is to test that a change in the ordering of the data points within a dataset has no impact on the quality of detection. Therefore, the hypothesis is that PYRAMID is not dependent on that order because it conducts its clustering detection at the cell level rather than at the data points. In Fig. 42.10, the

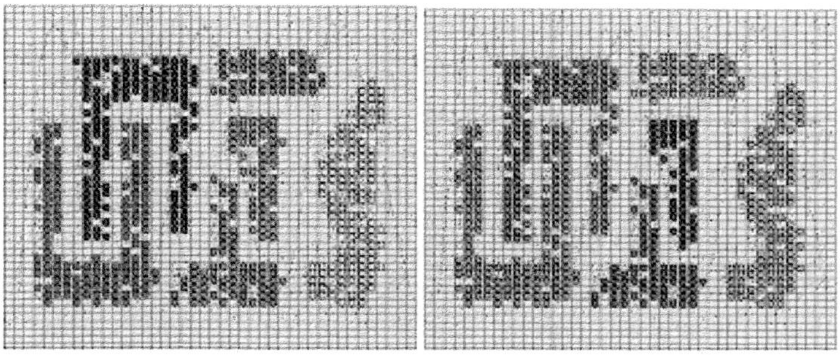

Fig. 42.10 PYRAMID detection of DS1 with original versus shuffled order

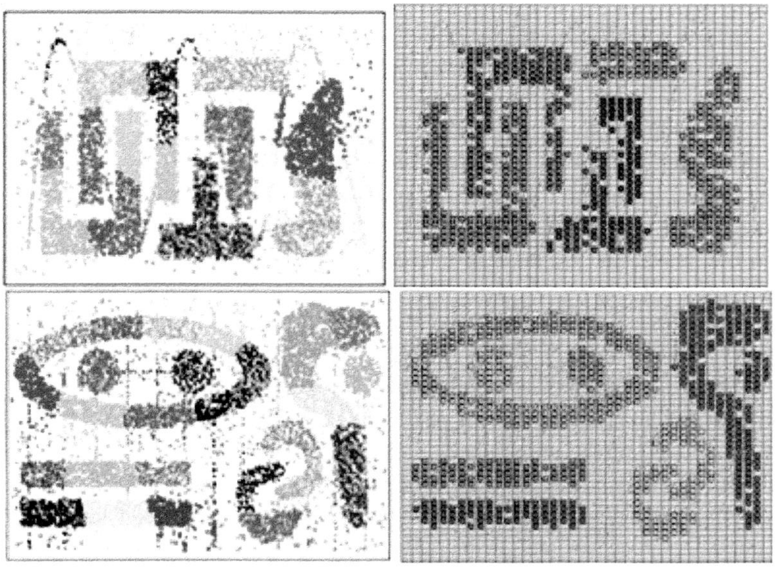

Fig. 42.11 Outliers by CURE [12] versus PYRAMID

left-hand side shows the result of running PYRAMID on the original dataset DS1, and the right-hand side depicts the results on a randomly shuffled version of DS1. It is noticeable how the detection is almost identical in both cases.

Experiment C: Built-In Handling for Outliers

The objective of this experiment is to test whether PYRAMID detects outliers. Therefore, the hypothesis is that PYRAMID's density-based fitness function and genetic operators would aid PYRAMID in discarding outliers. This experiment validates this hypothesis by running PYRAMID on DS1 and DS2, comparing them against CURE based on the figures published by Karypis et al. [12]. As demonstrated by the different shades in Fig. 42.11, in a scenario where CURE has a shrink factor of 0.3 and 10 representatives, it detected the clusters in DS1 and DS2 but it counted some outliers as part of the clusters as pointed out by Karypis et al. [12].

42.4.4.2 Quantitative Experiments

This category concentrates on quantifiable measurements by means of a set of experiments that were conducted to measure the speedup achieved when the proposed data parallelism was utilized. Furthermore, runtime metrics were captured by PYRAMID-CT and presented in the form of charts. These were compared to the ones produced by Sarafis et al. [20].

Experiment D: Speedup Using Parallelism

The objective of this experiment was to assess the performance improvement, or speedup, achieved when the above data parallelism approach is utilized. Therefore, the hypothesis is that data parallelism would result in performance improvements manifested by speedup values greater than one. Speedup is the ratio of the execution time on one processor (serial run) over the execution time on multiple processors (parallel run; [23]).

This experiment validates the above hypothesis by executing serial and parallel runs of PYRAMID on datasets DS1, DS2, and DS3. Subsequently, it calculates the speedup for each of these and presents their comparisons in a tabular form.

As part of this experiment, a serial run and two parallel runs of PYRAMID are executed for each of the above three datasets. One of the parallel runs is referred to as Parallel-4. It includes a master processor and four PYRAMID slave instances each running on a slave processor. The other parallel run is referred to as Parallel-16. It has a master processor and 16 instances of PYRAMID distributed on 16 slaves. It is worth noting that an averaging approach was adopted for the latter case due to insufficient processors to conduct the experiments.

Considerable improvements were achieved using parallelism with each of the above datasets. These are outlined in the following figures, which display the execution times, rounded up to seconds, of these datasets for a serial run, the parallel run with four slave processors (Parallel-4), and the parallel run with 16 slave processors (Parallel-16).

As shown in Figs. 42.12 and 42.13 and the associated table, Table 42.2, the speedup gain between the serial and Parallel-16 is sometimes more than twice that of the speedup gain between the serial and Parallel-4. The table shows a considerable improvement in speedup, because the lowest of these is 1.80, whereas the highest is 6.43. In addition, different instances of PYRAMID exhibit different running times, which may be due to the randomness inherent within evolutionary algorithms. This study adopted an averaging approach to obtain a typical value.

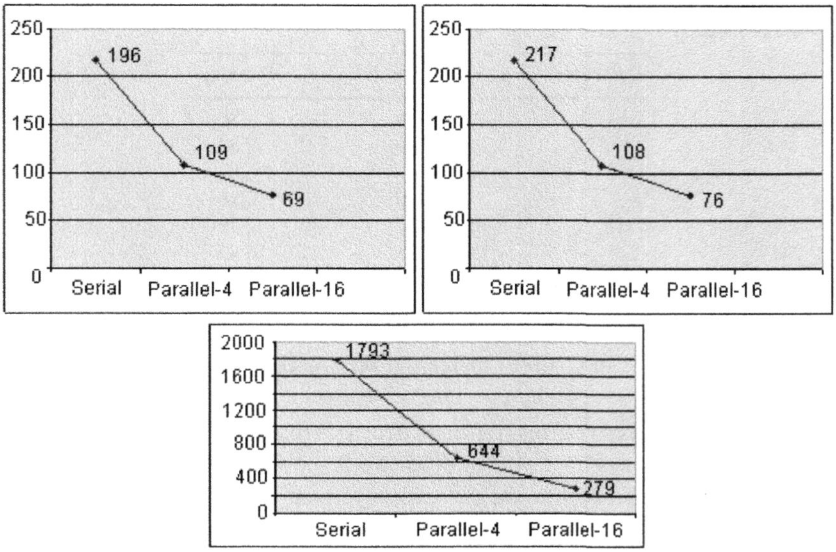

Fig. 42.12 Execution time (in seconds) achieved with datasets DS1, DS2, and DS3

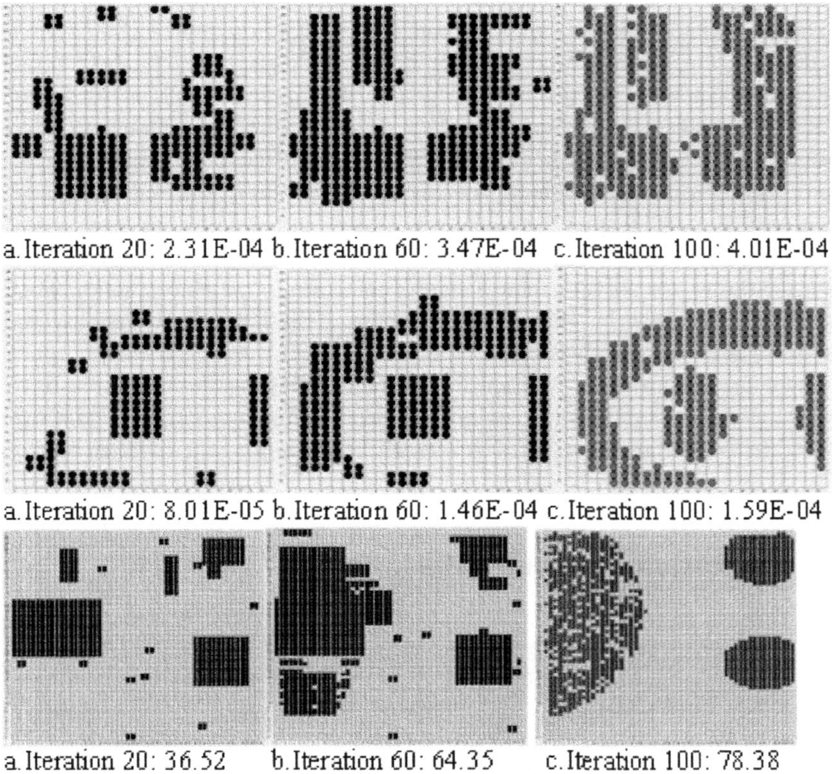

Fig. 42.13 PYRAMID detection and fitness values of DS1, DS2, and DS3 at various iterations

Table 42.2 Speedup for DS1, DS2, and DS3

Dataset	Serial against	Speedup
DS1	Parallel-4	1.80
	Parallel-16	2.84
DS2	Parallel-4	2.01
	Parallel-16	2.86
DS3	Parallel-4	2.78
	Parallel-16	6.43

Experiment E: Fitness and Cluster Detection Progression

Two of the fitness objectives, coverage and density, both in the numerator, are related to the data distribution. The coverage objective is the ratio of the discovered points over the total points. Therefore, the more discovered points, the higher the coverage is. The density objective is a product of the densities at the individual, rule, and cell levels. This objective is mostly used to circumvent outliers by reducing the fitness value if the discovered areas are sparse. As discussed earlier, the density objective has a considerably smaller value than the coverage objective. Therefore, an increase in coverage has a bigger impact on fitness than density.

During the experiments, it was observed that the majority of the cluster discovery, or coverage gain, was achieved in earlier iterations, that is, before the 60th iteration. Figure 42.13 shows how a quadrant assigned to one of the slave processors, running PYRAMID, detected a portion of the DS1 dataset at the 20th, 60th, and 100th iterations. Figure 42.13 indicates considerable cluster discovery, thus coverage gain, before the 60th iteration whereas the rest of the iterations up to the 100th show less coverage gain but smoother detection. This is supported by the other graphs of Fig. 42.13, which chart the fitness improvement of Fig. 42.12 throughout all iterations. It shows considerable fitness improvement up to the 60th iteration whereas later iterations exhibit less fitness improvement. A similar behavior is observed in Fig. 42.13 for DS2 as well as DS3.

The main conclusion of this section is that the fitness improvement appears to be steepest in earlier iterations where the algorithm still has considerable undiscovered cluster areas. Later iterations are spent mostly on smoothing the detected clusters.

42.5 Conclusion and Future Work

This chapter discussed a novel approach to clustering large datasets, called PYRAMID, which was introduced in Tout et al. [23]. PYRAMID leverages some of the concepts used in NOCEA [20] and improved over it by employing a hybrid combination of GP's global search and strong representational capabilities along with a powerful density-aware multiobjective fitness function. PYRAMID also employed data parallelism to achieve speedup. The experiments in Tout et al. [23] demonstrated that PYRAMID detects clusters of arbitrary shapes, is immune to outliers, and is independent of the order of input. In addition, its inherent data parallelism allows it to improve performance. This chapter added to Tout et al. [23] by exercising the ability of PYRAMID to detect a more challenging dataset, DS5, which was employed in previous well-known clustering research [21]. The results have demonstrated that, despite the challenges inherent within DS5, PYRAMID has exhibited relatively better detection than CURE [21]. Another experiment also attested to the independence of PYRAMID on user-supplied parameters.

One possible avenue for future research is to revisit the PYRAMID algorithm and explore the performance measurements through speedup with higher dimensions. Other avenues include: performing additional experiments to assess various aspects of cluster detection such as exploring the use of rules with variable shapes, not strictly rectangular, and using other datasets as well as other forms of parallelism.

References

1. Berkhin, P. (2002). Survey of clustering data mining techniques. *Accrue Software*. Retrieved February 28, 2005, from http://www.ee.ucr.edu/~barth/EE242/clustering_survey.pdf.
2. Berry, M.J. and Linoff, G. (1997). *Data Mining Techniques: For Marketing, Sales, and Customer Support*. New York: John Wiley and Sons.
3. Davis, L. (1991). *Handbook of Genetic Algorithms*. New York: Van Nostrand Reinhold.
4. Deb, K. (2001). *Multi-Objective Optimization Using Evolutionary Algorithms*. New York: John Wiley and Sons.
5. Dettling, M. and Bühlmann, P. (2002). Supervised clustering of genes. *Genome Biology, 3*(12), 39–50.
6. Dorai, C. and Jain, A.K. (1995). Shape spectra based view grouping for free-form objects. *Proceedings of the International Conference on Image Processing, Washington, DC*, 3, 340–343.
7. Ester, M., Kriegel, H.P., Sander, J., and Xu, X. (1996). A density-based algorithm for discovering clusters in large spatial databases with noise. *Proceedings of the Second International Conference on Knowledge Discovery and Data Mining, Portland, Oregon*, 226–231.
8. Guha, S., Rastogi, R., and Shim, K. (1998). CURE: An efficient clustering algorithm for large databases. *Proceedings of the 1998 ACM SIGMOD International Conference on Management of Data, Seattle, WA*, 73–84.
9. Han, J., and Kamber, M. (2001). *Data Mining, Concepts and Techniques*. San Francisco: Morgan Kaufmann.
10. Hinneburg, A., and Keim, D.A. (1998). An efficient approach to clustering in large multimedia databases with noise. *Proceedings of the Fourth International Conference on Knowledge Discovery in Databases, New York*, 58–65.
11. Jain, A.K., Murty, M., and Flynn, P. (1999). Data clustering: A review. *ACM Computing Surveys, 31*(3), 264–323.
12. Karypis, G., Han, S., and Kumar, V. (1999). Chameleon: A hierarchical clustering using dynamic modeling. *IEEE Computer: Data Analysis and Mining (Special Issue), 32*(8), 68–75.
13. Kaufman, L. and Rousseeuw, P.J. (1990). *Finding Groups in Data: An Introduction to Cluster Analysis*. New York: John Wiley and Sons, Inc.
14. Kolatch, E. (2001). *Clustering Algorithms for Spatial Databases: A Survey* (Technical Report No. CMSC 725). Department of Computer Science, University of Maryland, College Park, 1–22.

15. Koza, J.R. (1991). Evolving a computer program to generate random numbers using the genetic programming paradigm. *Proceedings of the Fourth International Conference on Genetic Algorithms, La Jolla, CA*, 37–44.
16. Ohsawa, Y. and Nagashima, A. (2001). A spatio-temporal geographic information system based on implicit topology description:STIMS. *Proceedings of the Third International Society for Photogrammetry and Remote Sensing (ISPRS) Workshop on Dynamic and Multi-Dimensional Geographic Information System, Thailand*, 218–223.
17. Rasmussen, E. (1992). Clustering algorithms. *Information Retrieval: Data Structures and Algorithms*, 419–442. Upper Saddle River, NJ: Prentice-Hall.
18. Ripley, B.D. (1996). *Pattern Recognition and Neural Networks*. Cambridge, MA: Cambridge University Press.
19. Sarafis, I., Zalzala, A., and Trinder, P. (2002). A genetic rule-based data clustering toolkit. *Proceedings of the 2002 World Congress on Evolutionary Computation, Honolulu*, 1238–1243.
20. Sarafis, I., Zalzala, A., and Trinder, P. (2003). Mining comprehensive clustering rules with an evolutionary algorithm. *Proceedings of the Genetic and Evolutionary Computation Conference, Chicago*, 1–12.
21. Sheikholeslami, G., Chatterjee, S., and Zhang, A. (1998). WaveCluster: A multi-resolution clustering approach for very large spatial databases. *Proceedings of the 24th International Conference on Very Large Data Bases, New York*, 428–439.
22. Solberg, A., Taxt, T., and Jain, A. (1996). A Markov random field model for classification of multisource satellite imagery. *IEEE Transactions on Geoscience and Remote Sensing, 34*(1), 100–113.
23. Tout, S., Sverdlik, W., and Sun, J. (2006). Parallel hybrid clustering using genetic programming and multi-objective fitness with density (PYRAMID). *Proceedings of the 2006 International Conference on Data Mining (DMIN'06), Las Vegas, NV*, 197–203.
24. Wang, W., Yang, J., and Muntz, R. (1997). STING: A statistical information grid approach to spatial data mining. *Proceedings of the 1997 International Conference on Very Large Data Bases, Athens*, 186–195.
25. Zhang, T., Ramakrishnan, R., and Livny, M. (1996). BIRCH: An efficient data clustering method for very large databases. *Proceedings of the 1996 ACM SIGMOD International Conference on Management of Data, Montreal*, 103–114.

Chapter 43
Transforming the Natural Language Text for Improving Compression Performance

Ashutosh Gupta and Suneeta Agarwal

43.1 Introduction

In the last 20 years, we have seen a vast explosion of textual information flow over the Web through electronic mail, Web browsing, information retrieval systems, and so on. The importance of data compression is likely to be enhanced in the future as there is a continuous increase in the amount of data that needs to be transformed or archived. In the field of data compression, researchers developed various approaches such as Huffman encoding, arithmetic encoding, Ziv–Lempel family, dynamic Markov compression, prediction with partial matching (PPM [1] and Burrows–Wheeler transform (BWT [2]) based algorithms, among others. BWT permutes the symbol of a data sequence that shares the same unbounded context by cyclic rotation followed by lexicographic sort operations. BWT uses move-to-front and an entropy coder as the backend compressor. PPM is slow and also consumes a large amount of memory to store context information but PPM achieves better compression than almost all existing compression algorithms.

In the recent past, [3, 4] developed a family of reversible Star-transformations which were applied to a source text along with backend compression algorithms. The basic idea of the transform module is to transform the text into some intermediate form which can be compressed with better efficiency. The transformed text is provided to a backend compression module which compresses the transformed text. However, execution time performance and runtime memory expenditure of these compression systems have remained high compared with the backend compression algorithms such as bzip2 and gzip.

The compression ratio achieved with compressing the transformed text is much better than compressing the text directly with well-known compression algorithms. Heap's law [5], an empirical law widely accepted in information retrieval, establishes that a natural language text of $O(u)$ words has a vocabulary of size $v = o(U^{\beta})$, for $0 < \beta < 1$. Typically, β is between 0.4 and 0.6 [6, 7], and therefore v is close to $O(\sqrt{u})$. An important conclusion of this law is that for a text of $O(u)$ words, the total

Oscar Castillo et al. (eds.), *Trends in Intelligent Systems and Computer Engineering.*

number of stop-words is u^* 40%. Stop-words are those words that occur frequently in text documents. Articles, prepositions, and conjunctions are natural candidates for a list of stop-words. We have identified a list of 520 stop-words.

Our work is based on idea that a natural language text of $O(u)$ words has $u^*0.4$ stop-words and the remaining words in a text are vocabulary. As the volume of stopwords is less as compared to the volume of vocabulary, we can transform all the stop-words in a text to some intermediate form and all the vocabulary in a text remains as it is.

We used an algorithm given in Franceschini and Mukherjee [4] for transforming text into an intermediate form suitable for compression. Compared with compress, gzip, and bzip2, the new transform achieves improvement in compression performance. Experimental results show that, for our test corpus, the average compression time using the new transform with bzip2, gzip, and compress is 11.2% slower, 36.35% faster, and 9.43% faster as compared with the original bzip2, gzip, and compress, respectively. The average decompression time using the new transform with bzip2, gzip, and compress is 42.21% slower, 15.63% faster, and 7.2% faster compared with the original bzip2, gzip, and compress, respectively.

We conducted experiments on our own test corpora. Results show that, using new transform, the average BPC (bits per character) improved 29.28% over bzip2, 28.71% over gzip, and 14.58% over compress.

43.2 Transformation: Stop-Word-Based Dictionary Approach

In this section, we first discuss stop-word-frequency distribution for which the transform algorithm [4] is used. After that a brief discussion about Heaps law in information retrieval is presented, followed by detailed description about the algorithm given by Franceschini and Mukherjee [4].

43.2.1 Stop-Word Frequency Distribution

First, we show the word frequency and length of words information, as given in Fig. 43.1. It is clear from the figure that words of lengths four and five have higher frequency as compared to other words of the English language.

43.2.2 Heap's Law

A natural language text consists of vocabulary and stop-words. Stop-words are those words which occur frequently in a text and do not contain valuable information in searching a pattern. The list of such words contains 520 stop-words. The example

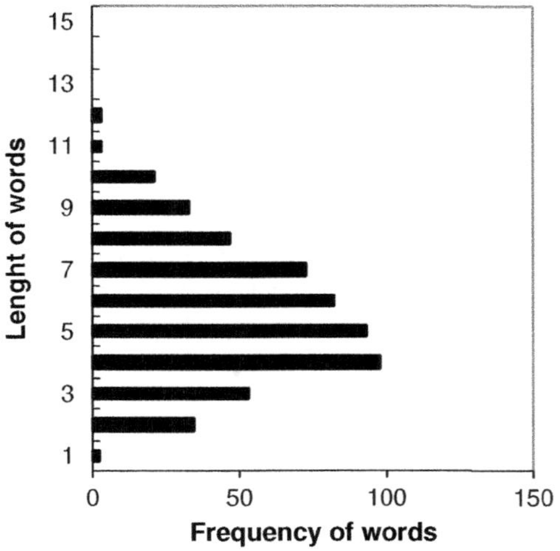

Fig. 43.1 Stop-word based frequency distribution

of stop-words includes a, i, an, an, so, by, and, but, the, from, and so on. The volume of stop-words is much less compared to the volume of the total vocabulary of the text but the appearance of the stop-words in the text are much higher than the appearance of the vocabulary. This is an important conclusion of Heaps law [5]. This will lead to conversion of original text into an intermediate form which contains $u^*40\%$ (transformed stop-words) and the remaining $u^*60\%$ of vocabulary. In this way we introduce an extra redundancy to the transformed text using the algorithm given by Franceschini and Mukherjee [4]. In the next section, we briefly discuss the transformation algorithm.

43.2.3 Transformation Algorithm

We are using the transformation algorithm as given by Franceschini and Mukherjee [4]. The set of stop-words are arranged in lexicographic order and by word length as shown in Fig. 43.1. This makes distinct dictionaries of stop-words. For each stop-word dictionary of length 1 to 15, we make a transformed stop-word dictionary. The dictionaries used in the experiment are prepared in advance, and shared by both the encoder module and decoder module. Currently the transform dictionaries only contain lower-case words. Dedicated operations were designed to handle the initial letter capitalized words and all letter capitalized words. The character "∼" appended to the transformed word denotes that the initial letter of the corresponding word in the original text file is capitalized. The appended character " ` " that all letters of the corresponding word in the original text file are capitalized. The character "\" is used as an escape character for encoding the occurrence of "*", "∼", "`", and "\" in the input text file.

The transformer reads a word from the input text file and checks the word into the corresponding stop-word dictionary. If the word is in the stop-word dictionary, then the transformer reads the transformed stop-word dictionary and emits the transformed word. Continuing in this way leads to a transformed output that contains redundant data. This introduced redundancy is helpful for any compression algorithms.

The transform decoding module performs the inverse operations of the transform encoding module. The escape character and special symbols ("*", "~", " ` ", and "\") are recognized, and transformed stop-words are replaced with their original stop-words.

43.3 Performance Evaluation

We evaluated compression performance as well as the compression time improvement using our own test corpus which consists of 22 files. All these test files are listed in Table 43.1. The experiment was carried out on a 1.6 GHZ Pentium IV machine housing Linux 9. We have chosen bzip2, gzip, and compress as a back-end compression tool.

Table 43.1 Test corpus

File	Size (bytes)
File1.txt	8833
File2.txt	48808
File3.txt	9432
File4.txt	11375
File5.txt	63826
File6.txt	18263
File7.txt	18988
File8.txt	56448
File9.txt	43460
File10.txt	39172
File11.txt	11025
File12.txt	29706
File13.txt	30726
File14.txt	2289
File15.txt	2932
File16.txt	2032
File17.txt	2968
Brief.rtf	52557
Project.rtf	7405686
Copying.txt	32874
Genesis	219118
Book1	312832

43.3.1 Timing Performance of New Transform

In the transform encoding module of the new transform, we create 14 dictionaries of stop-word of lengths 1 to 15 excluding stop-words of length 14 as there is no word of length 14. These 14 dictionaries are fixed to the encoding end as well as to the decoding end and because the number of stop-words is much smaller, there is no need to generate it every time. The strings in the 14 dictionaries are sorted lexicographically. The time complexity of searching a word of length m in an appropriate dictionary as a stop-word with $n$strings will require at most $O(\log n + m)$ time.

43.3.2 Timing Performance with Backend Compression Algorithm

The encoding/decoding time when the new transform is combined with the backend data compression algorithms (i.e., bzip2, gzip, compress) is given in Tables 43.2 and 43.3. The following conclusions can be drawn from these tables.

- The average compression time using the new transform algorithm with bzip2, gzip, and compress is 11.2% slower, 36.35% faster, and 9.43% faster compared with the original bzip2, gzip, and compress, respectively.
- The average decompression time using the new transform algorithm with bzip2, gzip, and compress is 42.21% slower, 15.63% faster, and 7.2% faster compared with the original bzip2, gzip, and compress, respectively

Table 43.2 Comparison of encoding speed (in ms)

	AVG
gzip	63.16
gzip + NT	40.2
compress	103.48
compress + NT	93.72
bzip2	550.44
bzip2 + NT	619.8

Table 43.3 Comparison of decoding speed (in ms)

	AVG
gzip	45.8
gzip + NT	38.64
compress	95.36
compress + NT	88.48
bzip2	54.8
bzip2 + NT	94.84

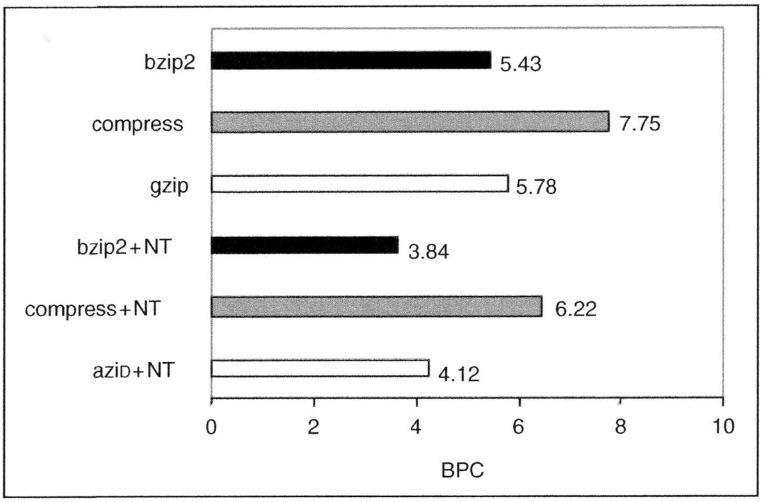

Fig. 43.2 Compression performance with/without transform

43.3.3 Compression Performance of New Transform

We show the compression performance (in terms of BPC) of the new transform, given in Fig. 43.2.

In our implementation the original stop-word dictionaries and transformed stop-word dictionaries are shared by both transform encoder and transform decoder. These dictionaries are generated independently. The dictionaries contain 520 entries of stop-words and 520 entries of transform stop-words. The size of dictionaries is nearly 5 KB. Figure 43.2 illustrates the comparison of average compression performance for our test corpus. The results are very clear.

- In our test corpus, facilitated with the new transform, bzip2, gzip, and compress achieve an average improvement in compression ratio of 29.28% over bzip2, 28.71% over gzip, and 14.58% over compress.

The compression performance of bzip2 powered by the new transform is superior to the original bzip2.

References

1. A. Moffat (1990) Implementing the PPM data compression scheme. *IEEE Transactions on Communications*, 38(11):1917–1921.
2. M. Burrows and D. Wheeler (1994) A block-sorting lossless data compression algorithm. Technical Report, SRC Research Report 124, Digital Systems Research Center, Palo Alto, CA.
3. F.S. Awan and A. Mukherjee (2001) LIPT: A lossless text transform to improve compression. In *Proceedings of International Conference on Information and Theory: Coding and Computing*, Las Vegas, Nevada, IEEE Computer Society.

4. R. Franceschini and A. Mukherjee (1996) Data compression using encrypted text. In *Proceedings of the Third Forum on Research and Technology*, Advances on Digital Libraries, 130–138. ADL.
5. J. Heaps (1978) *Information Retrieval—Computational and Theoretical Aspects*. Academic Press, New York.
6. M.D. Araujo, G. Navaaro, and N. Ziviani (1997) Large text searching allowing errors. In *Proceedings of the 4th South American Workshop on String Processing*. R. Baeza-Yates, Ed. Carleton University Press International Informatics Series, vol. 8. Carleton University Press, Ottawa, Canada, 2–20.
7. E.S. Moura, G. Navarro, and N. Ziviani (1997) Indexing Compressed text. In *Proceedings of the 4th South American Workshop on String Processing*. R. Baeza-Yates, Ed. Carleton University Press International Informatics Series, vol. 8. Carleton University Press, Ottawa, Canada, 95–111.

Chapter 44
Compression Using Encryption

Ashutosh Gupta and Suneeta Agarwal

44.1 Introduction

The basic objective of a data compression algorithm is to reduce the redundancy in data representation so as to decrease the data storage requirement. Data compression also provides an approach to reduce communication cost by effectively utilizing the available bandwidth. Data compression becomes important as file storage becomes a problem. In general, data compression consists of taking a stream of symbols and transforming them into codes. If the compression is effective, the resulting stream of codes will be smaller than the original symbols.

The decision to output a certain code for a certain symbol or set of symbols is based on a model. The model as described in Nelson and Gailly [1] is simply a collection of data and rules used to process input symbols and determine which code(s) to output. A program uses the model to accurately define the probabilities for each symbol and the coder to produce an appropriate code based on those probabilities.

Text compression can be divided into two categories, statistical-based and dictionary-based. In statistical models, the technique encodes a single symbol at a time by reading it in, calculating a probability, then outputting a single code. A dictionary-based compression scheme [2–4] uses a different approach. It reads in input data and looks for groups of symbols that appear in the dictionary. If a string match is found, a pointer or index into the dictionary can be output instead of the code for the symbol. Thus the longer the match found, the better is the compression ratio.

The dictionary-based approach is also divided into two parts, static dictionary methods and dynamic dictionary methods. The static dictionary is built before compression occurs and it does not change while the data are being compressed. This dictionary is common to both the encoder and decoder ends. In the dynamic dictionary-based methods, the meaning of dynamic is that we start out either with no dictionary or with a default baseline dictionary. As compression proceeds, the algorithm adds new phrases to be used later as encoded tokens. An example of dictionary-based text compression is LZW developed by Welch [1].

Oscar Castillo et al. (eds.), *Trends in Intelligent Systems and Computer Engineering.*

Researchers [3, 4] had proposed a word-based dictionary method which used a single dictionary. Most text compression algorithms perform compression at the character level. If the algorithm is adaptive, the algorithm slowly learns correlation between adjacent pairs of characters, then triples, quadruples, and so on. The algorithm rarely has a chance to take advantage of longer-range correlations before either the end of input is reached or the tables maintained by the algorithm are filled to capacity.

If text compression algorithms were to use larger units than single characters as the basic storage element, they would be able to take advantage of the longer-range correlations and, perhaps, achieve better compression performance. Faster compression may also be possible by working with larger units. We use a word as the basic unit. Here we adopted multiple dictionaries [2] with a dynamic approach so that the probability of hitting a word in a dictionary increases. Along with this, a new storage protocol and a back-search algorithm is used. To speed up the location search of a particular word, a hashing function [5] is used.

We present the results based on encrypting the compressed data. The basic idea of this algorithm is to define a unique encryption or signature of each word in the dictionary by replacing certain characters in the words by a special character "*" and retaining a few characters so the word is still retrievable. For any encrypted text the most frequently used character is "*" and the compression algorithms can exploit this redundancy in an effective way. For a compression algorithm A and a text T, we apply the same compression algorithm A on an encrypted text *T and retrieve the original text via a dictionary which maps the decompressed text *T to the original text T.

The above procedure is given in Fig. 44.1. In this figure, ND, CD, T, *T, and A stand for normal dictionary, cryptic dictionary, original text, encrypted text, and compression algorithm, respectively. Here we assume that the system has access to a dictionary of words used in all the texts. If a library or office organization were to use this technique, the availability of a in-house dictionary would be a one-time investment in storage overhead. If two organizations wish to exchange information using this technique, they must share a common dictionary.

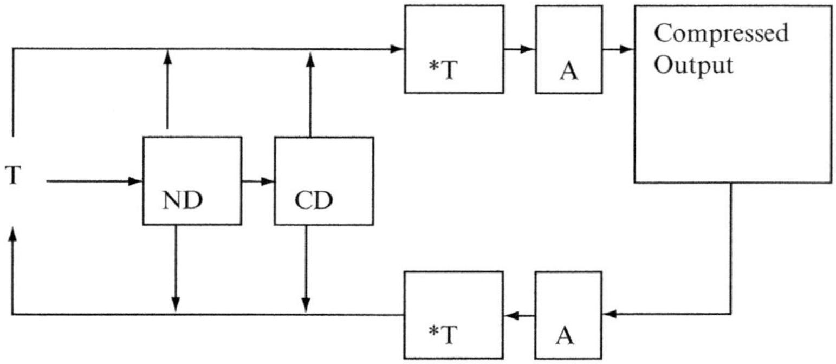

Fig. 44.1 Compression and decompression process

44.2 Encoding the Text Using Signatures

Whenever the text is presented to a reader, he reads the words not sequentially, but recognizing each word as a unique symbol. A signature is a cryptic word in which we replace certain characters in a word by a special character and retain a few characters so that the word is still retrievable. Consider an example of a set of six-letter words starting with letter c and ending with letter r in English: canker, canter, career, and carter. Denoting an arbitrary character by a special symbol "*", the above set of words can be unambiguously spelled as

canker	⟶	c**k**
canter	⟶	can***
career	⟶	ca*e*r
carter	⟶	*a****

An unambiguous representation of a word by a partial sequence of letters from the original sequence of letters in the word interposed by special characters; "*" is called a signature of the word. The collection of English words in a dictionary in the form of a lexicographic listing of signatures is called a cryptic dictionary and an English text completely transformed using signatures from the cryptic dictionary is called a cryptic text. For any cryptic text, the most frequently used character is "*". A dictionary is said to be optimal in which we maximize the use of the special character "*" for a given set of application texts. Here we assume that maximum word length is 15 letters.

Let F denote a finite string (or sequence) of characters (or symbols) $f_1 f_2 f_3 \ldots f_n$ over an alphabet Σ where $f_i = F[i]$ is the ith character of F, and n is the length of the sequence F. S is a subsequence of F if there exist integers $1 \le r_1 < r_2 \ldots < r_s \le n$ such that $S[i] = F[r_i]$, $1 \le i \le n$. Let D denote a dictionary of a set of distinct words. A cryptic word corresponding to F, denoted as *F, is a sequence of n characters in which $^*F[i] = ^*$ if $i \ne r_i$ and for all other i, $F[i] = F[r_i]$ as in S. For a dictionary D, the distinct shortest subsequence problem is the problem of determining a cryptic dictionary, *D such that each word in *D has maximal number of "*" symbols [6].

44.2.1 Constructing Cryptic Dictionary

A cryptic dictionary can be constructed in two ways. In the first method, Franceschini and Mukherjee [6] partitioned the dictionary D into n dictionaries D_i, $(1 \le i \le n)$, where n is the length of the longest word in the dictionary such that D_i contains the words of length i. Within each D_i, we computed a signature for each word as follows. For each word w in D_i, we determine whether w has a single letter (say a) that is unique in its position (say j) throughout the words of D_i. If so, then the signature of w is computed as $w =^{****} \ldots^{*} a^{*} \ldots^{****}$ where there are $j-1$ *s before

a and $|w| - j + 1$ *s after a. Once the signature of w is computed, w is removed from the continued processing for *D_i. Processing continues by considering pairs of letters, triples of letters, and so on, of words from D_i to find a unique signature for the words until signatures have been found for all words. Here we give an example that illustrates the procedure for constructing the normal and cryptic dictionaries. Finally we build the cryptic text by using the normal and cryptic dictionaries [6].

In the second method, Franceschini and Mukherjee [6] optimize the cryptic dictionary by ensuring that many more "*"s were used in the words. The cryptic dictionary is constructed by using the normal dictionary. In the construction of the cryptic dictionary, we partitioned the dictionary D into **n** dictionaries D_i, $(1 \leq i \leq n)$, where n is the length of the longest word in the dictionary such that D_i contains the words of length i. We assigned a signature to each word as follows. The first word received a signature consisting of i*s. The next 52i words received a signature consisting of a single letter (either lower-case or upper-case) in a unique position, surrounded by $i - 1$ *s. Here 52 comes from the fact that each position in a word contains either 26 lower-case letters or 26 upper-case letters. Similarly, the second word of length five received the signature a****. The next 52*52*$C(i,2)$ words received a signature consisting of two letters in unique positions as a pair surrounded by $I - 2$s (where C(i,2) represents the number of ways of choosing two positions from i positions). For example, one five-letter word received a signature of b*D**. It was never necessary to use more than two letters for any signature in the dictionary using this scheme [6].

44.2.2 Encryption and Decryption Process

To encrypt a text file, we first read a word from the file and find the word in the dictionary. If the word is found in the dictionary, its signature is obtained in the cryptic dictionary, and then the signature is output. Other characters such as punctuation marks and capital characters from the text file are not changed. Punctuation marks and spaces are handled by copying them from the input to the output directly; these nonletter characters are used as word delimiters because the dictionary does not contain the spaces and punctuation marks.

Because the dictionary contains only lower-case letters, we do not automatically recognize words with capitals that are in the dictionary. A naïve approach simply copies the unrecognized word to the encrypted file. A better approach is to append the capitalization information to the end of the encrypted word in the encrypted file. We append a special character (.) to the end of the word and then append a bit mask in which bit i is set if and only if position $i$is a capital in the original word. Because we are dealing with English text, we can make an optimization to improve the performance, as follows. The most likely capitalization patterns are initial capitalization of the word (e.g., at the beginning of a sentence) and all capitals. Instead of appending the bit-pattern, we append $\sim$ or $^\wedge$ to the end of the word to handle these cases. This saves us the storage of the bit patterns in the most common cases.

Finally, we used several special characters appear in the input file such as *,',:,;,-, ^ and `. We prepend an escape character (\) to them in the encrypted file. This adds one final special character to the encryption process (namely \), which we handle in the same way as the other special characters. To decrypt a text file, we read a signature from the encrypted file, look it up in the cryptic dictionary, obtain the corresponding word in the dictionary, and output the word. Again, other characters from the text file are not changed.

44.3 Compression Technique

We used the compression technique described in Ng et al. [2], which uses multiple dictionaries and each dictionary has a linked subdictionary. Along with this, a new storage protocol and a back-search algorithm are used. The codes which are used in this compression technique are divided into three types, namely, copy, literal and hybrid codes. The pseudo code for the back-search algorithm described in Ng et al. [2] is given below:

44.3.1 Back-Search Algorithm

The back-search algorithm is the most important part of the whole process. This algorithm described in Ng et al. [2] is used for reducing the number of literal codes. The algorithm generates the codes such as cascading copy and cascading literal by reducing the word character by character. The pseudo-code for the algorithm is given below.

```
loop until end of file
      n=word_length(word);
      If(matched occurs between word and word in n-table)
      {sent copy code;}
      else {
do {reduce one character at a time i.e. n - -
if(prefix in normal dictionary)
{sent cascading literal code; break;}
else {
if(prefix in subdictionary)
{sent cascading literal code; break;}
else{
if(prefix and suffix in normal dictionary)
                    {sent cascading copy code; break;}
                    else if(prefix in subdictionary and suffix in normal dictionary)
                    {sent cascading copy code; break;}
```

```
}
            }
        }(while n>=2)
    }
if (whole search fails)
{put literal code;}
end loop
```

44.3.2 Hashing Function

The hashing function used here was proposed by Peter K. Pearson [5]. This hashing function is used for variable-length text strings. This function takes as input a word W consisting of n number of characters such as C_1, $C_2, \ldots, C_n$, each character being represented by one byte, and returns an index in the range 0–255. The algorithm for the hashing function is given below.

```
h[0]:=0;
for i in 1.. n loop
h[i]:=T[h[i-1] xor C[i]];
end loop;
return h[n];
```

The hashing function used here serves two purposes: it is used to initialize the dictionary, and it is used as a quick search to a particular word. Words that are inside each dictionary are decided by using a hashing function. Figure 44.3 shows the algorithm flow of the coding process.

44.4 Experiments

This section shows the experiments that we have performed using the encryption approach. These are preliminary experiments, but we believe that they reflect the significant potential of this approach. We used different text files of different sizes. The compression ratio is calculated as:

$$\text{CR} = (\text{Compressed File Size/Input File Size}) \times 100$$

44.4.1 Implementation Results

In this section, we present the results of implementation of the compression algorithm on several text files. The results which are obtained by direct compression and by encrypted compression are also given in Table 44.1. The percentage savings in

Table 44.1 Performance result

Input file (in bytes) (I)	Direct compression (in bytes) (II)	Encrypted compression (in bytes) (III)	CR_1 (IV)	CR_2 (V)
119941	112087	97691	93.4	73.11
132313	117969	103821	89.15	70.90
148704	125499	111619	84.39	68.33
165017	133018	119395	80.6	66.29
181330	140537	127171	77.5	64.61
197643	148056	134947	74.91	63.21
218013	157231	144518	72.12	61.70
242531	168085	155853	69.3	60.13

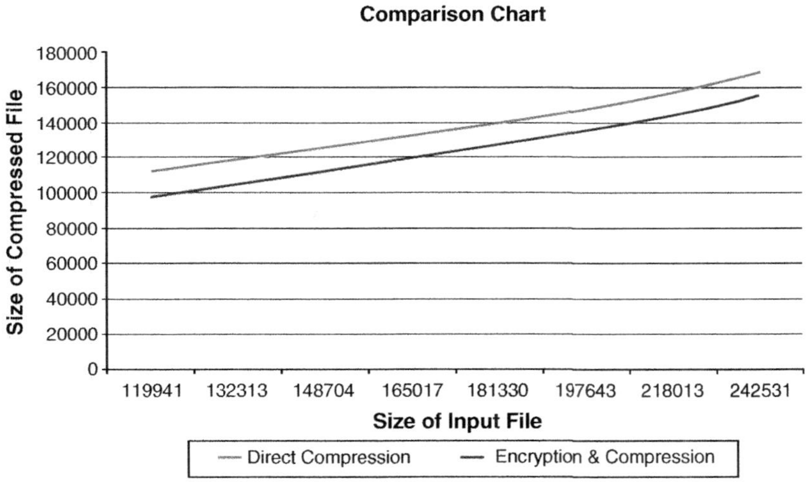

Fig. 44.2 Performance chart for compression after encryption

disk space is thus found. When we examine the data for compressed files, we see that all of the encrypted compressions yield uniformly better results than the direct compressions.

44.4.2 Performance Comparison

The improved performance of direct compression over encrypted compression can be better understood by the graphical means given in Fig. 44.2. As we study the graph, we can say that there is a constant rising in the values. The results obtained by encrypted compression are better than direct compression. The execution time for different files using direct compression and encryption is also given. Table 44.2 shows the execution time for direct compression and compression with encryption.

Table 44.2 Time execution

Input file (in bytes) (I)	Execution time for direct compression (II)	Execution time for encryption (III)	Execution time for compression after encryption (IV)	Total time T = (III) + (IV)
119941	1.96	1.9	1.59	3.49
132313	2	2	1.7	3.7
148704	2.04	2.2	1.83	4.03
165017	2.23	2.47	1.96	4.43
181330	2.37	2.7	2.1	4.8
197643	3.18	2.84	2.24	5.08
218013	3.39	3.11	3.02	6.13
242531	3.49	3.46	3.34	6.8

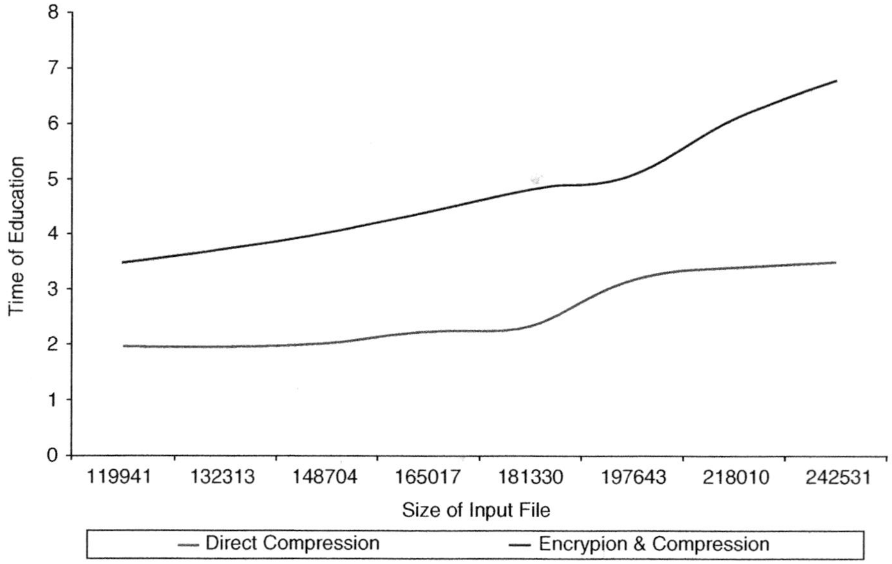

Fig. 44.3 Timing performance chart for direct compression and encrypted compression

The time execution chart for Table 44.2 is given in Fig 44.3. The execution time is larger for the encrypted compression as compared to direct compression. The reason is that the encryption process takes time depending upon the file size.

References

1. M. Nelson and J.L. Gailly (1996). *Data Compression Book*. 2nd edition, BPB Publication.
2. K.S. Ng, L.M. Cheng, and C.H. Wong (1997). Dynamic word based text compression, *IEEE Trans*. Proceedings of the Fourth International Conference on Volume 1, 18–20 Aug. 1997 Page(s):412–416 vol.1

3. G.V. Cormack, and R.N. Horspool (1992). Constructing word based text compression algorithms. *Proceedings of IEEE Data Compression Conference*, pp. 62–71.
4. J. Jiang and S. Jones (1992). Word based dynamic algorithms for data compression. *IEEE Proceedings-I*, Vol. 139, No. 6.
5. P.K. Pearson (1990). Fast hashing of variable length text strings. *Communications of the ACM*, Vol. 33, No. 3.
6. R. Franceschini and A. Mukherjee (1996). Data compression using encrypted text. *Proceedings of ADL*, pp. 130–138.

Index

D

G

H

Q

R

S

U

V

W

X